U0139909

Advances in Hydraulic Physical Modeling and Field Investigation Technology

— Proceedings of International Symposium on Hydraulic Physical Modeling and Field Investigation

Editors
Li Yun Wu Shiqiang

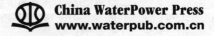

China WaterPower Press
www.waterpub.com.cn

Abstract

This book was the Proceedings of the International Symposium on Hydraulic Physical Modeling and Field Investigation, which was opened on September 13 – 15, 2010, in Nanjing. There were 4 topics included in the research field: New Technology of Physical Modeling of River, Coastal and Environmental Flows, Advancement in Field Investigation for Hydro and Environmental Engineering, Development of Instruments and Facilities for Hydraulic and Eco – hydraulic Measurement, Hybrid Model Approach and Combination of Physical Approaches with Numerical Simulation.

This book covered 114 papers, which were the newest research results in the world. It can be referenced by students, engineers and researchers.

图书在版编目（ＣＩＰ）数据

水工物理模型与原型观测技术进展 = Advances in
Hydraulic Physical Modeling and Field
Investigation Technology : 英文 / 李云，吴时强编
. -- 北京 : 中国水利水电出版社，2011.4
　ISBN 978-7-5084-8554-6

Ⅰ. ①水… Ⅱ. ①李… ②吴… Ⅲ. ①水工模型试验
－国际学术会议－文集－英文 Ⅳ. ①TV131.61-53

中国版本图书馆CIP数据核字(2011)第069073号

书　　名	**Advances in Hydraulic Physical Modeling and Field Investigation Technology** —Proceedings of International Symposium on Hydraulic Physical Modeling and Field Investigation	
作　　者	Editors Li Yun Wu Shiqiang	
出版发行	中国水利水电出版社 （北京市海淀区玉渊潭南路1号D座　100038） 网址：www. waterpub. com. cn E - mail：sales@waterpub. com. cn 电话：(010) 68367658（营销中心）	
经　　售	北京科水图书销售中心（零售） 电话：(010) 88383994、63202643 全国各地新华书店和相关出版物销售网点	
排　　版	中国水利水电出版社微机排版中心	
印　　刷	北京市兴怀印刷厂	
规　　格	184mm×260mm　16 开本　45.5 印张　1608 千字	
版　　次	2011 年 4 月第 1 版　2011 年 4 月第 1 次印刷	
印　　数	0001—1000 册	
定　　价	**168. 00 元**	

Welcome Address

Dear respected guests,
Delegates,
Ladies and gentlemen,

  On behalf of the Local Organizing Committee (LOC) and Nanjing Hydraulic Research Institute, I would like to extend my warmest welcome to all of you to attend the International Symposium on Hydraulic Physical Modeling and Field Investigation (ISHPF 2010) to be held on Sept. 13 – 15, 2010 in Nanjing, China.

  With the theme of *"Important role of physical modeling and field investigation technology in hydraulic engineering and research"*, the symposium is designed to help meet the challenges in hydraulic physical modeling and field investigation, and shall play a vital role in the process of developing solutions to these challenges. Along with the fast development of mathematical modeling in recent ten years and more, physical modeling, being a conventional tool for fellow hydraulic engineers, still plays an important role in hydraulic research and engineering today. This symposium will provide you, delegates from all over the world, with a forum to exchange you research results, experiences, new ideas and updated information in the fields related to the hydraulic physical modeling and field investigation.

  The symposium is organized by Nanjing Hydraulic Research Institute (NHRI) in close cooperation with the International Association for Hydro-Environment Engineering and Research (IAHR). NHRI, set up in 1935, is a multipurpose national hydraulic research complex in China, mainly dedicated to basic researches, applied researches and technological development, and undertaking directional, principal and comprehensive researches for water conservancy, hydroelectric power and waterway transportation projects as well as researches on soft science and macro decision making. During the symposium, you will be invited to visit the Water Experiment Center of the institute, the pivotal part of China's State Key Laboratory of Hydrology-Water Resources and Hydraulic Engineering, where more than 30 large-size physical models equipped with advanced testing and measuring devices are under operation. Through this symposium, we wish to develop new cooperation ties with fellow hydraulic engineers from home and abroad.

Located at the lower reaches of the Yangtze River, Nanjing is a beautiful city with a history of more than 2,500 years. There are many places of cultural and historical interests in Nanjing worth visiting. With its unique advantages in hi-tech industry, science, technology and education, Nanjing, capital of Jiangsu Province, is also a famous modern metropolis, attracting various investors and professionals to develop their new careers here.

I sincerely hope that all of you will enjoy your stay here professionally and culturally, and wish ISHPF 2010 a great success!

Thank you all.

Professor Zhang Jianyun

Chairman of LOC, ISHPF2010
Academician of Chinese Academy of Engineering
President of Nanjing Hydraulic Research Institute (NHRI)
Sept. 13, 2010, Nanjing, China

Forword

As Asian countries continue to be some of most important branches of the International Association of Hydro-Environment Engineering and Research (IAHR), the IAHR Asia and Pacific Division is playing an ever-increasing active role in the field of hydraulic research and its practical application. In order to meet such development, the International Symposium on Hydraulic Physical Modeling and Field Investigation (ISHPF 2010) was held on September 13 – 15th, 2010, in Nanjing, which provided a unique platform for technical exchange and demonstration of the development and new ideas of hydraulic physical modeling and field investigation technology.

The latest development of relevant technology in the field of hydraulic physical modeling and field investigation was presented via invited speeches and presentations in parallel sessions. The themes and sub-themes were as follows.

New Technology of Physical Modeling of River, Coastal and Environmental Flows

- Modeling of water flow over structures, energy dissipaters and dam breach
- Modeling of pollutants, salty, heated, multi-phase flow and ground flow
- Modeling of sediment transport and fluvial processes
- Modeling of estuarine, coastal processes and wave movement
- Similarity theory and scale effects of physical modeling

Advancement in Field Investigation for Hydro-Environmental Engineering

- Field investigation and experiment in inland waters and estuaries and oceans
- Field investigation and experiment for environment, eco-systems and river restoration
- Field investigation under extreme conditions: storm and tsunami, flooding, ice and snow and disasters

- Technology of long-term hydraulic monitoring and controlling of hydro-environmental engineering
- Comparison and integration of laboratory experiments and field investigation

Development of Instruments and Facilities for Hydraulic and Eco-hydraulic Measurement

- Facilities of indoor and outdoor measurement
- Sensor technology and instruments for hydraulic measurement
- New technology for hydro-environmental and eco-hydraulic measurement
- Data acquisition and data processing

Hybrid Model Approach and Combination of Physical Approaches with Numerical Simulation

- Coupling of physical models and numerical models
- Multi-process, data-driven modeling
- Other new approaches relating with physical modeling

The Editors
Nanjing，China
Nov. 2010

Sponsored by

Nanjing Hydraulic Research Institute, China

Co-sponsored by

International Association for Hydro Environment Engineering and Research

Chinese Hydraulic Engineering Society

China Society for Hydropower Engineering

Key Laboratory of Water Science and Engineering, Ministry of Water Resources of China

Key Laboratory of Navigation Structures, Ministry of Transport of China

State Key Laboratory of Hydrology Water Resources and Hydraulic Engineering

Hohai University

China Institute of Water Resources and Hydropower Research

Yangtze River Scientific Research Institute

Institute of Yellow River Hydraulic Research

Navigation Engineering Committee of Chinese Society of Hydropower Engineering

Estuarine & Coastal Science Research Center

Organization

Local Organizing Committee

Chairman	Zhang Jianyun	NHRI, China
Vice – Chairman	Li Yun	NHRI, China
	Tang Hongwu	Hohai Univ. , China
Members	Han Changhai	NHRI, China
	Cai Fulin	Hohai Univ. , China
	Yan Genhua	NHRI, China
	Huang Guobing	CRSI, China
	Xuan Guoxiang	NHRI, China
	Jin Hai	Ministry of Water Resources, China
	Dai Jiqun	NHRI, China
	Gao Jing	Chinese Hydraulic Engineering Society
	Ge Jiufeng	NHRI, China
	Wang Lianxiang	China Institute of Water Resources and Hydropower Research
	Wang Lingling	Hohai Univ. , China
	Luo Shaoze	NHRI, China
	Xu Shikai	NHRI, China
	Duan Xiangbao	NHRI, China
	Wu Xiufeng	NHRI, China
	Hu Yaan	NHRI, China
	Wu Yihong	China Institute of Water Resources and Hydropower Research
	Lu Yongjun	NHRI, China
	Li Zhonghua	NHRI, China

	Wang Zili	Yellow River Institute of Hydraulic Research, China
Secretary General	Wu Shiqiang	NHRI, China
Vice – Secretaries	Xie Xinghua	NHRI, China
General	Sun Feng	NHRI, China
	Feng Zhonghua	NHRI, China

Scientific Committee for ISHPF 2010

	Zhang Ruikai	NHRI, China
Chairman	Zhang Ruikai	NHRI, China
Vice – Chairman	Joseph Hun wei Lee	University of Hongkong, China
	Gao Jizhang	China Institute of Water Resources and Hydropower Research
Members	Alexander Sukhodolov	Institute of Freshwater Ecology and Inland Fisheries, Germany
	Zhang Changkuan	Hohai Univ., China
	Colin Rennie	IAHR HIS, Canada
	Cristobal Mateos	Spain
	David Zhu	University of Alberta, Canada
	Jiang Enhui	Yellow River Institute of Hydraulic Research, China
	Jin Feng	Yangtze River Scientific Research Institute, China
	Yan Genhua	NHRI, China
	Tan Guangming	Wuhan University, China
	Li Guifen	China Institute of Water Resources and Hydropower Research
	Wu Hualin	Ministry of Communications Research Center of the Yangtze, China
	Ichiro Fujita	Kobe University, Japan
	Jihn Sung Lai	Hydrotech Research Institute of Taiwan University, Taiwan, China

Contents

Topic Ⅱ Advancement in Field Investigation for Hydro-Environmental Engineering

Topic Ⅲ　Development of Instruments and Facilities for
Hydraulic and Eco-hydraulic Measurement

Topic Ⅳ   Hybrid Model Approach and Combination of Physical Approaches with Numerical Simulation

Topic I

New Technology of Physical Modeling of River, Coastal and Environmental Flows

Experimental Study of Flood Discharge through a High Arch Dam[*]

Weilin Xu

State Key Laboratory of Hydraulics and Mountain River Engineering (Sichuan University), Chengdu, 61000

Abstract: Flood discharge is one of the key problems in high dam projects. In China, many high dam projects have the following characteristics: high water head, large discharge and narrow river channel. In this paper, first, the 3-D flow fields in a plunge pool was simulated by the turbulence mathematical model and measured by the five-hole Pitot sphere combined with the pressure sensors and automatic data sampling and processing system; second, we developed a new type of flood discharge through a high arch dam, i. e., contracting dissipators with flaring gate piers in top spillways and slit-type buckets in middle orifices, and tested the design in a physical model of high-arch dam; finally, the influence of aeration on jet scour was investigated experimentally.

Key words: high arch dam; flood discharge and energy dissipation; contracting dissipators; plunge pool; aeration

1. Introduction

A lot of high arch dams are under construction in China. In these projects, the flood discharge is one of the key problems, and has the following characteristics: high water head, large discharge and narrow river channel. Duo to the large discharge and narrow river channel, both the top spillways and the middle orifices are needed to discharge flood through an arch dam into a plunge pool.

Under the condition of high water head, the plunging jets will hit strongly the slab of plunge pool and generate high impact pressure and strong pressure fluctuation, which will result in a rapidly varying cycle of loading and unloading of forces acting on each concrete or rock block slabs and damage them finally. This may cause dam foundations erosion and threaten the dam itself (Bollaert, 2003; Liu, 2007). In order to effectively reduce the impact pressure on the plunge pool slab, the jets should be dispersed as much as possible before entering the plunge pool. Motivated by this idea, in Ertan arch dam that is 240 meters high and constructed in 1990s, the jets from the top spillways and the middle orifices collide in air and are widely dispersed, so that the dynamic pressure on the slab of plunge pool is largely reduced. However, the collision of jets in air creates strong spray, which increases artificial rainfall intensity and rainfall area and may result in landslides. Therefore, new type of flood discharge through a high arch dam was researched in recent years.

Obviously, if the jets from the top spillways and the middle orifices do not collide in air, the excessive spray would be relieved significantly. In this case, however, the hydrodynamic pressure may be too high unless special design is developed to increase the effective area of jets impact zone in the plunge pool. In a narrow valley, one feasible way is to apply the contracting dissipators to the top spillways and the middle orifices in a high arch dam, such as flaring gate piers and slit-type buckets (Gong Z et al., 1983; Vischer et al., 1995). In this case, narrow jets can be obtained and the jet collision in air can be avoided, besides the jets can be dispersed strongly in the longitudinal direction to reduce the impact pressure on the plunge pool slab.

---

[*] Keynote lecture

2. 3-D flow featuer and energy dissipation characteristics in plunge pool

Investigating the flow field of plunge pool is the basis of an in-depth study of the energy dissipation problem of plunge pool. However, the flow features of plunge pool have not been understood fully, but also the 3-D flow field has not been previously measured in a physical model.

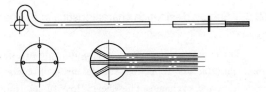

Figure 1　The sketch of the five-hole Pitot sphere

The intense turbulence and aeration make it very difficult to measure the flow field in the plunge pool. In view of the flow characteristics in the physical model, a five-hole Pitot sphere is used as velocity measurement instrument, by which the 3-D velocities can be measured at the same time. Figure 1 is the sketch of the five-hole Pitot sphere. The diameters of the Pitot sphere and its pressure holes are respectively 5mm and 0.35mm. In experiment, the pressure sensors are connected to each pressure hole by pipes. The hydrodynamic pressure in each hole is transformed into electric signal and transmitted to the automatic data sampling and processing system. These devices, together with a microcomputer, allow the measurement to be taken automatically.

Because the pressure holes of the Pitot sphere are connected to the pressure sensors, air cannot get into the piezometric tube and technical difficulty in measuring velocity of aerated flow is overcome.

2.1　3-D flow feature in plunge pool

Figure 2, Figure 3 and Figure 4 are the longitudinal, transverse and horizontal sections of the velocity vector field respectively in a plunge pool, with the calculated on the left-hand side and the measured on the right-hand side. The measured data are obtained from a physical model, in which the width of the pool floor and the height of the end sill are respectively 70cm and 35cm. The side slope is 1 : 1.1. When flood is discharged through five top spillways, the flow rate is 68.5L/s and the water depth in the plunge pool is 42.3cm. The nappe consists of five streams, with the incident angles 65°-69° and the profile as shown in Figure 4b. Because of symmetry, only a half of the flow field is calculated and measured, and the whole flow field is given when the velocity vectors are plotted by microcomputer (see Figure 3 and Figure 4). On the basis of the calculated and measured results, the typical features of the 3-D flow in plunge pool can be obtained.

The velocity in the upstream and impact region of the incident point is difficult to measure and no recirculation appears in the measured region. However, the calculated results show a little recirculation near the incident point, as shown in longitudinal section in Figure 2. It is also shown that the range of recirculation relates to the angle of incidence, decreasing as the latter tends towards zero. In the upstream region of the incident point, as is shown by the calculated results, there is a large vortex. It can also be seen in the simulated results that the high speed jet changes direction abruptly near the bottom plate, which is consistent with the result that the calculated and measured hydrodynamic pressure on the bottom plate rises abruptly near the impact point and drops to a gentler level on other part of the bottom plate.

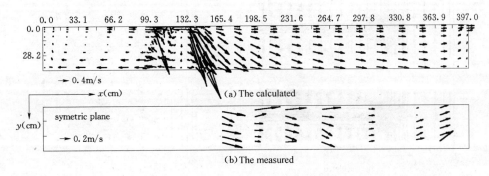

Figure 2　Longitudinal section of flow field in plunge pool

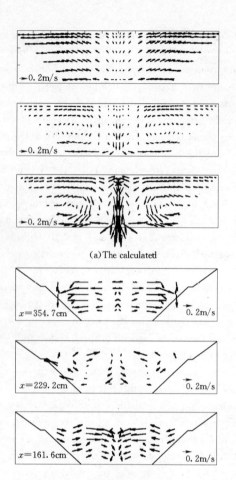

(a) The calculated

(b) The measured

Figure 3　Transverse sections of flow field in plunge pool

In the transverse sections (Figure 3), it can be seen from both the calculated and the measured results that there are two symmetric transverse circulations. It implies that the flow is spiral in the plunge pool. However, if the nappe were so wide as the width of the water surface, no space would be left for

water to form transverse circulations, then the transverse circulations seen in Figure 4 would not exist. In some transverse sections, there are two smaller secondary circulations near the water surface. In the region near the end sill, the velocity vectors have radial distribution in the transverse section.

A plane view is given in Figure 4, showing that the transverse distribution of longitudinal velocity component has several peaks. The positions of the peaks are correspondent with those of the incident points.

2.2　Energy dissipation characteristics in plunge pool

Turbulent energy and its dissipation rate are the important factors reflecting the features of energy transition and exchange. The energy loss of jet from its impinging into plunge pool to its flowing out of the plunge pool is firstly turned into the turbulent energy from the time-averaged energy, then is dissipated as the heat energy due to the fluid viscosity. In fact, the energy dissipation is the process of energy transferring and exchanging.

The distributions of turbulent energy in the horizontal planes are shown in Figure 5 and those of turbulent energy dissipation rate are shown in Figure 6. It can be seen in the figures that the turbulent energy and its dissipation rate mainly concentrate near the jet axis, especially the dissipation rate. Table 1 shows the percentages of the turbulent energy and dissipation rate in the regions near the jet axis over those in whole plunge pool. The results shown in Table 1 indicate again that the energy transition occurs mainly in the region near the jet axis and only a little of water mass in the plunge pool takes part in the energy dissipation. This reveals one of the reasons why the more concentrated the jets, the more difficult the energy dissipation. On the other hand, it can also be seen

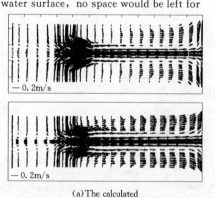

(a) The calculated

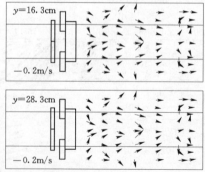

(b) The measured

Figure 4　Horizontal sections of flow field in plunge pool

in the distributions of turbulent energy and its dissipation rate that the maximum values of turbulent energy and dissipation rate appear at the position near the water surface. The above-mentioned characteristics, together with the flow feature, show that the shear of water flow is the main dissipation mode in plunge pool when the water depth is enough (the smaller the water depth, the less the shear dissipation of energy). In the region near the incident point, where the shear of water flow is very obvious, large amount of energy is dissipated. Then, in the region where the jet impacts the pool floor, the energy is dissipated secondly by impacting. Before reaching the end sill, the energy is dissipated thirdly by turbulent mixing of water flow in the plunge pool. On the basis of the above-mentioned characteristics, the whole water body in plunge pool can be divided into three energy dissipation regions, i.e., shear, impact and mixing dissipation regions of energy. The way of the optimization of plunge pool and dam outlets should be: expanding the range of the shear dissipation region (e.g. dispersing the nappes), reducing the impact intensity in the impact dissipation region (e.g. forming moving water cushion), and making use of the water body in the mixing dissipation region as much as possible.

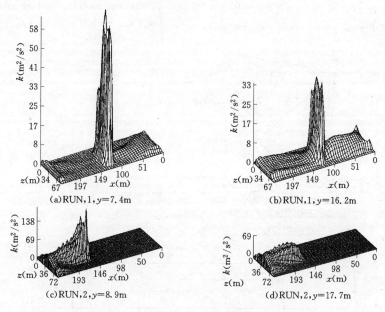

(a)RUN,1,y=7.4m   (b)RUN,1,y=16.2m

(c)RUN,2,y=8.9m   (d)RUN,2,y=17.7m

Figure 5   Distribution of turbulent energy

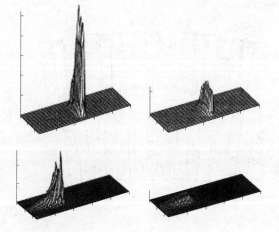

Figure 6   Distribution of turbulent energy dissipation rate

Table 1   Percentage of k and ε near the jet axis

Run No.	1	2	3	4	5	6
Percentage of volume of water body (%)	23	22	23	23	23	23
Percentage of k (%)	43	38	49	50	47	43
Percentage of ε (%)	90	65	89	88	85	88

It should be noted that not only the maximum pressure but also the gradient of the pressure distribution on the pool floor should be restricted in the design of plunge pools. The large pressure gradient means that the shear dissipation is not enough and the impact kinetic energy is concentrated too much. Besides, the large pressure gradient may cause a very large lift force on the pool floor. The 3-D model in this paper can be used to compute the forces on the upward surface of the concrete slab, but the forces on the downward surface are very complex. Therefore, the limitations of the maximum kinetic pressure and the maximum pressure difference need to be determined empirically. An example of the limitation is: the maximum time-averaged kinetic pressure head is limited under 15m (as in the designs of the Ertan, the Xiaowan and the Xiluodu plunge pools); the difference between the maximum and the minimum time-averaged kinetic pressure heads is also limited under 15m. In order to decrease the pool depth as much as possible under the limited condition, the water flow should be divided into multiple jets and each jet should be dispersed enough before entering the plunge pool. In addition, because the water body between the jet entry location and the dam is quite stable and has very little action in energy dissipation, the distance between the jet entry location and the dam should not be too long so as to decrease the plunge pool length and increase the energy dissipation rate per unit volume in the plunge pool.

3. Non-collision flood discharge from top spillways and middle orifices

3.1 Physical model

A physical model of a high-arch dam with top spillways and middle orifices was constructed at the

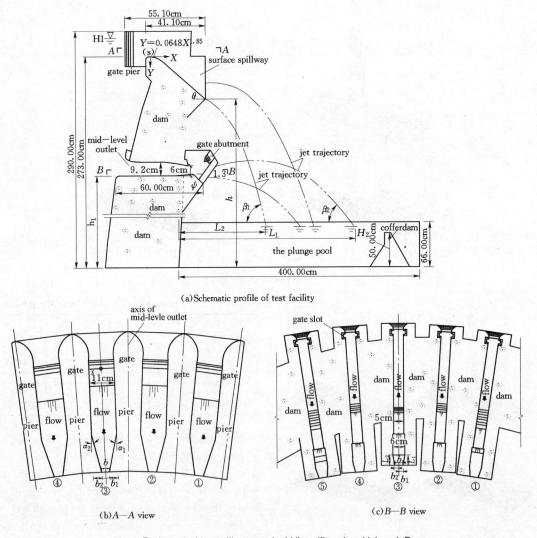

(a) Schematic profile of test facility

(b) A—A view

(c) B—B view

Figure 7 Layout of top spillways and middle orifices in a high arch Dam

State Key Laboratory of Hydraulics and Mountain River Engineering in Sichuan University, which simulates a prototype hydropower project with an arch dam in 305 meters high. The model includes the upstream reservoir, arch dam, plunge pool and downstream river. The model scale is 1 : 100 with the same Froude number between the model and the prototype.

The discharge system includes four top spillways and five middle orifices in the arch dam. All top spillways are of open style and free overflow, each of which has the width 11.0cm. The crest of overflow weir is 273cm high. The adjacent spillways are separated by gate piers 12.24cm wide. The overflow weir is of WES (Waterway Experiment Station) shape, with a water design H_d of 11.07cm neglecting the approaching velocity head. The pier type of top spillways is either normal gate piers or flaring gate piers,

whose configuration parameters and layout are schematically presented in Figure 7. The value of configuration parameters is list in Table 2. All spillway outlets are of rectangular shape. To solve the problem of radial concentration and unsymmetrical flow caused by the curvature of a high arch dam, asymmetrical flaring gate piers are applied to the two sides of the top spillways [see Figure 7 (b)].

The middle orifices are deployed below the gate piers of top spillways and the outlets are of either normal buckets or slit-type buckets shape. To control the landing of jets in the plunge pool and to avoid the collision between two adjacent jets, asymmetrical slit-type buckets are applied to the middle orifices [see Figure 7 (c)]. The parameters and layout are shown in Figure 7 and summarized in Table 3.

Table 2   **The configuration parameters of top spillways**

Body type		h (cm)	θ	α_1	α_2	b_1 (cm)	b_2 (cm)	b (cm)
Normal gate piers	①	255.34	$-30°$	0	$-3°$	0	-1.48	12.48
	②	262.00	$0°$	$-3°$	$-3°$	-1.48	-1.48	13.96
	③	257.50	$15°$	$-3°$	$-3°$	-1.48	-1.48	13.96
	④	255.34	$-30°$	$-3.0°$	0	-1.48	0	12.48
Flaring gate piers	①	252.40	$-30°$	$14°$	$16°$	3.00	4.50	3.48
	②	244.93	$-40°$	$15°$	$15°$	3.76	3.76	3.48
	③	244.93	$-40°$	$15°$	$15°$	3.76	3.76	3.48
	④	252.40	$-30°$	$14°$	$16°$	4.50	4.00	3.48

Note   The positive value of α represents the contraction, and the negative one represents the diffusion; the positive value of b_1 or b_2 represents broaden of piers and the negative one represents the contraction of piers.

Table 3   **The configuration parameters of middle orifices**

Body type		h_1 (cm)	h_2 (cm)	$\psi(°)$	b_1 (cm)	b_2 (cm)	b (cm)	l_1 (cm)	l_2 (cm)
Normal bucket	①	197.00	194.00	-12	3.0	3.0	6.0	0	0
	②	195.50	194.00	-5	3.0	3.0	6.0	0	0
	③	194.00	195.00	5	3.0	3.0	6.0	0	0
	④	195.50	194.00	-5	3.0	3.0	6.0	0	0
	⑤	197.00	194.00	-12	3.0	3.0	6.0	0	0
Slit-type bucket	①	197.00	194.00	-12	2.80	1.20	4.0	1.40	8.00
	②	195.50	194.00	-5	2.40	1.60	4.0	2.60	6.00
	③	194.00	195.00	5	2.00	2.00	4.0	4.00	4.00
	④	195.50	194.00	-5	1.60	2.40	4.0	6.00	2.60
	⑤	197.00	194.00	-12	1.20	2.80	4.0	8.00	1.40

Note   The positive value of ψ represents ski-jump angle, and the negative one represents the plunge angle.

The plunge pool is 400cm long and 112cm widen with a trapezoidal cross – section. In order to measure the dynamic pressure on the slab caused by the plunging jets, totally 507 piezometric tubes of diameter 1cm were installed on the right of the pool slab. The overall arrangement of physical model is shown as Figure 8.

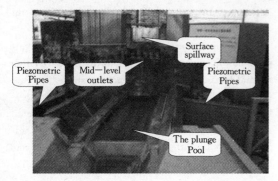

Figure 8   Experimental Setup

Two tests were conducted as Case 1 jets with collision and Case 2 jets without collision. In Case 1, the normal gate piers (Table 2) were on the top spillways and the normal buckets (Table 3) on the middle orifices. In Case 2, the flaring gate piers (Table 2) were on the top spillways and the slit-type buckets (Table 3) on the middle orifices. As a result, jets will collide in air in Case 1 and not collide in Case 2. In both cases, the upstream water level is $H_1 = 287.60$cm and tail-water level is $H_2 = 63.69$cm, which is the extreme flood condition with 5000 year return period.

3.2   Jets in air

In Case 1, normal gate piers were on the top spillways and normal buckets on thre middle orifices. In this case, the jets from the top spillways spread in the lateral direction and mixed with each other at the edges. Further, the jets from the middle orifices collided strongly with the jets from the top spillways, which resulted in strong spray on both sides of the plunge pool [see Figure 9 (a)].

In Case 2, flaring gate piers were on the top spillways and slit-type buckets on the middle orifices. By using the flaring gate piers which contract the discharge channel on the top spillways in the lateral direction, the jets were confined laterally and elongated fully in vertical and longitudinal directions. As the results, the jets from the top spillways did not mix with each other, and there was a rather

(a)Case 1, jets with collision   (b)Case 2, jets without collision

Figure 9   Flow pattern

large space between the jets. Since slit – type buckets were applied to the middle orifices, the jets were also narrow and there was a rather large space between the jets too. With the particular layout of the top spillways and the middle orifices (Figure 7), the free jets from the top spillways crossed over the jets from the middle orifices without collision in air and caused no strong spray [see Figure 9 (b)]. The careful plunging jets sketches measured in experiment are shown as Figure 10 (b).

To further quantify the jet pattern, the parameters of plunging jets measured for these two cases are presented in Table 4. The length of jet-impact zone in the plunge pool, which is defined as $L = L_1 - L_2$ (see Figure 7), is 139.40 cm in Case 2. This is larger than that of Case 1 ($L = 97.0$cm) because the jets from the top spillways with flaring gate piers and the middle orifices with slit-type buckets increased in longitudinal distance as narrow, high and long jets and had no collision. The area of jet-falling zone is 0.769m² in Case 2, which is much larger than that of Case 1 ($A = 0.623$m²). The angle of jet impingement is smaller than that of Case 1.

3.3   Dynamic pressure on the slab of plunge pool

A jet plunging into the plunge pool can be considered as an oblique submerged jet (Jeffrey, 1998), the velocity of which reduces rapidly with the increasing depth of water cushion. This process in turn reduces the dynamic pressure on the plunge pool slab. The water in a plunge pool can be divided into three zones: shear zone, impact zone, and mixing zones. The jets interact with each other in the plunge pool and form a very complex turbulent flow pattern. In this highly turbulent flow, the mean dynamic

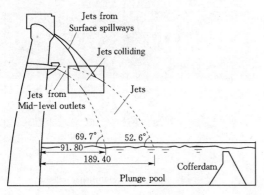

(a)Case 1 : Jets with collision

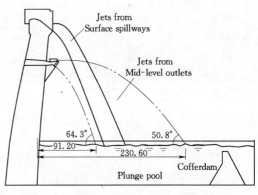

(b)Case 2 : Jets without collision

Figure 10   Sketches of plunging jets

Table 4   **The parameters of plunging jets**

Parameter	L_1 (cm)	L_2 (cm)	L (cm)	W (cm)	A (m²)	β_1 (°)	β_2 (°)
Case 1	189. 40	91. 80	97. 60	63. 80	0. 623	69. 6	52. 6
Case 2	230. 60	91. 20	139. 40	55. 20	0. 769	64. 3	50. 8

Note $L = L_1 - L_2$, W = the width of jet-falling zone; A = the area of jet-falling zone, $A = L \times W$.

pressure caused by jets hitting on the pool slab can be written as

$$\overline{p} = \frac{p_{max} + p_{min}}{2} - \rho g h_1 \qquad (1)$$

where $\overline{p}$ = the mean dynamic pressure on the pool slab; p_{max} and p_{min} are respectively the maximum and the minimum total pressure measured on the pool slab; h_1 is the still water depth in the plunge pool; g

is the gravitational acceleration; ρ is the density of water.

Further, $\Delta p = p_{max} - p_{min}$ reflects the pressure amplitude. The measured maximum $\overline{p}$ and Δp on the slab of plunge pool in two cases are listed in Table 5, and the contour lines of $\overline{p}$ is shown in Figure 11, the contour lines of Δp in Figure 12.

Table 5   **The maximum $\overline{p}$ and Δp on the slab**

Parameter	$\overline{p}_{max}$ (kPa)			Δp_{max} (kPa)		
Case 1	$x = 205$	$y = 10$ or -10	1. 09	$x = 205$	$y = 10$ or -10	2. 40
Case 2	$x = 180$	$y = 19$ or -19	0. 60	$x = 185$	$y = 43$ or -43	1. 58

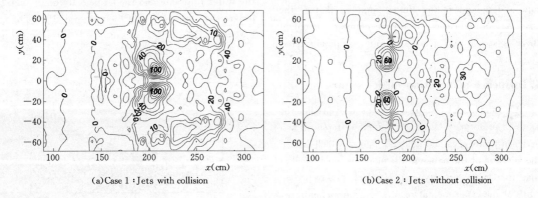

(a)Case 1 : Jets with collision   (b)Case 2 : Jets without collision

Figure 11   Contour lines of $\overline{p}$ on the slab

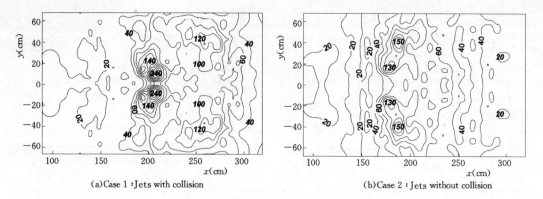

(a)Case 1 :Jets with collision　　　　　　(b)Case 2 : Jets without collision

Figure 12　Contour lines of Δp on the slab

The mean dynamic pressure is related to the discharge per unit area on the plunge pool and the jet impingement angle. The larger the discharge per unit area and the jet impingement angle are, the stronger the dynamic pressure will be. In this paper, the overall discharge and the impingement angle are about the same for both cases, and thus the larger the area of jets-falling zone in the plunge pool is, the smaller the dynamic pressure will be. From Table 4, the area of jet-falling zone is 0.769m² in Case 2, which is larger than that in Case 1. Therefore the maximum mean dynamic pressure $\bar{p}_{max}$ in Case 2 is only 0.60kPa, which is smaller than that of in Case 1, 1.09kPa (Table 5). In Case 2, the mean dynamic pressure on plunge pool slab is significantly reduced when compared to Case 1, due to the wide spread of jet-falling area.

In Figure 12, the pulsating pressure Δp in Case 2 is smaller than that in Case 1, and the maximum pulsating pressure Δp_{max} in jet impact zone is only 1.58kPa in Case 2, however the Δp_{max} is 2.40kPa in Case 1. So there is less surface turbulence observed in the pool in Case 2 than that of Case 1 in test. The zone of large Δp is the almost same as that of the large $\bar{p}_{max}$ in two cases, where the action of turbulent shear is strong and a large energy dissipation takes place.

4. Influence of aeration on jet scour

Can the model test results be used to predict the scour in prototype? Obviously, some differences xist between prototypes and models, one of which is that the aeration of the nappe in the model is much weaker than that in the prototype. This affects the comparability of the experimental results of scour. Therefore, the influence of aeration on scour needs further study.

4.1　Experimental equipment

The experimental study was conducted in a glass flume which is 3m length, 0.22m width, and 0.8m depth (see Figure 13). A set of nozzle systems was used to produce the plane jets. The discharge from a polymethyl methacrylate pipe with a rectangular cross section was aerated before reaching the nozzle. Aeration was through many fine pores so as to produce a uniformly aerated jet. The water flow rate and the air flow rate were measured by a weir and an air flow meter, respectively. A valve was used to control the water flow rate. The jet angle was set at 60°, and the jet velocity (mean mixture velocity) was varied from 1.4m/s to 2.1m/s. The bed material had particle sizes between 10 and 15mm and particle density of 2600kg/m³. The air concentration varied from 3% to 30%. The tailwater depth was controlled at 0.26, 0.36, and 0.45m, respectively. First, the experiment on the nonaerated jet was made, and then the nozzle was changed for the aerated jet so that both the water flow rate and the jet velocity of the aerated jet are the same as that of the nonaerated jet. The wa-

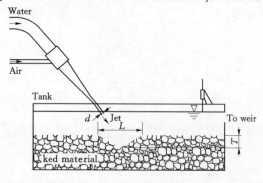

Figure 13　Experimental equipment of aerated jets

ter flow rate and the jet velocity were controlled by adjusting the valve opening and the head drop, respectively. The nozzle was submerged slightly in water to avoid additional air being entrained at the water surface by the jet. In experiment, the scour lasted 8h so that a stable scour hole can be formed and equilibrium scour depth is measured.

4.2　Scour depth of aerated jet

The aerated and the nonaerated jets are compared under the same conditions of jet velocity, tailwater depth, and discharge per unit width, which is different from the case in Mason's experiments (Mason, 1989) where the air concentration increases with increasing jet velocity. The main experimental results are shown in Table 6, where d is jet thickness; C_a is air concentration of jet [air flow rate/ (water flow rate air flow rate)]; V_0 = mean mixture velocity at the outlet of the nozzle (calculated by the flow rate and the outlet area); and L is length of the scour hole.

Table 6　　　　　　　　　　　　　　**Experimental results of scour**

h=0.26m					h=0.36m					h=0.45m				
d (mm)	C_a (%)	V_0 (m/s)	T (m)	L (m)	d (mm)	C_a (%)	V_0 (m/s)	T (m)	L (m)	d (mm)	C_a (%)	V_0 (m/s)	T (m)	L (m)
20.5	0	1.463	0.092	0.32	20.5	0	1.762	0.102	0.43	20.5	0	1.744	0.088	0.43
22.1	7.3	1.461	0.069	0.29	22.1	7.1	1.760	0.070	0.34	22.1	7.3	1.760	0.047	0.34
23.3	11.9	1.465	0.053	0.27	23.3	12.0	1.765	0.052	0.33	23.3	11.9	1.753	0.022	0.28
20.5	0	1.751	0.167	0.56	25.5	19.7	1.768	0.019	0.21	20.5	0	2.048	0.156	0.60
22.1	7.2	1.747	0.130	0.41	20.5	0	2.060	0.213	0.67	22.1	7.2	2.044	0.112	0.46
23.3	11.9	1.756	0.105	0.37	22.1	7.3	2.054	0.146	0.535	23.3	11.9	2.053	0.073	0.40
25.5	19.3	1.751	0.088	0.35	23.3	12.9	2.074	0.097	0.43	20.5	0	2.386	0.277	0.88
20.5	0	2.064	0.240	0.73	25.5	19.6	2.059	0.061	0.34	22.1	7.2	2.389	0.167	0.62
22.1	7.0	2.051	0.205	0.62	25.5	0	1.475	0.074	0.35	23.3	11.9	2.410	0.138	0.54
23.3	11.9	2.057	0.183	0.535	26.4	3.4	1.472	0.065	0.325	25.5	19.4	2.388	0.067	0.38
25.5	19.8	2.068	0.143	0.45	28.0	10.6	1.499	0.058	0.305	28.0	26.7	2.393	0.033	0.31
25.5	0	1.478	0.137	0.43	30.0	14.9	1.471	0.040	0.30	25.5	0	1.763	0.125	0.52
26.4	3.3	1.471	0.118	0.37	33.3	23.5	1.480	0.034	0.30	26.4	3.4	1.766	0.080	0.46
28.0	8.7	1.475	0.105	0.345	25.5	0	1.763	0.180	0.58	28.0	8.9	1.755	0.068	0.38
30.0	15.1	1.471	0.087	0.33	26.4	3.4	1.766	0.153	0.51	31.6	18.8	1.752	0.050	0.34
33.3	23.5	1.473	0.072	0.305	28.0	8.7	1.768	0.098	0.39	33.3	22.7	1.746	0.046	0.34
25.5	0	1.766	0.237	0.69	30.0	15.1	1.764	0.095	0.405	25.5	0	2.063	0.216	0.74
26.4	3.3	1.761	0.209	0.60	33.3	23.4	1.765	0.048	0.30	26.4	3.4	2.045	0.176	0.67
28.0	8.8	1.770	0.187	0.55	25.5	0	2.056	0.273	0.85	28.0	9.0	2.050	0.147	0.53
30.0	15.0	1.763	0.157	0.50	26.4	4.8	2.090	0.236	0.76	31.6	18.9	2.049	0.080	0.405
33.3	23.5	1.772	0.110	0.405	28.0	8.9	2.067	0.187	0.665	33.3	23.4	2.049	0.063	0.37
					30.0	14.6	2.055	0.154	0.525	36.5	30.1	2.050	0.025	0.29
					33.3	23.3	2.060	0.104	0.43					

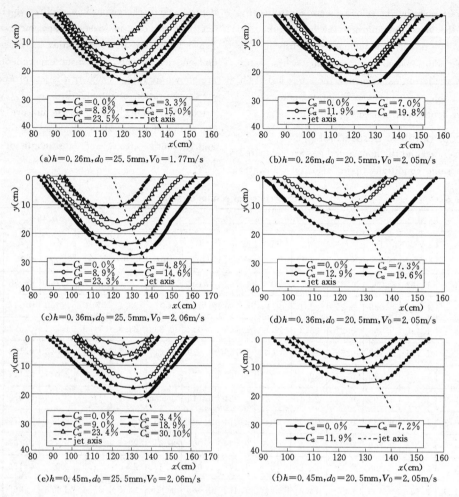

(a)h=0.26m,d_0=25.5mm,V_0=1.77m/s

(b)h=0.26m,d_0=20.5mm,V_0=2.05m/s

(c)h=0.36m,d_0=25.5mm,V_0=2.06m/s

(d)h=0.36m,d_0=20.5mm,V_0=2.05m/s

(e)h=0.45m,d_0=25.5mm,V_0=2.06m/s

(f)h=0.45m,d_0=20.5mm,V_0=2.05m/s

Figure 14　Measured scour holes

The profles of scour holes are plotted in Figure 14, where y is vertical distance from the original bed surface and the jet axis shows the direction of the nozzle. It can be seen that the difference between the scour holes caused by aerated jets and those by nonaerated jets is very obvious. In order to obtain the quantitative effect of aeration on scour depth, the relative scour depth T/T_0 is plotted against the air concentration of jet, C_a, where T and T_0 are depths of the scour holes caused by the aerated and the nonaerated jets, respectively, and measured from the original bed surface. Figure 15, Figure 16, and Figure 17 show the relationship of T/T_0 and C_a for the tailwater depth of 0.26, 0.36, and 0.45m, respectively. In these three fgures, the solid and the hollow points denote d_0=20.5mm and d_0=25.5mm, respectively, where d_0 is thickness of nonaerated jets

or the equivalent water thickness of aerated jets (i.e., the thickness of the nonaerated jet with the same water flow rate and jet velocity). From the experimental data points, the following relationship is obtained:

$$\frac{T}{T_0} = (1-C_a)^a \qquad (2)$$

In Eq. (2), the index a is equal to 2.5, 4.0, and 5.5, respectively, when the tailwater depth, h, is 0.26, 0.36, and 0.45 m. Figure 15-Figure 17 show little influence on the jet velocity, the equivalent water thickness of jets, and the water discharge per unit width on the relationship of T/T_0 and C_a. Under the conditions of the same tailwater depth and different jet velocity, equivalent water thickness of jets, and water discharge per unit width, the influence of aeration on the scour depth is nearly the same. It means that the index a in Eq. (2) is mainly related to the tailwater depth. On the basis of the experimental data, a can be expressed as a function of tailwater depth, i. e.,

$$a=\varphi\left(\frac{h}{h_0}-1\right) \qquad (3)$$

where h_0 is initial tailwater depth affected by aeration (i. e., when $h < h_0$, the influence of aeration on scour is neglectable) and approximately equal to 0.1 m; φ=coefficient and approximately equal to 1.57.

The results calculated by Eq. (2) and Eq. (3) are compared with the experimental data in Figure 18. The data measured by Rahmeyer (1990) are also compared with the calculated results in this figure. They are in reasonable agreement. Further, the effects of the particle diameter, the tailwater depth, and the jet velocity on the relationship of T/T_0 and C_a need to be analyzed. As mentioned above, the jet velocity has little influence on this relationship. The effect of the tailwater depth has been expressed by Eq. (3). The power of the particle diameter has been found inversely proportional to the scour depth, as shown by Mason (1985, 1989); therefore, the relative scour depth T/T_0 is independent of the particle diameter.

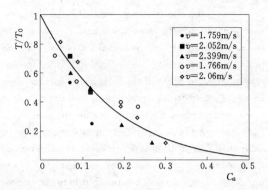

Figure 17   Relationship of $T/T_0 - C_a$ ($h=0.45$m)

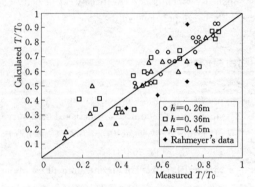

Figure 18   Comparison of measured and calculated T/T_0

5. Conclusions

The water body in plunge pool can be divided into three regions: shear, impact and mixing dissipation regions of energy. The way of the optimization of plunge pool and dam outlets should be to expand the range of the shear dissipation region (e. g. dispersing the nappes), reduce the impact intensity in the impact dissipation region (e. g. forming moving water cushion), and make use of the water body in the mixing dissipation region as much as possible.

A new type of flood discharge through a high arch dam was developed, i. e., contracting dissipators with flaring gate piers in top spillways and slit-type buckets in middle orifices, and tested the design in a physical model of high-arch dam. The experiment data show that the jets from the top spillways could cross through the jets from the middle orifices without collision, and thus significantly reduced the spray generation. Furthermore, this design could also make the jets diffuse fully in longitudinal direction and enlarge the total effective area of jets impact in the plunge pool so that the maximum mean hydrodynamic

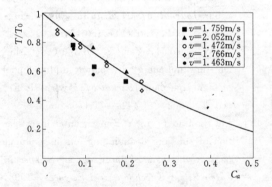

Figure 15   Relationship of $T/T_0 - C_a$ ($h=0.26$m)

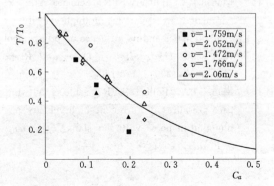

Figure 16   Relationship of $T/T_0 - C_a$ ($h=0.36$m)

pressure on the pool slab was reduced, and the pressure fluctuation was reduced as well. It proves that the new design of energy dissipator is feasible and provides a good alternative for flood discharge and energy dissipation for high arch dams in narrow river channels where strong spray should be avoided.

Aeration can reduce the scour depth. Therefore, the results of scour experiment in laboratory will be safer than that in prototype, because the aeration of the jets in the model is much weaker than that in the prototype. A quantitative relationship between the air concentration of jet and the relative scour depth was established, which is not affected by jet velocity and water discharge per unit width.

References

[1] Mason PJ. The choice of hydraulic energy dissipator for dam outlet works based on a survey of prototype usage. *Proc Inst Civ Engrs*, 1982, 72 (1): 209-219.

[2] Mason PJ. Effects of air entrainment on plunge pool scour. *Journal of Hydraulic Engineering-ASCE*, 1989, 115 (3): 385-399.

[3] Mason PJ, Arumugam K. Free jet scour below dams and flip buckets. *Journal of Hydraulic Engineering-ASCE*, 1985, 111 (2): 220-235.

[4] Merel T. W. The world's Major Dams and Hydro Plants. *Water Power and Dam Construction*, 1991: 67-77.

[5] Bollaert E, Schleiss A. Scour of rock due to the impact of plunging high velocity jets Part I: A state-of-the-art review. *Journal of Hydraulic Research*, 2003, 41 (5): 451-464.

[6] Bollaert E, Schleiss A. Scour of rock due to the impact of plunging high velocity jets Part II: Experimental results of dynamic pressures at pool slabs and in one-and two-dimensional closed end rock joints. *Journal of Hydraulic Research*, 2003, 41 (5): 465-480.

[7] Liu P, Li A. Model discussion of pressure fluctuations propagation within lining slab joints in stilling basins. *Journal of Hydraulic Engineering-ASCE*, 2007, 133 (6): 618-624.

[8] Ying D, Rao H, Su W. Flood Discharge and Energy Dissipation Design of Ertan Arch Dam. *Design of Hydroelectric Power Station*, 1998, 14 (3): 27-31.

[9] Liu Z. The prototype observation research report of discharge a atomization at Ertan hydroelectric station. *Beijing, China: China Institute of Water Resources and Hydropower Research*, 2000, 03.

[10] Yang M, Cui G. Study on control factors of stability for plunge pool slabs. *Journal of Hydraulic Engineering*, 2003, 34 (8): 6-10.

[11] Jeffrey G. Bohrer, Steven R. Fellow, and Rodney J. Wittler. Predicting plunge pool velocity decay of free falling, rectangular jet. *Journal of Hydraulic Engineering-ASCE*, 1998, 124 (10): 1043-1048.

[12] Xu W, Deng J, Qu J etc. Experimental investigation on influence of aeration on Plane Jet Scour. *Journal of Hydraulic Engineering-ASCE*, 2004, 130 (2): 160-164.

[13] Xu W, Liao H, Yang Y and Wu C. Turbulent flow and energy dissipation in plunge pool of high arch dam. *Journal of Hydraulic Research*, 2002, 40 (4): 471-476.

[14] Liu P, Gao J, Li Y. Experimental investigation of submerged impinging jets in a plunge pool downstream of large dams. *Science in China Series E-technological Sciences*, 1998, 14 (4): 357-365.

[15] Manso PA, Bollaert EFR, Schleiss AJ. Experimental investigations on high-velocity jet characteristics and its influence on plunge pool rock scour. *Hydraulics of Dams and River Structures*, 2004: 173-180.

[16] Manso PA, Bollaert EFR, Schleiss AJ. Impact pressures of turbulent high-velocity jets plunging in pools with flat slab. *Experiments in Fluids*, 2007, 42 (1): 49-60.

Design of Data Transmission Networks Based on CAN Bus for Physical Model Control System

Zhang Hongwei

Estuarine and Coastal Science Research Center, Ministry of Transport of People's Republic of China, Shanghai, 201201

Abstract: This paper analyzes the defects of the data transmission network for traditional physical model electronic system, and introduces the data bus technology in industrial automation. Based on CAN field bus, this paper presents a kind of hardware design scheme of data transmission network for physical model control and data acquisition system. The proposed scheme describes the overall architecture, distribution and interface circuit of the monitoring computer, interface circuit of the field nodes in detail. It is pointed out that the scheme of CAN bus network is an effective and efficient method of information transmission, and it can efficiently simplify system structure, raise reliability, and improve performance of whole system; it can meet the practical need of physical model control and data acquisition system.

Key words: physical model; electronic system; data bus; information transmission; CAN bus

With deeper development of the practice of waterway and harbor engineering, more and more measuring items will be completed during the corresponding physical model experiments. As a result, the number of electronic devices, specialized sensors, and executive equipments, will increase rapidly. If the traditional technology of signal cable layout is still adopted, much more cables will be used to construct an electronic system for a model and the whole cable networks will be more complicated. Thus, it will weaken the reliability of the system and will make it difficult to maintain. To solve this problem, this paper presents a kind of hardware design scheme of data transmission network based on CAN (Controller Area Network) bus for physical model control and data acquisition system. The proposed scheme describes the overall architecture, distribution and interface circuit of the monitoring PC, interface circuit of the field nodes in detail. The main aim of this paper is to make the physical model electronic system digitalized and become more intelligent.

1. Introduction of data bus technology

Data bus is a communicative system composed of bus control devices, physical media, topology of the network and the communication protocol. Information transmitted through the bus can be shared by every node connected to the bus, which make the whole system become more efficient. Essentially, data bus is computer local network technology applied in the field of automatic control and data acquisition[1]. The main characteristics of the data bus can be concluded as the following points.

1.1 High reliability

In order to prevent the network being paralyzed and make sure the data bus working reliably in bad conditions, the technology of reliability has been strengthened in design of hardware and communication protocol. Many practical technologies, such as isolation, redundancy, mechanism of error detection and fault recovery, have been taken to lower error probabilities of the bus.

1.2 Real time

The data bus generally have higher information transmitting speed and shorter message frames, so it takes less time to complete transmitting a message to meet the demands of real time control.

1.3 Instruction/Response communication protocol

The data bus controller controls the process of

instruction generating and transmitting and the relevant terminals responds to the instruction. The broadcast type topology is usually adopted to form the network so as to manage and supervise the data bus.

In present physical model, especially the tidal model, much more electronic control devices and sensors are applied, so the overall electronic system of physical model based on data bus should be researched and developed.

2. Design scheme of data bus transmission network

2.1 Choice of data bus

As a complete scheme, data bus includes hardware products, topology and communication protocol. The concrete indexes are data transmission bit rates, number of nodes, maximum transmission distance, transmission delay time, transmission media, topology structure and communication protocol. Within the above indexes, some real time parameters including data transmission bit rates can be decided by calculating the data transmission amounts of the relevant channel; the number of nodes and data transmission distance can be decided by the network layout. To design a proper network for physical model electronic system, we list several popular data buses applied in present industrial automation field in Table 1. MIL-STD-1553B bus has a high reliability, but its expensiveness limits it to military and aerospace engineering. Lonworks bus adopts all 7 layer communication model of ISO/OSI, but its flexibility is inferior to other data buses[2]. Due to the excellent performance comparable to the above two buses and the lower cost, CAN bus is used more widely than them. Therefore, we choose CAN bus to design data transmission network. As a data bus of short frame, CAN bus can be utilized flexibly by modifying its protocol easily.

Table 1 Typical data bus type applied in present industrial automation field

Data bus	Protocol mode	Transmission media	Typical speed (Mbit/s)
MIL-STD-1553B	Instruction/response	Twisted pair cables	1.0
Lonworks	Instruction/response	Coaxial cable/twisted pair cables	0.3
CAN	CSMA/CD	Coaxial cable/twisted pair cables	1.0

2.2 Data transmission network for physical model electronic system

The overall physical model electronic system can be divided into control sub-system and data acquisition sub-system. The former is made to control all borders of the physical model, such as the tailgate, underwater pump, dual direction pump. Most cables in this system carry feedback and control signal, so they are stem wires of information transmission. The latter is made to acquire data from all sensors placed in every pointed site of the physical model, including water level meter, flow velocity meter and other specialized instruments. It should be noticed that instruments of data acquisition sub-system are much more than that of the control sub-system and they spread all over the physical model.

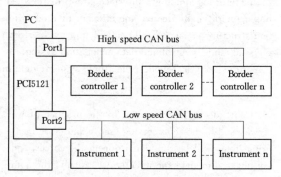

Figure 1   The sketch diagram of physical model data transmission network

Corresponding to the architecture of the physical model electronic system, the network is designed in two types with CAN bus, shown in Figure 1. The first type is a high speed data bus to transmit control information. All the individual electronic equipments in control sub-system are transformed with CAN interface circuit connected to CAN bus. To meet the requirements of real time control, data transmission rates of this branch is set to 250kb/s. The other type is low speed data bus with data transmission rates of 50kb/s. All the individual sensors in data acquisition sub-system are transformed into intelligent instruments with CAN interface circuit. Two bus branches with different data transmission rates are connected to a computer with a CAN communication card PCI5121 which has two CAN interface ports. In this way, high speed CAN bus is beneficial to guarantee the real time quality of the boarder controllers and low speed CAN bus extends the communicating distance of data acquisition instruments with better anti-interference and lower cost.

2.3 Design of CAN bus interface circuit

All electronic controllers and instruments connected to CAN bus form a local network demanding higher reliability. Hence, it is important to design CAN bus network of the system and the design of CAN interface circuit is the key, including choice of CAN controller, CAN trans-receiver and anti-interference devices[3].

2.3.1 *Design of interface circuit between computer and CAN bus*

PC is the monitoring node of the system. In order to add CAN communication function to PC, we select PCI5121 card produced by ZHIYUAN Company. Figure 2 shows the sketch diagram of PCI5121 and its application in the system.

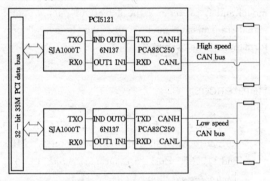

Figure 2 Hardware diagram of the monitoring PC node

The PCI5121 card is a CAN bus communication adaptor of high performance. It provides a good and convenient solution on connecting PC to the CAN bus. It adopts the standard PCI ports to adapt to the high speed data communications, and according with the PCI 2.1 standard it supports the PNP (Plug In Play) feature. It contains a double electronic isolated CAN channels. In each channel, SJA1000T is selected as CAN controller, 6N137 as high speed electro-optical coupler, and PCA82C250 as CAN transceiver. Each channel of PCI5121 is optical isolated to prevent the PC from the damages caused by the eddy current of ground collars, and this improves its reliability in the bad working environments.

The adaptor also adopts the 4 layer PCB designs and the SMT manufacturing technique to improve the reliability for long period workings. Applied in the system of this paper, one port of PCI5121 is connected to control sub-system designed as high speed data bus network, with the speed of data transmission up to 250kb/s. The other one is connected to data acqui-

sition sub-system designed as low speed data bus network, with the speed of data transmission up to 50kb/s. There are two terminal resistors added to two endpoints of each CAN bus to improve the reliabilities of the CAN communications. The value of the resistor is 120Ω for twisted pair cables.

2.3.2 *Design of interface circuit between field nodes and CAN bus*

There are many kinds of field nodes distributed in physical model electronic system. Although they serve as different functions, their interface circuit connected to CAN bus should be the same. In this paper, we select AnyCAN5401 module to design every field node regardless its belonging to control sub-system or to data acquisition sub-system.

AnyCAN5401 is an intelligent module produced by ZHIYUAN Company embedded with CAN Communication circuit. This module adopts 32-bit microcontroller as its control core, with 16 I/O channels and 4 ADC channels, and supports CAN2.0 protocol, which makes the designer flexibly realize control and data acquisition scheme based on CAN bus. There are many advanced technologies applied in the module, including multi-layer PCB routine, high resolution low noise power components, high speed signal isolating chips, ESD bus protection devices, and stand-alone hardware watchdog, which guarantee the module run stably in bad conditions of physical model.

The hardware sketch diagram of field node is shown in Figure 3. Besides the core chip, the node includes CAN interface, AD input, PWM output, serial port and reset circuit. CAN transceiver has already been integrated into AnyCAN5401 and can be directly connected to the network. There are 4 AD input of the module and its resolution is 10 bits. Reset circuit can be realized by connecting a button to the configuration pin of AnyCAN5401. In order to configure parameters of AnyCAN5401 or read parameter information, it is necessary to design serial port communication terminal connected to the relevant pins of AnyCAN5401. The serial communication baud rates of AnyCAN5401 is set to 9600 bit/s, through which every field node should be configured with configuration tool installed in PC before its running. System based on CAN bus can autonomously switch off defect nodes, so it is unnecessary to stop the whole network while adding or reducing a single field node and it is easy to maintain and expand the system.

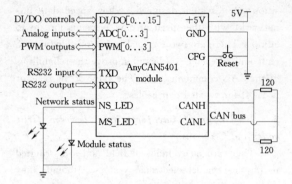

Figure 3   Hardware diagram of the field nodes

3. Analysis of bus transmission and its influence to electronic system

The essence of system based on data bus is its real time network. Compared to most computer centralized electronic control and data acquisition system, system based on bus transmission network distributes tasks of control and data acquisition to every controller and instrument connected to the bus, such as D/A, A/D, DI and DO that are executed by computer in the former system. Although digitalization of information transmission in bus is realized in this way, time of transmission delay increases too. Bus transmission delay is influenced by hardware performance and communication protocol that are reflected by indexes as response speed, transmission bandwidth and transmission efficiency. Accelerating as the bus transmission speed is, it still has time delay compared to point-to-point transmission with specialized wires. For example, more frequent data acquisition of sensors will lead to heavier data transmission load and influence of transmission delay will be obvious. The same problem can also be caused by suddenly emerged data transmission, and seriously data jam will happen[4-6]. Hence, before the scheme of system is decided we must make a thorough consideration of the bus transmission capability and select proper data acquisition frequency. During computing a period of sample and control, bus transmission should be paid enough attention to. An effective method is to enhance the computing speed, which can shorten the computing time delay and compensate for bus transmission delay. Moreover, a reasonably designed architecture of network can also balance the data flow of the bus.

4. Conclusions

In this paper, we introduced a scheme of data trans-

mission network based on CAN bus for a new type distributed electronic system of physical model. Due to less cables and connectors used in the system, it is easier to construct the network to control and coordinate every sub-system more comfortably. For CAN bus, it is also easy to adopt redundant data bus technology with low data transmission attenuation and high anti-interference. All these make it easy to maintain the system and enhance its reliability.

Data transmission network based on CAN bus is also an open system. When some new measuring items are required in physical model, relevant instruments can be included in the system quickly as long as they are equipped with CAN interface and compatible with the communication protocol.

In general, data bus transmission network is an important portion for the electronic system of physical model and it overcomes the defects of point-to-point signal transmission mode, enhancing the overall performance of the system.

References

[1]   Lian Baowang, Li Yong, Zhang Yi. The Performance Analysis and Comparison on CAN Bus and 1553E Bus. *Measurement and Control Technology*, 2000, 19 (6): 47-49.

[2]   Zhu Zhengli, Yin Chengliang, Zhang Jianwu. Application of CAN Bus System in Hybrid Electric Vehicle. *Journal of Traffic and Transportation Engineering*, 2004, 4 (3): 90-94.

[3]   Li Guangzhong, Wu Shitao. The Design of Intelligent Nodes Based on CAN Bus. *Microcomputer Information*, 2009 (26): 72-73.

[4]   Long Zhiqiang, Liu Shusheng, and Cao Chengkan. The Research about Communicatiao Network of Maglev Train Based on CAN Bus. *Electrical Drive*, 2002 (2): 44-47.

[5]   Yao Maode, Chen Jinping. Design of Data transmission Networks Based on CAN Bus for Automobile Electronic System. *Journal of Chang'an University*, *Natural Science Edition*, 2006, 26 (1): 86-89.

[6]   Hu Yonghong. Study of UAV Network System Based on CAN Bus. *Computer Measure & Control*, 2009, 17 (12): 2479-2481.

Role of Physical Models in River Development and Protection

Liu Tonghuan[1,2]   Yao Shiming[2]

[1]*State Key Laboratory of Water Resources and Hydropower Engineering Science*,
Wuhan University, *Wuhan*, 430072;
[2]*Changjiang River Scientific Research Institute*, *Wuhan*, 430010

Abstract: For a long time, physical model has been taken as an important measure to realize the target of river regulation and development. Based on a summary for the development history of physical models in China and abroad, this article briefly discusses the role that physical model has been playing in construction of large hydraulic projects, river regulation, utilization and development, river outlet and seashore development, and addresses the critical simulation techniques to be further studied and solved, including distorted models, similarity of bed resistance, model sediment, model measuring techniques and so on.

Key words: physical model; river regulation; river protection

1. Introduction

The evolvement of river bed is resulted from the interaction between water sand and boundary of river way. The movement process of water sand and the boundary of river way both change per the space time, which is how the different types of river course evolved. This is true either for the different rivers or for different sections of the same river. Up to now, the cognition on the evolvement regularity of river course is basically limited to qualitative description only and there is little achievement in quantitative research. Due to the complicity of riverbed evolvement, we cannot completely reply on the analytic and mathematic method to determine how much the river regulation project is affected by the bed load and in large hydropower project and the evolvement trend of a particular river section but need resort to physical model for in-depth research in order to solve the related technical difficulties we encountered in river regulation, development and protection. Since the founding of PRC, especially after China's opening and reform, the physical model test has been playing a significant role in regulation, development and protection in river like Yangtze River, Yellow river, Zhujiang River and Huaihe River. Based on a summary for the development history of Physical Model, this article focuses on the description for the role that physical model plays in construction of large hydraulic-power water conservancy and hydraulic power project, regulation, development and protection of river course and utilization of river seashore. The further technical difficulties to be solved by physical model are also addressed here.

2. Development history of physical model

Physical model test is a method to study the water flow structure, evolvement regularity of river bed and the outcome of engineering solution using the knowledge of river dynamics to simulate the boundary and dynamic conditions similar to the real model according to the principle that the water flow is similar to the movement of bed load from the mechanical perspective[1]. Physical model test has been started more than one century ago. Early in 1686, Newton had discussed the condition for flow movement similarity and predicted the establishment of similarity theory in his book "Mathematic Principle of Natural Philosophy". In 1875, a French scientist, Fargue (L. J. Fargue) conducted the earliest physical model test in order to clean Garonne River in Bordeaux City. The model stands for a river bed sized like 1 : 100 horizon to vertical, the seashore is fixed and with sand put onto the river base, so there is no geometrical deformation. In

1885, an British scholar, O. Reynolds, carried out the tide influence test for Mersey River outlet in Liverpool City. During the tests, he used a distorted model which was triple-sized in horizon to vertical. In 1891, Dr. Hubert Engels from Dresden Technology College in Germany established the first hydraulic test lab in the world and did the physical model test in the glass water slot. In 1913, Engels built a much larger hydraulic test lab and had been doing the model test concerning the comprehensive renovation over decades including the physical model test for Yellow River commissioned by the Chinese government. Early 2000s saw the rapid development of hydraulic power projects and that of water conservancy research. physical model, as an effective method to solve the problems of water conservancy project, was highly recognized all over the world due to its excellent application merit and hence prevailed around the world quickly from Germany, and the research institute and hydraulic testing lab have been founded in some countries[2].

From Jun. to Oct. 1932, invited by Chinese Relief Committee for Flood Disaster, Engels conducted a model test in Germany assuming the constricted levee space for Yellow River. The test aimed to study the possibility to apply the theory of sand-dredging by constricting water into the construction practice. This test promoted the progress for Chinese establishment of hydraulic lab. For Yangtze River, the 3rd largest river in the world, the largest physical model test has been started since 1935. The central hydraulic testing institute has carried out the model test for improvement projects in Madang section along lower Yangtze River and that for Zhenjiang watercourse. From 1944 to 1945, the institute also conducted the model test for improvement projects in Xiaojibei and Xiaonanhai sections along the branch of upper Yangtze River. Since the founding of People's Republic of China, with the rapid development of hydraulic construction, the physical model test and research has been put in the agenda in order to provide the scientific basis for the planning and design for Yangtze River[3-4].

In 1950, Yangtze River Water Conservancy Committee cooperated with Wuhan University and performed a physical model test for flood division project along Jingjiang River for the research on washing test. In 1950s, Yangtze River Academy of Sciences separately established the distortion model for San-douping section to Chenglinji section, undistorted model for Sandouping section to Yichang section, cut-off model for lower Jingjing River.

In 1960s, coupled with flood control and regulation in critical sections, model tests were conducted for silt-releasing project in north Jingjiang section and cutting-off project for lower Jingjiang section and Hanjiang section. A reservoir head test was done for Three Gorges Project and a natural physical model test is conducted to study the evolvement regularity and cause for the sinuous shape of Jingjiang River.

In 1970s, with the construction of Gezhouba Dam hydraulic project, the domestic research institutes and colleges together built 4 dam models and 1 reservoir model in order to solve the problems of hydro-project layout, sediment treatment, waterway regulation and sediment passing through turbine. Since 1980s, many new questions about sediment have been raised around the verification and construction of Three Gorges Project. The domestic institute and colleges separately built 3 hydroproject models, 9 models of reservoir head and models for different sections in lower reaches of Three Gorges Dam per different scale ratio using different model sand, which achieved substantially scientific research and provided the sufficient input for the construction of Three Gorges Project. Beside Gezhouba Dam Project and Three Gorges Project, with the rapid development of the social economy, the domestic institute also built other hydro-project model, river restoration model, channel regulation model, bridge model, hydropower station intake model, dock and harbor model, utilization model of shoals and beaches, regulation model for entrance of Yangtze River, fundamental research model and flood control model etc., which gained the substantial research achievement and hence provided the scientific basis to ensure all the project running smoothly and prompted the economic development along Yangtze River.

The research and application of model sand of different weight ratio and particle size as well as the mass application of advanced model test instrument, control device and computer automation further improved and developed the model theory and practice. Now many physical model tests are undergoing in many project of development and protection of rivers, this takes a leading position in the world and many works have been published which plays a significant role in economic construction.

3. Role of physical model in river development and protection

3.1 Role of physical model in large hydro-project construction

China has the most abundant hydropower reserve in the world, which would be 526 kW. The energy. even if being exploited, would be equal to the energy from 700 million tons of standard coal annually. Before the founding of PRC, 42 sites of power station have been built or partially built to provide 360MW installation power. Since the founding of PRC, the hydropower construction get a rapid development and a lot of key power stations have also been built, including the Three Gorges Project which is the largest one along Yangtze River and rich in installation power up to 18.20 million kW·h. Cascade reservoirs are being formed by the multiple reservoirs built or being built along the main and side streams on Yangtze River, Yellow River and other rivers. During the verification and construction process, a lot of sediment problems showed up which directly affect reservoir life, reservoir inundation, evolution of reservoir head channel, normal operation of dam lock and power station as well as the river bed evolution and shipping safety etc.; all these factors are critical technical questions for the construction and operation of the large hydro-projects. Incoming with sediment problem of key water control project, Yangtze Scientific Research Institute has built the physical models for Gezhouba Dam conjunction area, dam area of Three Gorges Project and reservoir head, Xiaonanhai water control conjunction area along Yangtze River, Tingzikou water control conjunction area along Jialing river, Xinglong conjunction area along Han river, etc., per the scale and using the different model sand.

China is rich in hydroelectric potential, but the distribution is quite irregular and the precipitation and runoff amount differ much in different area. There is above 82% water resources but only 36% farmland along and in the south of Yangtze River. There is less than 18% water resource but up to 64% farm land in the north of Yangtze River, and the farmland over the Huang-Huai-Hai Watershed, which is most potential in grain production increase, takes 41.8% of the farmland over the country but only shares less than

5.7% water resource. To solve the tough problem of irregular distribution of water resource, we have constructed lots of long-distance and large scale water transfer project, including the projects diverting the Yellow River water to serve Qingdao, Tianjin, Shanxi and the middle route and eastern route of South-to-North Water-Transfer Project. Such long-distance projects often encounter some technical difficulties like river crossing, railway crossing and barrier crossing etc.

3.2 Role of physical model in practice of river course realignment

Under natural conditions, the evolution of river course, especially the sedimentation, is very complicated as distortion for river happens in both vertical and horizontal direction. To fully develop the function of river course including flood control, shipping and comprehensive service, the river regime needs to be controlled and the river course need to be realigned comprehensively. The main measures of river course-realignment related to flood control is comprehensive realignment measures including bank paving, cutting-off, river branch closure, regime re-alignment etc. In the practice of river course re-alignment, no matter Yangtze River, Yellow River or other rivers, they all adopt the method of physical model test to study the realignment solution and technique. For example, the physical model test has been adopted in regulation and development for the key section along the main stream of Yangtze River including downtown section in Chongqing, Upper Jingjiang section, Lower Jingjiang section, Yueyang section, Wuhan section, Jiujiang section, Anqing section, Tonglin section, Wuyu section, Ma On Shan section, Nanjing section, Zhenyang section, Yangzhong section, Chengtong section, and Yangze head section. Through the model test, some evolution and developments within a certain space and time range for these key sections as well as the effect and influence level have been discovered, which provides the scientific basis for the planning, design and comprehensive regulation on these river sections.

3.3 Role of physical model in practice of river development

Shoreline, shoals and beaches of river-way are all limited precious resources and they play an increasingly important role in the economic development of riverside areas. With the construction and development of Yangtze Economic Zone and the execution of national strategy to develop western regions, the higher and newer requirements for development and utilization of river resources have been raised by all departments and areas along Yangtze River. The development and utilization of shoreline mainly covers levee revetments, river-crossing bridges, tunnel, water intake, water outlet, dock and harbor, river-crossing underwater cable, etc. During river development and utilization, the physical model test method is often used to study the flow characteristics, Scour-Sedimentation characteristics before and after planned construction on certain river section and the influence on the flood control, river regime and navigation channel. Meanwhile, the model testing results provide bases for optimization of utilization of shoreline resources and project design plan and also propose corrective preventive actions accordingly for the adverse impact that may be resulted from river development and utilization. Surrounding river development and utilization, Yangtze Academy of Science has conducted the physical model test to study the shoreline utilization for some sections of Sichuan river in upper Yangtze stream, Wuhan section, Huangshi section, Jiujiang section, Tongling section and Wuhu sections as well as river-crossing bridges, river-crossing tunnel in Wuhan and Nanjing, river crossing channels, Cuntan and Peiling dock, which provides the technical support for the project design and assessment on the flood control influences.

3. 4   Role of Physical Model in Ecological Conservation

Besides the role that physical model plays in traditional river development and utilization, it also helps in the study on the related ecological and environmental problems, like river concentration field, temperature field, contaminant diffusion, prevention of Oncomelania Spread and Ecological Rehabilitation Project, etc. Surrounding the critical technical questions being raised with the emerging fields, some scientific institutes from home and abroad have conducted model test for cooling water discharge respectively from Guodian Taizhou Power Plant, Nanjing Madu Power Plant, Yueyang Power Plant, Guodian Tongling Power Plant, Huadian Wuhu Power Plant, Ma On Shan' Third Power Plant, model test for Sewage

Outfall Project for Panzhihua Coal Corporation, model test to study the impact to the flow characteristics in Floodgates of Water Diversion Works from the Yangtze River to the Hanjiang River, and model tests to study the Ecological Rehabilitation Project of Rhine in German and Mississippi River in the US. The achievements have been greatly used in scientific research, planning, project design and decision managements.

3. 5   Role of physical model in estuarine and coasts utilization

Since the founding of PRC, we have achieved a great development in utilization of estuarine and coasts, Speaking in details, the Deep Water Channel and port newly built or under construction includes Yangtze River deep channel, Shenhua Huanghua Port, Shanghai Yangshan Deep Channel, Hebei Caofeidian Port and Jiangsu Rudong Port. All these ports are complicated in natural conditions and hence the sediment problem comes as the biggest to be solved. Although the Silt-Discharge Movement in estuaries and coasts is more complicated than that in rivers, using physical model for critical problems, we can get to a more practical conclusion even than mathematical model. Our country have built physical model respectively for deep port in Yangtze River outlet, Sheyang River outlet, Yongjiang River outlet, Qiantang River outlet, Shenzhen river outlet, Shanghai Yangshan River outlet, Tangshan Caofeidian port, Tianjin deep port and Shamen Deep port. With these models, the test and research have been conducted to study the key technical problems including navigation, seashore utilization and adverse impact of sediment accumulation in harbor dock and provides the reference for solving seashore problems. The above work provides the precious experience for physical model of estuarine and coasts and contributes a lot to the development of estuarine and coast science.

3. 6   Solid model for flood control along Yangtze river

After the completion of reservoirs in upper stream of Yangtze River with Three Georges Reservoir at the core, the flood control capacity will be greatly improved. However, the contradiction between safe channel capacity and upper stream inflow still exists, considering the regulation of reservoir which has changed the natural condition of incoming sediment, as a result, the correlation between water and sediment in middle and lower stream of Yangtze Rivers as well as in lake has changed a lot, which brings new challenge to flood control in middle and lower stream of Yangtze River. To cope with impact of reservoir groups to flood control, navigation, ecology, environment, etc. , in 2003, after approval by State Planning Commission, Yangtze River Academy of Science under Yangtze River Water Conservancy Commission have organized and implemented a crucial research project—Yangtze River Flood Control Modeling Project with total investment up to 200 million RMB. At present, a model testing hall up to 60,000 has been built and equipped with the state-of-art test-measuring control system and digital simulation system, and a physical model has also been established to cover the range between Zhicheng (a main stream of Yangtze River) to Luoshan section (about 400 km long) and Dongting Lake area (including 3-outlet water splitting channel) . Up to now, with this model, lots of tests have been done with this model to study the water flow motion characteristics in Jingjiang River, flood evolving characteristics in typical food year and regularity of river channel scouring and sedimentation in Jingjiang Section in the early operation period of Three Georges Project. The achievements have been widely used in design of emergency project, waterway regulations. The next step is to use the model for study on changes of correlation between rivers and lakes and evolution and improvements of rivers and lakes in order to provide strong support to the improvements and flood control on rivers and lakes located in middle stream of Yangtze River.

4. Critical technical problems to be further studied and solved for physical model

With the development of water conservancy construction of Yangtze River and the continuous deepening of the research in hydraulic and river dynamics and river bed evolution subject, the physical model test of Yangtze River got a positive development and same for model similarity theory and testing technique[3 ,10-15]. The physical model test of Yangtze River has experienced a process from general river improvement test to sediment study for large and integrated utilization of water conservancy hubs as well as study for estuaries and coasts, and also the process from the simple model test based on clear water and static river bed to a more complicated model test based on muddy and arenaceous water. At present, the achievement from Physical Model test is firstly due to the study and analysis for various sediment issues by macro-exercise of geology, landform, geography and hydro-science knowledge, analysis for cause of river form and silt movement characteristics for river sections of distinctive features with the back up of hydraulics, flow dynamics, river dynamics and river bed evolution theory based on the fundamemtal phenomena of river, which enabled a great improvement in both theory and application of simulation for physical model and provides the sufficient reference for the successful construction of various project and eventually promote the economic development along Yangtze River.

Due to the complicity of sediment issue of natural rivers, there are still couples of key technical points affecting the accuracy of physical model test, including the followings.

(1) Distortions of physical model. The distortions include geometric distortion and time distortion. At present, only an approximate similarity can be achieved in theory for the similarity of distorted physical model, so the result from model test is only an approximate forecast and surely deviates more or less from the natural river.

(2) Similarity of bed resistance. A similar flow pattern and velocity distribution is the precondition to reach similarity in silt movement for physical model , however, the similarity in resistance is the vital factor to ensure the similar distribution of flow pattern and velocity.

(3) Sand filtering difficulty. To ensure that the model testing result is similar to the natural condition, the supreme important pre-condition for selection of model sand is to guarantee the similarity in silt movement and hence the selection of model sand is a critical technique which directly relates to the similarity of model silt movement and river bed distortion and the forecasting accuracy of model test and even the success of model test finally.

(4) Technical difficulty for model measurement. In physical model test, as the requirements for control accuracy of flow, sand content and water level is very high, and surely lots of work load is involved to measure water level, flow velocity, sand content and wave height and other hydraulic factors. A complete set of measure and control instruments coupled with

the state-of-art measuring technique is definitely needed to guarantee the accuracy and quality of the measurement and meanwhile shorten the model testing period which finally helps to promote the scientific research and better meets the demands of modernization program.

5. Conclusion

The evolution process of natural river is very complicated due to the influence of Incoming Runoff and Sediment Runoff, boundary condition of river bed and base level of corrosion. Up to now, we cannot fully reply on analytical and mathematical model calculation to solve the impact that the sediment problem caused to river evolution or improvement project for large water conservancy project in river improvement, development and protection, but we have to resort to physical model test for further research in order to solve the critical technical problems encountered during river improvement, development and protection. The new development of economy along Yangtze River brings the new opportunities for the development of Physical Model and will bring the simulation precision and testing technique to a new level. With the rapid development of hydraulic and hydropower construction in our country, physical model design theory and testing technique gain remarkable achievements and comes with their own feature, and now they have become a common but effective tool to solve the problems for hydropower projects and they are playing a more and more important role.

References

[1] Qian Ning, Wan Zhaohui, Sediment Dynamics [M], Science Press, 2003.

[2] Tan Xuming, The Past and Today of Chinese First Hydraulic Laboratory-History Trace of Hydro-Technical Institute, Website of Beijing Hydro-Technical Institute, Apr 25, 2008.

[3] Yu Wenchou, Lu Jinyou, Evolution and Improvement of Yantze river [M], China Water Conservancy & Hydroelectric Power Press, Aug 2005.

[4] Lu Jingyou, Simulation Technique Study Report for Distorted Physical Model , Changjiang Water Conservancy Committee, Changjiang Science Institute, Mar 2007.

[5] Li Baoru. Design Method of Dynamic River Bed [J], Journal of Water Conservancy, 1993, (12): 18-25.

[6] Zhang Hongwu, Research Progress for Similarity of Dynamic River Bed [J], Development of Water Science Development, 2003, 3 (12): 124-127.

[7] Mao Ye, Wang Yonghua, Study and Review for Dynamic River Bed [J], Journal of Hehai University (Natural Science Edition) , 2003, 31 (2): 124-127.

[8] Li Yuanfa, Chen Junjie, Zhu Chao etc. Discussion for Simulation Technology of Physical Model [J], People's Yellow River, 2005, 27 (12): 18-19.

[9] Chen Cheng, Tang Hongwu, Chen Hong, Huang Jiantong, Study and Summary of Topographic Mapping Method for Domestic Physical Model [J], Development of Water Conservancy and Hydro Power, Water Science Development, 2009, 29 (2): 76-79, 94.

[10] Zhang Hongwu, Feng Shunxin, Existing Problems and Solutions for Dynamic River Bed Model, Development of Water Science, 2001, 12 (3): 418-423.

[11] Yin Yamin, Study on Design and Fabrication of Static River Bed Model [J]. Experiment Science & Technology, 2004, 2 (7): 12-13.

[12] Yao Shiming, Zhang Yuqin, Li Jinyun, Variability Study for Solid Model [J], Journal of Changjiang River Scientific Research Institute, 1999, 16 (5): 1-4.

[13] Qu Yin, Guo Xilin, Long Chaoping, Sun Guizhou, Study on Tim Distortion Issues of Solid Sediment Model [J] . Journal of Water Conservancy, 2007, 38 (11): 1318-1323.

[14] Hu Yanfen, Yu Minghui. Wu Weiming, Distortion Impact for Water Flow Similarity in River bend of Physical Model [J]. Development of Water Science, 2007, 18 (11): 63-67.

[15] Lv Lieming. Newfasional Automatic Measurement and Control System for Physical Model [J] . People's Yellow River, 2005, 27 (8): 12-13, 39.

Design and Application of the Automatic Instrumentation for Dispatching Used in Hydraulic Model Test of Cascade Hydropower Stations

Liu Shanyan[1] Yang Wei[1] Xie Di[2] Liu Wenzhong[2]

[1] Department of Hydraulics, Yangtze River Scientific Research Institute, Wuhan, 443002
[2] Department of Control Science and Engineering, Huazhong University
of Science and Technology, Wuhan, 430074

Abstract: In this paper, we introduce modern information technologies as digital instrumentation and scientific computational utilities to the daily regulation model of the Three Gorges-Gezhouba Cascade Hydropower Stations (TGGCHS). From the perspective of the popular Digital River model used in hydrological features analysis, the practical model provides evaluable real-time information which may be used to reduce the ill-posed property of numerical calculations. Therefore, information obtained in the daily regulation model of TGGCHS helps to decrease the uncertainty of solution in Digital River. The hydrological parameters as flow velocity and flow pattern of the channel between the two dams in the model can be recorded simultaneously in the automatic instrumentation system. So the coupling relationships of different hydrological parameters, as well as different dispatching schemes are accurately recorded and then studied by using the linear system theory and their inversion method, which provides us means to study the regulation and anti-regulation dispatching of cascade hydropower stations. The informatization of the daily regulation model of the TGGCHS is realized in the following two levels: (1) Up to 9 observation points of flow velocity and water level are used in the measurement system, with time tags for all data obtained from the daily model of TGGCHS; (2) Efficient management of the mass data obtained by synchronous acquisition of hydrological parameters by using database technology, including the software functions of storage and synchronous display of the data with time marks, historical data repetition and evaluation of arrival time of the flow.

Key words: cascade dispatching; hydraulic model; hydrological parameters; automatic measurement and control system

1. Background

With the completion of Three Gorges Dam and the Three Gorges power plant's fully operational, the Three Gorges Dam on the upper and lower reaches of the system gradually run deep-seated problems and some key issues of pressing need to be solved. The unsteady flow problem formed in the TGGCHS's operation during flood period and the Power Plant's operation is very prominent. Unsteady flow caused the river's flow; slope and flow are in constant flux, which is between the Three Gorges Dam and Gezhouba, especially during flood period traffic flow, have very large variety in flow range. So that river flow rate, such as navigation and foam swirling flow have also been enhanced. The Shun-up and Shun-off waves generated from the Three Gorges power plant's discharge rolls down from the dam to the downstream. And the waves generated by the resistance of the river may be constantly decayed in the TGGCHS[2-3]; the Gezhouba power station up against the anti-regulation produced by wave and reverse wave down passing to the upstream process, are also subject to the impact of the resistance along the river and are constantly decaying; smooth, reverse wave around the central encounter between two dams, colliding with each other, with wave's energy to be further consumption. Water near the dam is impacted by discharge of the two peaking power plants and the water level fluctuation and amplitude are large, as the water level amplitude between the two largest dam sites, hours variable rates are the same as amplitude level. How to make full use of existing mature, advanced information technology and automation technology to solve water quality problems in large complex river network of information access is one of the problems which research community needs to solve[2, 4-5].

This article focuses on building the hardware equipment and software platform which is needed in upgrading the automation control system of the Three Gorges-Gezhouba daily regulation model in order to solve the bottleneck problem when getting real-time hydrological information. The building of hardware-software platform is based on real-time data in the daily regulation of the various tests. The use of network sharing and large scale computing technologies have been gradually carried out through follow-up of research projects in conjunction with relevant units in common.

2. Real-time hydrological information collection platform upgrade

The upgrading of synchronous acquisition platform for real-time hydrological information includes the upgrading of the access flow measurement and control systems and civil works of Three Gorges Reservoir model[6]. The reconstruction project of Cascade Hydropower Joint scheduling model test platform needs to complete four major works of transformation and upgrade. The main tasks are as follows:

1) The replacement and upgrades of the Three Gorges and Gezhouba water outlet system hardware in daily regulation of hydraulic model. The transformation of the hardware system makes continuous flow measurement and control of between the Three Gorges and Gezhouba more accurate, timely and accurately simulate the prototype system hydraulics characteristics.

2) Upgrading the control system into the outflow stream and control software for various purposes in the daily regulation test pilot study.

3) Add 8 to 10 point velocity and water level synchronization acquisition systems between the two dams to downstream, which includes single-point velocity flow of information and critical two-dimensional plane channel flow of the hierarchical profile information.

4) The efficient management of massive data and applications of hydrological parameters of the synchronous acquisition, which includes data storage and synchronization showed recurrence of historical data, stream time of the evaluation of different scheduling options and hydrological parameters of coupling time, providing basic research data for digital-depth study of cascade hydropower stations.

3. Hydrological information collection hardware platform

3.1 Replacement and upgrades of hardware in access to water system

Among the Three Gorges and Gezhouba's inlet and outlet systems' hardware[7-9] in the daily regulation model of TGGCHS, the main components of the Three Georges's $\phi300$ into the flow system such as electromagnetic flow meter and electric valve have been totally aging damaged due to the use life of aging damage, so they need to be replaced. Gezhouba Power flow measurement is achieved through the turbine flow meter, because of the same reason as long service life and the work environment, turbo rust and cause serious traffic measurement error. In addition, there is no flow measurement equipment installed in Gezhouba sluice, and Gezhouba Power Station has been in lack of a continuous flow regulation function, which often resulted that daily flow test cannot be balanced. Therefore we need to add $\phi300$ electromagnetic flow meter and electric valves to the Into the flow system of the Three Gorges of the daily regulation model of TGGCHS; installed electromagnetic flow meter in the 21 units of Gezhouba, and installed electric control valves in two of the units. While taking into account the dam of the complexity of fluid flow, we install the multi-channel ultrasonic open-channel flow meter and gate position sensor in Gezhouba sluice gate, in order to achieve flow measurement and control of the sluice. After transformation of the hardware system, we can achieve the Three Gorges and Gezhouba out of the sluice flow measurement and control, and maximize the simulation prototype hydraulic characteristics.

3.2 Out flow control system

Based on the transformation of the above hardware, upgrades the corresponding control system and control software, specifically including the purchase of the Three Gorges $\phi300$ into the flow system's PLC control module with the corresponding software module, and the separation of control and joint control algorithm issues of the $\phi300$ into the flow system and the existing $\phi200$ system. In Gezhouba controller, the analog input signal's acquisition task will be in an unprecedented increase, the analog input will increase by 25 points, with analog output by 2 points. Flow measurement, such as 21-point analog input node, at least 3 points above the sluice flow measurement of

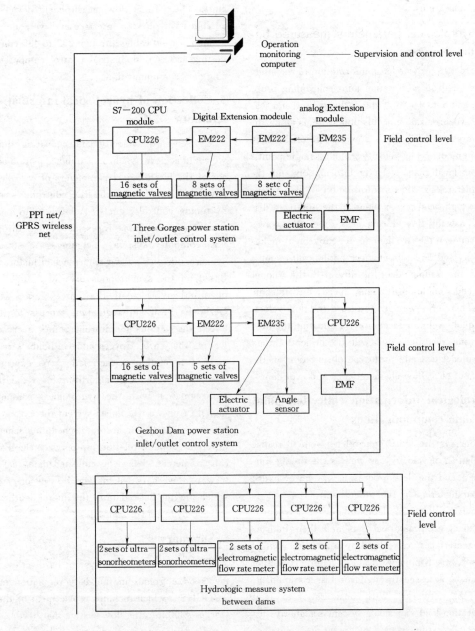

Figure 1   Block diagram of the cascade hydropower stations between two dams digital

transformation of the joint scheduling model for distributed network control system

(Doppler ultrasonic flow meter and ultrasonic profiler with specific arrangement

between the two channel models and Gezhouba Dam downstream)

multi-channel analog nodes, one point as gate position analog input nodes, two points as solenoid valve control analog output. Since these data are required to collect real-time synchronization, it presented some challenges to the control module for the PLC and the corresponding software module's computing power and communications.

3.3 Hydrological parameters measured between the two dams

In order to provide basic information to the study of cascade hydropower station joint scheduling problem, we add 8 to 10 three-dimensional flow and water level synchronization acquisition systems along the Three Gorges to Gezhouba's downstream of the daily regulation model. The velocity includes single-point's two-dimensional plane velocity information and the section hierarchical flow's information of key river. Since the high head power plant in the upstream discharged unsteady flow during peak time, the waves formed between the two dams in the movement makes the velocity and direction changes periodically, and it impacted the sailing fleet of aircraft, the rudder effect, rudder angle, drift angle, speed, etc. to some extent. Therefore, we need to record the flow velocity and liquid level more accuratey. Following the flow changes and recorded, we will do in-depth study through model test and prototype observation on the formation of such flow characteristics and evolution.

4. Hydrological information collection point for all data communications

In our system, PLC controller's help to realize the acquisition of velocity, as well as the timing control signals and the data processing system of hydro logical parameter's On the other hand, function as real time data collecting, data storage, data synchronization display of historical data and their human-computer interfaces are realized by a domestic configuration software Kingview. Data in remote collection points can be accessed through wireless communications or cable networks. With consideration to small communication load and small system scalability, as well as consideration to ease of operations, wireless networking solution is finally chosen in our system. Besides, the environment of the analog scene is very complex and the possibility of failure is high. Wired communication links along the road maintenance

needs check failure and it's usually difficult to identify the failure point in time, while setting up dedicated wireless data transmission through the use of wireless module allow us only to maintain data transmission module, and we could find out the reason of failure quickly as well as restore normal operation of circuits. The various flow measurement devices are all equipped with RS232 interface and we could lay 485 communication cables through 232 to 485 converters, getting access to industrial control computer, realizing data communications.

5. Collection of hydrological information system data

Based on the velocity data acquisition equipment, we could achieve synchronization acquisition and storage of the hydrological parameters of cascade hydropower stations through the PLC controller's collection by timing control signals. Achieve hydrological parameters of the data processing system through the host computer system, including data storage and synchronous display and recurrence of historical data. Based on the synchronous data with time marker, quantitatively calculate the Three Gorges Dam and Gezhouba Peak anti-regulation process to flow between the various monitoring points over time, as well as the Three Gorges and Gezhouba's regulation and anti-regulation characteristics. We expect to find an optimal scheduling algorithm for regulation and anti-regulation with the hydrological parameters of digital. And we use linear system theory and the inversion method to do accurate scheduling scheme with different coupling relationship between the hydrological parameters. Based on collected data, assess waterway between two dams and flow velocity changes which provide basic data in the digital-depth study of cascade hydropower stations.

6. Summary

The inlet and outlet system hardware's replacement and upgrades in the daily regulation model of TGGCHS, which includes replacement of the Three Gorges $\phi 300$ into the flow system that have been completely damaged, upgrading the flow measurement system of Gezhouba's flow (including sluice) system and the addition of a continuous adjustment function to Gezhouba's flow system. After transformation of the hardware system we can achieve the flow's measurement and control in the Three Gorges sluice and Gezhouba's out flow, timely and accurate simulating prototype hydraulic characteristics.

References

[1] Krebs D., Runte T., Strack G.. Planung für das Schiffshebewerk am Drei-Schluchten-Staudamm in China. Bautechnik, 2006, 83 (2): 73-84. (English translation: www. kuk. de. publications).

[2] Chinese Design Code: Specifications for seismic design of hydraulic structures (DL 5073—2000).

[3] Akkermann J., Hewener A.. Erdbebenbemessung des Schiffshebewerks am Drei-Schluchten-Staudamm, China. Bauingenieur 2006, 81 (4): 171-180.

[4] German RC Committee (DAfStb). Code of Practice, Manufacture and Application of Cement-Bound Grouting Concrete and Grouting Mortar, June 2006.

[5] Wright C. Funding the Three Gorges. Asia Money. 2003, 3: 20-25.

[6] Xie, Ping. Three Gorges Dam: Risk to Ancient Fish. Science, 2003, 302.

[7] Xing C., et al.. Human impacts on the Changjiang (Yangtze) River basin, China, with special reference to the impacts on the dry season water discharge into the sea. Geomorphology. 2001, 41: 111-123.

[8] Xuemin C. Rapid Growth Combats Acute Shortages. Modern Power Systems. 1995, 3: 33-36.

[9] Yeng S. L., et al.. Delta response to decline in sediment supply from the Yangtze River: evidence of the recent four decades and expectation for the next half century. 2003.

A New Arithmetic of Ship Motion by Draw-wire Length

Meng Xiangwei[1,2]   Gao Xueping[1]   Jiang Yunpeng[2]

[1] *School of Civil Engineering, Tianjin University, Tianjin, 300072*
[2] *Tianjin Research Institute for Water Transport Engineering, Tianjin, 300456*

Abstract: In this paper a measuring device of draw-wire type was designed for ship motion measurement in physical model test on ship mooring. And then a new arithmetic was proposed for the calculation of six-freedom ship motion. Six draw-wire displacement sensors of the device were fixed above the ship model and pulled when the ship model was moving. According to the position of ship model and wire length variation of the sensors, six-freedom ship motion can be calculated by the arithmetic. It is indicated that calculated result of six-freedom accords well with measurement result by the device.

Key words: ship model; wave; six freedom; experimental study

1. Introduction

Six-freedom ship motion is an important issue of harbor engineering research, which has an important significance for structure design and operation days for wharf. With development of port construction, safety of mooring ship, like LNG vessel, was raised higher and measurement of six freedom was attached much more importance than past.

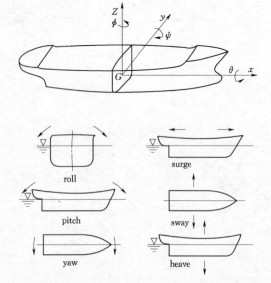

Figure 1   Sketch of ship motion of six degrees of freedom

Six-freedom ship motion in physical model test is usually expressed by roll, pitch, yaw, surge, sway, and heave, shown in Figure 1[1], and decompounded into rolling angle θ, ψ, ϕ around X, Y, Z axial and replacement x, y, z along X, Y, Z.

Generally speaking, measurement of six-freedom in physical model test is divided into non-contact and contact type. The non-contact type is mostly used by video[2] or electromagnetic measure and the contact type is used by acceleration or machine sensor, etc. Video measure has little interfering with ship motion but was sensitive to environment light. Acceleration sensor has certain accumulative error and machine device is difficult to process.

Draw-wire measuring device is of contact type. In this paper a new measurement method with draw-wire displacement sensors was proposed. It could measure six-freedom ship motion and was not sensitive to environment light and without accumulative error. In addition, it was easy to process and featured strong practicability.

2. System design

The sketch of the measurement system was shown in Figure 2.

In Figure 2, the number of 3 stands for rope sensors. The number of 41 stands for measurement device, connected with power supply and computer. The number of 42 stands for computer, the number of 43 stands for

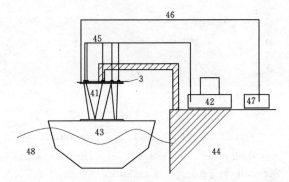

Figure 2　Sketch of measurement system

ship model, the number of 44 stands for wharf and bank, the number of 45 stands for signal line of rope sensor, the number of 46 stands for power supply line, and the number of 47 stands for power supply. The number of 48 stands for water.

　　Draw-wire displacement sensors were fixed with a shelf which was over ship model. And six ropes were pulled out from the sensors and fixed with ship body. When the ship had six-freedom motion induced by loading of wind, wave and current, the ropes were pulled and the sensor measured variation of the ropes. Then the ship position of at each moment could be computed by computer according to data from the sensor.

　　The method proposed in this paper was enlightened from simulation of operating platform of plane, car and ship[3]. The simulation platform was usually divided into upper platform and the base, between which six hydraulic cylinders were equipped. The motion of upper platform was controlled by six hydraulic cylinders, which were controlled by the computer. The hydraulic cylinders were not needed for ship motion in water waves, thus six ropes were proper to realize the purpose. In other words, the method in this paper was, to some extent, similar with traditional drawing puppet. For the puppet, it was moved by control of ropes; while for ship model, the ropes was pulled by ship motion under waves.

　　The device was mainly used for sea ship model test. Pulling force of single rope was usually less than 2N, and composition force was upwards. Influence of rope on ship motion could be neglected when the ship model was large enough (for example, more than 100kg in weight). That's to say that replacement measured by the rope reflected actual ship motion. The sketch of measurement was shown in Figure 2 and Figure 3, in which the upper platform was fixed

with the bank and the lower platform was fixed with the ship model. The rope, from sensor of upper platform, has a constant force which was made by permanent force spring. In ship model test, ship motion frequency was usually less than 2Hz. In other words, it was of low frequency and small scope, thus response requirement of rope was not high. And it was observed that the rope was keeping straight state during test.

　　In Figure 3, the number of 1 stands for upper platform. 31-36 stand for rope sensor, fixed with upper platform. The number of 11-16 stands for hole of rope. 2 stands for lower platform. 22-26 stand for connection point of rope and lower platform.

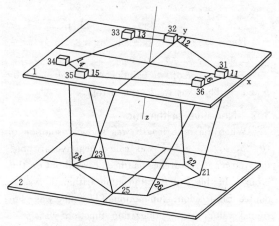

Figure 3　Stereogram of measurement device

　　The designed scope of six-freedom ship motion was shown as following:
　　roll, $-20°-+20°$; pitch $-10°-+10°$; yaw; $10°-+10°$;
　　sway $(-100+100)$ mm; surge $(-100-+100)$ mm; heave $(-50-50)$ mm.
　　sensitivity: angle$<0.02°$, length<0.01mm;
　　error: $<0.1°$; length<0.1mm.

3. Description of issue

3.1　Positive arithmetic

　　The system has six unknown variables. And six freedoms had corresponding relationship with lengths of six ropes during measurement scope. If the ropes were lie properly, six freedoms could be solved by equation group with rope length. More than six ropes would bring more bondage to ship model, although the measurement precision would be higher.

　　Platform 1 was of global coordinate, platform 2 was of ship coordinate. The ship motion could be reflected with six variables $(\theta, \psi, \phi, x, y, z)$. The coordinate of each point on platform at each time could be obtained by space transform, and length of six ropes could be calculated by original coordinate[4]. This process was called positive solution of the issue.

Supposed that original coordinate was $p_i = [x_i \; y_i \; z_i \; 1]$, and the transformed coordinate $p = p_i \times A$.

$$A = [cy \times cz, \qquad -cy \times sz,$$
$$sy, \qquad 0$$
$$sx \times sy \times cz + cx \times sz, \quad -sx \times sy \times sz + cx \times cz,$$
$$-sx \times cy, \qquad 0$$
$$-cx \times sy \times cz + sx \times sz, \quad cx \times sy \times sz + sx \times cz,$$
$$cx \times cy, \qquad 0$$
$$x, \qquad y,$$
$$z, \qquad 1];$$

Here, $sx = \sin\theta$; $sy = \sin\psi$; $sz = \sin\phi$; $cx = \cos\theta$; $cy = \cos\psi$; $cz = \cos\phi$。

The distance of two points could be calculated by p and p_i. That was the length of rope.

3.2 Negative arithmetic

When the rope length was known, solution of $(\theta, \psi, \phi, x, y, z)$ was called negative arithmetic. And many researches had been carried out on this issue[5]. In this paper optimization arithmetic was adopted. Six-freedom ship motion of last time was as original value, and optimization function was $f = \sum_{i=1}^{6} (l_i - l_{gi})^2$, in which l_{gi} was rope length calculated from new six freedoms.

The optimization arithmetic was as follows.

A single peak will exist as long as the original six freedoms were close to actual value. In Figure 4, function $y = f(x)$ has a single peak at point $x = m$ during $[a, b]$. The extent $[a, b]$ was divided into three parts.

When $f(x_1) > f(x_2)$, m was during (a, x_2) and (x_2, b) could be abandoned. When $f(x_1) \leqslant f(x_2)$, m

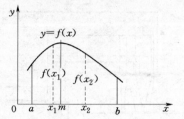

Figure 4　Sketch of optimization of single variable

was during (x_1, b) and (a, x_1) could be abandoned. If the process was repeated, $m = (x_1 + x_2)/2$ could be obtained with abs $(x_1 - x_2)$ less than enactment error.

According to single variable optimization principle, each variable $(\theta, \psi, \phi, x, y, z)$ was optimized with the same process until difference of six freedoms of two times was less than enactment error.

4. Programming Environment

The whole system was divided into upper platform, lower platform, draw-wire displacement sensors and computer[6,7].

The software environment was as follows.

Operating system: Winxp;

Program language: Delphi6.5, Matlab6.0;

Communication protocol: RS485.

5. Arithmetic validation

First, time series of ship motion were given and length of six ropes was calculated with positive arithmetic. Secondly, ship motion was calculated by negative arithmetic according to rope length. If calculated ship motion was consistent with the given value, it was concluded that the arithmetic was reliable. The result of validation was shown in Table 1.

Table 1　　Arithmetic validation for ship motion of six degrees of freedom

N	Rx_1 (°)	Rx_2 (°)	Dz_1 (mm)	Dz_2 (mm)	dRx (°)	dRz (mm)
1	0.00	0.00	125.00	125.00	0.00	0.00
2	−0.27	−0.33	120.45	120.54	0.06	−0.08
3	−0.54	−0.51	115.94	115.89	−0.04	0.05
4	−0.81	−0.81	111.51	111.51	0.00	0.00
5	−1.07	−1.07	107.19	107.19	0.00	−0.01
6	−1.32	−1.36	103.01	103.07	0.04	−0.06
7	−1.56	−1.52	99.02	98.96	−0.04	0.06
8	−1.79	−1.84	95.24	95.32	0.06	−0.08

N	Rx_1 (°)	Rx_2 (°)	Dz_1 (mm)	Dz_2 (mm)	dRx (°)	dRz (mm)
9	−2.00	−2.06	91.71	91.80	0.06	−0.08
10	−2.19	−2.12	88.46	88.36	−0.07	0.10
11	−2.37	−2.41	85.51	85.56	0.04	−0.05
12	−2.53	−2.59	82.88	82.97	0.06	−0.09
13	−2.66	−2.71	80.61	80.67	0.05	−0.06
14	−2.78	−2.84	78.70	78.79	0.06	−0.09
15	−2.87	−2.83	77.17	77.11	−0.04	0.06
16	−2.94	−2.96	76.05	76.08	0.03	−0.04
17	−2.98	−3.04	75.32	75.41	0.06	−0.09
18	−3.00	−2.98	75.01	74.98	−0.02	0.03
19	−2.99	−3.01	75.12	75.14	0.01	−0.02
20	−2.96	−2.98	75.63	75.66	0.02	−0.02
21	−2.91	−2.83	76.56	76.45	−0.07	0.11

Six points distributed equally on the upper and lower platform. The circum radius of upper platform was 250mm and angles of six points were 0°, 90°, 120°, 210°, 240°, and 330°. The circum radius of lower platform was 200mm and angles of six points were 30°, 60°, 150°, 180°, 270°, and 300°. Supposed $Rx1$ was rolling angle of lower platform around x axis, $Dz1$ was replacement of lower platform along z axis. Length of six ropes was computed by known ship motion with positive solution. And ship motion was computed by rope length with negative solution. The known ship motion and computed ones were thought to be accorded with each other when the error between them was less than 0.01.

Random error 1mm was placed on length of first rope for validation of solution stability. The result was shown in Table 1, in which $Rx2$ was computed rolling of lower platform around x axis, $Dz2$ was replacement of lower platform along z axis, dRx was rolling error of lower platform around x axis, dRz was replacement error of lower platform along z axis. The measurement error of rope sensor was 0.1mm, less than 1mm. Thus the system satisfied with error requirement of angle and length.

Because the computation time is larger than that between two measurements during the physical test,

the rope length was measured firstly. When the measurement was completed, six freedoms were computed. The result of validation indicated that the solution was reliable.

6. Measurement result

Here is an example of physical model test on ship mooring, in which length scale was 1 : 80 and the main dimension of ship was shown in Table 2. Result of empty ship with beam wave was shown in Figure 5.

It was seen from the result that scope of roll was maximum with obvious period and scope of pitch was small with obvious period. Yaw period was rather larger than roll and pitch. Surge has long period while sway and heave have short period. The maximum value measured from the test was: roll 1.8°, pitch 0.2°, yaw 0.5°, surge 10mm, sway 18mm, heave 3mm. Sample frequency was larger than 50Hz. Finally the measurement of ship model test was rational.

7. Conclusion

(1) Draw-wire measurement device has characteristics of simple principle, easy process, etc.

(2) Optimization arithmetic of six freedoms was realized in the system and the device could be used to measure ship motion under loading of wave, wind and current.

Table 2   **Main dimensions of ship**

	total length (m)	column length (m)	with (m)	depth (m)	draft (m)	displacement (t)
Real ship	332	320	58	30.4	23	357479.98
Model ship	4.15	4	0.72	0.38	0.29	0.68

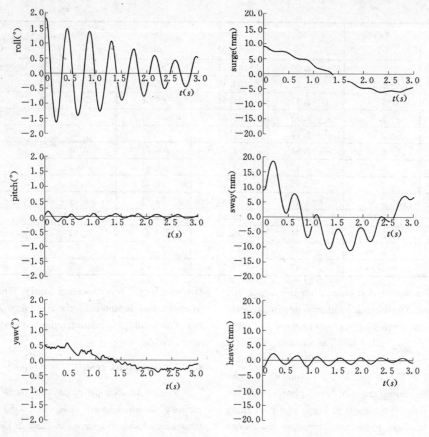

Figure 5   Six degrees of freedoms measurement results of mooring ship

(3) It was indicated by the mooring ship model test that the precision of the device was high enough and reached the anticipative purpose.

References

[1] Wu Xiuhuan. Ship maneuverability and seaseeking [M]. Beijing, Peoples' Communication Press, 1991: 25-31.

[2] Jiang Hairong. The Design of Untouched Float Six-Free-Parameter Testing-System Based on Double CCD [D]. Master's dissertation of Dalian Technology&Science University, 2005: 52-58.

[3] Hao Yining. A Study on a Six-Degree-of-Freedom Platform [J]. Transaction of Beijing Technology & Science University, 2002, 22 (3): 331-334.

[4] Hao Jianbin. Application of Matlab in Forward Displacement Analysis of Stewart Platform. Mechanical Engineering. 2005 (11): 92-94.

[5] Li Shujun. A Self-modified Successive Approximation Method for Forward Position Analysis of 6-3 Structure In-parallel Manipulator. Mechanism Science and Technology, 2002, 21 (1): 81-85.

[6] Luo Qiang. The Programming Method for TeeChart Control in Delphi4.0. Transaction of Changsha University, 1993, 13 (4): 50-52.

[7] Li Zhongming. A study of programming design methods in serial communication based on Delphi 6. Transaction of Lanzhou University, 2004, 40 (4): 41-44.

Calibration Test of Doppler Ultrasonic Flow-meter

Zhao Jianping   Yan Genhua   Gu Hua

State Key Laboratory of Hydrology-Water Resources and Hydraulic Engineering,
Nanjing Hydraulic Research Institute, Nanjing, 210029

Abstract: This paper mainly introduces the calibration situation of the model test of Doppler ultrasonic flow meter. The calibration is carried out in the standard glass water tank, and the output of the test instrument is evaluated by the actual measured flow. With regard to the influence of the cross section form, the test calibrations are implemented for different cross sections. Comparison of the instrument's output flow and the measuring weir's measured flow shows, in the current layout method of instruments, for different cross section forms, and on the basis of comparison of its test result and the measuring weir's measured result, the instrument's flow test accuracy is higher, and has the best regularity of the test result for the rectangle cross section. This provides a reliable basis for the application of such model of flow meter, when the simplified formula for simplified measurement in application is proposed in the paper.

Key words: calibration; flow meter; correlation coefficient; flow

1. Introduction

To rationally allocate, transfer and use the water resource, and to integrate the water rate into the production cost, it's essential to measure the flow of used water when accurate measurement is in favor of scientifically and rationally using water according to need. Especially for the regions lack of water, the effective use of the limited water resource can urge the region's economic and cultural development.

In selection of instruments, the instrument shall be able to guarantee the measured data can meet the current use demands on one hand, and not increase too much cost on the other hand. The Doppler ultrasonic flow meter is a type of flow meter used widely, which automatically calculates the flow value on the basis of the given area of the cross section by measuring the velocity of flow, the water depth and the temperature, and accumulates the total flow over time.

2. Introduction of instrument

Thanks to its small volume and light weight, the Doppler ultrasonic flow meter is convenient for field work, and it won't block the river course even if it's installed in the small canal for measuring flow for long-term; its installation, use and data analysis only need simple operations, which are easy for the beginners to understand and grasp; it's capable of measuring the flow velocity through the acoustical principle and accumulating the flow, when it has no intervene to the measured section current or the user's save, record or calculation, as it's of high automated performance; it's merely has dangerous contamination to the measured object; it's capable of working continuously and capturing the changes of parameters in short time interval; it can remotely sense and measure through the appropriate wireless network; and it doesn't need high-cost auxiliary building and just requires a few auxiliary work for the on-site measurement.

The water depth and temperature is directly measured by the pressure and temperature sensor, whose operational principle is excluded in this paper. The key flow measurement is through the principle of acoustical Doppler effect, and the flow is calculated by the frequency difference between the emissive frequency and the measuring sound wave's echo signal reflected by the moving suspended solids.

3. Calibration test

3.1 Design of test device

The test is carried out in a glass water tank with 60cm wide bottom, 0.000 bottom slope, and measuring weirs at both upstream and downstream, for testing the rectangle and trapezoid cross sections, whose dimensions are shown in Figure 1. The flow meter

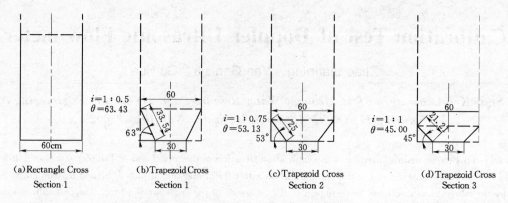

Figure 1   Different models of cross sections

testing any cross section shall be place in the middle of the bottom of the cross section.

3.2   Analysis of flow formula

In consideration of the instrument's actual use effect, the positions of the measured objects shall be of these features: the upstream landform shall be open, flat and hard, without obstacles or pollutions; within the measuring range, the current is stable and flat, with small surface waves; the cross section is in good geometric shape, and the material of flow is representative. On such premise, the basic calculation formula of the instrument's flow meter has been analyzed as follows, and the accumulation process of flow is shown in Figure 2.

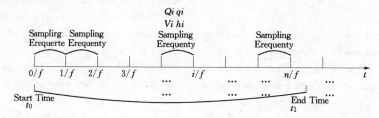

Figure 2   Schema of accumulation of flow

$$Q_t = \sum_{i=1}^{i=n} \left(q_i \frac{1}{f} \right) \tag{1}$$

In which:

Q_t is the total water consumption (m^3);

$$n = \left[\frac{t_1 - t_0}{f} \right] \tag{2}$$

n is the number of samples in the measurement period, which is an integer;

t_0 is the start time of measurement (s);

t_1 is the end time of measurement (s);

f is the instrument's collecting and recording frequency (1/s);

$$q_i = \int_A v_i dA = \xi_i \overline{V}_i A_i \tag{3}$$

q_i is the flow in each sampling period (m^3/s);

$\xi_i = f_1(\xi)$ ξ_i is the integral influence coeffi-

cient of boundary, flow velocity and water depth;

$\overline{V}_i$ is defined as the cross section's average flow velocity (m/s);

$$\overline{V}_i = \frac{1}{A} \int_A v_i dA \tag{4}$$

v_i is the flow velocity at any point on the cross section;

$$A_i = f_2(h_i, B, \theta) \tag{5}$$

A_i is the area of flow-passing cross section calculated by the water depth (m^2);

h_i is the average water depth of the cross section (m);

B, θ are the geometric information of the cross sections;

The cross section's water depth used in the formula is substituted in the formula by the instrument's measured value; the calculation of area needs to consider the deviation of measured water depth and cross section's actual water depth. The main factors influencing the measurement of water depth are the actual flowing status (such as the surface waves) and the instrument itself; for simplicity, h_i in the above formulas is replaced with $(\xi_h h_i + c_h)$, when ξ_h and c_h are the instrument's influence coefficient of water depth. The cross section's flow velocity distribution is related to the materials forming boundary, the cross section's geometric shape, the kinetic energy of liquid surface gas, and the physical characteristics of the liquid itself, so the distribution is very complicated. Also for simplicity, with the same model of cross section, it shall be replaced with $(\xi_v v_i + c_v)$, when ξ_v and c_v are the instrument's influence coefficient of flow velocity.

The cross section's average flow velocity used in the flow formula is calculated by substituting the instrument's measured flow velocity, when it's necessary to consider the deviation of flow velocity in the instrument's measuring range on one hand, and the correlation of the value and the whole cross section's average flow velocity needed in the calculation on the other hand.

During calibration, in consideration of the correlation of substituting the average flow velocity with the measured value, these 2 methods can be used: (1) consider the accurate flow velocity distribution on the whole cross section, and get the correlation coefficient of flow velocity through the comparison of the instrument's measured flow and the cross section's accurate flow velocity distribution; (2) as the instrument's measurement of flow ignores the correlation of single flow velocity, the correlation analysis can be carried out directly by the measured flow and the flow calculated by the instrument's measured flow velocity. The former method can get the clearer relation between the cross section's average flow velocity and the measured value, which is easy to adapt to the influence of measuring position on such correlation coefficient, and is in favor of partial calculations; yet, due to the cross section's flow velocity distribution is hard to be measured accurately, and the consideration of the correlation of flow shall be more practical, there will be different results from different measuring positions.

For simplicity, during each period, given it is the linear correlation (this is just for example, when

it's not linear, other correlation coefficients shall be considered), the flow formula is corrected to:

$$q_i = \rho_i q_{ce} + c_i''$$

$$\Rightarrow q_i = \rho_i \{ V [(\xi_v v_i + c_v]_i A [(\xi_h h_i + c_h]_i \} + c_i''$$

ρ_i and c_i'' are the correlation coefficients of the instrument's measured flow and the actual flow.

When a group is defined, the value is also defined. V, A are used for representing the functional relations of calculating values and measured values. It's to be remarked that, such correlation coefficient is mainly for "representing the whole cross section's average flow velocity by the measured flow velocity". Once such value is determined, the correct flow can be calculated by the measured water depth and flow velocity.

In the light of the influence of the measuring position on the measured flow velocity, although the measuring position is the same in calibration and actual measurement, when the coefficient is allowed not to be considered, the further consideration of the correlation of flow and different positions will be a big help for improving the accuracy of the discrete measurement for the cross section.

The said correlation coefficients will continuously change with the variation of the models of cross sections, the water depth, the actual measurement situation and the calibration situation, as well as the changes of flow. Even when it's linear in a certain measuring period, it would not be the same as that in analysis in the actual state, which usually isn't a constant.

In view of the utility of the instrument, it's necessary to further analyze the influence of different factors on such correlation coefficient, on the basis of getting such correlation coefficient (using the measured flow velocity to represent the capacity of the whole cross section's average flow velocity).

3.3　Analysis of test result

During the whole test, the water line in the tank slightly reduces on way in observation; particularly, the effluent at rear reduces apparently. Within the instrument's measuring range, the water doesn't contents a large quantity of find sand and bubbles, the current is slow and uniform, the surface wave is not large (what need to be explained is, along with the increase of test flow, the surface wave will become larger), so, in general, the test doesn't have large interference.

Figure 3 shows the linear fitting correlation of the instrument's measured value for the rectangle cross section 1 and the rectangle weir's flow through the least square method. Figure 4-Figure 6 show respectively the linear fitting correlations of the instrument's measured values for the trapezoid cross sections 1-3 and the rectangle weirs' flows through the least square method.

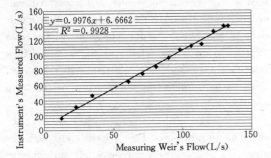

Figure 3　Fitting correlation curve of
instrument's measured value for rectangle
cross section 1 and weir's flow

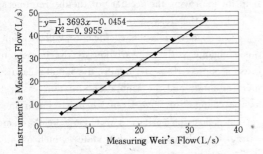

Figure 4　Fitting correlation curve of
instrument's measured value for
trapezoid cross section 1 and weir's flow

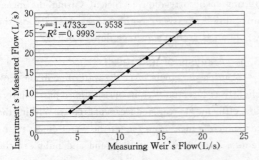

Figure 5　Fitting correlation curve of instrument's
measured value for trapezoid cross
Section 2 and weir's flow

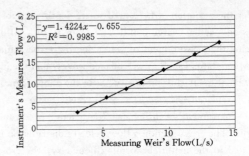

Figure 6　Fitting correlation curve of
instrument's measured value for
trapezoid cross Section 3 and weir's flow

The test result shows, in the above test conditions (flow control, instrument's layout position, method, etc.), the instrument's measured results for different models of cross sections are within the high accuracy range, comparing to the test results of the measuring weir's.

The test result of the rectangle cross section is of good regularity, and the instrument's measured value is of high accuracy, whose correlation coefficient with the measuring weir's test result reaches 0.99; however, there are large relative deviations at the points with small flow, which have surpass the instrument's sphere of application.

For the trapezoid cross sections, due to slow slopes at both sides, which increase the flow passing ranges at both sides, the whole cross section's flow velocity will be hard to keep being uniformly distributed on the whole section; particularly, the flow velocity at the measured part of the instrument placed in the middle position is much larger than the whole section's average flow velocity, so it can not represent the average flow velocity of the flow-passing cross section. The side slopes [(1 : 0.75)—(1 : 0.5)—(1 : 1)] of 3 different trapezoid cross sections used in the test respectively have corresponding correlation coefficients 1.3690, 1.4731 and 1.4226 of the instrument and the measuring weir. And the instrument's measured value is larger than the measuring weir's flow also states the instrument's measured flow velocity is larger than the average flow velocity of the flow-passing cross section.

The above calibration test shows, the measured flow Qm by the instrument at the middle bottom of the flow-passing cross section is different from the flow Qa; using the calibration coefficient defined in Formula 6.

$$Qa = K \times Qm \qquad (6)$$

For rectangle cross section: $K = 0.920\text{-}0.994$

For trapezoid cross section: $B = 30\text{cm}$, Slope Angle $\theta = 63.435°$, $K = 0.707\text{-}0.747$

$B = 30\text{cm}$, Slope Angle $\theta = 53.13°$, $K = 0.697\text{-}0.776$

$B = 30\text{cm}$, Slope Angle $\theta = 45°$, $K = 0.725\text{-}0.806$

4. Conclusion

The preliminary calibration test above states: the flow calibration coefficient K is a regular and stable scale-free coefficient, which would be influenced by the flow passage's cross section shape, the area of flow-passing cross section, the water depth, the roughness of side walls, the status of incoming flow and the water quality; and there will be deeper systematic test researches to further improve the instrument's measuring accuracy and simplify the work amount of measurement and calculation.

The current laboratory test shows, the instrument's measured flow is of smaller error than that of the measuring weir's measured value, so it can meet the demands of water metering in most occasions. As a highly automated instrument, it can be used in many places, and particularly fit for remote measurement. And application of this instrument can save the cost and improve the measuring accuracy.

References

[1] Zhao Jianping, Hu Qulie, Yan Genhua, etc. Ministry of Water Resources "948" project-the fourth report of Irrigation water measurement technology and research on equipment, Nanjing Hydraulic Research Institute.

[2] Editor: Nanjing Hydraulic Research Institute, Approved by: The People's Republic of China Ministry of Water Resources. Hydraulic (conventional) model test (SL 155—95). China Water Power Press.

[3] Main authors: Li Yun, Tao Xiuzhen, editor: Nanjing Hydraulic Research Institute, Approved by: The People's Republic of China Ministry of Water Resources, dynamic flow and flow standard device calibration method (SL/T 232—1999). China Water Power Press.

[4] Liang Guowei, Cai Wuchang. Flow Measurement Technology and Instruments. Machinery Industry Publisher.

Sonic Characterisation of Water Surface Waves

A. Nichols[1]   K. Horoshenkov[1]   S. Shepherd[1]   K. Attenborough[2]   S. Tait1

[1]*School of Engineering, University of Bradford, UK*
[2]*Acoustic Research Group, Open University, UK*

Abstract: This paper discusses the development of an acoustic device to estimate the dynamic surface profile of channel flow. The work builds on previous observations by Cooper et al. (2006) which suggest that the response of the scattered acoustic signal from a water surface is related to hydraulic conditions by energy losses within the flow. Temporal analysis of reflected acoustic intensity is used to estimate the scale of dynamic surface features, while analysis of excess attenuation spectra is developed as a potential method of describing the statistical shape of time-averaged spatially-variant surface features. The temporal analysis shows a strong trend between the range of surface movement, and the range of reflected acoustic intensity. Analysis of excess attenuation spectra is shown to estimate the size of static surface features with little error for a given range of surface features per unit length with geometry analogous to a water surface. This suggests that the method can be used to capture the static component of water surfaces. Such an achievement will allow for remote and inexpensive measurement and monitoring of real channel flows, and may facilitate deductions of flow parameters such as flow rate, depth, turbulence and bed topography among others.

Keywords: water surface; remote sensing; flow measurement; flow monitoring; acoustics

1. Introduction

Monitoring of flow conditions of free-surface flow is essential in a number of applications, such as river flood monitoring, and flow control in water and waste treatment plants. Current measurement techniques can be costly, time-consuming, and invasive. A non-invasive acoustic device would allow cost-effective real-time monitoring of water systems.

In the absence of free surface shear stress caused by wind, the shape of the free surface is a product of the internal flow structures caused by any number of factors including bed shape, wall roughness and flow rate. Tamburrino and Gulliver[1] used a moving bed flume and particle tracking velocimetry to show that the surface velocity vector field was related to vortices formed in the flow due to the relative motion of the bed. They attributed upwellings on the surface to vortices in the flow, and downwellings to areas of increased streamwise velocity. They conclude that vortices within the internal flow field form a "significant transport mechanism" which carry mass and energy to the free surface. Shvidchenko and Pender[2], when investigating the structure of open-channel flow over

gravel beds, saw similar vortices and showed them to span the entire depth of flow. Their results also suggest that different sized bed gravel gave rise to different scales of flow structures. These findings suggest that free-surface features relate directly to the internal flow structures, which in turn relate to the boundary conditions present.

Empirical relationships have been shown to exist between the energy losses of shallow channel flows, and the dynamics of an acoustic signal scattered from the water surface. Cooper etal[3]. showed that an increase in k_s value of the bed would result in a decrease in standard deviation of intensity of an acoustic signal reflected from the surface. Although the mechanisms behind this relationship are not understood, it suggests that an acoustic solution is plausible.

Water surfaces in shallow flow are formed from two groups of superimposed wave-shapes: standing waves which maintain their general shape and position over time; and quasi-cyclic dynamic waves which appear to travel downstream and whose shape fluctuates over time. Observations of shallow channel flows show that both the dynamic and static components of a water surface are affected by the channel bed shape and flow conditions. As such, both the static and dynamic components may yield information about the flow, and hence both should be captured by the remote device if possible. Whilst the method employed by Cooper et el[3]. detects temporal variation in the surface, it cannot detect the static wave patterns that were also observed.

This work investigates one method of measuring the dynamic components of a water surface, and one for measuring the static component. The dynamic technique, adopted from Cooper et al[3]. involves analysis of the temporal variation of acoustic intensity of a reflected spherical wave. It is thought that a surface whose shape changes with time, will cause a proportional change in acoustic reflection with time.

Analysis of the static component of a water surface is facilitated by the assumption that time-averaging of data will filter out the dynamic component and allow the static component to be isolated. Standard theory for acoustic surface measurement can then be applied. Several techniques have been developed for quantifying roughness parameters of various static surfaces. Oelze[4] for example has examined the use of acoustic backscatter to measure soil surfaces. Backscatter is likely to prove unfeasible over larger distances such as rivers and wide channels due to the energy content required & corruption from environmental noise.

2. Analysis of water surface components

In order to understand the surfaces to be measured acoustically, it was necessary to quantify the general shape and size of water surface features in free surface flows through rough boundary channels. From observations of water surfaces it was hypothesised that the surface would contain a dynamic component which varied in time and space, and also a static component which varied in space but was fixed in time.

2. 1 Methodology

An 11m long, 0. 46m wide tilting flume was used. The bed condition used was gravel with a mean grain size of 6mm, and the gradient was maintained at a constant value of 0. 002. Flow rate was varied using an inlet valve, and a height adjustment at the outlet allowed the water surface gradient to be held equal to the bed gradient, ensuring uniform energy loss along the length of the flume. The test conditions for three test runs are shown in Table 1.

Table 1   Conditions for hydraulic tests

Test run	Flow rate (m³/s)	Uniform depth (mm)	Mean vel. (m/s)
A	0. 0273	94. 6	0. 628
B	0. 0235	85. 0	0. 600
C	0. 0184	76. 4	0. 522

Two Churchill 2-core 2mm diameter wave probes were positioned in fixed locations in the flow, while two others were moved through a 1m length of flow in a streamwise direction at a constant rate of 2mm/s using an automated traverse. The probe gives a voltage output proportional to the amount of probe that is submerged. By fixing the probe in the flow, and applying a conversion, it directly outputs the instantaneous depth of the water. The fixed probes would provide a measure of the dynamic nature of the water surface, while data from the moving probes would be filtered to detect any static component that didn't vary with time. Spatial spectral analysis was then used to gain insight into the length scales of the static and dynamic components. For the dynamic component, mean flow velocity was used to convert frequency to wavelength, while for the static component, the traverse velocity was used.

2. 2   Results & discussion

Average spatial spectra from the two stationary probes are shown in Figure 1. It can be seen that the dominant dynamic wavelengths are around 5 to 10mm and 40 to 60mm, and that the amplitude of these

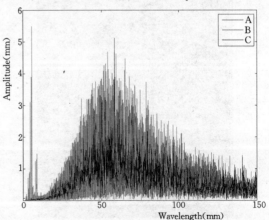

Figure 1   Spatial spectra of dynamic
surface-wave components

dominant waves is approximately 3 to 5mm.

It can be seen that as the flow rate decreases, the amplitudes and wavelengths of dominant dynamic surface waves decrease also. The average spatial spectra from the two moving probes at each flow rate are shown in Figure 2.

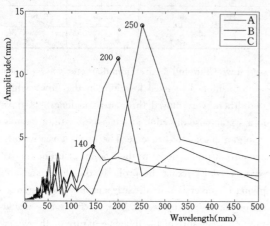

Figure 2　Spatial spectra of static
surface-wave components

These show that a static surface wave component does indeed exist and that the dominant wavelengths and amplitudes of these stationary waves decrease as the flow rate decreases. The approximate length scales of these temporally stationary waves are 100 to 250mm wavelength (1.8 to 2.6 water depths) and 3mm to 13mm amplitude (0.05 to 0.15 water depths). It can be seen that in test run A, the dominant static wave had a wavelength of 250mm. This is verified in Figure 3, where a metre long scale is attached to the side of the flume. Five peaks of the static

Figure 3　Visual estimation of static
surface-wave length-scale

wave pattern are indicated by the black lines. These lines can be seen to divide the scale into 4 segments, giving a wavelength of 25cm. It is also of note that the waves form a predominantly 2-dimensional pattern, with wave-fronts perpendicular to the direction of flow. Dominant static wavelengths for test runs B and C were verified in the same fashion.

3. Temporal variation of reflected acoustic intensity

The time-variant aspect of a water surface has been shown to relate to the conditions of the flow[3]. In this work the acoustic response is compared directly to the dynamic component of the water surface. It was thought that the presence of roughness would affect the intensity of reflection, and therefore a changing roughness would cause a changing intensity.

3.1　Methodology

For this experiment a 15m long, 0.68m wide channel with a fixed 0.006 bed-slope was used. The original bed of the channel was used which consisted of 10cm square tiles with grout between. This gave a small roughness to generate surface structures. The flow rate was adjusted to twelve different values between 0.0065m³/s and 0.03m³/s (see Table 2) in order to create different surface dynamics. At the outlet a gate was used to control outflow to ensure steady uniform flow along the length of the channel, and hence constant energy loss along the channel. At the test area, two large glass panels are placed at the edge of the channel in order to simplify acoustic boundary conditions. On one glass surface a 15mm dome tweeter was mounted. This has been tested to act as a point source for the frequency range used. On the opposite side a microphone was attached.

Table 2　　　Conditions for acoustic tests

Test number	Flow rate (m³/s)	Uniform depth (m)	Mean velocity (m/s)
1	0.0008	0.0054	0.2205
2	0.0014	0.0108	0.1870
3	0.0033	0.0164	0.2992
4	0.0065	0.0260	0.3680
5	0.0117	0.0367	0.4687
6	0.0173	0.0480	0.5295
7	0.0212	0.0534	0.5829

Test number	Flow rate (m^3/s)	Uniform depth (m)	Mean velocity (m/s)
8	0.0246	0.0606	0.5978
9	0.0265	0.0614	0.6341
10	0.0279	0.0633	0.6469
11	0.0288	0.0651	0.6505
12	0.0295	0.0658	0.6584

Continued

Due to the restrictions of the experimental setup a continuous signal was not an option due to multiple reflections. A short pulse was designed to be only 2ms long so that any multiple reflections could be windowed out. For each flow rate, 50 pulses were sent over the surface at random intervals. The standard deviation of the received intensity of each of these 50 pulses is calculated for each flow rate.

The actual dynamics of the water surface were recorded using a wave probe 1m downstream of the acoustic equipment. Probe readings were recorded for one minute for each flow rate. The wave probe was held in a constant location and therefore recorded the time-varying surface component at a point. It did not therefore record the spatially-varying temporally-static component. The statistics of these dynamic readings were then compared to that of the acoustics.

3. 2   Results & discussion

The standard deviation of the instantaneous water depth as recorded by wave probe is taken to be a

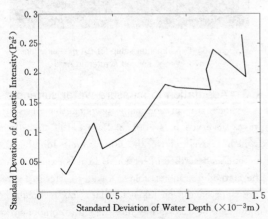

Figure 4 Standard deviation of acoustic response vs. standard deviation of water depth

measure of the dynamic range of the water surface. This is plotted against the standard deviation of the acoustic intensity in Figure 4. It can be seen that a general trend occurs. As the height of dynamic surface waves increases, this causes an increased variation in the reflected acoustic signal. This is thought to be due to larger surface structures behaving more like individual scattering elements which cause a more complex and diverse intensity field above the surface. This appears to give a reasonable method of calculating surface wave height from the reflected acoustic signal, however more work is required as the trend shown is not perfectly linear and may be related to other factors such as wavelength and surface velocity. In comparison to the work of Cooper et al[3], this would suggest that a smaller k, value results in larger surface waves. This would need further investigation to verify.

4. Analysis of excess attenuation spectra

An excess attenuation (EA) spectrum represents the frequency dependent signal received from a point source over a solid boundary. A key feature is the series of troughs that occur when the direct and reflected acoustic signals interfere destructively[5]. Relating this acoustic feature to the time-averaged water surface topography would allow for remote acoustic measurement of the static component of water surface roughness.

4. 1   Methodology

Much previous work has built on theories by Tolstoy[6] and Twersky which use a "boss" model. This allows a rough surface to be thought of as a flat hard surface, upon which are several 2-dimensional semi-cylindrical rods or scatterers[7]. The roughness is categorised by the volume of rod per square metre, and the centre-to-centre spacing between rods.

Empirical relationships have been derived to calculate the excess attenuation spectrum from a given surface acoustic admittance. This has been adapted to allow calculation of surface admittance from a fitting to a measured EA spectrum[8]. This fitting relies on accurate measurement of source and receiver separation, and their respective heights above the surface. Expressions have also been derived to calculate surface admittance from a given surface geometry[7]. Whilst this relationship has proven to be reliable, solving the inverse problem is not trivial. A simplification of the relationship is shown in Equation (1).

$$\beta = af^b + cf^d \qquad (1)$$

where β is the complex surface admittance, f is frequency, and a, b, c, d are variables related so surface geometry.

For this work a 100cm×100cm acoustically hard surface has been used as a base on which a number of 1cm radius semi-cylindrical scattering rods have been attached. The 1cm height of these rods are within the 0.3 to 1.3cm amplitude static waves observed in section 2.2, and produce a 2-d roughness akin to the static-waves observed. For this reason the setup used is thought to be a good analogy to a water surface. A tannoy speaker attached to a 100cm long, 2cm diameter copper pipe was used as a point source which generated white noise in a frequency band from 0 to 20kHz while a microphone was positioned 70cm away. Both source and receiver were set at a height of 7cm above the surface. Rod arrangements were organised by the mean separation between adjacent rods, which is proportional to the percentage of surface containing rods. Separations were chosen to include the wavelengths observed in Figure 2. Readings of EA were taken for 10 random arrangements at each coverage percentage. Random spacings were used to minimise the potential for coherent scattering, although it was noted that uniform periodic arrangement yielded the same results. The ten EA spectra were then averaged for each coverage percentage and the variables a, b, c and d from equation 1 were deduced using a curve fitting routine.

4.2　Results & discussion

It was seen that variables "a" and "d" are always very small, and in fact the volume of scatterers per unit area of surface linearly related to variable "c", for a reasonable range of surface coverages. A linear correction was empirically derived. Figure 5 shows the percentage error in scatterer volume estimation against percentage of surface covered. It can be seen that up to a coverage of around 32% the method gives an estimation of scatterer volume accurate to within 4%. 32% coverage corresponds to scattering elements having an average of 6cm separation between them.

It is thought that above 32% coverage the scattering elements became too close and began to interact, causing a significant number of multiple scattering paths to exist and adversely affect the results. This cov-

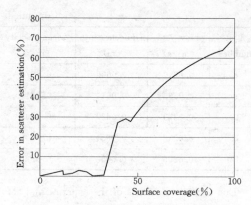

Figure 5　Error in scatterer volume estimation
vs. percentage of surface containing scatterers

erage corresponds to a mean spacing of 60mm. This spacing is analogous to water surface wavelength.

Similar results were achieved when using wedge shaped rods instead of semi-cylinders. An example of the experimental setup is shown in Figure 6, with the tube from the tannoy speaker on the left, and the microphone on the right.

Figure 6　Experimental setup for EA measurement
over wedge shaped scattering rods

4.3　Adaptation to measure water surfaces

Since the predominant wavelengths of static waves observed in section 2.2 were 100 to 250mm (which is well above the 60mm limit for the EA method as described in section 4.1) it is expected that the method is transferrable to a water surface. Therefore steps have been taken to adapt this technique to allow it to be used to measure the static component of water surfaces. The method would output a variable proportional to the volume of surface waves. A method would then be developed to infer surface wavelength and amplitude from this.

Due to the acoustic boundary conditions described in section 2.1, the continuous white noise signal could not be used as it would cause multiple reflections which would obscure the signal. Instead a short 2ms pulse has been designed to provide the acoustic signal. A modulated sinc pulse was used as a carrier, to contain frequency components from 1kHz to 30kHz.

The key element is that the frequency spectrum of this pulse has been designed to be absolutely flat in the band specified to allow for clear definition of the received EA spectrum. Figure 7 shows the time signature of the pulse, while Figure 8 shows the resulting frequency spectrum.

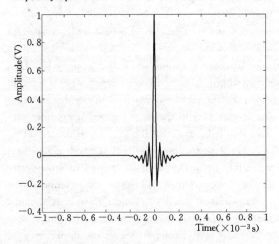

Figure 7　Time series of engineered pulse

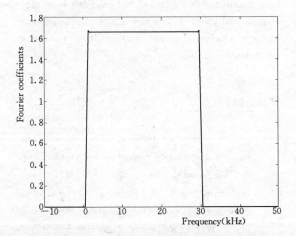

Figure 8　Frequency spectrum of experimental pulse

The imperfect frequency dependant response of the speaker has also been accounted for by adjusting the frequency spectrum to be the inverse of the speaker response, and then taking an inverse Fourier transform to obtain the corrected pulse time series.

Since the fitting of the EA spectra relies heavily on accurate measurement of source and receiver separation, and their respective heights above the surface, a water surface which may vary in height could cause significant problems, not to mention the fact that positioning the two components with a fixed separation in a real-life scenario would be equally problematic. To combat this, a method has been developed for calculating the relative positions of the components based on the acoustic signal alone. The method requires 3 microphones to be mounted in a vertical array with a known separation between the microphones (a, and b) as shown in Figure 9.

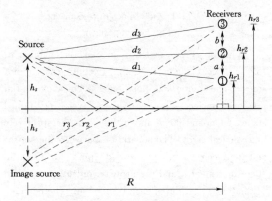

Figure 9　Experimental setup for determination of equipment position

By emitting a short spherical-wave pulse from the source, each receiver will detect two pulses, one which has travelled directly from the source, and one which has reflected from the surface. By measuring the time between the source emitting the pulse, and each receiver detecting the direct and reflected pulses, and multiplying by the speed of sound in air, the direct and reflected path-lengths (d_1, d_2, d_3, r_1, r_2, r_3) can be determined. Using the cosine rule, the internal angles of the triangle formed by lines b, d_2, d_3 can be calculated, and this can be used to calculate the horizontal separation between the source and the receiver array, R. The same process can be applied to triangles a, d_1, d_2 and ($a+b$), d_1, d_3 to calculate a total of three values for R. These are then averaged to minimise error. Trigonometry can then be used to calculate h_{r1} from d_1, r_1 and R. Two more values for h_{r1} can then be found by calculating h_{r2} and subtracting "a", and by calculating h_{r3} and subtracting "$a+b$". These values for h_{r1} can then be averaged to minimise error and hence h_{r2} and h_{r3} can be found using "a" and "b". To calculate h_s the vertical components of d_1, d_2 and d_3 can be subtracted from h_{r1}, h_{r2} and h_{r3} respectively, and the individual results averaged.

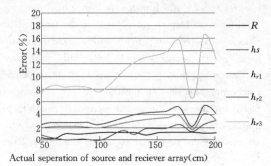

Figure 10   Error in acoustic estimation
of equipment positions

The theory was tested in an anechoic chamber using a 15mm diameter dome tweeter as a point source. The source was positioned 20cm above a flat plywood surface. Three receivers were positioned 20cm, 60cm and 80cm above the surface respectively in a vertical array. The separation between the source and the receiver array was varied from 0.5m to 2m in 10cm increments, and the acoustic pulse was used to estimate the position of each component. The percentage error in estimation of each distance for each separation is shown in Figure 10.

It can be seen that errors are predominantly lower than 5% except for h_{r1} where the actual value is small, causing the percentage error to be high. The limiting factor is the sampling rate of the data acquisition hardware. For these tests data was sampled at 96kHz, meaning that the time that each pulse is received is accurate to the nearest 1.0417×10^{-5} seconds. This means that each path-length could only be calculated to within 3.5mm. This is the cause of the error seen in Figure 10. By sampling at a higher frequency, error can be vastly minimised.

This technique has proven successful and will allow for the measurement equipment to be installed in arbitrary positions whilst still facilitating surface measurement using the excess attenuation method, and will eliminate problems caused by equipment moving due to wind, slippage etc.

These developments will allow the EA method to be investigated as a way of measuring the static component of water surface roughness by emitting multi-ple pulses and averaging the EA spectra over time to eliminate the effect of the dynamic surface component.

5. Summary & Conclusions

Methods have been described to measure acoustically both the static and dynamic components of water surface roughness.

The temporal variation of reflected spherical acoustic waves has been shown to relate to the temporal dynamics of the moving water surface. Further work is required to refine this technique, and developing a method of using a continuous signal would allow for a continuous series of intensity values to be recorded against time, which would facilitate spectral analysis.

Analysis of excess attenuation spectra has been shown to give accurate measurement of wave size for an idealised static surface. It has been shown that this surface is analogous to the static component of water surfaces so it is believed that the technique is transferable.

The promise shown by both methods shows that it feasible for a device to be created to acoustically measure the dynamic and static components of the water surfaces found in channel flow. It is thought that this information can then be used to infer flow details and bed characteristics so the device may be used for remote monitoring of water systems in rivers, treatment plants and elsewhere. Such a development would give rise to a new range of sensing technologies which would be easy to install, non-invasive, low-maintenance and low-cost.

References

[1]   A. Tamburrino, J. Gulliver. Free-Surface Turbulence and Mass Transfer in a Channel Flow. *AIChE Journal*, 2002, 48 (12): 2732-2743.

[2]   B. Shvidchenko, G. Pender. Macroturbulent structure of open-channel flow over gravel beds. *Water Resources Research*, 2001, 37 (2): 709-719.

[3]   J. R. Cooper, S. J. Tait, K. V. Horoshenkov. Determining Hydraulic Resistance in Gravel-bed Rivers from the Dynamics of their Water Surfaces. *Earth Surface Processes and Land-*

forms, *John Wiley & Sons*, *Ltd.*, 2006.

[4] M. L. Oelze, J. M. Sabatier, R. Raspet. Roughness Measurements of Soil Surfaces by Acoustic Backscatter. *Soil Sci. Soc. Am.* 2003, 67: 241-250.

[5] K. Attenborough, S. Taherzadeh. Propogation from a Point Source over a Finite Impedance Boundary. *J. Acoust. Soc. Am.* 1995, 98: 1717-1722.

[6] I. Tolstoy. Smoothed Boundary Conditions, Coherent Low Frequency Scatter, and Boundary Modes. *J. Acoust. Soc. Am.* 1984, 75: 1-22.

[7] P. Boulanger, K. Attenborough, S. Taherzadeh, T. Waters-Fuller, K. M. Li. Ground Effect Over Hard Rough Surfaces. *J. Acoust. Soc. Am.* 1999, 104: 1474-1482.

[8] S. Taherzadeh, K. Attenborough. Deduction of Ground Impedance from Measurements of Excess Attenuation Spectra. *J. Acoust. Soc. Am.* 1999, 105: 2039-2042.

47

Study on the Digital Camera Measuring System based on Digital Camera and PTV Technology

Mingxuan Ren   Jun Zheng   Huan Feng Zhaosong Qu

Beijing Sinfotek Technology Co., Ltd., Beijing, 100085

Abstract: The Digital Camera Measuring System based on digital camera and Particle Tracking Velocimetry (PTV) technique is introduced in this paper. It is applied to measure a wide range of surface velocity distribution of steady or unsteady flow in real time. The powerful post-processing analysis module in the system makes users analyze measure data efficiently and conveniently. The seeding Particles are scattered in the Hydraulic model, and the system will control cameras to capture images continuously. The system uses image processing technique to recognize particles, and the error is within 0.5 pixels. The particles are matched by the PTV technique, thereby the system can be used to measure wide range distribution of velocity. The velocity noise can be identified by automatic filtering, and the error vectors of the calculated velocity distribution could be excluded by the method. The velocity vector is interpolated by combining moving least squares method with inverse distance weighted method, so the system can generate the velocity distribution with uniform grid. The system can sustain 16 cameras to collect images in real time. And a study on the collecting image mode improves the synchronization of multi-camera collecting image greatly. In order to measure wider range distribution of velocity, the system is extended by using Local Area Network (LAN) technology. The system is widely applied to measure the velocity distribution in hydraulic, river and harbor model experiment.

Key words: velocity distribution measurement; PTV; model experiment; vector interpolation; automatic filter

1. Introduction

In the Hydraulic model experiments, it is usually needed to measure the distribution of velocity. In the early time, it is generally to use the instrument of current meter or the methods of calculating time of the surface tracer to measure velocity, but the measurement range is limited, and it often can't provide the information of the wide range distribution of velocity in the same time. Because of the lack of the measured data from the whole velocity distribution, there is a great impact on the study of hydraulics.

In the 1960s, the technology of the Particle Image Velocimetry (PIV) that can be used to simultaneously measure the instantaneous velocity of the whole velocity distribution, and make the particles with a high resolution, has been developed. In the same time, the technique is widely used in the flume experiments. In 1998, Fujita used the PIV technique to measure the wide range distribution of velocity. Since the algorithm of PIV is complicated, the amount of time used to calculate the measured wide range distribution of velocity is great, which makes the real-time measurement of wide range distribution of velocity hard. In the reality, the technology of PIV is not widely used in Hydraulic model experiments.

A branch of the PIV technology is the Particle Tracking Velocimerty (PTV). Because of its simple algorithm and fast computation, it is more economical to make use of the technology of PTV to measure the wide range distribution of velocity in the Hydraulic model experiments. Velocity distribution measuring system that has been developed based on the technology of PTV, has been used in model experiments (Xingkui Wang, 1996). The principle is: first, to throw tracer particles in the flow; the second, to record the particles' motion through the digital cameras; finally, to calculate the surface velocity distribution by using PTV technology. But, the images recorded and stored by video tape, are processed after they have been collected, and the calculation and analysis are done by the system, which makes the real-time measurement hard. As the development of digital camera technology, the technology of real-time measuring wide range distribution of velocity is realized (Yu Mingzhong, 2002). However, cameras collect images in turn, which means that the synchronization of collecting images is very poor. So, the method of collecting images will introduce cumulative error to the result, and lead to incorrect results for measuring the wide range distribution of velocity.

This paper introduces the Digital Camera Measuring System that has solved the problem of multi-camera collecting images simultaneously, and improved the synchronization of collecting images greatly. (Beijing Sinfotek Technology Co., Ltd. Open Patent No.: CN100487372C). In order to measure wider range distribution of velocity, the system is extended by using LAN technology.

The system has been being used widely in the Hydraulic model experiments in many scientific research institutes and universities.

2. System principles

2.1 Hardware system

The system uses high-resolution CCD camera for image acquisition. Each system contains 16 cameras. Cameras are installed above the Hydraulic model. The vertical distance from camera to the model is determined by the range of measurement. When camera is installed on a fixed height above model, the number of cameras can be calculated by measuring the size of model and the vertical distance from camera to model.

The signal of camera is the national standard PAL video signal. The system uses the Image Board to sample images, and then transfers the images to computer for processing. System structure diagram is shown in Figure 1.

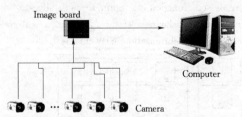

Figure 1　Hardware structure diagram

2.2 Software system

The software system mainly implements the function of recognizing particle by using the technology of image processing and extracting particles' coordinate. And then the system processes sequences of images for particle matched. The coordinate of the particle that has been matched is used in calculating particle velocity.

The system uses the method of adaptive threshold to achieve the function of image segmentation, the purpose of which is to recognize particles. The pixels belong to different particles will be divided into groups, one group per particle. Each group contains a few pixels. Calculating the center coordinate of a particle bases on calculating all the pixels' coordinate of a group. The average algorithm is used to calculate the center coordinate of a particle.

The method of image segmentation and particle recognition is used for calculating the center coordinate of a particle. After all the particles of each image frame have been recognized, the next important step is to match particle. At last, the trajectory and velocity of each particle will be obtained. In this paper, the algorithm of matching particle is based on the nearest neighbor particle matching algorithm. (Xingkui Wang, 1996).

Because of camera lens distortion, it is necessary to correct the distortion. In order to ensure the accuracy of measurement data, the system makes use of a formula of correcting camera lens distortion to calibrate it (Mingzhong Yu, 2002). The calibration error is less than 0.4 pixels.

After calculating the whole distribution of velocity, it is necessary to deal with the data that is unreal. The system has been added some functions to resolve the problem, such as deleting the error vectors, interpolating new vectors and so on. An automatic filter is applied to smooth out the velocity noise, hence, the error vectors of the calculated velocity distribution could be removed. The velocity vectors are interpolated by combining moving least squares method with inverse distance weighted method, so the system can generate the velocity distribution with uniform grid.

2.3 The synchronization of the system

The system uses multiple cameras for collecting images. Whether or not cameras collect images synchronously, the synchronization will have a direct effect on the accuracy of velocity.

If many cameras collect images in turn, synchronized delay time is increased greatly, which is due to the time each camera spends on collecting one frame image. Because of the delay, accuracy of the data will be greatly affected. Two formulas of synchronized delay time are shown [Equation (1) and Equation (2)].

$$T_{lag}=(T_g+T_s)(N_c-1) \qquad (1)$$
$$T_g=tN_f \qquad (2)$$

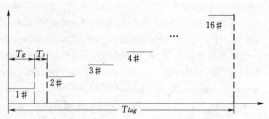

Figure 2　The collecting timing diagram

Where T_{lag} is synchronized delay time, T_g is the single-channel collecting time, and t is the time required for collecting one image, and here is 0.04s. N_f signifies the number of frames, and here is set to 8 and T_s is the channel switching time, and here is 0.01s with N_c, the total number of channels which is set to 16 here. By using the above formulas, the calculated synchronized delay time is 6.3s. The collecting timing diagram is shown in Figure 2.

It is clear that above-mentioned program still have some shortcomings. So, the system uses a new collecting program to achieve multi-channel simultaneously collecting images.

A kind of image board, which has two collecting modes that is gray mode and color mode, is used in the system. The image board includes 4 sub-boards, and the total frame rate is 100f/s. Furthermore, quick switching channels in each sub-board results that the 16 channels almost collect images at the same time. According to the characteristics of the image board, the 16 video channels are divided into four groups, and each group contains 4 video channels. Channels of a group are distributed in 4 sub-boards. It means that each group of 4 channels can synchronously collect images. The course of collecting images is shown in Figure 3. 1, 2, 3, 4 channels are considered as the same group. The four channels can

simultaneously collect images. 1, 5, 9, 13 channels are considered as the same sub-board. When 16 channels collect images at the same time, it only needs to switch between the four groups instead of switching between the 16 video channels. When switching groups is done, giving off a frame of image is a good way to avoid collecting the wrong image. It will take 0.04s to switch channels. The four groups will be switched three times. By using Formula 2 to calculate the synchronization delay time, the result is 1.08s. The system uses the new program to collect images, which can improve the synchronization of collecting images greatly, and obtain more accurate data of velocity.

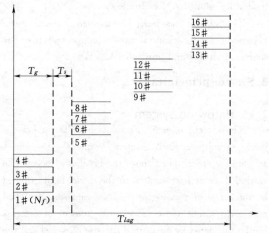

Figure 3　Sample Sequence chart in new system

When only some of the cameras are collecting images, the system will optimize current video group to obtain minimum synchronization delay time.

As shown in Figure 4 (a), there are 16 channels, and 1, 2, 4, 5, 7, 12, 14, 15 channels are selected. If there is no function to optimize and allocate video groups, all the channels of the four groups will collect images. The synchronization delay time is 1.08s.

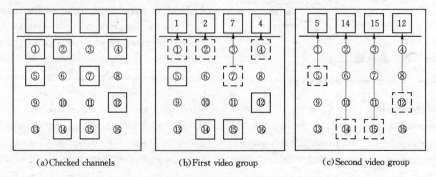

(a)Checked channels　　(b)First video group　　(c)Second video group

Figure 4　Sketch map of video group optimization

The system provides a best optimization way to collect images and reduce delay time. The principle of optimization is: if there is a channel of each group that is not selected, the channel will be replaced by the nearest channel that is selected in the same sub-board. After the channels shown in Figure 4 (a) are optimized and allocated, there are only two groups to collect images. The first group contains 1, 2, 7, 4 channels [Figure 4 (b)]. The second group contains 5, 14, 15, 12 channels [Figure 4 (c)]. And it only needs to switch one time. Through using Formula 1 to calculate the synchronization delay time, the result is 0.36s. The synchronization of the system is further improved.

3. LAN extension

With the development of model experiments, the wider range model is getting more attention, which can provide better experimental data. It is very difficult for standard system (16 cameras, 7 m installation height, covering an area of 300 m² or so) to meet the needs of wide range model. In this regard, the system provides LAN extension.

By adding the expansion program to the system, the system can measure wider range distribution of velocity. The system is consisted of a number of standard systems. The standard systems communicate with each other via Gigabit LAN. All the standard systems aggregate one network edition system. Any one of standard systems can be the role of "server" or "client". The structure is shown in Figure 5.

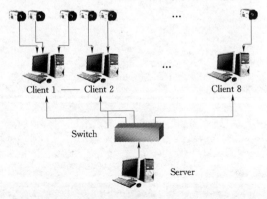

Figure 5　System extension structure diagram

The "server" sends instructions to "client", and receives the velocity data from "client", and then merges and saves the velocity data. The "server" also has other functions, such as, real-time monitoring each channel of the "client", adjusting threshold of each channel, setting parameters and so on.

The "client" receives the instructions from "server", and collects images in accordance with parameters from "server". And then, the "client" calculates the velocity data. Last, the "client" will send the velocity data to "server".

The system includes more than one computer. If computers are housed, to install the system will become complicated, and maintenance of the system will become difficult. In order to make the system more integrated, the "client" and switch will be packaged into a vertical cabinet. Operator only needs control a "server", which will control the whole system to perform simultaneous measurement. The system extension structure diagram is shown in Figure 5. The cabinet structure diagram is shown in Figure 6.

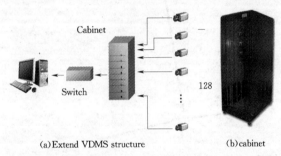

(a) Extend VDMS structure　　　(b) cabinet

Figure 6　The program using cabinet structure diagram

4. Application Cases

The system has been widely used in Hydraulic model experiments, River model experiments, Harbor model experiments and Flume experiments. Many universities and research institutes, such as Tsinghua University, Changjiang River Scientific Research Institute, Nanjing Hydraulic Research Institute and Wuhan University and so on, have used the system for measuring the distribution of velocity in River model experiments and Harbor model experiments.

The wider range of model using the network edition system for measuring the distribution of velocity is harbor model of Yangtze Estuary in Nanjing Hydraulic Research Institute. The system that is being used in the model consists of 4 standard sub-systems, and includes 60 cameras. Figure 7 and Figure 8 show the distribution of velocity obtained by the system with and without a project built on the model. Through comparing the two diagrams, it is clear

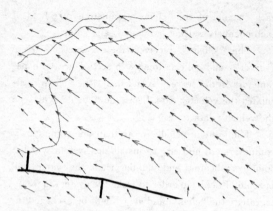

Figure 7   Distribution of velocity without project

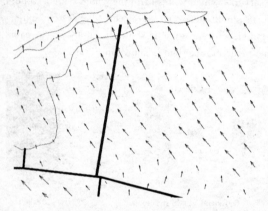

Figure 8   Distribution of velocity with project

that the distribution of velocity is changed because of building the project. So, the system can provide a visual, accurate experimental data.

In the experiment, it has measured the high tide and down emergency flow velocity in different location of the model that belongs to pre-project or different programs.

5. Conclusion

The technology of Particle Image Velocimetry, which is a non-contact and high-precision velocity measurement technique, is widely used in Hydromechanics.

The system that is developed based on the technology of PTV applies to the wide range real-time measuring distribution of velocity. The system has resolved the problems of synchronization and expansion, which improves the accuracy of velocity greatly and expands the scope of application of the system. The system is made better use in hydraulic measuring industry.

References

[1]   Fujita L, Muste M, and Kroger A. Wide Range Panicle Image Veloaimetry for Flow Analysis in Hydraulic Engineering Applications. Journal of Hydraulic Research, 1998, 36 (3): 397-414.

[2]   Wang X. K. , Pang D. M. , Wang G. . Application of Image processing technology in the physical model experiments velocity distribution measurement.

[3]   Yu. M. ZH. Study on the PTV Technique and 3D Movement of Particles. Beijing: Tsinghua University, 2002.

Remote Water Level Measurement System Based on Ultrasonic Ranging and Wireless Data Transmission Technology

Ji Hongjun   Xia Lijuan   Ren Mingxuan   Zheng Jun

Beijing Sinfotek Ltd. , Beijing, 100085

Abstract: Flow released by hydropower station in daily regulation mode results in rapid water level change in downstream waterway, and thus brings the safety problems of shipping. This paper presents a remote water level measurement system, which meets the need of real-time water level monitoring and early-warning for shipping based on ultrasonic ranging and wireless data transmission technology. The ultrasonic water level gauge is used to measure the water level in waterway, eliminating the problems associated with the manual water level gauge and measuring well, and proven to be more suitable for field water level monitoring. Through the wireless data transmission module via GPRS/CDMA network, the measured water level data can be transmitted to data center timely and accurately. The water level database is used to store and manage the real-time water level data, and provides the basis for further applications. The water level data can be queried and analyzed for a variety of applications in the system, and information of the waterway can be accessed via the network. This paper also describes the automatic early-warning function of the measurement system. Once meeting the early-warning criteria of the waterway, an early-warning message of water level will be automatically sent to the certain persons in the waterway management department, which provides decision support for shipping scheduling. The system with real time water level monitoring and early-warning function provides powerful information support for shipping scheduling and management, and has been applied in the real-time water level monitoring project in the waterway of Lancang River. The system has shown a good prospect in a wide range of applications.

Key words: remote water level measurement; ultrasonic ranging; wireless data transmission; water level early-warning

1. Introduction

The fluctuating discharge arising from the hydropower station's daily regulation may cause rapid water level change and shipping accidents in downstream channel (Yue, 2004). Thus the real-time water level measurement system is built to provide real-time data to enhance the navigation safety.

Traditionally the observation well is used to monitor the water level by manual and automatic methods. The method using the manual gauge is convenient and straightforward, but has to be operated manually. Auto-record gauge can record water level data automatically, while it works in the measuring well. The riverine terrain along most of the waterway is fairly steep, therefore the method using the manual gauge is infeasible and of low precision. Meanwhile, the measuring well cannot be dug if the bank is composed of

rugged stones. On the contrary, the ultrasonic ranging technology appears to be reliable and stable. Due to its advantage in the field water level measurement, the ultrasonic gauge has been applied to hydrologic survey recently (Zhang, 1998; Li, 2009).

Remote data transmission is an important step for real time measurement. Special network with high cost was commonly adopted to transmit the measured hydrologic data. Because of the development of the mobile communication technology and the commercialization of the data service in recent years, it is practical to transmit the hydrologic data via GPRS/CDMA network (Wang, 2005; Cao, 2009). By using national public telecommunications infrastructure, it can save the cost and improve the system stability greatly for long-running.

This paper presents a remote water level measurement system, which is based on ultrasonic ranging and wireless data transmission technology. The system provides the function of real-time water level monitoring in waterway. Also, the function of water level early-warning is developed to provide decision support for waterway safety management.

2. The structure of system

The remote water level measurement system comprises hardware and software subsystems. The hardware subsystem, which is based on dependent solar energy photovoltaic electricity, ultrasonic ranging and wireless data transmission technology, mainly consists of the device of solar panel, storage battery, controller, ultrasonic gauge and wireless data transmission module. The software subsystem, which is based on database and application programming technology, mainly consists of the water level database, water level data query and analysis module, water level information early-warning module.

Figure 1 illustrates the structure of this system.

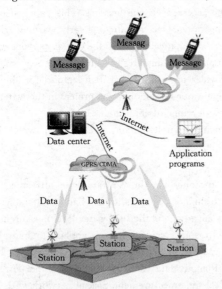

Figure 1   The structure of the system

The water level measuring stations along the waterway are equipped with the solar panel, storage battery, controller, ultrasonic gauge, wireless data transmission module and other equipments. As the power supply, the solar panel, storage battery and controller provide the power to the ultrasonic gauge and wireless data transmission instrument. The ultrasonic gauge, which is installed at a certain height above the water surface, measures the distance between the device and the water surface by sending and

receiving the ultrasonic wave. The water level can be calculated by the gauge elevation minus the measured distance. The wireless data transmission instrument transfers the water level data to the data monitoring center in real time via the GPRS/CDMA network.

The data received by the data monitoring center is stored in the water level database which can be used for further applications such as water level query, analysis and early-warning. The database system consists of B/S and C/S modes. The B/S mode provides the query function of real-time water level information on the web, and the C/S mode provides further analysis and early-warning of water level.

The remote real-time water level measurement system provides the functions of water level query and analysis. Besides, the function of water level automatic early-warning is developed to provide timely data information and decision support for navigation management department.

3. Key technology of the system

3.1   Ultrasonic ranging

The ultrasonic ranging gauge measures the distance based on the principle of ultrasonic reflection. The gauge sends ultrasonic wave and receives the wave reflected by the target object, and then calculates the distance. The speed of ultrasonic wave is influenced by the environment temperature, humidity, pressure, and etc, among which the temperature is an important factor. When the environment temperature rises by one degree, the speed of the ultrasonic wave will increase by 0.6m/s approximately. The gauge can compensate the variation caused by the change of temperature through the temperature sensor, and assure the measurement accuracy.

The ultrasonic gauge has the long service life, and is suitable for field work for its convenient installation. The gauge can meet the demand of water level monitoring with high frequency and precision. The gauge, which is installed above the water surface vertically, measures the distance between the instrument and the surface by sending and receiving the ultrasonic wave. Afterwards, the water level can be calculated from the distance and the elevation of the gauge. The ultrasonic ranging gauges with different ranges have different absolute precision. In the measurement stations, the gauges with proper ranges are deployed which meet the different water level and improve the measurement precision.

3.2 Wireless data transmission

The water level data of all measurement stations needs to be transmitted to the data center timely and accurately. In so doing, the system can overcome the limitation of manual and other measurement methods.

The GPRS and CDMA nationwide network has been established and the internet becomes very accessible, therefore the system adopts the wireless data transmission technology via GPRS and CDMA network. The water level data at the measurement stations can be transmitted to the fixed IP address in the internet. Afterwards, the data is received and stored to the database by the monitoring center server with the fixed IP address.

Every data transmission instrument has a unique ID number, which corresponds to the measurement station. When the computer receives the water level data transmitted by the instrument, the measurement station is identified by its ID number, and then the data is stored correspondingly. If the data transmission is disrupted when the instrument is working because of network jam and power failure, the system can automatically reconnect in fixed time interval until connected.

3.3 Water level automatic early-warning

If the massive data is not fed back to the navigation decision maker, it would be a great waste of resources, and the efficiency and level of navigation management would not be improved significantly.

There has to be the staffs on duty in the control center throughout the day for the traditional warning system. According to the warning information from the monitoring spots, the information will be reported to the relative departments and the relative persons in charge. To improve the warning efficiency and save the manpower, a module designed to send short message is installed in the monitoring center in this system. The mobile phone numbers of all the departments and persons in charge are set according to the type and authority of the information. When the early-warning information emerges, the system automatically sends the messages to the responsible phone numbers, and thus provides the data support for navigation management and decision-making timely.

There are several types of water level early-warning information, such as lowest navigable stage, flood or stage abrupt change. Specifically, the early-warning information is sent when the water level in the waterway is lower than the lowest navigable stage or higher than a threshold value in flood season, or the stage changes very rapidly.

4. Application case

The Lancang waterway, an international golden waterway connecting China with Southeast Asian countries, plays an important role in the economic development. The Lancang River in Yunnan province, China, is a mountainous river, and the water level rises and falls drastically. The fluctuating discharge by the electricity generation station along the river in daily regulation causes the water level changing rapidly, which has a significant influence on the shipping. To ensure the safety of the waterway shipping, the water level in the waterway needs to be measured in real time.

The water level measurement system is installed in Jinghong, Ganlanba, Guanlei stations downstream of Jinghong power station in Lancang River. Figure 2 shows the equipments in the measurement station of Ganlanba. The ultrasonic gauge in the tip of the stent is used to measure the water level of the waterway. The solar panel provides the power by collecting and transforming the solar energy. The wireless data transmission module in the device box transmits the data in real time.

Figure 2　The equipments in the measurement station of Ganlanba

The data monitoring center in Kunming, Yunnan Province, receives the real-time water level data from the measurement stations. In the system, the data can be queried and analyzed using a number of methods. The water level automatic early-warning module is developed to send the warning information to the waterway management department and the persons in charge automatically, according to the water level in Lancang River. The entire system greatly improves the efficiency in management and decision making.

5. Conclusion

The remote water level measurement system is developed based on the ultrasonic ranging and wireless data transmission technology and presented in this paper. The ultrasonic gauge is used to measure the water level in the waterway in this system. The data is transmitted by wireless data transmission module via GPRS/CDMA network. The water level information can be queried and analyzed using a number of methods, and the water level automatic early-warning function is developed as well. The system with the function of real-time monitoring and early-warning provides great support for shipping scheduling and management, and has a wide range of applicability.

References

[1] Cao Y. J., Cao N. H., Zhou H., Xiong G. Y., and Cheng X.. A Novel GPRS/CDMA Hydrological Telemetry System. Hydropower Automation and Dam Monitoring, 2009, 33 (2): 65-67. (In Chinese)

[2] Li Z., Chen X. H.. Application of ultrasonic depth detection technology in hydrological measurement of Xiluodu Hydropower Station. Yangtze River, 2009, 40 (7): 22-23. (In Chinese)

[3] Wang J. X.. GPRS Practice in Hydrological Information Inquiring. Automation in Water Resources and Hydrology, 2005 (1): 14-17. (In Chinese)

[4] Yue P. J., Wang Y. C.. Influence of release from daily storage plant on the downstream navigation and its preventive measure. Journal of Waterway and Harbour, 2004, 25 (3): 52-58. (In Chinese)

[5] Zhang Z. T., Hu H. Hu F. X., Gu G. C.. Application of Fathometer to Submerged Landform Measurement. Ocean Technology, 1998, 17 (4): 39-43. (In Chinese)

Design of Multi-point Concentration Measurement System in the Model Test of Low Level Radioactive Cooling Water Released from Nuclear Power Plants

Ruan Shiping[1]   Xu Shikai[1]   Song Zhenbo[2]   Zheng Jianchun[2]
Tong Zhongshan[1]   Wang Yong[1]

[1] *State Key Laboratory of Hydrology-Water Resources and Hydraulic Engineering,*
Nanjing Hydraulic Research Institute, Nanjing, 210029
[2] *Jinan Baud Electronic Science & Technology Co., Ltd., Jinan, 250021*

Abstract: The low level radioactive cooling water released from the nuclear power plant brings out the problem of radionuclide dispersion. Research on the pollution characteristics of the low level radioactive cooling water is very important for the environmental protection and working safety of nuclear power plants. This paper puts forward the principle and structure of a multi-point concentration measurement system (MCMS) which can be used to test the radioactive pollutant field of low level radioactive cooling water.

Key words: nuclear power plant; radionuclide; low level radioactive cooling water; pollution characteristics

1. Introduction

It's well known that nuclear power is safe, clean and efficient energy. However, the impact of radioactive leaks will do harm to the public health and ecological environment severely. Nuclear safety is the key problem of the development of nuclear power. During the normal operation of nuclear power plants, water discharge to the environment with low suspended solids, and low-level radioactive waste will cause nuclides pollution (137Cs, etc.). In the feasibility study of a nuclear power plant project, environmental impact assessment must be made through the model test and calculation to study the law of nuclides diffusion and dilution in receiving water.

In the traditional model test, Rhoda mine-B is generally used as a tracer to simulate the motion of nuclides pollution, and the concentration field is deduced from the concentration of water sampled from measuring points. Fluorescence spectrometer luminosity is used as the major instrumentation for Rhoda mine-B concentration detection, and the amount of work involved is simply too great. For example, given a model of 100 measuring points, if we take the water sample at the interval of 30 minutes (prototype), there will be about 72000 water samples to be ana-lyzed in a tide cycle. Moreover, the sampling and analysis process is very difficult to be automated.

A multi-point concentration measurement system (MCMS) was developed to detect the radioactive pollutant field of low level radioactive cooling water by Nanjing Hydraulic Research Institute and Jinan Baud Electronic Science & Technology Co., Ltd. This system can be used to measure the fluoride ion concentration of 200 points at the intervals of 5 seconds at the same time automatically.

2. Detection method and principle

2.1 Detection method

In the model test of low level radioactive cooling water from nuclear power plants, it is extremely difficult to find a reagent whose half-life can fulfill the requirement of model similarity, so the decay of nuclides usually do not be considered and this predigestion makes the results more secure. Moreover, the density difference between the solution and water cannot be too significant to avoid the density current so that the flow field can be simulated correctly.

The radioactive elements density of the near outfall zone and the far zone maybe differ by five orders of magnitude. Security threshold of this density is too small, so the instrument must has a high resolution and a wide measure range. We think electrode method is the best option.

In MCMS, NaF was used as tracer to simulate nuclides motion in receiving water and the analysis of fluoride was accomplished by ion selective electrodes (ISE) to measure the concentration field by means of multi-point measurement. The basic idea is to measure electrode potentials (voltages) and then to determine analyte (F-ion) concentrations in solution.

2. 2 Detection principle

The ion selective electrode (ISE) usually contains an inner reference electrode and a membrane which providing the interface between the sample solutions and the ISE. The potential develop across the membrane, E_{meas}, depends on the activity difference of a specific ion on each side of the membrane, E_{meas} can be expressed as[1-2]

$$E_{meas} = E_{outer\ ref} + E_{inner\ ref} + E_{junc} + E_{ISE} \quad (1)$$

where $E_{outer\ ref}$ is the potential of the outer reference electrode; $E_{inner\ ref}$ is the potential of the inner reference electrode; E_{junc} represents the various junction potentials that develop at liquid junction in the cell; E_{ISE} is the potential developed across the ion-selective membrane.

The cell potential is related to the logarithm of the concentration of the measured ion by the Nernst equation

$$E_{meas} = constant - \frac{2.303RT}{nF} \log \frac{a_{ioninner}}{a_{ionouter}} \quad (2)$$

where R is the gas constant, T is the temperature (K), F is the Faraday constant, n represents the charge on the analyte ion and a is the activity of the analyte ion.

For an ideal fluoride ISE, the cell potential is linearly related to the logarithm of the fluoride ion concentration and should increase 59.16 mV for every 10-fold decrease in the $[F^-]$. When the ionic strength of all standards and samples is constant, the response of a real fluoride ISE is described by a similar relationship

$$E_{meas} = constant - \beta(0.05916) \log[F^-]_{sample} \quad (3)$$

where β is the electromotive efficiency and typically has a value very close to unity (> 0.98).

So if we have known the relationship of E_{meas} and $\log[F^-]_{sample}$, we can deduce the F-icon concentration in water sample.

3. Hardware design

The measurement system includes electrode array, drive module array, transmitter module array, conversion module array, transmission module and software. The design block diagram is shown in Figure 1.

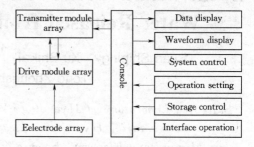

Figure 1   Design block diagram of MCMS

In this system, the electrode array which is made up of 200 fluoride sensitive electrodes and 200 saturated calomel electrodes, is designed to pick up analog signals of fluoride concentration. Drive module array which is made up of 200 drive modules, is designed to amplify and filter analog signals of fluoride concentration. Transmitter module array which is made up of 200 transmitter modules, is designed to data collection, conditioning, A/D conversion and protocol conversion. Transmission module is designed to transport signals by the data transmission line and our own protocol. Conversion module array is designed to implement the conversion of communication port and communication protocol. Software and console implement data control, data acquisition, display and operation setting.

The hardware topology is shown in Figure 2 and the function of each module is described on the right side. Figure 3 and Figure 4 are the pictures of ISE and drive & transmitter module respectively.

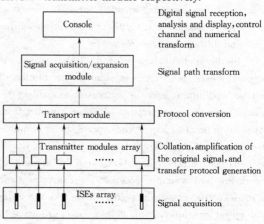

Figure 2   Hardware topology

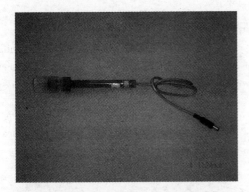

Figure 3　Ion selective electrode（ISE）

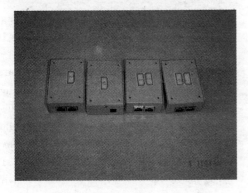

Figure 4　Drive & transmitter module

Figure 5 is the installation master plan. Through the bracket slot on the BNC connector, the electrodes are put vertically into the receiving water, and its detection signals will be transmitted through driver module and transmitter module after amplification. The data transmission line which connected with 485 extension module and ultimately the control terminal PC and the conversion modules are placed in the bracket slot.

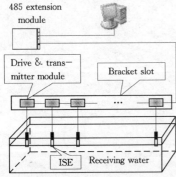

Figure 5　Installation master plan

The resolution of this system is $10^{-1} - 10^{-6}$ mol/ L, the applicable temperature is $5 - 45℃$, and its minimum time interval is 5 seconds.

4. Software design

As shown in Figure 6, console display interface involves data area, displaying content of the control system; wave area, showing the waveform in data area timely; control system area, controling the relationship between test points and inquiries test; operation area, managing test points, test intervals, alarm threshold and other basic settings; storage & control area, saving and echoing the display data; interface plan, completing interface refreshment, exit and initialization.

Data area	Control system area	
	Openration area	
Wave area	Storage & control area	Interface plan

Figure 6　Console display interface

The main function of the software is to aid the hardware systems to finish the fluoride ion concentration test and it involves: (1) test point selecting (all or part of 200 test points can be selected and controlled); (2) data access; (3) test data and waveform display; (4) test interval setting; (5) test process control.

The system consists of 6 models: control module, communication module, display module, operating module, basic setting module, and data access module. The functions of each module are listed in Table 1. Figure 7 and Figure 8 are the system's interface and data access module's interface respectively.

Table 1　Description of system module functions

Modules	Functions
Control module	The whole system (including software and hardware) control
Communication module	Data communication (including control signals and measurement data)
Display module	Measurement data or waveform display
Operating module	Operation interface
Basic setting module	System setting
Data access module	Access to measurement results

Figure 7   System's interface

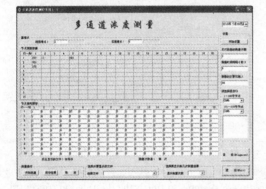

Figure 8   Data access module's interface

5. Conclusions

The system test has been completed successfully and will be used in the model test of Xudabu nuclear power plant. Comparing with traditional method, it features simple structure, easy installation and maintenance as well as wide range and high resolution. All these advantages illustrate a good prospect of application and extension for this system.

References

[1]   KORYTA J.. Ion-selective electrodes. *Annual Review of Materials Science*, 1986, 16 (1): 13-27.

[2]   APOSTOLAKIS J.C., GEORGIOU C.A., and KOUPPARIS M.A.. Use of ion-selective electrodes in kinetic flow injection: Determination of phenolic and hydrazino drugs with 1-fluoro-2, 4-dinitrobenzene using a fluoride-selective electrode. The Analyst, 1991, 116 (3): 233-237.

Ultrasonic Devices Designed for Concentration Measurements in Reservoirs

Y. J. Huang[1]   C. C. Sung[2]   J. S. Lai[3]   Y. G. Tan[4]   F. Z. Lee[5]

[1] *Postdoctoral Fellow, Hydrotech Research Institute,*
National Taiwan University (NTU)
[2] *Professor, Dept. of Engineering Science and Ocean Engineering, NTU*
[3] *Research Fellow, Hydrotech Research Institute, NTU*
[4] *Professor, Dept. of Bioenvironmental Systems Engineering, NTU*
[5] *Ph. D. Student, Dept. of Bioenvironmental Systems Engineering, NTU*

Abstract: Ultrasonic spectroscopy is a rapid, on-line, non-invasive measurement technique for suspension characterization of different particle sizes and wide-ranging concentrations. This study presents the design and fabrication of two experimental systems to simultaneously measure the sediment concentration of ultrasound wave propagation in liquid. Each monitoring system includes design and manufacture for an ultrasonic transducer, an electric transmitting and receiving circuit, and data log. This work also integrates these components into a portable type system operating underwater at 100 meters depth, and a fixed chamber type able to draw water from various depths. Measurement results show that ultrasonic attenuation variations are driven by concentration. The chamber type and portable type ultrasonic measurement system took successful measurements in the Shihmen and Tsengwen reservoirs during several typhoons.

Key words: sediment concentration; ultrasonic; attenuation

1. Introduction

The most landform of Taiwan is precipitous thus the rivers are all very short and the current of rivers is very rapid. Especially in the typhoon or torrential rain period, the flow from upstream watershed carries high concentration sediment into the reservoir. The highly-concentrated sediment not only reduces the life-span of the reservoir but also causes the waterworks not being able to deal with such muddy water. A real-time and long-term monitoring of the concentration of water flow becomes significantly important in Taiwan. In recent years there has been growing interest in monitoring of concentration in reservoir by ultrasonic diagnostic methods based on attenuation spectrometry[1]. The attenuation of sound waves through suspensions of a dispersed phase of one material in a continuous second phase could be used to characterize suspensions[2]. The measuring concentration range of the commercial instruments using traditional light scattering methods are usually up to 30000 ppm and are too narrow. The ultrasonic methods have the advantage that they can penetrate optically opaque mixtures[3]. Although ultrasonic signals can be strongly affected by particle size[4-7], temperature[8] and trapped air bubbles[9] which bring about errors in spectrometer results, it can be overcome by the design of the measuring system.

Several studies have investigated sound wave propagation through sediment suspensions. Sewell[10] in 1910 treated the fog problem by considering small, rigid, fixed spheres in viscous medium. Epstein[11] in 1941 improved on this by allowing the spheres to move and giving them elastic modulus or viscosity. Urick's measurements[12] in 1948 supported Epstein's results. Epstein and Carhart[13] in 1953 included heat conduction effects and showed that heat conduction and viscosity effects were both important for fogs but were simply additive to a good approximation. Chow[14] in 1964 researched surface tension effects showing that they are negligible for systems with solid particles or liquid droplets. Surface tension effects are important when bubbles are involved. These studies concentrate on isolated particle scattering. Interactions can be important in viscous systems, though. Batchelor[15] in 1974 stated that experiments on the simple shearing flow of a rigid sphere suspension show significant deviation from dilute suspension theory concentrations as low as two or three percent. A comment by Birkhoff in 1950[16] is illuminating in this context. He showed that a sphere falling slowly in a cylindrical tube of viscous liquid, has 100 times the cross section of the sphere, and encounters 20% more resistance than if there were no walls. This gives an idea of the extant to which particles may be expected to drag each other along. This effect is not the same as multiple scattering considered by Davis[17] in 1979, who merely applied the standard multiple-scattering formulae by Waterman and Truell[18] in 1961.

All ultrasonic methods intended for liquid flow measurements include several processes: transmission, propagation and reception of ultrasonic waves, signal conditioning, and data processing. During the processes, determining ultrasonic velocity in a movable liquid is the ultimate goal. Relative velocity is mostly measured. Ultrasonic methods can be successfully implemented in a liquid flow measurement because they have satisfactory stability, a wide dynamic range, and allow velocity measurement of electrical conductive and non-conductive liquids both in clean and impure liquids, mentioned by Bobrovnikov[19] in 1985, Hamidullin[20] in 1989 and Magori[21] in 1994. Establishing a correct and uniform dependence between liquid flow velocity and certain ultrasonic wave parameters accepted by receiving electronics is the major ultrasonic flowmeter problem. Measurement methods for principles applied in liquid flow velocity estimation can be divided into three groups based on: (a) Doppler's effect, (b) ultrasonic time propagation measurement and (c) cross-correlation technique by Sanderson and Hemp[22] in 1981.

The chamber type and portable type ultrasonic measurement system were designed and manufactured for real-time monitoring sediment concentration in the Shihmen and Tsengwen reservoirs during typhoon periods.

2. Concentration and flow velocity measurement

The parametric measurements of attenuation for sediment concentration, using spectral centroid and zero-crossing estimates of center frequency, generally yielded high attenuation coefficients than the ampli-

tude decay techniques. The variability of parametric techniques has been noted in clinical measurements by other researches[23]. Theoretically, if attenuation is well described by a power law model

$$\alpha(f) = \beta f^n \quad (n \neq 1) \tag{1}$$

where $\alpha(f)$ is the attenuation function of frequency, f is the frequency, n is the power of frequency and β is the attenuation coefficient. Specifically, a Gaussian spectrum is downshifted, but the downshift is related to β and n and the bandwidth of the spectrum decreases with depth instead of remaining constant.

Using Equation (1), the power law model is established to obtain the calibration function for concentration and attenuation. Once the power law model established for a sediment sample, we can use it to estimate the attenuation in any frequency and distance between probes.

Flow velocity measurements are performed by two acoustic transducers, based on the estimation of acoustic wave propagation time. In principle, wave traveling along the direction of flow requires less time than that traveling along the opposite direction. The acoustic wave propagates at an angle ψ with respect to fluid flow direction. Propagation time of the acoustic wave transmits from probe 1 to probe 2 denoted as t_1, transmitting from probe 2 to probe 1 denoted as t_2. With flow velocity V, the corresponding propagation times in both directions are expressed as

$$t_1 = L/[c + V\cos\psi] \tag{2}$$

$$t_2 = L/[c - V\cos\psi] \tag{3}$$

where L is the distance between probe 1 and probe 2 and c is sound wave velocity in the fluid. From Equation (2) and Equation (3) we obtain

$$c = \frac{L}{2}\left[\frac{t_1 + t_2}{t_1 t_2}\right] \tag{4}$$

$$V = \frac{L}{2\cos\psi}\left[\frac{t_1 - t_2}{t_1 t_2}\right] \tag{5}$$

3. Measurements in laboratory and field

The flow mixture of water and sediment was pumped through the chamber type measurement system, while the sample run through the gap between transmitter and receiver and the reading of sediment concentration could be obtained. Figure 1-Figure 4 show the experimental setup for the chamber type measurement system in the Shihmen and Tsengwen reservoirs. As shown in Figure 1 and Figure 3, the stratified withdraw of three-layer and six-layer system installed in the Shihmen and Tsengwen reservoirs, respectively. Sediment concentration at the selected layer along the vertical pipeline in reservoir water was measured by switching the pumping position.

Figure 1　The stratified withdraw of three-layer system installed in house of water intake in the Shihmen reservoir

Figure 2　The chamber type measurement system measured in house of water intake in the Shihmen reservoir

Figure 3　The stratified withdraw of six-layer system installed along the inclined shape of the intake structure in the Tsengwen reservoir

Figure 4　The chamber type measurement system measured in the Tsengwen reservoir

Figure 5　The prototype measurement system measured in the Shihmen reservoir

Before operating the devices in the field, experiments were conducted in the laboratory for calibration. The experiments were conducted in a 1m×0.5m ×0.5m plastic rectangular tank equipped with a speed controllable motor driving a propeller to maintain homogeneous particle suspension in liquid. The circulating flow was driven by a submerged pump. To avoid producing air bubbles, two vents were installed before the flow into the chamber. Also the experiment technique maintained a smooth and steady flow by adjusting the inlet and outlet valve. The central frequencies of transducers are designed with 1MHz and 10MHz together. The system switches to a suitable frequency for different sediment concentration automatically. The distance between transmitter and receiver is 20cm. The energy loss between transmitter and receiver was measured. Figure 5 shows the portable type of measurement system when on the boat in the reservoir.

4. Results and discussion

There are four sets of calibration data measured by four kinds of transducers. Figure 6 shows the calibration data for kaolin and reservoir sediment samples, using 1MHz, 3MHz, 5MHz and 10MHz probes. The temperature was controlled at 27℃ for both samples. Findings show a linear attenuation dependence on kaolin and reservoir sediment samples within the investigated concentration range. One can find the linear relationship of concentration and attenuation. The attenuation apparently increases as concentration increases. All the value of R^2 of regression data for both sediment samples are around 0.97-0.99. It is obvious that the higher frequency the better resolution of attenuation while concentration increases. Figure 6 plots the calibrating data for kaolin samples and Figure 7 plots the calibrating data for reservoir sediments.

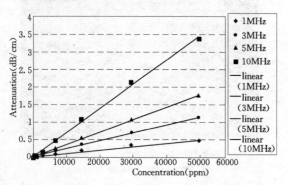

Figure 6　Attenuation vs. concentration for kaolin samples

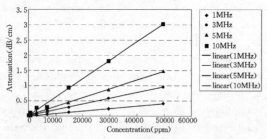

Figure 7　Attenuation vs. concentration for reservoir samples

Figure 8-Figure 10 show the results of concentration measurement by the portable measurement system at the dam, Lungchu bay and Amuping in the Shihmen reservoir during Sepat typhoon (2007). It was found that measured concentration has good agreement with the sampled data which was obtained by oven drying. The difference between results by portable type measurement system and by oven drying method would be the concentration in reservoir during

Sepat typhoon was too low to be far away from the best accuracy range of measurement system.

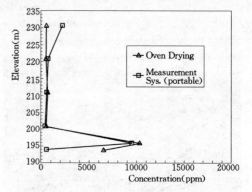

Figure 8　Measurement results by the portable measurement system at the dam in the Shihmen reservoir during Sepat typhoon, compare with those by oven drying

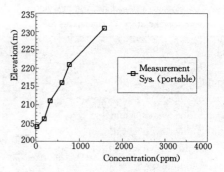

Figure 9　Measurement result by the portable measurement system at Lungchu bay in the Shihmen reservoir

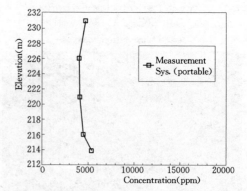

Figure 10　Measurement result by the portable measurement system at the Amuping in the Shihmen reservoir

In practice, it was difficult to measure concentration by system and sample by electromagnetic sampling machine at the same time. Figure 11 shows the results of flow velocity measurement by the portable measurement system at the Lungchu bay in the Shihmen reservoir during Sinlaku typhoon (2008).

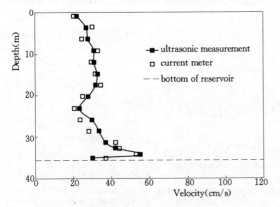

Figure 11　Measurement result by the portable measurement system at Lungchu bay in the Shihmen reservoir during Sinlaku typhoon

Figure 12 shows the chamber type measurement system that installed and gathered the different depth of samplings in house of water intake at Shihmen reservoir. Several water pumps at different level pumped water up. The water samples were directly measured by the chamber type measurement system and collected by Specific Gravity Bottle to derive the rudimentary concentration result at the same time. After a typhoon event, the measured data by ultrasonic system compared with oven drying method using the collected samplings in typhoon period. These experiments demonstrate that the total error in full scale concentration is within ±10%.

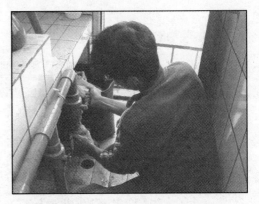

Figure 12　Photo of sampling sediment in house of water intake of Shihmen reservoir

This good tendency agreement was also found by using portable measurement system for sediment concentration measurements in turbidity currents during typhoon floods, which shows that the portable measurement system is operative and trustworthy in field.

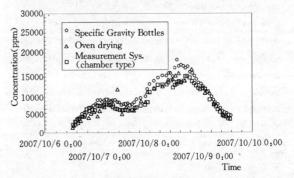

Figure 13　Measurement result by chamber type measurement system at water intake in the Shihmen reservoir during typhoon Krosa

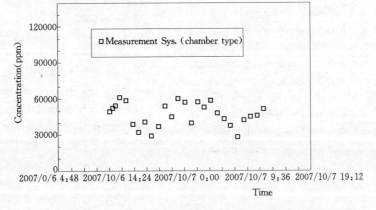

Figure 14　Measurement result by chamber type measurement system in Shihmen Power Company during typhoon Krosa (pipe route 1)

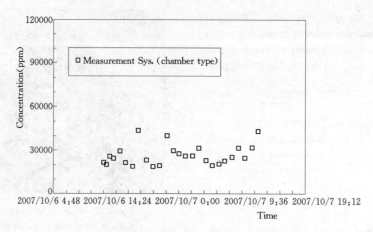

Figure 15　Measurement result by chamber type measurement
system in Shihmen Power Company during typhoon Krosa（pipe route 2）

Figure 13-Figure 15 show the results by chamber type measurement system in house of water intake at Shihmen reservoir and Shihmen Power Company during typhoon Krosa（2008）. In Figure 13, good agreement between measurement data and results by oven drying can be found. The concentration derived by Specific Gravity Bottle method is a quick and simple method. It can be used to examine the results immediately. However, the data obtained from SGB method have more error than those by oven drying. These results show that the chamber type measurement system is operative and trustworthy.

5. Conclusion

Two kinds of ultrasonic systems are designed and manufactured for real-time velocity and sediment concentration monitoring of the reservoir. Ultrasonic spectroscopy has been applied to measure velocity in many applications. It is a rapid measurement technique for suspension characterization over a wide range of particle size and concentration. The relationships of ultrasonic sediment attenuation and concentration for the Shihmen and Tsengwen reservoirs have been established. Field measurements for real-time velocity and sediment concentration monitoring demonstrate that the fixed chamber type and the prototype ultrasonic measurement system are workable and applicable during several typhoons.

Acknowledgment

We are grateful to the Water Resources Agency, Ministry of Economic Affairs, Taiwan, for data providing.

References

[1] Stolojanu V. , Praksh A. . Characterization of slurry systems by ultrasonic techniques. Chemical Engineering Journal, 2001 (84): 215-222.

[2] Harker A. H. , Temple J. A. G. Velocity and attenuation of ultrasound in suspensions of particles in fluids. J. Phys. D: Appl. Phys. , 1988 (21): 1576-1588.

[3] Judith A. Bamberger, Margaret S. Greenwood. Measuring fluid and slurry density and solids concentration non-invasively. Ultrasonics, 2004 (42): 563-567.

[4] Andrew K. Holmes, Richard E. Challis, David J. Wedlock. A wide-bandwidth ultrasonic study of suspensions: the variation of velocity and attenuation with particle size. Journal of Colloid and Interface Science, 1994 (168): 339-348.

[5] Stakutis V. J. , Morse R. W. , Dill M. , T. Bever R. . Attenuation of ultrasound in aqueous suspensions. The Journal of the Acoustical Society of America, 1955, (27): 539-546.

[6] James C. Austin, Richard E. Challis. The effects of flocculation on the propagation of ultrasound in dilute kaolin slurries. Journal of Colloid and Interface Science, 1998 (206): 146-157.

[7] Judith Ann Bamberger, Margaret S. Green-

wood. Using ultrasonic attenuation to monitor slurry mixing in real time. Ultrasonics, 2004 (42): 145-148.

[8] Ratjika Chanamai, John N. Coupland, D. Julian McClements. Effect of temperature on the ultrasonic properties of oil-in-water emulsions. Colloid and Surfaces A: Physicochemical and Engineering Aspects, 1998 (139): 241-250.

[9] James C. Austin, Richard E. Challis. Ultrasonic propagation through aqueous kaolin suspensions during degassing. Ultrasonics, 1999 (37): 299-302.

[10] C. J. T. Sewell. On the extinction of sound in a viscous atmosphere by small obstacles of cylindrical and spherical form. Phil. Trans. Roy. Soc. London, 1910, A210: 239-270.

[11] P. S. Epstein. Contributions to Applied Mechanics. Thedore Von Karman Anniversary Volume, California Inst. Tech., 1941: 162-188.

[12] R. J. Urick. The absorption of sound in suspensions of irregular particles. The Journal of the Acoustical Society of America, 1948, 20: 283-289.

[13] P. S. Epstein and R. R. Carhart. The absorption of sound in suspensions and emulsions. The Journal of the Acoustical Society of America, 1953, 25: 553-565.

[14] J. C. F. Chow. Attenuation of acoustic waves in dilute emulsions and suspensions. The Journal of the Acoustical Society of America, 1964, 36: 2395-2401.

[15] G. K. Batchelor. Transport properties of two-phase materials with random structure. Ann. Rev. Fluid Mech., 1974, 6: 227-254.

[16] G. Birkhoff. Hydrodynamics, a study in logic, fact, and similitude. Princeton University Press, 1950.

[17] M. C. Davis. Coal slurry diagnostics by ultrasound. transmission. The Journal of the Acoustical Society of America, 1978, 64: 406-425.

[18] P. C. Waterman and R. Truell. Multiple scattering of waves. J. Math. Phys., 1961, 2: 512-537.

[19] G. N. Bobrovnikov, B. M. Novozilov and V. G. Serafanov. Non-contact Flow-Meter. Plenum Publishing Corporation, 1985.

[20] V. K. Hamidullin. Ultrasonic control and measurement equipments and systems. St Petersburg University, 1989.

[21] V. Magori. Ultrasonic sensors in air Proc. IEEE Int. Ultrasonics Symp., 1994: 471-481.

[22] M. L. Sanderson and J. Hemp. Ultrasonic flowmeters - a review of the state of the art Proc. Conf. on Advances in Flow Measurement Techniques, September, 1981: 157-178.

[23] R. C. Waag. A review of tissue characterization from ultrasonic scattering. IEEE Truns. Biomed. Eng., 1984, V31: 884-893.

Development of Swirl Meter for Sewage Pumping Station Model of the Harbour Area Treatment Scheme in Hong Kong (Stage 2A)

Jiuhao Guo   Colin K. C. Wong   Joseph H. W. Lee

Croucher Laboratory of Environmental Hydraulics, Department of Civil Engineering, The University of Hong Kong, Pok Fu Lam, Hong Kong

Abstract: In the physical model study of Stonecutters Island main pumping station of the Hong Kong Harbour Area Treatment Scheme (HATS), a key design consideration is the avoidance of strong vortices and pre-swirl in the approach flow-which could impair the efficiency of pump operation. A swirl meter has been successfully developed to measure the swirl angle in the pump suction pipes. The performance of the swirl meter is tested in a swirling pipe flow with known angular momentum. LDA measurements of flow velocity are carried out to verify the swirling flow pattern. The rotational speed of the swirl meter is shown to be linearly proportional to discharge and depends on the inlet vane angle and associated velocity profile. The predictions of the rotational speed are in excellent agreement with data over a wide range of vane angles and discharges.

Key words: vortices and swirls; swirl meter; rotation; pump station design

1. Introduction

The Hong Kong Harbour Area Treatment Scheme (HATS) was implemented in 2001 to improve the water quality in Victoria Harbour and surrounding waters. Stage 1 of HATS involves the construction of a 23.6 km long deep tunnel sewerage system to collect sewage from the densely populated urban areas of Hong Kong to a centralized sewage treatment plant at Stonecutters Island. A sewage flow of 1.4 million m^3/d receives Chemically-Enhanced Primary Treatment (CEPT) followed by discharge via a 1.2km long outfall in the harbour[1]. The advanced primary treatment removes about 70%, 80%, and 50% of the BOD, suspended solids, and bacteria from the sewage effluent respectively, and has resulted in notable improvements in the water quality. In Stage 2 of HATS, the remaining urban areas of Hong Kong will be connected and the treatment capacity will be upgraded to 2.8 million m^3/d. The Stonecutters Island main pumping station Stage 2 (SCIMPS 2) is proposed to handle the increased flows.

To assist with the hydraulic design, a physical model of the pumping station has been designed and constructed. A key design consideration in the layout of the pump sump chamber and suction pipe arrangements is the avoidance of strong vortices and pre-swirl in the approach flow – which may affect pump efficiency and cause vibrations and cavitations[2-5]. On the other hand, tight space constraints for the HATS Stage 2 project precluded the provision of ideal approach flow conditions to the pump intakes, and the optimal sump chamber design and inlet arrangements need to be tested experimentally.

The swirl angle – defined by the ratio of tangential velocity to axial velocity, is an important indicator of the degree of pre-swirl. In the hydraulic design of pump stations, a generally accepted guideline is to limit the swirl angle to below $5°$[2,4]. The swirl angle can be indicated by using a swirl meter – a four-bladed vane usually mounted near the pump inlet to indicate the vortex strength and swirl condition of the flow. Studies of the performance of swirl meter have been carried out by several investigators. Hattersley introduced the swirl meter for indicating the swirl condition[3]. Experiments have also been conducted by Lee and Durgin[6] using a specially designed swirl generator. By adopting certain assumptions based on limited velocity measurements, a theoretical framework was developed to correlate the rotational speed with discharge and swirl angle.

In the hydraulic model test of SCIMPS 2, a swirl

meter has been developed to measure the swirl angle. A swirling pipe flow is generated in the laboratory using a swirl generator. The cross-section velocity field is measured using Laser Doppler Anemometry (LDA). A theoretical relation between the rotational blade speed, discharge and vane angle is developed. The application of the swirl meter to measure the flow field in the SCIMPS 2 model is illustrated.

2. Theory

Consider a steady uniform pipe flow with a known angular momentum injection at the entrance to the pipe (Figure 1). A steady flow enters the pipe from an inlet tank. The angular momentum is generated by a swirl generator that consists of a set of circumferentially mounted guide vanes that directs the flow at a preset uniform angle to the radial direction. For short lengths, angular momentum is conserved, and a swirling flow with axial symmetry is created.

The swirl angle at any point can be defined as:

$$\theta = \tan^{-1}(v_t/v_a) \tag{1}$$

where v_t = tangential velocity, and v_a = axial velocity.

For the bulk flow, an average measure of swirl can be indicated by the ratio of the velocity at the blade tip to the axial velocity:

$$\theta = \tan^{-1}\left[\frac{\pi d N}{Q/A}\right] \tag{2}$$

where d = tip-to-tip diameter of the swirl meter blade ($d = 0.75$ of pipe diameter D), N = rotational speed of swirl meter, Q = discharge, and A = pipe cross-sectional area.

The angular momentum flux across a control surface can be expressed as follows:

$$\vec{L}\int_{A} \rho\, \vec{rU}(\vec{nU})\mathrm{d}A \tag{3}$$

At the inlet, the flow enters the pipe radially through the swirl generator and guided by the vanes inclined at angle α from the radial direction. The angular momentum at inlet can be shown to be

$$L_{in} = \rho \frac{R_{in}}{A_{in}} Q^2 \tan\alpha \tag{4}$$

where Q = discharge, and R_{in}, A_{in} = radius and inflow surface area of the swirl generator respectively (Figure 2). In the pipe flow, the angular momentum is given by:

$$L_p = 2\pi\rho\int_{0}^{R_p} u(r)v(r)r^2\,\mathrm{d}r \tag{5}$$

Our velocity measurements show that the flow resembles a combination of a forced vortex near the centre of the pipe, and a free vortex beyond a certain radius r_l. The tangential velocity can be written as:

$$v(r) = \omega r \qquad \text{for } 0 \leqslant r \leqslant r_l, \text{ and}$$

$$v(r) = \frac{\omega(r_l)^2}{r} \qquad \text{for } r_l \leqslant r \leqslant R \tag{6}$$

where $v(r)$ is the tangential velocity and ω is the rotational speed of the vortex, and R_p = pipe radius. Adopting a uniform axial velocity (valid for small vane angles), we have:

$$u(r) = U = \frac{Q}{\pi R_p^2} \tag{7}$$

Using Equation (4), Equation (5), Equation (6) and Equation (7), by angular momentum conservation, the rotational speed of swirl meter can be related to the vane angle and discharge flow as:

$$N = \frac{R_{in}}{KA_{in}R_p^2}Q \tan\alpha \tag{8}$$

where K is a velocity coefficient that depends on the velocity profile and hence the vane angle. For large vane angles, the axial velocity is not uniform, but Equation 8 can be shown to retain the same form.

3. Experiments

Figure 1 shows the experimental setup. A recirculation system is driven by a 4.6kW pump connected to a frequency inverter. A 3.6m long pipe ($D = 2R_p = 120$mm) is connected to an inlet tank. The initial part of the pipe is made of Perspex for the purpose of flow

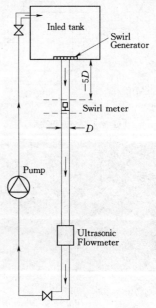

Figure 1　Swirling pipe flow

observation. The inlet from tank to pipe is fitted with a swirl generator. A swirl meter is installed at a distance of $5D$ downstream of the inlet. The discharge is controlled by a frequency inverter and monitored by a Controlotron 1010 ultrasonic flow meter.

The design of the swirl generator follows the work of Lee and Durgin (1987). A total of 16 aluminum guiding vanes are uniformly mounted around the periphery of the generator. Each vane can be adjusted to obtain different vane angles α (Figure 2).

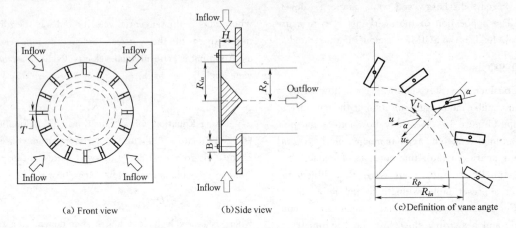

(a) Front view   (b) Side view   (c) Definition of vane angte

Figure 2   Swirl generator

Figure 3 shows the swirl meter which consists of four aluminum blades that are mounted on an axially aligned aluminum shaft supported on two stainless steel ball bearings. The tip-to-tip blade diameter is $d = 93\,\text{mm}$ ($1.5R_p$) and the blade length is $74.4\,\text{mm}$ ($1.25R_p$). Figure 4 shows a photo of the swirl generator and the swirl meter.

For each vane angle, the rotational speed of the swirl meter was measured for discharges $Q = 0$ to $9\,\text{L/s}$. The tested vane angles include $\pm15°$, $\pm10°$, $\pm7.5°$, and $\pm5°$, Negative values of vane angle lead to clockwise rotational flow and vice versa for positive angles.

In addition, the detailed velocity distribution is measured by using Laser Doppler Anemometry (LDA). A two-component Dantec LDA fibre-optic system in backscatter configuration with a 2W Argon-ion laser is used. Both the axial and tangential components are measured at a number of locations along a radial transect. By enclosing the measurement section in a Perspex square box, laser light distortion near the curvature wall is minimized.

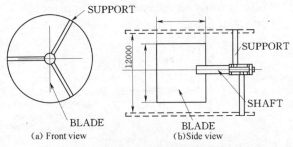

(a) Front view   (b) Side view

Figure 3   Swirl meter

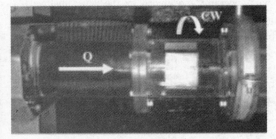

(a) Swirl generator   (b) Swirl meter

Figure 4   View of swirl generator and swirl meter in an experiment

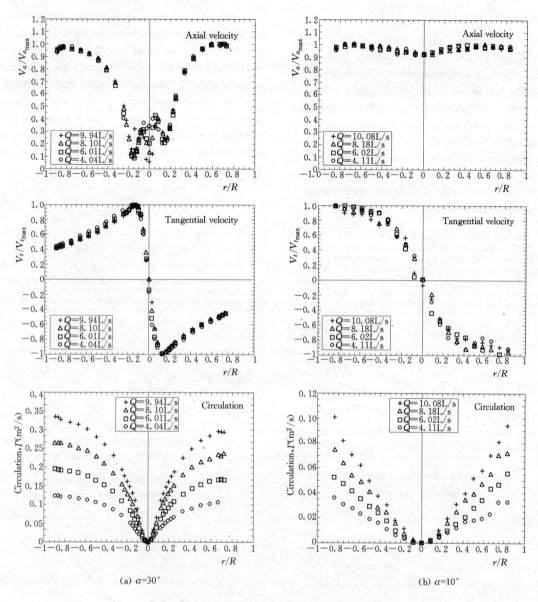

(a) $\alpha = 30°$ (b) $\alpha = 10°$

Figure 5   Measured velocity profile in swirling pipe flow

4. Results and Analysis

4.1 Velocity distribution

    Figure 5 (a) and Figure 5 (b) show the measured velocity profiles for two vane angles. In general, the flow field is complex. For a strong vane angle of $\alpha = 30°$, a Rankine vortex type of flow is indicated [Figure 5 (a)]. Near the center, the tangential velocity varies linearly with radius and reaches the maximum at a certain radius, beyond which the velocity falls off like a free vortex. The vortex behavior is indi- cated more clearly by the variation of circulation (bottom plot) with radius; the circulation increases with r up to $r_l \approx 0.6 R_p$ beyond which it is conserved. Furthermore, the strong swirl and pressure gradient effect lead to an axial velocity defect, with a decrease of axial velocity in the center. On the other hand, for a weaker swirl, $\alpha = 10°$, the axial velocity is rather uniform, and the tangential velocity and circulation is indicative of a flow dominated by a forced vortex [Figure 5 (b)]. For the discharges tested, self similarity is observed for profiles of both components, which suggests that swirl angle is independent of the discharge.

4.2 Swirl meter test

Figure 6 shows that the rotational speed (N) increases linearly with discharge (Q) for each vane angle. The best fit curves are found to have negative y-intercept. This indicates that some "initial discharge" is needed to force the swirl meter to overcome the inertia before rotation; alternatively the y-intercept can be interpreted as the bias if there is no flow. This y-intercept should be constant for all vane angle settings if the same swirl meter is used; the results suggest an approximate value of -3.98.

Figure 7 shows the comparison between inlet vane angle and swirl angle defined by v_t at $r = d$ [Equation (2)]. A linear correlation can be established between vane angle and swirl angle. The swirl angle is about 2 times of the vane angle. If half of the blade diameter is used, the swirl angle is found to be approximately equal to vane angle. This relationship suggests that the swirl meter provides an effective and accurate measurement of the average swirl in the pipe flow.

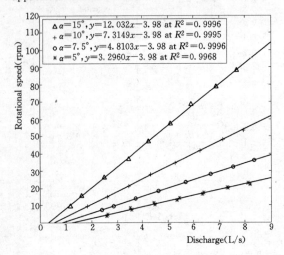

Figure 6　Variation of rotational
speed with discharge

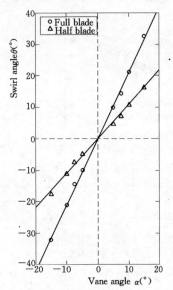

Figure 7　Comparison of inlet vane angle
and measured swirl angle

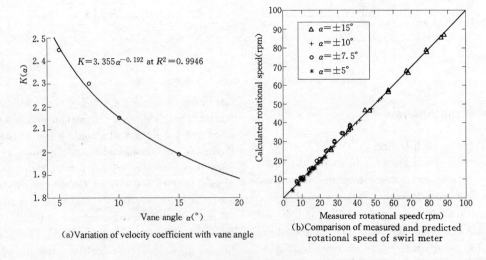

(a)Variation of velocity coefficient with vane angle

(b)Comparison of measured and predicted
rotational speed of swirl meter

Figure 8　Correlation of rotational speed of swirl meter with discharge
and inlet vane angle

To correlate the rotational speed of swirl meter with inlet vane angle and discharge, the velocity coefficient K in Equation (8) needs to be determined. Results show that K is a function of the vane angle, as shown in Figure 8 (a). It generally decreases as the vane angle increases. A best-fit curve can be obtained as:

$$K = 3.355\alpha^{-0.192} \qquad (9)$$

By substituting Equation (9) into Equation (8), the rotational speed of swirl meter can be predicted. Figure 8 (b) shows that the prediction is in excellent agreement with the data.

5. Application

Figure 9 shows the isometric view of the proposed design of SCIMPS 2 and some key dimensions in prototype scale. The station includes two rectangular pump sumps each of dimensions 40.7m×7.5m×37.5m. The pumping station handles a combination of flows from Stage 1 and Stage 2 of HATS. The inflow to each chamber is pumped to elevated wastewater treatment units by a set of 4 pumps each of 3.95m³/s capacity; the system peak flow is 31.6m³/s.

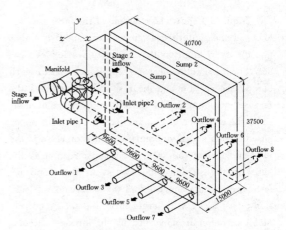

Figure 9 Isometric view of the pump sump (not to scale; unit: mm)

A 1 : 12.8 Froude scale physical model is constructed. The swirl meter is installed in each suction pipe at a distance of 620 mm (5D) from the suction-pipe inlet. In each experiment, the rotational speed is measured by the swirl meter. In addition, the velocity distribution inside each sump is measured using Acoustic Doppler Velocimetry (ADV).

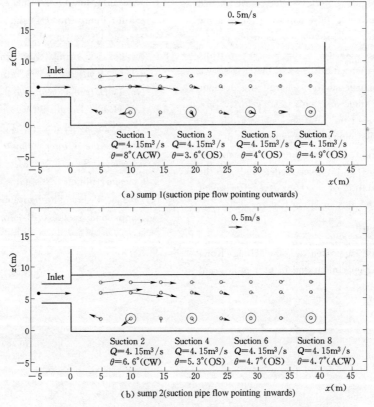

(a) sump 1(suction pipe flow pointing outwards)

(b) sump 2(suction pipe flow pointing inwards)

Figure 10 Measured flow field and swirl angle for SCIMPS 2
("ACW" =anticlockwise, "CW" =clockwise, "OS" =oscillation)

Figure 10 shows an example of the measured velocity in the centerline section of the sump chamber and the swirl angle in the suction pipes for two cases. The swirl angle measurement shows that for each sump, the suction pipe nearest to the inlet has the largest swirl angle. The velocity measurement indicates the existence of two circulation cells with opposite direction, with a dominant circulation close to the inlet and having the same rotational direction as the swirl in the suction pipe closest to the sump inlet. These results suggest that the swirl angle in suction pipe 1 and 2 with larger magnitude and relatively stable direction are governed by the stronger circulation near the inlet. In contrast, the swirl angles measured in the downstream suction pipes are generally weaker and more oscillatory in character.

6. Concluding remarks

A swirl meter has been designed for the measurement of swirl in hydraulic models of pumping stations. The performance of the swirl meter is tested in a swirling pipe flow with known angular momentum. LDA measurements of flow velocity are carried out to verify the swirling flow pattern. Results show that the swirl meter can provide a reasonable indication for the average rotation in the complex swirling flow. The rotational speed of the swirl meter is shown to be linearly proportional to discharge and depends on the inlet vane angle and associated velocity profile. The predictions of the rotational speed are in excellent agreement with data over a wide range of vane angles and discharges. The swirl meter has been successfully applied in the physical model study of the sewage pumping station of Harbour Area Treatment Scheme (HATS) Stage 2 in Hong Kong.

Acknowledgement

This study is supported by the Hong Kong Drainage Services Department and in part by a grant from the Hong Kong Research Grant Council (HKU 714309E).

References

[1] Choi K. W., Lee J. H. W., Kwok K. W. H., Leung K. M. Y.. Integrated stochastic environmental risk assessment of the Harbour Area Treatment Scheme (HATS) in Hong Kong. *Environmental Science and Technology*, 2009, 43 (10): 3705-3711.

[2] Clark P. B.. Some observations on the hydraulics of pumping stations. *Proceedings of the 2nd International Conference on Water and Wastewater Pumping Stations*, BHR Group, 2002.

[3] Hattersley R. T.. Hydraulic Design of Pump Intakes. *Journal of the Hydraulic Division, Proceedings of the American Society of Civil Engineers*, USA, March, 1965: 223-249.

[4] Hydraulic Institute. American National Standard for Pump Intake Design. Parsippany, New Jersey, USA, 1998: 28.

[5] Knauss J. Swirling flow problems at intakes. *IAHR Hydraulic Structures Design Manual No. 1, Balkema, Rotterdam*, 1987.

[6] Lee H. L., Durgin W. W.. The performanceof cross-vane swirl meters. *Vortex flows: presented at the Winter Annual Meeting of the American Society of Mechanical Engineers*, Chicago, Illinois, November, 1987: 16-21.

[7] Prosser M. J.. The hydraulics design of pump sumps and intakes, *British Hydromechanics Research Association*. Cranfield, Bedford, UK, 1977.

Topic Ⅱ

Advancement in Field Investigation for Hydro-Environmental Engineering

The Tai Hang Tung Storage Scheme for Urban Flood Control: Model Study and Field Performance[*]

Joseph Hun-wei Lee

Department of Civil engineering, The University of Hong Kong, Hong Kong

Abstract: The Tai Hang Tung Storage Scheme (THTSS) aims to solve the flooding problem in Mongkok-a highly densely populated commercial and industrial district of West Kowloon, Hong Kong. For a 50 - year design storm, a maximum upstream inflow of $Q=110$ m³/s must be conveyed within acceptable water levels. The heart of the flood detention scheme is a 100000 m³ underground storage tank, a triple side weir system, and transitions to the upstream inlets and downstream channels. The award winning THTSS design was developed from a combination of heuristic reasoning, physical and numerical model studies. The field performance of the THTSS during heavy rainstorms since commissioning in 2005 is reviewed. The complex hydraulics in the triple side - weir system is first studied via a 2D numerical solution of the shallow water equations using a shock - capturing finite volume method; the flow behavior and weir overflow distributions are compared with experiments. Measured flood levels in the surcharged system are studied using a 1D unsteady model for key storm events and different weir level settings. The post - operation audit offers insights into the hydraulic control and optimal operation of the system.

Key words: urban Flood Control; model Study; field Performance

---

[*] Keynote lecture

Research on Eco-Hydro-Morphological River Processes by Combining Field Investigations, Physical Modeling and Numerical Simulations[*]

K. Blanckaert[1,2,3,4]   W. van Balen[3,5]   A. Duarte[2]
V. Dugué[2]   X. F. Garcia[4]   W. Ottevanger[3]   M. T. Pusch[4]
A. Ricardo[2]   I. Schauder[4]   A. Sukhodolov[4]   W. S. J. Uijttewaal[3]   R. Wilkes[4]

[1] *State Key Laboratory of Urban and Regional Ecology, Research Center for Eco-Enviro-nmental Sciences (RCEES), Chinese Academy of Sciences (CAS), Beijing, China*
[2] *Laboratory of Hydraulic Constructions (LCH-ENAC), Ecole Polytechnique Fédérale Lausanne (EPFL), CH-1015 Lausanne, Switzerland*
[3] *Faculty of Civil Engineering and Geosciences, Delft University of Technology, Delft, The Netherlands*
[4] *Department of Ecohydrology, Institute of Freshwater Ecology and Inland Fisheries, Berlin, Germany*
[5] *HKV Consultants, Postbus 2120, 8203 AC Lelystad, The Netherlands*

Abstract: Hydrodynamical, morphodynamical and ecological river processes and their multiple linkages occur in an infinity of different configurations and over a wide range of spatial and temporal scales. This paper illustrates a research methodology that consists in combining field investigations, physical modeling in the laboratory, and numerical simulations in order to develop generic insight and tools for engineering and management of the river environment. This combined methodology is illustrated in research on (i) the macroscale characteristics of the velocity distribution and their relation to long-term and large-scale planimetric river processes, (ii) the flow field in the vicinity of the river bank and its importance with respect to bank erosion and, (iii) linkages between the characteristics of the mean flow and the turbulence on the one hand, and the behavior of invertebrates on the other. The reported research strongly relies on the use of state-of-the-art measuring instruments as well as numerical techniques.

Key words: field investigation; physical model; numerical simulation; meander; hydrodynamics; morphology; ecohydraulics

1. Introduction

Hydrodynamical, morphodynamical and ecological river processes and their multiple linkages occur in an infinity of different configurations and over a wide range of spatial and temporal scales. This complicates the development of generic insight and tools for engineering and management of the river environment.

The present paper illustrates a research methodology that consists in combining field investigations, physical modeling in the laboratory, and numerical simulations. Field investigations allow identifying and observing the processes of relevance in the natural environment. But their analysis and interpretation is complicated by the lack of control on the natural river environment and by interferences between all simultaneously occurring processes. Laboratory experiments in simplified schematized configurations allow isolating and/or accentuating certain processes under controlled condition. Moreover, they allow investigating the dependence of these processes on the control parameters as well as defining guidelines and require-

---

* Keynote lecture

ments for the numerical modeling of these processes. Field investigations and laboratory experiments are inherently limited to a small number of configurations. Numerical models allow broadening the parameter space. But they are only reliable after validation by means of the experimental data provided by the field and laboratory investigations.

The objective of the present paper is to illustrate the application of this combined methodology to investigate processes occurring on different temporal and spatial scales. Section 3 focuses on the macroscale characteristics of the velocity distribution and their relation to long-term and large-scale planimetric river processes. Section 4 highlights the flow field in the vicinity of the river bank and its importance with respect to bank erosion. Section 5 reports research on linkages between the characteristics of the mean flow and the turbulence on the one hand, and the behavior of invertebrates on the other. The reported research strongly relies on the use of state-of-the-art measuring instruments (Section 2) as well as numerical techniques (Section 3, Section 4).

2. Measuring instruments

Enhancing the understanding of hydrodynamical, morphodynamical and ecological river processes and their linkages requires accurate knowledge of the three-dimensional flow field as well as the large turbulent structures.

The laboratory experiments reported in this paper made use of an Acoustic Doppler Velocity Profiler (ADVP, see Figure 1) developed at Ecole Polytechnique Fédérale Lausanne (EPFL, Switzerland). The ADVP's working principle and estimates of the accuracy in themeasurements have been reported by Lemmin and Rolland[1], Hurther and Lemmin[2], Blanckaert and Graf[3], Blanckaert and de Vriend[4], Blanckaert and Lemmin[5]. Blanckaert[6] discusses the methodological advantages of the ADVP over commercially available velocimeters. The ADVP measures profiles of the three-dimensional velocity vector $\vec{v}(t) = (v_s, v_n, v_z)(t)$. The velocity vector is decomposed in an orthogonal reference system with curvilinear streamwise s-axis along the river centerline, transverse n-axis pointing towards the left bank, and vertically upward z-axis. The velocity range to be measured imposes a lower bound on the measuring frequency, whereas the maximum flow depth imposes

an upper bound to the measuring frequency[1]. The measuring frequency typically adopted in the laboratory experiments is about 30 Hz, which is sufficient to resolve the large scale turbulent structures. The profiling capacity of the ADVP enables measurements at high spatial resolution and resolving turbulent coherent structures.

Figure 1　Acoustic Doppler Velocity Profiler (ADVP) in the laboratory experiments (the horizontal size is about 25cm)

The upper limit on the measuring frequency imposed by the flow depth does not allow resolving the large scale turbulence structures in field configuration with flow depths larger than about 1m by means of acoustic velocity profilers. Therefore, Nortek Acoustic Doppler Velocimeters (ADV) that measure the three-dimensional velocity vector with high temporal resolution in one single point were applied in the field investigations. In order to perform measurements with high spatial resolution, simultaneous measurements were made with an array of ADV's mounted on a bridge (Figure 2).

Figure 2　Array of Nortek Acoustic Doppler Velocimeters (ADV) in the field investigations

3. The velocity distribution and its relation to long-term and large-scale river planform processes

Planform river processes occur on geological time scales and large spatial scales. This section focuses on the migration of meandering rivers, as illustrated in Figure 3. Meander migration occurs through erosion of the outer bank and corresponding accretion at the opposite inner bank.

Figure 3　Illustration of the migration of meandering rivers (picture from internet)

Field observations[7] indicate that the meander migration is largely driven by the hydrodynamics and can be modeled as:

$$M \sim \Delta U_s \qquad (1)$$

Here M is the rate of erosion at the (concave) outer bank, which is equal to the accretion rate at the (convex) inner bank because the width does not significantly change in the process of meander migration. ΔU_s is the excess of velocity at the outer bank with respect to the average velocity in the cross-section, for conditions corresponding to the bankfull discharge. Hence the knowledge of the velocity field on the spatial scale of a single meander bend for one specific flow condition can enhance the understanding of long-term and large-scale meander migration.

A field investigation was performed on a bend on the Ledra River (Figure 4 and Figure 5). IGB conceived the experimental set-up and provided the instruments and logistics of this field campaign that was coordinated by A. Sukhodolov. Flow depth and discharge varied by less than 5% during the measuring campaign, allowing for field measurements with an unprecedented spatial resolution on grids with more than 500 measuring points per cross-section. Six cross-sections were measured in detail in the bend illustrated in Figure 4 and Figure 5. The investigated reach has a nearly constant width of about 15m, an overall-averaged water depth of about 1.5m, and a discharge of about 18m³/s. Flow is subcritical, $Fr \approx 0.3$. The river is relatively narrow with an aspect ratio of about 10 and rather strongly curved with a ratio $R_{max}/B = 3$, where R_{max} is the centerline radius of curvature in the apex.

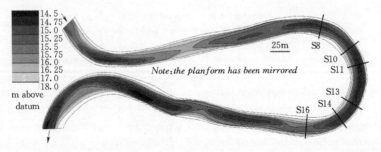

14.5
14.75
15.0
15.25
15.5
15.75
16.0
16.25
17.0
18.0
m above datum

25m　S8
S10
S11

Note: the planform has been mirrored

S13
S14
S16

Figure 4　Planform and bathymetry of the meandering Ledra River (Courtesy A. Sukhodolov)

Figure 6 illustrates the distribution of the depth-averaged velocity vectors (U_s, U_n), computed from measurements in about 30 vertical profiles with a vertical spacing of 0.1m between points. The velocity distribution is slightly inwards skewed in the cross-section S8 in the beginning of the bend, which may be attributed to the change in curvature. The velocity distribution is clearly outwards skewed in the cross-section S10, which requires outwards mass transport between S8 and S10. The maximum velocity excess at

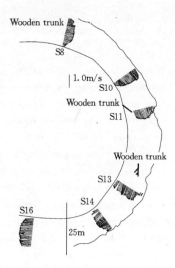

Figure 5　Depth-averaged velocity
vectors (U_s, U_n) measured in the
Ledra meander bend

the outer bank clearly occurs in this cross-section S10 situated upstream of the bend apex. The core of maximum velocities has shifted inwards in cross-section S11. Surprisingly, the velocities are nearly uniformly distributed over the width in the second part of the bend. A slight outwards shift of the velocities occurs at the exit of the bend, which may be attributed to the decreasing curvature.

The analysis and interpretation of the velocity distribution in this meander bend are complicated by various factors typical of natural rivers. Wooden trunks that affect the hydrodynamics are present in the outer part of the bend upstream of cross-section S8, at the water surface in the inner part of the bend upstream of cross-section S11, and in the middle of the cross-section upstream of cross-section S13 (Figure 5). Moreover, vegetation, outer-bank benches and local bank erosion created an irregular geometry of the outer bank that considerably affects the flow.

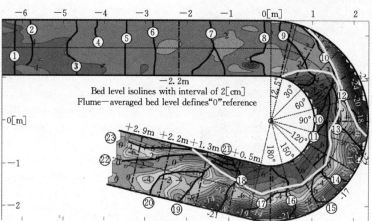

Figure 6　Isolines of the bed level with an interval of 0.02m derived from echosounder measurements [Additional ADVP measurement with higher spatial resolution are available in the indicated cross-sections. The position of dunes is based on photographs. The flume-averaged bed level defines the reference level. The white lines delineate approximatively the point bar and pool (bottom).]

Experiments in a physical laboratory model allow investigating the velocity distribution and the underlying processes under controlled conditions, without parasitical influences of wooden trunks, vegetation or irregular geometries. Moreover recent progress in instrumentation allows measuring the bed and water surface topographies (echosounders) and the 3D flow field (Acoustic Doppler Velocity Profiler) with unprecedented spatial and temporal resolution. Laboratory investigations concerning the bed topography have often been carried out with small flow depths[8],

or under dynamic conditions with migrating bed-forms[9] that did not allow for detailed velocity measurements whereas investigations of the flow field have often been carried out over schematized bed topographies[10-11]. Laboratory investigations including detailed measurements of the flow and the bed topography are extremely scarce[12-14].

The reported laboratory experiment was performed in the flume shown in Figure 6, which consists of a 193° bend of constant centerline radius of curvature, $R=1.7$m, preceded by a 9m long straight inflow reach and a 5m long straight outflow reach. The width was constant at $B=1.3$m. The bed consisted of mobile sand, with a mean diameter $d=2$mm. The flow and sediment discharges of $Q=0.089$m³/s and $q_s=0.023$kg/(m·s), respectively, resulted in a flow field and a bed topography that were in equilibrium and representative of their counterparts in natural sharp meander bends. The flume averaged velocity and flow depth were $U=0.49$m/s and $H=0.141$m, respectively, yielding a Froude number of $Fr=0.41$ The curvature ratio was $R/B=1.31$ and the aspect ratio was $B/H=9.2$.

Figure 7 shows the distribution around the flume of the normalized depth-averaged streamwise velocity. U_s/U. Moreover, it illustrates the location of the first moment (center of gravity) of the U_sh pattern as well as the vector ($\langle\!\langle U_sh\rangle\!\rangle,\langle\!\langle U_nh\rangle\!\rangle$) ($\langle\!\langle\rangle\!\rangle$ represents cross-sectional averaged values). Similar to the Ledra bend, the maximum velocities are not systematically found over the deepest part of the cross-section. A potential vortex velocity distribution establishes just downstream of the bend entry with maximum values near the inner bank. The maximum velocities are subsequently found over the pool which requires strong outwards mass transport. The maximum value of $U_nh/UH\approx0.7$ indicates that the flow seems to go straight on and to collide with the outer bank at an oblique angle at about $60°$ in the bend. A horizontal recirculation zone is observed over the shallow point bar, similar to recirculation zones observed in natural rivers[15-17]. Flow does not separate at the bend entry, but only at about $40°$ in the bend, which can be attributed to the discontinuity in curvature at the bend entrance that leads to pronounced local accelerations/decelerations in the inner/outer-half of the cross-section and corresponding inwards mass transport that opposes flow separation. Therefore, this behavior is characteristic of zones of pronounced curvature increase. The flow recirculation zone reaches its maximum width in the cross-section at $80°$, where it spans about $2/3$ of the total width. The maximum velocities move again inwards downstream of the cross-section at $90°$, accompanied by inwards mass transport. A pronounced outwards shift of the maximum velocity occurs at the bend exit, with corresponding strong flow deceleration over the shallow at the inside of the bend. Velocities tend

to uniformise in the straight outflow reach. The pattern of depth-averaged streamwise velocities (Figure 7), leaves a clear footprint on the observed dune pattern (indicated in Figure 6), which is representative for the migration speed of the dunes and the related sediment transport rate. The overall features of the depth-averaged velocity pattern in the Ledra bend (Figure 5) and the laboratory bend (Figure 7) are quite similar. Gradients are more pronounced in the latter, however, which is due to the more pronounced curvature.

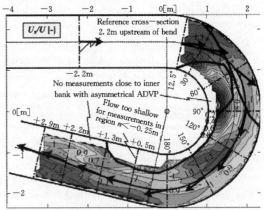

Figure 7　Normalized depth-averaged streamwise velocity, U_s/U [-] [Isoline pattern based on high-resolution measurements in the indicated cross-sections. The black lines delineate approximatively the point bar and pool. The black line with vectors indicates the first moment of the distribution of U_sh/UH, and the vectors ($\langle\!\langle U_sh\rangle\!\rangle$, $\langle\!\langle U_nh\rangle\!\rangle$) /$UH$]

Blanckaert[6] has made a detailed analysis of the mechanism underlying the velocity redistribution by means of a term-by-term analysis of the depth-averaged momentum equation. This analysis revealed that the water surface gradient is the principal mechanism with respect to velocity (re) distribution. Inertia and curvature-induced secondary flow were also found to be processes of dominant order of magnitude, whereas turbulence was found to play only a minor role. The water surface gradient is mainly determined by the curvature and by changes in curvature. Curvature gives rise to a transverse tilting of the water surface (commonly called superelevation) which is inversely proportional to R. Hence, changes in the curvature throughout a bend lead to corresponding changes in the transverse water surface slope, and changes in the downstream water surface slope.

The dominant processes with respect to the velocity redistribution can be captured by means of a one-dimensional (1D) numerical model. Despite the rapid evolution of computational power, 1D models are required for the prediction of long-term and large scale river processes.

Since the seminal models of Engelund[18] and Ikeda, Parker and Sawai[19], mathematical models of meander hydrodynamics and meander migration have been continuously further developed and refined. Recent developments were mainly made by Seminara[20], Camporeale et al[21]., Crosato[22] and Pittaluga et al[23]. The latter summarizes and compares existing 1D models for meander migration. All of these models are somehow based on the assumption of mild curvature. Blanckaert and de Vriend[24-25] have proposed a 1D model for meander hydrodynamics that remains valid in the high curvature range and encompasses the mild-curvature model of Johannesson and Parker[26]. Their model can formally be written in the form of a linear relaxation equation in the variable α_s that parameterizes the transverse distribution of the streamwise velocity, with adaptation length λ_{as}/R and driving mechanism F_{as}/R:

$$\frac{\alpha_s}{R} = (1+n/R)\frac{1}{U_s}\frac{\partial U_s}{\partial n} \quad (2)$$

$$\lambda_{as/R}\frac{\partial}{\partial s}\left(\frac{\alpha_s}{R}\right)+\frac{\alpha_s}{R}=F_{as/R} \quad (3)$$

$$\lambda_{as/R}=\frac{1}{2}\frac{H}{\psi C_f}\left\{1-\frac{1}{12}\frac{\alpha_s}{R}+\frac{1B^2}{R}\right\} \quad (4)$$

$$F_{as/R}=\frac{1}{2}\frac{S_n Fr^2 + A - 1}{R}$$
$$-\frac{1}{2}\frac{H}{\psi C_f}\frac{\partial}{\partial s}\left(\frac{1}{R}\right)\left(1-\frac{B^2}{6R^2}\right)$$
$$+\frac{4\chi}{\psi C_f}\frac{H^2}{B^2}\frac{\langle f_s f_n\rangle}{R}\left[1+\frac{1}{12}\frac{(S_n Fr^2 + A+3)B^2}{R^2}\right]$$
$$+\frac{1}{24}\frac{H}{\psi C_f}\frac{B^2}{R^2}\frac{\partial}{\partial s}\left(\frac{S_n Fr^2 + A}{R}\right) \quad (5)$$

The coefficient ψ parameterizes curvature-induced energy losses and S_n is a coefficient of order 1. This equation clearly indicates the processes underlying the velocity redistribution. The first term (I) in Equation (5) represents the effect of the transverse slopes of the bed and the water surface, parameterized by the coefficient A and the Froude number, respectively. The second term (II) represents local flow accelerations/decelerations due to the adaption of the transverse water surface slope (superelevation) to changes in curvature. The third term (III) represents

velocity redistribution by the secondary flow. It is parameterized by $\langle f_s f_n\rangle/R$, which is determined from the relation[24]:

$$\langle f_s f_n\rangle = fct(C_f)\,fct\left[C_f^{-0.275}\left(\frac{H}{R}\right)^{0.5}(\alpha_s+1)^{0.25}\right] \quad (6)$$

The second functional relation in Equation (6) accounts for the non-linear interaction between the horizontal flow distribution (α_s) and the vertical flow distribution ($f_s f_n$). The fourth term (IV) represents velocity redistribution by the cross-flow U_n resulting from streamwise variations in the transverse slopes of the bed and the water surface. When assuming that B/R is small, $\psi \approx 1$, the coefficient $\chi = 1.5$, and the second functional relation in Equation (6) is identical to one, the model reduces to the widely used mild-curvature model of Johannesson and Parker[26].

Figure 8 compares the evolution of α_s/R around the bend in the laboratory experiment, obtained from: (i) the experimental data; (ii) Blanckaert and de Vriend's 1D model without curvature restrictions; (iii) Blanckaert and de Vriend's 1D model in its asymptotic formulation for mild curvature. Obviously the complex 3D velocity pattern can only be parameterized by means of the single variable α_s/R [cf. Equation (2)] in an approximate way. Therefore, Figure 8 includes an uncertainty range (gray area) for the estimations from the experimental data. The mild curvature model significantly overestimates the outwards velocity distribution. Although significant deviations occur locally, the proposed model without curvature restrictions agrees globally satisfactorily with the data. Obviously a depth-integrated and width-integrated 1D description can only account for processes that occur on a spatial scale of the order of magnitude of the river width. The description of processes on a smaller spatial scale requires a 2D or 3D description and modeling approach. The resolution of the 1D approach is coherent, however, with the objective of investigating long-term and large scale river planform processes.

The transparency of the 1D model is exploited in Figure 9, which shows the evolution around the bend of the mechanism that drive the velocity redistribution according to Equation (5). According to the model, all four mechanisms are of dominant order of magnitude. Remarkably, variations in curvature [Term (II)] are the dominant contribution. Differences be-

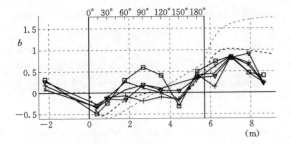

Figure 8 Evolution of α_s/R around the flume in the mobile-bed experiment [obtained from: (i) the experimental data (labeled lines, full lines and gray area); (ii) the proposed 1D model without curvature restrictions (long-dashed line); (iii) the proposed model in its asymptotic formulation for mild curvature (short-dashed line)]

tween the model without curvature restrictions and the mild-curvature model are considerable and include: (i) a reduction of the effect of the secondary flow (Term Ⅲ) due to non-linear interactions between the horizontal and vertical structures of the flow; (ii) velocity redistribution due to streamwise changes in the bed and water surface topography (Term Ⅳ) which are not accounted for in the mild-curvature model; (iii) reduction of the driving mechanisms due to the curvature-induced increase in energy losses parameterized by ψ (Terms Ⅱ, Ⅲ and Ⅳ).

More in general, the proposed model identifies $C_f^{-1}H/R$ as the main control parameter with respect to the velocity redistribution in curved open-channel flow. The driving mechanisms related to streamwise variations in curvature and the transverse bed and water surface slopes scale with it and the normalized adaptation length, $\lambda_{as/R}/R$ [cf. Equation (4)] is proportional to it. de Vriend[27] identified $C_f^{-1}H/R$ as an

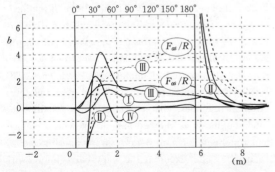

Figure 9 Evolution around the flume of the mechanisms that drive the velocity redistribution according to Equation (5) for the model without curvature restrictions (full lines) and the mild-curvature model (dashed lines) [The labels on the curves correspond to the terms in Equation (5).]

important control parameter. Because it represents a ratio between forcing by curvature (H/R) and dissipation by boundary-friction generated turbulence (C_f), he called $C_f^{-1}H/R$ the Dean number, similar to its definition in curved laminar flow. Blanckaert and de Vriend[24] identified $C_f^{-1}H/R$ as a major control parameter with respect to the vertical structure of the flow field and its interaction with the transverse flow structure. Johannesson and Parker[26] identified a similar parameter, $2\pi C_f^{-1}H/\lambda_m\,\lambda_m$ is the meander wavelength which they called the reduced wavenumber. The ratio B/R is traditionally the major scaling parameter used in field studies on meandering rivers. With respect to the velocity redistribution, B/R does not play a major role in mildly curved bends, but may be significant in sharp bends [cf. Equations (5)]. The parameter set $C_f^{-1}H/B$ and B/R is equivalent to the parameter set $C_f^{-1}H/R$ and B/R. $C_f^{-1}H/B$ has the advantage that it characterizes a river reach and is independent of the curvature of individual bends.

The hydrodynamic model without curvature restrictions[24-25] has been validated by means of the here reported laboratory experiment, which constitutes an extremely demanding case due to its very sharp curvature and its very pronounced variations in curvature. The model constitutes a powerful tool to investigate the hydrodynamics in meandering rivers as well as its relation to meander migration. Ottevanger et al[28]. have further validated this model by means of field measurements and they have applied it to investigate the processes responsible for the velocity redistribution in various natural meander bends.

4. The flow field near the bank and its relation to bank erosion processes

According to Equation (1), large-scale and long-term river planform processes are largely driven by the global distribution of the velocity. Understanding river processes on shorter time scale and a smaller spatial scale requires more detailed knowledge of the flow field. Bank erosion processes on the time-scale of a flood event and on a local spatial scale, for example, depend on the geotechnical behavior, the flow field near the bank and its interaction with groundwater flow.

Equation (1) does not account for the complex near-bank hydrodynamics, such as additional cells of secondary flow near the bank[29-37] or the flow separation and recirculation at the convex banks[38-40] or the formation of benches at concave banks accompanied by flow recirculation[41-44]. This section will focus on the flow pattern near the outer bank in meander bends.

Figure 10 illustrates patterns of the 3D flow field measured in the cross-section S10 of the Ledra meander bend (cf. Figure 4, Figure 5). Streamwise velocities are rather low over the shallow inner part of the cross-section, but characterized by a considerable transverse gradient. The velocities increase in out-wards direction, but the core of maximum velocities stays at about 2m, corresponding to about once the local flow depth, from the outer bank. Vertical profiles of the velocities are rather flat in this region, and characterized by important near-bed gradients and a maximum value that occurs slightly below the water surface. The secondary flow, defined as the projection of the velocity vector in the cross-section, shows in-wards mass transport over the entire flow depth close to the inner bank over the shallowest par of the cross-section. The typical curvature-induced secondary flow occurs in the central part of the cross-section.

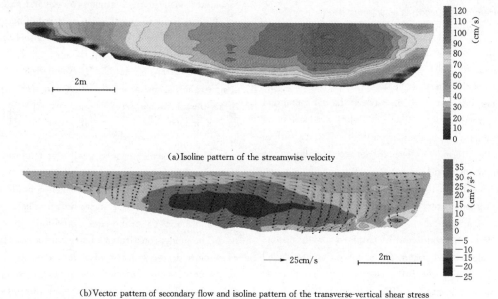

(a) Isoline pattern of the streamwise velocity

(b) Vector pattern of secondary flow and isoline pattern of the transverse-vertical shear stress

Figure 10   Patterns measured in the S10 cross-section (cf. Figure 4, Figure 5) on the Ledra meander bend

The most important feature with respect to bank erosion processes is the occurrence of a counter-rotating outer-bank cell of secondary flow in the upper part of the flow depth adjacent to the outer bank. It occupies a zone with a width of about once the local flow depth. The core of maximum streamwise velocities seems to coincide with the separation between both secondary flow cells. In the same region, advection of flow towards the toe of the outer bank occurs, but it does not result in a discernable increase of near-toe streamwise velocities. The pattern of both secondary flow cells is very well reflected in the pattern of cross-sectional (transverse-vertical) shear stress.

As aforementioned, vegetation, outer-bank benches and local bank erosion created an irregular geometry of Ledra's banks, which complicates the analysis of the near-bank flow patterns and the outer-bank cell of secondary flow. The influence of the inclination and roughness of the outer bank has therefore been investigated in a series of nine experiments in the laboratory flume shown in Figure 6. The bed was horizontal and consisted of immobilized sand. Three different bank inclinations (30°, 45° and 90°) and three different roughness characteristics (smooth, the same sand roughness as the bed and riprap with a equivalent roughness of $k_s = 30mm$) were investigated by Duarte[45]. The experiments were carried out with similar overall flow depth of $H = 0.16m$ and overall velocity of $U = 0.43m/s$.

The outer-bank cells are known to be relevant with respect to the stability of the outer bank and the adjacent bed. According to Bathurst, Thorne and

Hey[32] they endanger bank stability, whereas Blanckaert and Graf[46] found that they protect the outer bank and the adjacent bed by forming a buffer layer that protects the outer bank and adjacent bed from any influence of the center-region cell: the central secondary flow cell redistributes the velocity and causes it to increase towards the outer bank, whereas the outer-bank cell prevents this increase to continue through to the bank and keeps the core of maximum velocity a distance from the bank at the separation between both cells. The measurements in the Ledra comply with Blanckaert and Graf's results[46].

Figure 11 illustrates the measured patterns of the secondary flow in the cross-section at 90° in the bend in experiments with smooth vertical outer bank, rough vertical outer bank, and rough 30°-inclined vertical outer bank. The secondary flow is quantified by means of the streamwise component of the vorticity, defined as $\omega_s = \partial v_z / \partial n - \partial v_n / \partial z$. Outer-bank cells occurred in all nine investigated conditions. An increase in outer-bank roughness widens and strengthens considerably the outer-bank cell, whereas inclining the outer bank weakens the outer-bank cell and changes its pattern. But its protective effect by forming a buffer layer between the outer bank and the center-region cell is conserved. More details are reported in Duarte[45].

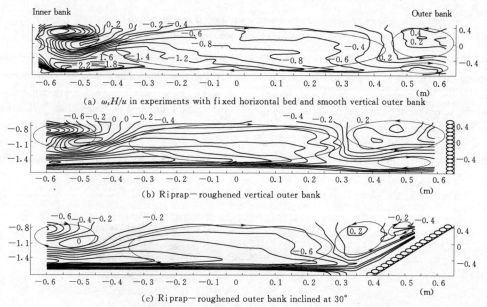

(a) $\omega_s H/u$ in experiments with fixed horizontal bed and smooth vertical outer bank

(b) Riprap—roughened vertical outer bank

(c) Riprap—roughened outer bank inclined at 30°

Figure 11　Measured patterns of the secondary flow
(quantified by the normalized streamwise component of the vorticity)

Blanckaert and de Vriend[4] have analyzed the mechanisms underlying the near-bank secondary flow cells by means of a term-by-term analysis of the vorticity equation and the kinetic energy fluxes between the mean flow and the turbulence. Based on their results they have postulated that the outer-bank cell can only be resolved by turbulence models that can account for the backscatter of kinetic energy from the turbulence to the mean flow. This hypothesis was confirmed in numerical simulations by Van Balen et al[47]. , who accurately resolved the outer-bank cell in the reported laboratory experiments with horizontal bed and smooth vertical banks (cf. Figure 11) by means of a Large Eddy Simulation (LES) numerical

model, but did not succeed to resolve them by means of a Reynolds Averaged Navier Stokes (RANS) model with two-equation turbulence closure (see Figure 12 which is modified from Van Balen et al.[47]) .

The results from the validated numerical LES model allow gaining further insight in the meander hydrodynamics. Contrary to the experimental data, the simulated data are not limited by a low spatial resolution in streamwise direction or by experimental scatter and uncertainty. Hence they are particularly appropriate for term-by-term analyses of the flow equations. Van Balen et al.[47] have analyzed the mechanisms underlying the near-bank cells of secondary flow and highlighted the important role played by

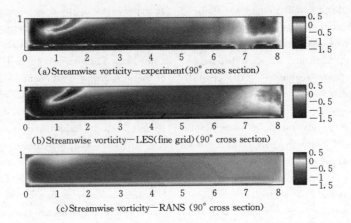

(a) Streamwise vorticity—experiment (90° cross section)

(b) Streamwise vorticity—LES (fine grid) (90° cross section)

(c) Streamwise vorticity—RANS (90° cross section)

Figure 12   Patterns of the secondary flow（quantified by $\omega_s H/U$ in the experiment with fixed horizontal bed and smooth vertical banks）

turbulence. The LES simulations also provide data in flow regions that are not accessible by the velocimeters, such as the flow regions close to the solid boundaries and the water surface, and variables that cannot easily be measured, such as the pressure distribution or the boundary shear stress. The letter plays a dominant role with respect to morphodynamical processes, and especially with respect to bank erosion processes. The distribution of the shear stress on the outer bank, shown in Figure 13, which is modified from Van Balen et al.[48], is clearly affected by the outer-bank cell of secondary flow. A more detailed description can be found in Van Balen et al.[48] The validated LES code also allows a broadening of the parameter space in order to obtain generic results on the bank shear stress that are valid for a wide range of natural rivers.

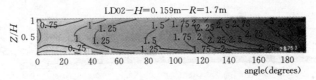

Figure 13   Patterns of the normalized shear stress on the outer bank （$\tau_{bank}/\rho U^2$ in the experiment with fixed horizontal bed and smooth vertical banks obtained from LES simulations）

5. Invertebrate dynamics and their relation to the flow field and the turbulence

Understanding invertebrate assemblages' dynamics in relation to flow dynamics, morphodynamic and sedimentological processes is a key approach in understanding river ecological functioning. Changes in flow dynamics have direct consequences on invertebrate individuals, i. e. drift (Gibbins et al.[49], Gabel et al.[50]) or changes in prey-predator relationships (Hart and Finelli[51], Gabel et al.[52]) Flow dynamics also influence habitat structure and food resource availability, two main components shaping invertebrate assemblages. For example, increases in flow velocities will trigger erosive processes, removing local silt deposits and organic debris as well as fine sand particles. The physical habitat for invertebrates switches then from fine instable organic enriched substrate to coarse more stable organic-poor substrate. As a result, invertebrate assemblages will change from communities dominated by sediment eaters or collector-gatherer taxa to communities dominated by filterer taxa.

This mechanism is however more complicated since each taxa has a range of tolerance to changes in environmental conditions, allowing a certain resilience of the communities in the ecosystem. Flow variations also differently influence environmental conditions at local scale according to local geomorphology or the presence of macrophytes. Hence, far to be simple and linear, the influence of flow dynamics on invertebrates is complicated by the dense network of interactions linking all components of the river ecosystem at different spatial and temporal scales.

Our ongoing research links field scales, laboratory and numerical approaches to examine the linkages between complex flow, morphology and ecology in meander bends, and especially their effect on aquatic invertebrate distributions and the use of food resources. Questions of interest that are addressed include:

(1) How and to what extend, daily and seasonal natural changes in hydrodynamic conditions influence the density of benthic invertebrates?

(2) What are the hydrodynamic parameters of relevance with respect to the drift of invertebrates?

(3) Do revitalized rivers provide better ecological conditions for invertebrates and how can this be quantified?

The field investigation was designed to describe and quantify benthic invertebrate assemblages, habitat structure and food availability at representative locations in meanders, in strict correspondence to local hydraulic conditions. The survey was performed along riffle and pool transects (Figure 14). A detailed description and quantification of the habitat structure (size fraction of the upper 5 cm of sediment, macrophytes mapping, and record of the presence of hardstones or coarse woody-debris substrates) as well as food availability (thickness and organic fraction of silt cover, biofilm and seston concentrations) was done on a 2m × 2m grid at each transect (Figure 14). Benthic invertebrates were collected at all differing meso-habitats—as characterized by a combination of specific flow and habitat conditions—identified at the date of the survey.

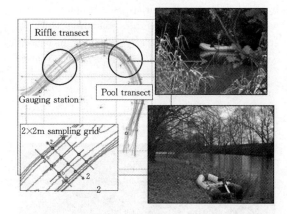

Figure 14 Illustration of the field sampling strategy used to survey benthic assemblages, habitat structure and food availability in strict correspondence to hydraulic characteristics on the Spree River meander at Neuenbrück, NE Germany

In order to link ecological parameters and hydraulics conditions, each sampling location was geographically referenced in the geographical system set up to establish topography and flow fields of the bend, and water level was recorded at referenced gauging stations installed upstream and downstream the meander. The survey was performed 4 times a year in order to count for seasonal differences in invertebrate assemblages.

Invertebrates were collected using a Surber net sampler of 250 μm mesh size and 0.05m^2 sampling surface (Figure 15) operated under water by direct diving. A metal stick, positioned from a boat at the desired sampling location along the transect served as guide to locate the exact sampling location when diving. Once the Surber sampler in place on the river bottom, the upper 5 centimeter of sediment (or any other sampled substrates, i.e. macrophytes, stones, dead wood) are pushed into the net where they stay trapped

(a)Surber net   (b)Brushing

Figure 15 Surber net device and brushing technique used to collect benthic invertebrates

by the flow. Back at the surface, the samples are fixed with 70% ethanol. Invertebrates colonizing hard substrates were detached by gently scratching the collected pieces with a brush (Figure 15). The scratched surface areas were recorded in order to calculate densities. At the laboratory, invertebrates were separated from the substrate, counted and identified till species level under stereo-microscope (Figure 16).

Figure 16　Typical working station to process invertebrate samples in the laboratory

Beside recording invertebrate densities for various flow conditions, identifying the origin of food ingested by the individuals of each taxa for these various flow conditions is of key importance aspect. For that purpose we used the stable isotope analysis approach. Conceptually, each food resource exhibits a specific ratio of C13/N15 isotopes which stays unchanged in invertebrate tissues when individuals mainly feed on that food resource. δN15 values increase through trophic levels but δC13 values do not change. Hence, it becomes easy to link food availability with the trophic regime of benthic individuals. Moreover, by correlating this information to quantitative information on food distribution, it becomes possible to estimate at what thresholds changes in food resources trigger changes in invertebrate assemblages.

Figure 17 illustrates possible mechanisms of shift in community composition due to changes in flow condition. The two graphs represent the food-web situation at two locations on the River Mulde (North-east Germany) that have different flow characteristics: plot A concerns high flow velocities (0.72m/s) whereas plot B concerns low flow velocities (0.23m/s). The sediment structure was comparable at the two locations, with the exception of the presence of a slightly thicker silt cover in location B due to lower flow velocities. Isotope analysis shows that communities of filterer taxa characterized by *Hydropsychae contubernalis*, *H. pelucidulla*, *Pisidium* sp (Figure 18) and *Sphaerium* sp dominate at high flow velocities. These taxa feed on seston transported through the water column as shown by the comparable values of δ13C found in the seston and the body tissues of these species.

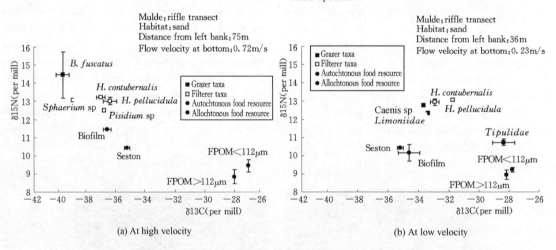

Figure 17　Plots of δ13C/δ15C ratios for benthic invertebrates and food resources collected in sand habitat at high and low flow velocities, respectively (FPOM: Fine Particulate Organic Matter)

At low flows, obligatory filterers like *Pisidium* sp or *Sphaerium* sp have disappeared. Collector gatherer taxa like *Caenis* sp, *Limoniidae* or *Tipulidae* (Figure 19), collecting fine deposited organic particles dominate the assemblages. The case of the *Tipulidae* is especially illustrative. The individuals of this taxa directly feed on the fine particulate organic material (FPOM) deposited at the surface of the sand as shown by the comparable values of δ13C found in the FPOM and the body tissues. *Caenis* sp and *Limoniidae* individuals probably feed more on the organic particles deposited or trapped in the biofilm. It is also interesting to see that non-obligatory filterer like *H. contubernalis* and *H. pellucidula* shift for a food regime more composed of deposited organic material as suggested by the shift, of 4 units of δ13C towards characteristic values for FPOM in their body tissue.

Figure 18　Specimen of *Pisidium* sp,
a typical filterer taxa

Figure 19　Specimen of *Tipulidae*,
a typical collector－gatherer taxa

Hence, it is easy to imagine that a drop in flow velocity on location A will favor the presence of the collector gatherer taxa, similarly to location B. Conversely, increasing flow at location B will favor the colonization of sand by true filter taxa like *Pisidium* sp or *Sphaerium* sp. This result exhibit how hydrodynamics trigger modification of invertebrate assemblages by controlling food availability.

Species distributions, however, do not only depend on such processes, but also on direct drag forces exerted by the flow on species individuals, and on how species with different body form and gripping strategies cope with these forces (see Gabel et al[50]., Schnauder et al[53]) . Where the study of the links between food and species distribution typically requires field investigations, the quantification of thresholds in drag forces for drifting species individuals and the study of the ongoing processes in invertebrate drift can be investigated under controlled laboratory conditions. Both approaches are complementary to achieve a realistic model of species distribution according to flow dynamics.

In orer to investigate the hydrodynamic parameters, and especially the role of turbulence, implicated in invertebrate drift, a set of experiments was performed that expose three benthic invertebrate taxa (*Aeshna cyanea*, *Calopteryx splendens* and *Cordulia* sp.) to three different levels of turbulence intensity. Each test was replicated three times, resulting in a set of 27 experimental cases. The experiments were performed in a laboratory flume that represents an alpine confluence. The bed topography and sedimentology in the main channel are the result of the interplay between flow and sediment transport, leading to pronounced transverse bed slopes and sediment sorting. Invertebrates were successively exposed to flow at locations differing in turbulence levels. The lowest level of turbulence was obtained upstream of the confluence where the flow is quasi-uniform. A higher turbulence level was found in the confluence zone due to the inclined bed topography and the coarser sediments, even without discharge in the tributary. The turbulence level was further increased due to 3D flow effects when the tributary contributed to the discharge. For each test, the discharge was increased until invertebrate drift occurred.

The investigation required the use of state-of-the art experimental techniques. The turbulent velocity fields in the vicinity of the invertebrate were measured by means of the Acoustic Doppler Velocity Profiler (ADVP, cf. section 2) . The ADVP's profiling capacity and its flexibility to follow the moving invertebrates were essential in this investigation. Simultaneously, the invertebrate behavior was recorded using a high acquisition rate video camera operated at 5 Hz. This allowed determining accurately the drift moment.

Figure 20 illustrates the results for an experiment carried out with the invertebrate species "*Aeshna cyanea*" . ADVP measurements provided with a temporal resolution of 31.25Hz the velocity field in the vertical profile located above the invertebrate. The Taylor hypothesis of frozen turbulence allows converting the temporal series into a spatial pattern of turbulent structures in the vertical plane that containing the invertebrate and that is oriented along the flow direction [Figure 20 (d)]. These preliminary experiments suggest that invertebrate drift is related to turbulence events. Figure 20 (d), for example, shows considerable turbulent accelerations near the invertebrate at the drift moment.

(a) One of the invertebrate species investigated

(b) Close-up of the invertebrate sitting on the gravel bed in the flume with indication of the vertical flow profiles measured above the invertebrate(1)

(c) Close-up of the invertebrate sitting on the gravel bed in the flume with indication of the vertical flow profiles measured above the invertebrate(2)

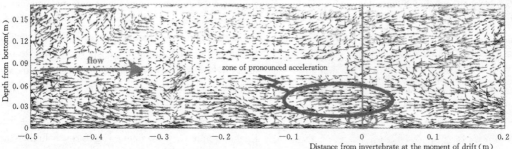

(d) Vector representation of the turbulent streamwise-vertical velocity fluctuations in a vertical plane through the invertebrate and parallel to the flow at the moment of drift

Figure 20   Experiment carried ont with *Aeshna cyanea*

The objective of this research is to implement the new knowledge in a numerical model that accounts for the linkages between invertebrate dynamics, the hydrodynamics, the sediment transport and the morphodynamics. Such a tool would allow a better evaluation of the ecological value of rivers and a quantification of the gain obtained in river revitalization projects.

6. Conclusions

Hydrodynamical, morphodynamical and ecological river processes and their multiple linkages occur in an infinity of different configurations and over a wide range of spatial and temporal scales. This paper has illustrated a research methodology that consists in combining field investigations, physical modeling in the laboratory, and numerical simulations in order to develop generic insight and tools for engineering and management of the river environment.

The first illustration of this methodology concerned the investigation of the velocity distribution in meander bends, which is known to be a predominant

forcing mechanism for the long-term and large-scale planimetric evolution of meanders. Detailed velocity measurements performed in a natural meander bend were presented. They revealed the dominant hydrodynamic processes, but their interpretation and analysis were hindered by the presence of floating wood, vegetation, outer-bank benches and local bank erosion. These dominant processes were subsequently investigated under controlled conditions in a physical model in the laboratory. Based on the analysis of the laboratory investigation, a numerical model for the velocity redistribution was developed, which computational requirements allow its application on a large spatial scale and/or a long temporal scale.

The second illustration concerned the investigation of the hydrodynamics near the river bank, which are relevant for bank erosion processes on the timescale of a flood event and on a local spatial scale. Field investigations in a meander bend revealed that the near-bank hydrodynamics are conditioned by the existence of a so-called outer-bank cell of secondary flow. This outer-bank cell of secondary flow was subsequently investigated under controlled laboratory conditions. Attention was focused on its dependence on the roughness and inclination of the outer bank. Based on the insight provided by the laboratory experiments, requirements for the numerical simulation of the outer-bank cell were defined, resulting in the successful simulation of the near-bank hydrodynamics by means of a three-dimensional Large Eddy Simulation model. This model provided the shear stress on the outer bank. This is the most important parameter with respect to the flow attack on the bank, but it could not directly be obtained from the laboratory experiments.

The third illustration concerned the investigation of the linkages between three-dimensional flow patterns, turbulence, morphology and ecology in meander bends. Field measurements were performed that mapped the local hydraulic and morphologic conditions, the habitat structure of invertebrates and the availability of food resources. Complementary laboratory experiments investigated thresholds in mean drag forces and turbulent coherent structures for drifting species individuals, and on how species with different body form and gripping strategies cope with these forces. The insight provided by the field and laboratory investigations will be implemented in a numerical model that accounts for the linkages between invertebrate dynamics, the hydrodynamics, the sediment transport and the morphodynamics.

Acknowledgments

This research was sponsored by the Swiss National Science Foundation under grants 2100-052257, 2000-059392. 2100-066992, 20020-103932 and 2000-20-119835, by the Deutsche Forschungsgemeinschaft (DFG) and the Netherlands Organization for Scientific Research (NWO) under grants SU 405/3-1 and DN66-149 in the framework of their bilateral cooperation program. The first author was partially funded by the Chinese Academy of Sciences fellowship for young international scientists under Grant No. 2009YA1-2.

References

[1] Lemmin U. , Rolland T. . Acoustic velocity profiler for laboratory and field studies. *J. Hydraul. Eng.* , 1997, 123 (12), 1089-1098.

[2] Hurther D. , Lemmin, U. . A constant beam-width transducer for three-dimensional Doppler profile measurements in open channel flow. *Measurement Science and Technology* , 1998, 9 (10), 1706-1714.

[3] Blanckaert K. , Graf W. H. . Experiments on flow in an open-channel bend. Mean flow and turbulence. *J. Hydraul. Eng.* , 2001, 127 (10), 835-847.

[4] Blanckaert K. , De Vriend H. J. . Secondary flow in sharp open-channel bends. J. Fluid Mech. , 2004, 498: 353-380.

[5] Blanckaert K. , Lemmin U. . Means of noise reduction in acoustic turbulence measurements. *J. Hydraul. Res.* , 2006, 44 (1), 3-17.

[6] Blanckaert K. Topographic steering, flow recirculation, velocity redistribution and bed topography in sharp meander bends. *Water Resour. Res.* , doi: 10. 1029/2009WR008303, 2010, in press.

[7] Pizzuto J. , T. Meckelnburg. Evaluation of a linear bank erosion equation. *Water Resour. Res.* , 1989, 25: 1005-1013.

[8] Whiting, P. J. , and Dietrich, W. E. . Experimental studies of bed topography and flow patterns in large-amplitude meanders. I : Observations. *Water Resour. Res.* , 1993, 29 (11): 3605-3614.

[9] Abad J. D. , M. H. Garcia. Experiments in a high-amplitude Kinoshita meandering channel: 2. Implications of bend orientation on bed morphodynamics, *Water Resour. Res.* , 2009, 45, W02402, doi: 10. 1029/2008 WR007017.

[10] Yen C. , B. C. Yen BC. Water surface configuration in channel bends. *J. Hydraul. Div.* , 1971, 97 (HY2): 303-321.

[11] De Vriend H. J. , Koch F. G. . Flow of water in a curved open channel with a fixed uneven bed. T. O. W. Rep. R657-VI/M1415-II , Delft Hydraulics Lab. , Delft Univ. Techn. , The Netherlands, 1978.

[12] Hooke R. L. . Shear-stressand sediment distribution in a meander bend, Tech. Rep. UNGI RAPPORT 30, Dep. of Phys. Geogr. , Univ. of Uppsala, Uppsala, Sweden, 1974.

[13] Kikkawa H, Ikeda S and Kitagawa A.. Flow and bed topography in curved open channles. *J. Hydraul. Div.*, 1976, 102 (9): 1327-1342.

[14] Odgaard A. J., Bergs M. A.. Flow processes in a curved alluvial channel. *Water Resour. Res.*, 1988, 24 (1): 45-56.

[15] Leeder M. R., Bridge P. H.. Flow separation in meander bends. *Nature*, 1975, 253: 338-339.

[16] Frothingham K. M., Rhoads B. L.. Three-dimensional flow structure and channel change in an asymmetrical compound meander loop, Embarras River, Illinois. *Earth Surface Processes and Landforms*, 2003, 28 (6): 625-644.

[17] Ferguson R. I., Parsons D. R., Lane S. N., Hardy R. J.. Flow in meander bends with recirculation at the inner bank. *Water Resour. Res.* 2003, 39 (11): 1322.

[18] Engelund F.. Flow and bed topography in channel bends. *J. Hydraul. Div.*, 1974, 100 (HY11): 1631-1648.

[19] Ikeda S., Parker G., Sawai K.. Bend theory of river meanders. Part 1. Linear development. *J. Fluid Mech.*, 1981, 112: 363-377.

[20] Seminara G.. Meanders. *J. fluid mech.*, 2006, 554: 271-297.

[21] Camporeale C., P. Perona, A. Porporato, L. Rodolfi. Hierarchy of models for meandering rivers and related morphodynamic processes, *Rev. Geophys.*, 2007, 45, RG1001, doi: 10.1029/2005RG000185.

[22] Crosato A.. Analysis and modelling of river meandering. PhD thesis, Delft University of Technolgy, Delft, The Netherlands, 2008.

[23] Pittaluga M. B., Nobile G. and Seminara G.. A nonlinear model for river meandering. *Water Resour. Res.* 2009, 45, Article Number: W04432.

[24] Blanckaert K., De Vriend H. J.. Nonlinear modeling of mean flow redistribution in curved open channels. *Water Resour. Res.*, 2003, 39 (12): 1375-1388.

[25] Blanckaert K., De Vriend H. J.. Meander dynamics: a nonlinear model without curvature restrictions for flow in open-channel bends. *J. Geoph. Res.*, 2010, doi: 10.1029/2009JF001301, in press.

[26] Johannesson H., Parker G.. Velocity redistribution in meandering rivers. *J. Hydraul. Eng.*, 1989, 115 (8): 1019-1039.

[27] De Vriend H. J.. Velocity redistribution in curved rectangular channels. *J. Fluid Mech*, 1981, 107: 423-43.

[28] Ottevanger W., Blanckaert K., UittewaalW. S. J. Analysis of the mechanisms responsible for streamwise velocity redistribution in natural rivers. *Geomorphology*. (submitted for publication).

[29] Hey R. D., Thorne C. R.. Secondary flow in river channels. *Area*, 1975, 7 (3): 191-195.

[30] Bridge J. S., Jarvis J.. Velocity profiles and bed shear stress over various bed configurations in a river bend. *Earth Surface Proc.*, 1977, 2: 281-294.

[31] Bathurst J. C., Thorne C. R., Hey R. D.. Direct measurements of secondary currents in river bends. *Nature*, 1977, 269: 504-506.

[32] Bathurst J. C., Thorne C. R., Hey R. D.. Secondary flow and shear stress at river bends. *J. Hydr. Div.*, 1979, 105 (10): 1277-1295.

[33] Thorne C. R., Hey R. D.. Direct measurements of secondary currents at a river inflexion point. *Nature*, 1979, 280: 226-228.

[34] Dietrich W. E., J. D. Smith. Influence of the point bar on flow through curved channels. *Water Resour. Res.*, 1983, 19 (5): 1173-1192.

[35] De Vriend H. J., Geldof H. J.. Main flow velocity in short and sharply curved river bends. 1983, Rep. 83-6, Lab. Fluid Mech., Dept. Civil Engng, Delft University of Technology.

[36] Thorne C. R., Abt S. R., Maynord S. T.. Prediction of near-bank velocity and scour depth in meander bends for design of riprap revetments. In: River, coastal and shoreline protection; Erosion control using riprap and armourstone, Eds. C. R. Thorne, S. R. Abt, F. B. J. Barends, ST., Maynord and K. W. Pilarczyk, Wi-

ley, 1995: 115-133.

[37] Markham A. J. , Thorne C. R. . Geomorphology of gravel-bed river bends. Dynamics of gravel-bed rivers, P. Billi, Hey, R. D. , Thorne, C. R. and Tacconi, P. , ed. , Wiley, 1992: 433-456.

[38] Leeder M. R. . Bridges P. H. . Flow separation in meander bends. *Nature*, 1975, 253 (5490): 338-339.

[39] Frothingham K. M. , Rhoads B. L. . Three-dimensional flow structure and channel change in an asymmetrical compound meander loop, Embarras River, Illinois. *Earth Surface Processes and Landforms*, 2003, 28 (6): 625-644.

[40] Ferguson R. I. , Parsons D. R. , Lane S. N. , Hardy R. J. . Flow in meander bends with recirculation at the inner bank. *Water Resour. Res.* , 2003, 39 (11): 1322-1333.

[41] Hickin E. J. . Hydraulic factors controlling channel migration. In R. E. Davidson-Arnott & W. Nickling (editors), Research into Fluvial Systems, GeoAbstracts, Norwich, 1977: 59-66.

[42] Jackson A. D. . Bedload transport and sorting in meander bends, Fall River, Rocky Mountain National Park, Colorado. PhD , Colorado State University, Fort Collins, colorada, 1992: 280.

[43] Andrle R. . Flow structure and development of circular meander pools. *Geomorphology*, 1994, 9: 261-270.

[44] Hodskinson A. , Ferguson R. I. . Numerical modelling of separated flow in river bends: Model testing and experimental investigation of geometric controls on the extent of flow separation at the concave bank. *Hydrol. Processes*, 1998, 12: 1323-1338.

[45] Duarte A. . An experimental study on main flow, secondary flow and turbulence in open-channel bends with emphasis on their interaction with the outer-bank geometry. PhD-thesis Nr 4227, Ecole Polytechnique Fédérale Lausanne, Switzerland, 2008.

[46] Blanckaert K. , Graf W. H. . Momentum transport in sharp open-channel bends. *J. Hydraul. Eng.* , 2004, 130 (3): 186-198.

[47] Van Balen W, Blanckaert K, Uijttewaal WSJ. Large-eddy simulations and experiments of single-bend open-channel flow at different water depths. *J. Turbulence*. 2010, 11 (12) .

[48] Van Balen W. Large Eddy Flow Simulation for the prediction of bank erosion and transport processes in river bends. PhD dissertation, Delft University of Technology, The Netherlands, 2010.

[49] Gibbins C, Vericat D, Batalla RJ. When is stream invertebrate drift catastrophic? The role of hydraulics and sediment transport in initiating drift during flood events. *Freshwater Biology* 2007, 52: 2369-2384.

[50] Gabel F, Garcia X. -F, Brauns M, Sukhodolov A, Leszinski M, Pusch MT. Resistance to ship-induced waves of benthic invertebrates in various littoral habitats. *Freshwater Biology*, 2008, 53: 1567-1578.

[51] Hart DD, Finelli CM. Physical-biological coupling in streams: The pervasive effects of flow on benthic organisms. *Annu. Rev. Ecol. Syst.* , 1999, 30: 363-395.

[52] Gabel F. , Stoll S. , Fischer P. , Pusch M. , Garcia X. -F. Waves affect predator-prey interactions between fish and benthic invertebrates. *Oecologia* (Submitted) .

[53] I. Schnauder, S. Rudnick, X. -F. Garcia, J. Aberle. Incipient motion and drift of benthic invertebrates in boundary shear layers. River Flow proceedings, 2010.

Comparison of Model Test and Prototype Observation on Flood Controlling of Town Housing

Huang Shuyou[1]   Ren Yushan[1]   Yin Zhigang[1]   Zhou Jinghai[2]   Liu Yongjun[2]

[1] *School of Water Conservancy and Environment Engineering,*
Changchun Institute of Technology, Changchun, 130012
[2] *School of Civil Engineering, Shenyang Architecture University, Shenyang,* 110168

Abstract: The paper analyzes the damage cause and effects of town housing through wash-out resistance tests of housing model (using the same construct material with different mortar) in varying flow. Compared with prototype observation, Test results show the important damage of town housing is under water, especially foundation. Based on the analysis of the necessity of consolidation, the consolidation measures combined with the characters of town housing is put forward. The research is good reference to flood region.
Key words: town housings; housing model; flood controlling of town housing.

1. Introduction

With the climate change and the extreme weather increasing, it is frequent of the flood damaged. Especially with the town development of the rural, it is becoming heavier increasingly that the expense of the town housings is suffered by flood damage. Therefore it is necessary to research the collapse mechanism of the town housings under the flood damaged. The purpose is in order to improve the flood-fighting ability of the town housings, and find measures of the practical reinforcement to improve the Flood-fighting ability of the town housings on this basis.

2. The town housings' collapse mechanism in the flood

The construction of the town housings is destroyed by the flood, it is main matter that the external force suffered more than the building itself support the stress limited. The action force of the flood to buildings is mainly including: the hydrodynamic pressure, the hydrostatic pressure and the immersion force. It is interacting that the three external force of the flood, and it is variant that the time of occurrence and the role of the force.

Early in the flood, it is the hydrodynamic pressure and the hydrostatic pressure that the building suffered major action force of the flood, then the largest pressure appears the face of the building to the

flood and the maximum is up to the sum of the two. For the brittle materials or the low strength construction, at this point, the most easily damaged is due to insufficient strength. When the flood enter the building, the hydrostatic pressure is reduced gradually, then the hydrodynamic pressure is mainly force around the building. When the flood erupt, the immersion force starts strengthening slowly, the flood immerses the walls of building and the building's foundation especially. The deluge and building materials interact, it made the building becouse it change the mechanical properties of the colloidal materials, and made its cohesive force descend or the foundation become deformed.

3. The test results and the experimental analysis of the housing model in clay

The actual situation for the rural architecture, the first test is on he clay building modeling.

The model was arranged in a 10m long, 3m wide, and 1.2m high artificial sink, according to the actual size of housing, the size of the experimental model was reduced to the length × width × height is 180cm × 100cm × 60cm. In test; the largest wall was faced the flood according to the building put in the most unfavorable position in the flood, in the wall height of 5cm above the ground, the measuring devices was put in the wall on the horizontal position. It was used in model test that the actual housing materials constructed, the test material that the length × width × height was 12cm × 6cm × 5cm.

In test, the flow rate was tested by the electromagnetic flowmeter, the water level was measured with a pin for surveying, the Time-average pressure was measured by the piezometer tubes and the fluctuating pressure measured by the pressure sensors.

Owing to the clayey characteristic, especially the mortar blended with the clay and sand, the clay and sand is separated into granular quickly and loss their bonding characteristics in the flood and the impact.

When the unit discharge was 127L/s and the test was carried on only 9 minutes and 30 seconds, the first piece of brick droped from the wall of the building model, 11 minutes 22 seconds, the model building was in collapse, the model should be considered to have been destroyed in this situation.

The flood will directly damage to the buildings with clay mortar or the clay buildings from the test simulation and the field observation of the scene.

Analysis the damage reasons for the following:

(1) The clay mortar buildings immersed in the flood, the clay mortar begins to grow soft, loss of bond strength, this time under the impact of the flood, and the current in the main area and the site of the mutation, where the most mortar of the building among suturas was washed away by the current first. The clay mortar was lossed, so that the constructions loss bond strength of the masonry buildings, some masonries begin to fall off, followed by others linked sites, and thus the entire building break down.

(2) From the test scene, the clay mortar building damaged is in the underwater in flood, to protect the clay mortar buildings from flood damage, protecting the building should from foundation begin to the flood could reach the wall of the maximum height.

4. Reinforcement measures and results analysis

Due to the clay mortar of the town houses, that do not resist the impacts in flood, take the three types of the reinforcement following:

(1) Clay mortar masonry wall, cement pointing.

(2) Clay mortar masonry wall, cement mortar (1 : 3) plastering.

(3) Cement mortar (1 : 3) masonry wall.

4.1 Clay mortar masonry wall, cement pointing

In order to test it impact effect that the rein-forced clay mortar masonry model. In the test process, the unit discharge was increased gradually, when the unit discharge reached to 157.1L/s, the pointed building model in cement, the welcoming wall began to crack. This shows that the buildings appear destruction. Figure 1 is the pressure changes and trends in the middle of the wall that face the flood and the border area of that wall. It can be to some extent increase the building capacity to the flood, from the test results only that the existing clay mortar town housings reinforced in this way.

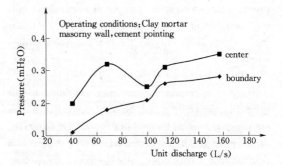

Figure 1　The characteristic points of the wall relate between pressure and flow curve in the program one

The town housings use this protective measure by cement pointing, a certain extent, although it can resist the impact of the flood, but it does not a long soak in the flood that put to use such reinforcement of the buildings. Due to the factors of the external, the position that masonry cement pointing appear leakiness or off phenomenon. Soaking in the flood, the wall will occur collapse, thereby causing the collapse of buildings, this destruction can be seen from the model experiment.

4.2 Clay mortar masonry wall, cement mortar (1 : 3) plastering

After the model experiment that in clay mortar walls was reinforced by the cement mortar (1 : 3) plastering, when the unit discharges over 157.1L/s, the model did not appear any form of damage. Figure 2 is the pressure changes and trends in the middle of the wall that face the flood and the border area of that wall.

From the testing process, the building can resist the larger flood flows through the existing clay mortar masonry residential construction reinforced by cement mortar Plastering. However, the clay mortar was soft and has lost gelation through observation of

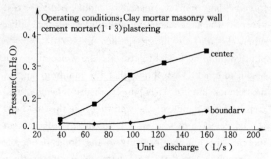

Figure 2   The characteristic points of the wall relate
between pressure and flow curve in the program two

the masonry can be seen. Once the cement mortar falls off, the building may collapse, even though the building was supported fully by the outer layer of cement mortar.

Town housings building plastered by cement mortar, to a certain extent, they can soak in the flood. In a period of time, the buildings collapse does not occur as long as the building foundation immersed in flood can not be destroyed. However, this way of reinforcement, requires a lot of cement, the project cost will be high; at the same time the protective measures similar to in the soft soil outside surround a hard shell, it can not be long-term immersed in the flood. This reinforcement, the actual value is difficult to promote in the vast rural area.

4. 3   Cement mortar masonry construction

After the town housings was constructed by the cement mortar, As the integrity of the building is better, so it can withstand more pressure in the flood. when the unit discharge over 220.0L/s in test, on the test model surface without any signs of damage. Figure 3 is the pressure changes and trends in the middle of the wall that face the flood and the border area of that wall.

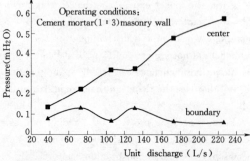

Figure 3   The characteristic points of the wall relate
between pressure and flow curve in the program three

It is known from the test data in measured, the cement mortar masonry buildings, it enhances buildings from the impact of the flood and it makes the buildings withstand a larger unit discharge. For the whole building, the collapse of the building does not occur normally as long as the basis of the building immersed in water will not be damaged. However, the foundation of the modle experiment and the basis of the actual are differential. In model experiment, the model is constructed on the cement mortar floor directly, it is solid foundation; In reality the foundation is ever-changing, in the process of the construction will take all kinds of reinforcement. And even take such measures as the strengthening, but in the actual damage of the buildings, it is main factor that due to the deformation of the foundation damage caused.

5. Prototype observation and Conclusion

In 1998, the catastrophic flood befell in Northeast region of China, the flood made the western region of Heilongjiang and Jilin provinces and the eastern Inner Mongolia Autonomous Region suffer severe disaster. 918, 490 houses collapsed and nearly 50 billion yuan of the direct economic lossed. After the flood, in the subsequent survey found that the collapse of the houses is collapsed in part of the collapsed houses in the flood period; also part of the houses is collapsed during the flood and after the flood subsided, the collapse of the majority houses caused by the foundation settlement.

From the inspection of the Huaihe River, the flood destroyed the western region of the rural houses in Anhui Province in 1991, in the majority collapse houses, caused by the foundation settlement mostly in the flood. This is consistent with the experimental results.

6. Conclusion

Study and analyze the test results from the several operating conditions on the model experiment of the town housings, combination of the survey available datas in flood. Its destruction is mainly caused by the parts of damage building in underwater, especially damage of the foundation, from the illustrations of the town housings building in the flood. According to the current actual situation in the vast rural areas, those should being reinforced that the building foundation and the wall which the flood may reach the height of the building in the process of the reinforcement of existing town housings.

References

[1] Industry standard of the People's Republic of China. hydraulic (conventional) model experiment (SL 155—95). China Electric Power Press, 1998, 11.

[2] Ge Xueli, Wang kaishun. The capacity analysis and improvement of the town housings in flood in west of Anhui [J]. Building Science, 1991, 4: 47-48.

[3] Ge Xueli, Wang kaishun. The destruction and defense of the town building in the natural disasters [J]. Disaster Reduction In China 2001, 2, 11 (1) : 38-42.

[4] Zhonggui Hui, Liu Shuguang. Experimental Study of Pressure on Village Houses Hit by Floods [J]. Urban Roads Bridges & Flood Control, 2008, 12: 92-95.

[5] Cui Qinghai, Tianlixuan. The flood and disasters of the Songhua River in 1998. Water resourse & hydropower of northeast China 2000, 18 (186): 41-43.

[6] Wisner B, Blaikie P, Cannon T, et al. At risk: natural hazards, people's vulnerability and disasters. 2nd ed. London: Routledge, 2004.

[7] Birkmann J. Risk and vulnerabiltiy indicators at different scales: applicability, usefulness and policy implications [J]. Environmental Hazards, 2007, 7 (1): 20-31.

[8] Hanfeng. The capacity of town housing, safety planning and security mechanisms in China [J]. Theorists, 2009, 8: 207-208.

Analysis on the Free Vibration Characteristics of Gate Rubber Seal Considering Fluid-solid Coupling

Xiong Run'e　Yan Genhua

State Key Laboratory of Hydrology-Water Resources and Hydraulic Engineering,
Nanjing Hydraulic Research Institute, Nanjing,　210029

Abstract: With the development of hydraulic construction, many high water-head, long-span gates have been constructed, which demand gate seal with higher performance. Due to the self-excited vibration of the gate seal, hydraulic gates always damages, which makes the importance of studies on gate seals increasingly outstanding. Triple nonlinear effect of the rubber material is considered by software ANSYS 10. 0. Finite element analyses are carried out, in order to discuss the influence of the seal' length, the bolt pre-stress and fluid-solid coupling on the seal free vibration characteristics. Based on the practices, some feasible scheme will be proposed to avoid the seal free vibration. This research can provide valuable reference for the study on the free vibration of the gate seal.

Key words: gate seal; ANSYS software; rubber; fluid-solid coupling; free vibration characteristics

1. Introduction

Gate seal is an important composition part of hydraulic gate. With the development of hydraulic construction, many high water-head, large-size gates have been constructed, which demand gate seal with higher performance in structure design, manufacture and installation. Engineering practice indicates that, the sealing effect is bad in a considerable amount of practice by which the self-excited vibration of the seal can be induced[1]. Furthermore, sharp vibration of the gate can always be evoked, affecting the project operation. The arch service gate of bottom outlets of the early Jiaokou Reservoir in China vibrates excessively just because of the seal free vibration, arms destroyed with dynamic instability. Flood gate of Wantan hydropower station of Miyi in Panzhihua in Sichuan vibrates strongly as the top seal surge self-excited vibration owing to leakage. Bottom seal on the radial gate in upper lock head of Anhui Mengcheng County sluice also surge self-excited vibration, which leads to the gate's excessive vibration[2]. Therefore, it is important in the process of gate construction to solve the problem of seal.

It is a waste of manpower and material to study the features of seal rubber with the traditional experiment method, involving the problems such as material nonlinearities, geometric nonlinearities and boundary nonlinearities, which make it hard to make numerical analysis and study systematically of seal rubber. Software ANSYS 10. 0 is applied to carry out finite element analyses, based on the top seal of a large width-to-height flat gate, in order to discuss the influence of the seal' length, the bolt pre-stress and fluid-solid coupling on the seal free vibration characteristics, revealing the change law. And some advices on the study of the seal vibration are given as well.

2　The triple nonlinearities of water seal in finite element analysis

The materials used for gate seal such as rubber and polymers, are mostly hyperelastic. Finite element analysis of the watertight rubber is complicated nonlinear calculation, the nonlinearities reflect in three aspects: the material nonlinearities, geometric nonlinearities and boundary nonlinearities [3-4].

2. 1　The material nonlinearities of watertight rubber

The material nonlinearities mean that material constitutive relationship (namely stress-strain relationship) is nonlinear. Mooney-Rivlin modal is always used in finite element analysis, which can vividly describe the mechanical properties of rubber material and can completely meet the engineering application requirements. The form of the strain energy potential is:

$$W = C_{10}(I_1 - 3) + C_{01}(I_2 - 3) + \frac{1}{d}(J - 1)^2 \quad (1)$$

where W is the strain energy potential; $I_1 = \lambda_1^2 + \lambda_2^2 + \lambda_3^2$ is the first invariant; $I_2 = \frac{1}{\lambda_1^2} + \frac{1}{\lambda_2^2} + \frac{1}{\lambda_3^2}$ is the second invariant; λ_1, λ_2, λ_3 are the principal stretch ratios in three spindles; C_{10}, C_{01} is material constants determining the partial deformation[4-5]; $d = 2\frac{(1-2\nu)}{(C_{10}+C_{01})}$ is the material incompressibility parameter, where ν is Poisson Ratio; J is determinant of the gradient of elastic deformation.

2.2 The geometric nonlinearities of watertight rubber

The geometric nonlinearities manifested as nonlinear strain and displacement relationship. Full Lagrangian method is adopted, and the thinking is based on a virtual power principle, configuration at time 0 is referenced by all variables, the second Piola-Kirch-hoff stress and Green strain relationship at time $t + \Delta t$ are obtained as follows:

$$\int_V S_{ij} \delta \varepsilon_{ij} \, dV = W \quad (2)$$

According to the Expressions of the second Piola-Kirch-hoff stress and Green strain at time 0 and time $t + \Delta t$, finite element equation can be derived then, the matrix form is:

$$([K]_D + [K]_\sigma + [K]_\varepsilon)\{\delta_q\} = \{F\} + \{T\} - \{P\} \quad (3)$$

where $[K]_D$ is tangent stiffness matrix, characterizing the relationship between load increment and displacement; $[K]_\sigma$ is initial stress matrix or geometric stiffness matrix, characterizing influence of initial stress on structure with large deformation; $[K]_\varepsilon$ is stiffness matrix with initial displacement or large displacement, characterizing the change of structural stiffness caused by large displacement; $\{\delta_q\}$ is node coordinates increment vector; $\{F\}$ is body load vector; $\{T\}$ is surface load vector; $\{P\}$ is equivalent resultant force vector when strain applies on nodes.

2.3 The nonlinear contact between watertight rubber and steel plate

Seal rubber deforms and contacts with steel plate, under water pressure, which makes the structural stiffness of the seal changed suddenly, meanwhile it presents nonlinearities. Lagrangian method in incremental form is applied, and the basic idea is total

potential energy and the multiplication of contact boundary condition and Lagrange multiplier constitute modified potential energy, and then get the stationary value to solve the final control equation. Get the variation of total potential energy, and assume it to 0:

$$\delta \Pi = \delta E + \delta W + \delta Q = 0 \quad (4)$$

where E is the internal force potential energy of the system; W is the external force potential energy; Q is the contact force potential energy[3-4].

3. Introduction of the modal

The top seal of a large width-to-height flat gate will be studied in this paper, which is fixed on the foundation of breast wall, with a "P" shape in section. The seal structure consists of the rubber seal component, the seat bang of breast wall, upper and lower mental platen, upper auxiliary platen and gate faceplate, which is shown in Figure 1. When retaining water, the plate gate moves downward and holds down the "P" head of the seal, outstretching the seal rubber head to press against the gate faceplate with certain compression which makes water retained. When the gate opens, the upper auxiliary platen moves upward and the P-type head of the seal retracts off the gate faceplate. Then, the gate opens smoothly.

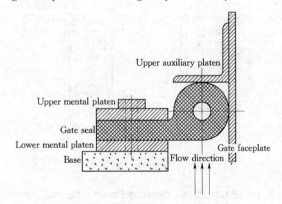

Figure 1　Section of the seal structure

4　Study on the seal self-excited vibration

The self-excited vibration of hydraulic gate is harmful heavily, which exists widely in hydraulic and hydroelectric engineering. It is usually induced by the leakage or the large deformation of the gate, and the vibration intensity is determined by gate stiffness, materials and installation precision of seals, smoothness of the gate faceplate and the work head etc[5]. When the installation precision of gate and conveyance structure or preloading is inadequate, someplace is compressed excessively and some other inadequately. Under high water pressure, it is probable to induce leakage by the latter.

Figure 2 shows the situation when high-speed jet gets through the slit between the seal and the panel. Because of the negative pressure on it while the jet passes by, the head of the seal is sucked to the panel, with the gap blocked. Then the seal rebounds, while the negative pressure disappears, gap forming repeatedly. The cycling appearance of the phenomenon causes the seal vibrating in a certain frequency[6]. The harmonic force generated by the seal free vibration motivates gate in a specified dominant frequency. When the frequency of the seal free vibration is the same or similar to that of the gate at certain step, the gate sets up resonance and the vibration levels is significantly increased. While there is always little structural damping of the steel gate, the resonance amplification factor is huge, which is one of the reasons why many gates damage when resonance happens[5]. Therefore, it is significant to study the dynamic characteristics of the seal to the normal operation of hydraulic structures.

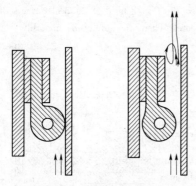

Figure 2 Gap flow phenomenon

4.1 Fluid-Solid Coupling Basic Theory

Hydraulic gates are typically elastomeric. The interaction between the flow and structure means that the elastomeric move and deform, suffering the fluid force, which affect the flow at the same time. Then, the flow field distorts which results in the change of the fluid force further. This is a typical fluid-structure coupling problem[5]. Establish finite element equation for the coupling system. In the structure domain, it is given by:

$$[M_s]\{\ddot{u}\} - [Q]\{\dot{\varphi}\} + [K_s]\{u\} + [Q_1]^T\{\varphi\} = \{F_s\} \quad (6)$$

Converted to the fluid domain, considering the structure motion, the formula is:

$$[M_p]\{\ddot{u}\} - [Q]^T\{\ddot{u}\} + [Q_2]\{u\} + [K_p]^T\{\varphi\} = \{0\} \quad (7)$$

Combining the two above equations, general coupling equation is obtained as follows:

$$\begin{bmatrix} M_s & 0 \\ 0 & M_p \end{bmatrix}\begin{Bmatrix} \ddot{u} \\ \ddot{\varphi} \end{Bmatrix} + \begin{bmatrix} 0 & -Q \\ Q^T & 0 \end{bmatrix}\begin{Bmatrix} \dot{u} \\ \dot{\varphi} \end{Bmatrix}$$
$$+ \begin{bmatrix} K_s & Q_1 \\ Q_2 & K_p \end{bmatrix}\begin{Bmatrix} u \\ \varphi \end{Bmatrix} = \begin{Bmatrix} F_s \\ 0 \end{Bmatrix} \quad (8)$$

where

$$[M_p] = \sum_{e_f}(M_{AB}) = \frac{\rho_f}{C^2}\sum_{e_f}\int_{V_f^e} \overline{N}_A \overline{N}_B \, dV$$

$$[K_s] = \sum_{e_s}(K_{ab}^{ij}) = \sum_{e_s}\int_{V_s^e} V_s^e C_{ijkl} N_{a,k} N_{b,l} \, dV$$

$$\{F\} = \sum_{e_s}\left[\int_{V_s^e} f_i N_a \, dV + \int_{S_\sigma^e} \overline{T} N_a \, ds\right]$$

$$[Q_2] = \sum_{e_c}(Q_{2aA}^i) = \sum_{e_c}\int_{S_c^e} \rho_f U_{\tau_a} \frac{\partial N_a}{\partial \tau_a} n_i \overline{N}_A I_i \, ds$$

$$[K_p] = \sum_{e_f}(K_{AB}) = \sum_{e_f}\int_{V_f^e} \overline{N}_{A,K} \overline{N}_{B,K} \, dV$$

$$[Q] = \sum_{e_c}(Q_{aA}^i) = \sum_{e_c}\int_{S_c^e} \rho_f N_a n_i \overline{N}_A \, ds$$

$$[Q_1] = \sum_{e_c}(Q_{1aA}^i) = \sum_{e_c}\int_{S_c^e} \rho_f U_{\tau_a} \frac{\partial \overline{N}_A}{\partial \tau_a} n_i N_a \, ds$$

where C_{ijkl} is elastic constant tensor, related to structural material characteristics; $[M_s]$, $[K_s]$, $[F_s]$ are mass matrix, stiffness matrix and load matrix, related to structure, respectively; $[M_p]$, $[K_p]$ are mass matrix and stiffness matrix, related to fluid, respectively; $[Q]$ is coupling matrix at the fluid-structure interface; $[Q_1]$, $[Q_2]$ are coupling matrixes related to fluid.

Equation (8) predicts fluid-structure coupling mechanism that the fluid forces the structure move and deform, which change the fluid field simultaneously [7].

4.2 The Analysis of Influence of the seal's length on the seal structure Mode

In order to research the seal free vibration characteristics, assume that prestressing force in bolts is 0 and bolt pitch is 100mm, avoiding the eliminate other interference. Table 1 lists the natural frequency of the first five orders. Figure 3 draws the frequency curve and the frequencies of different orders distributions along the length of the seal is shown in Figure 4.

Table 1   The natural frequencies of the first five orders of different length of the seal （Hz）

Length （mm）	775	1150	1550	1950	2350	2750
The 1st order	21.1	20.9	20.7	20.6	20.2	20.1
The 2nd order	22.8	21.7	21.1	20.9	20.4	20.2
The 3rd order	27.8	24.1	22.4	21.7	21.0	20.7
The 4th order	35.1	27.8	24.7	23.2	22.0	21.4
The 5th order	44.1	32.7	27.6	25.1	23.4	22.5

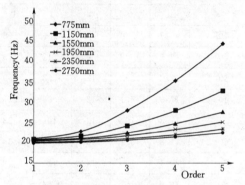

Figure 3   Frequency Curve along the orders

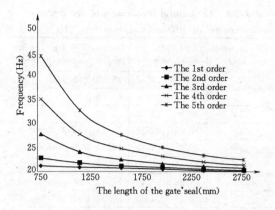

Figure 4   Frequency Curve along the length of seal

Figure 5 shows the first three modes of 1950mm seal.

The above chart data indicates that: (1) frequency of each order decreases with the increase of the length of the seal; (2) frequency of each order increases in a increasing amplitude with the increase of the order; (3) vibration modes of low orders of different length of seal are basically similar.

Considering the structural characteristics, the results are reasonable. With the increase of the seal's length, the seal displays the characteristics of beams, stiffness decreases and mass increases, which leads

(a) Mode 1   (b) Mode 2   (c) Mode 3

Figure 5   The first three modes of 1950mm seal

the decrease of the natural frequency. But for hydraulic structures, it is easy to induce self-excited vibration due to the low-frequency of structures, and gates strong vibration, which is the phenomenon to avoid in hydraulic structure design.

Based on a large width-to-height flat gate, some vibration protection should be adopted while long seal fixed. As for the top water seal, two seals would be suggested. Make sure the two seals work while the gate opens, and strong vibration due to the seal leakage will be avoided basically.

4.3   The analysis of influence of the bolt prestress on the seal structure mode

In order to study the influence of the bolt prestress on the seal free vibration characteristics, many different length of seals grouping, consistent conclusion is obtained. Here, take a typical length (775mm) to describe as an example. Table 2 lists the natural frequency of the first five orders in the length of 775mm of seal. Figure 6 draws the frequencies of different orders distributions along the length of the seal, respectively.

Table 2 The natural frequencies of the first five orders of different bolt prestress（Hz）

The prestress (kN)	The 1st order	The 2nd order	The 3rd order	The 4th order	The 5th order
0	18.41	20.19	25.19	32.28	41.1
10	20.09	21.76	26.68	33.75	42.57
20	20.88	22.56	27.46	34.58	43.46
30	21.58	23.28	28.19	35.37	44.34
40	21.7	23.43	28.33	35.5	44.49
50	21.69	23.43	28.34	35.52	44.53
60	21.69	23.43	28.35	35.54	44.54
70	22.11	23.82	28.72	35.84	44.8
80	21.47	23.23	28.28	35.45	44.48
90	21.11	23.02	28.02	35.29	44.36
100	21.03	22.94	27.96	35.25	44.35

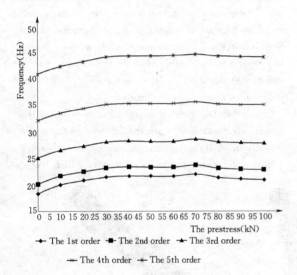

Figure 6　Frequency curve along the bolt prestress

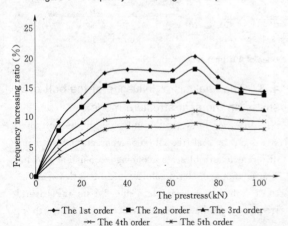

Figure 7　Frequency increasing ratio curve along the bolt prestress

The above data shows that frequency of each order increases to varying degrees. In order to analyze the laws that the natural vibration frequencies change with the prestress, compare the frequencies under different prestress with under no prestress, get the growth rate and list it in Table 3. Figure 7 draws the frequencies growth rate of different orders distributions along the length of the seal.

Table 3 The natural frequencies increase ratio of the first five orders of different bolt prestress（%）

The prestress (kN)	The 1st order	The 2nd order	The 3rd order	The 4th order	The 5th order
0	0	0	0	0	0
10	9.17	7.79	5.9	4.56	3.58
20	13.5	11.8	9	7.12	5.73
30	17.2	15.3	12	9.58	7.89
40	17.9	16.1	12	9.98	8.25
50	17.8	16	13	10	8.33
60	17.8	16.1	13	10.1	8.38
70	20.1	18	14	11	8.99
80	16.6	15	12	9.84	8.21
90	14.7	14	11	9.32	7.94
100	14.3	13.6	11	9.21	7.9

It is shown that: （1）the frequencies change with the prestress in a similar law; when the prestress increases among 0 - 30kN, the frequencies grow greatly; the prestress increases among 30 - 60kN, the frequencies grow slightly; the prestress increases among 60 - 100kN, there is a peak at 70kN, the first frequency increase by 20% correspondingly, as the prestress is greater than 70kN, the frequencies decrease; （2）the first frequency is 20Hz around and the five-order mode is similar.

Thus, the bolt prestress has staggered effect on the stiffness of the seal structure, leading a small increase of natural frequencies. The bolts just play a role as a join and affect the watertight structure of the stiffness. It is common that the bolts are cut off when the gate opens or closes, which results the seal loose, destroys the sealing and Stability of the structure, and then it is likely to induce the seal free vibration. Therefore, it is very necessary to do repair and maintenance for the bolts regularly.

4.4 The Analysis of Influence of fluid field on the seal structure Mode

Different lengths of seals are grouping in the analysis of the seal free vibration to study the influence of fluid field on the structure. The corresponding fluid-solid coupling models are shown in Figure 8 and Figure 9 respectively. List the natural frequencies of the first five orders in different length of fluid field in Table 4. The corresponding relation curve is shown by Figure 10.

Figure 8　Geometric model of flow-solid coupling

Figure 9　Finite Element Model of flow-solid coupling

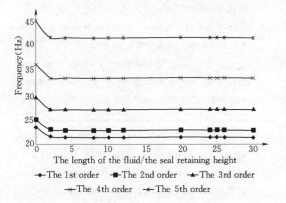

Figure 10　Frequency Curve along the fluid field's length

The results reveal that: when the length of the fluid field is twice the seal retaining height, the first main frequency decreases from 23.36 Hz to 21.45 Hz, with the decrease ratio of 8.18%; when the length of the fluid field is ten times as the seal retaining height, the first main frequency decreases to 21.35 Hz, with the decrease ratio of 8.60%; when 25, 26, 30 times, the first main frequency keeps almost unchanged at 21.36 Hz, with the decrease ratio of 8.56%. List the results in Table 5, and the curve of the first main frequency along the length of the seal is drawn in Figure 11.

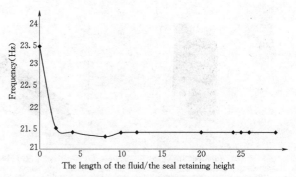

Figure 11　The first frequency curve along the fluid field' length

Table 4　The natural frequencies of the first five orders of different length of the fluid field (Hz)

Times	The 1st order	The 2nd order	The 3rd order	The 4th order	The 5th order
0	23.4	24.9	29.4	36.1	44.8
2	21.5	22.9	27.0	33.3	41.4
4	21.4	22.8	27.0	33.3	41.3
8	21.3	22.8	26.9	33.2	41.3
10	21.4	22.8	27.0	33.2	41.3
12	21.4	22.8	27.0	33.3	41.3
20	21.4	22.8	27.0	33.3	41.4
24	21.4	22.8	27.0	33.3	41.3
25	21.4	22.8	27.0	33.3	41.3
26	21.4	22.8	27.0	33.3	41.3
30	21.4	22.8	27.0	33.3	41.3

Table 5　The low-orders frequencies when the length of fluid field is 0, 10, 30 times as the seal retaining height (Hz)

Order	1	2	3	4	5
without water	23.4	24.9	29.4	36.1	44.8
10 times	21.4	22.8	27.0	33.2	41.3
corresponding decrease ratio (%)	8.6	8.4	8.2	8.0	7.7
30 times	21.4	22.8	27.0	33.3	41.3
corresponding decrease ratio (%)	8.6	8.4	8.2	8.0	7.6

It is predicted that: (1) frequency of each order decreases with the increase of the length of the fluid field; (2) when the length of the fluid field is ten times as the seal retaining height, the first main frequency keeps basically stable. When it is 30 times, the first main frequency keeps almost unchanged. So 10 times as the height can meet the precision requirement considering the influence of fluid-solid coupling; (3) fluid-solid coupling has little influence on the vibration modes of the seal structure, which are basically similar in the first six orders as with or without the fluid field, and Figure 12 displays the first three modes of free vibration in water.

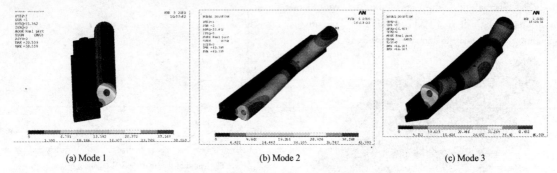

(a) Mode 1   (b) Mode 2   (c) Mode 3

Figure 12　The first three modes of flow-solid coupling

5. Conclusions

This paper applies ANSYS 10.0 to research the modal characteristics of the gate seal, it is concluded as the follows.

(1) Long seals display the characteristics of beams, the free frequencies decrease with the increase of the length of the seal which increases the possibility of the seal structure to vibrate and results in the resonance of the gate under certain condition, destroying the projects. Therefore, some vibration protection is suggested for the gate with long seals. As for the top water seal, two seals should be adopted. Make sure the two seals work while the gate works, and strong vibration due to the seal leakage will be avoided basically.

(2) The bolt prestress has staggered effect on the stiffness of the seal structure, leading a small increase of natural frequencies. The bolts play a role as a join and affect the watertight structure of the stiffness. The failure of bolts always results in the seal loose, destroying the sealing and stability of the structure, and then it is likely to induce the seal free vibration. Therefore, it is very necessary to do repair and maintenance for the bolts regularly.

(3) Considering the influence of fluid-solid coupling, frequency of each order decreases, low-step frequencies do with a high decrease ratio; when it is ten times as the seal retaining height frequencies keep almost unchanged. So, when the length of the fluid field is 10 times as the height can meet the precision requirement considering the influence of fluid-solid coupling.

References

[1] Luo Miaotong. Leakage Treatment for the gate of the west-south sluice in north-river levee of Guangdong province [J]. Guangdong Water Resources and Hydropower, 2008 (2): 67-69. (in Chinese)

[2] Yan Genhua. Examples of hydraulic gate free vibration and prevention measures [C]. Papers in the national hydropower engineering metal-structure major information network congress in 2009. (in Chinese)

[3] Liu Lihua, Lv Niandong, Zhao Jinping, YAO Changjie, HU Wei, ZHENG Zhenqin. Optimization calculation of adjustable seal in some high-head radial gate [J]. Engineering Journal of Wuhan University, 2007, 40 (1): 96-100. (in Chinese)

[4] Shang Xiaojiang, Qiu Feng, Zhao Haifeng, Li Wenying. Advanced Analysis methods and Its Application in ANSYS Finite Element [M]. Beijing, China WaterPower Press, 2006. (in Chinese)

[5] Yan Genhua. Hydrodynamic load and gate's vibration [J]. Hydro-Science and Engineering, 2001 (2): 10-15. (in Chinese)

[6] Yan Shiwu. Development trend of the structure of the hydraulic gate seals home and abroad [J]. Collection of translation of Hydro-Science and Engineering, 1982, 2.

[7] Lai Xiujin. Study on Fluid-Structure Interaction's FEM Dynamic Simulation of Dragon Washbasin Phenomenon [D]. WuHan: Hydraulic and Hydropower Engineering in Huazhong University of Science and Technology, 2005. (in Chinese)

Numerical Study on the Siltation in Harbor Area of the Hongqili Waterway, Pearl River Delta

Chengwei Xu[1]   Liqin Zuo[2]   Huaixiang Liu[2]

State Key Laboratory of Hydrology-Water Resources and Hydraulic Engineering,
Nanjing Hydraulic Research Institute, Nanjing, 210029

Abstract: The Hongqili waterway, one of the main navigation channel in the Pearl River estuarial network, locates in the southeastern part of the Pearl River Delta, where is right adjacent to the Lingding Ocean. Thus this waterway is under the strong influence of both runoff and tides, resulting in sophisticated hydrological environments and erosion & sedimentation variation. A harbor project named Dagang is going to be carried out in the Hongqili waterway in order to enhance the local transportation performance. More than twenty docks of 1000t class (can be upgraded to 3000t class) will be constructed according to the design so that large amounts of excavation and dredging will be required in the harbor district and its navigation channels. This paper calculated the future siltation amounts in the excavation and dredging areas after the accomplishment of Dagang harbor. A two-dimensional numerical model was established to achieve this purpose. In this model integrated datasets including hydraulic, tidal and sediment data were all put into consideration. The model verification was based on the data series of both flood seasons (July – August in 2004) and non-flood seasons (April in 2008, January in 2009). The results indicate that only 1 year after the 1000t class docks' construction the average siltation depth in the harbor area will reach 0. 41 – 0. 66m, and the figure will be 0. 90 – 1. 05m in the inlet channel. If the docks are all upgraded to the 3000t class, the average siltation depth will be 0. 60 – 0. 96m in the harbor area, 1. 6 – 1. 7m in the inlet channel, respectively. Therefore, in this study 1000t docks are suggested and 3000t upgrade can be expected in future after further research.

Key words: harbor construction; siltation; numerical model; Pearl River Delta

1. Introduction

The hongqili waterway locates at the southeast part of the Pearl River Delta. It stretches in the direction of northwest to southeast, running from Banshawei (in northwest) to the cities of Shunde, Fanyu and Zhongshan, with the total length of 41km (Figure1). Connecting with the Ronggui waterway in the upstream and with the Hengmen waterway in the downstream, it passes through the Hongqimen estuary, and finally pours into the Lingding Ocean (Xu et al. , 2009) . As a typical branching reach, there are both wide and narrow sections in the Hongqili waterway. Many sand bars locate not only in the channel but also outside the Hongqimen estuary. This waterway is under the mutual influence of both runoff from upstream and the Lingding Ocean tide from downstream. Thus the water and sediment environment is very sophisticated. And the runoff is the main factor which provides the most water and sediment.

A harbor is going to be built in the Hongqili waterway upstream (Figure 1). Totally 21 1000t class docks (and reserves the possibility of upgrade to 3000t class) were planned in this project, forming a 1443m length coastline. A 2D tidal-sediment numerical model was developed in this study. The flow field and sedimentation before and after the project construction was compared and analyzed, including the choice of 1000t class boat or 3000t class boat. The back-siltation amounts in the channel and harbor area were calculated according to that, providing technical supports to this project.

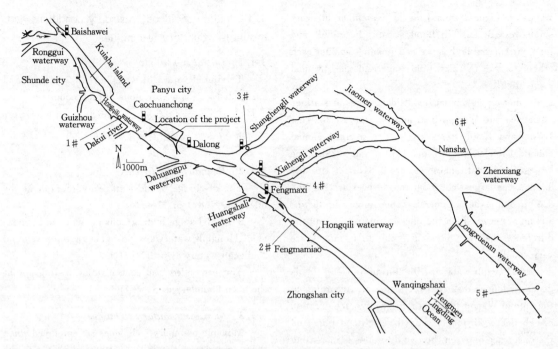

Figure 1 Sketch of Hongqili waterway and layout of vertical lines

2. Hydrological characteristics of Hongqili waterway

2. 1 Water and sediment conditions

The water in the river network of Pearl River Delta is supplied by 3 main tributaries called the Xijiang River, the Beijiang River and the Dongjiang River, respectively. And this river network connects ocean with 8 main estuaries. The runoff flowed through Hongqimen estuary was 20. 9 billion m³ in 1980s and 37. 2 billion m³ in 1990s, ranking the 5th among all the 8 estuaries. The water and sediment ratios between these estuaries changed a lot during this time. Due to the impact of both natural morphology and human activity, all southeast-heading channels in the Pearl River Delta incised deeper than the southwest-heading channels. Therefore the water proportion of the 4 east estuaries increased from 53. 4% in 1980s to 63. 5% in 1990s, and 47. 7% to 56. 8% for sediment proportion. The proportions of all the 4 individual estuaries also increased only except the Jiaomen estuary. And the increment of Hongqimen estuary was the largest, since the water & sediment proportions increased from 6. 4% and 7. 3% in 1980s to 11. 3% and 12. 2% in 1990s (Mo et al., 2009).

According to the analysis of the Pearl River Committee, Ministry of Water Resources of China, the April-September runoff measured at the Wanqinshaxi station in Hongqili waterway takes 79. 7% of that of whole year, while the runoff of the other 6 months only takes 20. 3%.

Among all the 7 largest Chinese rivers, the suspended load concentration of Pearl River is the lowest, only 0. 27kg/m³ for annual average. In flood season (April – September) the suspended load concentration varies between 0. 14 – 0. 53kg/m³, while in the non-flood season this value is only 0. 02 – 0. 07kg/m³. The sediment transported by Hongqili waterway is mainly from the Xijiang River and the Beijiang river, and some from the flood tide current. Sediment yield in the range of Hongqili waterway is small, only comprising some sediment scoured from the riverbed and sediment yield of several small rivers.

2. 2 Tide and tidal current

The tide in Hongqili waterway is the same with that of the whole Pearl River estuaries. Featured as a typical irregular semidiurnal tide, there are two flood tides and two ebb tides in a single day and the diurnal inequality is evident. The highest tidal stages ever at stations of Banshawei, Wanqinshaxi and Nansha are 3. 21m, 2. 62m and 2. 30m respectively, while the lowest being − 1. 31m, − 1. 77m and − 1. 82m. The tidal stages are highest from June to September in every single year. Because in flood season the runoff and tidal current are both large and pushing against each other, so both the high tidal stage and the low tidal stage are lifted.

The average annual flood tidal ranges at stations of Banshawei, Wanqinshaxi and Nansha are 0. 97m, 1. 77m and 1. 30m respectively. The average annual ebb tidal ranges are 0. 97m, 1. 17m and 1. 29m. The largest flood tidal ranges ever are 2. 84m, 2. 94m and 3. 50m. The largest ebb tidal ranges ever are 2. 87m, 3. 12m and 3. 36m. Generally the ranges in downstream are larger than the ones in upstream, and the differences between flood and ebb tidal ranges are very small.

In Hongqili waterway the duration of ebb tide is generally longer than that of flood tide. According to the data of Wanqinshaxi station, the average duration of ebb tide is 2h longer than that of flood tide. And the difference between the two durations is increasing from downstream to upstream, that is, the more upstream the shorter flood tide duration and the longer ebb flood tide duration. The average duration of flood tide is the longest in winter and shortest in summer, while the average ebb tide duration is just the opposite. The flood tide duration in the dry season is longer than that of the flood season, while the ebb tide duration is shorter.

3. ZD flow-sediment mathematical model and its verification

To overcome difficulties in computation for natural water bodies, such as irregular boundary figures, great disparity between length and width of a calculated area, etc. , the boundary-fitting orthogonal coordinate system is employed in 2D flow-sediment mathematical model in this paper. Basic equations include the flow continuity equation, flow momentum equation, the non-equilibrium transport equations of suspended load, the non-equilibrium transport equations of bed load, the gradation equation of bed materials and the bed deformation equations. About this model, its basic governing equations and numerical solutions could be found in the reference (Lu et al. , 1998; Lu et al. , 2002a; Lu et al. , 2002b) . The treatment methods of some key problems in the mathematical model are presented in the following sections.

3. 1 Initial conditions, boundary conditions and movable boundary techniques

3. 1. 1 *Initial conditions*

The initial values of water levels, velocities and suspended load concentrations in each node are

$$H(\xi,\eta)\mid_{t=0}=H_0(\xi,\eta) \qquad u(\xi,\eta)\mid_{t=0}=u_0(\xi,\eta)$$
$$v(\xi,\eta)\mid_{t=0}=v_0(\xi,\eta) \qquad S(\xi,\eta)\mid_{t=0}=S_0(\xi,\eta)$$

(1)

3. 1. 2 *Boundary conditions*

The discharge or tidal stage hydrograph at the inlet is $Q=Q(t)$ or $H=H(t)$.

The tidal stage hydrographs at the open boundary of Hongqili waterway, Shanghengli waterway and Xiahengli waterway are $H=H(t)$.

The suspended load concentration hydrograph at the open boundary is $S=S(t)$.

3. 1. 3 *Movable boundary techniques*

Movable boundary techniques are employed when the boundaries of shoals and battures change with the rise and fall of water levels. According to the seabed elevation of water depth (water level) node, the grid can be judged whether it is above the water surface or not, if it is not, roughness n takes normal values; otherwise, n takes an infinite positive number (for example, 10^{30}) . When the velocities of the four sides of the above grid are calculated by use of momentum equations, the mean value of roughness of adjacent nodes is employed. In any case, whether the adjacent nodes are above the water surface or not, the mean drag force is still an infinite value. Therefore, the ratio between various items and the drag forces in the momentum equations is infinitely small. The above calculated velocities must be an infinitely small number tending to be zero. In order to make the calculation continue, a tiny water depth (0. 005 m) is given for the grid node above the water surface.

3. 2 Treatment of some key problems in numerical simulation of channels influced by both tidal currents and runoff

3. 2. 1 *Critical conditions for erosion and deposition of bed*

The comparison between sediment concentration and sediment transport capacity is employed. If $S_L >$ S_L^*, the sediment concentration is larger than the sediment transport capacity, and deposition will oc-

cur. If $S_L \leqslant S_L^*$ and $V \geqslant V_{CL}$, the sediment concentration is smaller than the sediment transport capacity and the velocity is larger than the incipient velocity V_{CL}, erosion will occur. Considering cohesive sediment with the density changed with time, Tang Cunben's equation is employed (Tang, 1963)

$$V_{CL} = \frac{m}{m+1} \left(\frac{h}{D_L} \right)^{\frac{1}{m}} \sqrt{3.2 \frac{\gamma_s - \gamma}{\gamma} g D_L + \left(\frac{\gamma'}{\gamma'_0} \right)^{10} \frac{C}{\rho D_L}}$$

(2)

in which $m = 6$, $C = 2.9 \times 10^{-4}$ g/cm, $\rho = 1.02 \times 10^{-3}$ gs²/cm⁴; γ'_0 is the stable wet unit weight, and it usually takes 1.6 g/cm³; γ' the actual wet unit weight of mud, and γ'_0 has a relation with deposition duration and gradually increases with time.

3.2.2 *Sediment transport capacity*

Zhang Ruijin's formula $S_* = k \left(\dfrac{V^3}{gh\omega} \right)^m$ is usually employed for river channels' sediment transport capacity without tidal currents' effect. However, the Hongqili waterway is under the mutual influence of both runoff and tidal currents. The sediment transport capacity in this reach has some features of tidal reaches, though runoff is the main factor of sediment transport. It can be known that in Zhang Ruijin's formula, the sediment transport capacity is 0 when flow velocity is 0. In tidal reaches, because of the flow's strong unsteady characteristic, the sediment transport capacity could remain a value when flow velocity is 0. The principle of sediment transport capacity by tidal currents is thus employed in this paper (Lu, et al., 2002c), based on that the concept of background concentration S_0 is introduced, its formula resulted from the regression of the field data is as follows:

$$S_* = k_0 \frac{V^2}{gh} + S_0$$

(3)

In Hongqili waterway, the tidal currents are various with different runoff, and the sediment transport capacity's characteristic is various too. Based on some research, the different is shown in background concentration S_0. So, the sediment transport capacity in Hongqili waterway is

$$S_t^* = k_0 \frac{V^2}{h} + S_0$$

(4)

where $k_0 = 1.19$, $\begin{cases} S_0 = 0.1925 & dry\ season \\ S_0 = 0.6049 & flood\ season \end{cases}$

3.3 Model verification

The calculated region covers a length of about 25km, from Banshawei to Wanqingshaxi, including Shanghengli waterway and Xiahengli waterway. Some other small tributaries are treated as free flow. There are 388×95 grid points in this area. After orthogonal calculation, a orthogonal curvilinear grid is got, which is 30 - 100 m in length and 20 - 30m in width (Xu et al., 2009). Grid in project area is encrypted with grid spacing of about 10m. According to the hydrometrics data in Jul. to Aug. 2004 (flood season), Apr. 2008 (dry season) and Jan. 2009 (dry season), tidal level, flow velocity and direction hydrographs, discharge process and suspended load concentration were verified. The calculated values were in good agreement with the measured ones (Xu et al., 2009). Verification of hydrometrics data in Jul. to Aug. 2004 is given as follows.

3.3.1 *Verification of tidal currents*

Figure 2 shows comparison between the calculated and measured mean velocities and directions of vertical lines 1# and 3#, and discharge of the cross section corrspoingding to vertical lines 1# and 3# during the spring tidal period in Jul. to Aug. 2004. Vertical lines' location are shown in Figure 1. It is seen that the calculated velocities and directions of flood and ebb tides are in good agreement with the measured data in both magnitude and phase, which represents the motion chacteristics of the tidal currents.

The flood current is effect greatly by runoff in Hongqili waterway. During the neap tide period in Jul. to Aug. 2004, there was no flood current because of strong runoff, with the discharge of 7500 - 9500m³/s of cross section 1#. While during the spring tide period, with the discharge of 5800m³/s of cross section 1#, flood current happened.

Figure 3 shows the current filed during spring tidal period. It could be seen that under restrain of riverbed terrain, the currents are of obvious reversing characteristics. During the flooding period, the currents come from estuaries go through Shanghengli waterway, Xiahengli waterway and downstream of Hongqili waterway, and meet together in upstream of Hongqili waterway. During the ebb tides, the currents come from upstream of Hongqili waterway flows into Shanghengli waterway, Xiahengli waterway and downstream of Hongqili waterway. The computation may reflect gradual submergence and occurrences of shoals as well as local flow regime induced by irregular topography with the increase and decrease of water levels.

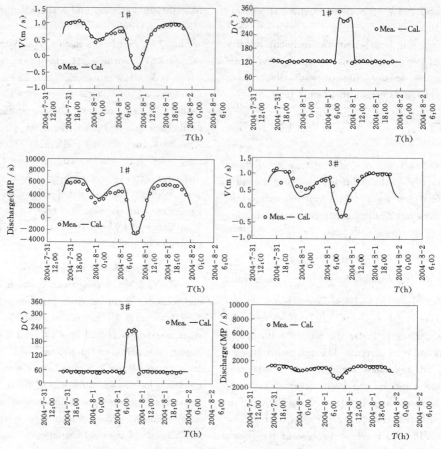

Figure 2   Verification of flow velocity, direction and discharge
process during the spring tidal period in Jul. to Aug. 2004

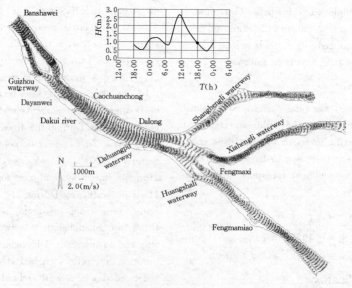

Figure 3   The calculated region and current field of Hongqili waterway
during the spring tidal period (18: 00, Aug. 1, 2004)

3.3.2 *Verification of suspended load concentration*

Figure 4 shows the grain size of suspened load and bed materials in Hongqili waterway. The median diameter of bed materials was 0. 03mm in Jul. 2004 and Jan. 2009, and 0. 016mm in dry season of 2008. The median diameter of suspended load was about 0. 008mm in Jul. 2004, 0. 013mm in dry season of 2008 and 0. 01mm in Jan. 2009. It could be seen that the suspended load came from upstream reaches was finer in flood season (Jul., 2004) with large runoff. While in dry season of 2008 and 2009, diameter of suspended load and bed materials had little difference, which revealed that the suspended load mainly came from riverbed.

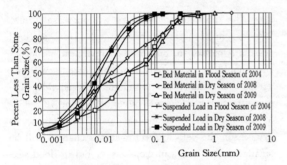

Figure 4　Grain size curve of suspended load and bed materials in Hongqili waterway

The suspended load concentration hydrographs of spring and neap tides are verified of Jul. 2004, Apr. 2008 and Jan. 2009. Figure 5 shows comparison between the calculated and measured suspended load concentration hydrographs of vertical line 1# during the spring tidal period in 2004. The verification shows that the calculated values are in agreement with the measured ones in magnitudes and phases, and it can

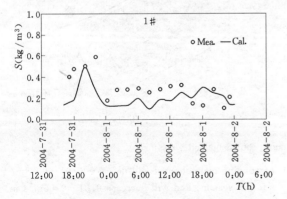

Figure 5　Verification of suspended load concentration process during the spring tidal period in Jul. 2004

reflect the general law of suspended load transport by tidal currents in Hongqili waterway. In flood season (Jul. 2004), the measured suspended load concentration was 0. 1 – 0. 6kg/ m³ during spring tide period, and 0. 4 – 1. 4 kg/m³ during neap tide period with larger runoff discharge. In dry season of 2008, the measured suspended load concentration was about 0. 1 kg/ m³, and only 0. 01 – 0. 06kg/ m³ in dry season of 2009. It shows that, the suspended load greatly influenced by incoming flow and sediment of upstream reaches. The suspended load concentration is higher with finer size in flood season, and is lower with coarser size in dry season.

3.3.3 *Verification of deposition and erosion of riverbed*

Evolution of Hongqili waterway was affected greatly by sand excavation, which was unsuitable to use to verify the mathematical model. Hutiaomen waterway located in the Pearl River Delta, has similar hydrological and sediment conditions with Hongqili waterway. Evolution of Hutiaomen waterway thus was chosen to verify deposition and erosion of riverbed. The 2D mathematical model has been used in Hutiaomen waterway in 2001 (Lu et al., 2001) to study navigation regulation projects. The projects have been implemented, and the predicted results have been tested by measured data (Guangdong Waterway Bureau et al., 2007).

4. Sediment back-siltation prediction

4.1　Construction plan

A harbor is going to be constructed in the left bank of Caochuanchong to Dalong section, upstream of Hongqili waterway. The site is shown in Figure 1 and the harbor layout is plotted in Figure 6. Totally 21 1000t class docks were planned and all the structures of them were able to upgrade to 3000t class. The length of the whole coastline will add up to 1446m. Some relative design parameters are listed in Table 1.

4.2　Flow field variation

After the harbor construction, the tidal currents in front of the docks will be reciprocating flow and the flow direction will be parallel to the dock frontline. The velocity of tidal currents in dry season will be 0. 1 – 0. 2m/s, and in normal flood season this value will rise to 0. 3 – 0. 6m/s. The docking area in front of docks will be excavated from land and thus

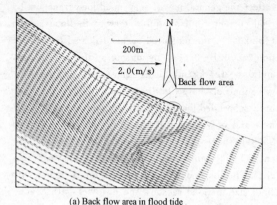

(a) Back flow area in flood tide

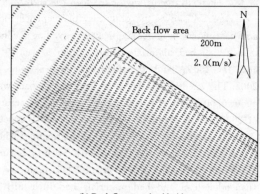

(b) Back flow area in ebb tide

Figure 6　Back flow area in different tides

controlled by the landform. There will local back flow in the east and west side of this docking area (Figure 6). The back flow will appear in the east side when flood tide, with its diameter being dozens of meters, and will appear in the west side when ebb tide, with its diameter being dozens of meters under the 1000t class case or 200 – 300 meters under the 3000t class case. Consequently severe siltation will occur in this field.

The flow velocity will rise a little if the harbor basins were excavated. When excavated to the 1000t class design depth, the docking area's tidal current velocity will increase from 0m/s to 0.05 – 0.15m/s in dry season and to 0.1 – 0.5m/s in flood season (Figure 7). For the boat-turning area inside harbor ba-

sins, the tidal current velocity will increase 0.00 – 0.09m/s in dry season and 0.0 – 0.14m/s in flood season. A part of the approach channel near harborbasins should be excavated, too. The currents will run across this channel and the angle between tidal current direction and channel direction will be 42° – 65°, 39° – 62° in west part and east part, respectively. In dry season the velocity vector across channel will be smaller than 0.3m/s in most cases, but in flood season this vector will be larger than 0.3m/s, even larger than 1m/s sometimes. The velocity in corresponding main channel will not change in dry season, while a slight decrease of 0.01 – 0.03m/s will appear in flood season. That means the effect of this harbor project to the main channel is small.

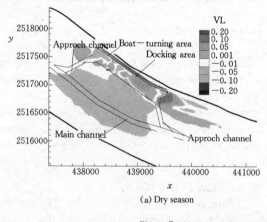

(a) Dry season

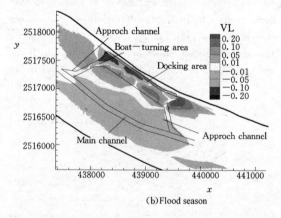

(b) Flood season

Figure 7　Variation of velocity during ebb tidal period（m/s）

Compared with the 1000t class plan, if excavated to the 3000t class design depth, the influence sphere on the harbor basin velocity will be enlarged and the velocity increment will be similar. The velocity in cor-

responding main channel will decrease 0.01 – 0.03m/s in dry season, and will decrease 0.01 – 0.06m/s in flood season. The minus degree is about 1% – 5%.

4.3 Back-siltation prediction

The velocity will increase after the excavation of the harbor basin. But due to a large dredging degree will lead to a large increase in water depth, the sediment load capacity will be reduced and there will be back-siltation in the harbor basin and navigation channel.

The combinations of the average annual runoff process and the spring tide or neap tide were adopted as the calculation condition. The boundary tidal stages at the inlet and outlet were supplied by a 1D numerical model of the Pearl River network (Xu, et al., 2009). Generally one typical tide was chosen for each month in the dry season, and two typical tides were chosen for each month in the flood season.

The excavation of harbor basin may cause back-siltation. In order to achieve the design depth of 1000t class, the docking area need to be excavated to -4.6m height, the back-siltation amounts will be 16 thousand m^3 each year, about 0.41m depth; the boat turning area need to be excavated to -5.2m height, the back-siltation amounts will be 124 thousand m^3 each year, about 0.66m depth; the approach channel near harbor basin need to be excavated 16 thousand m^3, the back-siltation amounts will be 11 thousand m^3 each year, about 0.90 – 1.05m depth (Table 1).

Table 1   **The back-siltation each year after the project construction**

Class	Location		Dredge length (m)	Dredge width (m)	Bottom height (m)	Excavation depth (m)	Excavation amounts (10^{3}m^3)	Siltation amounts (10^{3}m^3)	Siltation depth (m)	Siltation rate (%)
1000t	Docking area		1446	26	-4.6	6.66	254	16	0.41	—
	Turning area		1446	116	-5.2	3.13	585	124	0.66	21.2
	Approach channel	West	190	40	-5.2	1.36	11	7	0.90	66.7
		East	80	40	-5.2	1.48	5	4	1.05	70.8
		Total	270	—	—	1.39	16	11	0.95	68.0
	Total		—	—	—	—	871	162	—	—
3000t	Docking area		1446	36	-7.5	8.88	475	32	0.60	—
	Turning area		1446	216	-8.1	5.20	1642	305	0.96	18.5
	Approach channel	West	300	70	-8.1	4.00	118	49	1.65	41.2
		East	100	70	-8.1	3.19	25	13	1.70	53.2
		Total	400	—	—	3.83	143	62	1.66	43.2
	Total		—	—	—	—	2403	460	—	—

In order to achieve the design depth of 3000t class, the docking area need to be excavated to -7.5m height, the back-siltation amounts will be 32 thousand m^3 each year, about 0.60m depth; the boat turning area need to be excavated to -8.1m height, the back-siltation amounts will be 305 thousand m^3 each year, about 0.96m depth; the approach channel near harbor basin need to be excavated 143 thousand m^3, the back-siltation amounts will be 62 thousand m^3 each year, about 1.6 – 1.7m depth. The back-siltation at different sites of the docking area will be similar, indicating a trend of fully sedimentation. While in the boat turning area the back-siltation mainly locate at middle and downstream section where excava-

tion depth will be large.

5. Conclusions

The hongqili waterway is under the mutual influence of both runoff from upstream and the lingding ocean tide from downstream. Thus the water and sediment environment is very sophisticated. A 2D tidal-sediment numerical model was developed in this study. The equation for calculating the sediment carrying capacity was given. The 2004 flood season (July – August), 2008 dry season (April) and 2009 dry season (January) were used for the verification of this model. For the processes of water stages at every station, flow velocities, directions and sediment concentrations in the vertical lines and discharge at cross sections, the verification showed that calculation fitted well with the measured value.

A harbor is going to be constructed in the left bank of Caochuanchong – Dalong section, upstream of Hongqili waterway. The flow field and sedimentation before and after the project construction was compared and analyzed under different cases, including the choice of 1000t class boat or 3000t class boat. The sedimentation volumes in the channel and back-siltation amounts were calculated according to that. It is suggested that the 1000t class design depth should be excavated first and then monitor the back-siltation in the harbor basin. Then the 3000t class design depth may be expected after the further research.

Acknowledgment

This work was financially supported by the National Natural Science Foundation of China (Grant No. 50879047, No. 50779037 and No. 50909065) and China's "Eleventh Five-Year" National Science and Technology Support Program (Grant No. 2008BAB29B08).

References

[1] Guangdong Waterway Bureau and Guangdong Consulting Service Center of transport. Summary of technologies and experiences of navigation regulation projects in downstream of Xijiang River. 2007. (in Chinese)

[2] Mo Siping, et al. Summary and analysis of regulation projects in Hongqili waterway. Nanjing Hydraulic Research Institute. 2009. (in Chinese)

[3] Lu Yongjun and Yuan Meiqi. 2-D mobile-bed turbulent model for tidal estuary. *Advances in Water Science*, 1998, 9 (2): 151-158. (in Chinese)

[4] Lu Yongjun, Luo Zhaosen. Study on 2D flow-sediment mathematical model in Hutiaomen waterway of Xijiang River. Nanjing Hydraulic Research Institute. 2001. (in Chinese)

[5] Lu Yongjun, Li Haolin, Dong Zhuang, Two-Dimensional tidal current and sediment mathematical model for Oujiang Estuary and Wenzhou Bay. *China Ocean Engineering*, 2002a, 16 (1): 107-122.

[6] Lu Yongjun, Chen Guoxiang. A Non-uniform sediment transport model with the boundary-fitting orthogonal coordinate system. *Journal of Hydrodynamics*, 2002b, 14 (1): 64-68. (in Chinese)

[7] Lu Jianyu, Lu Yongjun, Li Haolin. A primary research for sediment carrying capacity in Oujiang Estuar. *The Ocean Engineering*, 2002c, 20 (1): 46-51. (in Chinese)

[8] Tang C. B.. Incipient motion of sediment. *Journal of Hydraulic Engineering*, 1963 (2): 1-12. (in Chinese)

[9] Xu Chengwei, et al. Study on 2D mathematical model for the Dagagn harbor project. Nanjing Hydraulic Research Institute. 2009. (in Chinese)

Physical Model Testing and Validation of Long-Throated Flumes

Robert Keller

*Department of Civil Engineering, Monash University,
Clayton, Victoria   3800, Australia*

Abstract: Long-throated Flumes (LTFs) are widely-used structures for monitoring flow rates in open channels. They may be built in a variety of different shapes and are generally very accurate when operated under unsubmerged flow conditions. This paper deals with the application of LTFs to monitor the flow distribution within a complex open channel system in a sewage treatment system. Augmentation of the capacity of the system required the construction of new channels and associated long-throated flumes to control and to monitor the various flow distributions. The preliminary design of the LTFs was carried out using the HEC-RAS backwater program coupled with the WINFLUME program for the flumes themselves. The preliminary design was tested within a large physical model, operated under the Froudian similarity law. It is shown that the accuracy of the flow distribution predicted by the HEC-RAS program requires a very precise determination of the energy losses, and that this determination is not possible with standard analytical models. On the other hand, a high degree of correlation is shown between theory, physical model, and prototype measurements of the designed LTFs.

Key words: long-throated flumes; theoretical analysis; physical model study; field measurements

1. Introduction

Long-throated Flumes (LTFs) are widely-used structures for monitoring flow rates in open channels. They may be built in a variety of different shapes and are generally very accurate when operated under unsubmerged flow conditions. When operated in this mode, LTFs may be used to control the distribution of flow through an upstream bifurcation in addition to monitoring the flow within each channel.

In this paper, a case study is presented in which LTFs are used to control and monitor the flow distribution within a complex open channel system in a sewage treatment system. Augmentation of the capacity of the system required the construction of new channels and associated LTFs to control and to monitor the various flow distributions.

The particular application presented is of the design of a new bifurcation within an existing channel. An LTF is present on the existing channel and has been subjected to extensive in-situ field calibration. The new channel from the bifurcation is also to have an LTF built on it. Despite the channels having different hydraulic characteristics, a principal design requirement was that the two LTFs should provide control such that the flow is distributed as evenly as possible throughout the design flow range.

The preliminary design of the new LTF was carried out using the HEC-RAS backwater program[1] coupled with the WINFLUME program[2] for the flumes themselves. This preliminary design was tested within a large physical model, operated under the Froudian similarity law. The fact that extensive field calibration data are available for the LTF on the existing flume provided an opportunity to test the WINFLUME program for LTF design, the physical model tests, and the prototype calibration against each other.

In this paper, the theoretical analysis of the LTF is first briefly reviewed. The design of the new LTF, utilizing HEC-RAS and WINFLUME, is then described. The physical model is described and the results presented.

Comparative results of analysis, physical modeling, and field measurements of the existing LTF are then presented and discussed. Conclusions complete the paper.

2. Analysis of long-throated flume

The analysis of the LTF has been presented previously[3]. It is included herein for completeness.

The major property of an LTF is that it is designed to create a constriction in the flow area sufficient to produce critical flow over the full range of expected flow rates. In addition, the head loss across the structure should not be excessive and afflux should be kept to a minimum.

The general profile of flow through an LTF is shown schematically in Figure 1, which also shows the notation for the theoretical analysis of the flume. In particular, it is noted that the energy level, H, and the stage height, h, are referenced to the invert level in the throat.

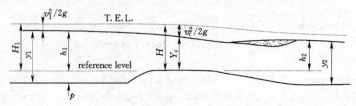

Figure 1   Flow profile through a long−throated flume

The control section is the approximate location of critical flow within the throat of the flume. It is not necessary to know precisely where this occurs because the developed head-flow rate relationship is expressed in terms of the head upstream.

With reference to Figure 1, application of the energy equation yields:

$$H_1 = y_c + \frac{v_c^2}{2g} \qquad (1)$$

where subscript, c, refers to critical conditions.

To proceed further, the shape of the control section must be known. For a rectangular cross-section, the properties of critical flow are such that:

$$y_c + \frac{v_c^2}{2g} = \frac{3}{2} y_c = \frac{3}{2}\sqrt{\frac{q^2}{g}} \qquad (2)$$

where q is the flow rate per unit width within the control section and g is acceleration due to gravity.

Substitution of Equation (2) into Equation (1) and expanding yields:

$$H_1^3 = \left(\frac{3}{2}\right)^3 \frac{q^2}{g} \qquad (3)$$

from which:

$$q = \frac{3}{2}\sqrt{\frac{2}{3}g}\, H_1^{3/2} \qquad (4)$$

In terms of the width of the control section, b_c, Equation (4) is written as:

$$Q = \frac{2}{3}\sqrt{\frac{2}{3}g}\, b_c H_1^{3/2} \qquad (5)$$

where Q is the total flow rate.

The development of Equation (5) has assumed ideal flow conditions—in particular, that there is no energy loss between the location of the upstream head, H_1, and the critical control. Secondly, Q is expressed as a function of H, the total energy level, whereas it is much more useful to express Q as a function of the measured upstream head, h.

These are taken into account by introducing a discharge coefficient, C_d, and a velocity coefficient, C_v such that

$$Q = C_d C_v \frac{2}{3}\sqrt{\frac{2}{3}g}\, b_c H_1^{3/2} \qquad (6)$$

It can be shown[3] that C_d and C_v are given by:

$$C_d = \left(1 - \frac{0.006L}{b_c}\right)\left(1 - \frac{0.003L}{h}\right)^{3/2} \qquad (7)$$

and

$$C_v = \left(1 + \frac{Q^2}{2gh_1 A_1^2}\right)^{\frac{3}{2}} \qquad (8)$$

Equation (5) to Equation (8) can be generalized for non-rectangular cross-sections once the relationship between the critical depth, y_c, and the upstream energy level, H_1 is known. These equations, or their equivalent for non-rectangular cross-sections, represent the computational heart of the WINFLUME program.

3. Design of new long-throated flume

The basic layout of the upstream bifurcation is shown in Figure 2.

The existing LTF is shown on the lower channel and the new LTF on the upper channel. These two flumes provide the downstream control on the flow distribution at the bifurcation.

A HEC-RAS model of the bifurcation down to and including the LTFs was established. The design

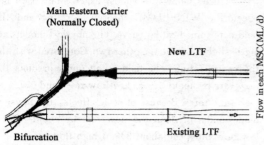

Figure 2　Basic layout of upstream bifurcation

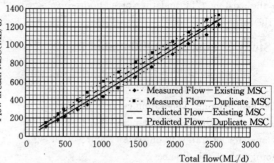

Figure 4　Theoretical and experimental
flow distribution at bifurcation

of the new LTF was then carried out iteratively by running the HEC-RAS model with progressively different geometries for the new LTF until the desired flow distribution was achieved. This flow distribution is shown in Figure 3.

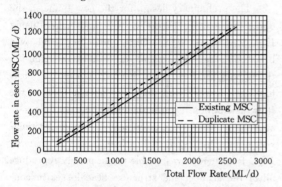

Figure 3　Designed flow distribution at bifurcation

The results of the analysis and design leading to Figure 3 were then tested in the physical model, as described in the following section.

4. Physical model study and results

The physical model was designed according to Froudian principles, requiring equality of the Froude number at homologous points in both model and prototype. A scale ratio of 1 : 20 was chosen as being the largest possible, consistent with available space and water supply.

The preliminary design for the new flume, developed theoretically as outlined in the previous section, was reproduced in the physical model, together with the existing flume shown in the lower channel in Figure 2. The flow split at the bifurcation was determined over the full flow range and the results are presented in Figure 4, together with the flow split established during the design phase.

The physical model results in Figure 4 show a significantly greater bias towards the new LTF in the upper channel than the predicted results, obtained from the use of the HEC-RAS and WINFLUME models. These results were considered to be sufficiently different to prompt a full review of the methodology of the predictions and of the physical modelling. This review highlighted some significant difficulties, unique to this site.

In particular, it was noted that a hydraulic jump formed right at the bifurcation and that its precise location depended on the magnitude of the total flow rate.

Two issues arise here. Firstly, the HEC-RAS computer modelling of the bifurcation has to accurately model the energy losses across the hydraulic jump, in addition to accurately model the bifurcation losses themselves, in order to predict an accurate flow split at the bifurcation. Secondly, the location of the hydraulic jump is not necessarily accurately predicted by the physical model. Even very small changes in flow depth can result in a significant shift in location of the hydraulic jump.

On balance, it is considered that the physical modelling issues impose a lesser effect than the theoretical issues. Accurate theoretical modelling requires more than the ability to predict the rating curve – the HEC-RAS component of the model must allow correctly for the head losses at the bifurcation, and across the transitions in both channels upstream of the LTFs.

From the discussion above, it is concluded that the flow split results from the physical model are to be trusted ahead of the theoretical HEC-RAS model, but that even the physical model results should be treated with some caution.

One important issue is to examine the correlation between physical model results, theoretical analysis, and prototype measurements. These are all available for the existing LTF and the correlation is examined in the following section.

5. Study of existing long-throated flume

The existing LTF in the lower channel of Figure 2 has a known rating curve which has been developed from field gauging. This rating curve has been used by the system operators to monitor the flows in the channel for operational purposes. Figure 5 shows a photograph of this flume.

Figure 5   Photograph of existing
long-throated flume

Because the present study requires the application of this rating curve as one of the boundary conditions of the HEC-RAS modelling, the opportunity was taken to check the field-determined rating curve against that predicted from the WINFLUME computer program and that determined from the physical model study. The results of this comparison are presented in Figure 6.

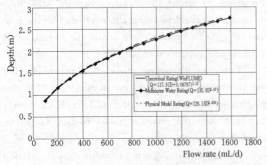

Figure 6   Comparisons of correlations
for existing long-throated flume

Figure 6 shows an almost exact comparison between the WINFLUME theoretical rating and the field-determined rating curve. The physical model rating is slightly above the other two. Compared with the theoretical rating, the field-rating under-predicts the flow rate by about 3% at low flow rates and 0.7% at high flow rates. Because of scale effects, the physical model under-predicts the flow rate by about 8% at low flow rates and about 3% at high flow rates.

Notwithstanding these comments, the obvious closeness of the ratings lends confidence to the use of the WINFLUME program in the present flume design study and also to the modelling procedures in the associated physical model study.

6. Conclusions

The preliminary design for the new flume in the upper channel of Figure 2 was undertaken using HEC-RAS and WINFLUME modelling and was based on the requirement of a flow distribution throughout the flow range that was as even as possible. The physical modelling was shown to bias the flow distribution towards the upper channel. In assessing the reason for the discrepancy between the theoretical and physical modelling it is concluded that the HEC-RAS model was unable to adequately reproduce all of the channel and structure head losses in conjunction with a mixed flow junction analysis. A further complicating factor was the occurrence of significant turbulence and an associated hydraulic jump at the bifurcation.

Through comparisons with physical modelling results and field rating of an existing flume the WINFLUME modelling and the physical modelling were shown to be very accurate.

Acknowledgement

The physical modeling was undertaken in the Hydraulics Laboratory at the Department of Civil Engineering, Monash University. Grateful thanks are expressed to Frank Winston and Eduardo Daly for assistance with the experimental work.

References

[1]   US Army Corps of Engineers. *HEC-RAS River Analysis System*, Version 4.0, March 2008.

[2] Clemmens A. J. , Wahl T. L. , Bos M. G. , Replogle J. A. . *Water Measurement with Flumes and Weirs*. International Institute for Land Reclamation and Improvement, Wageningen, The Netherlands, 2001.

[3] Bos M. G. （Ed. ）. *Discharge Measurement Structures*. International Institution for Land Reclamation and Improvement, Publication No. 20; also Delft Hydraulics Laboratory, Publication No. 161; also Laboratory of Hydraulics and Catchment Hydrology, Publication No. 4; Wageningen/Delft, The Netherlands, 1976.

Air Concentration Distribution at Lower Nappe of a Spillway Aerator

M. R. Jalili Ghazizadeh[1]   S. Esfandiari[2]   A. Ahmadi[3]   A. R. Zarrati[4]

[1] *Assistant professor, Power and Water University of Technology, Tehran, Iran*
[2] *M. Sc. Student, Shahrood University of Technology, Shahrood, Iran*
[3] *Assistant Professor, Shahrood University of Technology, Shahrood, Iran*
[4] *Professor, Amirkabir University of Technology, Tehran, Iran*

Abstract: Forced aeration on high speed flow in chute spillways has been widely used as an economical and technically feasible solution for prevention of cavitation attack. Study of air concentration distribution across the lower nappe is of high importance for aerator design. Although many studies have been carried out on aerators, measurement of air concentration distribution across the lower layer of aerator jet is rare. In the present study, vertical distributions of air concentration along lower layer of a jet were measured on a physical model of an aerator. On the basis of the measurements results, a new equation for air concentration distribution at the lower nappe was developed. Comparison of this equation, with data of previous researchers indicated a good agreement. The proposed equation can be found useful for analysis of air-water flow and optimal design of spillway aerators.

Key words: hydraulic structures; chute spillway; cavitation; air-water flow; aerators; air-water layers; air concentration distribution

1. Introduction

Air entrainment of high speed flow occurs in hydraulic structures. One of the advantages of air entrainment is to prevent or reduce cavitation damage[1]. If the flow velocity approaches 20 to 30 m/s there is a danger of cavitation attack. Flow aeration on chute spillways is known as an economical and technically feasible method for prevention of cavitation attack.

Peterka (1953)[2], Russel and Sheehan (1974)[3] performed experiments in Venturi test sections and showed that 5% to 10% of air was enough to protect concrete specimens against 10 to 20 MPa compressive strength. In China, Deng (1988)[4], Zhou and Wang (1988)[5], Zhang (1991)[6] tested concrete blocks on prototype spillways. They showed that the presence of 4% to 8% of air, in the flow layers close to the spillway invert, prevented cavitation damage for velocities up to 45 m/s.

If the amount of air, entrained by self-aeration, is not enough to avoid cavitation, air can be artificially introduced by aerators. Forced-aeration is possible by installation of aerators on the spillway surfaces.

Aerators are designed to deflect high-speed flows away from the chute surface. The aerator system can be consists of a ramp, groove, offset or combinations of those. The important factor in the mechanism of aeration in the lower and upper surfaces of the free nappe is turbulence. Undulations are formed along the outer nappe surfaces due to turbulence fluctuations. Air is then trapped inside the undulations and are then dragged and diffused into the nappe by turbulence. Reason for amplification of turbulence fluctuations over an aerator is sudden changing of flow direction and pressure distribution in the flow over the aerator ramp[7-10].

Due to aeration of flow, air-water layers are formed at the lower as well as the upper boundary of the nappe and air-water layer thickness grows in the flow direction. Amount of air entrained into the lower layer of the nappe is an important factor in aerator design.

Ervine and Flavey[9], Zarrati[11] and Lima et al. [2] showed that air concentration distribution in the nappe lower layer follows a Gaussian distribution. Chanson used an Error Function to show the air concentration distribution in air-water lower layer of the nappe[13].

In the present work, flow characteristics of nappe were studied in a physical model. By measurement of air concentration at different sections of a nappe in different hydraulic conditions, nappe aeration was studied. Based on measurements, a new equation was developed for air concentration distribution at the lower nappe. This equation can be used to calculate air discharge into the lower nappe.

2. Regions of flow over an aerator

The regions of flow over an aerator are shown in Figure 1 consisted of: (1) the approach flow zone, (2) the transition zone, (3) the aeration region, (4) the impact region, (5) the downstream flow region and (6) the equilibrium region.

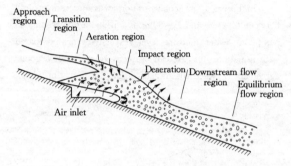

Figure 1   The regions of flow for a typical aerator[13]

The approach zone is a region not affected by the aerator. Due to existence of ramp, pressure distribution changes in the transition zone and shear stresses on the spillway floor increases. These changes alter the turbulent field and have a strong influence on the lower air-water layer aeration.

Without a ramp, there is still a pressure change at the lip of the aerator from almost a hydrostatic pressure to a uniform pressure distribution across the nappe. In the aeration region air is entrained by high intensity turbulent eddies close to the air-water interfaces of the nappe particularly in the absence of high pressure gradient in the nappe[9, 14]. In spite of many studies on aerators, due to complexity of flow behavior and required instrumentations, experimental data on nappe aeration is scares.

3. Characteristics of air-water mixture layers along the aerator nappe

As can be seen in Figure 1, by entraining air into the nappe, air-water layers grow from nappe boundaries to the nappe core along the aerator nappe. Since the majority of transported air downstream of the aerator is supplied from the entrained air into the lower nappe, recognition of effective parameters on

mechanism of air-water layer growth and air discharge entrainment along the lower nappe is essential.

Details of the lower and upper air-water layers in aeration zone are shown in Figure 2. Pan et al.[15] defined outer boundary of air-water nappe at concentration 60%. Some researchers used 90% or 95% concentration for definition of the outer boundary of the nappe[12-13,16-17]. Chanson[13] and Lima et al[12] used 10% and 5% air concentration for definition of non-aerated core boundary, respectively. In this study inner and outer boundaries of air-water layer were considered as air concentration of $C=1\%$ and $C=90\%$ respectively (Figure 2).

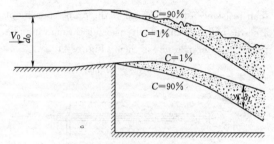

Figure 2   Definition of nappe layers characteristics in the aeration zone

4. Experimental set-up

In the present work a 270cm length perspex flume with 20cm width and 25% longitudinal slope, showed in Figure 3, was used. A head tank with dimensions of 80cm height, 50cm length and 50cm width was located in outset of the flume. A slide gate was installed just downstream of the head tank for regulation of flow and water depth. In order to reduce the energy loss and flow disturbance, the walls of head tank were designed as transition curves. The flume used in the present work included three parts. The first part had 100cm length from the slide gate. This part of flume was designed for formation of a fully developed turbulent flow. Using empirical formulas, thickness of boundary layer was calculated and it was shown that formation of a fully developed turbulent flow forms along the first part of the flume. The second part had a 50cm length and included the aerator. The length of the third part of the flume was 120cm. The third part was made for de-aeration region of the nappe (Figure 3).

In order to measure air concentration, a resistivity probe that had been designed and calibrated by the authors was used. The probe works based on the high difference in conductivity between liquid and gas. The resistance of water is about one thousand times lower

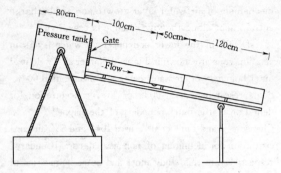

Figure 3　Definition sketch of the
physical model and aerator device

than resistance of air. The tip of the probe consists of a needle with an isolated wire passing through it and a potential difference across them (Figure 4).

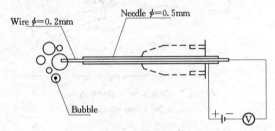

Figure 4　Configuration
of needle probe tip

In air-water mixture, when the probe is in water, the tip of the wire and the needle are connected through the water media and the current passes through the circuit. When the wire tip pierces a bubble, the resistivity increase sharply until the bubble is swept away[18].

The flow discharges were in range of 20 L/s to 40 L/s and Froude numbers 6.7 to 8.2. The air concentration distributions at different sections of the nappe and air discharge value in the duct were measured in all experiments. The hydraulic flow conditions for the present experiments are presented in Table 1.

Table 1　Hydraulic characteristics of flow in the present experiments

Test No.	Water Discharge (L/s)	Air Discharge in the Duct (L/s)	Flow Depth Approach to Deflector (mm)	Average Flow Velocity Approach to Deflector (m/s)
1	38.5	4.08	42.0	4.63
2	28.0	1.81	35.5	3.94
3	25.8	2.37	32.0	4.04
4	21.5	2.18	26.0	4.13

5. Profiles of air concentration at air-water layers

A gaussian equation for air concentration distribution in the nappe is suggested by Pan et al[15]. as:

$$c = c_m e^{-4.7(y/\delta_L)^2} \tag{1}$$

where y is vertical coordinate axis (Figure 2), δ_L thickness of lower air-water layer, c_m is air concentration at the outer boundary of the nappe (60% based on Pan et al. definition) and c is air concentration at distance y from the lower nappe boundary. Characteristics of parameters are shown in Figure 2.

In Figure 5 air concentration distribution measured between $C=1\%$ and $C=60\%$ as inner and outer boundaries of air water mixture layer respectively are compared with Equation (1). As can be seen, there is not a good agreement between the measured data and Equation (1).

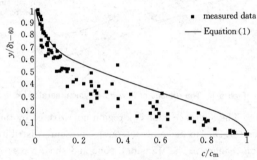

Figure 5　Comparison of the measured data and
Equation(1) for air concentration distribution in
the lower layer of the aerator nappe

To improve the agreement between the data and the Gaussian equation, with defining the outer layer as 90% air and modifying the coefficient of the equation 1 results in the following new equation:

$$c = c_m e^{-4.16(\frac{y}{\delta_L})^{1.57}} \tag{2}$$

In Figure 6 the measured data and the proposed equation [Equation (2)] are shown.

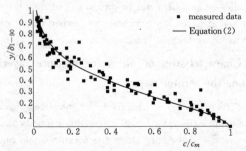

Figure 6　Comparison of the measured data and
Equation (2) for air concentration distribution in
the lower layer of the aerator nappe

6. Validation of the new air distribution equation with other resear-chers' data

Chanson conducted number of experiments on air-water flow and measured air concentration distribution at lower air water layer along the nappe[19]. In Figure 7, the data measured data from Chanson are compared with Equation (1). As can be seen in Figure 7, there is not a good agreement between the data and Equation (1).

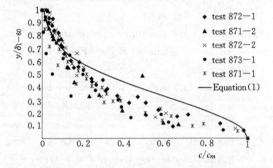

Figure 7   Comparison of a number of Chanson's experimental data and Equation(1)

In Figure 8 the data measured by Chanson and the proposed equation [Equation (2)] are shown. As Figure 8 shows, the agreement between the data and Equation (2) is satisfactory.

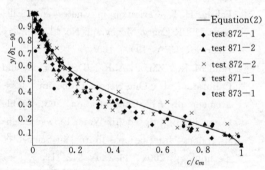

Figure 8   Comparison of a number of Chanson's experimental data and Equation(2)

Lima et al. also carried out a number of tests to determine the air concentration distribution at the lower layer aerator nappe[12]. Figure 9 shows a comparison between the data of Lima et al. and Equation (1). As can be seen in this Figure, there is not a good agreement between the data and Equation (1).

In Figure 10 the data measured by Lima et al. and the proposed equation [Equation (2)] are

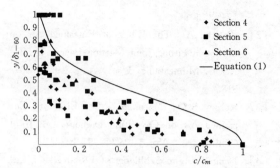

Figure 9   Comparison between a number of Lima et al. 's experimental data and Equation(1)

shown. Figure 10 shows a general agreement between the data and Equation (2). As can be seen, Equation (2) agrees better with the experimental data. It should be mentioned that the data in Figures 9 and 10 is measured at various sections along the nappe before it is disintegrated.

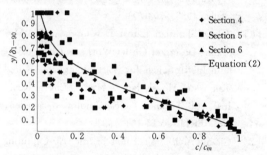

Figure 10   Comparison between a number of Lima et al. 's experimental data and Equation(2)

7. Conclusion

In the present study some experiments on air-water mixture flow over an aerator were performed. Based on experimental data, a new equation was developed for air concentration distribution at lower air-water layer along the nappe. Comparison between the suggested equation, as well as data of previous researchers indicated a good agreement. The proposed equation can be found useful for analysis of air-water flow and analyze the entrained air along the aerator nappe.

References

[1]   Ervine D. A., Falvey H. T., Khan A. R.. Turbulent Flow Structure and Air Uptake at Aerators. Hydropower & Dams. September,

1995: 89-96.

[2] Peterka A. J. The Effect of Entrained Air on Cavitation Pitting. Joint Meeting Paper, IAHR/ ASCE, Minneapolis, USA, Aug. 1953: 507- 518.

[3] Russell S. O. , Sheehan G. J.. Effect of Entrained Air on Cavitation Damage. Can. Jl of Civil Engrg. , Vol. 1, 1974: 97-107.

[4] Deng Zhenghu. Problems of Flood Relief/Energy Dissipation and High-Velocity Flow at Wujiangdu Hydropower Station. *Intl Symp. on Hydraulics for High Dams*, IAHR, Beijing, China, 1988: 230-238.

[5] Zhou Lintai, Wang Junjie. Erosion Damage at Fengman Spillway Dam and Investigation on Measures of preventing Cavitation. *Intl Symp. on Hydraulics for High Dams*, IAHR, Beijing, China, 1988: 703-709.

[6] Zhang Shoutian. Latest Developments in Hydraulic Design of Outlet Works in China. Bulletin Institutionen for Vattenbyggnad, Kungl Takniska Hogskolan (Hyd. Engrg. Royal Inst. of Technol.), Stockholm , Sweden, No. TRITA-VBI-154. 1991.

[7] Glazov A. T.. Calculation of the Air-Capturing Abiliti of a Flow behind an Aerator Ledge. Gidrotekhnicheskoe Stroitel'stvo, 1984, 11: 37-39 (Hydrottechnical Construction, 1985 , Plenum Publ. , 554-558) .

[8] Cain P.. Measurements within Self-Aerated Flow on a large Spillway. Ph. D. Thesis, Ref. 78-18, Dept. of Civil Engrg. , Univ. of Canterbury Christchurch, New Zealand. 1978.

[9] Ervine D. A. , Falvey, H. T.. Behaviour of Tubulent Water Jets in the Atmosphere and in Plunge Pools. Proc. Instn Civ. Engrs. , Part, 83: 295-314. Discussion: Part 2, Mar. -June 1988: 359-363.

[10] DeFazio F. G. , Wei, C. Y.. Design of Aeration Devices on Hydraulic Structures. American Society of Civil Engineers Hydraulics. 1983.

[11] Zarrati A . R. Studies of Air-Water Mixtures For Spillway Aerators. Thesis Submitted to University of London for degree of Ph. D. 1991.

[12] Lima A. C. , Schulz H. E. , Gulliver J. S.. Air uptake along the lower nappe of a spillway aerator. Journal of Hydraulic Research 2008, 46 (6): 839-843.

[13] Chanson H.. Air Bubble Entrainment in Free Surface Turbulent Shear Flows. Academic Press. 1996.

[14] Chanson H.. Study of Air Entrainment and Aeration Devices. J of Hyd. Res. , IAHR, 1989, 27 (3): 301-319.

[15] Pan S. B. , Shao Y. Y. , Shi Q. S. , Dong X. L.. Self-Aeration Capacity of a Water Jet over an Aeration Ramp. Shuili Xuebao (JI of Hyd. Engng.), Beijing, China, 1980 (5): 13-22 (in Chinies) . (USBR translated No. 1868, book No, 12, 455)

[16] Vischer D. L. , Hager W. H.. Dam Hydraulics, Willey, Chichester, U. K. 1998.

[17] Falvey H. T.. Cavitation in Chutes and Spillways. USBR Engrg. Monograph No. 42, Denver, Colorado, USA, 1990: 160.

[18] Jalili M. R. , Zarrati A. R.. Development and Calibration of a One-Needle Probe for Measurement of Air Concentration and Bubble Count in High-Speed Air-Water Flows. Scientia Iranica, International Journal of Science &. Technology, 2004, 11 (4): 312-319.

[19] Chanson H.. A Study of Air Entrainment and Aeration Devices on a Spillway Model. Ph. D. thesis, Ref. 88-8, Dept. of Civil Engr. University of Canterbury, New Zealand. 1988.

Determination of Discharge Coefficient for Oblique Side Weirs Using the Partial Least Square Method

Ali Parvaneh[1]   Seyed Mahmood Borghei[2]   Mohammad Reza Jalili Ghazizadeh[3]

[1]*Graduate Student, Department of Civil Engineering,*
Sharif University of Tech, Tehran, Iran
[2]*Professor, Department of Civil Engineering,*
Sharif University of Technology, Tehran, Iran
[3]*Assistant. Professor, Department of Water Engineering,*
Power and Water Univ. of Tech, Tehran, Iran

Abstract: In this paper, oblique side weir with asymmetric geometry has been experimentally studied. Modification to the geometry of ordinary oblique side weirs causes an increase in the effectiveness of the length of weir as being more in line with the streamlines. Based on the results of 200 model tests, a non linear relation has been presented for discharge coefficient in terms of geometric and hydraulic parameters through using the partial least square method. The results show that this kind of weir is more efficient than the ordinary oblique and normal rectangular side weirs.

Key words: side weir; discharge coefficient; oblique side weir; PLS method

1. Introduction

It is likely that the flow discharge exceeds the capacity of a channel or river and, thus, control structures such as side weir, should be employed to protect the system from overflowing. Usually side weirs, with different geometries, are installed in the channel side wall, parallel to the flow direction and at a desired height so that when the water level rises to the weir height, some portion of the flow would be deviated laterally over the side weir. A schematic view of the hydraulic of side weir is illustrated in Figure. 1.

The question of discharge capacity of side weirs has lead to ample study to obtain a relation for discharge capacity of the side weir due to different circumstances. These studies originated in the early twentieth century. However, the study conducted by De-Marchi (1934) was the basis for many future works. De-Marchi assumed that the specific energy is constant in the channel, along the weir length. Most researches have accepted this assumption, as experienced by some (El-Khashab et al., 1976; Borghei et al., 1999). While the assumption of constant energy

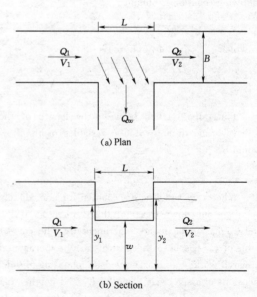

(a) Plan

(b) Section

Figure 1   A schematic view of side weir

is acceptable in many situations, especially for subcritical flow and low Froude numbers, other group of researchers believe that the specific energy is not constant along the side weir and, therefore, they have proposed alternative methods for flow analysis over

125

the side weir.

The De-Marchi approach, by taking $S_0 - S_f = 0$, where S_0 is the channel slope and S_f is the energy slope, leads to the differential equation of flow for a rectangular channel as (Chow, 1959):

$$\frac{dy}{dx} = -\frac{\left(\frac{Q}{gB^2y^3}\right)\left(\frac{-dQ}{dx}\right)}{1-\left(\frac{Q^2}{gB^2y^3}\right)} = \frac{Qy\left(\frac{-dQ}{dx}\right)}{gB^2y^3 - Q^2} \quad (1)$$

Where x is the flow direction, y is the water depth, Q is discharge in the main channel, B is the channel width and g is the gravitational acceleration. Also, the discharge outflow for an element of flow over the weir can be written as:

$$\frac{dQ}{dx} = -\frac{2}{3}C_M \sqrt{2g}(y-w)^{1.5} \quad (2)$$

Where C_M is the discharge coefficient and known as the De-Marchi coefficient, which is a function of channel geometry and flow variables. On the other hand, using the specific energy relation in a channel, flow rate is computed from:

$$Q = By \sqrt{2g(E-y)} \quad (3)$$

Substituting Equation (2) and Equation (3) in Equation (1), integrating the resultant equation, the following equation is obtained:

$$x = \frac{3B}{2C_M}\Phi(y,E,w) + cte. \quad (4)$$

In which $\Phi(y,E,w)$ is

$$\Phi(y,E,w) = \frac{2E-3w}{E-w}\sqrt{\frac{E-y}{y-w}} - 3\sin^{-1}\sqrt{\frac{E-y}{E-w}} \quad (5)$$

Thus, the relation between the length of the weir (L) and other hydraulic variables of the flow would be

$$L = \frac{3}{2}\frac{B}{C_M}(\Phi_2 - \Phi_1) \quad (6)$$

Where, Φ_1 and Φ_2 are accounted for immediately upstream and downstream of the weir respectively, and C_M has to be found experimentally. The correct value of C_M has been the goal of many experimental research works, as it is needed for choosing the dimensions of a side weir (length, L, and height, w) for the required diversion discharge. A dimensional analysis approach to the problem would suggest the most non-dimensional influential parameters affecting C_M in the form of: $C_M = f(Fr_1, w/y_1, L/B)$, where Fr_1 is the upstream Froude number at Section 1 in the channel (Figure 1). However, some researchers have found C_M to be only a function of Fr_1 (linear or

nonlinear), among them Subramanya and Awasthy (1972), Ranga Raju et al. (1979), Hager (1987), Cheong (1991), Agaccioglu and Yüksel (1998). Another group of investigators have included other non-dimensional parameters, i. e. , Singh et al. (1994), Jalili and Borghei (1996) have included w/y_1 in the function and Borghei et al. (1999) have included L/B as well as w/y_1. As the side weirs are very effective discharge controlling structures, many different weir geometry and channel cross sections are tested for evaluation of C_M values. For example, addition to rectangular channels, for circular channel the work of Ramamurthy et al. (2006), for triangular channel the study of Uyumaz (1992), for triangular side weir the paper by Ghodsian (2004) and for the influence of bed morphology the work of Rosier et al. (2005) are among them.

Many attempts have been done in order to improve the efficiency of the side weirs, i. e. decrease the length of the weir for the same weir height and lateral discharge requirement. One of the methods tested for ordinary sharp crested weirs would be the obliqueness effect (Borghei et al. , 2003 and 2006), yet not many studies have been performed on this effect for side weirs. Ura et al. (2001) carried out analytical and experimental studies on the side weirs to minimize disadvantages of conventional side weir. They believed that the normal side weirs have some disadvantages, such as flow segregation and decrease of discharge coefficient, with the increase of Froude number in the channel. Therefore, they employed an oblique side weir, as illustrated in Figure 2, and presented Equation (7) for discharge coefficient in terms of upstream Froude number (Fr_1) and angle of diversion (θ).

$$C_M = 0.611\left[\cos\theta\sqrt{\frac{3Fr_1^2}{2+Fr_1^2}} + \sin\theta\sqrt{1-\frac{3Fr_1^2}{2+Fr_1^2}}\right]\sin\theta \quad (7)$$

According to their findings, the influential oblique angle (θ) is between 60 and 80 degrees with 70 degrees being optimized. Their design included guide wall with gradual opening (Figure 2) and inclined apron between channel and side weir (Figure 3) to omit the dead zone and flow segregation, decrease sedimentation effects and increase the discharge coefficient. They concluded that discharge coefficient for oblique side weir with gradual opening is greater than those for oblique side weir with inclined apron between channel and weir and this coefficient is in turn greater than normal oblique side weir.

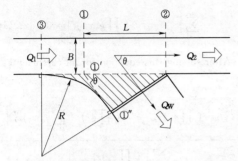

Figure 2　Flow at oblique weir (Ura et al., 2001)

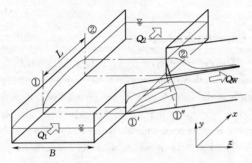

Figure 3　Oblique side weir with slope between channel and weir slope (Ura et al., 2001)

Parvaneh and Borghei (2009) modified oblique side weir geometry to improve the efficiency of the side weir. The proposed modified oblique side is shown in Fig. 4b. Their experimental works on an

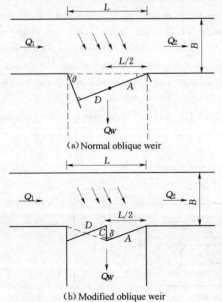

(a) Normal oblique weir

(b) Modified oblique weir

Figure 4　Plan view

asymmetric geometry of oblique side weir showed that the effect of obliqueness is an important parameter to increase the efficiency of the side weir and should be accounted for by the designer engineers. While Figure 4 (a) is the normal oblique side weir, similar to the one proposed by Ura et al. (2001), the modified oblique side weir (Figure 4b) is the one which was used by Parvaneh and Borghei (2009).

In other words, while length "A" of the oblique weir is fixed, length "D" is displaced and becomes closer to the main channel and, therefore, the flow. As a result, this led to an increase of the orthogonal effect of the weir with larger effective length and less occupying of the diversion channel. As it can be concluded from Figure 4 (b) the weir crest length (L') is $A+C+D$.

The linear equation proposed by Parvaneh and Borghei (2009) was in the form of

$$C_M = -0.153 + 0.252\left(\frac{L'}{L}\right)$$

$$-0.544\left(\frac{Fr_1}{\sin\frac{\delta}{2}}\right) + 0.510\left(\frac{y_2 - w}{y_1 - w}\right) \quad (8)$$

This equation shows that the length of the weir as well as the angle, are influential for this kind of weir. The other kinds of side weirs, which recently some works have been published on, are the labyrinth side weir. This is a kind of oblique side weir with equal oblique sides. Emiroglu et al. from about 2500 test runs have obtained equations for C_M by using normal data analysis (Emiroglu et al., 2010a) and soft computing (ANFIS) analysis (Emiroglu et al., 2010b). Their results showed that this kind of weir is much more efficient than the normal side weir.

Because only a few studies have been conducted on oblique side weirs and also the efficiency of oblique side weir is by far greater than normal side weir in most hydraulic circumstances, it is anticipated that oblique side weir will be the focus of a significant amount of analytical and experimental studies in the coming years.

This paper presents a new equation for discharge coefficient of labyrinth side weir using the Partial Least Square method.

2. Partial least square method

Partial least square (PLS) is a robust method to estimate and fit multivariable statistical data. In effect, PLS is used to determine a dependent variable in terms of independent variables for nonlinear problems. In the current paper, this method was employed to estimate C_M for modified oblique side weir.

For the first time Ramamurthy et al. (2006) used this method in hydraulic engineering. As far as this method has received less attention in hydraulic engineering practice in the current paper, this method was employed to estimate CM for modified oblique side weir. In this section, the PLS method is explained. To show the relation between variables, the following function was used:

$$f = g_1(x_1) \cdot g_2(x_2) \cdot \cdots \cdot g_n(x_n) = \prod_{i=1}^{n} g_i(x_i) \tag{9}$$

$$g_i(x_i) = a_{io} + a_{i1}x_i + a_{i2}x_i^2 + \cdots + a_{im}x_i^m$$

$$= \sum_{j=0}^{m} a_{ij}x_i^j \tag{10}$$

$$f = \prod_{i=1}^{n} \left(\sum_{j=0}^{m} a_{ij}x_i^j \right) \tag{11}$$

and SSE can be defined as

$$\delta^2 = \sum_{k=1}^{s} (f - f_k)^2 \tag{12}$$

Through using the least square method and taking derivative with respect to a_{ij}, we may have

$$\frac{\partial}{\partial a_{ij}} \left[\sum_{k=1}^{s} (f - f_k)^2 \right] = 0 \tag{13}$$

This equation can be reduced to Eq. (14).

$$\sum_{k=1}^{s} \left[(f - f_k) \frac{\partial f}{\partial a_{ij}} \right] = 0 \tag{14}$$

By using Equation (9) and Equation (10):

$$\frac{\partial f}{\partial a_{ij}} = \prod_{i=1}^{n} g_i \frac{\partial g_t}{\partial a_{ij}} = \prod_{\substack{i=1 \\ i \neq t}}^{n} g_i x_t^j \quad (j = 1, 2, \cdots, n) \tag{15}$$

Substituting Equation (15) in Equation (1) yields:

$$\sum_{k=1}^{s} \left[(f - f_k) \left(\prod_{\substack{i=1 \\ i \neq t}}^{n} g_i x_t^j \right) \right] = 0 \tag{16}$$

After solving this equation for all coefficients:

$$a_{tj} = $$

$$\frac{\sum_{k=1}^{s} \left\{ (f_k) \left[\left(\prod_{\substack{i=1 \\ i \neq t}}^{n} g_i \right) x_t^j \right]_k - \left[\left(\prod_{\substack{i=1 \\ i \neq t}}^{n} g_i \right)^2 \left(\sum_{\substack{p=0 \\ p \neq j}}^{m} a_{tp} x_t^p \right) x_t^j \right]_k \right\}}{\sum_{k=1}^{s} \left[\left(\prod_{\substack{i=1 \\ i \neq t}}^{n} g_i \right) x_t^j \right]_k^2} \tag{17}$$

3. Experimental setup

The experimental model involves an inclined rectangular section flume with the 40 cm width and 11m length is shown in Figure 5. The channel wall is made of glass and has a high of 66 cm. Also, the channel bottom is color coated metal. Water circulation system is a closed system in which water is conducted from main reservoir to glass flume using pump. Prior to reaching to the side weir, some portion of the flow in the glass flume deviate laterally and discharges into the lateral reservoir. After measuring the amount of lateral discharge QW by V-Notch weir of the reservoir, it flows into the main reservoir. The remainder flow in the flume discharges into an end reservoir at the end of flume and after measuring the amount of discharge (QW), it flows into the main reservoir. The used side weirs in this research are made of Plexiglass. Altogether, 200 tests were conducted with opening lengths of 30cm, 40cm and 60cm, heights of 5cm, 7.5cm, 10cm and 15cm and oblique angles of 20°, 30°, 45° and 60°. Test variables are presented in Table 1.

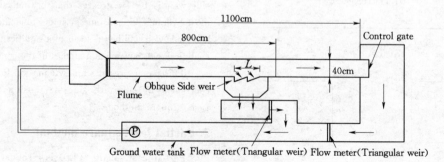

Figure 5   Plan view of the experimental set-up

Table 1   The test variables

L (cm)	w (cm)	δ	Fr_1	Q_1 (L/s)	Number of tests
60	5, 10, 15	140, 120, 90	0.24-0.87	24.6-29.3	45
40	5, 7.5, 10, 15	140, 120, 90, 60	0.19-0.87	20.0-30.0	75
30	5, 7.5, 10, 15	140, 120, 90, 60	0.20-0.91	19.1-30.0	80

4. Primary tests

In order to compare the performance of the present oblique with normal rectangular side weir, a series of comparison tests have been performed with the same experimental values. The measured sections for the water depth in the two cases along the center line of the main channel and at the downstream side of the weirs are shown in Figure 6.

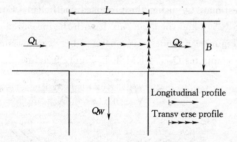

(a) Rectanguar weir

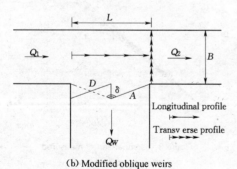

(b) Modified oblique weirs

Figure 6   Plan of considered side weirs

Figure 7 shows an example of the variation of the water depth for the normal and the presented oblique weirs for the same Q_1. Also, as shown for a test sample in Table 2, for the same upstream discharge and depth (Q_1 and y_1), the efficiency of oblique weir re-

spect to the normal side weir increased from 20% to 90% when δ were from 140 to 60 degrees. These results clearly reveal that the proposed oblique side weir is an efficient structure.

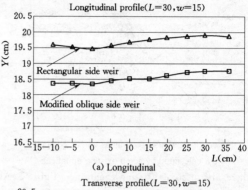

(a) Longitudinal

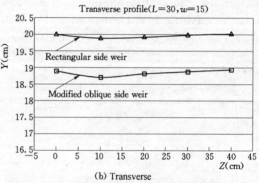

(b) Transverse

Figure 7   Variation of water depth for the two weirs for $Q_1 = 27.3$ L/s, $\delta = 60°$

Table 2   Primary sample test variables for normal and oblique side weir ($L = 40$cm, $w = 15$cm)

δ	y_1 (mm)	Q_1 (m³/s)	Fr_1	Q_w (m³/s) Oblique	Q_w (m³/s) Normal	Discharge increase (%)
140	186.1	0.0267	0.265	0.0046	0.0038	20
120	186.6	0.0244	0.241	0.0055	0.0042	30
90	187.7	0.0232	0.227	0.0075	0.0045	66
60	184.8	0.0237	0.238	0.0098	0.0052	90

5. Results and Discussion

With accepting the constant energy assumption along the weir in the main channel (with a mean value reduction about 1.3% reported by Parvaneh and Borghei 2009 for the same experimental set-up), De-Marchi's assumption and, thus, Equation (6) can be used for present oblique side weir.

For present tests, the observed discharge coefficients were calculated by Equation (6) and compared

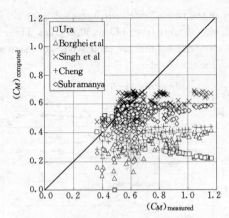

Figure 8　Observed C_M values versus computed
C_M calculated by some existing equations

with results of some existing equation in figure 8. As it can be concluded from Figure 8 none of the relations is suitable for evaluation of the present oblique side weir. Thus, a new relation for discharge coefficient of the oblique weir would be needed.

In the conducted tests, one of the important variables was weir effective length determined as follows:

$$L' = \frac{L}{\sin\left(\frac{\delta}{2}\right)} + \frac{L}{\tan\left(\frac{\delta}{2}\right)} \tag{18}$$

Beside the popular dimensionless parameters of normal side weirs, i. e., Fr_1 and w/y_1, other variables such as δ and L'/L were considered for oblique weirs. No doubt that with many different functions of these variables, a wide range of combination in the form of equation could be obtained. Preliminary function to estimate C_M was chosen similar to that used by Ramamurthy et al. (2006) as follows.

$$C_M = g_1\left[\frac{Fr_1}{\sin(\frac{\delta}{2})}\right] g_2\left[\frac{w\,\sin(\frac{\delta}{2})}{y_1 - w}\right] \tag{19}$$
$$g_3\left(\frac{L'}{L}\right) g_4\left(\frac{L'}{B}\right)$$

$$g_1\left[\frac{Fr_1}{\sin\,(\frac{\delta}{2})}\right] = a_{10} + a_{11}\left[\frac{Fr_1}{\sin\,(\frac{\delta}{2})}\right]$$
$$+ a_{12}\left[\frac{Fr_1}{\sin\,(\frac{\delta}{2})}\right]^2 + \cdots \tag{20}$$

$$g_2\left[\frac{w\,\sin(\frac{\delta}{2})}{y_1 - w}\right] = a_{20} + a_{21}\left[\frac{w\,\sin(\frac{\delta}{2})}{y_1 - w}\right]$$
$$+ a_{22}\left[\frac{w\,\sin(\frac{\delta}{2})}{y_1 - w}\right]^2 + \cdots \tag{21}$$

$$g_3\left(\frac{L'}{L}\right) = a_{30} + a_{31}\left(\frac{L'}{L}\right) + a_{32}\left(\frac{L'}{L}\right)^2 + \cdots \tag{22}$$

$$g_4\left(\frac{L'}{B}\right) = a_{40} + a_{41}\left(\frac{L'}{B}\right) + a_{42}\left(\frac{L'}{B}\right)^2 + \cdots \tag{23}$$

$$f = f(x_1, x_2, \cdots, x_n) \tag{24}$$

After several trial and error attempts and determination of all a_{ij}, the following equation was yielded:

$$C_M = \left[0.772 - 0.605\left[\frac{Fr_1}{\sin\frac{\delta}{2}}\right] + 0.559\left[\frac{Fr_1}{\sin\frac{\delta}{2}}\right]^2\right.$$
$$\left. - 0.358\left[\frac{Fr_1}{\sin\frac{\delta}{2}}\right]^3 + 0.084\left[\frac{Fr_1}{\sin\frac{\delta}{2}}\right]^4\right]$$
$$\times \left[0.693 + 0.622\left[\frac{w\,\sin\frac{\delta}{2}}{y_1 - w}\right] + 0.104\left[\frac{w\,\sin\frac{\delta}{2}}{y_1 - w}\right]^3\right]$$
$$\times \left[1.843 + 0.676\left(\frac{L'}{L}\right) + 0.563\left(\frac{L'}{L}\right)^3\right]$$
$$\times \left[0.499 - 0.13\left(\frac{L'}{B}\right) + \left(\frac{L'}{B}\right)^2\right] \tag{25}$$

In order to study the effect of C_M, it is required to estimate Q_W using the proposed coefficient [Equation (25)] and then compare it with the measured data. Figure 9 provides a comparison between measured and computes Q_W, NRMSE in Figure 9 defined as:

$$\mathrm{NRMSE} = \sqrt{\frac{\sum[F(x) - f(x)]^2}{\sum[f(x) - f\overline{X}]^2}} \tag{26}$$

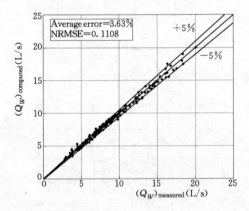

Figure 9　Comparison of measured
and computed Q_W

The lower value for NRMSE indicates lower data scattering and higher accuracy of the relation (Gonzales et al. 2001). Where, $F(x)$ is estimated value, $f(x)$ is measured value and $\overline{f}$ is mean measured value. As it can be seen in Figure 9 good results were achieved using Equation (25) for estimation of C_M.

6. Conclusion

In this study, a modified geometry for oblique side weirs was proposed. This modification on normal oblique side weir caused increase in both orthogonality length and effective length of side weir and consequently increased the discharge coefficient and side weir performance. Then using the PLS method a nonlinear relations for discharge coefficient were obtained. The estimated discharge was within the range of $\pm 5\%$ of measured discharge. Although the test variables cover a wide range for measured data, it is suggested that other experimental set-ups can be used to obtain a more precise result.

Acknowledgment

This research was partially financed by the Water Research Council, affiliated with the Ministry of Energy of Iran.

References

[1] Agaccioglu H. , Yüksel Y.. Side weir flow in curved channels. ASCE Journal of Irrigation and Drainage Engineering, 1998, 124 (3): 163-175.

[2] Borghei S. M. , Jalili M. R. , Ghodsian, M.. Discharge coefficient for sharp crested side weirs in subcritical flow. ASCE Journal of Hydraulic Engineering, 1999, 125 (10): 1051-1056.

[3] Borghei S. M. , Vattania Z. , Ghodsian M. , Jalili M. R.. Oblique rectangular sharp-crested weir. ICE Journal of Water and Maritime Engineering, 2003, 156 (2): 185-192.

[4] Borghei S. M. , Kabiri-Samani A. R. , Nekoee N.. Oblique weir equation using incomplete self-similarity. Can. Jou. Of Civil Eng. , CSCE, 2006, 33 (10): 1241-1250.

[5] Cheong H. F.. Discharge coefficient of lateral diversion from trapezoidal channel. ASCE Journal of Irrigation and Drainage Engineering, 1991, 117 (4): 461-475.

[6] Chow V. T.. Open channel hydraulics, New York, McGraw-Hill Book Compa-ny, Inc. 1959.

[7] De-Marchi G.. Essay on the performance of lateral weirs. L, Energia Eletterica, Milano, Italy, 1934, 11 (11): 849-860.

[8] El-Khashab A. , Smith K. V. H.. Experimental investigation of flow over side weirs. ASCE Journal of Hydraulic Engineering, 1976, 102 (9): 1255-1268.

[9] Emiroglu ME, Kaya N, Agaccioglu H.. Discharge capacity of labyrinth side-weir located on a straight channel. ASCE Journal of Irrigation and Drainage Engineering, 2010, 136 (1): 37-46.

[10] Emiroglu M. E. , Kisi O. , Bilhan, O.. Predicting discharge capacity of triangular labyrinth side weir located on a straight channel by using an adaptive neuro-fuzzy technique. Journal of Advances in Engineering Software, 2010, 41 (2): 154-160.

[11] Ghodsian M. Flow over triangular side weir. Journal Scientia Iranica Transaction on Civil Engineering, 2004, 11 (1): 114-120.

[12] Gonzalez J. , Rojas I. , Poamares H. , Orteag J.. A new clustering techinque for function aproximation. IEEE Transactions on Neural Networks, 2002, 13 (1): 132-142.

[13] Hager W. H.. Lateral out flow over side weirs. ASCE Journal of Hydraulic Engineering, 1987, 113 (4): 491-503.

[14] Jalili M. R. , Borghei S. M.. Discussion of Discharge coefficient of rectangular side weirs by R. Singh, D. Manivannan and T. Satyanarayana. ASCE Journal of Irrigation and Drainage Engineering, 1996, 122 (2): 132.

[15] Novak P. , Cabelka J.. Modls in Hydraulic Engineering. Pitman Publishin-g Inc. 1981.

[16] Parvaneh A. , Borghei S. M.. Oblique side weir. In Proceedings of the 33th Congress of the International Association for Hydraulic Research, Vancouver, 9-14 August. 2009 (11310): 1046-1053.

[17] Ramamurthy A. S. , Junying Qu. , Di-ep Vo.. Nonlinear PLS method for side weir flows. ASCE Journal of Irrigation and Drainage Engineering, 2006, 132 (5): 486-489.

[18] Ranga Raju K. G. , Prasad B. , Gupta S.. Side

weirs in rectangular channels. ASCE Journal of Hydraulic Engineering, 1979, 105 (5): 547-554.

[19] Rosier B., Boillat J. L., Schleiss A.. Influence of side overflow induced local sedimentary deposit on bed form related roughness and intensity of diverted discharge. Proceedings of the XXXI IAHR Congress, Theme C, Seoul, Korea, 11-16. September, 2005: 1639-1650.

[20] Singh R., Manivannana D., Satyan-arayana T.. Discharge coefficient of rectangular side weirs. ASCE Journal of Irrigation and Drainage Engineering, 1994, 120 (4): 814-819.

[21] Subramanya K., Awasthy S. C.. Spatially varied flow over side weirs. ASCE Journal of Hydraulic Engineering, 1972, 98 (1): 1-10.

[22] Uyumaz A.. Side weir in triangular ch-annel. ASCE Journal of Irrigation and Drainage Engineering, 1992, 118 (6): 965-970.

[23] Ura M., Kita Y., Akiyama J., Moriyama H., Kumar Jha A.. Discharge coefficient of oblique side weirs. JSCE Journal of Hydro-science and Hydraulic Engineering, 2001, 19 (1): 85-96.

Notation

B	width of main channel
C_M	discharge coefficient for weir
E	specific energy
Fr	Froude number in the channel
g	acceleration due to gravity
L	length (width) of side weir
L'	weir crest length
Q	discharge in the main channel;
Q_W	discharge over side weir
V	velocity in main channel
w	weir height
x	distance along side weir
y	depth of flow
δ	vertex angle
θ	oblique angle

Long-Term Scheduling of Large-Scale Hydropower Systems for Energy Maximization

Systems for Energy Maximization

Zhang Ming   Fan Ziwu   Guo Yongbin

*State Key Laboratory of Hydrology-Water Resources and Hydraulic Engineering,
Nanjing Hydraulic Research Institute, Nanjing, 210029*

Abstract: A procedure called k-criterion method is introduced to solve the large-scale hydropower scheduling problem. It can reflect the internal operating principle of the hydropower system directly. In order to avoid the difficulties arising from calculating the equations in the process, the coordinates rotation method is adopted to transfer the decision-making process for multi-reservoirs simultaneously to single reservoir gradually. Utilizing the routine operation chart of hydro plant, the k-criterion method is improved and the result derived from it is more reasonable. Taking for an example of the large-scale hydropower system which consists of eleven hydro plants in the middle-upper Yangtze River basin, the model of maximizing the energy with fixed firm power is tested in the paper.

Key words: large-scale mixed hydropower system; k-criterion method; operation chart; coordinates rotation method; hydropower scheduling problem

For the hydropower system, the general methods applied to solve the optimal scheduling problem include dynamic programming, progressive optimization arithmetic[1-3], linear programming, net flow method, genetic algorithm, etc. Basd on those optimal procedures, objective functions are established to improve the energy generation. Though methods mentioned above are advantageous to solve the problem, there are some shortcomings of them still: for the linear programming, linear disposal are needed for the objective function and the constraints, which will lead to the deviation from the original problem; the calculation speed of the net flow method is fast, but the structural requirements of the constraints is too rigid. The genetic algorithm has especial advantage to solve the complex objective function, but it is not effective for the disposal of multitudinous constraints; the dynamic programming is limited by the separability requirement of the objective function and the computer memory capacity limitation. More correlative commentary can be consulted in the references[4-7].

K-criterion method improved by the routine operation chart is introduced in the paper. Not only can it optimize the hydropower system scheduling, but also can reflect the internal operating principle directly and meet the synthesis application requirements of the hydro plants. This procedure can give direct guide for the mixed hydropower system and the result derived from it can be analyzed manifestly.

1. The *k*-criterion method

For the large-scale hydropower system, some reservoirs are always needed to supply water to increase the generation or store water to decrease the generation to fulfill the power balance during the real-time operation. How to allocate the water needed efficiently among the reservoirs is the optimal scheduling of hydropower system. The k-criterion method adopted in this paper is one of the methods which are always applied to solve this problem.

Each hydro plant needs to make certain its position in the system and its upriver or downriver plants with hydraulic correlation should be distinguished synchronously. The serial number of each reservoir is signed from top down according to the master-slave relationship among the reach in the basin. If the first downriver reservoir with hydraulic correlation is decided, all of the upstream or downstream stations with hydraulic correlation of the current reservoir can be made certain dynamically. The hydropower system will be a dynamic open one and new hydro plant can be added to the system dynamically when needed.

1.1 The principle of the k-criterion method

For the hydropower system, if the power generation by the natural inflows can't fulfill the power balance of the electric power system, it will have to supply water to generate the difference. On the other hand, if the power generation by the natural inflows exceeds the demand, the hydropower system will store water correspondingly [6].

The hydraulic head will lower if the reservoir supply water and this will cause its natural generation losing of the latter inflows and the storing energy losing of its upstream reservoirs. Assuming the difference power is generated by reservoir i at time t, the additive power can be described as: $d\theta_{it} = 0.00272\eta_i F_{it} dH_{it}(H_{it} + \cdots + H_{nt})$. The losing power caused by the reduced hydraulic head dH_{it} can be regarded as: $\Delta\theta_{it} = 0.00272\eta_i(W_{tpi} + V_{ei}) dH_{it}$. The losing power of per additive supplying is:

$$K_{it} = \frac{\Delta\theta_{it}}{d\theta_{it}} = \frac{W_{tpi} + V_{ei}}{F_{it}(H_{it} + \cdots + H_{nt})} \qquad (1)$$

where i, t = index of reservoir and time; K_{it} = k value of the i reservoir at time t; F_{it} = the water surface area value of i reservoir at time t; W_{tpi} = summation of latter natural inflows of i reservoir; V_{ei} = summation of operating storage of the upstream reservoirs of reservoir i. $H_{it} \cdots H_{nt}$ = hydraulic head series of the downstream reservoirs of i reservoir.

On the other hand, if the reservoir increases its hydraulic head through storing water, it will improve its natural generation of the latter inflows and the storing energy of its upstream reservoirs. Analyzed as the supplying case approximately, the contribution power of per storing is:

$$K_{it} = \frac{\Delta\theta_{it}}{d\theta_{it}} = \frac{W_{tpi} + V_{ei}}{F_{it}(H_{it} + \cdots + H_{nt})} \qquad (2)$$

As is shown in Equation (1) and Equation (2), the equation of k value is uniform for either storing or supplying. Supposing the real inflows of i reservoir form the current t time to the end of storing or supplying period is W_{in}, the equation of k value can be described as:

$$K_{it} = \frac{W_{in}}{F_{i,t}(H_{it} + \cdots + H_{nt})} \qquad (3)$$

In the conditions of the power balance being fulfilled, maximization of contribution or minimization of losing is the objective that should be pursued. So it is advantageous for the reservoirs with smaller k value to supply water and the reservoirs with larger k value to store water. This is the guideline of k-criterion method to direct the operation of the hydropower system.

1.2 K-criterion improved by the routine operation chart

In real-time operation, if there are great differences among the k values and the operating decision is made by the k-criterion rigidly, some reservoirs will have to supply or store water overmuch. For one reservoir, that will lead to its empty or fullness in a short time and cause the generation fluctuating too hard consequently. The comprehensive utilization requirements of the reservoir will be negatively affected and the probability of spillage or discontent will increase too. The downstream reservoirs will store water prior to the upstream ones generally according to the k-criterion in the flood season. When the reservoirs storages are full, they will lose the regulation capacity and have to discard water if the quantity of the following inflows is large. For the upstream reservoirs, which have no chance to store water before the fullness of the downstream ones fulfilled, there is large risk of discontent for there will perhaps be no enough water to store because of the flood season being over when they begin to store water. Proper measures must be adopted to improve the guideline of k-criterion to avoid the negative occurrence. Routine operation chart of hydro plant is suggested here to improve the decision-making process. Running directed by the routine operation chart can avoid the excessive concentration of storing or supplying and ensure the achievement of the comprehensive utilization requirement of the reservoir. The visibility and controllability of the scheduling can be improved too.

For the compensative regulation of the hydropower system, applying the k-criterion method amended by the operation chart can decrease the spillage and fulfill the comprehensive utilization requirement of the reservoirs. The following control policies are taken in this paper.

(1) The generation should be controlled equally to the reduction power of the hydro plant to fulfill the comprehensive utilization requirement or ecological water demand of the river lower reaches when the operating level is within the reduction power domain.

(2) The generation should be controlled between firm power and twice of it when the operating level is within the firm power domain.

(3) The generation should be controlled between firm power and installed capacity of the hydro plant when the operating level is within the increscent power domain.

(4) The initial water level of the period or the end water level of the period can be taken as the operating control variable in the real time operation. The latter is adopted in this paper. The control strategy can be modified and adjusted according to the difference of the hydropower systems.

2. Mathematic model

The optimal decision model can be described as: With the initial reservoir water levels and inflow sequence known, how to realize the maximal generation of the hydropower system through arranging the generation properly is the pursued objective. During the process, all kinds of constraints should be met and as large as possible firm power of the system with which there is no destruction case during the simulation period should be pursued.

2.1 Objective function

In the model, the objective is expressed as:

$$E^* = \max \sum_{t=0}^{T-1} \sum_{i=1}^{N} p_{it} \Delta t \qquad (4)$$

where $N=$ number of hydro plants; $p_{it} =$ generation of i reservoir at time t; $\Delta t=$ number of hours of the period.

2.2 Constraint conditions

Water conservation for each reservoir must be observed:

$$v_{i,t+1} = v_{it} + I_{it} + \sum_{j \in \Omega_i} q_{jt} - q_{it} \qquad (5)$$

Reservoir minimum and maximum content limits are:

$$\underline{v_{it}} \leqslant v_{it} \leqslant \overline{v_{it}} \qquad (6)$$

Minimum and maximum release limits are:

$$\underline{q_{it}} \leqslant q_{it} \leqslant \overline{q_{it}} \qquad (7)$$

Chart-based minimum and maximum power limits are:

$$\underline{p_{it}} \leqslant p_{it} \leqslant \overline{p_{it}} \qquad (8)$$

Plant-based minimum and maximum power limits are:

$$p_{it}^{\min} \leqslant p_{it} \leqslant p_{it}^{\max} \qquad (9)$$

Generator output characteristic must be satisfied:

$$p_{it} \leqslant f_{it}(h, q_{it}) \qquad (10)$$

Firm output constrain of the system:

$$\sum_{i=1}^{N} p_{it} \geqslant P_t \qquad (11)$$

Where the variables are defined as follows: i, $t=$ index of reservoir and time; $v_{it} =$ storage of i at t; $I_{it} =$ local inflow of i at t; $\Omega_i =$ set of immediate upstream reservoir of i; $q_{it} =$ the release of i at t; $\underline{q_{it}}$, $\overline{q_{it}} =$ the lower and upper release bounds of i at t; $\underline{p_{it}}$, $\overline{p_{it}} =$ the lower and upper output bounds of i according to the routine operation chart based on the water level at t; $p_{it}^{\min} =$ minimum of technical power, $p_{it}^{\max} =$ installed capacity of i; $f_{it}(h, q_{it}) =$ generator output function of outflow discharge and hydraulic head; $P_t =$ the given firm power of the system at t.

3. Solution of the model

3.1 The coordinate rotation method

In real-time operation, when the natural generation can't fulfill the power demand in the low water season, the reservoir with the least k value will supply water first according to the k-criterion. With the lowering of the hydraulic head, the k value of the first reservoir keeps lessening simultaneously. As the first reservoir's k value has decreased as equally as the second reservoir's k value and the power generation can't meet the power demand still, these two reservoirs should supply water together trying to keep the k values equal to each other. Even more reservoirs should attend to supply water if needed there. Similar analysis can be given in the flood season when the natural power output exceeds the demand. For the current period, all of the supplying or storing water quantity of the attending reservoirs should be figure out simultaneously by solving the equations and all of the k values should be kept tending to be similar to each other at the end of the period. During the process, all of the constrains should be considered. As is seen from the k equation, it is quite difficult to solve the equations for the complicated coupling relations which come from the hydraulics relations among the reservoirs. The coordinates rotation method is adopted to transfer the decision-making process for multi-reservoirs simultaneously to single reservoir gradually. The specific process can be described as follows.

3.2 The Specific Process of Solution

Step 1: Calculate the natural generation of each hydro plant with the inflows.

Step 2. For the current water levels, check to make certain each generation is within the output domain decided by the operation chart. If not, modify it by the upper or lower generation bound correspondingly and calculate the summation of generation then.

Step 3: Contrast the summation of generation to the load demand of the system. If the difference meets the accuracy requirements, iterative process looking for better decision may stop and turn to the next period. If not, storing or supplying water is needed for the power balance. Sample the supplying case, the iterative process looking for the optimal scheduling can be illuminated as follows. The storing case can be analyzed approximately. The release is taken as the decision-making variable here.

(1) Calculate the k value of each hydro plant and arrange them in order from small to large. Proposing the reservoir i with the least k value should supply water first according to the k-criterion.

(2) Increase the release quantity of i reservoir with a given step, which is variable during the iterative process.

(3) Calculate the generation of i reservoir with the current release and check to make certain it is within the domain decided by the operation chart. If not, modify it by the upper or lower generation bound and account the release correspondingly.

(4) Check to make certain each hydro plant output fulfills all of constrains. If not, modify the decision-making variable. Calculate the generation summation then. Contrast it to the given load demand of the system. Iterative process should stop and turn to the next period if the difference meets the precision requirement. Otherwise, the k value of each hydro plant should be recalculated.

Arrange all k values in order and recalculate the iterative from the step again.

4. Case study

The model is tested for a large-scale hydropower system consisting of 11 reservoirs to show its efficiency. Besides the Three Gorges hydro plant and the Gezhouba hydro plant, there are other 5 hydro plants locating in the Wujiang River basin including Hongjiadu,

Dongfeng, Wujiangdu, Goupitan, Pengshui from the top down and 4 other hydro plants locating in the Jinshajiang River basin including Wudongde, Baihetan, Xiluodu, Xiangjiaba from the top down. The following Figure 1 shows the schematic layout of the reservoirs system. Table 1 shows the characteristic parameters of the hydro plants. Forty and two years of historical inflows data of period of ten days are adopted for simulation in the example. For the long-term scheduling, the lag time between the plants is neglected. The flood period consists of months from June to October and the low water period consists of months from November to May of the next year.

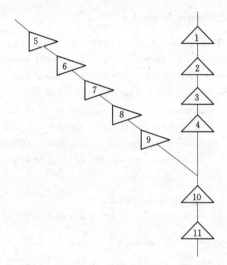

Figure 1　Schematic layout of the system

In order to testify the usefulness of the k-criterion method improved by the routine operation chart, the routine scheduling method was adopted in the paper and two series of results were gained through simulation by the two methods respectively. The results are shown in the following Table 2. Where 1 = index of the routine operation chart method; 2 = index of the k-criterion method; P_f = firm generation requirement; $\bar{E}$ = average generation per year; $\overline{E_l}$ = average generation of low water period per year; $\overline{E_f}$ = average generation of flood period per year; $\overline{W_s}$ = average spillage per year; $\overline{H}$ = average hydraulic head; R_{atio} = water use ratio.

As is shown from the Table 2, contrasting with the routine scheduling method, the average summation of generation per year of the system increases 1.6 percents and the average summation of generation of the low water period increases 6. 4 percents by the k-criterion method. What should be demonstrated is that the generation by the k-criterion method meets 22, 750 MW firm power without destruction instance during the simulation period and the routine method with a firm power of 20, 160 MW only.

Table 1   **Characteristic parameters of the reservoirs/hydro plants**

	Normal Water Level (m)	Dead Water Level (m)	Regulation Volue ($10^6 m^3$)	Capacity Coefficient	Installed Capacity (MW)	Average Inflows per Year ($10^6 m^3$)
Wudongde	950	925	16. 6	0. 014	7400	1200. 03
Baihetan	820	760	100. 32	0. 078	12000	1290. 66
Xiluodu	600	540	64. 6	0. 045	12600	1423. 16
Xiangjiaba	380	370	9. 03	0. 0062	6000	1447. 12
Hongjiadu	1140	1076	32. 23	0. 672	540	45. 17
Dongfeng	970	936	4. 9	0. 044	510	105. 06
Wujiangdu	760	720	13. 5	0. 084	1220	151. 69
Goupitan	630	585	36. 6	0. 156	3000	223. 28
Pengshui	293	278	5. 6	0. 014	1750	424. 78
The Three Gorges	175	145	222. 9	0. 051	18200	4333. 79
Gezhouba	64. 5	64. 5		—	2715	4333. 79

Table 2   **Two series of results by the two methods**

	Method	Wu dongde	Bai hetan	Xi luodu	Xiang jiaba	Hong jiadu	Dong feng	Wu jiangdu	Gou pitan	Peng shui	Three Gorges	Ge zhouba	Total
P_f(MW)	Pro1	1270	3520	4290	2210	144	250	349	860	449	5615	1200	20160
	Pro2	1270	3520	4290	2210	144	250	349	860	449	5615	1200	22750
$\overline{E}$ (10^8 kW·h)	Pro1	290	554	605	308	15. 6	29	43. 6	96	70	820. 3	150	2980. 3
	Pro2	288	550	617	314	14. 6	31	44. 6	100	69. 7	846. 2	153	3028
$\overline{E_l}$ (10^8 kW·h)	Pro1	72	163	200	108	6. 23	12	15. 8	38	30	281. 3	59. 8	985. 2
	Pro2	70. 6	156	209	110	9. 24	16	20. 7	47	37. 4	309. 6	63. 2	1048
$\overline{E_f}$ (10^8 kW·h)	Pro1	218	391	405	201	9. 39	17	27. 9	59	40	539	89. 8	1995
	Pro2	217	393	408	204	5. 34	15	23. 8	53	32. 2	537. 6	90. 1	1980
$\overline{W}$ ($10^8 m^3$)	Pro1	153	159	157	185	0. 86	11	3. 97	4. 4	22. 2	511. 9	1044	2253
	Pro2	151	131	123	162	0. 12	4. 9	1. 76	3. 2	22. 4	415	939	1954
$\overline{H}$ (m)	Pro1	116	204	199	101	151	129	126	188	74. 1	92. 73	19. 9	—
	Pro2	115	198	199	102	139	131	127	193	73. 9	93. 44	19. 8	—
R_{atio} (%)	Pro1	87. 2	87. 7	89	87. 2	98. 1	90	97. 4	98	94. 8	88. 19	75. 9	—
	Pro2	87. 2	89. 7	91. 2	88. 7	99. 7	95	98. 8	99	94. 7	90. 29	78	—

From the results of the water level sequences of the simulation, conclusions can be drawn as follows: for the upstream reservoirs such as Baihetan and Hongjiadu, they keep low water level during the sim-ulation operation process in an average sense and the fluctuation is hard relatively. The water levels keep almost invariable or increase some for other reservoirs whereas. In the low water period, the upstream reservoirs are always prior to the downstream ones to supply water to fulfill the power demand and the downstream reservoirs always store water with priority to keep high hydraulic head in the flood period. All of the water utilization ratios increase in a different degree.

The hydrologic year that from Jun. 1968 to May. 1969 adopted is a normal year, where Pro1 = generation by the routine method and Pro2 = generation by the k-criterion method.

The programming was implemented in C++ on a computer. The whole time spent was about 30 seconds.

5. Conclusions

K-criterion method improved by the routine operation chart is introduced to solve the optimal scheduling of large-scale hydropower system in the paper. It can reflect the internal operating principle of hydropower system directly. Utilizing the routine operation chart of hydro plant, the k-criterion method is improved and the result derived from it is more reasonable. The coordinates rotation method is adopted to avoid the difficulties of calculation of equations. A large-scale hydropower system with 11 hydro plants is sampled to test the usefulness of the model and the simulation generation is analyzed in the end. From the results of the operation, conclusion can be drawn that the k-criterion method amended by the routine operation chart can improve the generation of the hydropower system effectively and the model adopted in the paper has excellent direction value in practice.

Acknowledgement

This paper was founded by ministry of water Research special public service sectors (200901067); National water pollution Control and Treatment of major science and technology (2008Z × 07105008—03); Surface Scientific Research Funds of Naning Hydraulic Research Institate Foundation (Y109001); Youth Scientific Research Founds of Nanjing Hydraulic Research Institute (Y10808)

References

[1] Zong Hang, Zhou Jianzhong, Zhang Yongchuan. Research and Application for Cascaded Hydroelectric Optimized Scheduling Based on Modified Adaptive POA. Computer Engineering. 2003, 29 (17): 105-109.

[2] Luo Qiang, Song Chaohong, Lei Shenglong. Saturated-unsaturated unsteady seepage analysis of porous medium. Engineering Journal of Wuhan University. 2001, 34 (3), 22-26.

[3] Wu Zhengyi, Tang Jie, Zou Jianguo. Cascaded Hydroelectric Optimized Scheduling Studying of WuJiang basin. Automation of Hydropower Plants. 2005, 1: 109-113.

[4] Wu Y G. HoCY, Wang D Y. A. Diploid Genetic Approach to Short-term Scheduling of Hydro-thetmal System. IEEE Trans on Power Systems, 2000, 15 (4): 1268-1274.

[5] Yeh W W G. Reservoir Management and Operation Models: A State of the Art Review. Water Resources Research, 1985, 21 (12): 1797-1818.

[6] Zhang Yongchuan. the Economic Operation Principle of the Hydroplant. BeiJing: the Water Resources and Hydropower Publishing Company of China, 1998.

[7] Cervellera C, Chen VCP and Wen A. Optimization of a large-scale water reservoir network by stochastic dynamic programming with efficient state space discretization, Eur. J. Oper. Res., 2006, 171 (3): 1139-1151.

Experimental Study and Numerical Verification of 3D Thermal Stratified Wake Flow

Xing Linghang   Huang Guobing

Hydraulic Research Laboratory, Changjiang River Scientific Research Institute, Wuhan, 430010

Abstract: In this paper, the experiment of thermal shear stratified cylinder wake flow was carried out in a straight flume. The inlet section is divided into two layers including the upper warm water and lower cold water. The measurement section is equipped with Micro ADV and digital thermometer and the automatic tail gate is installed at the end of outlet section to control the water level. The distribution of 3D flow structure and temperature around the cylinder are analyzed based on experimental data. In addition, a modified explicit algebraic Reynolds stress and flux model was used to simulate the complex turbulent flow accounting for buoyancy on the three-dimensional unstructured grids. The computational region was distributed by arbitrary hexahedron grids to fit the complex boundary, and the local grid refinement was also adopted on the interface of warm and cold water, as well as those around the cylinder. The calculations show that the model yields better results.

Key words: explicit algebraic Reynolds stress model; thermal stratified wake flow; 3D; unstructured grids

1. Introduction

The phenomena of stratified wake flow are so general in nature. For example, when the environmental pollutants in water get across the obstacles, such as ait, piers, shipping, etc., the pollutant structure in stratified flow will be disturbed by the resistant effect of these impediments and varying degrees of tail mixing will be induced, which changes the distribution of pollutant in water. Another case is oil platforms at sea, the density stratification of temperature and salt water can have a relative large effect on the drag and lift to the platforms as well as the vortex shedding, this reduces the operational performances of equipments, even seriously threatens the operation safety of the system, so potentially decreases development of offshore oil and gas resources. One more example is heat conduction problem, such as heat exchangers, cooling tower, etc., if well controlling the full of characteristics of thermal stratified flow or wake flow, the rate of heat exchange can be increased and the efficiency of heat exchange equipment can be improved. In addition, by using the ocean thermocline and its acoustic properties along with its deep-sea channel, submarine can be reduced the opportunity to be listened by surface naval ships and increase greatest distance to listen to the underwater ships. Thus, further studying characteristics of thermal stratified wake flow has far-reaching significance on environmental protection and industrial or agricultural production. However, three-dimensional thermal stratified wake flow would induce strong pressure gradients and buoyancy effects, which results in asymmetry distribution of water density, pressure and temperature. In this paper, both experiments and numerical methods are used to study or discover its phenomena.

2. Experimental equipments

The experiment was carried out in a straight rectangular flume composed of transparent plexiglass. The whole system is equipped with two independent inlet systems and a backwater system. The two formers include a hot water system partly from the guiding pool heated up by a heater and a cold water system mostly from the flux of the guiding pool. Schematic layout of test equipment can be seen in Figure 1. The section dimension of the test flume is 34cm × 30cm, the bottom slope is 1‰. The flume consists of inlet section, measurement section and outlet section (see Figure 2). The inlet section is 300cm long, divided into the upper and lower layers by a 0. 2cm -

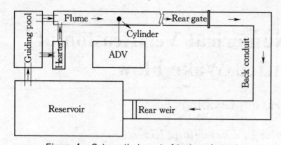

Figure 1　Schematic layout of test equipment

thin glass plate equipped at 20 cm above the bottom

of the flume. The measurement section is 200cm long, equipped with measuring instruments including Micro ADV and digital thermometer. The automatic tail gate is installed at the end of outlet section to control the water level. A 5cm diameter cylinder obstacle is placed in the center of the cross-section, which is 40cm downstream of the end of two separated flows. Detailed layout profile can be seen from Figure 2 and Figure 3. In present study, the upper flow is warm water and the lower flow is cold water.

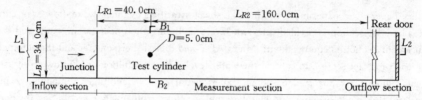

Figure 2　Plane schematic diagram of test flume

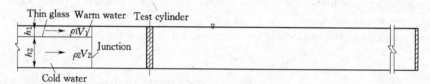

Figure 3　Vertical schematic diagram of test flume

3. Layout of observed velocity and temperature

In order to capture the main features of the thermal stratified cylinder wake flow, the velocities are measured along the vertical center section $L_1 L_2$ and the temperature are observed along the horizontal section $B_1 B_2$ (see Figure 2) plus the section $L_1 L_2$, each section can get across the centre of the cylinder barrier. The detailed observed vertical lines and their corresponding points for velocity and temperature are arranged as follows.

Velocity: along the section $L_1 L_2$ of the flume, 19 observed vertical lines for velocity are laid out at 20cm, 28cm, 32cm, 34cm, 35cm, 46cm, 47cm, 48cm, 49cm, 50.5cm, 52cm, 54cm, 57cm, 60cm, 65cm, 75cm, 87cm, 110cm and 180cm downstream the primal junction section of warm water and hot water. On each vertical line, 14 monitoring points are set from 0.5cm to 14.5cm subsurface the water at the pace of 1cm.

Temperature: along the section $L_1 L_2$ of the flume, the observed vertical lines for temperature are arranged at 20cm, 36.5cm, 46cm, 49cm, 52cm,

54cm, 57cm, 65cm, 75cm, 80cm, 87cm, 95cm, 110cm, 120cm, 130cm, 140cm and 180cm downstream the primal junction section of warm water and hot water. Along the horizontal section $B_1 B_2$, the monitoring lines are set in the order of 1cm, 6cm, 9cm, 11cm, 12.5cm, 13.5cm, 14.5cm, 19.5cm, 20.5cm, 21.5cm, 23cm, 25cm, 28cm, and 33cm from the up bank side. The observed points on every line including section $L_1 L_2$ and section $B_1 B_2$ are laid out at the same way, this means the observed points start from 0.5cm under the surface water to the bottom of the tank at every 1cm.

4. Test parameters

In the present study, the fluxes of the upper warm water (31.53℃) and the down cold water (20.65℃) are respectively controlled as 0.3L/s and 3.883L/s. Thickness of the warm water layer is 3.00cm, depth of the cold water layer is 20.00cm, average velocity is 5.73cm/s based on geometric mean depth, velocity ratio v_1/v_2 is 0.48, relative density is 0.0029. Some key experimental hydraulic parameters are listed in Table 1.

Table 1   **Test parameters**

Upper layer parameters				Lower layer parameters			
v_1	h_1	ρ_1	T_1	v_2	h_2	ρ_2	T_2
2.94cm/s	3.00cm	0.9952g/cm^3	31.53℃	6.151cm/s	20.00cm	0.9981g/cm^3	20.65℃

5. Numerical model

In recent years, the explicit algebraic stress model (EASM) is so flourishing and become an important branch of turbulence model. EASM model has similar numerical efficiency to that of the two-equation turbulence model and can partly overcome the shortcomings of the linear eddy viscosity model, reflect the action of Reynolds stress anisotropy. In addition, it can effectively avoid numerical singularity and improve the model stability and convergence, which is much superior to traditional algebraic Reynolds stress model.

5.1 Explicit algebraic Reynolds stress model

In present study, the Reynolds stress will be broken into two parts under the assumption of the principle of linear superposition. The first part is caused by strain stress, the second is controlled by the buoyancy impaction and its mathematical formula is available on the relationship of implicit algebraic stress model by Rodi. The final form of Reynolds stress is

$$\overline{u_i u_j} = \overline{u_i u_j}\big|_s + \overline{u_i u_j}\big|_b \qquad (1)$$

where the non-buoyancy part $\overline{u_i u_j}\big|_s$ contains isotropic part and the remaining part of the anisotropy, its expression is referenced to Wallin & Johansson model[1]

$$\overline{u_i u_j}\big|_s = k[2/3\delta_{ij} - 2C_\mu^{eff}S_{ij} + a_{ij}^{(an)}] \qquad (2)$$

C_μ^{eff} is effective coefficient, $a_{ij}^{(an)}$ is anisotropy part and can be written as

$$a^{(an)} = \beta^2(s^2 - \mathrm{II}_s/3\,\mathrm{I}) + \beta^3(\omega^2 - \mathrm{II}_\omega/3\,\mathrm{I})$$
$$+ \beta^4(s\omega - \omega s) + \beta^5(s^2\omega - \omega s^2) + \beta^6(\omega^2 s$$
$$+ s\omega^2 - 2/3\,\mathrm{IV}\,\mathrm{I} - \mathrm{II}_\omega s) + \beta^7(\omega^2 s^2$$
$$- s^2\omega^2 - 2/3\,\mathrm{VI}) + \beta^8(s\omega s^2 - s^2\omega s)$$
$$+ \beta^9(\omega s\omega^2 - \omega^2 s\omega) + \beta^{10}(\omega s^2\omega^2 - \omega^2 s^2\omega)$$

The coefficients in detail can be seen from Wallin & Johansson[1].

The buoyancy part is

$$\overline{u_i u_j}\big|_b = (1 - c_3)\frac{k}{\varepsilon}G\left(G_{ij}/G - \frac{2}{3}\delta_{ij}\right)/$$
$$[c_1 + (p+G)/\rho\varepsilon - 1] \qquad (3)$$

where $c_3 = 0.55$, $c_1 = 0.22$, $G_{ij} = -\rho\beta g_i\,\overline{u_j\theta}$, $G = -\rho\beta g_i\overline{u_i\theta}$.

5.2 Modified explicit algebraic active scalar flux model

A modified active explicit algebraic scalar flux model by Wikström, Wallin and Johansson[2] (A^{WWJ} model) is proposed while abandoning the usual passive explicit scalar transport model. The partly explicit algebraic active scalar flux can be derived as follows

$$\overline{u_i\theta} = -(1 - c_{\theta_4})A_{ij}^{-1}\frac{k}{\varepsilon}\overline{u_j u_k}\frac{\partial\Theta}{\partial x_k}$$
$$- \frac{c_{\theta_6} - 1}{c_\theta}A_{ij}^{-1}\frac{k^2}{\varepsilon^2}\beta g_j\overline{u_k\theta}\frac{\partial\Theta}{\partial x_k} \qquad (4)$$

Where, $c_\theta = 0.13$ (Launder 1975, 1976), some other parameters can be found in A^{WWJ} model by Wikström et al[2].

The present model inherits many advantages of A^{WWJ} model and is regarded as an extension for the passive explicit algebraic scalar flux model.

5.3 Numerical method

A method incorporating 3D unstructured grids along with a modified QUICK scheme and a pressure-correction equation is developed to solve incompressible flows. This approach can be fitted to arbitrary geometric shape easily. Generalized Minimum Residual (GMRES) method with the Incomplete LU (ILUT) precondition is used to accelerate the convergence speed for linear equations. The computational region can be divided in arbitrary hexahedral mesh to fit the complexity of the border. In order to fully capture the stratified wake flow information in detail, very fine grids are laid out nearby the cylinder. The computational grids are illustrated in Figure 4. Total number of nodes is 50, 568 and total number of elements is 45, 245. The under-relaxation factors for each discrete equation, namely the momentum equation, the turbulent kinetic energy equation, the turbulent dissipation rate equation, the scalar transport equation and the pressure correction equation, are 0.4, 0.4, 0.4, 1.0 and 0.1.

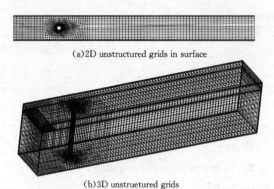

(a)2D unstructured grids in surface

(b)3D unstruetured grids

Figure 4　Computational grids

6. Experimental and numerical results

The experimental velocities are illustrated along L_1L_2 direction in Figure 5 and the distributions of temperature along L_1L_2 and B_1B_2 directions are depicted in Figure 5 and Figure 6. It can be seen from Figure 7 that a stable two-layer flow can keep well upstream the obstructive cylinder unless a little rising flow close ahead of the barrier due to resistance effect. While downstream the impediment, the stable state is disturbed and a recirculation zone is appeared with significant loss in velocity. On account of the po-

tential energy difference, the upper mixing water at the close rear of the cylinder barrier flows to the lower cold layer. In addition, the plotted separatrix of the recirculation zone shows that the length of recirculation is about 2cm at water surface and gradually enlarges to approximate 8cm at almost 7.7cm below the surface the water. To this end, the length of recirculation can maintain till to the waist of the cylinder as far as present monitoring scope. After the recirculation zone, the velocities recover slowly and the weak stratified flow can keep till to the export section. From Figure 5, it can be seen that the two-layer thermal stratification can mostly keep well upstream the cylinder. While the flow closing up the cylinder, the lower cold water rises quickly ahead of the obstruction accompanying with energy exchange between the upper hot water and lower cold water. This results in great reduction of water temperatures around the upper cylinder (more clearly seeing from Figure 6). After the flow bypassing the obstacles, the upper low-temperature areas are supplied by surrounding hot water, so that the water temperature gradually rises along the vertical cross-section, the corresponding flow state is also transformed to a weak thermal stratified flow, until the export section.

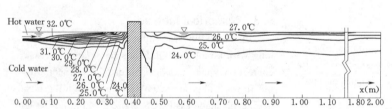

Figure 5　Illustration of contours of temperature along L_1L_2 direction

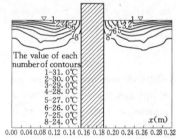

Figure 6　Illustration of contour of temperature along B_1B_2 direction

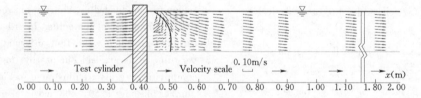

Figure 7　Illustration of velocity along L_1L_2 direction

The comparisons of experimental and computational data are depicted in Figure 8, Figure 9 and Figure 10. It can be seen that EASM model can fit measured values well wherever upstream or downstream

the cylinder. Thus, it gives a good reflection of anisotropy characteristics of the shear thermal stratified wake flow. It should be noted that the present EASM model can not give good forecasts in the recirculation

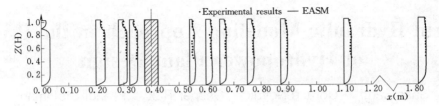

Figure 8　Comparisons of experimental and computational velocity along L_1L_2 direction

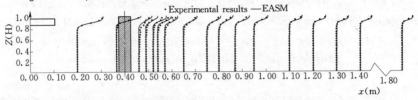

Figure 9　Comparisons of experimental and computational temperature along L_1L_2 direction

zone. Thus, more work would be done to better handle the simulation of the above—mentioned shortcomings and deficiencies.

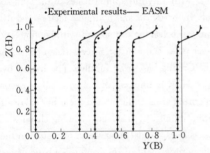

Figure 10　Comparisons of experimental and computational temperature along B_1B_2 direction

7. Conclusion

The experimental study of a shear thermal stratified wake flow is successfully designed and tested in laboratory, features of distributed velocity and temperature in the stratified flow are captured by experiments.

A modified EASM model is proposed including explicit algebraic Reynolds stress model accounting buoyancy and explicit algebraic active scalar flux model. The present EASM model was validated by experimental data and can mostly reflect the anisotropy characteristics of the shear thermal stratified wake flow. The current EASM model was successfully applied on 3D unstructured grids and can fit arbitrary geometric shape.

Acknowledgment

This research was sponsored by 2009 National Public Research Institutes for Basic R & D Operating Expenses Special Project （ No. YWF0901, No. YWF0905 and CKSF2010011）; National Basic Research Program of China （No. 007CB714106, Subproject of No. 2007CB714100）.

References

[1]　Wallin S. , Johansson A. V.. Modeling of streamline curvature effects on turbulence in explicit algebraic Reynolds stress turbulence models. In Proceedings of Turbulence and Shear Flow Phenomena II, 2001: 223-228.

[2]　Wikström P. M. , Wallin S. , Johansson A. V.. Derivation and investigation of a new explicit algebraic model for the passive scalar flux. Physics of Fluids, 2000, 12 (3): 688-702.

Hybrid Hydraulic Modeling Approach in the Process of Hydropower Plant Design

Sašo Šantl[1]   Gorazd Novak[2]   Gašper Rak[1]   Franci Steinman[1]

[1]*Chair of Fluid Mechanics, Faculty of Civil and Geodetic Engineering,*
University of Ljubljana, Slovenia, Hajdrihova 28, 1000 Ljubljana, Slovenia
[2]*Institute for hydraulic research-hydraulic laboratory,*
Hajdrihova 28, 1000 Ljubljana, Slovenia

Abstract: The paper presents a hybrid approach to the hydraulic modeling of the run-of-river hydropower plant on the Sava River in the Republic of Slovenia, where a physical model and a 2D mathematical model are combined. The objectives of hydraulic modeling are, at first, to define the current retention capacity, run-off regime and flood hazard situation in the discussed area. In the second phase, the objectives are to optimally design the hydropower plant, including full infrastructure, with the emphasis on keeping the same run-off regime downstream and to mitigate the present flood and erosion hazards. The main reasons for choosing hydraulic modeling with a hybrid approach lies in the complexity of this high flood-prone area which is defined by complex morphological, hydrological and hydraulic factors and it is burdened with different anthropogenic land use and activities. In addition, a nuclear power station with cooling water demands is also situated in the influential area of the planned hydropower plant and a strict agreement on conservation of the run-off flow regime with the downstream neighboring Republic of Croatia has to be fulfilled. The hybrid hydraulic model approach combines benefits from both hydraulic modeling approaches. A physical model can still more precisely assure data where the influence of turbulence and other 3D flow phenomena are significant. On the other side, a mathematical model upgrades the physical model in the phase where significant number of analysis, planning changes and data presentations for decision support occur. This two-model approach also allows mutual comparison of results in the phase of calibration and verification of both models which ensures significant improvement of the confidence band of measured or calculated quantities. Because of the uncertainties related to discharge and water level data and the above mentioned complexity of the area in the process of model calibration and verification a brief sensitivity analysis of all significant parameters was conducted.

Key words: hybrid hydraulic model; flood hazard; hydropower plant; calibration; sensitivity analysis

1. Introduction

In Slovenia two run-of-river hydropower plants on the Sava River are in the process of spatial planning and design. The analyzed area is a flood-prone area with significant retention flood water volume. It is also morphologically, hydrologically and hydraulically very complex with different land use and activities, where also a nuclear power plant with cooling water demands is situated.

The design of the HPPs also requires a designing process of dikes with flood overflow structures which will preserve the existing flood retention areas also in the future. According to the EU Directive on Floods, those areas are planned to be designated as significant flood areas and have to be preserved as such. Moreover, a strict agreement on conservation of the existing run-off flow regime with the downstream neighboring Republic of Croatia has to be fulfilled.

Because of the above mentioned elements and because of a too wide confidence interval regarding the existing hydrological and hydraulic data, it was decided to analyze and search for an optimal design solution with the hybrid modeling approach[7].

The entire modeling area is divided into three separate overlapping parts (Figure 1). The first upstream part of the Sava River (the area of constructed HPP Krško) was analyzed with a hybrid model using an undistorted physical model (PM) and a 1D mathematical model. The main purpose was to determine the discharges of previous significant flood events and to support spatial planning of transport infrastructure in the influential area.

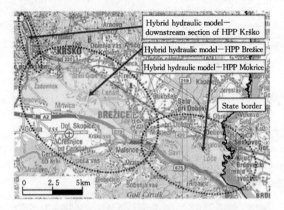

Figure 1　Overview map of design area of HPPs with flood warning map and presentation of extent of hydraulic models

The mid and downstream sections were divided with regard to the influential area of each planned HPP. In the mid section a HPP Brežice, and in the downstream section, which borders Croatia, a HPP Mokrice are planned.

These areas were modeled with hybrid models where PMs were built with distorted dimensions with the aim to assure adequate results. A mathematical (numerical) model (MM) was developed with MIKE Flood software[3] where coupling of a 1D model for river streams and a 2D model for flooded areas was done.

Besides the benefits gained by development of two complementary models for the same area (PM can more precisely simulate turbulence related phenomena and MM can simulate wider areas and supports much faster optimization process), hybrid modeling also assures mutual verification of both models.

After the verification of the hybrid model, optimization of HPP infrastructure was done mainly with MM with final verification of the optimized HPP design on the PM (Figure 2).

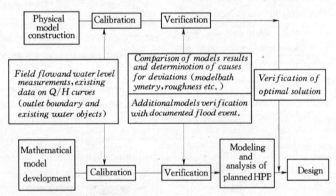

Figure 2　Parallelism of hybrid model building, calibration, HPP design process and verification

Further on, the process of hybrid model building for the mid section of the Sava River (HPP Brežice) is presented with the emphasis on the process of model development, calibration, sensitivity analysis and verification. At the end, some results of the optimization process are presented.

2. Hybrid model building

As it was mentioned, a confidence interval for higher discharges can be narrowed with the use of a hybrid model. Since PM has a limited reach of high discharges simulation, mainly due to the spatial capacities of the laboratory, MM provides also the results for the highest discharges which can occur in the area. This is very important for the analyzed area, because in this area a nuclear power plant is situated, which must be protected against Probable Maximum Flow (Figure 3).

The process starts with the building of PM and development of MM, for which detailed geodetic data on terrain, objects, river cross sections and land cover were obtained. The area was fully scanned with LIDAR (LIght Detection And Ranging) method supported with geodetic survey, ortho-photo, etc. To support model calibration, besides the existing flood event data in the past decades, additional intensive water level measurements at higher discharges of the Sava River were performed.

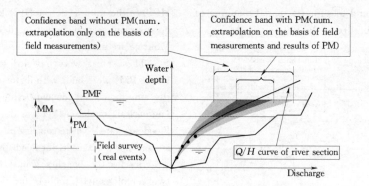

Figure 3　Schematic presentation of certain cross section with hybrid model approach for minimizing a confidence interval and extension of model applicability up to probable maximum flood（PMF）

2.1　Building and operation of distorted physical model

The modeling of a prototype area that includes 11 km long river reach and 23 km^2 of flood plains dictated the use of vertical distortion. This approach allowed to avoid laminar flow on the wide flood plains and assured a well measurable mean water level difference. The choice of distortion factor and of length scales was limited mainly by the capacities of the laboratory, the modeling accuracy and the following hydraulic limitations[5] :

$$n \leqslant 0,1 \left(\frac{b}{h} \right)_n \qquad (1)$$

$$n_{\max} = \left[\frac{0,2}{\left(\frac{k}{h} \right)_n} \right]^{\frac{1}{3}} \qquad (2)$$

$$L_{r\max} = C \cdot n^{\frac{10}{3}}, \ C = 5,65 \cdot 10^{-3} \cdot Re_n^{\frac{2}{3}} \left(\frac{k}{h} \right)_n^{\frac{7}{9}} \qquad (3)$$

Where b is water surface width, m; C is coefficient, h is water depth, m; k is equivalent sand roughness according to Nikuradse, mm; L_r length ratio; n distortion factor; Q is discharge, m^3/s, and Re is Reynolds number. Indices are represented with n as prototype quantity and r as relative quantity = prototype / model quantity.

The open space available at the laboratory allowed the realization of a length scale number of $L_r = 125$, which was well within the limitation of length scales, while the distortion factor was chosen as $n = 2$, as it turned out that $n_{\max} = 2$, 2 for the $Q_n = 5000 \mathrm{m}^3$/s. This led to the discharge scale factor $Q_r = L_r^{5/2} / n^{3/2} = 61763, 4$.

Construction of the model was done with the aid of hardboard profiles. Concreting of the 5 cm thick upper layer of 1500 m^2 model was completed in several stages demanding considerable effort. The model was equipped with 40 piezometric cylinders for the measurements of water levels and finally coated with cement milk.

The model calibration was achieved by adjusting the river bed and flood plain surface roughness with mostly uniformly distributed patches of sand with various grain diameters, representing the land use (forests, orchards etc.)

The calibrated model was most of all used to investigate water levels at flood discharges, but also to determine inflow and outflow from flood plains. Red dye was injected at different locations to determine pathlines of water flowing over inundations (Figure 4) . It turned out that these model flows could be

Figure 4　Photography of physical model with main flood streams determination

separated by construction of auxiliary walls that ran along analyzed pathlines. The quantities of these isolated flows were then measured using V-notch weir.

2. 2　Mathematical model development

A hydrodynamic mathematical (numerical) model of the area was developed with MIKE Flood software, which combines 2D modeling (MIKE 21) with 1D modeling (MIKE 11). The main decision for the application of this type of modeling arises from the needs for efficient modeling of hydraulic structures (weirs, gates, spillways). MIKE 11 is also equipped with a control structures module which models the dynamic operation of movable gates dependent on flow[3]. This software capability was a necessary requirement for modeling the dynamic operation of the gates of the run-off river HPP Brežice where final design, which will assure efficient electricity production on one hand and will preserve the same run-off regime with improvement of flood risk on the other hand, has to be optimized.

The developed MM, which covers about 32 km², is presented in Figure 5. The Sava River channel, the Krka river channel as the main tributary and Potočnica creek channel were modeled with 1D model. The rest of the area was modeled with 2D model with a cell size 20m×20m. Such a cell size was selected mainly to reduce computational time of optimization process of design state, where calculations with dynamic operation of movable weir gates are time consuming. As it will be described in the next paragraph, additional full 2D model was also developed with a cell size of 10m×10m in the process of verification. Deviations between DMR and model bathyme-

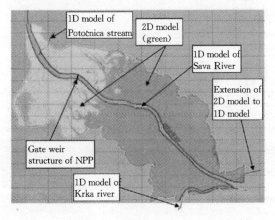

Figure 5　Design of mathematical model—coupling of 1D stream model with 2D flood prone areas

try, which could have an influence on the hydraulic flow regime, were briefly analyzed and corrected (elevation of roads, walls, etc.).

The MM has three inlet boundary conditions and two outlet boundary conditions (one is extension of the 2D model at left bank of the Sava River into 1D model). Besides boundary conditions, inner condition of the Nuclear Power Plant Krško (NPP Krško) weir structure was modeled by using a standard weir formula[3] (Villemonte).

3. Calibration and verification

Both models (PM, MM) were calibrated on the basis of field measurements in 2007 (this is also the year of geometry data acquisition) of water levels in periods of higher discharges ($Q = 1230\text{m}^3/\text{s}$, $1595\text{m}^3/\text{s}$ and $2466\text{m}^3/\text{s}$) which were measured in 13 locations along the river reach of the planned HPP Brežice (Figure 6, measurement locations from 1 to 13). The calibration process was done separately for PM and MM.

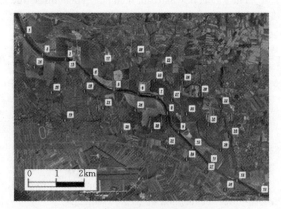

Figure 6　Ortho photo of the planned area with measurement locations

In the first phase of verification process of hybrid model, measurements (PM) and calculations (MM) for high discharges were performed ($Q = 3000\text{m}^3/\text{s}$, $4000\text{m}^3/\text{s}$ and $5000\text{m}^3/\text{s}$). The results were then mutually compared and a brief sensitivity analysis of different influential parameters was conducted. A comparison of the measurements on PM and the results of MM was performed at 38 measurements locations where locations from No. 14 to No. 38 represent locations on flood plains. This phase was then followed by an additional calibration phase of each model.

In the verification process, a comparison with the measured water levels of flood event in 1990 was crucial ($Q = 4000\text{m}^3/\text{s}$) since this data was not used for models calibration. Additionally, a full 2D MM with a cell size $10\text{m} \times 10\text{m}$ for the upper half of the analyzed area (including the NPP Krško weir structure) was also developed. In the process of verification and sensitivity analysis, influence of this weir structure (with deepening of the river bed for more than 1 m at the downstream section in the last 20 years) on the flooding regime was determined.

Further on, a process of the MM calibration, sensitivity analysis and verification is described more briefly.

3.1 Mathematical model calibration

The process of calibration of the MM was divided into three areas:

(1) calibration of model downstream (outlet) boundary condition,

(2) calibration of weir structure parameters of the NPP Krško (as inner boundary condition) and

(3) calibration of roughness coefficients of the Sava River mainstream channel and flood areas.

The model downstream boundary stage discharge curve was constructed on the basis of the field measurement and a downstream water gauge station. The results from the downstream MM of HPP Mokrice were also used in the verification process.

As it was mentioned, the Sava River bed changed in the last 20 years so that the experimentally determined stage discharge curve (determined on the basis of a physical model built in 1976) was not relevant any more for the calibration. Therefore, the MM cross sections of the Sava River from 1986 were included in the first stage of calibration. This provided more proper calibration of weir parameters.

In the calibration process of roughness coefficients, initial values were selected at first (Table 1).

Table 1　Initial values for Manning roughness coefficients for 1D and 2D model

Land cover	Roughness coefficient (Manning)
Main river channel (1D)	0.033
River banks (1D)	$1.5 n_G$ of main river channel
Grassland (2D)	0.0286
Vegetation (2D)	0.058
Objects (2D)	0.1

The sensitivity analysis where roughness coefficients of flood plains (2D model) were changed ($\pm 25\%$) showed that water levels varied in range ± 0.08 m at higher discharges. Since those variations are insignificant in comparison to other initial parameters, roughness coefficients for flood plains remained as initial values.

Roughness coefficients for 1D model were calibrated for sections divided in compliance with the measurement locations of the Sava River channel (Figure 6) and separately for each discharge of field survey.

In the calibration process of roughness coefficients it was found out that they varied with discharge. The sensitivity analysis showed that changes of water level caused by varying of roughness coefficients were about the same amplitude as those caused by varying of certain discharge (both parameters were varied $\pm 5\%$). Figure 7 shows differences in the water level with variation of $Q = (2466\text{m}^3/\text{s} \pm 5\%)$, at which also significant flooding starts.

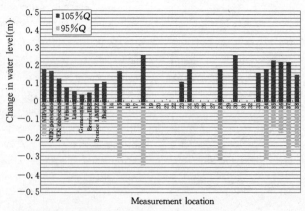

Figure 7　Sensitivity analysis of water level caused by changes in discharge $Q \pm 5\%$

On the other hand, precisely calibrated roughness coefficients for different discharges vary up to 25%. According to the above mentioned and because these phenomena can appear with the increase of the discharge (where water velocity is a significant parameter), it was decided to model roughness coefficients depending on the cross-section mean velocity[1,4,6]. This modeling approach is fully supported by MIKE 11 software. This change is also noticed in the phase when more intensive sediment transport process starts to develop (bank-full discharge of the Sava River in this river reach is around 2000m³/s). It is also expected that after sediment transport is fully developed and flood plains start to take a significant part in water conveyance, roughness coefficient change will converge.

Figure 8 shows an example of precisely calibrated roughness coefficients of the river section at measurement location No. 7 (Figure 6) and a roughness coefficient depending on the cross-section mean water velocity with upper and lower limit definition.

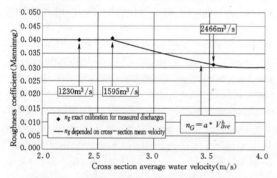

Figure 8   Calibration of roughness coefficient in the river section around measurement location No. 7 for different discharges depending on water cross section average velocity

The lower limit of $n_g = 0.03$ and final definition of equation parameters for roughness coefficient depending on the cross-section mean velocity were determined in the process of model verification.

With that approach, the differences between field survey of water levels and the calculated values are higher [± 0.16m; at highest measured flood discharge ($Q = 2466$m³/s) this difference is up to ± 0.09m] than in the case of exact calibration of roughness coefficients for each discharge separately (± 0.05m). However, since these differences in water levels are smaller than the differences, which were determined in the process of sensitivity analysis,

which included varying of measured discharges (differences in water level were ± 0.18 m in the area of 1D model—points from 1 to 13), the modeling of roughness coefficient depending on the cross-section water mean velocity can be applied.

3. 2   Verification

After the calibration phase, water levels for higher discharges ($Q = 3000$m³/s, 4000m³/s and 5000m³/s) were measured (PM) and calculated (MM). After mutual comparison of the results and comparison with the past significant flood event, an analysis of the main reasons for differences was conducted. The main reasons for the differences which were determined with an additional sensitivity analysis are:

(1) accuracy of topology (bathymetry) of the river channel and flood plains, especially deepening of the Sava River bed,

(2) the fact that PM does not capture the entire upstream right flood plain,

(3) sensitivity of weir structure of the NPP Krško and

(4) influence of coupling 1D model with 2D model.

To provide an acceptable level of model agreement, results modifications of each model were performed. PM needed better accordance of topography of flood plains and MM needed modification of connection between 1D and 2D model (it is modeled as weir where hydraulic parameters can be modified). For more precise modification of models parameters, additional full 2D mathematical model of the upstream half of the analyzed area was developed.

Figure 9 presents water depth comparison between field survey, PM and MM for in situ measured

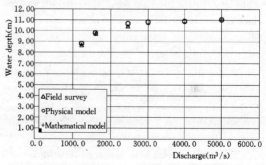

Figure 9   Measured and calculated water depths for field survey, PM and MM for different discharges at measurement location No. 7

discharges 1230m³/s, 1595m³/s and 2466m³/s and between PM and MM for discharges 3000m³/s, 4000m³/s and 5000m³/s.

The differences in water levels at this measurement location are in the range of ±0.10m for each data, what is acceptable tolerance of accuracy for flood events.

4. Objectives and results

Besides determination of the current flood hazard and risk, the main objective of this project and hybrid model application was to provide a decision support tool for optimal design of water structures of HPP Brežice, where besides efficient electricity production:

(1) current flood and erosion hazard and risk should be mitigated or at least sustained and

(2) current run-off regime with retention water capacity has to be preserved.

In the process of design, the left and right bank dikes location, flood water overflow structures dimensions and location and HPP weir structure dimension were optimized.

As it was mentioned, preservation of the same run-off regime must be assured and also proved in the process of design. Thus, a comparison between the current state hydrograph and the design state hydrograph at the outlet boundary location was carried out in the optimization process.

The developed MM was also used for flood hazard and risk mapping, according to the EU Directive on Floods. Due to the vicinity of NPP in the influential area, flood hazard and risk in the case of Probable Maximum Flood occurrence was also analyzed.

The hybrid hydraulic model also gives a brief insight into different hydraulic related phenomena which can occur in such wide flood areas. For example, it would take much more time to examine and verify the situation where despite the increase in the total discharge, water velocity and discharge in the main river channel start to decline. Such example is presented in Figure 10 for the river section at measurement location No. 7 when average water velocity at some discharge starts to decline.

This phenomenon was observed by PM as well by MM. It is caused when water flow at flood plains in the downstream section starts to dam the flow in the main river channel and when the entire cross

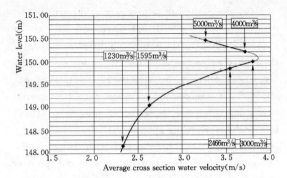

Figure 10　Presentation of water level depending on average water velocity at certain river cross section（measurement location No. 7）with the phenomena of slowing down of water velocity and discharge in the main river channel

section, including flood plains, starts to work as one river flow.

5. Conclusions

In the complex spatial cases where there is no sufficient or accurate data about discharge stage curves and where structures with significant influence on the run-off regime are planned, hybrid hydraulic models provide an efficient decision support tool. The application of two different types of hydraulic models combines advantages from both models and also provides mutual comparison of results in the phase of verification of the models.

Different influential hydraulic parameters have to be considered in the process of hydraulic model development, from topography, land cover, upstream, downstream and inner boundary conditions to proper roughness coefficient determination which can change with water depth and velocities. The accuracy of this data always has a certain confidence interval, therefore it is of a great importance to conduct a correct sensitivity analysis along the process of model calibration and verification.

Since many different land uses and activities are present in that design area, the developed mathematical model with updates and periodic verification will provide an efficient tool for further future spatial planning.

References

[1]　H. T. Nguyen, and J. D. Fenton. Identification of Roughness in Compound Channels. MODS-

IM 2005 International Congress on Modelling and Simulation. Modelling and Simulation Society of Australia and New Zealand. 2005.

[2] G. Anderson, I. D. Rutherfurd, A. W. Western. An analysis of the influence of riparian vegetation on the propagation of flood waves. Environmental Modeling and Software 2006, 21: 1290-1296.

[3] MIKE by DHI. MIKE 11 - A modeling system for Rivers and Channels - User Guide, MIKE by DHI, 2009.

[4] P. P. Nayak, K. K. Khatua, K. C. Patra. Variation of Resistance Coefficients in a Meandering Channel, National Conference Advances in Environmental Engineering, 14-15 Nov. 2009, at NIT, Rourkela.

[5] H. Kobus (editor). Hydraulic Modelling. Verlag Paul Parey, Hamburg. 1980.

[6] D. H. Yoo, T. H. Lee. Friction Factor of Smooth Turbulent Open Channel Flow. 32th Congress of IAHR, July 1-6, 2007, Venice, Italy.

[7] J. Mlačnik et. al.. Inception report - project "Hybrid hydraulic model development for downstream section of HPP Krško, section of HPP Brežice and section of HPP Mokrice." Institute for hydraulic research - hydraulic laboratory. 2010.

Swimming Behavior of Isolated Ayu, Plecoglossus Altivelis Altivelis, in Running Water

Kouki Onitsuka[1]   Juichiro Akiyama[1]   Hikaru Takeuchi[1]   Atsushi Ono[1]

[1]*Department of Civil Engineering, Kyushu Institute of Technology, Kitakyushu, 804-8550, Japan*

Abstract: Swimming behavior of isolated fish in static water has been investigated. Onitsuka et al[3,4]. investigated on the swimming behavior of isolated ayu in static water. However, swimming behavior of isolated fish in running water has not been investigated. In this study, the swimming behavior of isolated ayu in the running water was recorded with a digital video camera with changing the velocity for five cases between 0-0.70m/s. The area where the number of branch is more than 3 from starting point of ayu and distance from sidewall is more than one time of body length of ayu was defined as the universal area. Swimming along the sidewall after ayu arrived where distance from sidewall is less than one time of body length of ayu was defined as the wall area. Only a universal area is targeted in the analysis. The result showed that the swimming trajectory can be described by nodes and branches. The frequencies of the ground distance, swimming distance, ground speed, swimming speed, turning angle between each branch are modeled by the gamma distribution. It was found that the ground distance, swimming distance, ground speed and swimming speed in the streamwise direction increases with an increase of the flow velocity, because the ayu has positive rheotaxis. In contrast, the ground distance, swimming distance, ground speed and swimming speed in the spanwise direction is constant, irrespective of the flow velocity. The turning angle of swimming trajectory decreases with an increase of the flow velocity.

Key words: isolated ayu; node; branch; swimming distance; angle; running water

1. Introduction

Research of swimming behavior of fish is divided roughly into three: (1) schooling behavior of fish; (2) fish's relative behavior in school of fish; (3) behavior of fish that does not belong to school of fish. Inoue[1] showed that the swimming speed of the school of fish was 0.3m/s using a scanning sonar. Hasegawa & Soeda[2] pointed out that the each individual in the school interacts with each other so that its swimming speed fluctuates with time. Recently, the behavior of the fish that does not belong to the school of fish was researched. Onitsuka et al.[3-4] showed swimming behavior of isolate or a couple of ayu described by nodes and branches, and swimming speed and the swimming distance were quantitatively evaluated. Therefore, fish's behavior in the statistic water is being clarified. On the other hand, behavior of the fish in the running water has been hardly experimentally investigated. The examination by the parametric analysis is early. Takagi et al.[5] showed that behavior of ayu was calculated by

random walk model. However, neither the model constant used to calculate nor the assumptions validate the evidence. Therefore, it is necessary to investigate the swimming behavior of fish in running water. In this study, the swimming behavior of isolated ayu in the running water with changing the flow velocity was analyzed.

2. Experimental set up and hydraulic conditions

2.1 Experimental set up

Figure 1 shows the open-channel used for experiment. The length L, width B and height H are 2.4m, 0.8m and 0.2m, respectively. x, y and z are the coordinates of the streamwise, vertical and spanwise directions, respectively.

2.2 Hydraulic conditions and experimen-tal fish

Table 1 shows hydraulic conditions. Water depth was set to 0.04m. Flow velocity divided by averaged body length of ayu was set to five patterns within the

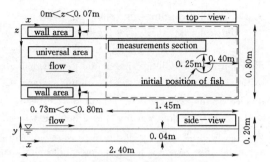

Figure 1　Sketch of experimental open-channel

range from 0 to 10. The case name shows the flow velocity divided by averaged body length of ayu. For instance, flow velocity shows ten times the averaged body length of ayu in C10.

Ayu was used for the experiments. The number of fish was 100. Averaged body length $\overline{B_L}$ is about 70mm.

Table 1　　Experimental conditions

case name	C0	C1	C3	C5	C10
$\overline{B_L}$ (mm)	70				
h (m)	0.04				
U_m (m/s)	0	0.07	0.21	0.35	0.70
$U_m/\overline{B_L}$ (1/s)	0	1	3	5	10

2.3　Experimental Method

A circular wire net of 0.25m in diameter is set up 2m downstream from upstream edge and isolate ayu is inserted. After it is confirmed that the ayu settled down, the circular wire net is taken up. Further, trajectory of ayu was recorded with a digital video camera set up the foreside of the open-channel. Recording speed of video camera is 30 frames per second and the number of pixels is 1440×1080. It experimented 100 times in each case, and 500 times in total. Three components of flow velocities, i.e., $\widetilde{u} = U + u$, $\widetilde{v} = V + v$, $\widetilde{w} = W + w$, in the open-channel were measured at 70 points (10 points in x direction, 7 points in z direction) with a 3-D electromagnetic current meter after removing the fish. In witch $\widetilde{u}$, $\widetilde{v}$ and $\widetilde{w}$ are the instantaneous velocity component in the streamwise, vertical and spanwise coordinate, respectively. The capital letter denotes the time-averaged value and small one denotes the velocity fluctuation. The sampling frequency and sampling time were set to 0.05s and 25.6s, respectively.

Onitsuka et al. [3] pointed out that the swimming trajectory of isolated ayu in statistic water can be described by nodes and branches. In this study, a similar tendency was confirmed as shown in Figure 2. After recording, the turning position and elapsed time in swim trajectory of ayu were identified. The ground distance L_G and the turning angle θ are calculated from a consecutive turning position in Figure 3. Turning angle θ that right rotation and left rotation were defined as plus and minus. The ground distance L_G is resolved the component and the ground distance in the x direction L_{Gx}, the ground distance in the z direction L_{Gz} were calculated. Moreover, swimming distance L, swimming distance in the x direction L_x, swimming distance in the z direction L_z were calculated from three kinds of ground distance and flow velocity. On the other hand, ground speed V_G, ground speed in the x direction V_{Gx}, ground speed in the z direction V_{Gz} were calculated from three kinds of ground distance and swimming time. Further, swimming speed V, swimming speed in the x direction V_x, swimming speed in the z direction V_z were calculated from adding flow velocity to these ground speeds.

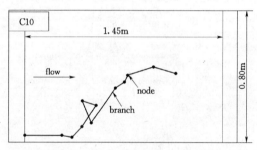

Figure 2　Example of swimming trajectory (C10)

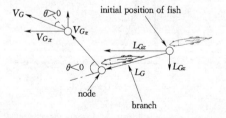

Figure 3　Modeling of swimming trajectory

3. Experimental results and considerations

3.1　Separation of universal swimming and wall effect swimming

Onitsuka et al. [3] pointed out that the swimming

trajectory of isolated ayu in statistic water has influence of the swim beginning by 2 branches. A similar tendency was seen also in this study. On the other hand, ayu did not arrive at the upstream edge directly. So ayu swims along the sidewall after it approaches in the neighborhood of the sidewall. As a result of the observation, when the distance from sidewall is less than one time of body length of ayu, it was judged that the swim characteristic changed. Therefore, the area where the number of branch is more than 3 from starting point of ayu and distance from sidewall is more than one time of body length of ayu was defined as the universal area. Swimming along the sidewall after ayu arrived where distance from sidewall is less than one time of body length of ayu was defined as the wall area. Only a universal area is targeted in the analysis.

3. 2 Ground distance in universal area

Figure 4 (a) - (c) show the frequency distribution of the value in which the ground distance in the x direction L_{Gx}, absolute value of the ground distance in the z direction $|L_{Gz}|$, and the ground distance L_G divided by the averaged body length $\overline{B_L}$, respectively is shown in each flow velocity. The ground distance in the spanwise direction is assumed to be an absolute value. Because, there is no physiological meaning that ayu select in the right bank direction and in the left bank direction. The ground distance in the x direction L_{Gx} shown in Figure 4 (a) shows a symmetric distribution type in each flow velocity. Therefore, distribution was shown by normal distribution and it showed by the curve in Figure 4 (a). The normal distribution is defined as follows:

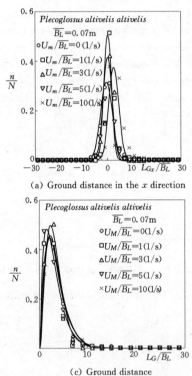

(a) Ground distance in the x direction

(c) Ground distance

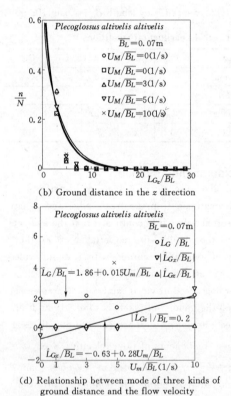

(b) Ground distance in the z direction

(d) Relationship between mode of three kinds of ground distance and the flow velocity

Figure 4　Ground distance according to change in flow velocity in universal area

$$f(L_{Gx}/\overline{B_L}) = \frac{1}{\sqrt{2\pi}\left(\frac{L_{Gx}}{\overline{B_L}}\right)'}\exp\left\{-\frac{\left(\frac{L_{Gx}}{\overline{B_L}}-\overline{\frac{L_{Gx}}{\overline{B_L}}}\right)^2}{2\left(\frac{L_{Gx}}{\overline{B_L}}\right)'^2}\right\} \quad (1)$$

The average $\overline{L_{Gx}/\overline{B_L}}$ and variance $(L_{Gx}/\overline{B_L})'^2$ were calculated from the least squares method. Mode of $L_{Gx}/\overline{B_L}$ increases with an increase of the flow veloci-

ty. Absolute value of the ground distance in the z direction $|V_{Gz}|$ shown in Figure 4 (b) takes the maximum value in the vicinity of $|L_{Gz}|/\overline{B_L}=0$ in all flow velocity and the distribution that decreases in the high value direction is shown. Therefore, chi-square distribution was adopted and it showed by the curve in Figure

4 (b) . The chi-square distribution is defined as follows:

$$f(|L_{Gz}|/\overline{B_L}) = \frac{(1/2)^{\left[\left(\frac{|\overline{L_{Gz}}|}{\overline{B_L}}\right)/2\right]}}{\Gamma\left[\left(\frac{|\overline{L_{Gz}}|}{\overline{B_L}}\right)/2\right]}$$

$$\times \left(\frac{|\overline{L_{Gz}}|}{\overline{B_L}}\right)^{\left[\left(\frac{|\overline{L_{Gz}}|}{\overline{B_L}}\right)/2-1\right]} e^{\left(-\frac{|\overline{L_{Gz}}|}{\overline{B_L}}\right)/2} \quad (2)$$

The average $|\overline{L_{Gz}}|/\overline{B_L}$ and variance $2 \times (|\overline{L_{Gz}}|/\overline{B_L})$ were calculated from the least squares method. Absolute value of the ground distance in the z direction $|L_{Gz}|$ is constant, irrespective of the flow velocity [see Figure 4 (b)] . The ground distance L_G shown in Figure 4 (c) has a low value in high frequency and the distribution that decreases in the high value direction is shown. Therefore, gamma distribution was adopted and it showed by the curve in Figure 4 (c) . The gamma distribution is defined as follows:

$$f(L_G/\overline{B_L}) = \frac{1}{\Gamma(\lambda)} \alpha^\lambda (L_G/\overline{B_L})^{\lambda-1} e^{-\alpha L/\overline{B_L}} \quad (3)$$

$$\Gamma(\lambda) = \int_0^\infty e^{-x} x^{\lambda-1} \, dx \quad (4)$$

Coefficients α and λ were calculated from the least squares method. The ground distance L_G increases with an increase of the flow velocity [see Figure 4 (c)] . Because, the ground distance in the x direction L_{Gx} increases [see Figure 4 (a)] . Figure 4 (d) shows the relationship between mode of three kinds of ground distance $\hat{L}_{Gx}/\overline{B_L}$, $|\hat{L}_{Gz}|/\overline{B_L}$ and $\hat{L}_G/\overline{B_L}$, respectively and the flow velocity $U_m/\overline{B_L}$. $\hat{L}_{Gx}/\overline{B_L}$ and $\hat{L}_G/\overline{B_L}$ have increase with an increase of the flow velocity. $|\hat{L}_{Gz}|/\overline{B_L}$ is constant, irrespective of the flow velocity. Therefore, it was found that ayu did not change the ground distance in the z direction, and

increase of the ground distance in the x direction with an increase of the flow velocity.

3.3 Swimming distance in universal area

Figure 5 (a) and Figure 5 (b) show the frequency distribution of the value in which the swimming distance in the x direction L_x, and the swimming distance L divided by the averaged body length $\overline{B_L}$, respectively is shown in each flow velocity. The swimming distance in the x direction L_x shown in Figure 5 (a) shows a symmetric distribution. Therefore, normal distribution was adopted and it showed by the curve in Figure 5 (a) . The swimming distance in the x direction L_x increases with an increase of the flow velocity. Swimming distance L shown in Figure 5 (b) has a low value in high frequency and the distribution that decreases in the high value direction is shown. Therefore, gamma distribution was adopted and it showed by the curve in Figure 5 (b) . The swimming distance L increases with an increase of the flow velocity. Absolute value of the ground distance in the z direction $|L_{Gz}|$ shown in Figure 4 (b) is constant, irrespective of the flow velocity. It is thought that an increase of the swimming distance L was caused by an increase of swimming distance in the x direction L_x. Figure 5 (c) shows the relationship between mode of three kinds of swimming distance $\hat{L}_x/\overline{B_L}$, $|\hat{L}_z|/\overline{B_L}$ and $\hat{L}/\overline{B_L}$, respectively and the flow velocity $U_m/\overline{B_L}$. $\hat{L}_x/\overline{B_L}$ and $\hat{L}/\overline{B_L}$ have increase with an increase of the flow velocity. $|\hat{L}_z|/\overline{B_L}$ is constant, irrespective of the flow velocity. It is thought that swimming distance in the upstream direction increases with an increase of the flow velocity and the swimming distance increased with an increase of the flow velocity.

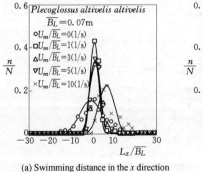

(a) Swimming distance in the x direction

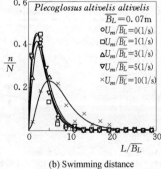

(b) Swimming distance

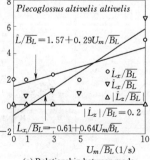

(c) Relationship between mode of three kinds of swimming distance and flow velocity

Figure 5   Swimming distance according to change in flow velocity in universal area

3.4 Ground speed in universal area

Figure 6 (a) - (c) show the frequency distribution of the value in which the ground speed in the x direction V_{Gx}, absolute value of the ground speed in the z direction $|V_{Gz}|$, and the ground speed V_G divided by the averaged body length $\overline{B_L}$, respectively is shown in each flow velocity. The ground speed in the x direction V_{Gx} shown in Figure 6 (a) shows a symmetric distribution type in each flow velocity. Therefore, normal distribution was adopted and it showed by the curve in Figure 6 (a). Mode of the ground speed in the x direction V_{Gx} increases with an increase of the flow velocity. It is thought that the ground speed in the x direction V_{Gx} increases with an increase of the flow velocity, because the ayu has positive rheotaxis. Absolute value of the ground speed in the z direction $|V_{Gz}|$ shown in Figure 6 (b) takes the maximum value in the vicinity of $|L_{Gz}|/\overline{B_L}=0$ in all flow velocity and the distribution that decreases in the high value direction is shown. Therefore, chi-square distribution was adopted and it showed by the curve in Figure 6 (b). Absolute value of the ground speed in the

z direction $|V_{Gz}|$ is constant, irrespective of the flow velocity. Therefore, it is thought that the ground speed in the spanwise direction is not affected by the flow velocity. The ground speed V_G shown in Figure 6 (c) has a low value in high frequency. Moreover, the distribution that decreases in the high value direction is shown. Therefore, gamma distribution was adopted and it showed by the curve in Figure 6 (c). The ground speed V_G increases, with an increase of the flow velocity, because an increase of the ground distance in the x direction L_{Gx}. Figure 6 (d) shows the relationship between mode of three kinds of ground speed $\hat{V}_{Gx}/\overline{B_L}$, $|\hat{V}_{GZ}|/\overline{B_L}$ and $\hat{V}_G/\overline{B_L}$, respectively and the flow velocity $U_m/\overline{B_L}$. $\hat{V}_{Gx}/\overline{B_L}$ and $\hat{V}_G/\overline{B_L}$ have increase with an increase of the flow velocity. $|\hat{V}_{GZ}|/\overline{B_L}$ is constant, irrespective of the flow velocity. It is thought that ayu did not change the swimming speed in the spanwise direction, irrespective of the flow velocity. However, the swimming speed in the streamwise direction increases, because ayu has positive rheotaxis.

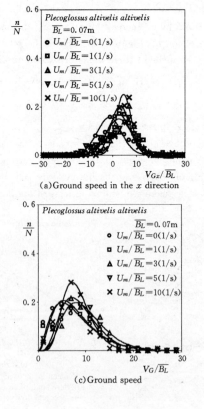

(a) Ground speed in the x direction

(c) Ground speed

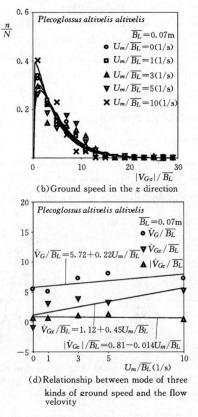

(b) Ground speed in the z direction

(d) Relationship between mode of three kinds of ground speed and the flow velocity

Figure 6   Ground speed according to change in flow velocity in universal area

3.5　Swimming speed in universal area

Figure 7 (a) and Figure 7 (b) show the frequency distribution of the value in which the swimming speed in the x direction V_x and the swimming speed V divided by the averaged body length $\overline{B_L}$, respectively is shown in each flow velocity. The swimming speed in the x direction V_x shown in Figure 7 (a) shows a symmetric distribution type in each flow velocity. Therefore, normal distribution was adopted and it showed by the curve in Figure 7 (a). It is confirmed that mode of the swimming speed in the x direction V_x increases with an increase of the flow velocity. The swimming speed V shown in Figure 7 (b) has a low value in high frequency. Moreover, the distribution that decreases in the high value direction is shown. Therefore, gamma distribution was adopted and it showed by the curve in Figure 7 (b). Mode of the swimming speed $\hat{V}$ increases with an increase of the

flow velocity. The increasing tendencies were more remarkable than ground speed V_G shown in Figure 6 (c). Therefore, it is thought that ayu increases remarkably the swimming speed, because ayu increases the ground speed with an increase of the flow velocity. Figure 7 (c) shows the relationship between mode of three kinds of swimming speed $\hat{V}_x / \overline{B_L}$, $|\hat{V}_z| / \overline{B_L}$ and $\hat{V} / \overline{B_L}$, respectively and the flow velocity $U_m / \overline{B_L}$. $\hat{V}_x / \overline{B_L}$ and $\hat{V} / \overline{B_L}$ have increase with an increase of the flow velocity. $|\hat{V}_z| / \overline{B_L}$ is constant, irrespective of the flow velocity. Moreover, when flow velocity $U_m / \overline{B_L}$ is 10, $\hat{V}_x / \overline{B_L}$ and $\hat{V} / \overline{B_L}$ were reached 15. However the burst speed of the fish is generally known as ten times body length. Recently, Izumi et al. [6] obtained the data that the burst speed of the fish is more than ten times body length. Therefore, it is thought that this research results are appropriate.

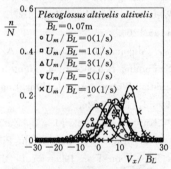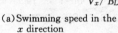

(a) Swimming speed in the x direction

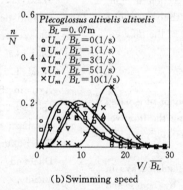

(b) Swimming speed

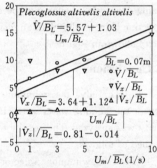

(c) Relationship between mode of three kinds of swimming speed and flow velocity

Figure 7　Swimming speed according to change in flow velocity in universal area

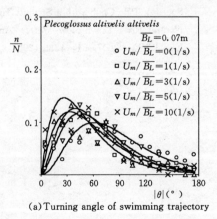

(a) Turning angle of swimming trajectory

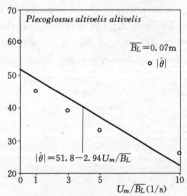

(b) Relationship between mode of absolute value of the turning angle and the flow velocity

Figure 8　Turning angle of swimming trajectory according to change in flow velocity in universal area

3.6 Turning angle of swimming trajectory in universal area

Figure 8 (a) shows the frequency distribution of absolute value of the turning angle $|\theta|$ in each flow velocity.

Absolute value of the turning angle $|\theta|$ has the distribution that decreases in the high value direction is shown. Therefore, gamma distribution was adopted and it showed by the curve in Figure 8 (a) . Absolute value of the turning angle $|\theta|$ decreases with an increase of the flow velocity. Figure 8 (b) shows the relationship between mode of absolute value of the turning angle $|\hat{\theta}|$ and the flow velocity $U_m/\overline{B_L}$. Mode of absolute value of the turning angle $|\hat{\theta}|$ decreases with an increase of the flow velocity. It is thought that positive rheotaxis increases remarkably, because swimming speed in the spanwise direction is constant. However, swimming speed in the streamwise direction is increasing with an increase of the flow velocity shown in Figure 7.

4. Conclusions

In this study, the swimming behavior of isolated ayu in the running water with changing the flow velocity was analyzed. The results of this study are described as follows.

(1) The swimming trajectory of isolated ayu in the running water can be described by nodes and branches.

(2) The ground distance, swimming distance, ground speed, swimming speed in the streamwise direction increases with an increase of the flow velocity, because the ayu has positive rheotaxis. In contrast, the ground distance, swimming distance, ground speed, swimming speed in the spanwise direction is constant, irrespective of the flow velocity.

(3) The turning angle of swimming trajectory decreases with an increase of the flow velocity, because ayu has positive rheotaxis.

References

[1] Inoue Y.. Size and Moving Behaviour of Fish School Around The Set-net. Japanese Society of Scientific Fisheries, 1987, 53: 1307-1312.

[2] Hasegawa E. , Soeda H.. Mutual Relationship among Individuals Composing a Fish School. Japanese Society of Scientific Fisheries, 1985, 51: 1921-1926.

[3] Onitsuka K. , Akiyama, J. , Yamamoto, A. , Waki, T.. Swimming Behavior of Isolated Ayu in Statistic Water. Ann. J. Hydr. Eng. JSCE. , 2008, 52: 1195-1200.

[4] Onitsuka K. , Akiyama J. , Yamamoto A. , Waki T.. Swimming Behavior of a Couple of Ayu in Statistic Water. Ann. J. Hydr. Eng. JSCE. , 2009, 52: 1219-1224.

[5] Takamizu K. , Kurihara T. , Aoki M. , Uchiyama F. , Fukui Y.. Hydraulic Functions of the Pile Dile and its Impact on Fish Behavior in the Vicinity. Ann. J. Hydr. Eng. JSCE. , 2007, 51: 1273-1278.

[6] Izumi M. , Yataka K. , Azuma N. , Kudo A. , Kato K.. On-Side Swimmng Experiment of Burst Speed of Pale Chub with a Stamina Tunnel Using Natural River Down-Flow Water. Ann. J. Hydr. Eng. JSCE. , 2007, 51: 1285-1290.

Overbank Flow Estimation using ANFIS and Genetic Programming

Lai SaiHin   Mah   DarrienYauSeng

River Engineering and Urban Drainage Research Centre (REDAC),
Engineering Campus, Universiti Sains Malaysia (USM),
14300, Nibong Tebal, Pulau Penang, Malaysia

Abstract: The estimation of discharge capacity in river channels is complicated once the river is flowing out-of-bank due to complex 3D turbulent structure and interactions at the interface region between main channel and flood plain. These interactions can significantly reduce the discharge capacity of the river or channel. When flooding, traditional flow equations and methods in estimating the discharges are found to be not very accurate and may lead to over or under estimation of discharge capacity. Therefore, this study is carried out to estimate the discharge for overbank flow in compound channels and natural rivers using traditional methods, empirical method and soft computing tools. It is found that Genetic Programming (GP) is the most accurate among all the methods being tested with a smallest average error of 5.76%, 5.29% and 1.02% in overbank flow estimation for River Senggi, River Senggai and River Main, respectively.

Key words: ANFI; discharge estimation; flood; genetic programming; overbank flow

1. Introduction

In analyzing flow through river channels, one of the most common tasks of a river engineer is to make estimates of discharge based on an estimated, recorded or simulated water level. This is very important not only to ensure sufficient water supply and waste disposal etc. during low flow, but also for practical purposes such as flood forecasting and flood mitigation during overbank flow or extreme water level.

For inbank flow the theoretical determination of the stage-discharge relationship at a given cross-section of a river is a straightforward issue. It is sufficient, in general, to use the overall hydraulic radius as the parameter, which characterizes the properties of the cross section. It is then possible to calculate the discharge through the channel from one of a range of well-known uniform flow formulas (such as Chezy, manning or Darcy-Weisbach) in terms of the channel roughness, slope and depth.

However, once the river is in flood, and flowing out-of-bank, it becomes much more difficult due to the complex 3D turbulent structure and momentum transfer at the interface region between main channel and flood plain (Figure 1 and Figure 2). These interactions,

i.e. momentum transfer and apparent shear can significantly reduce the discharge capacity of a river.

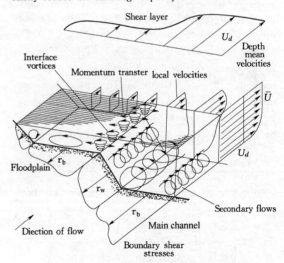

Figure 1   Mechanisms of overbank flow
in a straight compound channel[1]

Due to this reason, traditional flow equations and methods in overbank discharge estimation are found to be not very accurate, and may lead to very serious over-or-under estimation of discharge capacity[3-9]. This has become the subject of considerable researches in the past 30 years focusing on various as-

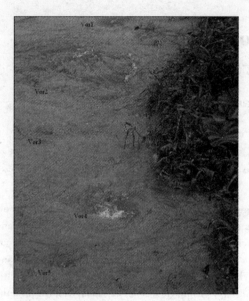

Figure 2　Momentum transfer characterized by
series of vortices at the interface region of
main channel and flood plain[2]

ga-tions, with certain idealized conditions, i. e. uniform channel cross-section, surface roughness and bed slope etc. Under such conditions, the equations derived will be representative of a certain channel type, and it is not applicable to channels/rivers with different geometrical shapes and boundary conditions.

In recent years, soft computing tools, such as ANN, ANFIS and GP have been used widely to solve various water related problems[26-36]. Therefore, this study is carried out to explore the possibility of using ANFIS and GP for overbank flow estimation.

2. Methodology

Extensive experimental data on overbank flow in compound channels as shown in Table 1 have been collected from various sources. Field data for 3 frequently flooded natural rivers, namely River Senggai, River Senggi[2, 17] and River Main[37] as shown in Table 2 and Table 3 have also been collected from river measurement for the analysis of this study.

By using the Manning equation and geometrical data such as flow depth, cross sectional parameters and roughness coefficient obtained from the respective sources. Various methods have been used to estimate overbank flow in the above mentioned compound channels and rivers. These methods include: (1) Traditional methods such as Single Channel Method (SCM), Vertical Divided Channel Method (VDCM), and Horizontal Divided Channel Method (HDCM); (2) Empirical method such as Weighted Divided Channel Method (WDCM[8]); (3) Soft computing tools such as Adaptive Neuro-Fuzzy Inference System (ANFIS) and Genetic Programming (GP).

pect of compound channel such as stage-discharge relationship, momentum transfer, flow resistance, apparent shear, discharge estimation, etc[8-17].

From the studies being carried out, various methods as well as empirical formulas have been proposed for overbank flow calculation. These methods can be divided into several groups, such as: (i) The apparent shear force methods[3-5, 17-18]; (ii) The correction factor methods[7, 19-21]; (iii) The numerical methods[22-25]; etc. Unfortunately, none yet commands wide spread acceptance. The main reason for this is that most of the previous researches are based on small, and some in large-scale laboratory investi

Table 1   Data for compound channels

Source	Dimension				Slope				Bed Material		Remark
	b (m)	B (m)	B/b	h (m)	S_0	S_1	S_2	S_3	MC	FP	
[8, 13]	0.75	5.00	6.67	0.15	0.001	1	0	∞	SC	SC	FCF, Phase A Series1 Smooth MC + FP
	0.75	1.65	2.20	0.15	0.001	1	0	1	SC	SC	FCF, Phase A Series3 Smooth MC + FP
[9]	0.75	3.00	4.20	0.15	0.001	1	0	1	SC	SC	FCF, Phase A Series2 Smooth MC + FP
	0.75	3.00	4.20	0.15	0.001	1	0	1	SC	RR	FCF, Phase A Series7 Smooth MC, Rough FP
[3]	0.076	0.152	2	0.076	0.000966	∞	0	∞	PP	PP	
	0.076	0.228	3	0.076	0.000966	∞	0	∞	PP	PP	Smooth MC+FP
	0.076	0.305	4	0.076	0.000966	∞	0	∞	PP	PP	

Continued

Source	Dimension				Slope				Bed Material		Remark
	b (m)	B (m)	B/b	h (m)	S_0	S_1	S_2	S_3	MC	FP	
[8]	0.300	0.080	4.75	0.080	0.00191 0.00109 0.00061 0.00037	∞	0	∞	PP	PP	Geometry 1
	0.180	0.080	3.25	0.080	0.00127 0.00097 0.00032	∞	0	∞	PP	PP	Geometry 2
	0.300	0.080	4.75	0.120	0.00178 0.00145 0.00105	∞	0	∞	PP	PP	Geometry 3
[4]	0.145	0.375	4.17	0.12	0.00043	∞	0	∞	PP	$n=0.011$	Smooth MC Roughened FP
	0.145	0.375	4.17	0.12	0.00043	∞	0	∞	PP	$n=0.014$	
	0.145	0.375	4.17	0.12	0.00043	∞	0	∞	PP	$n=0.017$	
	0.145	0.375	4.17	0.12	0.00043	∞	0	∞	PP	$n=0.021$	

Table 2　　　　　　**Geometrical properties for compound natural rivers**

Geometrical Properties	River Senggai	River Senggi	River Main
Bankfull depth, h (m)	1.060	1.306	0.900
Top width, B (m)	5.285	5.500	13.700
Aspect ratio, B/H_{bf}	4.986	4.211	15.222
Bed slope-MC, S_0	0.0010	0.0010	0.0030
Bed slope-left FP, S_L	0.0010	0.00085	0.0030
Bed slope-Right FP, S_R	0.0010	0.00085	0.0030

Table 3　　　　　　**Summary of data obtained from river measurements[17]**

River	Depth $(H-h)/H$	Total Q, (m³/s)	MC Q, (m³/s)	FP Q, (m³/s)
Senggai	−0.596 ↓ 0.361	0.2838 ↓ 5.0885	0.2838 ↓ 4.6877	0 ↓ 0.4008
Senggi	−1.023 ↓ 0.387	0.4100 ↓ 2.8773	0.4100 ↓ 2.4188	0 ↓ 0.4585

3. Discharge estimation—traditional and empirical methods

By comparing the results obtained from discharge estimation and observed data. the accuracy of each method in overbank flow estimation is summarized in Tables 4 and 5. For laboratory compound channels with smooth boundary conditions, it can be seen that even though the averaged errors for all the methods are ≤10%, the SCM which treated the whole cross section as single unit is found to give a maximum error of up to 52.51%. The commonly used HDCM and VDCM also give a maximum error of 32.39% and 44.25% respectively. This shows that the interactions between the main channel and flood plain is significant, and in any conditions, it should not be ig-

161

nored in discharge calculation.

Table 4　Discharge estimations in smooth compound channels using traditional and empirical methods

Statistical Calculation	Calculation Method			
	HDCM	VDCM	WDCM	SCM
Max error (%)	32.39	44.25	38.33	52.51
Average error (%)	5.70	9.72	6.53	7.78
RMSE (%)	7.43	11.81	8.85	11.98

For laboratory compound channels with roughened bottom surface, Table 4 shows that the VDCM overestimate the discharge of up to 83.18% with an average error of 24.52%. The WDCM and HDCM are also found to produce maximum errors of up to 68.23%, 53.33% and average errors of 15.72%, 9.35% respectively. While, the SCM is found to gives general better results among the methods with a maximum error of 17.01%.

Table 5　Discharge estimations in rough compound channels using traditional and empirical methods

Statistical Calculation	Calculation Method			
	HDCM	VDCM	WDCM	SCM
Max error (%)	53.33	83.18	68.23	17.01
Average error (%)	9.35	24.52	15.72	7.45
RMSE (%)	13.35	28.98	20.40	8.61

For all the rivers selected, the results obtained are shown in Table 6. It can be seen that all the methods are significantly in error, in extreme case (e. g. River Senggai), the maximum error and average error using VDCM can be as high as 89.46% and 51.76% respectively.

Table 6　Discharge estimation for rivers using traditional and empirical methods

River	Method	Max Error (%)	Ave Error (%)	RMSE (%)
Senggi	SCM	67.641	39.807	43.795
	HDCM	26.158	10.745	13.117
	VDCM	46.843	24.336	29.779
	WDCM	19.303	10.131	11.409

Continued

River	Method	Max Error (%)	Ave Error (%)	RMSE (%)
Senggai	SCM	71.610	46.666	49.274
	HDCM	28.492	12.815	15.370
	VDCM	89.457	51.756	53.902
	WDCM	52.320	28.372	30.076
Main	SCM	22.385	13.230	14.485
	HDCM	28.195	12.943	14.723
	VDCM	35.262	9.995	14.673
	WDCM	24.454	6.950	9.511

4. Discharge estimation-ANFIS

The input to ANFIS consists of 150 sets of training data and 40 sets of testing data arranged in 7 dimensionless groups.

$$Q_o \propto Q_T \left(\frac{(H-h)}{H}, \frac{n_f}{n_m}, \frac{A_f}{A_m}, \frac{P_f}{P_m}, \frac{R_f}{R_m}, \frac{S_f}{S_m}, \frac{V_f}{V_m}, \right) \cdots \quad (1)$$

Eq. (1) indicates several independent variables that could provide the adjustment needed to zonal calculation of discharge capacity, in which Q_o is observed discharge; Q_T is predicted discharge using traditional method; $(H-h)/H$ is depth ratio of flood plain and main channel; n_f and n_m are flood plain and main channel roughness coefficients estimated by assuming that there is no interaction occur at the interface region; A_f and A_m are wetted areas for floodplain and main channel; P_f and P_m are wetted perimeter for floodplain and main channel; R_f and R_m are floodplain and main channel hydraulic radius; S_f and S_m are bed slopes for floodplain and main channel; whereas V_f and V_m are flow velocities for floodplain and main channel estimated using Manning equation. This approach was preferred to allow the estimation of overbank flow in a predictive sense, i. e. without the need for measuring the velocities of the subsections.

Four schemes of ANN are tested namely, Back Propagation Networks (trainbfg and traincgf), Radial Basis Function (RBF), and Adaptive Neuro-Fuzzy Inference System (ANFIS). These ANN networks are employed by using computer codes in MATLAB with neural network and fuzzy logic toolboxes.

Based on the results obtained, ANFIS with a cluster radius of 0.5 is found to give the best performance among the ANN schemes. The correlation of determination (R^2) for this ANFIS architecture for training and testing data was 0.9867 and 0.9790 respectively. This implies that the agreement between the observations and predictions is good. Figure 3 and Figure 4 illustrate the correlation between the discharges estimated by ANFIS against observation data. However, Table 7 shows that the estimation of overbank flows for all the rivers are significantly in error. These results reflect that the ANFIS model being developed mainly based on laboratory data is not applicable to natural rivers.

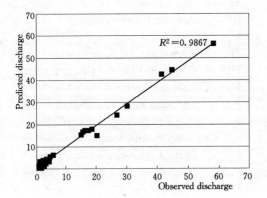

Figure 3　Predicted versus observed discharge by ANFIS for training data

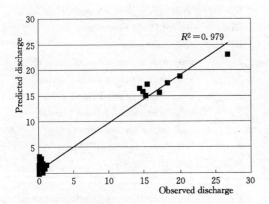

Figure 4　Predicted versus observed discharge by ANFIS for testing data

Table 7　Discharge estimation for rivers using ANFIS

River	Max Error (%)	Ave Error (%)	RMSE (%)
Senggi	84.298	28.31	35.550
Senggai	24.936	11.230	13.623
Main	26.698	8.860	11.228

5. Discharge estimation-genetic programming

The same training (150 sets) and testing data (40 sets) are also employed in a GP model for overbank flow estimation. The correlation of determination (R^2) obtained for training and testing data are both 0.9998. Figure 5 and Figure 6 illustrate the correlation for estimated discharge by GP model against observation data.

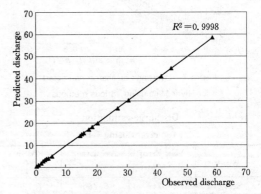

Figure 5　Predicted versus observed discharge by genetic Programing for training data

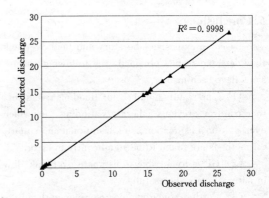

Figure 6　Predicted versus observed discharge by genetic programing for testing data

The accuracy for overbank flow estimation using GP is shown Table 8. These results show that GP is able to estimate the overbank flow in all the three natural rivers accurately with an average error of less than 6%, A sample of the discharge estimated for River Main using VDCM, HDCM, ANFIS and GP is plotted in Figure 7. Also plotted is the observed discharge for comparison. This figure further shows that among all the methods being tested, GP is the most accurate method for estimating overbank discharges in the rivers.

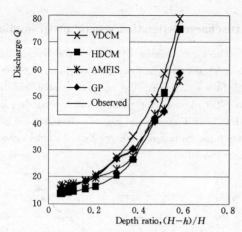

Figure 7 Overbank discharge estimated
for River Main using various methods

Table 8 Discharge estimation
for rivers using GP

River	Max Error (%)	Ave Error (%)	RMSE (%)
Senggi	18. 749	5. 760	7. 726
Senggai	8. 169	5. 290	5. 792
Main	2. 230	1. 020	1. 087

6. Conclusions

Based on extensive laboratory and field overbank flow data and results obtained, the following conclusions have been made.

(1) The results obtained for flooding rivers further confirm previous laboratory finding, in which overbank flow estimation using traditional and empirical methods are found to be significantly in error.

(2) GP is found able to estimate overbank flow in natural rivers successfully.

(3) Further study on flooding river with different scale, sinuosity, and geometrical conditions has to be carried out in order to provide more data to develop a reliable method for hydraulic analysis under overbank flow conditions.

Acknowledgement

The authors are grateful to the Government of Malaysia for the financial support from the Ministry of Science and Technology Innovation through the Escience grant No. 04-01-05-SF0415. Thanks to Dr. Myers from School of the Built Environment, University of Ulster at Jordanstown for providing the overbank flow data for River Main.

Notation

A_f	wetted areas for floodplain
A_m	wetted areas for main channel
ANFIS	Adaptive Neuro-Fuzzy Inference System
ANN	Artificial Neural Network
b	bottom width,
B	top width
FCF	Flood Channel Facility at Wallingford, UK.
FP	flood plain
GP	Genetic Programming
h	bankfull depth
H	flow depth
HDCM	Horizontal Divided Channel Method
MC	main channel
n	manning roughness coefficient
n_f	flood plain roughness coefficient
n_m	main channel roughness coefficient
P_f	wetted perimeter for floodplain
P_m	wetted perimeter for main channel
PP	perpeks
Q_o	observed discharge
Q_p	predicted discharge
R_f	hydraulic radius for floodplain
R_m	hydraulic radius for main channel
RBF	Radial Basis Function
RR	rod roughened
S_0	longitudinal bed slope of main channel
S_1	side bank slope of main channel
S_2	lateral bed slope of flood plain
S_3	side bank slope of flood plain
S_f	bed slopes for floodplain
S_m	bed slopes for main channel
SC	smooth concrete
SCM	Single Channel Method
V_f	flow velocities for floodplain estimated using Manning equation
V_m	flow velocities for main channel estimated using Manning equation
VDCM	Vertical Divided Channel Method
WDCM	Weighted Divided Channel Method

References

[1] Shiono K., Knight D. W.. Turbulent Open

Channel Flows with Variable Depth Across the Channel. Journal of Fluid Mechanics, 1991, 222: 617-646.

[2] Lai S. H. , Bessaih N. , Law P. L. , Aminuddin Ab. Ghani, Nor Azazi Z. , and Mah Y. S. . A study of hydraulic characteristics for flow in equatorial rivers. International Journal of River Basin Management, International Association of Hydraulic Research (IAHR), 2008, 6 (3): 213-223.

[3] Knight D. D. , J. D. Demetriou. Flood plain and main channel flow interaction. Journal of Hydraulic Engineering, ASCE, 1983, 109 (8): 1073-1092.

[4] Wormleaton P. R. , J. Allen, P. Hadjipanos. . Discharge assessment in compound channel flow. Journal of Hydraulic Division, ASCE, 1982, 108 (HY9): 975-994.

[5] Prinos P. , R. Townsend. . Comparison of methods for predicting discharge in compound open channels. Adv. in water resour. , 7 (Dec), 1984: 180-187.

[6] Myers W. R. C. . Velocity and discharge in compound channels, Journal of Hydraulic Engineering, 1987, 113: 753-766.

[7] Ackers P. . Hydraulic design of two-stage channels. Proceedings of the Institution of Civil Engineering, Water, Maritime and Energy, 1992, 96. 247-257.

[8] Lambert M. F. , W. R. Myers. Estimating the discharge capacity in straight compound channels. Proceedings of the Institution of Civil Engineering, Water, Maritime and Energy, 1998, 130: 84-94.

[9] Myers W. R. C. , J. F. Lyness, J. Cassells. Influence of boundary on velocity and discharge in compound channels, Journal of Hydraulic research, 2001, 39 (3): 311-319.

[10] Myers W. R. C. . Momentum transfer in a compound channel. Journal of Hydraulic Research, 1978, 16 (2): 139-150.

[11] Baird J. I. , D. A. Ervine. A resistance to flow in channels with overbank floodplain flow. Proceedings of the 1st International Conference on Hydraulic Design in Water Resources Engi-

neering: Channels and Channel Control Structures, University of Southampthon, 1984: 137-415.

[12] Myers W. R. C. , J. F. Lyness. Discharge ratios in smooth and rough compound channel, Journal of Hydraulic Engineering, 1997, 123 (3): 182-188.

[13] Myers W. R. C. , E. K. Brennan. . Flow resistance in compound channel. Journal of Hydraulic Research, 1990, 28 (2): 141-155.

[14] Myers WRC. . Influence of geometry on discharge capacity of open channels, Journal of Hydraulic Engineering, ASCE, 1991, 117: 676-680.

[15] Abril J. B. , Knight D. W. . Stage-discharge prediction for river in flood applying a depth-averaged model, Journal of Hydraulic Research, IAHR, 2003, 41 (6): 1-14.

[16] Myers W. R. C. , D. W. Knight, J. F. Lyness, J. B. Cassells. Resistance coefficients for in-bank and overbank flows. Proceedings of the Institution of Civil Engineers, Water, Maritime and Energy, 1999, 136: 105-115.

[17] Lai S. H. , Bessaih N. , Law P. L. Aminuddin Ab. G. , Nor Azazi Z. , and Mah Y. S. Determination of composite friction factor for flooded equatorial natural rivers. International Journal of River Basin Management, International Association of Hydraulic Research (IAHR), 2008, 6 (1): 3-12.

[18] Christodolou G. C. , W. R. C. Myers. . Apparent Friction Factor on the Flood Plain-Main Channel Interface of Compound Channel Sections, Proceedings of 28th Congress of the International Association for Hydraulic Research, Craz, Austria, SE3, A379. 1999.

[19] Radojkovic M. , S. Djordjevic. Computation of discharge distribution in compound channels, Proceedings of the 21st Congress of International Association for Hydraulic Research, Melbourne, Australia, 1985: 367-371.

[20] Wormleaton P. R. , D. J. Merritt. An improved method of calculation for steady uniform flow in prismatic main channel flood plain sections. Journal of hydraulic research, 1990, 28 (2):

157-174.

[21] Ackers P. . Flow foemulae for straight two-stage channels. Journal of Hydraulic Research, 1993, 31 (4): 509-531.

[22] Keller R. J. , W. Rodi. . Prediction of flow characteristics in main channel/flood plain flows, Journal of Hydraulic Research, 1988, 24 (4): 425-441.

[23] Tominaga A. , I. Nezu. Turbulent structure in compound open-channel flows, Journal of Hydraulic Engineering, ASCE, 1991, 117 (1): 21-41.

[24] Abida H. , R. D. Townsend. . A model for routing unsteady flows in compound channels, Journal of Hydraulic Research, 1994, 32: 145-153.

[25] Knight D. W. , K. Shiono. . Turbulence measurements in a shear region of a compound channel, Journal of Hydraulic Research, 1990, 128 (2): 175-196.

[26] Azamathulla H. M. , Deo M. C. , Deolalikar P. B. . Neural networks for estimation of scour downstream of a ski-Jump bucket. Journal of Hydraulic Engineering © ASCE. 2005.

[27] Azamathulla H. M. , Chang C. K. , Ghani A. A. , Ariffin J. , Zakaria N. A. , Hasan Z. A. An ANFIS-based approach for predicting the bed load for moderately sized rivers. Journal of Hydro-environment Research, 2009a: 1-10.

[28] Bateni S. M. , Jeng D. S. . Estimation of pile group scour using adaptive neuro-fuzzy approach, Technical Note: Ocean Engineering, 2007, 34: 1344-1354.

[29] Bateni S. M. , Borghei S. M. , Jeng, D. S. . Neural network and neuro-fuzzy assessments for scour depth around bridge piers. Engi-

neering Applications of Artificial Intelligence, 2007, 20: 401-414.

[30] Jain A. , Varshney A. K. , Joshi U. C. . Short-term water demand forecast modeling at IIT Kanpur using artificial neural networks. Water Resour. Manage. 2001, 15 (5): 299-321.

[31] Kumar J. , Jain A. , Srivastava R. . Neural network based solutions for locating groundwater pollution sources. Hydrol. J. , 2006, 29 (1-2): 55-66.

[32] Maier H. R. , Dandy G. C. . Neural networks for the prediction and forecasting of water resource variables: A review of modeling issues and applications. Environ. Modell. Software, 2000, 15: 101-124.

[33] Raman H. , Sunilkumar N. . Multivariate modeling of water resources time series using artificial neural networks. Hydrol. Sci. J. , 1995, 40 (2), 145-163.

[34] Sajikumar N. , Thandaveswara B. S. . A nonlinear rainfallrunoff model using an artificial neural network. J. Hydrol. , 1999, 216: 32-55.

[35] Singh A. K. , Deo M. C. , Kumar V. S. . Neural network-genetic programming for sediment transport, Maritime Engineering 160 Issue MA3, 2007: 113-119.

[36] Tay J. H. , Zhang X. . Neural fuzzy modelling of anaerobic biological waste water treatment systems. ASCE J Environ Eng 1999, 125 (12): 1149-1159.

[37] Martin L. A. , W. R. C. Myers. Measurements of overbank flow in a compound river channel, Proceedings of the Institution of Civil Engineers, London, Part 2, 1991, 91: 645-657.

Transmission Coefficient of Wave Permeable Breakwater

Yan Yixin[1,2]   Li Xi[2]

[1]*Key Laboratory of Coastal Disaster and Defence,*
Ministry of Education, Hohai University Nanjing, 210098
[2]*College of Harbor, Coastal, and Offshore Engineering,*
Hohai University, Nanjing, 210098

Abstract: The hydrodynamic characteristics of the wave permeable breakwater are described by new form of semi-empirical equation on transmission coefficient, based on available experimental data. The phenomenon of partial transmission of wave around permeable breakwater is discussed and the aim of the new form equation is to afford a convenient method for engineering design and practice. The equation was compared with the former experimental and theoretical work, e. g., the empirical expression by Van de Meer and by d'Angremond, it is concluded that in the range of the three expression applications, present formula shows a wider agreement with experimental data. Also it proves present expression is valuable to engineering practice.

Key words: permeable breakwater; wave transmission coefficient; breakwater design

1. Introduction

As the water waves propagate from deep water to shallow water, wave profile transforms from symmetrical to unsymmetrical, and this phenomenon is more intensified when the water depth is decreasing. Surface wave breaking happens when water depth reaches a threshold value. Thus a detailed description of nearshore hydrodynamics is of engineering values for wave and structure interaction. Though nonlinear wave theory has been developed to a wide range, it is still difficult to apply in coastal engineering practice. Linear theories and empirical equations are of practical advantage in analyzing wave transformation, such phenomena like breaking, run-up, overtopping. The main theoretical characteristics of wave-permeable breakwater is that wave can transmit through the breakwater and keep wave profile after the breakwater, while wave diffraction and refraction are not depending on the structure is permeable or not. Hydrodynamic characteristics of this permeable identity is analyzed on transmission coefficient in this paper with comparison of former works by Van der Meer and d'Angremond (Van der Meer et al., 2005; d'Angre-

mond et al., 1996).

2. Parameters influencing transmission coefficient of permeable breakwater

The factors affect transmission coefficient is directly related to the structural type of wave permeable breakwaters, as an example in Figure 1, which includes the follows.

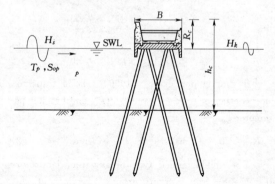

Figure 1   Governing parameters of permeable breakwater

H_s—Significant wave height of incident wave

H_k—Transmitted wave height behind permeable breakwater

T_p—Peak period of incident wave

S_{op}—Wave steepness, $S_{op} = H_s / g T_p^2$

R_c—Crest freeboard

K_t—Transmission coefficient H_k/H_s

h_c—Structure height

B—Structure width

ζ_{op}—Breaker parameter, e. g. , Iribarren number, $\zeta_{op}=h_x/(H_s/gT_p^2)^{1/2}$

h_x—Seaward slope of structure

Wave overtopping and transmission are main phenomena of energy transportation inside the permeable breakwater, partial energy is dissipated by wave breaking in the seaward side, and partial energy is reflected in front of structure. Since permeable breakwaters are presently applied in coastal protection, wave reduction, berthing stability, etc. , it is meaningful to engineering practice to predict transmitted wave height and transmission coefficient.

3. Existing formula and present formula of transmission coefficient

The transmission coefficient is directly related to type of wave permeable structures. Two main types are rubble mound and piles. For rubble mound breakwater, referred to Figure 2, based on extensive experimental data, Van der Meer and Deamen 1994 suggested the following expression using medium diameter of breakwater body stone to describe the transmission coefficient:

$$K_t=a\frac{R_c}{D_{n50}}+b \qquad (1)$$

where a and b are coefficient respectively

$$a=0.031\frac{H_s}{D_{n50}}-0.024$$

$$b=-5.42S_{op}+0.0323\frac{H_s}{D_{n50}}$$
$$-0.017\Big(\frac{B}{D_{n50}}\Big)^{1.84}+0.51$$

d' Angremond et al. (1996) related transmission coefficient to R_c/H_s and B/H_s, and breaker parameter ζ_{op}

$$K_t=-0.4\frac{R_c}{H_s}+0.64\Big(\frac{B}{H_s}\Big)^{-0.31}(1-e^{-0.5\zeta_{op}}) \qquad (2)$$

For rubble mound type breakwater, laboratory test results from various groups (Kramer et al. , 2005; Calabrese et al. , 2002; Gironella et al. , 2002; Hirose et al. , 2002; Melito etal. , 2002; Seabrook and Hall 1998), a wide range of data has been collected and was compared with the above two formula for a functional design of crest freeboard, R_c and structure width, B, about permeable breakwa-

ter. Eq. (1) is mainly applied to rubble mound breakwater, Eq(2) is not limited to rubble mound breakwater, but a wide deviation may appear when applied to other types of permeable breakwater, especially for large transmission coefficient.

Based on available data and above suggested equations, present investigation suggests the following expression:

$$K_t=\frac{1-\dfrac{R_c}{H_s}}{1-\dfrac{R_c}{H_s}}\exp\Big(-0.18\frac{B}{H_s}\Big)[1-\exp(-0.5\zeta_{op})] \qquad (3)$$

Eq. (3) applies to rubble mound permeable breakwater, application range can be extended to large value of transmission coefficient; the limitation is the crest freeboard should be less than incident wave height ($R_c<H_s$), in practice the value of R_c may be larger than H_s, such as a vertical cylinder piled breakwater, thus R_c may be neglected in transmission coefficient calculation since there is no overtopping.

For pile type permeable breakwater, the submerged depth of splash board t_0 plays an important role in defining the transmission coefficient K_t, and overtopping is controlled by crest freeboard R_c, by analyzing the experimental data (Li et al. , 2003; Chen et al. , 2008) . The following formula is introduced:

$$K_t=2.3\frac{1-\xi\dfrac{R_c}{H_s}}{1+\xi\dfrac{R_c}{H_s}}\exp\Big(-0.18\frac{B}{H_s}\Big)(1-e^{-0.5\xi_{op}}) \qquad (4)$$

where $\xi=0.23t_0/H_s$.

Referred to Table 1, and the calculated transmission by Eq. (4) is compared with the experimental data in flume of pile type breakwater, it is concluded Eq. (4) has an acceptable accuracy with experimental data and can be used for estimation of transmission coefficient by pile type breakwater.

Table 1 Experimental data of pile type permeable breakwater in flume under regular waves (Li et al. , 2005; Chen et al. , 2008)

Test No.	t_0/H_s	R_c/H_s	B/H_s	K_t (Calculated)	K_t (Measured)
1	2.21	0.93	3.25	0.42	0.42
2	2.08	1.25	3.45	0.44	0.44

Continued

Test No.	t_0/H_s	R_c/H_s	B/H_s	K_t (Calculated)	K_t (Measured)
3	1.78	0.93	3.25	0.54	0.39
4	1.62	1.25	3.45	0.56	0.43
5	2.17	1.17	4.55	0.34	0.37
6	1.59	1.79	4.59	0.47	0.45
7	0.00	4.63	6.29	0.74	0.59
8	−0.58	5.31	6.43	0.95	0.79
9	2.36	1.17	4.55	0.30	0.37
10	1.78	1.79	4.59	0.42	0.45
11	0.27	4.63	6.29	0.66	0.59
12	−0.31	5.31	6.43	0.83	0.79

4. Comparison of transmission coefficient formulae

The transmission coefficient is directly related to type of wave permeable structures, two main types are rubble mound and piles, in order to compare accuracy of above equations, Eq. (3) is plotted through available data sets in Van der Meer et al. (2005), referred to Figure 2, it shows, in the application range of Eq. (1) and Eq. (2), present investigation Eq. (3) has acceptable agreement with Eq. (1) and Eq. (2); and Eq. (3) has the better agreement with experimental data when ζ_{op} is ranged (0, 5) and H_s/D_{n50} is around ranged (0, 2). For large value of K_t, present expression shows better accuracy and application range.

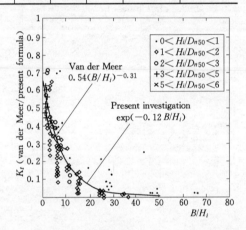 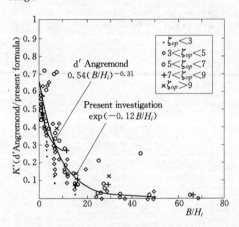

Figure 2   The comparison of expression by Van der Meer, d′Angremond and present investigation

5. Conclusions and discussions

The transmission coefficient is directly related to design parameters in this paper, a new formula is suggested based on former works by Van der Meer and Deamen (1994), d′Angremond et al. (1996), Van der Meer et al. (2005). It is concluded present formula is of a wider range of application in comparison with data sets available, and can be apply to different type of permeable breakwaters of engineering design reference.

Acknowledgment

The authors would like to thank the Fundamental Research Fund (Grant No. 2013B1020100) of Hohai University.

References

[1] Van der Meer J. W. , Briganti R. , Zanuttigh B. , Wang B. . Wave transmission and reflection at low-crested structures: design formulae, oblique wave attack and spectral change. Coastal Engineering, 2005, 52: 915-929.

[2] Van der Meer J. W. , Daemen I. F. R. . Stability and wave transmission at low crested rubble mound structures. Journal of Waterway, Port Coastal and Ocean Engineering, 1994, 1: 1-9.

[3] d′Angremond K. , Van der Meer J. W. , De Jong R. J. . Wave transmission at low crested structures. In: Proceedings of the 25th International Conference on Coastal Engineering.

ASCE, 1996: 2418-2427.

[4] Kramer M., Zanuttigh B., van der Meer J. W., Vidal C., Gironella F. X.. Laboratory Experiments on Low-Crested Breakwaters. Coastal Engineering, 2005, 52.

[5] Gironella X., Sa'nchez-Arcilla A., Briganti R., Sierra J. P., Moreno L.. Submerged detached breakwaters: towards a functional design. Proc. 28th Int. Conf. on Coastal Engineering ASCE, 2002: 1768-1777.

[6] Calabrese M., Vicinanza V., Buccino M.. Large scale experiments on the behaviour of low crested and submerged breakwaters in presence of broken waves. Proc. 28th Int. Conf. on Coastal Engineering, ASCE, 2002: 1900-1912.

[7] Melito I. Melby J. A.. Wave runup, transmission, and reflection for structures armoured with CORE-LOC. Coastal Engineering, 2002, 45: 33-52.

[8] Seabrook S. R., Hall K. R.. Wave transmission at submerged rubble mound breakwaters. Proc. 26th Int. Conf. on Coastal Engineering, ASCE, 1998: 2000-2013.

[9] Li Xi, Yan Yixin. Numerical simulations of nonlinear wave transformation around wave-permeable structure. Journal of Hydrodynamics, Ser. B, 2005, 17 (17): 699-703.

[10] Chen Dechun, Wang Yigang, LI Xi, et al. Experimental study of wave permeable breakwater in Sansha central fishingport, Xiapu, Fujian. Technical report, Hohai University, 2008. (In Chinese)

Mechanism of Sediment Transport in Uni-directional, Bi-Directional Flows

Shuqing Yang

School of Civil, Mining and Environmental Engineering,
Univ. of Wollongong, NSW2522, Australia

Abstract: The aim of the present study is to investigate the relationship between flow strength and sediment discharge. The appropriate definition of energy dissipation rate E in Bagnold's theorem is discussed and it is found that the sediment transport rate g_t in unidirectional and bi-directional flows can be well predicted when E is defined as the product of bed shear stress τ_o and near bed velocity u_*'. Then the linear relationship between u_*'E and the sediment transport rate is examined using measured data. The good agreement between measured and predicted values indicates that the phenomena of sediment transport can be reasonably described by the near bed flow characteristics.

Key words: sediment transport; river flow; tidal flow; breaking waves

1. Introduction

Sediment transport plays an important role in the evolution of river-beds, estuaries and the coast-lines; consequently it exerts a considerable influence on the evolution of the topography of the earth's surface. Therefore, the mechanism of sediment transport is of great interest to hydraulic engineers, coastal engineers, geologists, hydrologists, geographers and so on. Integral modeling of sediment transport from river to marine environment requires a quantitative and universally applicable law governing the motion of the transported sediment in all flow situations ranging from pure current to complex flow in the wave-current situation including irregular and sometimes breaking waves. The equation of sediment transport correlated with local flow parameters such as bed shear stress and near bed velocity is highly demanded by hydrodynamic modelers who are able to precisely determine these flow parameters by solving the Reynolds equations in rivers, estuaries and coastal waters.

For sediment transport in rivers, Chien and Wan (1999) provided a good summary of the well-cited equations, such as those proposed by Einstein (1942), Meyer-Peter and Muller (1948), Bagnold (1966), Yalin (1977), Engelund and Hansen (1972), and Ackers and White (1973). Chien and Wan show that all these equations can be expressed as $\Phi = f(\psi)$, i. e., the dimensionless sediment transport rate Φ

$$\Phi = \frac{g_t}{\gamma_s} \left(\frac{\gamma}{\gamma_s - \gamma} \right)^{\frac{1}{2}} \left(\frac{1}{g d_{50}^3} \right)^{\frac{1}{2}} \qquad (1)$$

is a function of the Shields shear stress parameter,

$$\Psi = \left(\frac{\gamma_s - \gamma}{\gamma} \right) \frac{d_{50}}{RS} \qquad (2)$$

where Φ = Einstein's sediment intensity parameter; g_t = sediment transport rate per unit width in kg/(m · s); d_{50} = median sediment size; γ_s and γ = specific weight of sediment and water, respectively; R = hydraulic radius; S = energy slope; g = gravitational acceleration.

There are other parameters developed to express sediment discharge or concentration, such as the $V^3/(gR\omega)$ proposed by Velikanov (1954), the dimensionless unit stream power, VS/ω by Yang (1996), and the transport-stage parameter, $T = (u_*'^2 - u_{*c}^2) / u_{*c}^2$ by van Rijn (1984), in which V = depth-averaged velocity, ω = sand fall velocity and $u_*' = (g^{0.5}/C')V$ = bed-shear velocity related to grains; $C' = 18\log(12R/3/d_{50})$ = Chezy-coefficient related to grains; u_{*c} = critical shear velocity.

Yang and Lim (2003) developed the following equation to express the sediment transport

$$\Phi = k \left(\frac{\gamma}{\gamma_s - \gamma} \right)^{\frac{1}{2}} T'_T \qquad (3)$$

where k is a dimensionless coefficien = 12.5, T' is Yang and Lim's (2003) dimensionless parameter.

The objectives of this paper are (1) to express Yang and Lim's (2003) empirical equation in a more general form; (2) to extend the basic equation to wave-current motion and longshore sediment transport; (3) to determine coefficients for different cases; and (4) to provide a general relationship between sediment transport and local flow characteristics to numerical modelers.

2. Mechanism of sediment transport

Bagnold (1966) developed a sediment transport function from the power concept. He considered the relationship between the available energy, E to an alluvial system and the rate of work being done by the system in transporting sediment. The total load, g_t is expressed in the following way from the power concept

$$g_t = \frac{\gamma_s}{\gamma_s - \gamma} k_1 E \frac{u_s}{\omega} \qquad (4)$$

where u_s = mean transport velocity of sediment, k_1 = efficiency coefficient; tan α = ratio of tangential to normal shear force.

Bagnold (1966) expressed the available energy as follows

$$E = \tau_0 V \qquad (5a)$$

Yalin (1977) notes that Eq. (5a) means the loss of potential energy of the flow per unit area and time because the "available power" termed by Bagnold can be expressed as follows

$$E = \tau_0 V \approx \gamma S \int_0^h u \, dy = \gamma S q \qquad (5b)$$

where q = water discharge per unit width, h = water depth.

Different from Bagnold's assumption, some researchers found empirically that the sediment transport is closely related to the energy dissipation near the bed (Yalin, 1977; Cheng, 2002), viz. the product of bed shear stress τ_0 and shear velocity u_* with the following form

$$E = \tau_0 u_* \qquad (5c)$$

where $u_* = (gRS)^{0.5} = (\tau_0/\rho)^{0.5}$. In Eq. (5c), the near bed velocity is represented by the over-all shear velocity, u_*. However, the Bagnold's assumption shown in Eq. (5b) indicates that the importance of any unit water volume is identical to sediment transport, regardless of its position. Obviously, this is not correct. Yalin (1977) noticed the shortcomings in Bagnold's assumption, and he expressed the near bed velocity using the shear velocity as shown in Eq. (5c). From the relationship of $u_* = (\tau_0/\rho)^{0.5}$, one can conclude that the Yalin's approach actually indicates that the energy supporting the sediment transport is proportional to $\tau_0^{3/2}$, in other words the bed shear stress is the sole parameter to express the energy

supporting the sediment transport. According to Einstein (Chien et al, 1999) the total shear stress, τ_o can be divided into two components, viz. bed-shear related to grains, $\rho u_*'^2$ and sand wave shear. The bed-shear velocity related to grains u_*' rather than the total shear velocity, u_* or mean velocity, V, plays a dominant role for sediment transport. Therefore, it is acceptable to express the "available energy", E in Eq. (4) in the following way

$$E = \tau_0 u_*' \qquad (6)$$

In Eq. (6), if the bed shear stress τ_0 is replaced by bed-shear related to grain $\tau'_0 (= \rho u_*'^2)$ and u_s is assumed to be proportional to the mean velocity V, then Eq. 4 indicates virtually that the total sediment discharge g_t is proportional to V^4/ω, or the sediment concentration c relies on $V^3/(hg\omega)$ that becomes the Velikanov's parameter.

The mean transport velocity of sediment, u_s in Eq. (4) is defined as follows (Yalin, 1977)

$$u_s = \frac{1}{\lim\limits_{\varepsilon \to 0} \int_\varepsilon^h c_y \, dy} \lim_{\varepsilon \to 0} \int_\varepsilon^h c_y u \, dy \qquad (7)$$

where y = distance from the bed, u = local velocity at level y, c_y = volumetric concentration of particles at level y. Bagnold (1966) assumed that $u_s = V = \int_0^h u \, dy / h$, this is equivalent to the assumption that the concentration c_y does not vary with y as commented by Yalin (1977). Therefore, it is rational to assume that the mean transport velocity of sediment is proportional to the near bed water velocity, i. e.,

$$u_s = a_1 u_*' \qquad (8)$$

in which a_1 = coefficient.

Substituting Eq. (6) and Eq. (8) into Eq. (4), one obtains a sediment transport formula

$$g_t = \frac{\gamma_s}{\gamma_s - \gamma} k u_*' \cdot \frac{E - E_c}{\omega} \qquad (9)$$

where $k = k_1 a_1$, E_c is introduced to express the critical condition for sediment motion. Eq. (9) states that the sediment transport rate depends on the energy dissipation on the bed, E and near bed velocity, u_*', the net transport will be always in the same direction as the movement of u_*' that is expressed as follows (Yang et al., 2003)

$$u_*' = \frac{V}{2.5 \ln \frac{11R}{2d_{50}}} \qquad (10)$$

3. Sediment transport in uni-directional flows

By simply assuming $u_*' (E - E_c) = \tau_o (u_*'^2 - u_{*c}^2)$, one can obtain Yang and Lim's (2003) equation from Eq. (9)

$$g_t = k \frac{\gamma_s}{\gamma_s - \gamma} \tau_o \left(\frac{u'^2_* - u'^2_{*c}}{\omega} \right) \qquad (11)$$

In order to test which hydraulic parameter is suitable for the expression of total sediment discharge, Barton and Lin's (1955) data set is particularly selected to compare the correlation coefficients of existing hydraulic parameters with measured sediment discharge. Figure 1 (a) shows the relationship between the sediment concentration c and the hydraulic parameters of VS/ω, $V^3/(gR\omega)$ and $T = (u'^2_* - u^2_{*c})/u^2_{*c}$, respectively. Yang's (1996) unit stream power parameter, VS/ω gives the correlation coefficient of 0.942; Velikanov's parameter $V^3/(gR\omega)$ provides correlation coefficient of 0.948; the dimensionless parameter, T developed by van Rijn (1984) is correlated with the sediment concentration for a coefficient of 0.943. Figure 1 (b) shows the measured

sediment discharge g_t versus the stream power $\tau_o V$ and $\tau_o u_*$. Figure 1 (c) exhibits that a one-to-one and linear relationship between g_t and T'_T can be provided by Yang and Lim's (2003) parameter, which gives the highest correlation coefficient of 0.973.

In Table 1, all Gilbert's observations are used in the calculation of correlation coefficients for avoidance of prejudice; Column 1 shows the number of experimental runs done by Gilbert and Column 2 indicates the median sediment size, Columns 3-5 list the calculated correlation coefficients between the measured sediment concentration c and VS/ω, $V^3/(gR\omega)$ and T developed by Yang, Velikanov and van Rijn, respectively; Columns 6-7 exhibit the correlation coefficients of the measured sediment discharge Φ or g_t with parameters Ψ and T'_T proposed by Einstein and Yang and Lim, respectively; the last Column shows the ratio

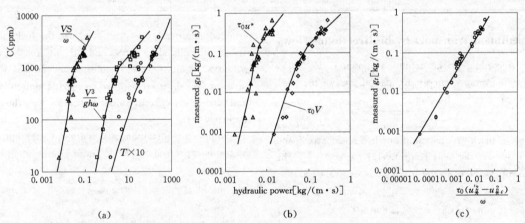

(a)   (b)   (c)

Figure 1   Relationship between the measured sediment discharge and various hydraulic parameters based on Barton and Lin's (1955) data

of difference between the values in Columns 3 and 7 to the values listed in Column 3. The last row of Table 1 shows the mean values, it can be seen that the parameter T'_T achieves obviously the highest correlation coefficient in all cases, in other words, sediment transport is strongly related to T'_T. Gilbert's data confirms the conclusion drawn from Figure 1.

In the experiments, bakelite ($\rho_s = 1.56\text{t/m}^3$ and $d_{50} = 1.05$, 0.67mm), sand ($\rho_s = 2.67\text{t/m}^3$ and $d_{50} = 0.7$mm) and Nylon ($\rho_s = 1.14\text{t/m}^3$ and $d_{50} = 3.94$mm) were used. The measured contact-load transport rate is shown in Figure 2. Figure 2 also includes the measured sediment discharge at high velocity (1-2m/s) from filed (Voogt et al. 1991), the estimated energy slopes for Krammer tidal chan-

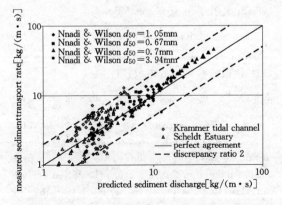

Figure 2   comparison of sediment discharge measured from high shear stress conduit flows and high velocity channel flows with computed results from Eq. (11)

nel and Scheldt Estuary are 0.18‰ and 0.13‰, respectively. Figure 2 demonstrates a good agreement between Eq. (11) and the measured sediment transport rate.

Table 1   **Correlation coefficients between sediment transport and hydraulic parameters based on Gilbert's (1914) data**

No. of runs (1)	d_{50} (mm) (2)	c & VS/ω (3)	c & $V^3/(gR\omega)$ (4)	c & T (5)	Φ & Ψ (6)	g_t & T'_T (7)	$[(7)-(3)]/(3)$
62	0.305	0.863	0.137	0.265	−0.384	0.972	12.6%
207	0.37	0.845	0.508	0.528	−0.800	0.921	9.0%
235	0.51	0.781	0.549	0.491	−0.768	0.956	22.4%
116	0.79	0.901	0.747	0.669	−0.659	0.988	9.6%
47	1.71	0.903	0.758	0.688	−0.699	0.973	7.8%
35	3.08	0.893	0.681	0.503	−0.895	0.972	8.8%
68	4.94	0.926	0.880	0.846	−0.808	0.989	6.8%
Mean		0.873	0.608	0.570	−0.716	0.967	11%

4. Sediment transport by bi-directional flows

According to the definition shown in Eq. (6), the time-averaged energy dissipation rate on the bed, E is expressed as follows

$$E = \overline{\vec{\tau}_{b(t)} \vec{u}_{b(t)}} \qquad (12)$$

where τ and u are instantaneous bed shear stress and velocity near the bed, respectively; t is the time; the subscript b refers to the bed. $u_{b(t)}$ is expressed as follows

$$\vec{u}_{b(t)} = \vec{u}_{w(t)} + \vec{u}_{cu} = \vec{u}_{um} \sin(\omega_1 t) + \vec{u}_{cu} \qquad (13)$$

where ω_1 is the angular frequency, $\vec{u}_{um}$ is the amplitude of harmonic near-bed orbital velocity near the bed. The widely accepted expression of instantaneous bed shear stress is shown as follows:

$$\vec{\tau}_{b(t)} = \frac{1}{2} f_w \vec{u}_{b(t)} |\vec{u}_{b(t)}| \qquad (14)$$

in which f_w is wave friction factor. Substituting Eq. (14) and Eq. (13) into Eq. (12) yields

$$E = \frac{f_w}{2T} \int_0^T |\vec{u}_{b(t)}| |\vec{u}_{b(t)}^2| dt = \frac{f_w}{2T} \int_0^T [u_{cu}^2 + u_{um}^2 \sin^2(\omega_1 t) + 2 u_{cu} u_{um} \sin(\omega_1 t) \cos \beta_1]^{\frac{3}{2}} dt \qquad (15)$$

where $\beta_1 =$ the angle between the current and the direction of wave propagation; T is the wave period. The integral in Eq. (17) cannot be solved analytically. For simplification it is assumed that the bed shear stress is superposed as follows:

$$\vec{\tau}_{b(t)} = f(\vec{\tau}_{w(t)} + \vec{\tau}_{cu}) \qquad (16)$$

where f is a function of u_{um}/u_{cu} and needs to be determined empirically, $\vec{\tau}_{w(t)}$ and $\vec{\tau}_{cu}$ are the bed shear stresses introduced by waves and current, respectively. Lodahl et al.'s (1998) experiments showed that the mean bed shear stress in wave-current motion may retain its steady-current value, indicating that the coefficient $f \approx 1$.

Thus, the instantaneous energy dissipation on the bed is expressed by use of Eq. (13) and Eq. (16) as follows:

$$\vec{\tau}_{b(t)} \vec{u}_{b(t)} = f(\vec{\tau}_{w(t)} \vec{u}_{w(t)} + \vec{\tau}_{cu} \vec{u}_{w(t)} + \vec{\tau}_{w(t)} \vec{u}_{cu} + \vec{\tau}_{cu} \vec{u}_{cu}) \qquad (17)$$

the averaged energy dissipation E over one wave period can be obtained:

$$E = f\left(\overline{\vec{\tau}_{w(t)} \vec{u}_{w(t)}} + \overline{\vec{\tau}_{cu} \vec{u}_{cu}}\right) = f\left(\overline{\vec{\tau}_{w(t)} \vec{u}_{w(t)}} + \vec{\tau}_{cu} \vec{u}_{cu}\right) \qquad (18)$$

Madsen et al. (1988) obtained

$$\overline{\vec{\tau}_{w(t)} \vec{u}_{w(t)}} = \frac{1}{4} \rho f_w \cos \varphi \left(\frac{2\pi A}{T}\right)^3 \qquad (19)$$

where A is the water particle semi-excursion near the bed, f_w is the wave friction factor, φ is the phase lead of the near bottom wave orbital velocity which is expressed as follows:

$$\tan \varphi = \frac{\pi/2}{\ln \dfrac{300 k u_{*um}}{\omega_1 2 d_{50}} - 1.15} \qquad (20)$$

in which $u_{*um} =$ the shear velocity driven by waves. The energy dissipation by current is:

$$\vec{\tau}_{cu} \vec{u}_{cu} = \tau_{cu} u'_{*cu} \qquad (21)$$

Thus the time-averaged rate of energy dissipation in

wave-current conditions is

$$E = f\left[\frac{\rho}{4}f_w\cos\varphi\left(\frac{2\pi A}{T}\right)^3 + \tau_{cu}u'_{*cu}\right] \quad (22)$$

Nielsen (1992) gives in the following form

$$f_w = \exp\left[5.5\left(\frac{2d_{50}}{A}\right)^{0.2} - 6.3\right] \quad (23)$$

The bottom orbital excursion amplitude A may be computed by use of

$$A = H/[2\sinh(2\pi h/L)] \quad (24)$$

where h = water depth, L = wave length, $H = H_s/\sqrt{2}$ for random wave, H_s being the significant wave height. Substituting Eq. (22) into Eq. (9) leads to

$$\vec{g}_t = k\frac{\gamma_s}{\gamma_s - \gamma}\frac{\vec{u}'_{*cu}}{\omega}\left\{f\left[\frac{\rho}{4}f_w\cos\varphi\left(\frac{2\pi A}{T}\right)^3 + \rho u'^3_{*cu}\right] - \rho u^3_{*c}\right\} \quad (25)$$

The procedure of calculation is shown as follows.

(1) Determine u^2_{*c} from the Shields curve based on d_{50}.

(2) Calculate the current shear velocity u'_{*cu} with Eq. 10 if the mean velocity is measured or if the local velocity u at elevation y is measured, the following equation is suggested for assessment of the bed-shear velocity u'_{*cu}.

$$\frac{u}{u'_{*cu}} = 2.5\ln\left(\frac{30y}{2d_{50}}\right) \quad (26)$$

(3) Calculate f_w and A using Eq. (23) and Eq. (24), using wave height, wave period and water depth.

(4) Calculate the sediment discharge g_t using Eq. (25) and assuming $k = 12.5$ and $f = 1$.

Van Rijn et al.'s (1993) experimental data is used for the verification. the current velocities varied in the range of 0.1-0.5m/s, irregular waves following and opposing current with a peak period of 2.5s were generated. The significant wave height varied in the range of 0.075m to 0.18m, the water depth was about 0.5m. The calculated fall velocities for $d_{50} = 100\mu$m and $d_{50} = 200\mu$m are 0.616cm/s and 1.92cm/s, respectively. The estimated critical shear velocities are $u_{*c} = 1$cm/s and 1.31cm/s, respectively. Figure 3 shows the comparison of measured and predicted sediment transport rate. The positive sign shown in Figure 4 represents sediment transport following the wave direction and the negative sign means the sediment transport against the wave direction. Without the knowledge of measured velocity profiles and ripple heights, Eq. (25) also shows reasonable results, a-

bout 52% of the predicted transport rates are within the discrepancy ratio of 2 of the measured values. Besides, it can be seen from Figure 4 that significant errors occur in the region of low transport rates, the discrepancy could be ascribed to two main reasons: 1) the measurement error is relatively larger in the low transport region, this is why researchers generally exclude the data with low sediment concentration or discharge for comparison (Yang, 1996); 2) the low transport region is actually the wave-dominated regime, as mentioned in such case f is less than 1, but for simplification $f = 1$ is employed to estimate the transport rate, this may introduce some errors. Nevertheless, the results of Eq. (25) are still encouraging.

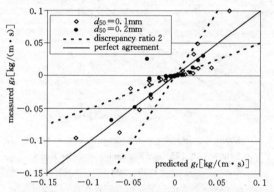

Figure 3 Comparison between the rate of total sediment transport in combined current and wave conditions from Eq. (25) and Van Rijn et al.'s data (1993)

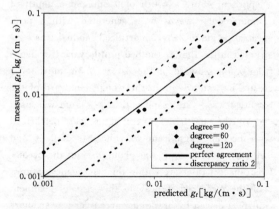

Figure 4 Comparison between total sediment discharge between Eq. (25) and Van Rijn and Havinga's data (1995) for different current-wave angles

Van Rijn and Havinga (1995) carried out experiments in a wave-current basin, in which the water depth was about 0.4m for all tests, and three different wave conditions were generated; significant wave height = 0.07m, 0.1m and 0.14m for three wave directions—60°, 90° and 120° between wave orthogonal and current direction. Irregular waves with a single-topped spectrum and a peak period of 2.5s were generated. Sediment of a median diameter of $100\mu m$ was used. Figure 5 displays good agreement between the measured sediment discharge and that calculated with Eq. (25), in which the coefficient $f = 1$ and $k = 12.5$ are applied. The fall velocity and critical shear stress are given as 0.616cm/s and 1cm/s, respectively.

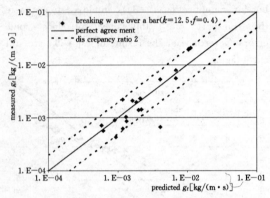

Figure 5 Comparison of sediment transport rate between Eq. (25) and Grasmeijer and Van Rijn's (1999) data

In 1999, Grasmeijer and van Rijn measured sediment transport by breaking waves over a near shore bar. The experiments were conducted in a flume with a length of 45m, a width of 0.8m and a depth of 1.0m. irregular waves were generated with a peak spectral period of 2.3s, an artificial sandbar was constructed in the flume, the bed profile varied in depth from 0.6m seaward of the bar to 0.3m at the bar crest. The water depth in the trough landward of the bar crest was 0.5m, sand with $d_{50} = 95\mu m$ was used to form the sand bar. Two test series were performed with incoming significant wave heights of 0.16 and 0.19m, respectively. It was about 95% of the predicted transport rates are within the discrepancy ratio of 2 of the measured values, in this case of sediment transport driven by breaking waves over a near shore bar no other methods are available for comparison in the literature, as far as the author knows.

5. Conclusions

To investigate the general relationship, different expressions of E in Bagnold's theorem are dis-

cussed. A new relationship has been examined and the following conclusions can be drawn from this study.

(1) In open channel flows, the total rate of sediment transport Φ is highly correlated to the proposed parameter $T'_T = \tau_o (u'^2_* - u^2_{*c})/[(\gamma_s - \gamma)\omega \sqrt{gd^3_{50}}]$, among the existing hydraulic parameters including the widely cited parameters, such as VS/ω, $V^3/gh\omega$ and T, the highest correlation coefficient is achieved by the new parameter when the same experimental data is used for comparison. A linear relationship exists between $\dot{\Phi}$ and T'_T, and the proportionality factor k is found to be 12.5.

(2) For breaking wave condition, or the wave-dominant regime, it is found that $k = 12.5$, but $f = 0.4$, providing good estimation for sediment transport under breaking waves over a nearshore bar.

(3) This study establishes the general expression of sediment transport for numerical modelers as shown in Eq. (9), It is suggested that the users calibrate the proportionality k using measured data because the definition of near bed velocity generally differs from each other.

References

[1] Ackers P., White W. R.. Sediment transport: new approach and analysis. *J. Hydr. Div. ASCE*, 1973, 99 (11), 2041-2060.

[2] Bagnold R. A.. An approach to the sediment transport problem from general physics. *Geol. Survey Professional paper*, 1966, 422-I, U. S. government printing office, Washington.

[3] Barton J. R. and Lin P. N.. A study of the sediment transport in alluvial streams. Report No. 55 JRB2, Civil engrg depart., Colorado Colleage, Fort Collins. 1955.

[4] Bodge K. R.. Short term impoundment of longshore sediment transport. Ph. D. Disertation, Gainesville; Univ. of Florida. 1986.

[5] Chien N. and Wan Z.. Mechanics of Sediment Transport. ASCE, Reston, Va. 1999.

[6] Einstein H. A.. Formulas for the Transportation of Bed Load. *Trans. Soc. Civ. Engrg.*, 1942, 107: 561-597.

[7] Engelund F., Hansen E.. A Monograph on Sediment Transport in Alluvial Streams.

Teknisk Forlag, Copenhagen, Denmark. 1972.

[8] Gilbert G. K.. The transportation of debris by running water. USGS Professional Paper 86. 1914.

[9] Grasmeijer B. T. , Van Rijn L. C.. Transport of fine sands by currents and waves. III: Breaking waves over barred profile with ripples. *J. Waterway, Port, Coastal, and Ocean Engrg.* , 1999, 125 (2), 71-79.

[10] Lodahl C. R. , Sumer B. M. , Fredsoe, J.. Turbulent combined oscillatory flow and current in a pipe. *J. Fluid Mechanics*, 1998, 373: 314-350.

[11] Madsen O. S. , Poon Y-K. , Graber, H. C.. Spectral wave attenuation by bottom friction theory. Proc. of 21 Intern. Coastal Engrg. Conf. , Malaga, Spain, 1988, 1: 492-504.

[12] Meyer-Peter E. , Muller R.. Formula for bed load transport. Proc. 2nd Meeting, IAHR, Stockholm, 1948, 6.

[13] Nielsen P.. Coastal Bottom boundary layers and sediment transport. World Scientific, Singapore. 1992.

[14] Van Rijn L. C. , Martin W. C. N. , Kaay T. V. D. , Nap E. , Kampen A. V.. Transport of Fine Sands by Currents and Waves. *J. Waterway, port and ocean engineering*, ASCE, 1993, 119 (2): 123-143.

[15] Van Rijn L. C.. Sediment transport Part II: suspended load transport. *J. of Hydr. En-grg. ASCE*, 1984, 110 (11): 1613-1641.

[16] Van Rijn L. C.. Two-dimensional vertical mathematical model for suspended sediment transport by currents and waves. Rep. S488 part IV, Delft Hydraulics, Delft. The Netherlands. 1985.

[17] Van Rijn L. C.. Handbook of sediment transport in currents and waves. Rep. H461, Delft Hydraulics, Delft. The Netherlands. 1989.

[18] Van Rijn L. C. , Havinga F. J.. Transport of fine sands by currents and waves. II. *J. Waterway, port and ocean engineering*, ASCE, 1995: 121 (2): 123-133.

[19] Velikanov M. A.. Gravitational theory for sediment transport. *J. of Science of the Soviet Union*, Geophysics, 1954, 4. (in Russian) .

[20] Voogt Leo. , Van Rijn L. C. , Betg J. H. V. D.. Sediment transport of fine sands at high velocities. *J. Hydr. Engrg. ASCE*, 1991, 117 (7) 869-890.

[21] Yalin M. S.. Mechanics of sediment transport. Pergamon Press, Oxford. 1977.

[22] Yang C. T.. Sediment transport: theory and practice. McGraw-Hill International Editions. 1996.

[23] Yang S-Q. , Lim S-Y. Total Load Transport Formula for Flow in Alluvial Channels. *J. Hydr. Engrg.* , ASCE, 2003 (1): 68-72.

Coupling Physical and Numerical Models: Example of the Taoussa Project (Mali)

Sébastien Erpicum[1]   Benjamin J. Dewals[1,2]   Jean-Marie Vuillot[3]
Pierre Archambeau[1]   Michel Pirotton[1]

[1]*Research unit HACH, ArGEnCo Department, University of Liege, Belgium*
[2]*Fund for Scientific Research - F.R.S.-FNRS*
[3]*Coyne et Bellier - Tractebel Engineering, France*

Abstract: Physical modeling and numerical modeling are two efficient analysis approaches in hydraulic engineering. The interactive application of both methods is obviously the more effective response to most flow problems analyses. Indeed, it enables combining the inherent advantages of both approaches, which are complementary, while being beneficial to the delays as well as the quality of the analysis. This paper presents the results of a successful application of such a combined numerical - physical study carried out by the Laboratory of Engineering Hydraulics of the University of Liège on behalf of Coyne et Bellier (Tractebel Engineering). It concerned, at the stage of detailed draft, the hydraulic study of the Taoussa Project on the Niger River in Mali. This project accounts for a 1.3km long embankment dam, a ship lock, a hydroelectric power plant and a gated spillway (10 gates) with a downstream stilling basin. The studies, performed in less than 6 months, focused on the flow characteristics at the scale of the reservoir and the river, using the numerical approach, as well as on hydrodynamic details in the spillway using a large scale factor physical model with boundary conditions defined on the basis of numerical modeling.

Key words: hydraulics; physical modeling; numerical modeling; coupled approach

1. Introduction

Physical modeling has always been widely used in hydraulic engineering for research as well as project design. It consists in building, with a degree of complexity dependent on the goals of the application, a model of the hydraulic systems to be studied[1]. This model enables to reproduce, in a controlled environment making a qualitative as well as quantitative analysis possible, the whole complexity of the flows occurring in these systems.

The main advantage of the physical models lies in their intrinsic capacity to reproduce, provided that adapted scale factor and similarity laws are applied, the complete flow features, even very complex. Moreover, they easily enable physical interactions with these flows and are very useful for promoting and communicating on a project. Indeed, they irrefutably demonstrate not only to the specialist engineer but also to each individual how a structure or a procedure will work[1].

The general use of numerical modeling in hydraulic engineering dates from no more than about twenty years. It has developed following the set up of increasingly representative mathematical models as well as more and more robust and accurate resolution schemes, coupled to a tremendous increase and popularization of the numerical computing potentialities.

Just like the efficiency and the pertinence of a physical model depend on a judicious choice of the scale factor and the similarity laws, carrying out representative and reliable flow numerical modeling depends on a number of parameters, sometimes still little recognized such as the choice of the mathematical model, of the resolution scheme, of their parameters, the quality and the adequacy of the input data, and in particular the boundary conditions, or the modeler experience.

The main advantages of the numerical models are their low application costs compared to the costs of a scale model building, especially for large study areas, their flexibility, in terms of geometry for example, or also the ease with which the evolution of the unknowns can be followed everywhere in the studied system.

Today, the coupled application of physical and numerical modeling is obviously the more effective response to most flow problems analyses. Indeed, it enables combining the inherent advantages of both approaches, which are complementary, while being beneficial to the delays as well as the quality of the analysis.

This paper presents the results of a successful application of such a combined numerical—physical study carried out by the Laboratory of Engineering Hydraulics of the University of Liège on behalf of Coyne et Bellier (Tractebel Engineering). It concerned, at the stage of detailed draft, the hydraulic study of the Taoussa Project on the Niger River in Mali.

The ability of the numerical models to consider large areas has been used to perform the project general hydraulic analysis and to precisely define the flow conditions near the spillway. This last one, which is a structure with complex hydrodynamics, has then been studied in details using a physical model with a large scale factor as the area to model has been drastically limited thanks to considering the numerical results.

2. Project features and goals of the study

The Taoussa hydraulic project (Figure 1) is mainly made of an embankment dam around 15m high and 1300m long across the valley of the Niger River, 130km upstream of the town of Gao in the eastern part of Mali.

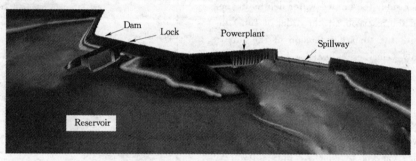

Figure 1   3D view of the Taoussa Project

The dam is equipped with a hydroelectric power plant counting for 5 groups with an equipment discharge of 75m³/s each, a 12m wide lock and a spillway with a release capacity of 3100 m³/s. The spillway is made of a Creager weir at level 252.75m and divided in 10 gated notches 8.5m wide, separated by 3.3m wide piers. A stilling basin 55m long is located downstream of the weir. Its bottom elevation is at level 244m and there is a downstream extremity step 2m high. The stilling basin is divided longitudinally in 5 identical parts by in between walls at level 255m, extending the weir piers.

The spillway has to control the reservoir level in between levels 254.50m and 258.75m in annual usual operation and normal conditions. During high flood events (i. e. discharge higher than 2500m³/s / 100 – year flood), the reservoir level could rise to a higher level, in particular up to level 260.80 m for the 1000 – year flood of 3100m³/s with 8 gates opened.

The hydraulic studies of the project have been carried out successively by two complementary approaches: a first step realized on the basis of numerical modeling to study the flows at the scale of the reservoir and the river, and a second experimental investigation, using a scale model, to analyze in details the spillway operation conditions.

The numerical study aims at analyzing the large scale flow conditions in the reservoir during the spillway operation and at looking to the hydrodynamic interactions between the dam, the lock and the hydroelectric power plant. It also has to help in defining the limited extension of the physical model of the spillway and, if necessary, in determining the water supply conditions to be reproduced in order to make the model representative of the real flow conditions. In the same framework, it has been used to compute the value of the water depth downstream boundary conditions to be prescribed in the physical model.

On another side, the goal of the experimental investigations was to validate and optimize the hydraulic design and the dimensions of the gated spillway and the stilling basin, to verify the flow and release conditions downstream of the project structures, including the temporary derivation stage, and to analyze the operating scheme of the spillway gates to optimize the river discharges release under varied reservoir levels and for normal operation conditions.

3. Large scale numerical model

3.1   Flow solver

The numerical study has to consider the flow conditions at large scale in the river and the reservoir as well as hydrodynamic details near the dam and the associated structures. The discretization of the numerical model has thus to cover a large area with a locally high accuracy.

On another side, the limited water depths in the river reach and the reservoir (in the order of 10m) in comparison with the surface extension of the flow, coupled to flow velocities essentially horizontal because of the geometry of the projected structures, do not require the use of a full 3D modeling approach.

On the basis of these considerations, the numerical studies have been carried out with the depth averaged flow solver WOLF2D, using a square mesh grid with a cell size varying by blocks. This software, part of the WOLF modeling system, solve by a finite volume method the depth – integrated Navier – Stokes equations, capable in mean values to reliably represent all flow specificities and taking explicitly into account turbulence effects with suited mathematical models[2-5].

3.2   Data, layout and validation

Two limnimetric stations are located on the river, upstream and downstream of the dam location, providing water level and discharge data.

On the basis of the topographic data provided by Coyne & Bellier, three digital elevation models (DEM) of the study area have been built. A DEM is indeed the basic data to perform flow numerical modeling. The first DEM was representative of the whole of the actual Niger River bed between the two limnimetric stations. The second one was limited to the future reservoir and the third one to the area downstream of the future structures down to the downstream limnimetric station.

In order to build these DEMs, the topography

and bathymetry data available have been grouped together in a single file as points defined by x, y and z coordinates. Interpolation using the adequate tool in WOLF2D allows generating a regularly distributed DEM on the whole area of interest. Superimposing the DEM and the digital drawings of the Taoussa project, the geometry of the structures has been accurately added to the topography.

The DEM of the Niger River in its present state covered an almost 7km long reach upstream of the downstream limnimetric station with more than 2550000 meshes (Figure 2).

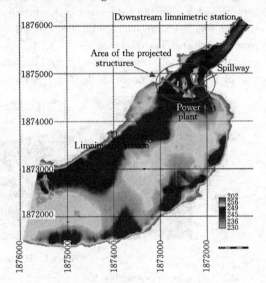

Figure 2   DEM of the Niger reach considered in the study

In order to model the flow on the different DEMs, a computation grid has been defined in each case with a regular square meshes size varying by blocks depending on the hydrodynamics features to model and of the modeling objective. For example, in the upstream and the downstream models (Figure 3), meshes sizes of 12m, 4m and 2m have been used with a progressive refinement of the grid close to the spillway, where the flow velocities are the most important.

The discharge is the upstream boundary condition of the numerical models. It is injected in the computation domain directly as a specific discharge at the downstream extremity of the stilling basin (downstream model) or on the whole of the river width using a numerical infiltration area (present river and reservoir models), i. e. without choosing a preferen-

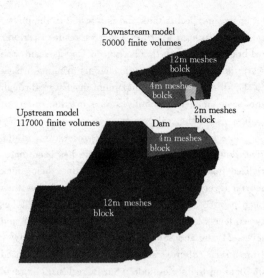

Figure 3　Computation grids and blocks for the upstream and the downstream models

real measured one with an accuracy in the order of 50 cm in the discharge variation range of the river, although the accuracy on the bathymetry can be locally only several meters (Table 1).

Table 1　　Real and computed Q/Z relations

Q (m^3/s)	Z down station (m)	Z up station		
		Real	Num model (m)	Δ num-real
415	251. 29	251. 42	252. 26	0. 84
1085	252. 43	252. 62	252. 99	0. 37
1500	253	253. 22	253. 46	0. 24
2484	254. 14	254. 42	254. 61	0. 19
3100	254. 74	255. 05	255. 24	0. 19

tial flow direction. The turbines discharge is removed from or injected in the computation domain using the same numerical infiltration artifice upstream and downstream of the hydroelectric power plant.

For each simulation, the free surface level is imposed as a downstream boundary condition, on the spillway crest or in the river cross section close to the downstream limnimetric station. For the reservoir simulations, the free surface level imposed on the spillway crest has been defined in order to reach in the reservoir the theoretical head corresponding to the injected discharge, depending on the predicted head/discharge curve of the weir. This enables to reach water depths and thus flow velocities in the computation domain representative of those which will be created in reality by the spillway. In all other cases, the free surface level imposed as the downstream boundary condition directly comes from the water rating curve of the downstream limnimetric station, provided by Coyne & Bellier.

Before the application of the numerical model for the large scale hydraulic study of the Taoussa Project, modeling of the flow in the present river state between the two existing limnimetric stations has been carried out for several discharges. These computations enable to assess the DEM accuracy without considering the project structures and to fit the roughness coefficient value in the river bed. The computed surface elevation Z discharge Q curve for the upstream limnimetric station is finally the same as the

3. 3　Numerical results

The upstream numerical simulations enable to characterize the flow in the reservoir for various operation conditions of the project structures. They are also at the basis of propositions to modify locally the bathymetry in order to improve the alimentation conditions of the power plant and the spillway and they allow quantifying the amplitude of the water level variations along the dam upstream face, because of head losses and velocity changes.

The downstream numerical simulations enable to determine the water depth/discharge relation at the downstream limit of the physical model (boundary condition). In particular, they showed that the water depth at this specific place cannot be directly interpolated from the data of the two limnimetric stations surrounding the project area for the same discharge. Indeed, at this specific location, clearly identified hydrodynamic situations lead to free surface levels lower than those of both limnimetric stations when the power plant is not in use. Similarly, the water levels at the lock and the power plant toe have been quantified for varied operation conditions of the spillway.

The water levels along the right bank cofferdam during the temporary derivation stage have been computed. They are significantly higher than those observed in the current situation, because of the river cross section reduction as a result of the temporary cofferdam.

Finally, the layout of the physical scale model and the location of the water alimentation wall in the model upstream basin have been defined on the basis of an analysis of the numerical results of the reservoir flow modeling. The flow fields in the reduced surface of the physical model and in the whole of the reservoir have been compared in order to validate the representativeness of the scale model regarding the spillway alimentation conditions.

4. Local physical model

4. 1　Features

A scale model representing the whole spillway and the power plant water intakes, a part of the reservoir (105m long and 300m wide) upstream of the gated weir crest and a 160m long reach of the natural river downstream of the stilling basin has been built with a scale factor of $1/40^e$ considering a Froude similarity (Figure 4 and Figure 5). It has been possible to consider this large scale factor thanks to the considerable restriction of the model area in the reservoir as well as downstream of the stilling basin, using the preliminary numerical modeling results.

Figure 4　General view of the scale model

Figure 5　Detail of the gated weir and the stilling basin

The water system is a closed circuit with two regulated pumps, a buried 400m³ storage tank and pressurized pipes. In order to gain a discharge and velocities repartition at the model entrance complying with the numerical results, the water is discharged in

an upstream specific tank and is injected in the physical model through a permeable screen made of perforated bricks and a synthetic fiber screen. Downstream, the water is collected in a specific channel bringing it back to the storage tank. The maximum discharge through the scale model is 350L/s, i. e. in the order of 3500m³/s for the real project.

The fixed bathymetry in the reservoir and the river downstream of the stilling basin has been materialized inside a waterproof tank with a 5cm thick mortar layer on a base of concrete bricks. It has then been painted with latex. It has been reproduced between levels 244. 00 and 262. 00 in the reservoir and levels 240. 00 and 256. 00 downstream of the spillway from a series of cross section profiles regularly spaced of 20cm on the scale model (8m in real size), drawn from the topography files provided by Coyne & Bellier. The spillway and the stilling basin have been built in several parts using PVC and aluminum. The different weirs have been profiled using synthetic resin. All these elements have been assembled on the floor of the waterproof tank delimiting the physical model, prior to the bathymetry realization. Each notch of the spillway is equipped with a mobile gate and can be blocked by a cofferdam. The five pipes of the hydroelectric power plant have been equipped with a valve in order to be able to represent varied operating configurations. All these valves discharge in a downstream concentration tank which is linked to a pipe with a discharge meter and a regulating valve. This last one enables to accurately regulate the head losses in the system and thus the amplitude of the discharge in the turbines.

The upstream boundary condition is the discharge injected in the scale model. The water level in the reservoir regulates automatically depending on the release capacity of the spillway and the hydro power plant operation conditions. The water level/discharge relation at the downstream extremity of the model is regulated by a specific weir. This water level is controlled at the river cross section where it has been computed using the numerical model.

4. 2　Experimental results

The scale model has been tested with discharges ranging from 75 to 3100m³/s. Except few minor modifications, these tests enable to validate the design of the structures, and in particular the dimensions of the stilling basin.

The head H/discharge Q relations of the weir have been measured on the scale model for varied opening configurations of the gates (10 and 2 gates opened, full or partial opening) for gated and free surface operating conditions (Figure 6). Operation conditions have also been verified for damaged or maintenance configurations (2 gates not available). Using the experimental data, a generic analytical $H - Q$ relation has been defined for each of the 2 operating configurations (free or gated weir) with parameters depending only of the number and opening degree of the gates.

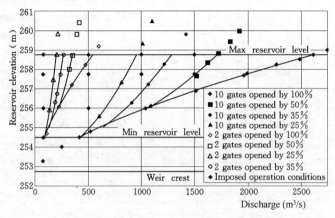

Figure 6   Analytical $H - Q$ relations and measurements

The water levels and the flow velocities have been measured in the stilling basin, on the weir, the piles and along the guiding wall on the left bank, for varied gates opening configurations and reservoir levels.

The power plant influence on the spillway operation has been analyzed through specific tests for varied discharges and opening configurations of the gates. In particular, the study of the downstream water levels variation when the turbines are in use showed no significant change in the flow characteristics in the stilling basin and the spillway. Identically, $H - Q$ relation of the weir is not affected for discharges in the reservoir up to 2500m³/s. It has been proved from these tests that, except a modification of the water levels in the stilling basin and the downstream river reach, the hydropower plant has no significant influence on the spillway operation.

Using the $H - Q$ analytical relations of the weir and following the observation on the scale model of the best sequence to open the gates, spillway management rules have been set up and secondly validated by specific tests on the scale model. They define the sequence and the opening rate of the gates to enable a secure control of the reservoir elevation between levels 254. 50 and 258. 75 for discharges up to the maximum released by the spillway under the considered

reservoir level.

Finally, the flows during temporary diversion stage have been analyzed in order to characterize the hydraulic efficiency of the spillway when the weir crest reaches the intermediate level 249. 00. The water levels along the structures have also been determined for varied successive constructions stages.

5. Conclusions

In hydraulic engineering, physical modeling and numerical modeling are two efficient analysis approaches. Their interactive application is obviously the more effective response to most flow problem analyses.

Carried out carefully, this combination enables a global improvement of the quality of the studies thanks to an increase in the considered areas, the objective determination of the critical part of the project and their detailed analysis. These advantages can be accompanied by a decrease in the costs and the delay of scale models building thanks to a preliminary study of variants with numerical modeling.

This paper presents the results of a successful application of such a combined numerical - experimental approach where the whole studies of the hydraulic features of a large project on the Niger River have been carried out in less than 6 months. The studies focused on the flow characteristics at the scale of the reservoir and the river, using the numerical approach, as well as on hydrodynamic details in the spillway using a large scale factor physical model, with boundary conditions defined on the basis of numerical modeling.

References

[1] ASCE (2000). *Hydraulic modeling - Concepts and practice*. ASCE Manuals and reports on engineering practice No. 97. Reston, Virginia: ASCE.

[2] Dewals B. J. , Erpicum S. , Archambeau P. , Detrembleur S. , Pirotton M. . Depth-integrated flow modeling taking into account bottom curvature. *J. Hydraul. Res.* 2006, 44 (6): 787-795.

[3] Erpicum S. , Dewals B. J. , Archambeau P. , Detrembleur S. , Pirotton M. Detailed inundation modeling using high resolution DEMs. *Engineering Applications of Computational Fluids Mechanics*, 2010, 4 (2), 196-208.

[4] Erpicum S. , Meile T. , Dewals B. J. , Pirotton M. , Schleiss A. . 2D numerical flow modeling in a macro-rough channel. *Int. J. Numer. Methods Fluids*, 2009, 61 (11), 1227-1246.

[5] Erpicum S. , Dewals B. J. , Archambeau P. , Pirotton M. . Dam-break flow computation based on an efficient flux-vector splitting. *J. Comput. Appl. Math.* , 2009: published online: 29 Aug 2009. DOI 10. 1016/j. cam. 2009. 8. 110.

Study of Transport Characteristics of Chinese Sturgeon's Eggs by Field Investigation and Numerical Simulation

Wei Zhangping[1]   Han Changhai[1]   Gao Yong[2]   Yang Yu[1]

[1] *State Key Laboratory of Hydrology-Water Resources and Hydraulic Engineering,*
Nanjing Hydraulic Research Institute, Nanjing, 210029
[2] *Institute of Chinese Sturgeon, China Three Gorges Corporation,*
Yichang, 443002

Abstract: Chinese sturgeon is one of the endangered animals in China, this paper initially studies their eggs' transport characteristics in Yangtze River. Based on the three dimensional hydrodynamic model and particle tracking method, the trajectories of drifting eggs are obtained by numerical simulation, further more, this method is used to predict the spawning grounds of Chinese sturgeon in 2009.

Key words: Chinese sturgeon; eggs; transport characteristics; field investigation; numerical simulation

1. Introduction

Chinese sturgeon is classified as R selection animal, because they lay a lot of eggs during the spawning period every year, and many drifting eggs and juvenile are found in the downstream of the spawning grounds. Nowadays, artificial incubation of the Chinese sturgeon's eggs is an effective way for us to protect them, so it is very important to study the drifting trajectories of their eggs or juvenile.

Since the drifting eggs and juvenile can be simplified as particles in the water, so this paper initially uses numerical simulation to study their transport characteristics and distributions in the downstream of Gezhouba Project, what's more, this method is revised to predict the spawning grounds of Chinese sturgeon in 2009; our research achievements are useful for protecting Chinese sturgeon spawning grounds and artificial incubation of their eggs.

2. Field investigation

In order to calibrate the numerical model, Chinese sturgeon's eggs and juvenile's distributions in the downstream of Gezhouba Project are investigated during the spawning period in 2009.

2.1 Investigation method

2.1.1 *Layout of monitor points.* According to some obtained eggs location in history records, 7 monitor points are fixed between the Gezhouba Dam and Miaozui reach, which are indicated in Figure 1.

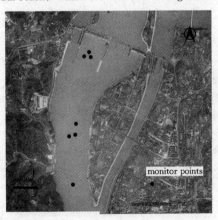

Figure 1   locations of monitor points

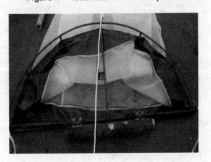

Figure 2   Device for collecting eggs

2. 1. 2 *Devices for collecting eggs.* It is very important to choose some special nets which can collect the drifting eggs and juvenile. In this investigation, the net used for collecting plankton is changed to collect eggs. It is showed in Figure 2.

2. 2 Investigation results

2. 2. 1 *Period of Chinese sturgeon spawning behavior.* 114 eggs and 127 eggs membranes are obtained in the investigation. According to their incubation status, we predict that Chinese sturgeon only had once spawning behavior and the time was between the midnight of 23rd and the early morning of 24th November in 2009.

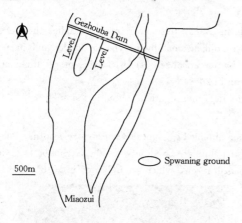

Figure 3   Location of spawning ground

2. 2. 2 *Numbers of spawning Chinese sturgeon.* According to the method[1] used to calculate the numbers of collected juvenile, we firstly calculate the total number of juvenile is 1.07×10^6. What's more, some studies show that the average number of eggs laid by Chinese sturgeon is 400000[2], and Chinese sturgeon's fertilization rate is 70%[2], incubation rate is 70%[2], 90% of their eggs are eaten by other fish[3], we predict that there are 54 spawning Chinese sturgeons in 2009.

2. 2. 3 *Spawning ground of Chinese sturgeon.* According to the locations and velocities in different monitor points, we predict that the spawning ground of Chinese sturgeon is between the Gezhouba Dam and the end of levee of the Dajiang section. It is showed in Figure 3.

3. Numerical simulation

3. 1 Hydrodynamic model
Continuity equation:

$$\frac{\partial u}{\partial x} + \frac{\partial v}{\partial y} + \frac{\partial w}{\partial z} = S \qquad (1)$$

Horizontal momentum equations:

$$\frac{\partial u}{\partial t} + \frac{\partial u^2}{\partial x} + \frac{\partial vu}{\partial y} + \frac{\partial wu}{\partial z} = fv - g\frac{\partial \eta}{\partial x} - \frac{1}{\rho_0}\frac{\partial P_a}{\partial x}$$

$$-\frac{g}{\rho_0}\int_z^\eta \frac{\partial \rho}{\partial x}\, dz - \frac{1}{\rho_0 h}\left(\frac{\partial s_{xx}}{\partial x} + \frac{\partial s_{xy}}{\partial y}\right)$$

$$+ F_u + \frac{\partial}{\partial z}\left(\nu_t \frac{\partial u}{\partial z}\right) + u_s S \qquad (2)$$

$$\frac{\partial v}{\partial t} + \frac{\partial v^2}{\partial y} + \frac{\partial uv}{\partial x} + \frac{\partial wv}{\partial z} = -fu - g\frac{\partial \eta}{\partial y} - \frac{1}{\rho_0}\frac{\partial P_a}{\partial y}$$

$$-\frac{g}{\rho_0}\int_z^\eta \frac{\partial \rho}{\partial y}\, dz - \frac{1}{\rho_0 h}\left(\frac{\partial s_{yx}}{\partial x} + \frac{\partial s_{yy}}{\partial y}\right)$$

$$+ F_v + \frac{\partial}{\partial z}\left(\nu_t \frac{\partial v}{\partial z}\right) + v_s S \qquad (3)$$

All of those parameters and variations are indicated in DHI Water & Environment (2007a).

3. 2 Principle of particle tracking method
The particle tracking method uses a Lagrange discretization, splitting all mass in the system into a number of particles with specific 3D coordinates and masses. Besides the basic process with drift and dispersion, this method also can include other processes: settling, buoyancy, and erosion.

The transport and dispersion of particles follows the Langevin Equation[5], which describes the trajectories of particles' movement in water. The results equation are written as

$$dX_t = a(t, X_t)dt + b(t, X_t)\xi_t dt \qquad (4)$$

Here a is drift term, b is the diffusion term and ξ is the random number.

In order to calculate the Y at certain time, we simply start from the initial value $Y_0 = X_0$ and precede recursively to generate the next value

$$Y_{n+1} = Y_n + a(t, X_t)Y_n \Delta_n + b(t, X_t)Y_n \Delta W_n \qquad (5)$$

Here, $\Delta W = W_t - W_s \in N \ (\mu = 0, \ \sigma^2 = \Delta_n)$ is the normal distributed Gaussian increment of the Wiener Process W, which is a continuous time Gaussian stochastic process with independent increments over $\tau_n \leqslant t \leqslant \tau_{n+1}$.

3. 3 Velocity calibration
The measured velocities by ADCP on 16th November, 2004 are used to calibrate our numerical model. The calibrated results show that our model is effective for hydrodynamic modeling.

3. 4 Processes and simplifications for modeling the transport characteristics of Chinese sturgeon' eggs

3. 4. 1 *Modeling time.* According to the field investigation, it is assumed that spawning time is between 23:00pm 23rd and 2:00am 24th November, and the stop time for modeling is set to 4:00am 24th November.

3.4.2 *Estimated resealed velocity of particles*. Since the numbers of spawning Chinese sturgeon and their spawning time are obtained, the resealed velocity of particles in our model can be estimated by the following method

$$v_e = \frac{\sum_{i=1}^{n=54} m_i}{T_{\text{total}}} \quad (6)$$

Here, m_i is the mass of one Chinese sturgeon egg, T_{total} is the spawning time. So the resealed velocity of particles is $v_e \approx 2.5 \text{g/s}$.

3.4.3 *Modeling zone*. The spawning ground is divided into 9 sub-zones, and it is assumed that all of the 54 Chinese sturgeons are uniformly distributed in every sub-zone, which is showed in Figure 4.

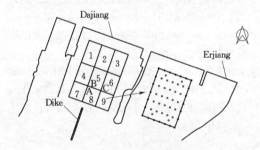

Figure 4   9 sub-zones and distributions
of Chinese sturgeons

3.4.4 *Modeling setting*. The whole modeling area is divided into 5 layers in vertical direction, because the Chinese sturgeon likes to swim in the bottom of river, so the model focuses on the fifth layer. Settling, dispersion, erosion and drifting are also considered, they are based on the results of the hydrodynamic modeling.

3.4.5 *Criterion for modeling results*. In the field investigation, 7 monitor points (three of them are indicated in Figure 4) are fixed in the downstream of Gezhouba Dam, so if the modeling trajectories of eggs pass through those points, the corresponding sub-zone is considered to be a possible spawning ground for Chinese sturgeon in 2009, and vice versa.

4. Results

The trajectories of particles resealed from the 9 sub-zones are obtained by numerical simulation; Figure 5 is an example of the simulation results.

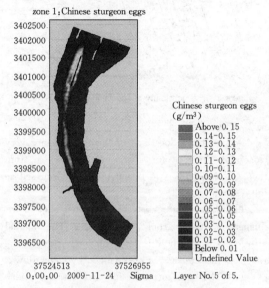

Figure 5   Trajectories of particles
resealed from sub-zone 1

All of the results can be classified as three groups; the first group's particles trajectories pass through all of the 7 monitor points, which include sub-zone 1, sub-zone 2, and sub-zone 5; the second group's particles trajectories pass through one or two monitor points in the upstream of the dike and the 4 monitor points in the downstream of dike, which include sub-zone 3, sub-zone 4, sub-zone 6 and sub-zone 8, the third group's particles trajectories do not pass through any of the three monitor points in the upstream of the dike, which include sub-zone 7 and sub-zone 9.

5. Conclusion

5.1   Prediction of spawning grounds

According to the numerical simulations, the trajectories of particles resealed from the assumed 9 spawning sub-zones are different, which means that those sub-zones may be the spawning grounds of Chinese sturgeon, but the possibility for them to be the spawning ground can be classified into three levels: Very likely (sub-zone 1, sub-zone 2 and sub-zone 5); Likely (sub-zone 3, sub-zone 4, sub-zone 5 and sub-zone 8); Unlikely (sub-zone 7 and sub-zone 9), which are indicated in Figure 6.

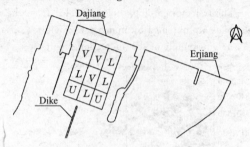

Figure 6   The possibility of spawning grounds in 2009

5. 2   Discussions

Although the locations of 9 sub-zones are very close, but the trajectories of particles resealed from them are different, two phenomena are discussed as follows.

(1) The topography of the river has a great influence on the particles' trajectories. The spatial differences cause the difference of water flow in the river, so the trajectories are changed by water flow.

(2) The transport time for the particles is also different. The results show that the particles' locations are total different in the downstream even at the same time, this is very important for us to decide when we should start to collect the eggs and juvenile.

Acknowledgment

This research is supported by NSFC for Youth (No. 50909064) and Ministry of Water Resources for Nonprofit Research (No. 200801105). The authors would like to extend our thanks to those Foundations.

References

[1] Yi Bolu et al.. Study of the relations between Gezhouba Project and the four major Chinese carps. Science & technology of Hubei Press, Wuhan, 1988.

[2] Wei Qiwei. Reproductive behavior ecology of Chinese sturgeon with its stock assessment. Institute of Hydrobiology, Chinese Academy of Science, Wuhan, 2003.

[3] Chang Jianbo. Structure and dynamics of the spawning stock of Chinese sturgeon in the Yangtze River. Institute of Hydrobiology, Chinese Academy of Science, Wuhan, 1999.

[4] DHI Water & Environment. MIKE21 & MIKE3 FLOW MODEL FM (Hydrodynamic and Transport Module Scientific Documentation) [M]. 2007a: 3-4.

[5] DHI Water & Environment. MIKE21 & MIKE3 FLOW MODEL FM (Particle Tracking Module Scientific Documentation) [M]. 2007b: 1-3.

Physical and Numerical Modelling of a Tangential Vortex Intake for Supercritical Storm Flow Diversion

Colin K. C. Wong   Joseph H. W. Lee

Croucher Laboratory of Environmental Hydraulics, Department of Civil Engineering, The University of Hong Kong, Pokfulam, Hong Kong, China

Abstract: In densely populated cities, urban flood control often requires the use of compact drainage structures. A compact bottom rack chamber with tangential vortex intake structure has recently been developed for diverting fast moving turbulent flow on steep catchment into a drainage tunnel for discharge into the sea. The flow in this newly developed intakes structure (generic design) is three-dimensional, turbulent and highly aerated. During the detailed design of the intake structure, it has been necessary to adopt a mirror-image orientation of the tangential inlet alignment so as to suit particular site constraints. A 1 : 16 physical model of the proposed tangential vortex intake has been carried out. The performance characteristics of the proposed design are compared with those of the generic design.

The complex 3D flow in the interception and tangential vortex intake structure are investigated by both the Froude-scale physical model and computational fluid dynamics (CFD) modeling using the FLOW3D software. In the numerical model, the air-water flow is simulated using the Volume of Fluid (VOF) method; turbulence closure is made via a two-equation k - ϵ model. Experimental results show the supercritical flow along the tangential vortex intake is characterized by flow impingement; it is highly non-uniform and the shockwave height reduces towards the downstream end. The air core size in the dropshaft is adequate for the design flow. The computed flow depth variation in the tangential inlet are in reasonable agreement with the experimental data; the major flow features can be simulated. Both the physical and numerical model results show insignificant differences in the drainage performance of the proposed and generic designs.

Key words: tangential vortex intake; urban hydraulics; physical model; aerated flow; numerical model

1. Introduction

In recent years rapid urbanization and climate change have resulted in more intense storm water runoffs in Hong Kong. However, the space available for upgrading urban drainage systems is very limited due to existing buildings and underground facilities. The Hong Kong West Drainage Tunnel (HKWDT) scheme has been developed to intercept and divert supercritical storm flows at hillside level to an underground tunnel via a system of vortex drop shafts (Figure 1). Central to this scheme is a compact intake structure that consists of a bottom rack chamber and a vortex intake. Figure 2 (a) - (c) shows the generic design of a representative intake. The storm flow passes through the bottom racks that serve to exclude debris from entering the downstream drainage system. The flow inside the bottom rack chamber is three-dimensional, highly turbulent and aerated. The supercritical outflow from the chamber is conveyed in a tapering inlet channel that leads the flow tangentially into a vortex dropshaft. A general theory or design guideline does not exist for such a complex hydraulic structure. The hydraulic design was developed based on a combination of heuristic reasoning, theoretical analysis and extensive experiments on physical models of different scales[5].

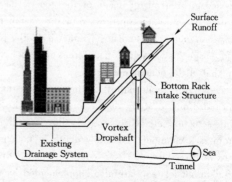

Figure 1   Hong Kong West Drainage Tunnel (HKWDT) scheme

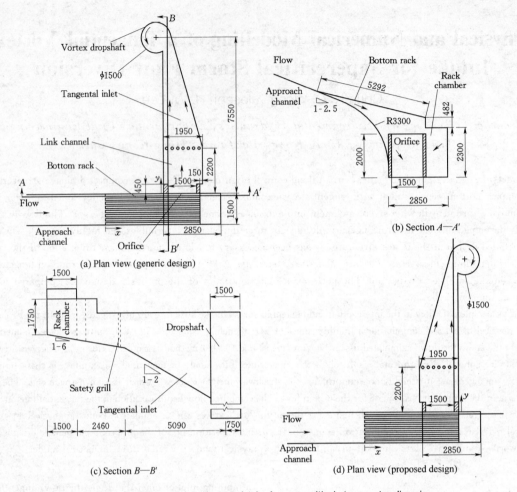

Figure 2   Bottom rack and vortex intake for supercritical storm water diversion

During the detailed design of the intake structures in the HKWDT scheme, it has been necessary to introduce variations of the generic intake design to suit particular site constraints. For one of the intake sites, the orientation of the tangential vortex intake is proposed to be a mirror-image of the generic design (Figure 2d). As the outflow from the chamber is highly non-uniform[7] and not symmetrical, the proposed design may result in a different flow behavior from that of the generic design.

A physical model study of the proposed intake was carried out to study the performance of the hydraulic structure in relation to: (i) flow stability in bottom rack chamber; (ii) flow depth in tangential vortex inlet; and (iii) air core in vortex dropshaft. The model tests are also supplemented by numerical computations using the FLOW3D software. The numerical predictions are compared with experimental data.

2. Physical model experiments

A 1 : 16 Froude scale model of the proposed design of the HKWDT scheme is carried out (Figure 2 d): the model is built of Perspex and comprises the approach channel with the bottom rack, the bottom rack chamber, a link channel and the tangential vortex intake. The supercritical inflow on the steep approach channel (1 : 2.5 in bottom slope) is intercepted by bottom racks inclined at slope 1 : 5. Based on extensive model tests[5], rack bars of circular cross-section with a void ratio of 0.43 are selected. The outflow from the bottom rack chamber enters a steep link channel (1 : 6 bottom slope) fitted with safety grills. The outflow is conveyed in a tapering inlet channel of 1 : 2 bottom slope into a vortex dropshaft. Experiments were carried out for four discharges corresponding to (i) nominal low flow; (ii) 1 in 50 year rainfall; (iii) 1 in 200 year rainfall; and (iv) maximum overflow conditions. The water surface profile in the approach flow channel and tangential intake are measured along with the bottom rack wetted length and the minimum air core area in the drop shaft. Corresponding experiments are also performed on a similar model of the generic design.

3. Experimental Results

Figure 3 (a) shows the observed flow for the maximum design flow of $Q=4.3$ m^3/s. The supercritical flow on the steep channel passes through the opening between the racks which present a resistance to the flow. After the rack interception, the main channel flow follows the channel bed; in addition the flow is also attached to the underside of the racks for a certain distance before leaving the rack as a sheet-jet beneath the rack, plunging onto the main channel flow. The entire flow enters the rack chamber. The inflow enters the chamber as a wall jet that plunges to the chamber bottom and the end wall, resulting in a three-dimensional spiral flow. The re-circulating flow is re-directed by the ceiling of the chamber to collide with the incoming flow, resulting in intense turbulent mixing and significant air entrainment. The flow in the chamber is surcharged. It appears that a continuous thin air layer dragged along by the inflow is formed on the jet surface and this air is captured into the foamy water in the chamber. Furthermore, an aerated roller is formed near the point when the inflow enters the chamber.

The outflow from the bottom rack chambers is highly non-uniform, and exits through an orifice into a link channel with significant flow contraction (Figure 2b); it resembles an outflow under a gate. The depth and flow are greater at the downstream end of the orifice. However, the flow that enters the tangential inlet also impinges on the tapering side wall, resulting in a piling up of the flow. Figure 4 (a) and (b) show the observed side and top views of the flow in the tapering inlet channel respectively. It is seen that the highly irregular flow swirls into the vortex dropshaft at the downstream end of the inlet channel (Fig-

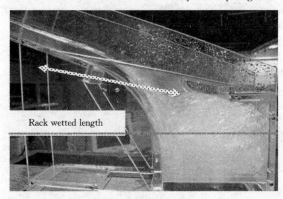

Rack wetted length

(a) Side view

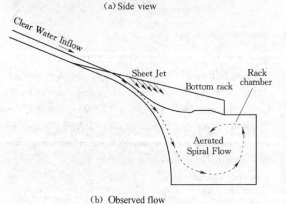

(b) Observed flow

Figure 3　Observed flow in bottom rack chamber
($Q=4.3$ m^3/s)

(a) Side view

Tangential inlet channel

(b) Top view

Figure 4　Observed flow in tangential vortex intake
(proposed design) ($Q=4.3$ m^3/s)

191

ure 4a); the lateral non-uniformity of the water surface an also be seen.

Figure 5 and Table 1 show that the measured depth profile along the straight side of the inlet channel for the proposed and generic designs for two representative flows. Figure 6 (a) and (b) show the measured air core in the vortex dropshaft for the design flow. In general the depth in the proposed tangential intake is somewhat less than that of the generic design; the measured air core area ratio (minimum air core area/dropshaft cross-section area) of 38 percent is notably larger than that of the generic design (29%; Table 1). However, both designs are able to meet the threshold criterion of 25% for the design flow.

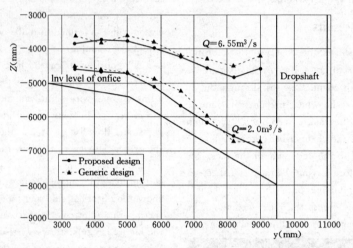

Figure 5   Measured flow depth along the tangential inlet
(Q=2. 0 and 6. 55m^3/s)

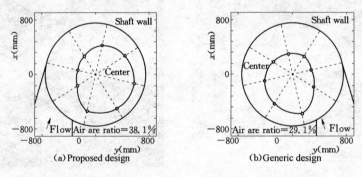

Figure 6   Measured air core in dropshaft (Q=4. 3m^3/s)

In summary, both the proposed and generic designs of the tangential vortex intake are able to meet the performance criteria with respect to smooth flow entry, flow stability and adequate vortex air-core size. It is interesting to note that the performance of the proposed design is somewhat better than the generic design.

4. Numerical modeling

CFD modeling of 3D free surface flows in hydraulic structure problems have been increasingly reported in recent years-e. g. the use of the volume of fluid (VOF) model to solve for the flow over a stepped spillway[1, 6]. In this study, the three-dimensional air-water flow is computed using the FLOW3D Computational Fluid Dynamics (CFD) software. The Reynolds-Averaged Navier Stokes (RANS) equations are solved with a two-equation k – ε model for turbulent closure. The air-water mixture flow is solved using the Volume of Fluid (VOF) method and a recently developed air entrainment model[3-4]. The one-fluid variable density version of the model is used - accounting for both the bulking of fluid volume by air and the associated buoyancy effects. The boundary and initial conditions are set in accordance with the experimental conditions in the physical model.

Figure 7 shows the numerical grid of the tangential intake; the $10.6m \times 11.8m \times 12.9m$ ($X \times Y \times Z$) computational domain is spanned by around 311, 280 cells. For the inflow boundary, the measured flow depth and velocity are specified. At the outlet boundary, the "outflow" boundary condition is adopted. The time step satisfies the Courant stability criteria $[u_{max} (\delta t / \delta x) < 0.4]$ while the maximum grid size is around 0.3m in the Z-direction. As previous studies showed that the bottom rack bars have an insignificant influence on the aerated flow, the rack bars are not included in the simulation.

Figure 8 shows the measured transverse depth profiles at different sections along the tangential inlet for three representative flows. It is seen that the depth of the supercritical outflow is greater at the downstream end (e. g. $y=3.4m$ section). As the flow impinges on the tapering side - resulting in higher

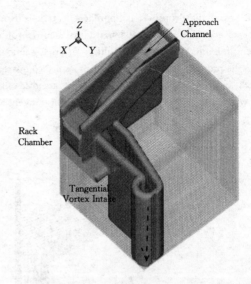

Figure 7　Grid arrangement (proposed design)

depths on both edges (e. g. $y=5.8m$ and $6.6m$ sections). The shock wave and lateral non-uniformity

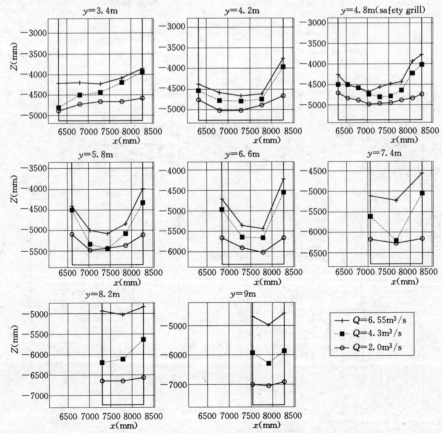

Figure 8　Measured transverse depth profile along the tangential inlet channel (proposed design)

reduces as the dropshaft is approached at the downstream end. The computed longitudinal depth profile along the inlet channel and at different cross-sections for the proposed design at $Q=4.3\,\mathrm{m^3/s}$ are compared with experimental data in Figure 9. A smooth outflow from the bottom rack chamber to the tangential vortex intake can be noted. In general, the numerical predictions can capture the shock wave height along the tangential intake and agree reasonably with the measurements in the physical model. A similar flow behavior in the tangential inlet channel has also been measured for the generic design (Figure 10).

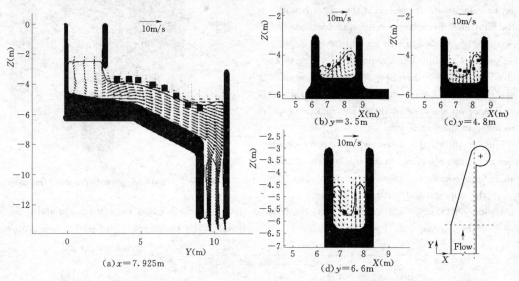

Figure 9　Comparison between predicted and measured flow depth（symbols）（proposed design; $Q=4.3\,\mathrm{m^3/s}$）

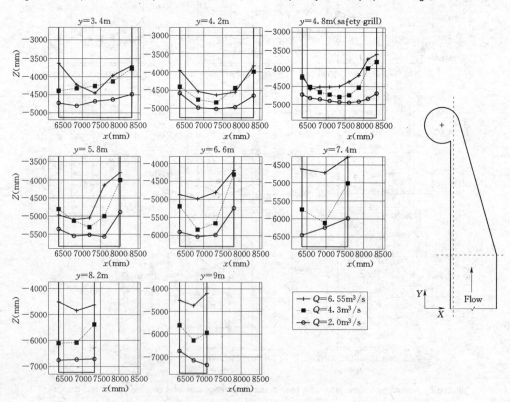

Figure 10　Measured transverse depth profile along the tangential inlet channel（generic design）

5. Concluding remarks

A comprehensive physical and numerical model investigation of a tangential vortex intake for supercritical storm flow diversion has been carried out. The steady three-dimensional air-water flow is computed by solving the RANS Equations with a two-equation turbulence model and using the Volume-of-Fluid method via the FLOW3D software. The computed flow characteristics in the bottom rack chamber and the tapering inlet channel are well-supported by the experiments.

Table 1 Measured flow depth and air core area ratio for tangential intake (in prototype scale)

	Rack wetted length (m)	Laterally-averaged depth (m) and (Froude no)								Air core area ratio (%)
		$y=3.4$	$y=4.2$	$y=4.8$	$y=5.8$	$y=6.6$	$y=7.4$	$y=8.2$	$y=9$	
Generic design	4.0	0.9 (0.83)	0.59 (1.57)	0.95 (0.76)	0.7 (1.41)	0.56 (2.26)	0.66 (2.12)	1.16 (1.13)	1.42 (1.11)	29.1
Proposed design	4.2	0.75 (1.09)	0.49 (2.03)	0.81 (0.96)	0.54 (2.06)	0.65 (1.83)	0.56 (2.69)	1.12 (1.19)	1.42 (1.10)	38.1

Acknowledgment

This work was supported by a grant from the Hong Kong Research Grants Council (Grant RGC HKU7143/06E).

References

[1] Chen Q., Dai G., Liu H.. Volume of fluid for turbulence numerical simulation of stepped spillway overflow. *Journal of Hydraulic Engineering*, 2002, 128 (7): 683-688.

[2] FLOW3D.. User's Manual Version 9.2 Flow Science Inc. 2007.

[3] Hirt, C. W., Nichols, B. D.. Volume of fluid (VOF) method for the dynamics of free boundaries. *Journal of Computational Physics*, 1981, 39 (1): 201-225.

[4] Hirt, C. W.. Modeling turbulent entrainment of air at a free surface. Technical note: TN61. Technical report, Flow Science Inc. 2003.

[5] Lee J. H. W., Yu D., Chan H. C., Gore L., Ackers J.. Bottom rack intake for supercritical storm flow diversion on steep urban catchment. *Proceedings of the 31st IAHR Congress*, Seoul, Korea, September 11-16, 2005.

[6] Souders D. T., Hirt C. W.. Modeling entrainment of air at turbulent free surface. In *World Water and Environmental Resources Congress 2004*, Salt Lake City, Utah, USA. 2004.

[7] Wong C. K. C.. *Hydraulics of bottom rack chamber for supercritical flow diversion*. M. Phil thesis, The University of Hong Kong, 2009.

Experimental Study and Numerical Simulation on the Dynamic Characteristics of One Large Plate Gate

Wang Xin   Luo Shaoze   Yuan Qiang

*State Key Laboratory of Hydrology-Water Resources and Hydraulic Engineering,
Nanjing Hydraulic Research Institute, Nanjing,* 210029

Abstract: Elasticity similitude model of one large plate steel gate was made and experiment mode analysis method was employed to study the dynamic characteristics of the gate structure. Meanwhile, numerical simulation means with three-dimensional finite element method was used to calculate the natural vibration characteristics of the gate. The results indicated that the calculated values and experimental values of the natural vibration frequencies of the gate were very close to each other and the vibration modes of the former several phases were consistent, which proved that analysis results of the dynamic characteristics were correct. By the way, the influence of the flow-structure interaction on the dynamic characteristics of the gate was researched by numerical simulation.

Key words: plate gate; dynamic characteristics; model test; numerical simulation

1. Introduction

Steel gate is the most common light water retaining structure. It is very important for the whole key water control project to keep gates in normal operation. Flow-induced vibration is the ubiquitous problem when it works, and the terrible vibration will affect the safety and reliability of the gate. There are a lot of examples that gates can not work or be damaged because of the serious vibration. At abroad, such as America, Japan and so on, the damage accident of the gate is also not uncommon. Therefore, the vibration problem has attracted more attention in the field of hydraulic engineering. A great deal of experience in the gate design, manufacture and operation about the gate vibration safety is accumulated from the long time engineering practice, and the technical level is enhanced enough. With the fast improvement of computer, dynamic signal measurement, data handling technique and so on, the flow-induced vibration problem research is developed enough as well.

Generally, it is starting from the external load forced on the structure and the natural characteristics of the structure to study the gate flow-induced vibration problem. The external load is the external cause of the gate vibration and the natural characteristics is the internal cause. The gate structure dynamic analy-sis need to compare the self-vibration frequency of the structure with the flow pulsation frequency and make them different as much as possible to insure the safety of the gate. In this paper, mode experiment and finite element method are employed to study the structure natural vibration characteristics for one large plate gate in initial design.

2. Natural vibration characteristics

2.1 Mode experiment

The structure dynamic equation is conducted La-place transmission and the transmission function of the vibration system is obtained:

$$H(j\omega) = \sum_{r=1}^{N} \left(\frac{A_r}{j\omega - \lambda_r} + \frac{A_r^*}{j\omega - \lambda_r^*} \right) \qquad (1)$$

where A_r, A_r^* are the residue matrix and its conjugation of the rth mode. The structure transfer function includes all the mode parameters. Therefore, the structure dynamic characteristics will be obtained after getting the transfer function and identifying the parameters.

One large steel plate gate 16.8m$\times$20.5m (width $\times$height) is used to control flood for one water control project. The dynamic characteristic model is designed by elasticity similitude law, and the model is made by plexiglass with the geometric scale $Lr = 20$. According to the structure feature of the gate, the gate is dispersed into 207 nodes, and every node includes three directions, the flow direction, side direction and vertical direction. Therefore, the model contains 621 freedom degrees, just as Figure 1 shows. The method that all nodes are excited one by one and the responses of one node are collected is employed in the experiment, that is, arbitrary forces are used to excite every node, meanwhile the exciting force and the vibration signal of the three directions are record, amplified by charge amplifier, filtered, then delivered to the computer. The time domain complex exponential fitting method is adopted to solve the transfer function, and then the mode frequency, damping and mode shape are obtained. Exciting hammer is used to excite the gate, and the exciting force is measured by force sensor in the head of the hammer. The vibration responses are measured by three-direction micro acceleration sensor.

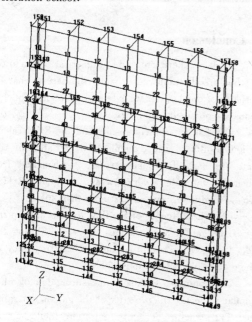

Figure 1　Node discretization

The self-vibration frequencies and damping ratios of the first ten steps are shown In Table 1, from which it can be found that the fundamental frequency is very low and the frequency and damping ratio of the first step are 4.68Hz and 5.15% respectively. From the first and second step mode shapes shown in Figure 2, it can be found that the vibration of the first two steps are all following the normal of the panel and the first step is anti-symmetric while the second step is symmetric.

Table 1　Self-vibration frequencies and damping ratios of the gate

Step	1	2	3	4	5
frequency（Hz）	4.68	24.37	38.65	52.62	58.67
Damping ratio（%）	5.15	3.65	3.56	4.05	3.77
Step	6	7	8	9	10
frequency（Hz）	76.58	101.69	118.42	139.03	165.32
Damping ratio（%）	4.83	5.03	5.1	3.89	5.26

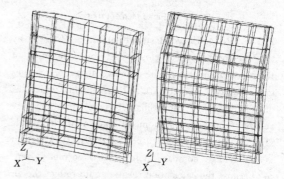

Figure 2　The first and second step mode shapes

2.2　FEM calculation

Mode analysis by FEM has been very mature. In order to compare with the experiment results, the FEM calculation is conducted and the influence of the fluid-solid interaction to the natural characteristics is study as well.

2.2.1　Dry mode

Three-dimensional finite element model of the gate is built. Because of the sheet structure of the gate, such as the panel, main girder, secondary beam, edge beam and so on, the space quadrilateral plate element with six freedom degrees is employed to discretize the whole model. The joint location of the panel and the upper flange of the main girder and secondary beam is considered as two elements to calculate each stiffness and add to the whole stiffness. The node number of the whole model after discretization is 17074 and element number is 20245. The model grid is shown as Figure 3.

The first and second step mode shapes of the gate obtained by calculation are shown as figure 4. The corresponding frequencies are 5.25Hz and 22.52Hz respectively. Result comparison between the experiment and the calculation is shown in table

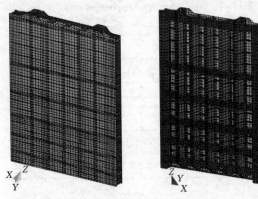

Figure 3　Finite element model

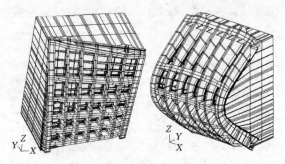

Figure 4　The first and second mode
shapes of the gate with water

2. The first and second step frequencies obtained by two methods are very close, and the differences of them are within 10%. The first two step mode shapes gotten by two methods are full agreement, which indicates that the dynamic characteristic parameters are reliable. The lower step mode shapes of the whole structure vibration agree each other well. However, it is very hard to compare their higher step mode, simply because the calculation results could reflect the vibration of the local weak structure and the frequencies are relatively dense. By the way, the numerical calculation cannot obtain the damping of the structure, but experiment could.

Table 2　Comparison between experiment results and calculation results

| step | frequency（Hz） | | Mode shape |
	experiment	calculation	
1	4. 68	5. 25	same
2	24. 37	22. 52	same

2.2.2　*Wet mode*

The most commonly used added mass method is adopted to consider the interaction of the flow and the gate. When the incompressible flow moves with small amplitude, the variation of the dynamic water meets the Laplace equation. The added mass matrix $[M_p]$ can be derived according to the boundary condition and the flow-structure interface condition, and the dynamic control equation is:

$$([M]+[M_p])\{\ddot{\delta}\}+[C]\{\dot{\delta}\}+[K]\{\delta\}=[R_0] \tag{2}$$

Taking the close gate condition for example, the first and second step self-vibration frequencies considered the interaction of the flow and structure are 2.556Hz and 10.86Hz respectively, which are reduced by about 50% comparing with the dry condition. It can be found that the influence of the flow to natural frequency of the gate is very obvious. The mode shapes are all consistent with no water condition.

3. Conclusion

Model experiment and numerical calculation method are all employed to study the natural characteristics of one large plate gate in detail. Some conclusions are obtained as follows.

(1) The experiment results and the FEM calculation results of the dynamic characteristics of the gate are basically consistent. The fundamental frequency of the gate is very low and at about 4-5 Hz. The second step frequency is more than 20Hz.

(2) The influence of the flow to natural frequency of the gate is very obvious. After considering the added mass the flow, the fundamental frequency of the gate is reduced by about 50%.

(3) Compared with some measuring data of other similar projects, the integrity of the gate is not satisfied. The structure design optimization should be conducted to increase the structure stiffness and reduce the vibration response of the gate.

References

[1]　G . H . Yan. Research development of flow-induced gate vibration [J]. Hydro-science and engineering, 2006 (1): 66-73.

[2]　Z. D. XU, L. H. DU, J. M. CAI. Study on the free vibration characteristics of liquid-solid cou-

pling of plane gate [J]. Journal of hydropower, 2001 (4): 39-43.

[3] Y. K. LIU, H. G. NI, Z. Q. YE. Analysis of flow-induced vibration of hydraulic radial gate [J]. Journal of Dalian University of Technology, 2005, 45 (5): 730-734.

[4] H. G. NI, Y. L. LIU, B. LIU. Prediction of tainter gate vibration using combined physical and numerical models [J]. Journal of Hydrodynamics, 2005, 20 (1): 24-32.

[5] J. F. WU, L. R. ZHANG, L. YU. Verification of perfect hydroelastic model test of flow induced gate vibration by prototype observation [J]. Journal of Yangtze River Scientific Research Institute, 2005, 22 (5): 62-64.

[6] G. H. YAN, S. W. YAN, B. K. FAN. Study of flow-induced vibration for the high-head and large dimension gate [J]. Journal of Hydroelectric engineering, 2001 (4): 65-75.

Unsteady Turbulent Flow Numerical Simulation of the Whole Channel Large Bulb Tubular Pumping Station

Wang Xin   Luo Shaoze   Yuan Qiang

State Key Laboratory of Hydrology-Water Resources and Hydraulic Engineering, Nanjing Hydraulic Research Institute, Nanjing, 210029

Abstract: The pressure pulsation caused by rotor-stator interaction in the large bulb tubular pump is the main hydraulic reason for vibrations of pumping station. In order to study the mechanics of flow induced vibration, incompressible continuity equation, Reynolds average Navier-Stokes and RN G $k-\varepsilon$ turbulent equations are used to simulate the 3-D whole channel unsteady turbulent flow of bulb tubular pump in this paper. Advanced sliding mesh technology are used to simulate the interaction between the rotor and the stator in the unit. Velocity and pressure distributions of the flow passage are analyzed. Some measuring points are set in the flow passage to obtain the pressure pulsations of every location especially near the blades. Using Fast Fourier Transformation, the spectrum characteristics of the pressure pulsations are studied, which will provide a good foundation for the structure vibration analysis of the pumping station.

Key words: bulb tubular pump; 3D unsteady turbulent flow; rotor-stator interaction; pressure pulsation

1. Introduction

Among low lift pump stations, large bulb tubular pump has been widely applied in eastern route of South-to-North Water Transfer Project for its high efficiency. 7 bulb tubular pump stations have been built in first period project of eastern route. High requirements are proposed because of their large-scale and long running time. Vibration in pump station is inevitable and the three most important and complex influence are mechanical, electrical and hydraulic factors. So it's necessary to study the hydraulic characteristics and vibration mechanism with CFD method. So far, the research of the flow in hydraulic machinery focuses on turbine and centrifugal pump and usually analyzes on unit paragraph. This paper gives unsteady turbulent flow simulation of the whole channel on new large tubular pumping station with Reynolds time-averaged model and sliding mesh technique and predicts pressure pulsation caused by rotor-stator interaction.

2. Numerical Simulation

2.1 Calculation model

The fundamental parameters of rear-mounted bulb tubular pump station: 5 blades, 8 guide vanes, wheel diameter of 3.14m, designed flow of 33.3m³/s, whole channel length (including inlet conduit segment, impeller chamber segment and outlet conduit section) of 34m. The whole channel model of bulb tubular pump station, which is built with unstructured tetrahedral mesh unit, is shown as Figure 1 and the external structure is shown as Figure 2. The grids is close in order to exactly simulate the guide vane. The total unit number of the model is 1290892 and the number of nodes is 292432.

2.2 Computing method and boundary conditions

2.2.1 Computing method. The flow in the pump station, whose flow rules could be described by Navier-Stokes equation, is 3-D unsteady incompressible tubular. Finite volume method is applied to discrete the control equation. In the differential format, the pressure and velocity are used to second order center and second order upwind difference scheme respectively. The turbulent kinetic energy and the turbulent dissipation rate is used to the second order upwind. We achieve the coupling solution of pressure and velocity with SIMPLEC algorithm and solve algebraic equations with under-relaxation iteration. Because of the rotation of the wheel in bulb tubular pump station, second order rotor-stator interaction is formed between the wheel and the upstream inlet section and the downstream blade respective-

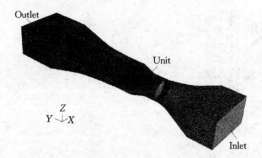

Figure 1   Perspective of the whole model

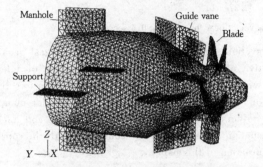

Figure 2   External structure perspective of bulb body

ly. Sliding mesh technique is used to simulate rotor-stator interaction flow field in unsteady analysis. 3 grid slip plane is set up in front of the inlet of the wheel, on the back of the outlet of the wheel and between the wheel and the wall. The grid of the water around the wheel rotate compared to the grid of the water in inlet and outlet. On both sides of interface the mesh nodes do not overlap. The speed component and turbulence is the same after interpolation cross the interface, and at the same time the pressure and flow flux is the same. At every unsteady step, SIM-PLEC is used to solve discrete equations; the position of the rotating grid is updated after computing convergence and next time step begin.

2.2.2 *Boundary conditions*

(1) The inlet boundary conditions. Compute the turbulence intensity of RNG $k-\varepsilon$ model with the default value of Fluent when the speed is given. (2) The outlet boundary conditions. Outflow is applied, the change rate of all the variables in the surface except pressure is zero. (3) The wall boundary conditions. Static wall is used in inlet passage, guide vane, the body of bulb, the outlet passage and blade chamber. Standard function is used in near wall region and ho-slip boundary condition is used in static wall. Moved wall is used to all the rotational wall with the blade in blade chamber and the moving speed and direction is the same with the blade.

3. Results and analysis

In the simulation on normal operation (with the up head of 13.6m, below head of 8.82m, evaluation of 4.40m) of large-scale bulb tubular pump station for unsteady tubular, it's speed is 125r/min, frequency 2.08Hz and designed time step 0.001s. The center of inlet section is reference pressure point. The 5 measuring point is set up to monitor the pressure of every step.

3.1   The distribution of flow velocity and pressure

The streamline at 0.5s is shown as Figure 3. We can know form the figure that the it's uniform flow in the contraction region of the outlet passage and the flow is nearly axial and has no circumferential velocity components. The flow in the chamber has large circumferential velocity components. The pressure distribution and velocity is shown as Figure 4. The circumferential velocity in chamber is changed into axial because of the static guide vane. The pressure is large at the upstream face and is small on the back face. The flow in the outlet passage has steady conditions because of the diversion effect.

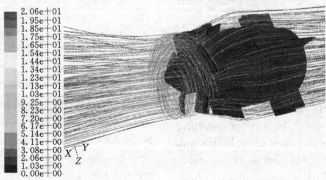

Figure 3   Stream line of the channel at 0.5 second

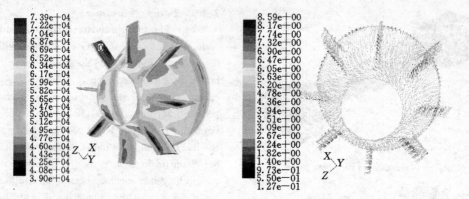

7.39e+04	8.59e+00
7.22e+04	8.17e+00
7.04e+04	7.74e+00
6.87e+04	7.32e+00
6.69e+04	6.90e+00
6.52e+04	6.47e+00
6.34e+04	6.05e+00
6.17e+04	5.63e+00
5.99e+04	5.20e+00
5.82e+04	4.78e+00
5.65e+04	4.36e+00
5.47e+04	3.94e+00
5.30e+04	3.51e+00
5.12e+04	3.09e+00
4.95e+04	2.67e+00
4.77e+04	2.24e+00
4.60e+04	1.82e+00
4.43e+04	1.40e+00
4.25e+04	9.73e-01
4.08e+04	5.50e-01
3.90e+04	1.27e-01

Figure 4　Pressure distributions and velocity vector near the vanes at 0.5 second

The pressure distribution of the blade on the pressure surface and suction surface at 0.5s is shown as Figure 5. We can see that: (1) The pressure is uniform on the pressure surface and is larger at the inlet of blade. Near the top of the blade the pressure is smaller and the cavitation is serious. (2) The pressure of some region on the suction surface is below zero and the largest negative pressure happens near the bottom of inlet where cavitation is possible. The cavitation is serious according to Huaian third bulb tubular pump station and happened at the top of blade and inlet suction surface, which is consistent with the computation of this paper.

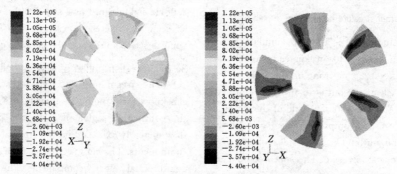

1.22e+05	1.22e+05
1.13e+05	1.13e+05
1.05e+05	1.05e+05
9.68e+04	9.68e+04
8.85e+04	8.85e+04
8.02e+04	8.02e+04
7.19e+04	7.19e+04
6.36e+04	6.36e+04
5.54e+04	5.54e+04
4.71e+04	4.71e+04
3.88e+04	3.88e+04
3.05e+04	3.05e+04
2.22e+04	2.22e+04
1.40e+04	1.40e+04
5.68e+03	5.68e+03
−2.60e+03	−2.60e+03
−1.09e+04	−1.09e+04
−1.92e+04	−1.92e+04
−2.74e+04	−2.74e+04
−3.57e+04	−3.57e+04
−4.04e+04	−4.40e+04

Figure 5　Pressure distributions on the pressure surface and suction surface at 0.5 second

3.2　The forecast of pressure pulsation

For studying the spectrum characteristics of pressure pulsation, 5 measuring points is set up respectively at the inlet chamber, before the wheel, between the wheel and guide vane, on the back of guide vane and at divergent section, as shown in Figure 6. Analyze the computation result of 1 - 2s.

The time-history and spectrum curve of pressure pulsation of the 5 measuring points are shown as Figure 7. We get spectrum curve with FFT. The sampling frequency is 1000Hz and the frequency is 500Hz after conversation. It is known from the picture that: (1) Point 1, wide frequency band of 50-150Hz with high frequency pulsation and low pulsation amplitude. (2) Point 2, in front of the chamber, with rotation frequency of 2.08Hz, 5 blades, the main fre-

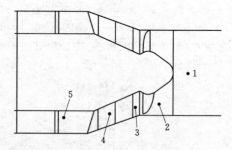

Figure 6　Sketch of the measuring point locations

quency of 10.74Hz which is the number of blade times. Other frequency is harmonic. (3) Point 3, between the wheel and the guide vane. The wheel influences the flow greatly and the main frequency components is similar to point 2. (4) We can see from Point 2 and Point 3 that the pulsation amplitude is

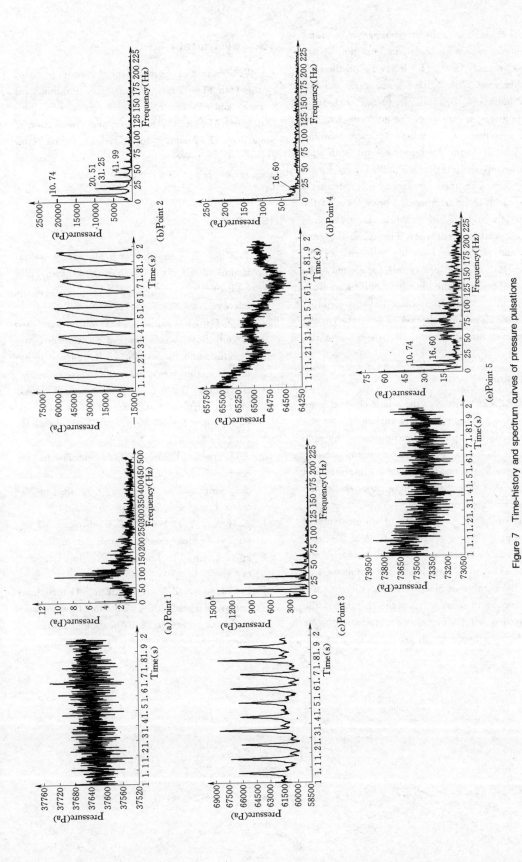

Figure 7 Time-history and spectrum curves of pressure pulsations

great in the chamber and the interference of the rotation and static section is the main pulsation source which lead unsteady flow. (5) Point 4, on the back of the guide vane. With 8 fixed guide vane and low frequency pulsation, the main frequency component of 16.6 Hz gives the impact of the guide vane on the flow. The pulsation amplitude become lower obviously. (6) Point 5. At the divergent section, with wide frequency brand, the interference passed to backward with the flow. The main frequency component is 10.74 Hz which is the number of blades times of the wheel and 16.60 Hz is the number of guide vanes times of the rotation frequency. The high frequency harmonic and pulsation amplitude reduced; of the pressure pulsation time-history and spectrum, the pulsation amplitude in the chamber, which is much larger than that at inlet and outlet flow passage, is the largest. So the main pulsation is generated in the unit and caused the vibration of the whole pump station.

4. Conclusion

(1) By the simulation of the whole channel large tubular pumping station unsteady turbulent, we obtain the velocity and pressure distribution at any time and find that the pressure is small on the top of pressure interface, in the middle of suction interface and the negative side water of the guide vane, where the cavitation may happened.

(2) By the FFT conversation of the pressure at the measuring point, we get the spectrum of the pressure pulsation. It can be seen that the pulsation amplitude in the chamber is largest and the dominant frequency is the number of blade times. So the pressure pulsation lead by interference near the chamber is the main excitation source. The proper prediction gives access to better vibration analysis.

Acknowledgment

We are grateful for the financial support provided by the 11th Five Years National Key Programs for Science and Technology Development of China (No. 2006BAB04A03), Research and Application of Some Key Technology for South-to-North Water Transfer Project. We would also like to thank anonymous reviewers for their useful comments.

References

[1] L. Xing, T. C. Li, X. Wang. The safety evaluation of vibration resistance of the Huayin Third bulb tubular pump station [J]. Hydroelectric energy, 2008, 26 (5): 116-118.

[2] Z. X. Gao, X. J. Zhou, S. D. Zhang, etc. 3-D viscous flow calculation and performance prediction inside the turbine [J]. Journal of hydraulic engineering, 2001, 32 (7): 30-35.

[3] S. H. Liu, Q. Shao, J. M. Yang, etc. Unsteady viscous calculation and pressure pulsation analysis [J]. Hydrauelectric Engineering, 2004, 23 (5): 97-101.

[4] S. Q. Yuan, C. Chen, M. Zheng, etc. The 3-D imcompressible flow field calculation in no-overloading centrifugal pump blade [J]. Journal of Mechanic Engineering. 2000, 35 (5): 31-34.

[5] Z. H. Xu, Y. L. Wu, N. X. Chen, etc. 3-D unsteady viscous excitation calculation in high speed pump [J]. Journal of Tsinghua University, 2003, 43 (10): 1428-1431.

[6] P. C. Guo, X. Z. Luo, S. Z. Liu. 3-D turbulence numerical simulation of coupling flow between blade and volute on centrifugal pump, Transactions of the CSAE, 2005, 21 (8): 1-5.

Rating Curve Assessment at the Yuansantze Flood Diversion Works Using Laboratory and Field Data

Jihn-Sung Lai[1]   Ming-Hsi Hsu[2]   Tsang-Jung Chang[2]

Fong-Zuo Lee[3]   Der-Liang Young[4]

[1] *Research Fellow, Hydrotech Research Institute and*
Adjunct Professor, Dept. of Bioenvironmental Systems Eng.,
National Taiwan University (NTU)
[2] *Professor, Dept. of Bioenvironmental Systems Eng., NTU*
[3] *Ph. D. Student, Dept. of Bioenvironmental Systems Eng., NTU*
[4] *Professor, Dept. of Civil Eng., NTU*

Abstract: The Yuansantze Flood Diversion Works (YFDW) was competed in 2005. The design flood discharge of YFDW is $1620m^3/s$ and flood diversion discharge is $1310m^3/s$. The main purpose of YFDW is to protect downstream urban areas of the Taipei City from flooding disaster. The design flood stage can be reduced by 1.5m on average at downstream reach under 200-year return period protection. Hydraulic physical model was constructed to verify and evaluate the engineering design. Various return-period inflow discharges were tested to find out flow mechanism in the YFDW. In physical modeling, the water levels and flow velocities at specific cross sections were measured. Using these water levels data and corresponding discharges, the rating curve of flood diversion discharge was then established. After completion of YFDW, field measurements of water levels have been recorded during typhoon and storm floods. However, the obtained flood diversion discharge adopting the rating curve obtained in the physical model did not consist with the discharge measured in the field. In this study, the assessment for applicability of the rating curves regressed from laboratory and field data is analyzed. The 3D numerical model is suggested to simulate flow field for detailed mechanism of the flow patterns in YFDW. The numerical results may be used to modify the rating curve for more accurate flood diversion discharge.

Key words: Yuansantze Flood Diversion Works; hydraulic physical model; rating curve

1. Introduction

Taiwan is situated at a geographical location with special climatic condition that brings to the island 3.6 typhoons per annum on the average. These typhoons often result in flood disasters and cause serious damage to properties and sometimes with severe casualties. For examples, Typhoon Zeb and Typhoon Babs in 1998 attack northern Taiwan and result in serious flood damage around Sijhih City of Keelung River Basin. Typhoon Xangsane in 2000 and Typhoon Nari in 2001also result in serious damage in economy and human life lost in Taipei area. Therefore, the Yuansantze Flood Diversion Work (YFDW) in 2004 was constructed to protect downstream urban area of Keelung River Basin from flood disaster [1-2]. The construction site of YFDW is located at the

end of the upper land catchment of the Keelung River. The Keelung River Basin has a drainage area of $501km^2$ with the mainstream length of 86km. The Keelung River, one of the three major tributaries of the Tanshui River in northern Taiwan, flows through the Keelung City, Taipei City and Taipei County of Taiwan. It is rather steep in the upstream and very flat in the lower downstream which near the sea; this is ideal terrain for the occurrence of frequent floods when heavy rain fall occurs in the basin. Rapid urbanization has resulted in the formation of highly developed and densely populated zones over the Keelung River Basin. However, the hydraulic facilities that existed before the project was initiated were unable to provide enough flood protection. Based on the design, the YFDW project can divert $1310m^3/s$ of water at the design peak discharge from the upper Keelung River

Basin into the East China Sea, while the remaining design flood of 310m³/s is discharged further downstream of the river weir in the Keelung River. The YFDW project provides an assurance of flood-carrying capacity (1620m³/s) and significantly increases the safety of the Keelung River Basin under 200-year return-period flood protection[3-4].

The YFDW consists of four major features: a river weir, a diversion inlet, a 2.48km-long tunnel and an outlet. The layout and field water level measurement system of the YFDW were presented in Figure 1. There is a side weir dividing the main channel of the Keelung River and a sediment settling basin, which traps sediment before being carried into the tunnel. And, there are two sluicing gate beside sediment settling basin to flush sediment after typhoon or heavy rainfall. An ogee-shaped diversion weir is designed to generate supercritical flows to prevent the 200-year return-period design discharge from filling up the entrance of the tunnel. In the diversion inlet which before the tunnel entrance, a chute contraction as a transition structure with a bottom slope of 10% connects from the ogee-shaped diversion weir to the tunnel entrance[5-7].

In YFDW project, the hydraulic model was built to investigate water level distribution and flow field. The results obtained from hydraulic model tests were used to establish the rating curve for diversion discharge. However, the diversion discharge measurement in the field is very difficult. As a result, this research collects hydraulic model results for establishing rating curves to estimate diversion discharge of inflow and outflow discharge, and apply those rating curves to calculate the measured discharge in the field. Through mass conservation analysis, the accuracy of rating curve can be assessed and improved.

2. Hydraulic model

During the design period of the YFDW, the measurements of the water levels in the chute contraction were considered the main issue in the tasks of hydraulic model at the same time for the approval of proper design. A hydraulic model of the YFDW was constructed to investigate hydraulic phenomena and water depth while flooding in a different return period. Based on the scale dimension, the geometric scale ratios of the prototype to the model were determined to be 100 in both vertical and horizontal directions, due to the limitations of the construction space.

The test area of hydraulic model arranged from 500m upstream to the river weir location, as shown in Figure 1. The Figure 1 shows that the hydraulic structures consist of a river weir, a side weir, an ogee-shaped diversion weir, two river flow outlet gates and two sluice gates. Figure 1 also displays measured positions of LS1 (for inflow), LS2 (for river outflow), LS3 (for sluing gate) and LS4 (for tunnel outflow). The test flow discharge ranged from 285m³/s to 1660m³/s, and the flow condition was kept steady. When flow phenomenon satisfies steady state, the water levels were measured from LS1 to LS4. Figure 2 shows the results of measured data of water elevation from LS1 to LS4. Because of the flow mechanism was steady state, the water elevations show the same tendency at various inflow discharge. Based on hydraulic model tests, the water elevation of LS1 has the lowest position and LS4 has the highest position. Besides, when inflow discharge from lower discharge to higher discharge, the water elevations have different water depth variation between LS1 to LS4, as shown in Figure 2. Because of flow direction and mechanism changing from left side of the YFDW to the middle area, the water elevation changes against with inflow discharge. Even though, the flow phenomenon changing against with inflow discharge and the water elevation distribution from LS1 to LS4 also maintain consistence.

Based on hydraulic model results of measured water elevations, the established rating curves for inflow discharge (LS1), outflow discharge at river flow outlet (LS2) and diversion discharge (LS3, LS4) are plotted in Figures 3, – Figures 6, respectively. The model -

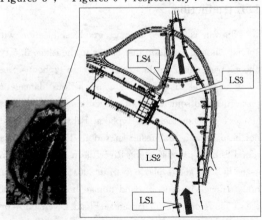

Figure 1 Location of YFDW and its layout
with water level gauge stations (LS1-LS4)

obtained data plotted in figures show that the accuracy R^2 of each rating curve reach over 0.95 to match the model test results quite well.

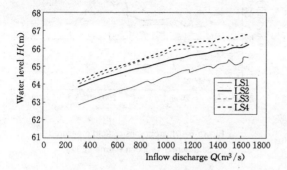

Figure 2   Water level variations at
various inflow discharges

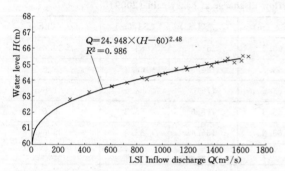

Figure 3   Rating curve based on LS1

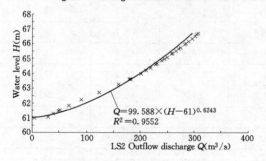

Figure 4   Rating curve based on LS2

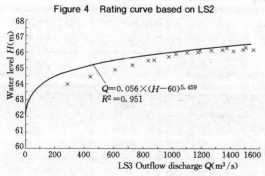

Figure 5   Rating curve based on LS3

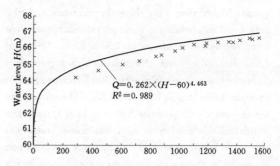

Figure 6   Rating curve based on LS4

3. Application of rating curve

Statistically, flooding is the worst natural hazard in Taiwan during the typhoon season, causing serious economic and social impacts. Data collected by the Ministry of the Interior, Taiwan, showed that 15366 people died from typhoon floods in Taiwan from 1960 to 2009, which was about 50% of total deaths due to natural hazards. In order to reduce the flood-related damages, the YFDW, one of the largest flood mitigation projects in Taiwan, had been built from 2001 to 2004. After the project was completed in 2004, it diverted $7.5 \times 10^7 \, \mathrm{m}^3$ discharge through the YFDW in 18 flood events from 2004 to 2009. The diverted discharge was calculated by the rating curves of LS3 and LS4 (see Figure 5 and Figure 6). The accuracy of the amount of the diverted discharge in the field has been questioned. Figure 7 and Figure 8 show the water elevation hygrograph during a heavy rain storm and Typhoon Parma in 2009. The figures show that the flow mechanism was unsteady, and hydrographs between these two events did not present its consistence. According to the data, these two events reached the diversion condition (i. e. the water elevation higher than 63m). In Figure 7, the water ele-

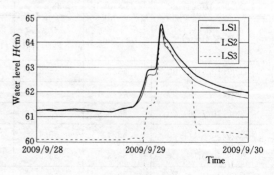

Figure 7   Hydrographs of field measured data
during the heavy rain storm (2009)

vations higher than 63m show the sequence in LS1, LS3 and LS2, while the water elevations higher than 63m show the sequence in LS1, LS4, LS2 and LS3. But, the water elevations which higher than 63m show the sequence at water level decreasing period were LS1, LS4, LS3 and LS2. Therefore, the water level rising behavior does not have consistence with those obtained from hydraulic model tests. On the other hand, this study emphasizes on diversion condition in these two flood events and uses rating curves plotted in Figure 3 – Figure 6 to check the accuracy of mass conservation.

Table 1 and Table 2 list the results of comparison. Theoretically, inflow discharge of LS1 is equal to outflow discharge of LS2 plus LS3 or LS2 plus LS4.

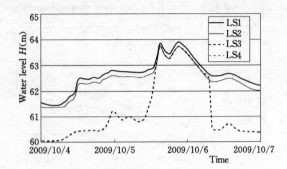

Figure 8　Hydrographs of field measured data during Typhoon Parma (2009)

However, the comparison results show that the relative error could be more than 60% in these two events.

Table 1　　Relationship between inflow and outflow discharge at heavy rain (2009)

Data (time)	LS1 (m)	Q_{in} (LS1)	LS2 (m)	Q_{out} (LS2)	LS3 (m)	Q_{out} (LS3)
2009/10/5 14:00	63.4	506.9	63.2	161.7	63.2	30.8
2009/10/5 15:00	63.9	707.2	63.7	184.9	63.7	73.6
2009/10/5 16:00	63.7	623.0	63.5	176.0	63.5	51.5
2009/10/5 17:00	63.5	563.2	63.3	168.6	63.3	39.6
2009/10/5 18:00	63.4	534.6	63.3	165.4	63.3	35.6
2009/10/5 19:00	63.6	598.0	63.4	172.2	63.4	45.5
2009/10/5 20:00	63.8	677.5	63.6	181.2	63.6	61.1
2009/10/5 21:00	63.9	734.8	63.7	185.4	63.7	72.2
2009/10/5 22:00	63.8	704.5	63.7	183.2	63.7	66.8
2009/10/5 23:00	63.7	659.9	63.6	178.9	63.6	56.5
2009/10/6 0:00	63.6	606.2	63.4	173.7	63.4	47.4
2009/10/6 1:00	63.4	536.9	63.3	166.0	63.3	36.7
2009/10/6 2:00	63.3	480.1	63.1	159.6	63.1	28.5
2009/10/6 3:00	63.1	426.7	63.0	152.4	63.0	22.8
2009/10/6 4:00	63.0	385.5	62.8	146.0	62.9	17.2

Data (time)	LS4 (m)	Q_{out} (LS4)	Q_{out} (LS2+ LS3)	Q_{out} (LS2+ LS4)	Q(LS2+LS3) (Error %)	Q(LS2+LS4) (Error %)
2009/10/5 14:00	63.2	45.8	192.5	207.6	62%	59%
2009/10/5 15:00	63.7	95.4	258.4	280.3	63%	60%
2009/10/5 16:00	63.5	73.0	227.6	249.1	63%	60%
2009/10/5 17:00	63.4	58.4	208.3	227.1	63%	60%
2009/10/5 18:00	63.3	53.5	200.9	218.9	62%	59%
2009/10/5 19:00	63.5	66.5	217.7	238.7	64%	60%

Data（time）	LS4（m）	Q_{out} (LS4)	Q_{out} (LS2＋LS3)	Q_{out} (LS2＋LS4)	Q(LS2+LS3) (Error %)	Q(LS2+LS4) (Error %)
2009/10/5 20：00	63.6	82.6	242.3	263.8	64%	61%
2009/10/5 21：00	63.7	94.0	257.6	279.4	65%	62%
2009/10/5 22：00	63.7	88.9	250.0	272.1	65%	61%
2009/10/5 23：00	63.6	77.4	235.4	256.3	64%	61%
2009/10/6 0：00	63.5	68.3	221.1	241.9	64%	60%
2009/10/6 1：00	63.3	54.0	202.7	220.0	62%	59%
2009/10/6 2：00	63.2	44.1	188.0	203.7	61%	58%
2009/10/6 3：00	63.0	36.5	175.1	188.8	59%	56%
2009/10/6 4：00	62.9	29.3	163.2	175.3	58%	55%

Table 2   Relationship between inflow and outflow discharge at Typhoon Parma (2009)

Data（time）	LS1（m）	Q_{in} (LS1)	LS2（m）	Q_{out} (LS2)
2009/9/29 4：00	64.7	1147.4	64.5	216.0
2009/9/29 5：00	64.2	871.7	63.9	194.3
2009/9/29 6：00	64.0	759.8	63.7	184.3
2009/9/29 7：00	63.7	625.5	63.4	171.3
2009/9/29 8：00	63.4	529.2	63.1	159.9
2009/9/29 9：00	63.2	451.7	62.9	150.8
2009/9/29 10：00	63.0	383.3	62.7	140.0

Data（time）	LS3（m）	Q_{out} (LS3)	Q_{out} (LS2+LS3)	Q (LS2+LS3) (Error %)
2009/9/29 4：00	64.6	224.5	440.5	62%
2009/9/29 5：00	64.0	107.6	302.0	65%
2009/9/29 6：00	63.7	72.9	257.2	66%
2009/9/29 7：00	63.4	45.1	216.4	65%
2009/9/29 8：00	63.1	29.1	189.0	64%
2009/9/29 9：00	63.0	20.7	171.5	62%
2009/9/29 10：00	62.7	13.3	153.4	60%

4. Numerical simulation

According to the error of calculating diverting discharge using the rating curve relationship did not consist with the measured data in the field. In the present study, the 3D numerical model CFX is employed to simulate flow field in the YFDW physical model. The CFX is one of the computational fluid dynamics software. The numerical model adopts volume of fraction (VOF) method to simulate interface location between water and air. And, the numerical model employed continue-continue phase module to calculate flow field of water and air at the same time. The boundary condition of bed and wall were set to be no slip condition, and the free surface was set by VOF. The out flow discharge of main channel was set specific mass flow and diversion weir was free overflow approximation. The governing equations are written as follows: Conservation of mass

$$\frac{\partial}{\partial t}(\gamma_a \rho_a) + \nabla(\gamma_a \rho_a u_a) = \sum_{\beta=1}^{N_P} (m_{a\beta} - m_{\beta a}) \quad (1)$$

Conservation of momentum

$$\frac{\partial}{\partial t}(\gamma_a \rho_a u_a) + \nabla \cdot \gamma_a \{\rho_a u_a u_a - \mu_a [\nabla u_a + (\nabla u_a)^T]\}$$

$$= \gamma_a (B - \nabla p_a) + \sum_{\beta=1}^{N_p} c_{a\beta}^{(d)} (u_\beta - u_a) + F_a + \sum_{\beta=1}^{N_p} (m_{a\beta} u_\beta$$

$$- m_{\beta a} u_a) \qquad (2)$$

where γ_a = VOF of phase α; ρ_a = density of phase α; u_a = velocity of phase α; $m_{a\beta}$ = momentum exchange rate; μ_a = viscosity coefficient of phase α; B = body force; p_a = pressure of phase α; N_p = phases number; $c_{a\beta}^{(d)}$ = drag force exchange coefficient; F_a = non-drag force of phase α.

The 5-year return-period discharge as an example was simulated and the results were presented in Figure 9. Based on simulation results [Figure 9 (a)], there was circulation area near the right bank of YFDW and main flow direction was closed to left bank. As a result, velocity distribution did not uniformly distribution in simulation area. Beside, Figure 9 (b) shows the water level distribution which near the crest of the ogee-shaped diversion weir. Flow field does not uniformly flow into the diversion tunnel, and the water surface also appears non-uniform distribution. The numerical results based on the measured inflow discharge can be used to compare with the laboratory data for obtaining more accurate rating curves.

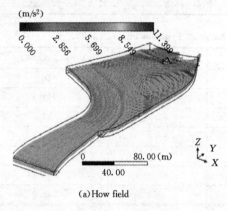

(a) How field

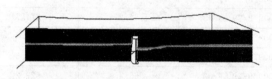

(b) Water level distribution near the crest of the oggee-shaped diversion weir in YFDW

Figure 9   Simulation results

5. Discussion and conclusion

To protect downstream urban areas of the Taipei City from flooding disaster, the Yuansantze Flood Diversion Works (YFDW) was competed in 2005. Many public hearings were held for this project to communicate with the public and get their support during the period of project design and construction[2]. The diverted discharge is one of the key issues for flood mitigation of the Keelung River basin to protect downstream reach. Therefore, the government was able to effectively run the project for public safety. Under 200-year return-period flood condition, the design flood stage can be reduced by 1.5m on average at downstream reach. Hydraulic physical model was constructed to verify and evaluate the engineering design. The water levels and flow velocities at specific cross sections were measured in physical modeling to establish the rating curves of diversion discharge.

In this study, the rating curves for obtaining di-version discharge in the YFDW project were investigated. The calibration results between inflow and outflow discharge by using rating curve showed that the relationship did not satisfy conservation of mass. It was found that the relative discharge error could be more than 60%. In Figures 7 and 8, the water level hydrographs in YFDW are analyzed to be under unsteady flow condition. Therefore, the diversion discharge calculated by the rating curve from steady state flow field of hydraulic model might not be suitable to estimate diversion discharge in unsteady state condition. Figure 10 (a) shows that the water elevation distribution under 200-year return-period condition from hydraulic model test. The water elevation is higher near the ogee-shaped weir than that at the inlet. Figure 10 (b) presents the water elevation distribution near the peak flow in Typhoon Parma. Beside, the simulated results of 5-year return-period also confirm that the flow field and water level distribution did not keep uniform flow state. In the present study, the 3D numerical model can be applied to simulate water elevation and flow field and improve accuracy of the rating curve assessment.

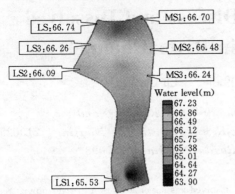

(a) Under 200-year return-period condition from hydraulic model test

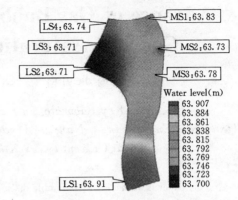

(b) Near peak flow in Typhoon Parma

Figure 10   Water elevation distribution in YFDW

Acknowledgment

The presented study was financially supported by the 10th River Management Office, Water Resources Agency Ministry of Economic Affairs. The writers would like to thank the Hydrotech Research Institute of National Taiwan University and the National Taipei University of Technology for their technical supports.

References

[1] Jan C. D. , Chang C. J. , Lai J. S. , Guo W. D. . Characteristics of hydraulic shock waves in an inclined chute contraction-numeral simulations. Journal of Mechanics, 2009, 25 (1): 75-84.

[2] Lai J. S. , Kang S. C. , Chang W. Y. , Chan Y. C. , Tan Y. C. . Development of a 3D virtual environment for improving public participation: case study-the Yuansantze Flood Diversion Works project. Advanced Engineering Informatics. 2010. (In press)

[3] Water Resource Planning Institute. Physical model studies of the Yuansantze flood diversion works of Keelung River 1st. Technical Report (in Chinese), Water Resources Agency, Taiwan, 2002.

[4] Water Resource Planning Institute. Physical model studies of the Yuansantze flood diversion works of Keelung River 2nd. Technical Report (in Chinese), Water Resources Agency, Taiwan, 2003.

[5] Water Resource Planning Institute. Physical model studies for detailed design of the Yuansantze flood diversion works of Keelung River, Technical Report (in Chinese). Water Resources Agency, Taiwan, 2005.

[6] The 10th River Management Office. Yuansantze flood diversion works-operation management and safety conservation project. Technical Report (in Chinese) . Water Resources Agency, Taiwan, 2006.

[7] The 10th River Management Office. Hydrological measurements and hydraulic analyses for the Yuansantze flood diversion works. Technical Report (in Chinese) . Water Resources Agency, Taiwan, 2009.

Advance of Gas Bubble Disease and Dissolved Gas Supersaturation in River Channels

Dong Jieying[1,2]   Chen Qihui[1,2]   Qi Liang[1,2]

[1] State Key Laboratory of Hydrology-Water Resources and
Hydraulic Engineering, Nanjing Hydraulic Research Institute, Nanjing, 210029
[2] Hohai University, 1 Xikang Road, Nanjing, 210098

Abstract: This paper is a review of gas bubble disease and dissolved gas supersaturation in river downstream of dams. High speed aerated flow arisen by discharge of dams can lead to dissolved gas supersaturation downstream, initiate gas bubble disease and threaten the security of water ecology downstream. The research of gas bubble disease and dissolved gas supersaturation in river was reviewed. Cause and symptom of gas bubble disease, mechanism, factors and harm of dissolved gas supersaturation were also introduced.

Key words: gas bubble disease; dissolved gas; supersaturation; ecology of water

1. Introduction

With the dams constructed one after another, the impact that dams have on the water environment downstream grasps more and more attention [1-2]. Flood discharge of reservoir dams often leads to total dissolved gas (TDG) supersaturation in downstream, which will cause gas bubble disease (GBD) of fish and they would suffer from death [3-5]. Then a significant reduction of fish stocks in downstream will emerge and the community structure and diversity will be alternated. The complex two-phase flow of water-vapor and TDG supersaturation under super atmospheric pressure [4] currently rarely described. The study of how dissolved gas attenuates [6] is not deep enough. A lot of experimental and theoretical researches ought to be taken on to understand and diminish the impact of TDG supersaturation on fish. In this paper, TDG supersaturation effects on fish are reviewed, and the mechanism of dissolved gas supersaturation, influencing factors and effects on fish are analyzed.

2. Fish gas bubble disease

Gas bubble disease [7] is a kind of fish environmental disease, related to over saturation of some gas in water, leading to many small bubble attached to the fish body surface and gill, even exist in intestinal canal. When floating or swimming it will make fish lose balance and sometimes cause mass mortality. Both juvenile and adult fish could be affected by this disease, and the damage to juvenile is more serious. Dissolved gas in water become supersaturated when dam begin to discharge flood, which directly lead to the occurrence of disease and cause massive death of fish, not only bring serious economic losses, but also cause recessions of fishery resources.

When dissolved oxygen in water become supersaturated, the oxygen can enter into fish through gill and blood cycle. Then a lot of free gas will stay in the blood vessel, and bubbles would form and expand to the heart with the increase of free gas, which cause blood block of the whole blood vessel [8], meanwhile, atrial and ventricular wall emphysema phenomenon also occurs, blood flows through the fins, fish skin capillaries, due to the pressure of oxygen at this time, oxygen dissociated from blood go into the tissues, excess oxygen remains in the organization and formats bubbles which are initially very small and then gradually increased. Fish swallows oxygen bubbles and bubbles can be formatted in the gut. when swallow more, it can form larger bubbles, which can make the fish float on the surface, lose balance, go round and round, struggle to swim, die at last.

Fish suffered from bubble disease, swimming slowly, weak, floating, welt, or distributed in the upper of more moderate water flow, severe fish swims with the abdomen upward against water because of a large number of bubbles in abdomen. Dissolved gas concentrates in normal tissues of fish and accumulates. The most common symptoms are proptosis and cornea, bubbles can be found on body surface, fins, gills[9]. Seriously, in front of the mouth, many bubbles can be found linear arrangement in both sides of the trench fissure Microscopically, pale gills, mucus increased between gill, there are many small bubbles, gill is complete and liver is white, some stomach have food, Intestine has yellow mucus and bubbles, there are no other symptoms expect dark appearance just like fish bleeding to death, some fish are entirely head congestion, mouth swelling around and cannot close.

The difference between total pressure and atmospheric pressure up to 400mm Hg[10-12] discharged flood caused massive fish, especially salmon die. This event draws people's concern about the phenomenon of gas supersaturation in river due to the high dam. So that research on the gas supersaturation in river and gas bubble disease was carried on. Ebel and others did research on the effect to salmon and steelhead trout by oversaturated solubility of nitrogen in Colombia rivers [11]. Beiningen and others carried research on the impact of the Snake River salmon by solubility of nitrogen [13]. In China, there are also reports about bubble disease because of the similar problem. In 1994, Zhejiang Province of China, because of Xin'anjiang reservoir twice flood discharge, the steelhead trout on the farm of Jiande which is 3km away from the power station suffered from bubble disease generally[14].

Research on fish bubble disease in the response of total dissolved gas in water or dissolved oxygen is mainly laboratory-based. Gunnarsli K S and others studied the adaptability of juvenile cods to the bubble disease [15], who points out that 103% saturation of dissolved gas is cod's lowest risk of bubble disease. By studying the migration of adult trout in the complex show the depth of choice behavior to get their adaptation of total dissolved gas supersaturation[16], Johnson EL and others point out that in Low saturation, steelhead trout can avoid the occurrence of bubble disease by hydrostatic compensation. However, when saturation is more than 130%, hydrostatic compensation can not stop the bubble disease.

3. The mechanism and advance of Dissolved Gas Supersaturation

3.1 The mechanism of Dissolved Gas Supersaturation

Dissolved Gas Supersaturation (DGS) in water generally happens in the tail water of the high dam or zones of complicated topography and water flow, however, water flow passing hydropower station has negligible effects in gas saturation in water.

Flood releasing outlet or spillway discharge flow can substantially increases gas saturation in water [5], moreover, among the dissipaters; ski-jump energy dissipater has greater impacts to gas saturation in tail water.

After discharge flow passed the surface of flip bucket, the discharge flow flies into air and entrains air by a large margin, as a result, typical aerated water or even atomized water is generated. Aerated flow falls after reaching the peak and then lashes the water cushion downstream [17]. When the nappe reaches at certain velocity, the interface of water and vapor becomes unsteady because of turbulence. Consequently, the surface of nappe tatters into spray and the flow run to downstream water cushion pool with abundance gas. The water column run deep into downstream water cushion pool with abundance gas, the deeper it goes the greater pressure it bears and the gas solubility is enhanced. Finally, the water column becomes saturation at certain depth. As the saturated water runs downstream and arrives in shoal water, there is not enough time for the dissolved gas to runoff and gas supersaturation water is generated.

3.2 Research Techniques of Dissolved Gas Supersaturation

There are generally three methodologies for the research techniques of Dissolved Gas Supersaturation: prototype measurements, numerical model and generalized test.

3.2.1 *Prototype Measurements*

The prototype measurements for Dissolved Gas Supersaturation has been well established at home and abroad. The Dissolved Gas Supersaturation in tail water of McNary Dam, which located at Snake River, Columbia, has been systemically analyzed [18]. Wilhelm's set forth the distribution pattern of general dissolved gas broadwise, endwise and streamwise respectively, he argues that the topography and depth of water in downstream channel are main factors affecting general dissolved gas, in addition, he establishes regression equation of flow per unit depth and general dissolved gas. Field measurements of Dissolved Gas Supersaturation have been done in Zipingpu tail water[6], Jiangliang indicates that the saturation of general dissolved gas in main stream zones is greater than non-mainstream zones and the variation trend of undershot flow discharge is similar to saturation of general dissolved gas, he concludes that the depth of water is the important factor affecting Dissolved Gas Supersaturation. Chen Yongcan compares the two sections in upper and lower reaches in The Three Gorges Reservoir[5], he analyzes the downstream variation of saturation of general dissolved gas, which caused by The Three Gorges flow discharge, he concludes that the ratio of orifice discharge to overall flow might influence downstream saturation of general dissolved gas by a large margin.

3.2.2 *Numerical Simulation*

The predictions of dissolved gas supersaturation experienced empirical formula, unidirectional flow module and current two-directional flow module [19-20]. Referring to the researches of U. S. army corps of engineers, Li Ran and Li Jia predict downstream saturation of general dissolved gas in several high dams (such as Zipingpu, Ertan and The Three Gorges) with prediction module of saturation of general dissolved gas in water cushion pool and bed scoured [21], the d-value between results of module calculation and prototype measurements is less than 5%. They also state that under the same flow discharge circumstance, flow released from deep outlet has less effect in the downstream saturation of dissolved gas than from surface and middle outlet. Learning from others' researches, Cheng Xiangju combines gas-liquid two-phrase mixture module [4] and scalar equation of diffusion dissolved oxygen saturation, then she get a numerical model which could extensively apply in stimulation of downstream saturation of dissolved gas in high dam. Through sensibility analysis of factors in saturation of dissolved oxygen, she believes that gas ratio, turbulence level and depth of downstream water are major factors in distribution of saturation of dissolved oxygen.

3.2.3 *Generalized Test*

With high velocity jet instrument and other instrument designed by themselves, Jiang Liang and Li Ran validated the cause of Dissolved Gas Supersaturation in lower reach of high dam [3], they figured that without entrained air, the discharge rate of supersaturated dissolved gas increases with the raise of turbulence level and decreases with the depth of water.

4. Prospect

From the available data, we can conclude that flow discharge from high dam induces Dissolved Gas Supersaturation, which bring fatal disaster to aquatic organism, specifically to fish. On the basis of Xu Xuzhong's investigation [22], the juvenile amount of four major Chinese carps in the Yangtze River has fall from 30 billion in 1950s to less than 1 billion at present, the fishery structure in Yangtze River has been significantly changed and biodiversity has been losing. Four major Chinese carps' proportion in catches experiences a decline from 20%-30% in 1960s to 5% now. Besides overexploitation, water pollution and degradation of habitat, the Dissolved Gas Supersaturation, which is caused by flow discharge from high dam, has become one major factor that undermines fishes in tail water of dams.

In the Snake River basin, the problem of dissolved gas supersaturation is especially serious, which directly lead to the endangerment of salmon and trout. In China, there are a number of high dams being or to be constructed, such as Xiluodu, Xiangjia Dam, Jinping and Baihetan. Those dams bring about economic benefits while they pose enormous threat to fishes and other aquatic organism; it is a big challenge for ecosystem preservation. Therefore, it is of great important to research on Dissolved Gas Supersaturation caused by flow discharge from high dam and its effects in gas bubble disease.

Based on the current research findings, the downstream saturation of general dissolved gas in tail water of dam increases with the raise of flow discharge, the depth of tail water and turbulence level [5-6, 23]. Besides, factors such as water temperature and salinity can affect it as well. More prototype measurements and module test are required to acquire the exact influence.

The topography is complicated in China, there are identical difference in the lower reaches of river bed and the design of flood releasing outlet and over-flow dam, moreover, reservoir regulations are not same, so the accuracy of predictions and numerical simulation of dissolved gas in the tail water of dam need to be tested and popularized.

There are insufficient research findings concerning different fishes' response to saturation of dissolved gas in water. In order to preserve aquatic organism and water eco-environment in different rivers and reaches, it is very important to acquire the least saturation of dissolved gas that leads to gas bubble disease and sensibility of different fishes, especially fishes under state protection. Researches on reservoir regulation and engineering remedial measure to reduce the rate of gas bubble disease and surveys on the proportion of fish dead from gas bubble disease in irregular death is also of significance.

Acknowledgment

This research is supported by NSFC for Youth (No. 50909064) and Ministry of Water Resources for Nonprofit Research (No. 200801105).

References

[1] QIANG W, XINBIN D, SHUYING X, et al. Studies on Fishery Resources in the Three Gorges Reservoir of the Yangtze River. Freshwater Fisheries, 2007, 37 (2): 70-75.

[2] WEIGANG S, MINGYING Z, KAI L, et al. Stress of hydraulic engineering on fisheries in the lower reaches of the Yangtze River and compensation. J. Lake Sci., 2009, 21 (1): 10-20.

[3] LIANG J, RAN L, JIA L, et al. The supersaturation of dissolved gas in downstream of high dam. Journal of Sichuan university (Engineering Science Edition), 2008, 40 (5): 69-73.

[4] XIANGJU C, YONGCAN C, XUEWEI C, Numerical simulation of dissolved oxygen concentration in the downstream of Three Georges Dam. Chinese journal of hydrodynamics, 2009, 24 (6): 761-767.

[5] YONGCAN C, JIAN F, ZHAOWEI L, et al. Analysis of the variety and impact factors of dissolved oxygen downstream of Three Gorges Dam after the impoundment. Advances in water science, 2009, 20 (4): 526-530.

[6] LIANG J, JIA L, RAN L, et al. A study of dissolved gas supersaturation downstream of Zipingpu dam. Advance in water science, 2008, 19 (3): 367-371.

[7] QINDONG W. Prevention of gas bubble disease in fish. Shandong Fisheries, 2007, 24 (2): 18-19.

[8] GRAHN B H, SANGSTER C, BREAUX C, et al. Case report: Clinical and pathologic manifestations of gas bubble disease in captive fish. JOURNAL OF EXOTIC PET MEDICINE, 2007, 16 (2): 104-112.

[9] SALAS-LEITON E, CANOVAS-CONESA B, ZEROLO R, et al. Proteomics of Juvenile Senegal Sole (Solea senegalensis) Affected by Gas Bubble Disease in Hyperoxygenated Ponds. MARINE BIOTECHNOLOGY, 2009, 11 (4): 473-487.

[10] CLARK, R M J. Environmental Protection Dissolved Gas Study: Data Summary-1977, 1977.

[11] EBEL, J W. Supersaturation of nitrogen in the Columbia River and its effect on salmon and steelhead trout. US National Marine Fisheries Service, Fishery Bulletin, 1969, 68: 1-11.

[12] WEITKAMP, E. D, KATZ M. A review of dissolved gas supersaturation literature. Trans. Am. Fish. Soc, 1980, 109: 659-702.

[13] BEININGEN, T. K, EBEL W J. Effect of John Day Dam on dissolved nitrogen concentrations and salmon in the Columbia River. Trans. Am. Fish. Soc, 1970, 99: 664-671.

[14] CHENGGEN W. Gas bubble disease of Steelhead. China Fishery, 1994, 10: 27.

[15] GUNNARSLI K S, TOFTEN H, MORTENSEN A. Effects of nitrogen gas supersaturation on growth and survival in larval cod (Gadus morhua L.). AQUACULTURE, 2009, 288 (3-4): 344-348.

[16] JOHNSON E L, CLABOUGH T S, CAUDILL C C, et al. Migration depths of adult steelhead Oncorhynchus mykiss in relation to

dissolved gas supersaturation in a regulated river system. Journal of Fish Biology, 2010, 76 (6): 1520-1528.

[17] XIANGJU C, YONGCAN C, QIANHONG G, et al. Supersaturated re-earation analysis on the Three Gorges Reservoir during discharge. Journal of hydroelectric engineering, 2005, 24 (6): 62-67.

[18] WILHELMS S C, SCHNEIDER M L. Near-field study of TDG in MaNary spillway tailwater. Fisheries and Structural Hydrodynamics Branch, 1997.

[19] C D H, L W L, Z G J. Numerical simulation of two-phase turbulent flow in hydraulic and hydropower engineering. Sci China Ser E-

Tech Sci, 2007, 50 (Suppl I): 79-89.

[20] HUANG H Q. Computational model of total dissolved gas downstream of a spillway. Iowa: University of Iowa, 2002.

[21] RAN L, JIA L, KEFENG L. Predictionfor supersaturated total dissolved gas in high-dam hydropower projects. Sci China Ser E-Tech Sci, 2009, 52 (12): 2001-2006.

[22] XUZHONG X. Yangtze aquatic challenge. Bussiness watch, 2010, 8: 40-41.

[23] XUEWEI C, XIANGJU C, WEI Z. Progress of Model Studies on total dissolved gas supersaturation downstream of Sluicing Dam. Science and technology, 2009, 27 (17): 101-105.

The Application of Two-phase Flow Numerical Simulation in Fengman Dam Overall Treatment Project

Yuan Qiang   Wang Xin   Luo Shaoze

State Key Laboratory of Hydrology-Water Resources and Hydraulic Engineering,
Nanjing Hydraulic Research Institute, Nanjing, 210029

Abstract: A new dam would be built 120m downstream of the old dam in accordance with the requirements of the overall treatment project of Fengman Hydropower Station. The water-air two-phase flow 2-D unsteady numerical simulation of ski-jump energy dissipation of the old dam from upstream to plunge pool is established by the combination of VOF method, k-ε turbulent model, PISO algorithm, structured and unstructured grid. The simulation vividly shows the impact situation on the new dam, which gives basis for the erosion experiment and supplies good guidance on the construction of the new dam during the flood season. What's more, it provides important reference value on the future renovation projects of more old dams.

Key words: ski-jump energy dissipation; two-phase flow; VOF method; numerical simulation; old dam renovation

1. Introduction

A new dam would be built 120m downstream of the old dam in accordance with the requirements of the overall treatment project of Fengman Hydropower Station. The new dam in construction period will be influenced deeply by high-speed flow discharged from the old dam in flood season. In order to study the impact on the new dam, such as impact velocity, angle and location, the numerical simulation of ski-jump energy dissipation of the old dam from upstream to plunge pool is established.

There have been many research results about ski-jump energy dissipation[1-2]. The numerical simulation of ski-jump energy dissipation is related to water-air two-phase flow. The water-air two-phase problem is very complicated itself. Meanwhile, boundaries of jet flow in air are unknown in advance, which brings trouble for making certain its boundaries. So it is difficult in computational fluid dynamics[3]. Based on the review and summarization of existing research achievements concerned at present, a 2-D unsteady simulation of the jet flow of the Fengman old dam from upstream to downstream is established by the combination of VOF method, k-ε turbulent model, PISO algorithm, structured and unstructured grid.

2. Computing Theory

2.1 VOF method

The basic idea of this method is: the functions $a_w(x, y, z, t)$ and $a_a(x, y, z, t)$ represent water and gas volume fraction (the relative volume ratio) in the calculation region. The sum of water and gas volume fractions is 1 in each unit:

$$a_w + a_a = 1 \qquad (1)$$

The VOF method[4] is designed to track the position of the interface between two or more immiscible fluids. A single momentum equation is solved and the resulting velocity field is shared by all phases. Turbulence and energy equations are also shared by all phases, if required. Surface tension and wall adhesion effects can be taken into account.

If the volume fraction of water and air in the flow field are known, all the other unknown quantities and features parameters of water and air can be represent by the weighted average of volume fraction of water and air. So in any given cell, these variables and parameters represent either water or air, or a mixture of both.

The control differential equation of the water volume fraction is:

$$\frac{\partial a_w}{\partial t} + u_i \frac{\partial a_w}{\partial x_i} = 0 \qquad (2)$$

Where t is time, μ_i and x_i is velocity components and coordinate components respectively. Water-air interface is tracked by solving the continuity equation above.

2.2 Stratified two-phase flow k-ε model

There are many turbulence models at present. But the most widely used in computational fluid dynamics is k-ε model[5]. In the VOF method, water and gas have the same velocity and pressure field, so the water-air two-phase flow field can be described by a set of equations as single-phase flow. In the k-ε turbulence model, the continuity equation, momentum equation and the k, ε equations can be expressed as follows.

The continuity equation:

$$\frac{\partial \rho}{\partial t} + \frac{\partial \rho u_i}{\partial x_i} = 0 \tag{3}$$

The momentum equation:

$$\frac{\partial \rho u_i}{\partial t} + \frac{\partial}{\partial x_j}(\rho u_i u_j) = -\frac{\partial p}{\partial x_i} + \frac{\partial}{\partial x_j}\left[(\mu + \mu_t)\left(\frac{\partial u_i}{\partial x_j} + \frac{\partial u_j}{\partial x_i}\right)\right] \tag{4}$$

The k equation:

$$\frac{\partial(\rho k)}{\partial t} + \frac{\partial(\rho u_i k)}{\partial x_i} = \frac{\partial}{\partial x_i}\left[\left(\mu + \frac{\mu_t}{\sigma_k}\right)\frac{\partial k}{\partial x_i}\right] + G - \rho\varepsilon \tag{5}$$

The ε equation:

$$\frac{\partial(\rho\varepsilon)}{\partial t} + \frac{\partial(\rho u_i \varepsilon)}{\partial x_i} = \frac{\partial}{\partial x_i}\left[\left(\mu + \frac{\mu_t}{\sigma_\varepsilon}\right)\frac{\partial \varepsilon}{\partial x_i}\right] + C_{1\varepsilon}\frac{\varepsilon}{k}G - C_{2\varepsilon}\rho\frac{\varepsilon^2}{k} \tag{6}$$

Where ρ and μ are the average density of volume fraction and molecular viscosity. p is correction pressure. μ_t is turbulent viscosity, which can be obtained by turbulent energy k and turbulent dissipation rate ε:

$$\mu_t = \rho C_\mu \frac{k^2}{\varepsilon} \tag{7}$$

Where C_μ is empirical constant and usually $C_\mu = 0.09$.

σ_k and σ_ε are Prandtl number of k and ε. $\sigma_k = 1.0$, $\sigma_\varepsilon = 1.3$. $C_{1\varepsilon}$ and $C_{2\varepsilon}$ are ε equation constant, $C_{1\varepsilon} = 1.44$, $C_{2\varepsilon} = 1.92$. G is the turbulent produced item caused by the average velocity gradient. It can be defined by the equation below:

$$G = \mu_t\left(\frac{\partial u_i}{\partial x_j} + \frac{\partial u_j}{\partial x_i}\right)\frac{\partial u_i}{\partial x_j} \tag{8}$$

With the introduction of VOF method, the form of k-ε turbulence model equations is as the same as that of single-phase flow model. But the specific expression of ρ and μ is different because they are given by the weighted average volume fraction. They are a function of volume fraction rather than a con-

stant. They can be expressed as follows:

$$\rho = \alpha_w \rho_w + (1 - \alpha_w)\rho_a \tag{9}$$

$$\mu = \alpha_w \mu_w + (1 - \alpha_w)\mu_a \tag{10}$$

Where α_w is the volume fraction of water, ρ_w and ρ_a is the density of water and air, μ_w and μ_a is molecular viscosity of water and air. ρ and μ can be obtained through the iterative solution of water volume fraction.

2.3 PISO algorithm

PISO (Pressure Implicit with Splitting of Operators) algorithm is proposed by Issa in 1986. It is created as a pressure speed computer program initially for unsteady compressible flow of non-iterative calculation, and then it is widely used in iterative calculation of the steady state problem. PISO is pressure-velocity coupling approximation algorithm which based on the highly similar relationship between pressure and velocity[6]. It is different from SIMPLE and SIMPLEC algorithm: SIMPLE and SIMPLEC algorithm are two-step algorithms, that is, a prediction step and a correction step; PISO algorithm adds a correction step and so includes a prediction step and two correction steps. After obtain (u, v, p) with the completion of the first amendment, it searches for second low-time improvement value, aiming to better satisfy the momentum equation and the continuity equation at the same time. With the use of predict - correction - revise, PISO algorithm can speed up the convergence speed of single iteration step[7].

3. Numerical simulation

3.1 Model establishing

The 2-D model of flip bucket is established using the actual size of Fengman old dam. The origin of Coordinate is set at the highest overflow weir. The simulation ranges from reservoir upstream to downstream river, including the whole weir section. The calculation model is divided by structured and unstructured grids and the calculation mesh division is shown in Figure 1.

3.2 Calculation conditions and boundary conditions

3.2.1 *Calculation conditions.* Choose the check flood level 267.70m as the calculation condition.

3.2.2 *boundary conditions.* A UDF (User-De-

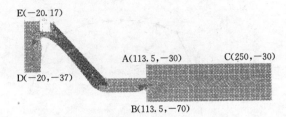

Figure 1　Calculation mesh division

fined-Function) is written by C language to make pressure change with the depth. In upstream, the inlet condition is defined by the UDF and pressure-inlet

is used above the water level. Besides, the boundary condition is pressure-inlet upper the model, pressure-outlet in the downstream and the other are solid wall. The downstream water level is not considered.

3. 3　Results and analysis

Figure 2 shows the volume fraction from the beginning of flow discharge ($t=0$) to discharge stability, in which blue is for water and red is for air. From the Figure we can clearly see the jetting process and pattern. Figure 3 shows water-air mixture velocity vector, Figure 4 and Figure 5 show the stream line and pressure contour respectively.

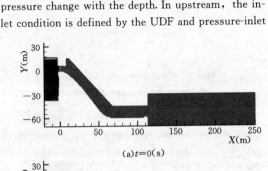

(a)$t=0$(s)　　　　　　(b)$t=5$(s)

(c)$t=7$(s)　　　　　　(d)Stability

Figure 2　Volume fraction

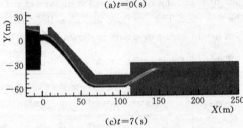

Figure 3　Mixture velocity vector

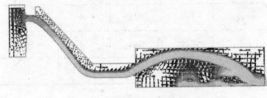

Figure 4　Stream line

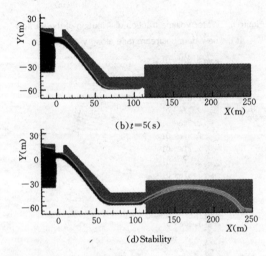

Figure 5　Pressure contour

Figure 6 shows water volume fraction distribution of the jet flow at the new dam upstream face along y-coordinate. We can see that the jet water tongue focus on vertical axis from-48m to-40m. Considering that the height of new dam foundation surface is 182. 96m, the jet flow

impacts on the new dam at the height from 204. 5m to 212. 5m. Figure 7 shows the mixed-phase velocity vector of the jet flow at the new dam upstream face. Figure 8 shows the mixed-phase velocity magnitude along the vertical axis and we can see that the maximum speed of about 33. 7m/s appears at y-coordinate of-43. 2m or the height of 209. 3m. Figure 9 shows the mixed-phase veloc-

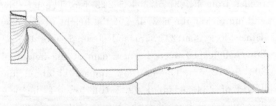

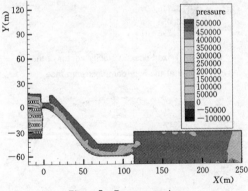

ity angle along the vertical axis at the new dam upstream face and we can see that the angle of the jet flow is about 18.7 degrees.

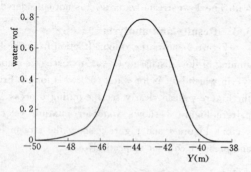

Figure 6　Water volume fraction distribution of the jet flow at the new dam upstream face along y-coordinate

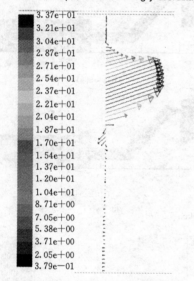

Figure 7　Mixed-phase velocity vector of the jet flow at the new dam upstream face

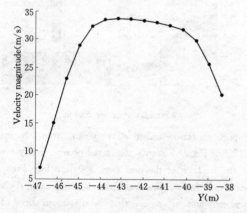

Figure 8　Mixed-phase velocity magnitude of the jet flow at the new dam upstream face

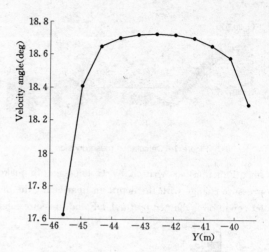

Figure 9　Mixed-phase velocity angle of the jet flow at the new dam upstream face

To research the impact of the jet flow on the new dam, we need know the construction process of the new dam and the relative position of the old dam and the new dam. The pre-flood and after-flood construction height of the new dam is shown in Table 1.

Table 1　Construction height of the overflow section

year	height (m)		model y-coordinate (m)	
	pre-flood	after-flood	pre-flood	after-flood
3	186	191	−66.5	−61.5
4	193	212	−59.5	−40.5
5	220	242	−32.5	−10.5
6	250	250	−2.5	−2.5

The jet tongue at the new dam upstream face ranges from height of 204.5-212.5m. The jet flow will jump over the new dam if the height of the new dam is less than 212.5m and some part of the jet flow will leaped over the new dam at the height of 204.5-212.5m. The jet flow impact at about the height of 234-241m or 99.5-106.5m off the new dam upstream face when the height of the new dam is 186m, at about the height of 230-236m or 230-236m off the new dam upstream face when 191m, at about the height of 200-205m or 65.5-70.5m off the new dam upstream face when 193m. The detail is in Figure 10.

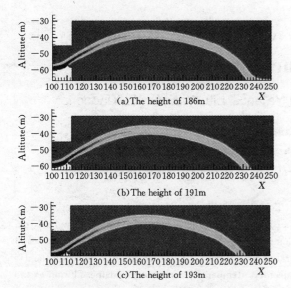

(a) The height of 186m

(b) The height of 191m

(c) The height of 193m

Figure 10　Throw distance of different height

4. Conclusion

The erosion problem of the new dam in construction is a key factor that affects the successful construction and needs to conduct systematic research. The water-air two-phase flow numerical simulation provides technical support for the safety of the new dam in construction in flood season. On the other hand, it supplies important reference value on the future renovation projects of more old dams. The water-air two-phase flow numerical simulation has good effect for the erosion research of new dam construction.

References

[1]　Xu Weilin, Liao Huasheng, Yang Yongquan, Wu Chigong. Computational and experimental investigation on the 3-D flow feature and energy dissipation characteristics of plunge pools [J]. Chinese Journal of Theoretical and Applied Mechanics, 1998, 30 (1): 35-42.

[2]　Chen Yongcan, Xu Xieqing. Numerical simulation of flow in flip scour pit [J]. Journal of Hydraulic Engineering. 1993 (4): 48-54.

[3]　Diao Mingjun, Yang Yongquan, Wang Yurong, Liu Shanjun. Numerical simulation of water-air two-phase jet flow from flip bucket to plunge pool [J] Journal of Hydraulic Engineering, 2003 (09): 77-82.

[4]　Hirt C W, Nichols B D · Volume of Fluid (VOF) Method For The Dynanics of Free Boundary [J]. J. Comput. Phys. 1981, 39: 201-225.

[5]　Lannder B E, Spalding D B · Mathematical Models of Turbulence [M]. Academic Press, London and NewTork, 1972: 90-110.

[6]　Qi Yuantian. Analysis on flood discharge and energy dissipation of high arch dam [D]. PhD thesis of Hohai University, 2006.

[7]　Wang Fujun, Computational Fluid Dynamics Analysis [M], Tsinghua University Press, 2005.

Study on Data Mining of Jet Field Based on Artificial Neural Network

Cao Xiaomeng[1]   Gu Zhenghua[1]   Hu Ya'an[2]   Liu Wang[1]   Xu Xiaodong[1]

[1] *Institute of Water Resources, Zhejiang University, Hangzhou, 310058*
[2] *Department of Hydraulic Engineering, Nanjing Hydraulic Research Institute, Nanjing, 210029*

Abstract: For the problems of single analysis mean of jet field data and inadequate knowledge discovery, we uses standard k-ε turbulence to found mathematical model of the vertical round buoyant jet in cross flow based on previous test data firstly, and then we obtains a large number of jet field data by numerical experiments, and further we introduces the data mining mode based on neural network theory (EDM-π&ANN). Through the data mining of jet field, non-linear formulas of velocity trajectory, temperature trajectory and the dilution at the highest position of temperature trajectory have been got finally, which have the features of harmonious dimension and unified form. And some new laws of vertical round buoyant jet in cross flow have also been founded. The investigation results provide an effective method to data analysis for knowledge discovery of jet and other similar scientific issues, and may have some academic reference value for further understanding of movement mechanism of jet.

Key words: jet; data mining; artificial neural network; CFD simulation; π theorem

1. Introduction

Jet is a fluid which enters into the surrounding fluid from a variety of discharge mouths or nozzles, and occurs strong mixing with the surrounding fluid (Shao Lei, 2007). With the rapid development of industry and agriculture, a large number of domestic sewage, industrial wastewater discharge into rivers, lakes, such discharges of emissions can be generally summarized into jet problems. Jet has been applied widely in water environment protection; e. g. , jet has been applied to the design of sewage discharge, flushing, dredging, and aeration to improve water quality, controlling lake eutrophication, and prevention of salt water intrusion and so on. In addition, jet has been also widespread in the ventilation and air conditioning, hydraulic mining, civil fire fighting, agricultural irrigation and other projects. Therefore, research on jet can not only deepen the understanding of jet, further enrich the jet theory, and also has important practical significance to project practice (Yang Zhonghua, 2004).

Currently, the primary research method of jet is combination of physical model test and numerical simulation. Such as: Lee and Balasubramanian et al (Balasubramanian et al. , 1978; Kuang et al. , 1999) researched on horizontal buoyant jets in quiescent shallow water and the stability of vertical plane buoyant jet; Johnston and Phillips et al (Volke, 1994; Johnston et al. , 1998) did some systematic experimental study on horizontal buoyant jets in static shallow water, and gave appropriate models of that jets; David, Habib et al. (Habib et al. , 1987; Dominique et al. , 1984) did lots of researches on surface buoyant jet, including experimental research, model research and numerical simulation; Huai Wenxin, Zeng Yuhong et al (Zeng Yuhong et al. , 2005; Huai Wenxin et al. , 2006) did lots of numerical simulation on horizontal and vertical jets with different environments, different mobility patterns, and obtained a series of laws and the empirical formulas.

There are many factors which affect the characters of jet, such as the relationship of data is complex, and the data are strong nonlinear and there are noisy data (Zhao Gaofeng et al. , 2008), so research data have not been fully utilized, this mainly show in two aspects (Gu Zhenghua et al. , 2006): (1) The research means of test data are still more dependent on manual analysis, because analytical tool is backward and single comparative, the mining of physical laws inherent in these data is inadequate; (2) a small part of biased data affect the understanding of physical laws, a way of knowledge discovery which is fault tolerance is needed eagerly. Data mining technology which birthed in 1980s provides a promising solution to use these data fully. Data mining is also known as knowledge discovery from databases (KDD), which is a complex process that mining unknown, valuable models or laws from large data base (Gu Zhenghua et al., 2006). In recent years, data mining is a hotspot at home and abroad, which is widely used in business, information and other fields, in water science, data mining is successfully applied to dam safety monitoring (Yin Zhengjie, 2006), hydrological forecasting (Zhang Chi, 2007) and flood control. Among that, Gu Zhenghua et al. (Gu Zhenghua et al., 2006) presented a test data mining mode which jointed π theorem and neural network, and had applied this theory to the understanding of greyhound's wake field successfully. For the problems of single analysis mean of jet field data and inadequate knowledge discovery, we give an example of vertical round buoyant jet which is a common phenomenon in practical project, and introduce EDM-π & ANN to data analysis of jet field to further understand the motion law of jet field.

2. Jet field data collection

2.1 Numerical experiment design

We consider the problem of vertical round buoyant jet in cross flow likes Figure 1 (Zeng Yuhong et al., 2005), that a round jet with outlet velocity V_j and outlet temperature T_j enters into the environmental fluid which velocity is u_a, temperature is T_a ($T_a < T_j$) from the nozzle of diameter D, and two fluids occur strong mixing.

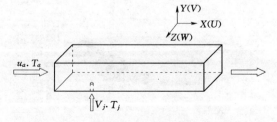

Figure 1 3-D round buoyant jet issued from the bottom in flowing environmental fluid

We use international more popular commercial CFD package-FLUENT to build numerical simulation model of jet. The condition designs of numerical experiment are shown in Table 1, which sets 13 kinds of conditions, in order to test the accuracy of parameters, conditions 1-3 (Zeng Yuhong et al., 2005) are selected reference (Zeng Yuhong et al., 2005)'s test conditions. Hexahedral meshes are used in calculation, the problem is considered to be steady in simulation, and pressure based solver is selected, turbulence model uses standard k-ε model, which is solved by SIMPLE algorithm. In the simulation process, fluctuation of density and transport coefficient are ignored, and Boussinesq assumption is set, which means density changes are only accounted in gravity items. The boundary condition of nozzle is velocity-inlet, $v = v_j$, $u = 0$, $w = 0$, $T = T_j$, and $k = 0.06v_j^2$, $\varepsilon = 0.06 \dfrac{v_j^3}{D}$; The boundary condition of upstream flow is also velocity-inlet, $u = u_a$, $v = 0$, $w = 0$, $T = T_a$, and $k = 0.06v_j^2$, $\varepsilon = 0.06 \dfrac{v_j^3}{D}$; The boundary condition of downstream exit is pressure-outlet; The boundary condition of wall is no slip and stationary wall; Near-wall treatment is solved by standard wall function, and free surface is solved by rigid lid hypothesis.

Table 1   Calculation conditions of CFD

$c\#$	D(mm)	v_j(m/s)	T_j(℃)	u_a(m/s)	T_a(℃)	H/D	R
1	8.1	0.6	34	0.1	13	24.7	6
2	8.1	0.8	34	0.1	14.5	24.7	8
3	8.1	1	34	0.1	13	24.7	10
4	8	0.4	34	0.1	13	25	4
5	8	0.5	34	0.1	13	25	5
6	8	0.7	34	0.1	13	25	7
7	8	0.9	34	0.1	13	25	9
8	6	0.6	34	0.1	13	33.3	6
9	10	0.6	34	0.1	13	20	6
10	12	0.6	34	0.1	13	16.7	6
11	8	0.6	30	0.1	13	25	6
12	8	0.6	40	0.1	13	25	6
13	8	0.6	45	0.1	10	25	6

Note R ($= v_j/u_a$) is velocity ratio, H/D is submergence depth.

2.2 Preliminary analysis of CFD simulation results

The key factors which impact the movement of jet in surrounding fluid are the characters of jet and environmental fluid. Jet trajectory can describe the movement of jet in environmental fluid clearly and vividly. The track line of maximum velocity on symmetry plane xoy is defined as velocity trajectory; the track line of maximum temperature on symmetry plane xoy is defined as temperature trajectory. Contrast the simulation results of condition 1-3 to the reference (Zeng Yuhong et al., 2005)'s equations of velocity trajectory (1) and temperature trajectory, (2) which are in the range of $x/D < 20$, the comparison results are shown in Figure 2 and Figure 3. From the figures we can see the two are in good agreement, and the development trend of trajectories is consistent.

$$\frac{y}{D} = 0.99 \left(\frac{\rho_i}{\rho_0}\right)^{0.47} R^{1.0} \left(\frac{x}{D}\right)^{0.3} \quad (1)$$

$$\frac{y^*}{D} = 0.8 \left(\frac{\rho_i}{\rho_0}\right)^{0.63} R^{1.1} \left(\frac{x}{D}\right)^{0.25} \quad (2)$$

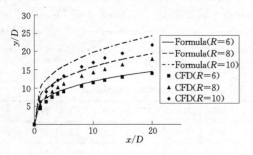

Figure 2　Velocity trajectory

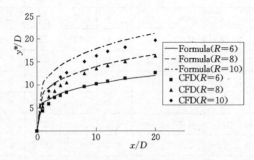

Figure 3　Temperature trajectory

The characters of vertical round buoyant jet in cross flow have certain complexity. According to previous scholars' studies in relevant aspects, it found that vertical round buoyant jet in cross flow had bi-furcation phenomenon and U-shaped structure (Zeng Yuhong et al., 2005). Figures 4, and Figure 5 give the temperature contours respectively which are at different sections along x direction and y direction under condition 1. From Figure 4, bifurcation phenomenon can be clearly seen along the cross sections of x direction ($x = 3D$, 5D); from Figure 5, U-shaped structure can be clearly seen at different heights sections above the nozzle ($y = 2D$, 4D, 6D, 8D). These phenomena are mainly caused by the interaction between cross flow and jet, which are accordant with described situation of the reference (Zeng Yuhong et al., 2005). Figure 6 show the temperature contours on symmetry plane xoy under condition 1, 2, 3 (denoted c1, 2, 3). From the figures we can see temperature declines rapidly near the nozzle, and in the downstream spread process, temperature shows a

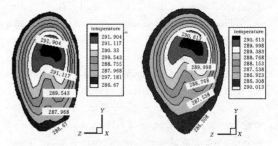

Figure 4　Contour lines of temperature ($x = 3D$, 5D)

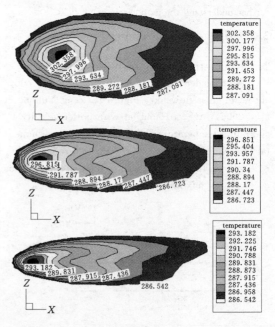

Figure 5　T contour lines above nozzle
($y = 2D$, 4D, 6D, 8D)

clear stratification phenomenon, and with velocity ratio increasing, the maximum height of jet also increases. These all laws are accordant with the test results of Zeng Yuhong et al. (Zeng Yuhong et al., 2005). Therefore, the established mathematical model and the model parameters in this paper are appropriate; it can be used in further analysis for the laws of vertical round buoyant jet.

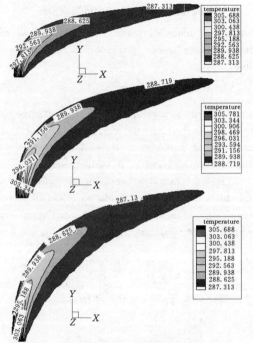

Figure 6   T contour lines on symmetry
plane xoy (c1, 2, 3)

Figure 7 shows the laws of V_m with the variation of x/D under conditions 4, 5, 6, from which can be seen that trend of the three curves is generai same. With the development of jet, V_m reduces gradually and tends to stabilize, and V_m declines fastest at the vicinity of nozzle.

3. Data mining of jet field based on EDM-π & ANN

3.1   EDM-π & ANN mode

The basic principle of EDM-π & ANN mode is dimensional analysis to researched problems using π theorem. And establish dimensionless function expressions of test data mining model, define inputs and outputs of neural network model, train the neural network model by applying the learning materials,

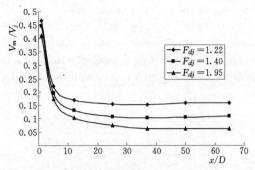

Figure 7 x/D affecting $\dfrac{V_m}{V_j}$ with different F_{dj}

estimate the π items based neural network model, define input and output of neural network over again, and train minutely until obtain the final mining model, and then begin data mining, and find potential and useful knowledge. The basic steps can be summarized as: 1) determine independent physical variables which impact the water flow system; 2) select basic variables; 3) according to π theorem, obtain (n-m) dimensionless function expressions which are composed by π items; 4) determine dependent variables and independent variables and establish artificial neural network model; 5) train the neural network model by the learning materials, obtain nonlinear mapping between dependent variables and independent variables; 6) based on non-linear relationship, evaluate the factors further, repeat 4), 5) until obtain the final test data mining model; 7) discover useful knowledge using the mining model, and make proper assessment (Gu Zhenghua et al. 2006).

3.2   Mining results

3.2.1   *Jet trajectory*

Studying condition 11 and condition 13, the different of two is only jet temperature and environmental temperature. Figure 8 gives the distribution of velocity V and temperature T along the y direction at $x/D=12$ on symmetry plane xoy under these two conditions. The figures show that the vertical coordinates of V_m and T_m are affected little by temperature, so velocity trajectory and temperature trajectory can neglect the effect of temperature. In addition, the viscous effect of buoyant jet is small comparatively. For simplifying the problem, the impact of Reynolds number is not considered here.

Therefore, the vertical coordinate y of maximum velocity V_m and the vertical coordinate y^* of maximum temperature T_m on symmetry plane xoy can be expressed by the following relationship:

$$(y, y^*) = f(\rho, g, H, D, x, v_j, u_a) \qquad (3)$$

where, ρ is water's density, g is gravitational acceleration, H is depth of environmental water, D is the diameter of nozzle, and x is the coordinate of x direction.

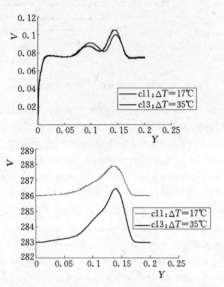

Figure 8 ΔT affecting V and T at xoy ($x/D = 12$)

According to π theorem, we choose ρ, g, D as the basic variables and obtain dimensionless expression

$$\begin{cases} \left(\dfrac{y}{D}, \dfrac{y^*}{D}\right) = F\left(\dfrac{H}{D}, \dfrac{x}{D}, F_{dj}, R\right) \\ F_{dj} = \dfrac{v_j}{\sqrt{gD}} \\ R = \dfrac{v_j}{u_a} \end{cases} \qquad (4)$$

$\dfrac{H}{D}$, $\dfrac{x}{D}$, F_{dj}, R as the inputs of neural network, $\dfrac{y}{D}$ and $\dfrac{y^*}{D}$ as the outputs of neural network, we obtain the topological structure of artificial neural network model which is shown in Figure 9.

Though numerical experiments we can find that when R is large (e. g., c3, c7), jet is up to the free surface of sink, so select the simulation results of c1-c2, c4-c6, c8-c11 in Table 1 as the learning materials. In addition, we can also find that jets are up to maximum height at $x/D = 70$ under various conditions, so we choose 80 as the maximum of x/D, and site 11 points according the former are dense and the latter are sparse. Learning materials are normalized by

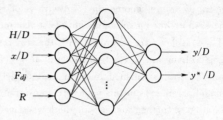

Figure 9　Computational model's topological structure of y/D and y^*/D based ANN

Formula 5 uniformly, which can prevent neurons from reaching super saturation (Yang Rongfu, 1998).

$$Y_i = \frac{Z_i - Z_{min}}{Z_{max} - Z_{min}} \beta + \alpha \qquad (5)$$

where, Z_i and Y_i is variables before and after normalized respectively; Z_{min} and Z_{max} is the minimum and maximum of variables correspondingly; β takes a number between 0 and 1, this paper uses 0.9.

Train the neural network model appropriately, and obtain the nonlinear calculation model of vertical coordinate y of maximum velocity V_m and vertical coordinate y^* of maximum temperature T_m on the symmetry plane xoy eventually.

$$\begin{cases} a = \left(\dfrac{9}{166}\dfrac{H}{D} - 0.86, \dfrac{9}{80}\dfrac{x}{D} + 0.05, \right. \\ \left. \qquad \dfrac{9}{11}F_{dj} - 0.87, \dfrac{3}{20}R - 0.55\right)^T \\ b = \dfrac{1}{1 + e^{-(\omega A - \theta)}} \\ \left(\dfrac{y}{D}, \dfrac{y^*}{D}\right) = \dfrac{1}{26_{1 + e^{-(vb - r)}}} \end{cases} \qquad (6)$$

Where, $\omega(4, 6) =$

$$\begin{bmatrix} -0.358 & 0.847 & 5.832 & 8.121 & -0.581 & -0.103 \\ -3.874 & -1.509 & 11.159 & 12.950 & 5.904 & 46.608 \\ 3.338 & -3.258 & 2.317 & 2.155 & 1.194 & 0.693 \\ 2.166 & 1.096 & 0.577 & -1.989 & 0.294 & -1.274 \end{bmatrix}$$

$$\theta(6, 1) = \begin{bmatrix} -1.626 \\ -2.282 \\ 9.270 \\ 10.956 \\ 1.688 \\ 2.638 \end{bmatrix},$$

$$v(6, 2) = \begin{bmatrix} 1.705 & 1.504 \\ -3.330 & -4.553 \\ -2.710 & -7.512 \\ 2.752 & -4.015 \\ 2.169 & 1.198 \\ 12.628 & 12.177 \end{bmatrix},$$

$$\gamma(2, 1) = \begin{bmatrix} 12.254 \\ 10.391 \end{bmatrix}.$$

Contrast the trained results of neural network with the results of CFD numerical simulation, the results are shown in Figure 10. The two agree well, which shows the learning effect of neural network model is satisfactory and they can be used for further analysis.

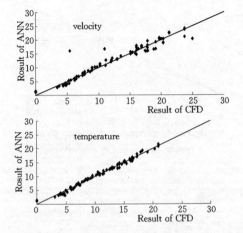

Figure 10   The comparison results of ANN and CFD on velocity trajectory and temperature trajectory

Taking Formula (6) as data mining model, we analyze the laws of trajectory which are affected by impact factors. Fixing $H/D=25$ and $R=6$, the forecasting results of velocity trajectories and temperature trajectories under $F_{dj}=1.3$, 1.65, 4.0 are shown in Figure 11. We can see that velocity trajectories and temperature trajectories are up with the increase of F_{dj}, this is mainly because the greater F_{dj} the larger

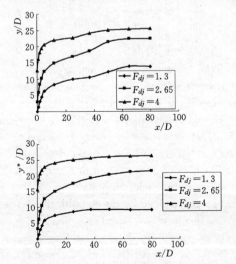

Figure 11   Velocity trajectory and temperature trajectory under different F_{dj}

jet initial momentum, so the jet rises higher, which is showed in track line is uplift. Fixing $R=6$ and $F_{dj}=2.65$, order H/D change from low to high, the corresponding velocity trajectories and temperature trajectories are shown in Figure 12. We can see that the change of H/D affects the law of velocity trajectories unclearly, while the temperature trajectories are affected by the change of H/D is obviously, it can clearly be seen that with H/D increasing, temperature trajectories are upward gradually, this mainly because the greater H/D the greater hydrostatic pressure at the location of nozzle, the repression effect to jet also increases, which hinders jet's moving in environmental fluid, so the height of jet rising decreases. Fixing $H/D=25$ and $F_{dj}=2.65$, order R change from low to high, the corresponding velocity trajectories and temperature trajectories are shown in Figure 13. It can be seen that the change of R affects the law of velocity trajectories also unclearly, while temperature trajectories can be clearly seen that with R increasing, trajectories are upward gradually, its mechanism is similar with F_{dj}. In addition, from the comparison of three figures above it can also find that the affected degree of H/D and R to trajectory is less than F_{dj}, so the development of jet trajectory mainly depends on F_{dj}.

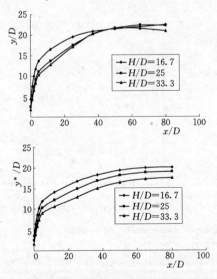

Figure 12   Velocity trajectory and temperature trajectory under different H/D

3.2.2   *Mixing of jet and environmental fluid*

The mixing of jet and environmental fluid can be characterized by dilution at characteristics loca-

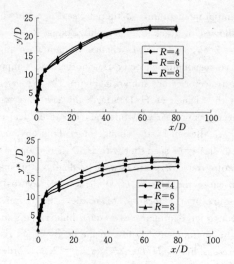

Figure 13　Velocity trajectory and temperature trajectory under different R

tions. The maximum rising height is an important reference parameter of buoyant jet, which is an important index to evaluate the scope of pollution (Yang Zhonghua, 2004), so the paper studies the dilution S_t at temperature trajectory's maximum rising height. The main affecting factors of S_t can be expressed by the following relationship.

$$S_t = f(\rho, g, H, D, v, j, u_a, T_j, T_a) \tag{7}$$

According to π theorem, choose ρ, g, D, T_a as the basic variables and obtain dimensionless expression

$$\begin{cases} S_t = F\left(\dfrac{H}{D}, F_{dj}, R, K\right) \\ F_{dj} = \dfrac{v_j}{\sqrt{gD}} \\ R = \dfrac{v_j}{u_a} \\ K = \dfrac{T_j}{T_a} \end{cases} \tag{8}$$

where, K is temperature ratio of jet and ambient fluid, $S_t = \dfrac{T_j - T_a}{T_t - T_a}$ (T_t is the dilution at temperature trajectory's maximum rising height).

H/D, F_{dj}, R, K as the inputs of neural network, and S_t as the output of neural network, establish the structure of artificial neural network model. Using the data results of numerical simulation under c2, c4-c6, c8-c10, c12-c13 in Table 1 as learning materials set, train the neural network model, and compare the trained results of neural network and results of CFD simulation, which is shown in Figure

14, the two are in good agreement, which shows that the learning effect of neural network model is satisfaction, they can be used for further analysis. Finally we obtain the nonlinear computational model of dilution S_t at maximum raising height.

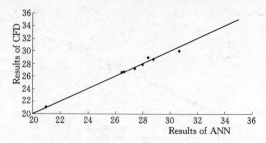

Figure 14　Comparison of dilution S_t of temperature trajectory's maximum rising height

$$\begin{cases} A = \left(\dfrac{9}{166}\dfrac{H}{D} - 0.86, \dfrac{9}{11}F_{dj} - 0.87, \dfrac{3}{20}R - 0.55, \dfrac{30}{73}K - 0.90\right)^{\mathrm{T}} \\ b = \dfrac{1}{1 + e^{-(\omega A - \theta)}} \\ S_t = \dfrac{1}{32}\dfrac{1}{1 + e^{-(vb - r)}} \end{cases} \tag{9}$$

where,

$$\omega(4,6) = \begin{bmatrix} 2.192 & 2.324 & 2.310 & 9.217 & 4.801 & 3.996 \\ 0.931 & 0.541 & 0.790 & -0.617 & 2.004 & 4.186 \\ 0.778 & -0.120 & 0.390 & -4.568 & 0.187 & 4.699 \\ 0.097 & 0.120 & 0.093 & 5.968 & 1.247 & -2.427 \end{bmatrix}$$

$$\theta(6,1) = \begin{bmatrix} 1.423 \\ 1.701 \\ 1.489 \\ 1.524 \\ 2.438 \\ 3.218 \end{bmatrix}, v(6,1) = \begin{bmatrix} -2.919 \\ -2.307 \\ -2.116 \\ 6.136 \\ 2.076 \\ 4.123 \end{bmatrix}, \gamma = 4.123.$$

Take Formula 9 as data mining model, analyze the laws of dilution S_t at temperature trajectory's maximum rising height which are affected by impact factors, the results are shown in Figure 15. Fixing $R = 6$, $H/D = 25$, order K change from low to high, under $F_{dj} = 1.3$, 2.65, 4.0 respectively, the corresponding laws of S_t are shown in Figure 15 (a). It can be seen that under the same F_{dj}, K has little effect on S_t; under different K, with the increase of F_{dj}, S_t also increases accordingly. Fixing $K = 3.65$, $H/D = 25$, order F_{dj} change from low to high, under $R = 4$, 6, 8 respectively, the corresponding laws of S_t are shown in Figure 15 (b). It can be seen that under different R, with the increase of F_{dj}, S_t also increases accordingly. This mainly because the greater F_{dj} the higher jet rising, jet mixes environmental fluid more fully, so S_t is greater. Fixing $K = 3.65$, $R = 6$, order H/D change from low to high, under $F_{dj} = 1.3$, 2.65, 4.0 respectively, the corresponding laws of S_t shown in Figure 15 (c). It can be seen that the laws of $S_t - H/D$ under different F_{dj} are consistent generally. With the increase of H/D, S_t first increases gradually, and reaches the peak points then drops gradually. The physical mechanism still needs further understanding. Fixing $K = 3.65$, $F_{dj} = 2.65$, order R change from low to high, under $H/D = 25$, 29.7, 33.3 respectively, the corresponding laws of S_t are shown in Figure 15 (d). It can be seen that with R increasing, the trend of S_t is also ascendant, and its mechanism is similar to F_{dj}.

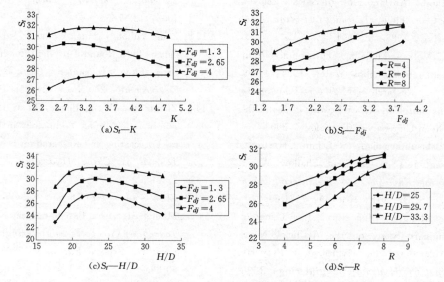

Figure 15　Laws of S_t under different factors

4. Conclusions

This paper attempts to apply data mining mode based on neural network (EDM-π & ANN) to the data analysis of jet field. Non-linear calculation formulas of velocity trajectory, temperature trajectory and dilution at temperature trajectory's maximum rising height are obtained finally. Comparing with the empirical formulas which are fitted by previous physical model experiments, these formulas take into account more affected factors, have broader scope of application, and have the features of harmonious dimension, unified form, which are convenient to build a knowledge base. In addition, some new laws of vertical round buoyant jet in cross flow are found. The results of this paper not only have some academic reference significance to the further research of jet, and also provide a new solution to high-level data analysis of jet fluid in the future.

Acknowledgment

This work was supported by grants from Open Foundation of State Key Laboratory of Hydrology-Water Resources and Hydraulic Engineering (No. 2008490411), National Natural Science Foundation of China (No. 50909085) and Open Foundation of Key Laboratory of Navigation Structures.

References

[1]　A. J. Johnston, N. Nguyen, and R. E. Volker. Round Buoyant Jet Entering Shallow Water in Motion. *Journal of Hydraulic Engineering*, 1998: 1364-1382.

[2]　C. P. Kuang, J. H. Lee. A Numerical Study on the Stability of Vertical Plane Buoyant Jet in Confined Depth. *Environmental Hydraulics*, Rotterdam, 1999: 205-210.

[3]　Gu Zhenghua, Tang Hongwu, Duan Zibing. A Pattern of Experiment Data Mining Based on π Theorem Combined with ANN. *The Ocean Engineering*, 2006: 86-90, 94.

[4] Huai Wenxin, Fang Shenguang. Numerical Simulation of Obstructed Round Buoyant Jets in a Static Uniform Ambient. *Journal of Hydraulic Engineering*, 2006: 428-431.

[5] N. Dominique, Brocard. Surface Buoyant Jets in Steady and Reversing cross Flows. *Journal of Hydraulic Engineering*, 1984: 792-809.

[6] O. Habib, Anwar. Flow of Surface Buoyant Jet in Cross Flow. *Journal of Hydraulic Engineering*, 1987: 892-904.

[7] R. E. Volke. Modeling Horizontal Round Buoyant Jets in Shallow Water. 1994: 41-59.

[8] Shao lei. Research *on Multiport Buoyant Jet of Thermal Water in* Co-flowing, Master thesis, Xi'an University of Technology, 2008.

[9] V. Balasubramanian, S. C. Jain. Horizontal Buoyant Jets in Quiescent Shallow Water. *J. of the Environmental Engineering Division*, ASCE, 1978, 104 (4): 717-729.

[10] Yang Rongfu, Ding Jing, Liu Guodong. Preliminary Study on the Artificial Neural Netwok Based on Hydrological Property. *Journal of Hydraulic Engineering*, 1998: 23-27.

[11] Yang Zhonghua. *Study on the Behavior of Negatively Buoyant Jets*, Ph. D thesis, Wuhan University, 2004.

[12] Yin Zhengjie, Wang Xiaolin, Hu Tiesong, Wu Yunqing. Water Supply Reserv oir Operating Rules Extraction Based on Data Mining. *Systems Engineering—Theory & Practice*, 2006: 129-135.

[13] Zeng Yuhong, Huai Wenxin. Numerical Analysis for a Round Turbulent Buoyant Jet in a Cross Flow. *Journal of Basic Science and Engineering*, 2005, 13 (2): 120-128.

[14] Zhang Chi, Wang Bende, Li Wei. Application of Data Mining Technology in Hydrological Forecasting and Research on Development. *Journal of China Hydrology*, 2007: 74-77, 85.

[15] Zhao Gaofeng, Bi Duyan, Sun Wei. A Data Mining Algorithm Based on Fuzzy Neural Net. *Journal of Air force Engineering University (Natural Science Edition)*, 2008, 9 (3): 63-66.

Topic Ⅲ

Development of Instruments and Facilities for Hydraulic and Eco-hydraulic Measurement

Similarity Criteria of Homogeneous Embankment Failure due to Overtopping Flow*

Li Yun　Wang Xiaogang　Xuan Guoxiang　Liu Huojian

State Key Laboratory of Hydrology-Water Resources and Hydraulic Engineering,
Nanjing Hydraulic Research Institute Nanjing, 210029

Abstract: The Process of dam breach is a strongly nonlinear process and very complex. The research of similarity criteria of dam breach has no breakthrough as yet. According to five highest prototype tests of the world and twenty laboratory tests conducted by Nanjing Hydraulic Research Institute, the mechanism of homogeneous embankment failure due to overtopping flow is "headcut". Based on this mechanism and results of tests, "headcut" migration rate similarity scare λ_R and the time similarity scare of flow process λ_t of homogeneous embankment failure due to overtopping flow were derived. According to the similarity criteria several tests were conducted for two field tests. The results verified the similarity criteria are available and it could be used to s simulate the breach process for homogeneous embankment failure due to overtopping flow.

Key words: dam break; model; similarity criteria

1. Introduction

The dam has brought enormous social and economic benefits to human, however for a variety of reasons, quite a lot of dams exist potential safety hazard. Once the dam wrecked and breached, it would cause a tremendous losses in the lives and properties downstream. Dam breach usually occurs in embankment dam[1-2], and is a gradually process. Forecast accuracy of the flow process in dam breach directly determines the accuracy of flood routing calculation. Therefore, it is great significance to master the practical process of dam breach development.

The previous research of dam-break model test and numerical simulation mostly concentrate on the flood routing [3-5]. The research of similarity criteria of dam breach is little as yet. The process of dam breach involves sediment transport in unsteady rapidly varied flow, and is a strongly nonlinear process. The mechanism involves hydraulics, soil mechanics, sediment transport mechanics and so on. Therefore, the research of dam breach has no breakthrough as yet[6]. Dam-break model test technology is still at the starting stage, Understanding of the core issue of similarity criteria still depends on lots of basic research work. For all that, some domestic and foreign scholars also have carried out some beneficial studies on similarity criteria. It can be divided into two types of methods. One is that the similarity criteria of dam breach derived from the theory of sediment transport, and the other is determined by a series of model tests.

The first method for establishing similarity criteria is based on the theory of gravity and sediment movement similarity criteria[7]. The same kind of soil for model and prototype were adopted to deduce the similarity criteria for dam-break tests. For example, DuPont obtained the dam break similarity criteria for non-cohesive homogeneous earth dam by adopting the same resistance coefficient of model and prototype and basing on the theory of gravity and Yalin sediment movement similarity[8]. These similarity criteria are mainly for non-cohesive homogeneous earth dam and the sediment transport similarity criteria adopted are based on plain river sediment transport formula. In fact it is different for sediment erosion between dam-break high velocity flow and river flow. So it is difficult for these similarity criteria to extend to cohesive homogeneous earth dam breach test.

The second method for establishing similarity

---

* Keynote lecture

criteria is based on a series of model tests. More than 30 times fuse-plug dam breach model test had done in Henan province[9]. The range of these model tests is 1 : 2 to 1 : 32. Through the series of model tests, Hydraulic Research Institute of Henan Province presented the formula of dam lateral growth rate, vertical incision rate and model similarity criteria of breach formation time. Jia Cuilan studied the dam erosion characteristics by a series of model extension method in 1996. She pointed out that flow similarity followed the gravity similarity criteria [10] and the material's characteristics of model adopted the same as prototype. Under the hypothesis, the dam lateral erosion rate was obtained by using a series of model extension method. The method usually requires data form a lot of different scale model. Moreover the model test results show that the experimental regularity needs to be improved according to the data of 30 times fuse-plug dam breach model test done in Henan province.

Some research on Gouhou face rockfill dam breach had been done in Nanjing Hydraulic Research Institute. A gradually approximation method had been adopted to solve similarity of model problem[11]. Recently five highest prototype tests of the world and twenty laboratory tests had been conducted under support of National Key Technology R&D Program in the 11th Five year Plan of China in Nanjing Hydraulic Research Institute. A very important mechanism of "headcut" erosion of homogeneous embankment failure due to overtopping flow has discovered [12]. The study of the article carried out is mainly based on this important mechanism and reference to the current international popular mathematical model theory for embankment dam breach. Particularly, sediment transport formula on high-speed water flow was used. Model similarity criteria for embankment dam breach had been presented in the article. Two verification tests for large-scale dam breach test conducted by Nanjing Hydraulic Research Institute had done which carried out base on the model similarity criteria presented in this article. The results proved the model similarity criteria were valuable.

2. Derivation of similarity criteria for homogeneous earth dam breach due to overtopping

Because dam-break flow belongs to gravity flow the flow similarity of dam breach model should follow the Froude similarity criteria. Then the velocity scale is $\lambda_v = \lambda_l^{1/2}$ and the discharge scale is $\lambda_Q = \lambda_l^{5/2}$. The difficulty of dam breach model test is how to guarantee the process similarity of dam breach (it should guarantee the soil mechanics similarity and sediment erosion similarity).

The process of "headcut" erosion in embankment dam overtopping breach is shown in Figure 1. In the process of embankment dam overtopping breach, step-like "headcut" in the downstream slope firstly form, then several "headcuts" merge into a big "headcut" and gradually erode upstream. High velocity flow on the "headcut" is the main power in the dam breach. So the migration process of headcut must be simulated in model test. An indirect method to simulate "headcut" migration speed is to simulate the sediment transport rate of the process of dam breach.

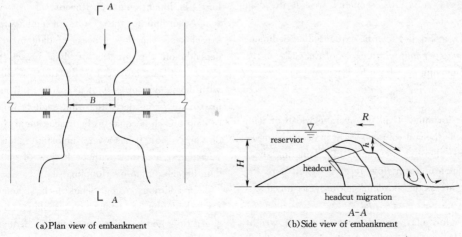

(a) Plan view of embankment   (b) Side view of embankment

Figure 1　Schematic layout of "headcut" development

Because the present sediment transport rate theory is mainly summed up by the data of sediment transport in Plain River and the flow in dam breach is very different from it, choosing the sediment transport rate formula suitable for dam breach become the key to forecast dam material erosion speed. The article adopted the Meyer-Peter and Muller formula which is suitable for calculating sediment transport of large Froude flow and is used wildly in current dam breach model (such as Breach, Ponce-Tsivoglou, Nogueira and so on).

According to Meyer-Peter and Muller sediment model, sediment transport rate q_s can be expressed as Eq. (1):

$$q_s = a(\tau - \tau_c)^b \qquad (1)$$

In the formula, a, b are coefficients, q_s is sediment quality for unit time and unit width [kg/(m·s)], τ is flow shear stress (N/m^2), τ_c is sediment starting shear stress (N/m^2), for big Froude number water flow $b = 1.5$[13].

Then sediment transport rate q_s can be expressed as:

$$q_s = a(\tau - \tau_c)^{3/2} \qquad (2)$$

Erosion rate of the sediment in the dam breach flow is rapid, assuming $\tau \gg \tau_c$, so:

$$q_s = a\tau^{3/2} \qquad (3)$$

According to mechanism of "headcut" erosion of embankment dam overtopping breach, migration rate of "headcut" can be set up as R (m/s), so:

$$R = \frac{q}{A} \qquad (4)$$

In the formula, q is sediment discharge for unit time (m^3/s), A is cross-sectional area of the dam (m^2).

Here q in Eq. (4) can be expressed as Sediment transport rate per unit width in use of Eq. (3), so the "headcut" moving speed can be expressed as:

$$R = \frac{q_s B}{\gamma_s A} \qquad (5)$$

In the formula, B is the average width of dam breach; γ_s is the soil volume weight in dam (N/m^3).

Because there is no formula of calculating shear stress for non-uniform flow for reference, here the flow shear stress on "headcut" τ can be assumed as:

$$\tau = \gamma_s h J \qquad (6)$$

In the formula, γ_s is the water volume weight (N/m^3); h is water average depth of the "headcut" flow (m); J is hydraulic gradient of the "headcut"

flow. Assuming there is a linear relationship between cross-sectional area and height of the dam, then $A = kH^2$; H is the height of the dam (m); k is dimensionless coefficient. Integrating Eq. (3), Eq. (6) and Eq. (5) the "headcut" migration rate R can be expressed as:

$$R = \frac{a\tau^{3/2} B}{\gamma_s k H^2} = \frac{a\gamma^{3/2} h^{3/2} J^{3/2} B}{\gamma_s k H^2} \qquad (7)$$

To avoid discussing the complex similarity rate of mechanical properties of dam materials (such as shear strength of soil; value of bond strength C; the value of internal friction angle φ; dry density; compaction and so on), the article assumes that model used the same material, compactions and water content as prototype. Then the similar scale of "headcut" migration rate can be expressed as:

$$\lambda_R = \frac{\left[\dfrac{a\gamma^{3/2} h^{3/2} J^{3/2} B}{\gamma_s k H^2}\right]_p}{\left[\dfrac{a\gamma^{3/2} h^{3/2} J^{3/2} B}{\gamma_s k H^2}\right]_m}$$

$$= \frac{\lambda_a \lambda_\gamma^{3/2} \lambda_h^{3/2} \lambda_J^{3/2} \lambda_B}{\lambda_{\gamma_s} \lambda_k \lambda_{H^2}}$$

$$= \lambda_l^{1/2} \qquad (8)$$

In the formula, $[\gamma_s]_p = [\gamma_s]_m$, then $\lambda_{\gamma_s} = 1$, For other dimensionless numbers, the scale is 1.

According to the similar scale of "headcut" migration rate the similar scale of time scale of erosion process of embankment dam can be derived. Assuming $A = kH^2$ and the equivalent width of dam cross-section $D = \xi A/H$ the time scale λ_t of erosion process of embankment dam can be expressed as:

$$\lambda_t = \frac{\lambda_C}{\lambda_R} = \frac{\lambda_\xi \lambda_A / \lambda_H}{\lambda_l^{1/2}} = \frac{\lambda_l^2 / \lambda_l}{\lambda_l^{1/2}} = \lambda_l^{1/2} \qquad (9)$$

here ξ is dimensionless coefficient.

According to the results of embankment dam breach tests the flow process of dam breach largely depends on the process of dam breach[14-15]. The process of dam erosion has the time synchronization with developing process of dam breach. It can be approximately considered that time scale of flow process of dam breach is equal to the time scale of the process of dam erosion. It has a very important value for studying the arrival time with peak flow of embankment dam breach due to overtopping.

In addition, using Schoklitsch sediment model for calculating the sediment transport under high-speed flow and carrying out a similar deduction, the same result can been got. The sediment transport rate with high velocity flow in Schoklitsch model can be expressed as:

$$G = 39.4J^{1.5}Qd^{-0.5} \qquad (10)$$

In the formula, G is sediment quality for unit time (kg/s), Q is water flow (m³/s), d is dam material diameter (m). Assuming $A = kH^2$, so the moving velocity of "headcut" can be expressed as:

$$R = \frac{G}{\gamma_s A} = \frac{39.4J^{1.5}Qd^{-0.5}}{k\gamma_s H^2} \qquad (11)$$

Assuming model use the same material and filling standard as prototype, it can been got that $\lambda_d^{-0.5} = 1$, $\lambda_{\gamma_s} = 1$. Then the migration rate scale of "headcut" can be expressed as:

$$\lambda_R = \frac{\left[\dfrac{39.4J^{1.5}Qd^{-0.5}}{k\gamma_s H^2}\right]_p}{\left[\dfrac{39.4J^{1.5}Qd^{-0.5}}{k\gamma_s H^2}\right]_m}$$

$$= \frac{\lambda_J^{1.5}\lambda_Q\lambda_d^{-0.5}}{\lambda_{\gamma_s}\lambda_H^2}$$

$$= \lambda_l^{1/2} \qquad (12)$$

Similarly, time scale of "headcut" moving (same as time scale of dam breach discharge process) can be expressed as:

$$\lambda_t = \lambda_l^{1/2} \qquad (13)$$

According to this, although the Meyer-Peter and Muller and Schoklitsch sediment model is based on different sediment transport data under high-speed, there is good consistency between them. This proved it is reasonable in some sense for using the sediment transport formula in high - speed water flow in this article.

The process of embankment dam breach is strong nonlinear development process under extremely dynamic condition, it is almost impossible to inverse the process accurately by model tests under the current technical condition. Grasping the core issue of dam breach and simplifying the secondary process is very useful for getting the similarity scale of migration rate of "headcut" λ_R, particularly the time of process of dam breach similarity scale λ_t. According to derivation process of similarity criterion it can been seen that the flow scale of dam breach follows gravity similarity, and the time scale of the process of dam breach flow was obtained from the similarity criteria of "headcut" migration rate. In addition, the time scale λ_t of the process of dam breach in this article is consistent with gravity similarity criteria. But they are based on different theory and have no inevitable connection.

3. Model test

3.1 Experimental procedures

To verify the reasonableness of similarity criterion two semi-global model tests were conducted. They had done based on two large—scale field tests of dam breach due to overtopping which were carried out by Nanjing Hydraulic Research Institute at Chuzhou, Anhui province. Test series numbers were F-A and F-B. To verify the applicability of similarity criterion the material viscosity of dam filling is relatively large in test F-A (the cohesion C of dam soil is large) and relatively small in test F-B.

During the process of dam-break, reservoir water level, releasing flow, water surface area and storage capacity of reservoir were changing with time, but in the model test it was difficult to simulate the whole topography of reservoir because of limited area in laboratory. In order to simulate discharge flow of the process of dam breach a method of equivalent storage capacity was adopted. The principle is to ensure the similarity of storage-capacity curve between model and prototype[16].

The arrangement of model was shown in Figure 2.

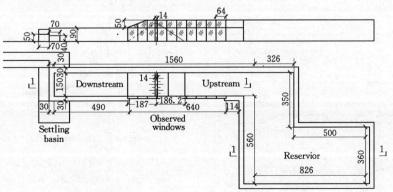

Figure 2 Schematic layout of the model

Model dam section was placed in Plexiglas flume for convenient observation. The data of falling process of reservoir water level was measured by water pressure transducer (the sampling frequency is 0.5Hz) and was real-timely recorded by the American multi-channel data acquisition system Wavebook. At the same time it was recorded by the camera for checking. The flow velocity was measured by Acoustic Doppler Velocimeter (ADV). High resolution CCD was set at the side of dam and in the front of dam downstream for recording the development of dam breach. Discharge process of dam breach was calculated by falling process of reservoir water level and water storage capacity curve. The model dam was filled with the same material as prototype dam in Chuzhou, Anhui Province.

3.2 Results

For F-A test, the prototype of dam is homogeneous cohesive soil dam and the maximum dam height is 9.7m, the crest width is 3.0m and the dam length is 120.0m. The clay content is 17.8%, the cohesion C is 13.0kPa, compaction of dam is 99.6%; water content is 15.5%. The formation of initial dam breach was set by the way of citation notching which was 1.0m×0.8m (depth×width). During the test there was no inflow from upstream. The design of model followed the similarity criteria obtained in article. The geometric scale of dam was $\lambda_l = 10.00$. The flow scale was $\lambda_Q = 316.20$. The scale of "headcut" migration rate was $\lambda_R = 3.16$. The time scale of the process of dam breach was $\lambda_t = 3.16$. Difference was found between the dam model and prototype because of construction. Average compaction of dam model was 100% and water content was 16.97%, but it would not obviously influent the results of tests.

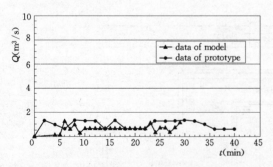

Figure 3   The flow process of dam breach

The comparison of predicted and measured value

of flow in dam breach was shown in Figure 3. As the cohesion of the dam material is large the dam is firmness. The flow in breach fluctuated and always maintained at a small values. The final average width and depth of dam breach could be seen in Table 1. According to the results it could be found the predicted values were consistent with measured values.

For F-A test, the prototype of dam is homogeneous cohesive soil dam and the maximum dam height was 9.7m, the crest width was 3.0m and the dam length was 120.0m. The clay content was 11.5%. The cohesion C was 7.5kPa. The compaction of dam was 97%. The water content was 19.4%. The size of initial dam breach was 1.0m × 0.8m (depth × width). There was no inflow from upstream during the test. The geometric scale of dam was $\lambda_l = 10.00$. The flow scale was $\lambda_Q = 316.20$. The scale of "headcut" migration rate was $\lambda_R = 3.16$. The time scale of the process of dam breach was $\lambda_t = 3.16$. Practical compaction of dam model is 96%; water content is 17.0%.

Table 1   Important parameters of test F-A

Item	The final average width of breach (m)	The final depth of dam crest breach (m)	Peak flow (m³/s)	Arriving time of peak flow (min)
Measured value of prototype	4.6	1.4	1.33	2
Predictive value of model	4.4	1.3	1.36	1

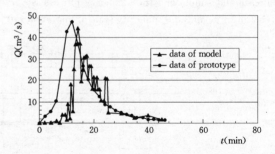

Figure 4   The flow process of dam breach

Compared with test F-A, the value C of dam filling reduces dramatically. After overtopping a large "headcut" soon formed and moved upstream quickly. The comparison of predicted and measured value of flow in dam breach was shown as Figure 4. The final value of average width of breach and depth of dam crest breach can see in Table 2. According to the Figure 4 it could be seen the predicted peak flow of model was very close to peak flow of the prototype (peak flow of the prototype is 47. 0m³/s and the predicted peak flow of model is 44. 0m³/s). The predicted arrival time of the peak flow of model delayed comparing to the prototype test. This may be due to the impact of scale effect.

Table 2   Important parameters of test F-B

Item	The final average width of dam breach (m)	The final depth of dam crest breach (m)	Peak flow (m³/s)	Arriving time of peak flow (min)
Measured value of prototype	18. 95	4. 6	47	12
Predictive value of model	16	4. 7	44	14

4. Conclusion and discussion

The article based on the results of large-scale field tests of dam breach due to overtopping in Chuzhou of Anhui province and laboratory tests carried out by Nanjing Hydraulic Research. According to the "headcut" mechanism of dam breach simplified the process of dam breach properly, deduced and got two important model similarity criteria for model test of homogeneous earth dam breach due to overtopping. One was the similarity criteria of "headcut" migration rate λ_R, the other was time similarity criteria λ_t of the flow process of dam breach. They were used for simulated two large-scale field tests. The results showed that these similarity criteria presented in this article can be used simulated the overtopping breaching of homogeneous embankments preferably and had certain practicality and maneuverability.

Meanwhile, in addition to the similarity criteria and test methods presented above it should be pointed out that the tests for overtopping breaching of homogeneous embankment must be satisfied some other basic requirements of hydraulic model tests. For example, supercritical and subcritical flow area of overtopping flow should be similar which requires the model has an enough big model scale. The similarity criteria in the article were only verified in the tests for homogeneous embankment. For the other types earth dams (such as core dam) the migration rate of "headcut" might be different and needed further study. In addition, there were certain differences in the theory between the sediment transport under high velocity flow and the "headcut" erosion of dam breach. So much more works should be done in future.

References

[1] Li Yun, Li Jun. Review of experimental study on dam break [J]. Advances in Water Science, 2009, 20 (2): 304-310.

[2] Zhu Yonghui. Review on oversea earth dam break modeling [J]. Journal of Yangtze River Scientific Research Institute, 2003, (2): 26-29.

[3] Zhang Dawei, Wang Xinkui, Li Danxun. Numerical modeling of dam-break flow under the influence of buildings [J]. Journal of Hydrodynamics (Ser. A), 2008, 23 (1): 48-54.

[4] Wang Xin, Cao Zhixian, Yue Zhiyuan. Numerical modeling of shallow flows over irregular topography. Journal of Hydrodynamics (Ser. A), 2009, 24 (1): 56-62.

[5] Fu Chuan jun, Lian Jijian. Three-dimensional numerical simulation of dam-break flow in complicated river sections. Journal of Hydraulic Engineering, 2007, 38 (10): 1151-1157.

[6] Wang Guangqian, Zhong Deyu, Zhang Hong wu. Simulation of flow process of the Tangjiashan Quake Lake at Wenchuan in China. Chinese Science Bulletin, 2008, 52 (24): 3127-3133.

[7] COLEMAN S E, Andrews D P, Webby M G. Overtopping Breaching of Non-cohesive Homogeneous Embankments [J]. Journal of Hydraulic Engineering, 2002, 128 (9): 829-838.

[8] DUPONT E, Dewals, B J. Experimental and numerical study of the breaching of an embankment dam [C]. Proc. 32nd Congress of IAHR Venice, IAHR, Madrid. 2007, 1

(178): 1-10.

[9] Yang Wuchen. Model experiment of erosion rate of fuse-plug dam. Water Resources and Hydropower Engineering, 1985, (3): 1-6.

[10] Jia Cuilan. Model design and instrument introduction on fuse-plug in Xinlin Reservoir. 1996 (3): 28-29.

[11] Hu Qulie, Yu Bo. Study of breach formation on face slab dam [R] . Nanjing: Nanjing Hydraulic Research Institute, 1996.

[12] Zhang Jianyun, Li Yun, Xuan Guoxiang, et al. Overtopping breaching of cohesive homogeneous earth dam with different cohesive strength [J] . Science in China Series E: Technological Sciences, 2009, 52 (10):

3024-3029.

[13] Singh V. Dam Breach Modeling Technology [M] . Dordrecht: Kluwer Academic Publishers, 1996.

[14] Visser P J. Breach growth in sand-dikes [D]. Ph. D Thesis. Netherlands: Delft University of Technology, 1998.

[15] Zhu Yonghui. Breach Growth in Clay-Dikes [D] . Ph. D Thesis. Netherlands: Delft University of Technology, 2006.

[16] Li Jun,Li Yun, Xaun Guoxiang, et al. Model design and preliminary study on dam breach due to overtopping [C] . International conference on dam safety management Nanjing University Press, Nanjing, 2008: 209-214.

An Elementary Study for the Mechanism of Earth Dam Break[*]

Xinghua Xie [1]   Jianjun Zhao[1]   Zhonghua Feng[1]   Yanqiong Geng[2]

[1] *State Key Laboratory of Hydrology-Water Resources and Hydraulic Engineering,*
Nanjing Hydraulic Research Institute, Nanjing, 210029
[2] *Lian Yungang Technical College, Lian Yungang, 222000*

Abstract: Based on the force balance fundamental of sand grain, it was analyzed that the force condition of the grain start velocity in water flow erosion as earth dam breaking process. Expressions of a rolling and a suspending start velocity of grain were deduced from that analysis. The relationships between grain start velocity and the grain diameter, flow direction, shear strength parameters were studied respectively, which were illustrated as curves. It was made known that there were different manners of grain start moving because of the sand grain diameter from small to large. If the grain diameter was smaller than 4.0cm, grains start moving as suspending manner. And other vice, grains might roll in water flow erosion happening. The suspending start velocity was not affected by the shear strength parameters, but the rolling start velocity was augmenting with the shear strength parameters go large as a nonlinear curve manner. The slope gradient could have a decisive action to force direction of grains. The larger the gradient, the easier the grains start moving. And rolling might be the start moving manner.

Key words: earth dam break; start velocity; force balance

1. Introduction

There were more than 8 million reservoirs in China, which played important social and economic roles in flood control, irrigation, power generation, water supply, shipping and other aspects. They were lifeblood for agriculture, economic and important infrastructure of social development in local area. They were also the fundamental guarantee of people's lives and safeties. As the potential factors, the possibility of earth dam break even evoked severe disasters may last for a long time. This situation was threatened the national public security, people's lives and property and socio-economical sustainable development. According to statistic results, there were 70 dam failures in 30 years from the mid-1950s to the late 1980s. The accident killed a total of 670 people, caused economic losses about $ 100 million. There were two most important reasons of the dam break. The first was that spillway facilities not working, the second was that the ability of the spillway facilities was not big e-nough. So the research about mechanism of dam break in complex conditions was very important. It was urgent demands to enhance defense capabilities against China's major natural disasters. It also strategically demands to protect the socio-economical sustainable development, people's stability and harmonious living.

The major reasons of dam break were divided into two types, infiltration break and overtopping dam failure deformation. As a result of overtopping dam occurred suddenly, break fast, the flood caused by dam break had a powerful destructive, so overtopping dam had got lots of attention.

In addition, researches focused more on the dam flood flow, flood mode, and the risk of loss of life. Nevertheless, the number of researches on overtopping dam break mechanism was rare. Generally, the simulation process of earth dam breach was an integrated process of hydraulics, hydrodynamics, sediment dynamics and soil mechanics. It was two-phase medium interaction process. (1) The water pour off

---

* Funded by the Non-profit Industry Financial Program of MWR (200801030, 201101005-03).

the dam breach from the reservoir, would lead the breach to expansion by erosion and collapse. This situation would last until the reservoir was empty or the dam could resist the water pour further. (2) Dike breach formation was largely depends on the source stream erosion on construction materials. This was related to dam height, construction materials, materials in dense and overtopping discharge status. (3) The dike breach was developed in horizontal and vertical direction at the same time. As time goes on, slope lost its stability, leading dam break, the top dam breach enlarge gradually. (4) Based on assumptions, the study of dike breach's development process abroad had several forecast model. prediction models of overtopping dam break are mainly DAMBRK, BEED, BREACH and Head cut erosion.

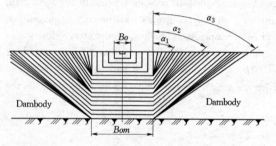

Figure 1   The erosion expand process of BREACH model

The modal DAMBRK was first proposed in 1988 by Fread. It was assumed that the dike breach start from a point at the bottom, its width grow by linear rate throughout the duration, until the final width of the dike breach. At the same time, the height of the bottom dike breach keep growing, until the final position. The modal SEED developed by Singh and Scarlatos in 1996, divided the dike breach into two parts, the horizontal part in the top of the dam and the slope part in the bottom of the dam. Sections in both parts were assumed to be trapezoidal, the horizontal dam part used as broad crested weir. The world's most popular modal was the BREACH designed by Fread in 1984-1988. Modal BREACH could simulate the dam break caused by overtopping or piping. It could also simulate the dam which was not built by the same material. So we could analysis the dams when their core and the external region were not build by the same material. Set the dike breach as rectangular at first, after the collapse in the dike breach of slope division, the section became a trapezoid. The collapse

of the dike breach appeared, when the depth of the breach get marginal value during its development, this marginal value was of a function of the dam body's material properties (such as the internal friction angle, bond and bulk density, etc).

On the basis of experiments and observation of instances, Ralston[6], Wahl[7], Hanson[8] and other researchers proposed a new mechanism of earth dam break, called head cut erosion. When water flow through the "head cut", the overflow water lash down the bed and produce swirling flow, swirling flow apply shear stress in a vertical or nearly vertical way, lash deep to the bed and corrosion the foundation of inner toe of slope, resulting in surface instability and collapse. The "head cut" keep upstream develop. The emergence and development of "head cut" was closely related to the nature of soil. The viscous the soil was, the easier the "head cut" appeared. On the contrary, cohesionless soil generally did not have the "head cut" phenomenon.

Cai Xiaoyu and other researchers divided the process of wave erosion slope into three stages, incident, climb and fall. Whatever stage, start from granular practical's force and motion two-pronged, analyze the break mechanism of wave scouring grains, explained the practical situation about model experiment of grain slope erosion damage. Tian Weiping[10] studied the mechanism about water flow erosion road bed, he did so many experiments in different conditions using artificial bend generalization modal. Combined field data with field data and test data, he established a relationship between maximum erosion depth and leading factors which include water depth, sediment particle size, river width, river bend radii.

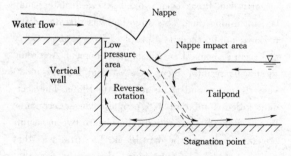

Figure 2   The sketch map of steep erosion

In the process of overtopping dam break, the sand grain force and motion were different from seepage piping dam break grains. The major reason about overtopping dam break was erosion destruction. The flow follows the Novier-Stokes equation, no longer the Darcy's law. Sediment grain start conditions increase with the velocity. The water flow's drag force destroyed grain force balance; make the sediment grain move with the flow. The flow of the earn dam break mainly washed down, scoured both sides of the river, made the dike breach wider and finally destroy the dam.

2. The mechanical condition of the surface grain's start

In normal conditions, the sand grains which compose dams were in mechanical equilibrium. After overtopping, flow floods the sand grain. Every sand grain which contacts with the water would be under hydrostatic pressure and acted by velocity. The original force balance was broken; it actually became erosion problem. With the ever-changing flow pattern, sand grain's mechanical status changed. If stability condition could not meeting, the grain started to move. Otherwise, the grain deposes. At the beginning of the overtopping dam break, it was only an erosion problem. The surface sand grain of the dike breach moves with the flow. When the depth of the dike breach got the particular position, the two sides of the drain collapsed; made the drain expand in the transverse way. The following was a derivation of the dam erosion mechanical procedure.

2.1 Mechanical analysis of horizontal flow grain

As shown in Figure 1, one sand grain was taken out off the dam surface, set the two-dimensional of dx and dy. The grain was acted by its own gravity G, so

$$G = \gamma_s V \quad \text{or} \quad dG = \gamma_s dV \qquad (1)$$

when the sand grain was in water, it would get its buoyancy. According to Archimedes' law, sand grain's buoyancy is equal to the gravity that the water in the sand particle's size had.

$$W = \gamma_w V \quad \text{or} \quad dW = \gamma_w dV \qquad (2)$$

It also affected by the drag force P which was made by the flow. The force P was caused by the pressure differential when the two flow regime changed. More information could be found in Duo Guoren (2003).

$$P = \gamma_w \lambda_x dz \frac{v_g^2}{2g} \qquad (3)$$

λ_x was resistance coefficient, v_g was the bottom ve-

locity which acts on the grain, g was gravitational acceleration.

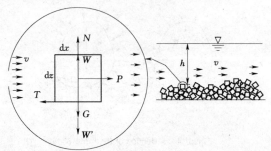

Figure 3　Grain force balance illustration in 2D

The grain had the trend of move with the flow. There were frictional forces between the sand grains. The force would prevent the movement of the grains.

$$T = \tau dx \qquad (4)$$

τ was shear stress, act on the sand grain. The shear stress was not only related to the sand grain's shear parameter but also related to its own normal stress.

$$\tau = C + \sigma \tan\varphi \qquad (5)$$

Where, C was sand's cohesion strength; φ was sand's internal frictional angle, normal stresses σ was in connection with the grain's depth.

$$\sigma = \gamma_w h + (\gamma_s - \gamma_w) dz \qquad (6)$$

Besides, according to the fundamental theory of hydraulics, flow velocity was not uniform distribution in vertical direction. The differential water pressure N, caused by the difference in flow velocity of lower and upper layer, also acted on the grain.

$$N = \gamma_w \lambda_z dx \frac{v_g^2}{2g} \qquad (7)$$

Where, λ_z was resistance coefficient, could be ignored when N was relatively small in normal condition.

When grains were immersed in water, as the studied grain had contacts with the adjacent grain, interface made the grain under the water bodies pressure W'.

$$W' = \gamma_w (h + H_a) \omega_K \qquad (8)$$

Where h was grain's water depth, H_a was hydraulic head values of atmospheric pressure. ω_K was a proportion about grain contact area in the total grain surface area.

$$W' = \gamma_w h \omega_K \qquad (9)$$

In fact, due to the contact area among grains bottom, there was no hydrostatic pressure. So, in the calculation of buoyancy, this buoyancy was redundant. When the grain starts to move, the pressure of this section goes derivation as follows:

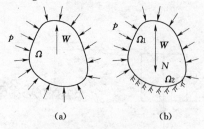

Figure 4   A deduce of the force N

As we all know, when solid was immersed in water, the solid' buoyancy was equal to solid surface water pressure in the area of solid surface's integral, as showed in Figure 4 (a). Or it could to say the buoyancy that the solid had, when it was completely immersed in the water, was equal to the gravity of the same volume water (Archimedes). Actually, the sand grains were not completely in floating condition when they're immersed in the water. The lower area of the grains always contacted with other grains, these part of area did not got the action of the hydro—static pressure. As Formula (2), it had to consider the hydro-static pressure of that part when calculated the buoyancy. For the buoyancy was upside, the pressure should be downside. As showed in Figure 4 (b).

2.2   Start condition of horizontal flow grain

According to the rule of force balance

$$\begin{cases} F_x = P - T \\ F_z = N + W - G - W' \end{cases} \quad (10)$$

where $F_x = 0$ was the critical condition. When $F_x > 0$, grains started to move. F_z decided the grains suspension or rolling. When $F_z > 0$, the grains suspend. When $F_z < 0$ and $F_x > 0$, the grains roll. As showed in Formula (1) – Formula (6).

$$F_x = P - T = \gamma_w \lambda_x dz \frac{v_g^2}{2g}$$
$$- \{C + [\gamma_w h + (\gamma_s - \gamma_w) dz] \tan\varphi \} dx \quad (11)$$

$$F_z = \gamma_w \lambda_z dx \frac{v_g^2}{2g} + \gamma_w dx dz - \gamma_s dx dz - \gamma_w h \omega_K$$
$$(12)$$

All the parameters were known except v_g. v_g was the flow velocity of grains' bottom. So, the forces in all directions were all proportional to the square of the v_g.

Assume $F_x = 0$, according to Formula (9)

$$v'_g = \sqrt{\frac{2g dx}{\gamma_w \lambda_x dz} \{C + [\gamma_w h + (\gamma_s - \gamma_w) dz] \tan\varphi \}}$$
$$(13)$$

Assume $F_z = 0$, according to Formula (10)

$$v''_g = \sqrt{\frac{2g}{\gamma_w \lambda_z} [(\gamma_s - \gamma_w) dz + \gamma_w h \omega_K]} \quad (14)$$

when $v''_g < v_g < v'_g$, grains started to roll, when $v_g > v'_g$ grains suspend and flow.

2.3   The relationship between grains' horizontal flow start velocity and the grains' diameter

Assume one kind of sand, volumetric weight $\gamma_s = 17.64 \text{kN/m}^3$, $C = 0.0 \text{kPa}$, $\varphi = 30°$, resistance coefficient $\lambda_x = 0.02$, $\lambda_z = 0.02$, the water depth above the sand $h = 0.05 \text{m}$, $dx = dz$, then the relationship curve of start velocity and grain diameter was illustrated in Figure 5.

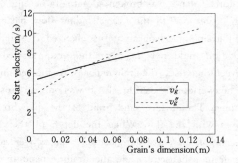

Figure 5   The relationship curve of start velocity and grain diameter

In Formula (11) and Formula (12), v_g was the velocity of the bottom of water flow. Based on the cross section's characteristic, the average velocity could be taken out.

In which, the reduction factor was 0.6.

$$v = \frac{v_g}{0.6} \quad (15)$$

3. The affection of shear strength parameters

Shear strength parameters only affected the grains' internal frictional angle. According to Formula (4) and Formula (5), the actions of the sand's C and φ acted on the T.

To cohesive soil, shear strength parameters were related to water percentage. According to the research made by Bishop (1960), Fredlund (1978, 1996), etc., when water percentage increased, the soil matrix suction reduced, cause effective shear strength reduced. But to sand, it did not have viscosity, it was to say $C = 0$. The size of internal frictional angle was decided by grain's size, shape and the relationship of the grains. The internal friction coefficient of the small grain piles was smaller than the bigger grain pile's, the globular grain pile's was smaller than the flaky grain pile's, the horizontal flaky grain pile's was smaller than the vertical flaky grain pile's. Shear strength parameters could get from inhouse experiment.

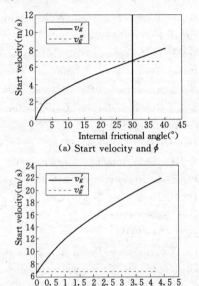

(a) Start velocity and ϕ

(b) Start velocity and C

Figure 6 The relation ship curve of start velocity and shear strength parameters

4 The influence of slope gradient

After overtopping, the water flow erosion in the downstream would become very violent. Along with the changes of the slope gradient, the flow pattern of the erosion area changed fiercely. The origin of the changes was that, when slope gradient changes, the mechanical state of the sand grains changed. The relationship could get from sand grains' mechanical equilibrium equations.

As the angular separation of water flow direction and slope gradient is α, so the balance equation was

$$\begin{cases} F_x = P - T + (G - W - N)\sin\alpha \\ F_z = (N + W - G - W')\cos\alpha \end{cases} \quad (16)$$

According to Formula (10) and Formula (11)

$$F_x = \gamma_w \lambda_x \mathrm{d}z \frac{v_g^2}{2g} - [C + (\gamma_w h + (\gamma_s - \gamma_w)\mathrm{d}z)\tan\varphi]$$
$$- (\gamma_w \lambda_z \mathrm{d}x \frac{v_g^2}{2g} + \gamma_w \mathrm{d}x\mathrm{d}z - \gamma_s \mathrm{d}x\mathrm{d}z - \gamma_w h\omega_K)\sin\alpha$$
$$(17)$$

$$F_z = \left(\gamma_w \lambda_z \mathrm{d}x \frac{v_g^2}{2g} + \gamma_w \mathrm{d}x\mathrm{d}z - \gamma_s \mathrm{d}x\mathrm{d}z - \gamma_w h\omega_K\right)\cos\alpha$$
$$(18)$$

The earth dam slope gradient was always lesser than $45°$. The slope gradient changes from $0°$ to $45°$, $\sin\alpha \in (0, \sqrt{2}/2)$, $\cos\alpha \in (\sqrt{2}/2, 1)$. So, on the slope, whatever F_x increased or F_z decreased, it was good to grains' start condition. The start velocity was

$$v_g' = \langle \frac{2g}{\gamma_w(\lambda_x \mathrm{d}z - \lambda_z \mathrm{d}x\sin\alpha)}\{[C$$
$$+ (\gamma_w h + (\gamma_s - \gamma_w)\mathrm{d}z)\tan\varphi]$$
$$+ [(\gamma_w - \gamma_s)\mathrm{d}x\mathrm{d}z - \gamma_w h\omega_K]\sin\alpha\}\rangle^{\frac{1}{2}} \quad (19)$$

$$v_g'' = \sqrt{\frac{2g}{\gamma_w \lambda_z \mathrm{d}x}[(\gamma_s - \gamma_w)\mathrm{d}x\mathrm{d}z + \gamma_w h\omega_K]} \quad (20)$$

According to the instance of last section sat the grains diameter as horizontal flow start diameter. The relationship curve of start velocity and slope gradient was shown in Figure 7. There was $v_g' \gg v_g''$, that was to say, if the grains start move with the flow, they would never roll.

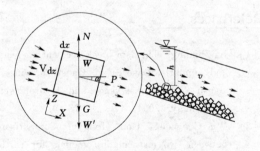

Figure 7 The force balance illustration of grain on slope

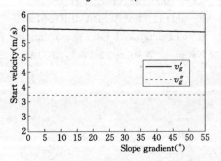

Figure 8 The relationship curve of start velocity and slope gradient

5 Conclusion

Based on the force balance fundamental of sand grain, it was analyzed that the force condition of the grain start velocity in water flow erosion as earth dam breaking process. Expressions of a rolling and a suspending start velocity of grain were deduced from that analysis. The relationships between grain start velocity and the grain diameter, flow direction, shear strength parameters were studied respectively. And which were illustrated as curves. It was made known that there were different manners of grain start moving because of the sand grain diameter from small to large. If the grain diameter was smaller than 4.0cm, grains start moving as suspending manner. And other vice, grains might roll in water flow erosion happening. The suspending start velocity was not affected by the shear strength parameters, but the rolling start velocity was augmenting with the shear strength parameters go large as a nonlinear curve manner. The slope gradient could have a decisive action to force direction of grains. The larger the gradient, the easier the grains start moving. And rolling might be the start moving manner.

References

[1] Jia Zhiming. The dam break accident analysis and support measures in Gansu province. Gansu Water Conservancy and Hydropower Technology. 2004, 40 (1): 18-19.

[2] Wu Congcong, Guo Hongyu. Comparison between two mathematical models for earth dam break. J. Wuhan Univ. of Hydr. & Elec. Eng, 2000, 33 (4): 17-20.

[3] Fread D L. The NWSDAMBRK Model: Theoretical Background/User Documentation [R]. National Weather Service, NOAA, Silver Spring, Md, 1988.

[4] Singh V P. Dam Breach Modeling Technology [M]. Kluwer, Dordrecht, The Netherlands, 1996.

[5] Fread D L. BREACH: An Erosion Model for Earthen Dam Failures [R]. National Weather Service (NWS) Report, NOAA, Silver Spring, MA, 1988.

[6] Ralston D C, Mechanics of Embankment Erosion During Overflow [A]. Proceedings of the 1987 ASCE National Conference on Hydraulic Engineering [C], Williamsburg, Virginia, 1987: 733-738.

[7] Wahl T L. Prediction of Embankment Dam Breach Parameters: A Literature Review and Needs Assessment [R]. DSO298 2004, Dam Safety Research Report, U. S. Bureau of Reclamation, 1998.

[8] Hanson G J, Temple D M, and Cook K R. Dam Over topping Resistance and Breach Processes Research [A]. Proceedings of the 1999 Annual Conference Association of State Dam Safety Officials [C], St. Louis, MO, 1999. CD - ROM.

[9] Cai Xiaoyu, Ling Tianqing, Tang Baiming, et. al. Mechanics of the wash-out of the particle bank slope by wave [J]. JOURNAL OF CHONGQING JIAOTONG UNIVERSITY, 2006, 25 (2): 73-76.

[10] Tian Weiping, Li Huiping, Gao Dongguang. Scour depth and mechanism of highway subgrade along rive [J]. Journal of Chang'an University (Natural Science Edition), 2002, 22 (4): 39-42.

[11] Dou Guoren. A study of silt incipient velocity [A], In: Dou Guoren memoir. Beijing: China WaterPower Press, 2003, 12: 54-69.

Experiences from Modifications of Curved Spillway Channels in Dam Rebuilding Projects

James Yang[1]   Patrik Andreasson[2]   Malte Cederström[3]

[1]*Vattenfall R&D, Älvkarleby & the Royal Institute of Technology (KTH), Stockholm*
[2]*Vattenfall R&D, Älvkarleby & Luleå Technical University (LTU), Luleå, Sweden*
[3]*Vattenfall Nordic, Hydro Power, Stockholm*

Abstract: To safely discharge higher design floods, a number of curved spillway channels need to be modified. The design flood of a dam is often released at higher reservoir level, thus giving rise to higher flow velocity in the channel. The combination of more water and higher velocity results in accentuated non-uniform channel flows. The unfavorably distributed flow into a stilling basin or plunge pool aggravates also the energy dissipation. The curved spillway channels of several existing dams are modified in physical model tests. The adopted countermeasures include heightened sidewalls, prolonged spillway piers, new partition walls, streamlining discontinuities in the waterway, the use of differential bottom elevations in cross section, addition of deflector on sidewalls or a combination of several measures. The project experiences are summarized in this paper, with emphasis on flow patterns at high discharges and countermeasures.

Key words: dam safety; curved spillway channel; model tests and countermeasures

1. Dam rebuilding background

In the light of the revised design floods and higher dam-safety requirements, many existing dams in Sweden have been rebuilt and many others are undergoing safety evaluations or an upgrading process. The updated damsafety guidelines, governing the dam-safety practice in the country, stipulate that dams should be maintained in such a way that a high level of safety is guaranteed and the risk of dam break is minimized to a practically non-existent level.

The rebuilding engineering measures may include raise of impervious core or/and dam crest, modification of spillway structures, foundation reinforcement and stabilizing measures of dam body, strengthening erosion protection, etc. The measures aim at higher dam safety margins, not at increasing the power production of a specific dam.

Many spillways were designed for discharges lower than the revised design floods. The increase in the flood magnitude, often by 20%-50%, gives rise to problem in the curved channels. Non-uniformly distributed flow builds up and the water may overtop the sidewalls. The energy dissipation becomes less sufficient due to the unfavorable distribution of flow into a stilling basin or plunge pool. No matter whether the stilling basin needs to be rebuilt or not, the channel flow patterns must be corrected.

A few examples are described in the paper to illustrate modifications of the spillway channels to deal with the higher design floods. Potential countermeasures include streamlining spillway piers and discontinuities, raising sidewalls, new partition wall in the channel, differential bottom cross-sectional profile, additions of deflector on sidewalls and a combination of several measures (Yang, 2009).

2. Letsi spillway channel

Letsi was constructed during 1967-1970. The dam is of rock-fill type, having a maximum height of 85m and a crest length of ~550m. It is grounded on solid rock and has a vertical impervious core of moraine, surrounded by filter and rock fill. The dam axis is given a slightly convex form in the upstream direction, Figure 1.

The spillway is placed on the right side of the rock-fill dam and consists of two 15m gated openings. The discharge capacity corresponded to some 1500m³/s at the full reservoir level. The 120m spillway channel bends to the left in plan and the elevation drop is about 15m. The channel bottom is partially in

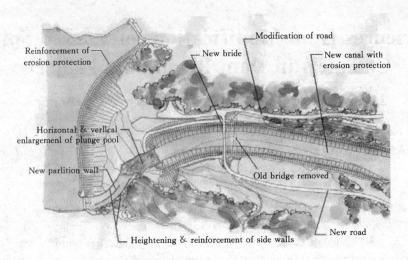

Figure 1　Layout of Letsi dam with curved spillway channel

rock and partially in concrete and is characterized by a higher bottom elevation on its right side, Figure 2 (a). From the channel, the spillway water is dis-

(a) The chanrel layout

(b) The flow pattern

Figure 2　Letsi spillway-existing channel layout and flow pattern at the design flood

charged into an energy dissipator some 40m below the downstream end of the channel. The water is then conveyed to the river in a canal with ripraps as erosion protection.

The spillway discharge capacity at the Letsi dam has to be increased by 25%-30% from the existing level that was only 1500m³/s. Without any structural modification in the channel, the flow distribution would be highly nonuniform in cross section throughout the channel, with shock waves propagating downstream and too much water on the right half of the channel. The water runs high against the right sidewall and becomes then reflected to the left, Figure 2 (b). The plunging water jet into the pool becomes too concentrated, rather than reasonably distributed along the pool (Bond et al. , 2003).

The major parts of both sidewalls of the channel are heightened. They are also strengthened to increase structural stability. There existed an end sill and one baffle block at the end of the channel. They are removed so as to reduce the amount of water cascading and spray outside the plunge pool. A partition wall, 2. 5-5. 0m high, is added and optimized in the middle of the channel in order to even out the flow. Limited by the existing bottom topography, it starts a few meters from the spillway piers and ends where the flow runs straight, Figure 3 (a). The prototype channel modified according to the model tests is shown in Figure 3 (b) (Yang et al. , 2006).

(a)

(b)

Figure 3   Letsi spillway rebuilding, with higher sidewalls and new partition wall

(a)Proto type

(b)The mode in scale 1 : 50

Figure 4   Ligga dam spillway channel, prototype and model in scale 1 : 50

3. Ligga spillway channel

Ligga was commissioned in 1954. The dam is constructed with an impervious core of moraine, having a height of 35m and a crest length of 350m. The active reservoir storage is $6 Mm^3$. The spillway is situated on the left side of the dam. It has three 20-m gated openings, the discharge capacity being $2200m^3/s$. The spillway channel, bending to right, is bounded by natural bedrock, Figure 4. The past operations have seen severe damages in the rock. Rock instability becomes visible even at moderate discharges.

The channel modifications are governed by several principles. (1) The spillway water is reasonably distributed in the channel; (2) The risk for erosion in the dam toe is minimized; (3) The right part of the channel is characterized by poor rock quality, implying rock excavation can only be done on the left side and (4) Concreting and use of concrete walls should be minimized as large costs are involved.

It is desirable that the channel modifications should provide about $24000 m^3$ excavated rock material, for both erosion protection upstream and a toe berm for the dam slope stability. This requirement affects even the way to excavate the channel. In order that large fragmentations of rock are obtained, the excavated depth should be kept as large as possible; excavation of less than one meter in depth is avoided. Several re-shaping options are examined in the model. Based on preliminary test results, two ways of excavation are chosen for further evaluation, Figure 5 and Figure 6.

Tests have shown that, with a moderate cross-sectional slope in option A, there is a good chance to procure reasonable flow patterns. However, the excavated volume amounts to only $5000m^3$. If the excavated depth in the left bank is increased to a nearly horizontal slope, the channel flow is satisfactory up to $1500m^3/s$. However, at higher discharges, too much water runs on the left side of the channel. Construction of a concrete sidewall downstream is needed. Otherwise, the water would flow into the forest to the left of the channel, causing floating debris for the dam downstream. Due to this, option A is abandoned.

By trial and error, satisfactory flow conditions

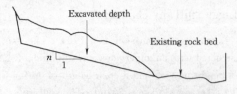

Figure 5　Ligga channel modification in a
scale model，option A

Figure 6　Ligga channel modification，option B

can be achieved if the channel is excavated deeper in the middle part, option.B. In this way, a "stepped" profile is given in cross section. The step deflects effectively part of the water and reduces the load imposed on the left bank. With a proper combination of cross-sectional slope and excavation depth, the spillway water is reasonably distributed in the channel. Up to the design flood, the main current follows the excavated channel and is directed away from the dam toe. The water is calm with little wave motions lon-

gest the whole dam toe.

The excavation in option B provides the required amount of rock material for the upgrading of the dam. The risk of rock erosion in the spillway channel, both before and after the rebuilding, is determined through numerical simulations (Ekström et al., 2007).

4. Höljes spillway chute

Höljes, completed 1961, is one of the highest embankment dams in Sweden. It has a height of 80m and a crest length of 400m. Two 14-m gated spillway openings and a log flume, Figure 7 (a), handle the flood. Their discharge capacity is 1300m³/s. The revised design flood the dam is to be rebuilt for amounts to some 55% higher.

The existing spillway chute bends slightly in plan to the left and sharply downward. The drop in elevation is some 60m. The upper part of the chute, some 80 m long from the gates, goes down almost linearly, with an average slope of about 27% (22m in 80m). The 50m long lower part drops 37. 5m in elevation.

One rebuilding layout involves a new spillway adjacent to the existing and the log flume is replaced. It is given a width of 17m and the same sill level as the existing. Its capacity is estimated at 740 m³/s, giving a total discharge of 2000m³/s at the full reservoir level. In proportion to the discharge, the existing chute, as well as the stilling basin, is widened to its right by 80%-90%. The chute width increases from 32 to 57m at the gates, and from 15 to 29m at the stilling basin. The extended part follows the existing channel curvature.

Originally, the widened part of the chute is given the same cross-sectional elevation as the existing. Tests showed that the chute flow is not uniformly distributed in cross section. Due to the inertia force, a large part of the water runs against the right sidewall and is reflected to the left, with unfavorable flow into the basin, Figure 7 (b).

By trial and error and also from previous engineering experiences, a layout with a differential bottom elevation in cross section is devised, Figure 8. Along the right edge of the existing chute, the widened chute part is given a higher bottom elevation. The bottom elevation difference starts at the downstream end of the spillway piers and increases to 3. 0 m that then runs all the way to the stilling basin. Due to the large increase in the flood magnitude, this bottom differentiation is however not enough to produce satisfactory flow distribution in the chute. A partition wall, placed along the upper bottom edge in the middle, is therefore added to improve the situation (Stenström et al. , 2009).

(a)The existing channel

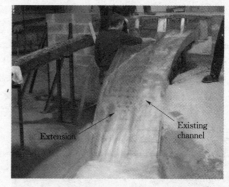
(b)The extension

Figure 7　Höljes, addition of a new spillway & extension of the existing channel

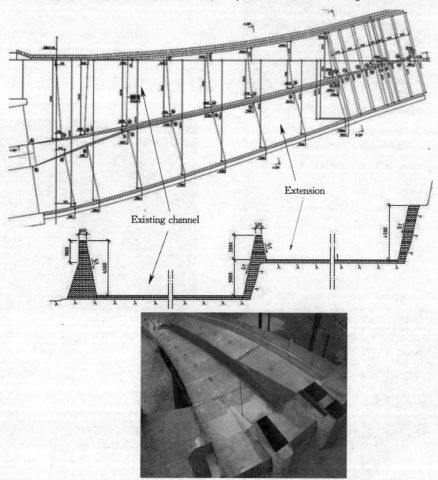

Figure 8　Höljes, channel with differential bottom
elevation & partition wall

The partition wall is 2m high and transits to the new spillway pier with the help of a slim wedge of 8 m in width. The pier is made thick due to structural considerations. The transition angle is optimized, so that no flow separation is created alongside of it.

The existing spillway pier, 4m wide, is extended downstream and given a streamlined shape. Its length and height are finalized at the design flood and the introduction of disturbance to the flow is kept to a minimum. A similar spillway pier extension can be found of the Alqueva dam, Portugal, Figure 9.

Figure 9   Spillway pier extension of the
Alqueva dam, Portugal

5. Storfinnforsen bottom outlet & canal

Storfinnforsen, completed in 1953, is 1200m long, of which the concrete dam accounts for 800 m. The buttress dam comprises 66 monoliths, the highest one being 39m.

There are two bottom outlets, with a size of 5.50m (w) by 3.65m (h) and a sill elevation 37m under the full reservoir level, Figure 10. The left

Figure 10   Storfinnforsen bottom outlet

outlet is sealed with a concrete wall, while the right one operates with an upward radial gate. The design flood is totally handled through the overflow spillway - the right bottom outlet does not need to share the flood. However, it does need to be functional and meet other dam-safety requirements, partly for reservoir lowering if required and partly for acting as reserve. The right outlet has never been used after the dam commissioning. It is not clear either for what hydraulic conditions the outlet was originally designed.

The outlets discharge into a canal that bends to the left. That is why a strongly non-uniform flow is formed in the canal when discharging from the right outlet, Figure 11. The flow velocity is as high as 20-21m/s. The main current runs along the right side of the canal and against the right sidewall. Irrespective of the outlet opening, the right sidewall is overtopped at high reservoir water levels. The overtopping water runs straight downstream outside the canal and undermines the sidewall from behind, which constitutes a safety risk if it discharges water for a long period of time. Necessary measures are required to prevent the overtopping and wall instability.

Figure 11   Storfinnforsen, flow pattern in existing
bottom outlet canal

In the model studies, several countermeasures in the form of baffle wall are tested against overtopping of the right canal wall, Figure 12 (a). The baffle wall reflects part of the water back to the canal and moved the beginning point of overtopping further downstream. Due to the fact that the canal bends to the left, it is not practical to try to prevent the overtopping along the whole sidewall. To overtop the downstream part of the sidewall can be allowed, as

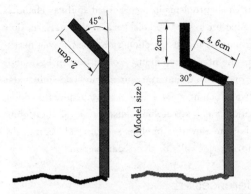

(a) Tested counter measures

(b) The feasible measure

Figure 12　Storfinnforsen outlet canal-countermeasure to prevent overtopping

the water over the wall runs basically into the river course. The use of the 30° baffle wall, with a vertical part that corresponds to supposed concrete thickness, seems to be a feasible measure, Figure 12 (b). The pressure acting on the baffle walls is also measured in different flow situations for the sake of design.

6. Halvfari discharge channel

Halvfari is an embankment dam of 43m in height. Its spillway consists, after proposed rebuilding, of a siphon outlet, a bottom outlet and two overflow openings. The plunge pool is located downstream of the powerhouse, Figure 13.

The right sidewall runs at an angle with the spillway piers. Due to this, the spillway water plunges obliquely into the downstream part of the plunge pool that has limited volume and water-surface area. The pool water volume is not effectively used. The plunging causes a plane-circulating flow pattern to

build up in the pool, with considerable water level difference. Due the insufficient energy dissipation, high flow velocity and strong wave motions accompany the flow in the watercourse downstream, Figure 14.

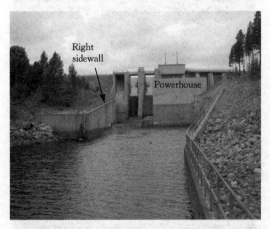

Figure 13　Halvfari dam with spillway structure

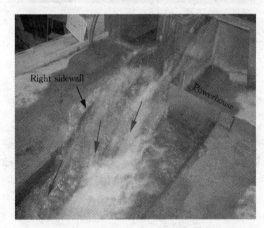

Figure 14　Halvfari, spillway water plunging unfavorably into the pool

The existing spillway piers end too early upstream, the water from both sides interrupts each other. It is found that if the spillway piers are extended, the water is directed further downstream, which is favorable. The extensions cause also the water level in the pool to rise about 2.0m downstream of the powerhouse.

The flow conditions and energy dissipation in the plunge pool is further improved by adding vertical deflectors at the end of the piers and on the right sidewall, Figure 15. Tests show that deflectors are needed in both spillway openings. The reasonable angle, measured in the flow direction, falls within 20°-30°.

Figure 15   Halvfari, improved energy dissipation
with streamlined pier extensions and added
deflectors at the end of the piers and on
the right sidewall

The suitable length of the deflectors is 2m (Yang et al. , 2007).

With the pier extension and the deflectors, the concentrated water jet into the plunge pool is avoided; the spillway flow is redirected in three separate plunging jets, reasonably distributed over the pool area. The pool water level becomes also somewhat higher. The jets plunge roughly into the middle of the pool, consequently resulting in less impact for the river course downstream. The proposed structural modification functions well for spillway discharges up to the design flood.

7. Conclusions and discussions

The increased design floods in many existing dams cause problems in their curved spillway channels designed for lower flood discharges. Non-uniform flow distributions in cross section and shock waves necessitate the need of rebuilding, often limited by practical conditions. Countermeasures may conclude extra sidewall height, extending spillway piers, adding partition wall, streamlining discontinuities, differential bottom elevation in cross section, addition of deflector on sidewalls or their combination.

References

[1] Bond H, Yang J, Cederström M, Halvarsson A. Letsi Dam-Refurbishment for Safe Passage of Extreme Floods, Hydraulic Considerations. WaterPower XIII, July 2003, Buffalo, NY.

[2] Ekström I, Yang J, Mörén L, Cederström M. Adapting Ligga to higher design flood, spillway channel modification through physical & numerical modeling. WaterPower XIV, July 2007, Chattanooga, TN.

[3] Stenström P, Yang J, Bond H, Sjödin A, Steiner, R. Increasing the discharge capacity at the Höljes dam in Klarölven, Sweden. 25th ICOLD Congress, May 2009, Brasilia.

[4] Yang J, Larsson P, Dath J, Löwén K-L. Halvfari Dam-hydraulic concerns and rebuilding for higher safety standard. International Symposium on Modern Technology of Dams, Oct. 2007, Chengdu, China.

[5] Yang J. Experiences from Hydraulic Model Tests in Dam Rebuilding Projects. Elforsk report, April 2009.

[6] Yang J, Halvarsson A, Bond H, Cederström M. Modification of flood discharge structure for updated design flood at Letsi dam. Dam Safety 2006, Sept. 2006, Boston.

Experimental Investigation on the Effect of Sluice Gate Distance on the Flow Structure and Amount of Sediment Entry to the Lateral Intake in Diversion Dam

Mahdi Esmaeili Varaki[1]   Javad Farhoudi[2]   David Walker[3]

[1]*Assistant Prof of Department of Water Engineering, University of Guilan, Rasht, Iran*
[2]*Professor of Department of Irrigation and Reclamation Engineering, University of Tehran, Tehran, Iran*
[3]*Associate Professor, School of Civil & Environmental Engineering, The University of Adelaide, Adelaide, Australia*

Abstract: A set of laboratory experiments has been carried out involving a layout that is common in irrigation diversion works. The layout involves a lateral intake channel with diversion angle 110° and diversion dam which are positioned to allow the flow to be controlled and managed. Based on experimental measurements and observations the effect of sluice gate distance from downstream corner of the intake entrance on the hydrodynamic behavior of diverging flow and amount of sediment entering the intake was investigated. The velocity profiles, both upstream of the intake in the main channel, in front of the intake in the sluiceway and a cross the intake entrance and amount of sediment entering into the intake were measured under various discharge combinations of the intake and sluice gate. Experimental observation and analysis of the velocity profile along the intake entrance showed that the structure of the diverging flow into the intake was a function of the sluice gate discharge. Comparison of the result showed that as the sluice gate distance from upstream corner of the intake increased, maximum return velocity increased. Analysis of the sedimentary measurements data showed the amount of sediment entering the intake increased with increasing intake discharge. Comparison of the results showed that with increase relative intake discharge (D_r) from 0.3 to 0.6, amount of sediment entry into the intake increase 3 timed approximately. In addition, an increase in the sluice gate discharge caused an increase of sediment entry to the intake under the same intake discharge. Comparison of the sedimentary measurements data under different sluice gate discharges showed that an increase the sluice gate discharge from closed gate condition to relative sluice gate discharge (S_r) 0.18, amount of sediment entry into the intake increase 59.81 percent approximately. Comparison of sedimentary result showed that amount of sediment entering into the intake increase 13.8 percent with increase of the sluice gate distance from the intake entrance.

Key words: river; lateral intake; sediment transport; flow structure; sluice gate; diversion dam

1. Introduction

According to river flow and water users demand, there are various options to divert water into off-take channels. Among them, diversion dams are commonly used to divert water into irrigation canals. The dams perform a number of functions including maintaining channel levels in times of low flow, and reducing the flow of sediment by allowing it to settle in calm slow-flowing waters. A typical layout of a diversion dam includes a spillway and stilling basin, a sluiceway, a sluice gate, and an intake structure and conveyance channel which convey the diverted water.

Operational experience has shown that sediment control for the diverted water is a significant problem. Sediment entry to lateral intakes and the downstream conveyance channel causes serious damage for users such as hydropower plants and cooling water systems. Furthermore, sediment deposited in conveyance channels brings the cost of dredging, the potential for plant growth, a decrease in channel capacity, an increase in the resistance to flow, and the potential for damage to the channel surface. This is of concern and it is important that the intake and sluiceway are designed and operated so that the amount of sediment entering the intake is minimized.

To reduce sediment entry to lateral intakes an in-depth understanding of the diverging flow structure and the interaction between the flow and sediment transport is required. Results of previous work on free lateral intakes in rigid bed flumes, which do not have any diversion works such as a diversion dam, have shown that there are complex 3D flow features including counter-clockwise secondary motion in the main channel downstream of the intake entrance, and clockwise secondary motion and a separation zone along the inner wall of the intake channel. Furthermore, the surface currents in the main channel are narrower than those at the bottom (Razvan, 1989; Neary et al., 1993; Neary et al., 1999 and Esmaeili Varaki, 2009). Also, the lateral distribution of velocity at the intake entrance was non-uniform and increased form upstream corner of intake entrance to downstream of it (Chen et al., 2004; Esmaeili Varaki, 2009). Barkdoll et al. (1999) and Esmaeili Varaki et al. (2009) found that an unsteady vortex caused sediment to be suspended and to enter the intake channel. Also, Esmaeili Varaki et al. (2009) found that the vortex was formed on the leeward side of the dune at the upstream side of the intake and its strength increased as the intake discharge increased.

Field and laboratory observations of hydraulic models of lateral intakes reveal that the height of the entrance sill, the angle of diversion and the rate of the intake discharge are important geometric and hydraulic parameters in determining the point of sediment entry to the intake (Raudkivi, 1993; Esmaeili Varaki, 2009). The entrance sill prevents entry of the bottom current that usually has a high concentration of sediment. Razvan (1989) recommended that the height of the entrance sill be greater than one third of the flow depth in front of the intake. Previous field and laboratory observations of lateral intakes showed that the diversion angle has an influence on the flow curvature and sediment transport into the intake. An optimum angle for minimum sediment diversion was found to be between 105° and 110° by UPIRI (1973), between 95° and 120° by Razvan (1989),

and between 95° and 120° by Garde and Ranga Raju (2000) and 110° by Esmaeili Varaki (2009).

In the case of lateral intakes in diversion dams, the structure of the approaching flow to the intake and the associated sediment transport are very different from those of free lateral intakes. In this situation there is an obstacle, the diversion dam weir and sluice gates, downstream of the diversion that causes the river morphology to change completely and to differ from those that have been studied previously. A laboratory investigation, performed by Esmaeili Varaki (2009), has shown that sluice gate discharge has a considerable influence on diverging flow structure and the amount of sediment entry to the lateral intake in diversion dam. However it is not clear what would happen if the distance of the sluice gate from the downstream corner of the intake entrance changes.

The aim of this study, therefore, was to investigate the effect of the sluice gate distance on both the flow structure and sediment transport in a lateral intake in a diversion dam.

2. Experimental Investigation

The experiments were carried out in a rectangular flume 18m long, 0.9m wide and 0.6m deep. Both the walls and bed were made of Plexiglas. The intake flume was 4.5m long, 0.4m wide and 0.6 m deep and was located 9 m from the inlet of the flume. Two centrifugal pumps with a total discharge of 150 L/s supplied flow to the flume. The inlet discharge was measured with a triangular weir. The diversion dam was located 9.4 m from the inlet. The dam was 0.6 m in length and 0.27 m high with a radial gate for adjusting the elevation of the water upstream of the dam and the overflow discharge of the dam weir. The sluiceway, located to one side of the dam, had a width of 0.3 m and included a dividing wall 0.3 m in length. The intake had an entrance sill 0.1 m in height and the outflow from the intake channel discharged into a tank 5 m in length and 1 m wide and 1 m deep. The discharge in the intake channel was measured in this tank using a rectangular weir. The outflow flowed into the main tanks made up of 8 tanks each 2 m long, 2 m wide and 1 m deep. These tanks were joined with pipes and flanges. The layout of the flume is shown in Figure 1. To study the effect of sluice gate distance on flow structure and the amount of

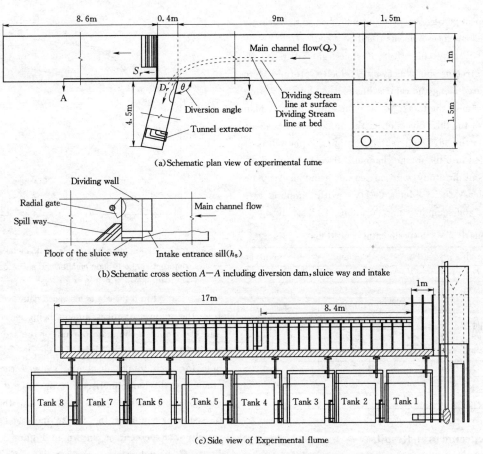

(a) Schematic plan view of experimental fume

(b) Schematic cross section $A-A$ including diversion dam, sluice way and intake

(c) Side view of Experimental flume

Figure 1　A schematic layout of experimental flume with definition of flow features.

sediment entering into the intake, two distances 0.05 m (the minimum distance that could be installed) and 0.2 m (half of the intake width) were selected.

The initial bed of the main channel was formed with an 0.08 m layer of sediment. In order to prevent scour near the inlet of the channel the first 3 m of the bed was formed with a coarse sediment. The rest was formed with approximately uniform sediment with 0.6 mm diameter. To manage the flow of sediment, a sediment injector was used that had a mechanism for regulating the intensity of the engine speed in such a way that sediment could be injected at selected concentration into the flow. The location of the injection point was less than 3 m from the inlet. A sediment extractor was constructed to collect sediment from the entry to the intake channel. It was installed at a location 3.5 m from the intake entrance.

Prior to starting the main experiments, it was necessary to carry out some preliminary work to ensure that the sediment was in dynamic equilibrium,

with equal sediment being injected and flowing into the intake channel After that, the elevation of the bed in the main channel was based on a rate of sediment injection such that the change of bed elevation was not significant.

In each experiment, outflows from the intake channel and sluiceway gate were adjusted with the tailgate, and the elevation of the water upstream of the dam was adjusted with the radial gate of the dam weir. The quantity of sediment entering the intake channel was measured over a set time interval, ranging from 10 to 15 minutes. Thus, it was possible to study the trend of sediment entry and the time taken to reach equilibrium. The experimental conditions for the runs are listed in Table 1. Parameters in Table 1 is defined at the notation section.

Each experiment was run in two phases. In the first a mobile bed was used that allowed sediment transport rates and patterns to be determined. Then, for the second phase focus shifted to the flow structure and a layer of cement powder was used to fix the bed so that velocity profiles and measurements could be taken.

To study the effect of sluice gate distance on flow structure in sluiceway, velocity profiles were measured along the central line of the sluiceway from upstream of the intake to near the sluice gate. A plan view of the flow diversion to the intake was visualized by injecting dye and taking continuous photos as it traveled into the intake channel. All velocity measurements in the study were carried out using an Acoustic Doppler Velocity Meter (ADV) with sampling rate 50 Hz and sampling time duration 10-15 second.

Table 1   Experimental conditions

Variable	Unit	Maximum	Minimum
Q_r	L/s	70. 2	70. 2
y_r/h_s	—	0. 32	0. 32
D_r	—	0. 65	0. 3
S_r	—	0. 21	0
Fr_r	—	0. 24	0. 19
Fr_{in}	—	0. 44	0. 15
C_s	—	7. 5	4. 8
X_{sluice}	m	0. 2	0. 05
D_s	mm	0. 6	0. 6

3. Experimental Results

3. 1   Effect of sluice gate distance on amount of sediment entered into the intake

In each mobile bed experiment, the amount of sediment entering the intake was measured over a time to reach equilibrium condition on a given time interval as both wet and dry weights. Results of the analysis in each sluice gate discharge showed that the amount of sediment entering the intake increased with increasing intake discharge. Analysis of the data for the amount of sediment entering the intake showed that as relative intake discharges increased from $D_r = 0. 3$ to $0. 65$, the amount of sediment entering the intake increase 2. 9 times. Also, for each intake discharge, the amount of sediment entering the intake increased with an increase in the discharge of the sluice gate as shown in Figure 2. This is because any increase of sluice gate discharge increases the discharge of the diverted flow and an increase of the diverted flow caused an increase in the total sediment being transported toward the sluiceway. Experimental observation showed that the sluice gate could not ex-

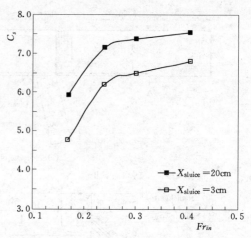

Figure 2   Curves of sediment entry to the intake under different intake Froude number and sluice gate discharge (S_r) 0. 3 (1−D_r) for two distances of the sluice gate

tract the amount of the sediment being carried in the sluiceway and so the main portion of the sediment entered the intake channel. Analysis of the sedimentary data for each intake discharge showed that as the sluice gate discharge increased from closed gate condition to 0. 3 (1−D_r) (that is 12. 6L/s), the amount of sediment entering the intake increased approximately 62. 8 percent as shown in Figure. 2. In this figure C_s is volumetric concentration, Fr_{in}, intake Froude number and X_{sluice} the distance of the sluice gate from the upstream corner of the intake, respectively. As shown in Figure. 2, the amount of sediment entering the intake increased approximately 13. 1 percent as the distance of the sluice gate from the downstream corner of the intake increased. Therefore it can be expected that as the distance of sluice gate increases, the effectiveness of the sluice gate discharge to extract sediment being transport in the sluiceway decreases.

3. 2   Effect of sluice gate distance on flow structure in the vicinity of the intake in the sluiceway

Experimental observations and analysis of the velocity profiles showed that the sluice gate discharge has an important effect on the pattern of the velocity profiles in the sluiceway. For all experiments where the sluice gate was shut all velocity profiles along the intake entrance had an inflection point and return velocities. The depth to the inflection points and the maximum return velocities were all functions of the intake discharges and increased with increases in it as shown in Figure 3 to Figure 5. Comparison of the velocity profiles for the two sluice gate positions demonstrated that values of maximum return velocity increased as the sluice gate distance increased.

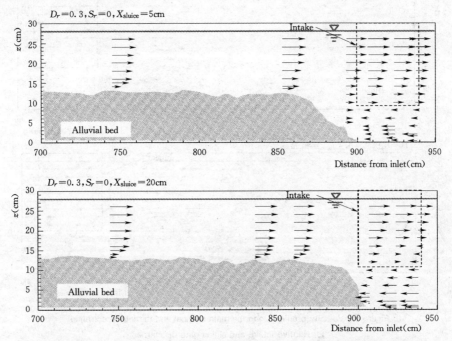

Figure 3　Velocity profiles along the main channel and the intake entrance for relative intake and sluice gate discharge

$D_r = 0.3$ and closed gate condition $(S_r = 0)$

(Note: Velocity vectors are scaled based on magnitude.)

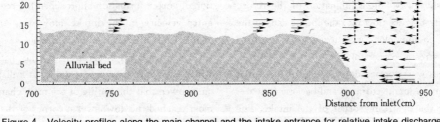

Figure 4　Velocity profiles along the main channel and the intake entrance for relative intake discharge

$D_r = 0.5$ and closed gate condition $(S_r = 0)$

(Note: Velocity vectors are scaled based on magnitude.)

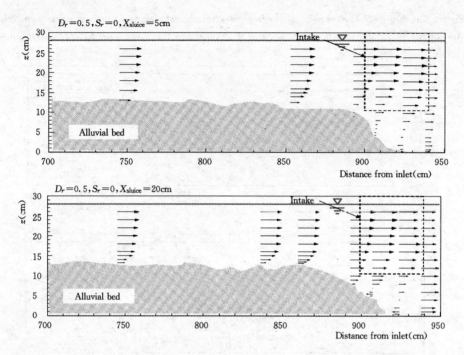

Figure 5   Velocity profiles along the main channel and the intake entrance
for relative intake and sluice gate discharge
$D_r = 0.5$ and $S_r = 0.3$
(Note: Velocity vectors are scaled based on magnitude.)

Analysis of the results showed that the inflection point in the velocity profile occurred near the sill level. Formation of the return velocity resulted from two effects: the adverse pressure gradient on the leeward side of the dune, and the closed sluice gate. The adverse pressure gradient on the leeward side of the dune is to be expected, but it is not believed to be the main cause of the return flow. That is believed to be due to the closed sluice gate which causes the part of the flow beneath the sill to be blocked and to return upstream. The combination of the return flow and the return velocity near the toe of the dune forms a zone of recirculation beneath the level of the sill of the intake in the sluiceway. The combination of the return velocity and transverse flow into the intake caused the formation of counter clock-wise tornado-like vortices near the upstream corner of the intake entrance. These vortices were formed downstream of the toe of the dune near the intake entrance and became fully developed near the upstream corner of the intake. The vortices picked up sediment from upstream of the intake and caused it to spill into the intake channel, as shown in Figure 6 (a). Observations and analysis of

movies that were made during the experiments showed that the intensity of these vortices, and the rate at which they formed and were then carried out into the intake by transverse flows, increased with an increase in the intake discharge.

Opening the sluice gate caused a change in the velocity profiles. In this situation, the return flow vanished and the dune shape of the bed profile of the main channel bed marched to the middle part of the intake entrance. In addition, the rates at which vortices formed and were then carried into the intake by the transverse flow increased due to the velocity of flow in sluiceway increasing. Furthermore, the location of vortex formation approached the region of high entry velocity in the intake. However, the vortices could not occur higher than the entrance sill. Experimental observation during conditions in which the sluice gate was installed at a distance of 20 cm showed that the profile of the main channel bed almost reached the downstream corner of the intake entrance and sediment entered the intake from the whole of intake entrance, as shown schematically in Figure 6 (b).

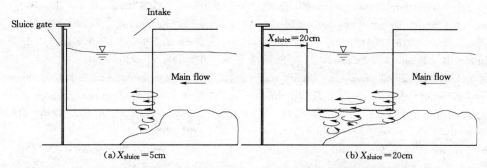

(a) $X_{sluice}=5$cm   (b) $X_{sluice}=20$cm

Figure 6   Formation of tornado-like vortex near the intake entrance

4. Conclusion

A lateral intake with a diversion dam is commonly used for diverting water. The most important problem for design and operation of such diversion structures is a reduction in the amount of sediment being carrying by the diverging flow. There is, therefore, a need to understand in some detail the hydrodynamics of diverging flow, the mechanism of transport of sediment into the intake, and the effect of important hydraulic and geometric parameters on the diverging flow structure and the amount of sediment entry into the intake.

Experimental observations and analysis of velocity measurements in a channel and in the vicinity of the intake showed that the diverging flow to the intake experiences a very complex 3D structure that is characterized by a return flow, and depends on the sluice gate discharge, the formation of tornado-like vortices, which picked up sediment from the toe of the dune on the bed upstream of the intake and caused it to spill into the intake, and a recirculation zone in front of the sluice gate.

Experimental observation showed that the intensity and rate at which vortices formed and were then carried into the intake was a strong function of both the intake and sluice gate discharges. From this result, it suggests that if it is possible to keep the toe of the dune away from the intake entrance it would be possible to achieve a reduction of sediment entry into the intake.

Results of the sedimentation investigation showed that any increase in the sluice gate discharge will increase the amount of sediment entering the intake. Comparison of result showed that as sluice gate discharge increase from closed gate condition to 0.3 $(1-D_r)$, 12.6 L/s, the amount of sediment ente-

ring the intake increased approximately 1.84 and 2.12 times for sluice gate distance 0.05 and 0.2 m, respectively. Also, comparison of the results reveals that an increase at sluice gate distance caused the amount of sediment entering the intake to increase approximately 13.1 percent.

Therefore, it is important that operators maintain calm slow flowing water upstream of the diversion dam and closed sluice gates, unless they are open for flushing. Also, they should close the intake gate or reduce the intake discharge to prevent entrance of sediment to the intake.

Notation

C_s = concentration of sediment diverted water in the intake channel (Q_{sin}/Q_{in})

D_r = relative discharge of intake rate (Q_{in}/Q_r)

D_s = mean diameter of sediment particle

Fr_{in} = Froude number in intake channel

Fr_r = Froude number in main channel

g = gravity acceleration

h_s = height of entrance sill

Q_d = discharge of sluice gate

Q_{sr} = volumetric sediment discharge of river

Q_{sin} = volumetric sediment discharge of intake

Q_r = Discharge of main channel

Q_{in} = Discharge of intake channel

S_r = relative discharge of sluice gate ratio(Q_d/Q_r)

X_{sluice} = Distance of sluice gate from downstream of intake entrance

y_r = Depth of flow in front of intake in sluice gate

References

[1]   Abbasi A.. Experimental study of sediment

control at free lateral intake in straight channel, PhD thesis, University of Tarbiat Modaress, Tehran, Iran, 2004.

[2] Barkdoll B. , Ettema R. , Odgaard, A. J. Sediment Control at Lateral Diversions: Limits and Enhancements to Vane Use. *J. Hydr. Engrg.* , *ASCE.* 1999, 125 (8): 855-861.

[3] Chen H. , J. Cao. Some 3D Hydraulic Features of 90 Lateral Water-Intake and Its Sediment Control. *Proc*, 9th *Symposium on River Sedimentation*. China, 2004.

[4] Esmaeili Varaki. M. Experimental investigation on the Effect of Diversion Angle on Flow Structure and Sediment Entry to a Lateral Intake in diversion dam, *PhD Dissertation in Hydraulic Structure*, *School of Soil & Water Eng. College of Agriculture & Natural Resources*, *University of Tehran*, *Iran*, 2009.

[5] Esmaeili Varaki. M. , J. Farhoudi, D. Walker. Experimental Investigation of the Flow Structure at a Right-angled Lateral Intake. *I C E*, *Journal of Water Management*, 2009, 162 (6): pages 379-388. DOI: 10. 1680/wama. 2009. 162. 6. 379.

[6] Garde R. J. , Ranga Raju K. G. Mechanics of Sediment Transport and Alluvial Stream Problem. Third edition. *New Age International Pub.* India, 2000.

[7] Hsu C. C. , Tang C. J. , Lee. W. J. , Shieh M. Y. Subcritical 90° Equal-Width Open-Channel Dividing Flow *J. Hydr. Engrg.* , *ASCE.* 2002, 128 (7): 716-720.

[8] Neary V. S. , Odgaard A. J. Three-dimensional flow structure at open channel diversions. *J. Hydr. Engrg.* , *ASCE*, 1993, 119 (11): 1224-1230.

[9] Neary V. S. , Sotiropoulos F. , Odgaard. J. Three dimensional numerical model of lateral intake inflows. *J. Hydr. Engrg*, *ASCE*, 1999, 125 (2): 126-140.

[10] Ramamurthy A. S. , J. Qu, D. Vo. Numerical and Experimental Study of Dividing. *J. Hydr. Engrg.* , *ASCE*, 2007, 133 (10): 1135-1144.

[11] Raudkivi A. J. . Sedimentation, exclusion and removal of sediment from diverted water. *A. A. BALKEMA*, Rotterdam, Netherlands, 1993.

[12] Razvan R. River Intake and Diversion Dams. Elsevier, 1989.

[13] Ruether N. , J. M. Singh, N. R. B. Olsen, E. Atkinson. 3-D computation of sediment transport at water intakes. *I C E*, *Journal of Water Management*, 2005 (158): 1-8.

[14] UPIRI. Hydraulic design of under sluice pocket at Lower Sadra Barrage including divide and excluder- Model study. T. M. 43 RR (H1-1), 1973.

Overall Geometry of a Submerged Inclined Jump Over a Sill

J. D. Demetriou [1]   D. J. Dimitriou[2]

[1] *School of Civil Engineering, National Technical University of Athens, Greece*
[2] *Davy Process Technology, London, UK*

Abstract: In this experimental study the overall geometry of a submerged inclined (angle φ with $0° \leqslant \varphi \leqslant 12°$) water jump over a sill is presented, analyzed and discussed. Two equations are given for the upstream and downstream dimensionless important lengths of these parts of the entire flow region. The overall flow's free surface profiles are also presented and pertinent equations are determined in dimensionless terms. The results may be useful to the hydraulic engineer when designing the stabilization of a weak jump through a sill.

Key words: Submerged Jump; Flow Over Sill

1. Introduction

The steady water flow over a sill is an interesting hydraulic problem. Two flow cases are important, the free hydraulic jump over the sill and the so called "submerged hydraulic jump" over the sill. Actually the last case does not constitute a hydraulic jump under the usual meaning of a supercritical flow of low depths followed by a subcritical flow of larger depths, but it is usually associated with the free hydraulic jump over the sill.

Figure 1 schematically shows the flow geometry within an inclined (angle φ, slope $J_o = \sin\varphi$) rectangular open channel, in which there are a rectangular opening (a) and a downstream thin sill (of w height), both perpendicular to the channel floor. If the flow comes out the opening with a depth a then a free hydraulic jump (broken line) is created, but if

the flow depth at the opening is d_s with $d_s > a$, then a submerged hydraulic jump is appearing over the sill. The flow submergence for any jump is due to a particular combination among all flow parameters, such as (a large) w, a, q ($=$ discharge per unit channel width) and the distance between opening and sill. The free surface (d) of the submerged jump is ascending until a maximum depth d_m ($> w$) is reached, followed by a fall until a small depth d_2-where the flow is supercritical since (in this investigation) it ends up to a drop. The length between opening and maximum flow depth is L, while between d_m and d_2 it is L', where $L + L' = L_t$. Along L, at any distance x the water depth is d, while along L' at any distance x' (from d_m) the water depth is d' ($> w$) -all perpendicular to the channel floor. Under the free surface (profile) along L a recirculating region (roller) of zero net discharge is created, simply increasing the underflow pressures.

In this experimental study the overall characteristics of the flow between Section 1 (depths a-of uniform flow-and d_s) and Section 2 (of uniform flow) at a total distance L_t is examined. Equations are given for L/d_m and L'/d_m, and the flow profiles along L_t are also investigated.

Since at Section 1 both a and d_s have the same discharge-which continues through Section 2 (d_2), the following characteristic Froude numbers are considered,

$$Fr_s = q/g^{1/2} \cdot d_s^{3/2}, Fr_2 = q/g^{1/2} \cdot d_2^{3/2} \qquad (1)$$

which are interconnected, $Fr_s = Fr_2 \cdot (d_2/d_s)^{3/2}$.

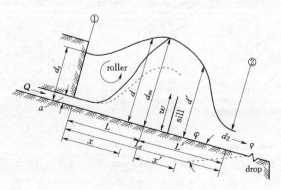

Figure 1   Flow geometry

Demetriou et al.[1-8], present a large number of initial laboratory measurements, a part of which are presented here after further elaboration.

2. The Experiments

All measurements were performed in a small tilting perspex rectangular flume. A large number of L and L' lengths and d_m, d_2, d_s, d and d' depths were measured in the L_t region, while w varied among 1-3-5 cm at various locations, and various q were volumetrically measured. Fr_2 varied between 1.25 and 5.9, Fr_s reached values up to 1.9 and all flows were fully turbulent flows. The inclination angles were $\varphi=$ $0°$, $-3°$, $-6°$, $-8°$, $-10°$, $-12°$ and in all cases the flows were organized to be steady and freely developing, while no air pockets were observed in touch with the downstream faces of the sills.

3. Results. Analysis And Discussion

Figure 2 presents a number of experimental measurements concerning L/d_m vs. Fr_s and all angles φ.

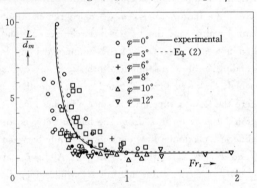

Figure 2 L/d_m vs. Fr_s and all angles φ

The measurements appear to be non systematic in relation to angle φ and thus a single curve is traced among the experimental points (solid line), showing that when Fr_s is increasing L/d_m is decreasing-independently of angle φ. An empirical equation was given to this experimental curve,

$$L/d_m = (L/d_2) \cdot (d_2/d_m) = 1.39 + 0.05 \cdot Fr_s^{-4.96}$$
(2)

holding for $0.35 \leqslant Fr_s \leqslant 1.9$ and $1.3 \leqslant L/d_m \leqslant 10$ and giving a (broken) curve on Figure 2 which almost coincides with the experimental curve.

Figure 3, Figure 4 and Figure 5 present the experimental results of L'/d_m vs. Fr_2 and angle

φ. Figure 3 and Figure 4 show that L'/d_m is angle φ dependent, while Figure 5 presents all the pertinent (solid) experimental curves which are very systematic in relation to angle φ. Along any curve ($\varphi=$const.) L'/d_m is falling when Fr_2 is increasing, while for $Fr_2=$const. L'/d_m are increasing with angle φ.

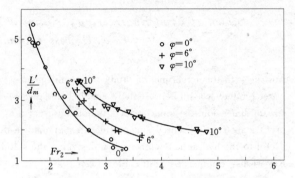

Figure 3 L'/d_m vs. Fr_2 for $\varphi=0°$, $6°$, $10°$

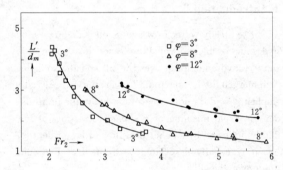

Figure 4 L'/d_m vs. Fr_2 for $\varphi=3°$, $8°$, $12°$

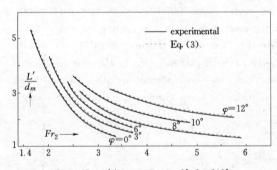

Figure 5 L'/d_m vs. Fr_2 for $0°\leqslant\varphi\leqslant12°$

To any curve of Figure 5 an empirical and practical equation of the form of

$$(L'/d_m) = (L'/d_s) \cdot (d_s/d_m) = a + b \cdot Fr_2^{-c} \quad (3)$$

was given for any angle φ at $1.7 \leqslant Fr_2 \leqslant 5.9$ and $1.2 \leqslant L'/d_m \leqslant 5.3$. Table 1 presents all a, b, c, arithmetic coefficients determined from the experimental curves of Figure 5, while along all experimental curves of the same figure a family of (broken) curves correspond to Eq. (3) in conjunction with Table 1. As it may be seen the experimental curves are almost coinciding to the curves given by the above empirical equations -with a, b, c, taken from Table 1.

Table 1   Geometrical coefficients

$L'/d_m = a + b \cdot Fr_2^{-c}$						
	$\varphi=0°$	$\varphi=3°$	$\varphi=6°$	$\varphi=8°$	$\varphi=10°$	$\varphi=12°$
a	0.23	0.94	0.35	1.21	1.38	1.72
b	15.13	43.41	16.21	68.64	15.97	28.04
c	2.15	3.51	1.93	3.54	2.16	2.49
$\bar{d}' = a' \cdot (\bar{x}')^{1.5} - b' \cdot (\bar{x}')^3$						
a'	1.97	2.08	2.19	2.31	2.42	2.53
b'	0.95	1.07	1.18	1.3	1.42	1.53

From Eq. (2), Eq. (3), and Table 1, the entire length L_t/d_m may be determined,

$$(L_t/d_m) = (L/d_m) + (L'/d_m) \qquad (4)$$

which depends on both Fr_s and Fr_2-interconnected through Eq. (1) -and varies between 2.5 and 15.3, i. e., L_t is much larger than d_m.

In order to determine the free surface profiles along L and L' all experimental measurements of d_s, d_m, d_2, d (at x) and d' (at x') were put in the dimensionless forms of

$$\left.\begin{array}{l} \bar{d} = (d - d_s)(d_m - d_s) \text{ and } \bar{x} = x/L \\ \bar{d}' = (d_m - d')/(d_m - d_2) \text{ and } \bar{x}' = x'/L' \end{array}\right\} \qquad (5)$$

and the elaboration of the experimental results gave the empirical equations

$$\bar{d} \equiv 1.29 \cdot (\bar{x})^{1.5} - 0.29 \cdot (\bar{x})^3 \qquad (6)$$

$$\bar{d}' = a' \cdot (\bar{x}')^{1.5} - b' \cdot (\bar{x})^3 \qquad (7)$$

where a', b', may also be found in Table 1 for $0° \leqslant \varphi \leqslant 12°$.

Eq. (6) shows that, in the L region, $\bar{d}$ depends only on $\bar{x}$ (for all angles with $0° \leqslant \varphi \leqslant 12°$), while Eq. (7) shows that d' depends on x' and angles φ (in the region L'). Although, the dependence of $\bar{d}$ on $\bar{x}$ actually means that $\bar{d}$ is indirectly associated to L, d_m and Fr_s (Figure 2), while d' is associated to x', i. e., it depends on L', d_m and Fr_2 (Figure 3 to Figure 5) —apart from angle φ. Both the above equations give $\bar{d} = 0$, $\bar{d}' = 0$, for $\bar{x} = 0$ and $\bar{x}' = 0$ (i. e. $d = d_s$ and $d' = d_m$), while for $x = 1$ and $\bar{x}' = 1$, $\bar{d} = 1$ (i. e. $d = d_m$) and $\bar{d}' \equiv 1$ (i. e. $d' \equiv d_2$).

A more practical expression for Eq. (6) is,

$$d/d_s = 1 + [(d_m/d_s) - 1] \cdot [1.29 \cdot (\bar{x})^{1.5} - 0.29 \cdot (\bar{x})^3] \qquad (8)$$

while a corresponding expression for Eq. (7) is,

$$d'/d_m = 1 + [(d_2/d_m) - 1] \cdot [a' \cdot (\bar{x}')^{1.5} - b' \cdot (\bar{x}')^3] \qquad (9)$$

Figure 6 and Figure 7 present two typical flow profiles along L_t (for $\varphi = 3°$ and $\varphi = 10°$ correspondingly) in terms of $\bar{d}$, $\bar{x}$, and $\bar{d}'$, $\bar{x}'$, while Figure 8 shows all present flow profiles along L_t for $0° \leqslant \varphi \leqslant 12°$. All profiles are shown according to Eq. (6), Eq. (7) and Table 1. For L region a unique water profile is holding for all angles φ (actually all pertinent curves for all φ were so dense that finally only one curve is fully representative), while for L' region a family of water profiles are holding for $0° \leqslant \varphi \leqslant 12°$. In general the profiles along L and L' are not symmetrical, and in

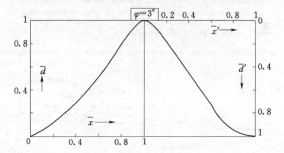

Figure 6   Flow profiles for $\varphi = 3°$

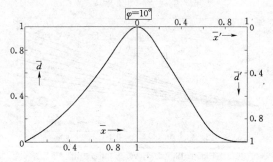

Figure 7   Flow profiles for $\varphi = 10°$

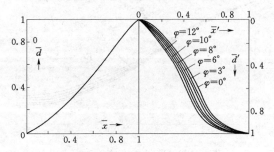

Figure 8   Flow profiles for $0° \leqslant \varphi \leqslant 12°$

the region L' the flow profile curves are lowering when angle φ is increasing. When the flow profiles are needed Eq. (6), Eq. (7) and Table 1 should be used. The position of the sill could not be shown in Figure 6 to Figure 8, since the relative location of d_m to the sill is not constant. In some flow cases d_m is formed upstream the sill and in other flow cases d_m appears downstream the sill—although d_m is steady in any particular flow.

Furthermore, Figure 9 and Figure 10 present a number of water profiles in the L region-but in terms of Eq. (8) -for $\varphi = 0°$ and $\varphi = 12°$ correspondingly, while Figure 11 and Figure 12 present a number of profiles in the L' region-but in terms of Eq. (9) -for $\varphi = 6°$ and $\varphi = 8°$ respectively. For φ=const. , the d/d_s vs. $\bar{x}$ profiles are lowering with Fr_s, while for φ=const. , the d'/d_m vs. $\bar{x}'$ profiles are lowering when Fr_2 are increasing. In both cases d/d_s and d'/d_m vary from 1 to (d_m/d_s) and from 1 to (d_2/d_m) correspondingly.

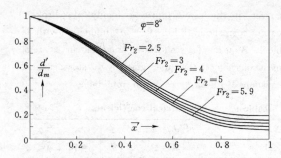

Figure 12 d'/d_m vs. $\bar{x}'$ for $\varphi = 8°$

4. Conclusions

In this experimental study the overall geometry of a submerged inclined (angle φ with $0° \leqslant \varphi \leqslant 12°$) jump due to the presence of a sill is presented, analyzed and discussed. Figure 1 presents the flow geometry, Figure 2 shows the $(L/d_m) - Fr_s$ curves for all angles φ, Figure 3 to Figure 5 explain the relationship between $L'/d_m - Fr_2 - \varphi$, while Eq. (2), Eq. (3), and Table 1, give practical expressions for those lengths. Figure 6 to Figure 8 give the dimensionless profiles ($\bar{d}$ vs. $\bar{x}$ and $\bar{d}'$ vs. $\bar{x}'$) accompanied by Eq. (6), Eq. (9), while Figure 9 to Figure 10 present the flow profiles in terms of d/d_s and d'/d_m vs. $\bar{x}$, $\bar{x}'$ and angles φ. The main conclusions are: (i) $L/d_m > 1$, $L'/d_m > 1$ and $L_t/d_m \gg 1$. (ii) The flow profile $\bar{d}$ vs. $\bar{x}$ for the L region is unique-while for the L' region depends on angle φ, where the overall profiles are not symmetrical among them. (iii) The flow profiles in the L region, in terms of d/d_s vs. $\bar{x}$ and Fr_s, vary from 1 to d_m/d_s, while in the L' region, in terms of d'/d_m vs. $\bar{x}'$ and Fr_2, vary from 1 to d_2/d_m. The results may be useful to the hydraulic engineer when designing the stabilization of a weak hydraulic jump though a sill.

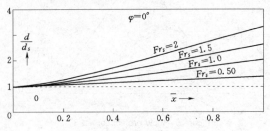

Figure 9 d/d_s vs. $\bar{x}$ for $\varphi = 0°$

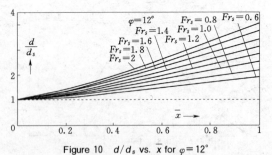

Figure 10 d/d_s vs. $\bar{x}$ for $\varphi = 12°$

Figure 11 d'/d_m vs. $\bar{x}'$ for $\varphi = 6°$

References

[1] J. Demetriou. Experimental Measurements on Local Hydraulic Flows Within Inclined Open Channels. *JD Research Hydrolab publications, Athens*, 2010 (2nd edition-in press).

[2] J. Demetriou, S. Hadjieleftheriou. Inclined Jumps Over Sills, Part 1, Upstream Geometry. *16th IAHR-APD and 3rd IAHR-ISHS Congress*,

Nanjing, China, 20-23, Oct. 2008: 8.

[3] J. Demetriou. Inclined Jumps Over Sills, Part 2, Downstream Geometry. *16th IAHR-APD and 3rd IAHRISHS Congress*, Nanjing, China, 20-23 Oct. 2008: 8.

[4] J. Demetriou. Overall Geometry of an Inclined Jump Over a Sill. *8th Int. Hydrogeological Congress*, Athens, Greece, 7-10 Oct. 2008: 8.

[5] J. Demetriou. Local Energy Loss Consideration in a Weak Inclined Jump Around a Sill. *7th Conf. Development and Management of Water and Energy Resources*, Bhubaneswar (Orissa), India, 4-6 Febr. 2009: 8.

[6] J. Demetriou. Prediction of Forces Along Inclined Jumps Over a Sill. *2nd Int. Conf. on Recent Advances in Experimental Fluid Mechanics*, Andhra Pradesh, India, 36th, March 2008: 8.

[7] J. Demetriou, M. Christia. Submerged Sloped Jump Over a Sill. *33rd IAHR Congress on Water Engineering for a Sustainable Environment*, Vancouver, Canada, August 14, 2009: 8.

[8] J. Demetriou. Theory and Geometry Comparison Among Inclined Free-Over Sill-Repelled Hydraulic Jumps. Flucome, *10th Int. Conf. on Fluid Control-Measurements and Visualization*, Moscow, Russia, Aug. 17-21, 2009: 8.

Laboratory Experiment on Reparation of Anti-Seepage Wall in Overburden Base of Dam[*]

Ning Bo[1,2]   Xie Xinghua[1]   Wu Shiqiang[1]   Geng Yangqiong[3]

[1] *College of Water Conservancy and Hydropower Engineering,*
Hohai University, Nanjing, 210098
[2] *State Key Laboratory of Hydrology-Water Resources and Hydraulic Engineering,*
Nanjing Hydraulic Research Institute, Nanjing, 210029
[3] *Lian Yungang Technical College, Lian Yungang, 222000*

Abstract: Through the laboratory experiment by grouting in sandy gravels, which carried on in the sandbox, the study on anti-seepage wall reparation in the overburden base of dam was conducted. Firstly, after thinking about the mixing ratio, the gel time, stone body intensity, grouting costs, grouting procedure etc., the cement slurry was finally selected as the grouting material and related parameters were determined. Secondly, on the basis of the test in relations among grouting pressure, grouting velocity and diffusing scale of slurry, connectivity of grouting calculus was observed and the vertical layout of the hole filling was determined, moreover how the changing of flow velocity under different water heads affecting diffusing of grout was studied. Finally, in terms of the reparation on anti-seepage wall, the effect of anti-seepage wall reparation by grouting was discussed. The results indicated that the parameters obtained in simulating experiment had the certain practical significance on the designing of reparation of anti-seepage wall and perfecting the anti-seepage theory of overburden dam foundation.

Key words: overburden base of dam; anti-seepage wall; reparation; the properties of slurry; grouting test

1. Introduction

Compared with other vertical anti-seepage measures including curtain grouting and high pressure jet grouting, the anti-seepage wall has significant advantage of reliable impermeableness, complete construction methods and inspection means, applicable to varied geological conditions. Therefore, the anti-seepage wall which is viewed as the most effective vertical anti-seepage form, is widely used in anti-seepage projects in overburden base of dam[1]. 5/12 Wenchuan earthquake has a greater influence on the reservoirs and dams located in seismic region, most of which are in danger of cracking and leaking. Therefore, it is an important task to reinforce dangerous reservoirs and dams before their operating well[2]. The anti-seepage wall which commonly used in dam foundation of many projects in seismic region, is a thin plate structure buried in loose alluvium layer with large area, and is more likely to suffer from destruction under the in-

fluence of earthquake loads. The main damage of the cut-off wall is cracking. Once under the situation, there will be strong leakage channel in the dam foundation, in which sand erosion damage caused by the high-speed water flow will further resulting in uneven settlement which is fatal injury for dam safety. As the anti-seepage wall is deeply buried in overburden base of dam, it is more likely to apply non-direct methods such as grouting rather than direct methods for its reparation. Grouting method which was used as measures for impervious curtain in the reinforcement of a number of projects has been widely used[3], but as reparation measures for cracking of the anti-seepage wall in overburden base of dam, fewer research paper and project application can be founded. The reason is that compare the grouting for anti-seepage wall reparation with grouting for curtain construction, the purposes of them are completely different, the former is the reparation of the damaged wall, and is a complementary measure, whereas the latter is the formation of the

---

* Funded by the Non-Profit Industry Financial Program of MWR (200801030, 201101005-3).

major anti-seepage structure. In this paper, laboratory experiment on reparation of anti-seepage wall in overburden base of dam was conducted. The main objective of the paper was proposing theoretical basis of recovering the performance of anti-seepage wall by grouting, determining related parameters in the grouting procedure, developing the solutions to repair the crack of the anti-seepage wall and investigating the feasibility and effectiveness of proposed solutions. The results indicated that the parameters obtained in laboratory experiment had the certain practical significance on the designing of reparation of anti-seepage wall and perfecting the anti-seepage theory of overburden dam foundation.

2. Experimental Setup

2.1 Design of the Model

Experimental model consists of grouting simulation system, water circulation system, slurry preparation and transportation system (Figure 1).

Figure 1 Experimental model design

Grouting simulation system was composed of sandbox and anti-seepage wall model, both of which were made of organic glass. There were reservoirs on either side of the sandbox, and a constant head was maintained in these reservoirs through water circulation system connected through inlets at the bottom of the reservoirs. The sandbox has outside dimensions of 70cm in length (reservoirs on either side is 10cm in length), 30cm in width, and 30cm in height. In order to meet the need of simulating the performance of anti-seepage wall with crack, the joint between its model and sandbox was sealed up with solid gum and a special hole was made. The anti-seepage wall model has outside dimensions of 30cm in width, 30cm in height. The hole which has dimensions of 1cm in width, 10cm in height, is 12cm far from the bottom of

sandbox, and is 14.5cm far from the front (back) of sandbox (Figure 2). The water circulation system consists of a water tank, drowned pump and inlet (outlet) plastic pipes. Water in the reservoirs enters the packed sand through perforated plates on either side of the box which allow water to pass through, but prevent sand with larger size from leaking into the reservoirs. Slurry preparation and transportation system consists of an air compressors, a pressure tank, a high pressure pipe and grouting gun. During the experiment, it can provide a steady power source, prepare and transmit the cement slurry for the grouting procedure.

Figure 2 Anti-seepage wall model

2.2 Experimental Equipments

The major experimental apparatus and equipments are as follows: an air compressor (discharge pressure is 0.8 MPa), a pressure tank (10L), a high pressure pipe, a pressure gauge (the maximum range is 0.25MPa), a pressure gauge (the maximum range is 0.1MPa), a grouting gun, precision electronic balance (the maximum range is 5000g, the minimum range is 0.1g, error is 0.5g), a measuring cylinder (1000ml), a stainless steel basin, 2 sandboxes, a water tank, a drowned pump, 6 pieces of plastic reinforced pipes, two valves, fine wire, solid gum, iron jaws and raw material belt.

2.3 Choose of the materials

Experimental material used in the experiment was aggregates in the construction site. The grading curve was shown in Figure 3.

When chosing the grouting material, the factors as follows were taken into account. Firstly, the grouting calculus were required to have relatively high strength because they were to be dug out and taken

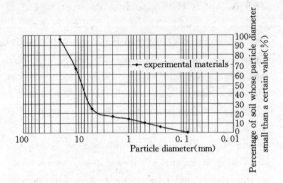

Figure 3   Grading curve of experimental materials

photograph. Secondly, the setting time of slurry was required to be as short as possible because the experiment was made in the situation of different mixing ratios, grouting pressures and modes. If the gel time was too long after grouting, the experiment schedule was bound to be affected. Thirdly, the cost of grouting material was required to control strictly on the premise of completion of the experiment because a great amount of slurry was needed in a trial-and-error process. Lastly, the grouting procedure was required to be as simple as possible because complicated grouting process could bring about something unexpected. After thinking about the factors above, the cement slurry was finally selected as the grouting material, of which P. O32. 5 cement produced by Ma-an Shan Conch Cement Co., Ltd and running water taken from the laboratory room were chose.

2. 4   Determination of Parameters

In the first place, the water-cement ratio of cement slurry was determined. Based on predecessors' experience[4-5], it was generally selected range from 0. 5 : 1 to 2 : 1. In order to choose the proper ratio,

the writer conducted numerous experiments. Results indicated that when the water-cement ratio was less than or equal to 0. 5 : 1, the slurry viscosity was relatively larger and the slurry flowed extremely slowly. For example, only when the pressure came up to 30kPa, the slurry can slowly drop out from the grouting tube. When the water-cement ratio was greater than or equal to 1 : 1, the slurry had good mobility, but grouting calculus had relatively lower intensity, which was usually broken down in the progress of excavating after 24 hours, resulting in affecting the measurement of the calculus shape and delaying the experiment schedule. Finally, the water-cement ratio of cement slurry in this experiment were four modes (0. 6 : 1, 0. 7 : 1, 0. 8 : 1, 0. 9 : 1).

In the next place, the time, the pressures and the depth for grouting were determined. Shorter time and lower pressures for grouting may result in difficulties in the formation of grouting calculus. Likewise, longer time and higher pressure for grouting not only cause a waste of materials and place greater demands on the equipments, but also prolong the experiment schedule. Therefore, the writer conducted a large number of tests to determine the grouting time and grouting pressure, according to the grouting volume corresponding to various time under the different water-cement ratio. Finally, the grouting time was 15s, the grouting depth was 10cm, and the grouting pressure were four modes (30kPa, 40kPa, 50kPa, 60kPa).

3. Method of Experiment

3. 1   Experimental Procedure

The concrete experimental procedure is shown in Figure 4.

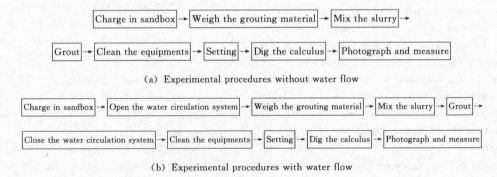

(a) Experimental procedures without water flow

(b) Experimental procedures with water flow

Figure 4   Experimental procedures with different conditions

3.2 Experimental Modes

Four concrete modes implemented during the experiment are shown in Table 1.

At the same time, another four groups of test were conducted for 2nd experimental mode, corresponding to Mode 2-1 and Mode 2-4, which is shown in Table 2. Another three groups of test were conducted for 3rd experimental mode, corresponding to Mode 3-1 and Mode 3-3, which is shown in Table 3.

Table 1   **Experimental modes**

Modes	Contents of the experiment	Grouting conditions				
		Water-cement ratio	Grouting pressure (kPa)	Grouting time (s)	Water head difference between upstream and downstream (cm)	Simulation of anti-seepage wall
1	Grouting test without water flow	0.6 : 1, 0.7 : 1, 0.8 : 1, 0.9 : 1	30, 40, 50, 60	15	—	—
2	Calculus connection test without water flow	0.7 : 1, 0.8 : 1	30, 40	15	—	—
3	Grouting test with water flow	0.7 : 1	30, 40, 50, 60	15	5, 10, 15	—
4	Leakage reparation by grouting with water flow	0.7 : 1	40	15	10	√

Table 2   **2rd Experimental mode**

Modes	Grouting pressure (kPa)	Water-cement ratio
2-1	30	0.8 : 1
2-2	40	
2-3	30	0.7 : 1
2-4	40	

Table 3   **3rd Experimental mode**

Modes	Upstream water head (cm)	Downstream water head (cm)	Water head difference between upstream and downstream (cm)	Outlet flow (mL/s)	Average velocity in outlet section (cm/s)	Grouting pressure (kPa)	Water-cement ratio
3-1	26.5	21.5	5.0	207.8	0.32	30, 40, 50, 60	0.7 : 1
3-2	26.5	16.5	10.0	349.5	0.71		
3-3	27.0	12.0	15.0	362.2	1.01		

4. Results

4.1 Results of grouting test without water flow

In the absence of water flow, the shape of calculus appears irregular ellipsoid. Figure 5 reflects that the dimensions of calculus in all directions various with the grouting pressures in different water-cement ratio. x-axis direction, y-axis direction and z-axis direction represent the length, the width and the height of the sandbox respectively.

4.2 Results of calculus connection test without water flow

Figure 6 reflects that the calculus connection corresponding to experiment mode 2-2 and mode 2-3 re-

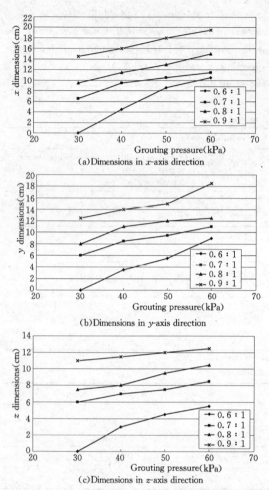

(a) Dimensions in x-axis direction

(b) Dimensions in y-axis direction

(c) Dimensions in z-axis direction

Figure 5　The size of calculus various with the grouting pressures in different water-cement ratio

(a) Calculus of experiment mode 2-2　(b) Calculus of experiment mode 2-3

Figure 6　The calculus connection

spectively. The measured z-axis directional (along the height of the sandbox) dimensions of the two calculus are in accordance with that of single calculus in grouting test without water flow.

4.3　Results of grouting test with water flow

In the conditions of water flow, Figure 7 reflects that the dimensions of calculus in all directions various with the water head difference between upstream and downstream in different grouting pressures. The meaning of x-axis direction, y-axis direction and z-axis direction are the same as before.

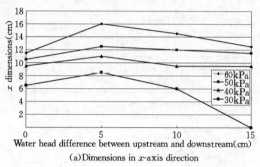

(a) Dimensions in x-axis direction

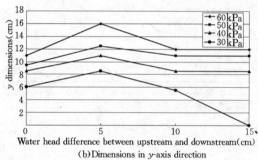

(b) Dimensions in y-axis direction

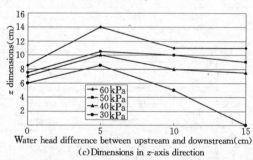

(c) Dimensions in z-axis direction

Figure 7　The size of calculus various with the water head difference between upstream and downstream in different grouting pressures

4.4　Results of leakage reparation by grouting with water flow

Before and after the leakage reparation, a constant water head was maintained in the upstream reservoir and other varying data were measured. Table 4 lists the comparison of the water head of downstream and the export seepage quantity. Table 5 lists the comparison of the location of saturation line, location represents the distance from the observing point to the perforated plate of the upstream reservoir.

Table 4   Comparison of the water head of downstream and the export seepage quantity

Water head \ Location	Upstream water head (cm)	Downstream water head (cm)	Export seepage quantity (mL/s)
Before the reparation	26.5	16.5	219.5
After the reparation	26.5	2.5	76.5

Table 5   Comparison of the location of saturation line (unit: cm)

Water head \ Location	0	12.5	25	25	37.5	50
Before the reparation	26.0	25.8	25.6	19.5	19.0	18.5
After the reparation	26.0	25.6	25.4	9.0	8.3	7.6

5. Conclusions and discussions

In this paper, a laboratory experiment on reparation of anti-seepage wall in overburden base of dam was conducted, the main conclusions were as follows.

In the absence of water flow, the shape of calculus appears irregular ellipsoid. When a constant grouting pressure is maintained, the dimensions of calculus in all directions increase with the increasing of slurry water-cement ratio. When a constant water-cement ratio is maintained, the dimensions of calculus in all directions increase with the increasing of the grouting pressure.

In the absence of water flow, after grouting in the same location but with two different depth, observe that the two calculus have the better connection. Meanwhile, the measured z-axis directional dimensions of the two calculus are in accordance with that of single calculus in grouting test without water flow, which is not only a validation of results of 1th experimental mode, but also illustrate the approach for selecting vertical dimension of a single calculus as the hole spacing is reasonable.

In the conditions of water flow, when the water head difference between the upstream and downstream is lower (5cm), the flow velocity is lower, compared with conditions without water flow, the distance of slurry diffusion increased significantly, and the dimensions of calculus in all directions increase significantly. With the increasing in water head difference between upstream and downstream (up to 10cm), flow velocity increases, but the distance of slurry diffusion decreases, and the dimensions of calculus in all directions decrease significantly. Meanwhile, compared with conditions without water flow, larger grouting pressure (50kPa, 60kPa) have the larger dimensions, while lower grouting pressure (30kPa, 40kPa) have the equivalent dimensions. When the water head difference between the upstream and downstream reach the largest (up to 15cm), under the lower grouting pressure (30kPa), lager flow velocity breaks up of the slurry and fails in formation of calculus. Under the proper grouting pressure (40kPa), flow velocity has little influence on the distance of slurry diffusion, and calculus dimensions in all directions have remained fundamentally unchanged. Under the larger grouting pressure (50kPa, 60kPa), the distance of slurry diffusion decreases, and the calculus dimensions in all directions decrease significantly. But the dimensions are still larger than that of without water flow.

In the conditions of water flow, after the reparation on anti-seepage wall by grouting, the downstream water head significantly decreases, and the export seepage quantity significantly reduces. Meanwhile, the location of saturation line upstream slightly decreases, and the location of saturation line downstream decreases noticeably. The finally results reflect the effect of the reparation by grouting is obvious, and indicate that the selection of grouting parameters is appropriate.

This paper did not consider the influence on the effect of reparation caused by performance of other slurry, and did not consider the influence caused by various water-cement ratios on the effect of reparation in the conditions of water flow. Moreover, cracking in the anti-seepage wall was assumed a simplified form. Yet the practical cracking was complicated, therefore, the simulation of the varying in locations of cracks, length of cracks, and the number of cracks should be further investigated.

Acknowledgment

This research was supported by the National Eleventh Five Science Developing Program under grant 2008BAB29B04-3-4, the National 973 Program under grant 2007CB714100 and NHRI Science Fund Program under grant Y10802.

References

[1] Song Yucai, Yan Qiao, Zhao Xianyong. Vertical anti-seepage on sand and gravel foundation [M]. Beijing: China WaterPower Press, 2009.

[2] Yang Baoquan, Zhang Lin, Wang Hua. Analysis on Consolidation of a Typical Earth-rockfill Dams Which Damaged in Wenchuan Earthquake [J]. Journal of Sichuan University (Engineering Science Edition), 2009, 41 (3): 146-150.

[3] Zhang Jingxiu. Seepage controling and grouting technology in dam foundation [M]. Beijing: China Water Power Press, 2002.

[4] Ruan Wenjun. Research on diffusion of grouting and basic properties of grouts [J]. Chinese Jounal of Geotechnical Engineering, 2005, 27 (1): 69-73.

[5] Wu Xiaoming. Drilling fluid and geotechnical engineering serous fluid [M]. Wuhan: China University of Geosciences Press, 2002.

Experimental Investigation of Beach Nourishment

Yang Yanxiong[1]   Xie Yaqiong[1]   Zhang Jiabo[1]   Qiu Ruofeng[1]

Zou Zhili[2]   Fang Kezhao[2]

[1]*Qin Huangdao Mineral Resource and Hydrogeological Brigade, Hebei Geological Prospecting Bureau, Qin Huangdao, 066001*
[2]*State Key Laboratory of Coastal and Offshore Engineering, Dalian University of Technology, Dalian, 116024*

Abstract: This paper presents a three-dimensional laboratory experiment on beach nourishment project in Beidaihe, China. The experiment is mainly designed to investigate the effects of alongshore sandbars and submerged breakwaters on the protection of nourished beach. The original erosive beach berm is firstly extended 50 meters after beach fill (Case A), then an alongshore sandbar (Case B), three submerged breakwaters (Case D) are considered to strengthen the protection to the nourished beach. After the reliability of the experiment is verified by comparing the experimental results to the available field data sets, the above three cases are conducted under specific wave conditions. In the experiments, wave elevation is recorded by the resistance type wave gages and seabed elevation is measured using URI-III bathymetry measurement system. Based on amounts of experimental data, the effects of each element in the above face protection discipline (beach fill + alongshore sandbar + submerged breakwaters with different planform arrangement) on the evolution of beach profile and shoreline are discussed in detail and Case D is recommended for practical use due to its better performance.

Key words: beach nourishment; breakwater; sandbar; wave

1. Introduction

Beidaihe China, with its beautiful beaches, is an excellent tourist resort. However, in recent years sandy beach in Beidaihe has been suffering from the sever erosion due to the combination of several factors, especially the reduction of sediment discharge from river into the sea and sand excavation near the coast. It is reported that most part of Beidaihe beach have been eroded by 20-50m in recent five years[1-2], which indeed affects the development of the beach tourism and inhabitation and etc. Thus the treatment and protection of the erosive beach in Beidaihe become the urgent task for engineers[2].

Traditionally, coastal structures such as seawalls, groins and breakwaters are developed to protect the erosive beaches. Today, these kinds of structures continue to be used for shoreline defense and this attitude of counteraction instead of working in concert with natural processes is referred to as "hard engineering". Lately in 1990s, the demanding request of the sustainable development of the coastal environment has led coastal researchers and engineers to recognize the importance of developing "soft engineering" approach[3]. Among these newly developed beach protection techniques, beach nourishment has emerged as a favored remedy for coastal erosion in many locations all over the world. Large quantities of beach-quality sand are placed to the nearshore zone of eroding area, shifting the mean waterline seaward. Nearshore sediment transport processes then redistribute the placed sand, typically to the benefit of adjoining sections of coast. It is recognized that beach nourishment is the most effective method. However, beach nourishment also has disadvantages, nourishment must be performed every few years to keep beaches from retreating after storms and thus make it extremely cost.

Some researchers hence suggested that the construction of lower-crest (or submerged) breaker water is necessary to enhance the protection to the filled beach. Most beach nourishment projects in America and Europe thus exhibit the typical characteristic of combination use of "soft engineering" (beach fill) and "hard engineering" (supporting structures like groins and breakwaters) and this has been proved to be the most effective way to protect or restore the eroded coasts[4]. Following this trend and considering the specific characteristics of Beidaihe beach, four designing schemes for beach nourishment have been proposed and a series of pilot projects have been conducted on Beidaihe beach since 2000[1].

These designing schemes have been investigated previously using numerical methods[1-2,5-6]. However cost-effective design and enhanced project lifetime are desired, which requires reliable predictive capabilities. The present study thus presents a three-dimensional laboratory experiment on beach nourishment project in Beidaihe, China. This experiment is mainly designed to investigate the effects of alongshore sandbars and submerged breakwaters on the protection of nourished beach. To our knowledge, this kind of experiment, designed specially for practical beach nourishment project, has never been reported before in China.

2. Experiments description

The experiments were conducted in the State Key Laboratory of Coastal and Offshore Engineering, Dalian University of Technology. The experiment layout is sketched in Figure 1 and the corresponding picture is presented in Figure 2.

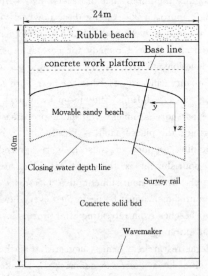

Figure 1   Experiment layout

The wave flume is 40m long, 24m wide and 1.0m deep with a flap-type, multi-direction wave maker equipped on one end and the rubble beach placed on the other end. A concrete platform, serving

Figure 2   Picture taken in the laboratory

as a workspace, is built near the end of the flume. The entire bathymetry, starting form the wavemaker to the bottom edge of the concrete platform, is built according to the given field water contour map and composed of two parts, one part adjacent to the wavemaker is the fixed bed made of concrete and the other part is the movable sandy beach resting on the solid concrete bottom. The interface of fixed and movable bathymetry is defined by the closure water depth $h_c = 5.0$m (prototype size). The movable sandy beach was placed opposite the wavemaker, as shown in Figure 1. In the experiments, wave characteristics obtained by analyzing the field wave data are used as the input signal to generate the corresponding waves, using a JONSWAP spectrum with peak parameter $\gamma = 3.3$. Though sand transportation induced by tide on Beidaihe beach is negligible based on the field observations, wave level change due to tide is still considered in the present investigation by changing water depth based on the fact that water level change affects the sand movement in intertidal zone.

A sketch map of Beidaihe beach is shown in the first panel in Figure 3. The entire beach is about 3300m long with a groin located at $x = 3000$m. It is worthy to note that a pilot beach nourishment project was conducted on Beidaihe beach in 2007. In the project, beach ranging from about $x = 300$m to 900m was filled and an offshore submerged breakwater and artificial sandbar were built to protect the filled beach, as shown in the figure. After the complete of the pilot project, the change of beach profiles have been surveyed for a long period and these valuable field data will be used to validate the experiments.

In the experiments, there are three designing

schemes are investigated, namely Case A, Case B and Case D. The schematic diagram for each case is plotted in Figure 3. The solid polygon is the approximate region of artificial sandbar and other sold lines denote the groins and breakwaters while thin lines are water contours. In case A, only beach fill is conducted to widen beach berm to 50m and no other hard constructions emerge, so this case is the typical application of pure "soft engineering". The rest two cases, i. e., Cases B and D are the applications of combination of soft protection (beach fill) and hard protection (construction of structures). Case B is the extension of Case A, with extending the sandbar to the entire beach and construction the low-crest breakwaters at the two ends of the beach to prevent the sand loss from the bay head. Based on Case B, two additional offshore low-crest breakwaters are constructed to enhance the protection to the filled

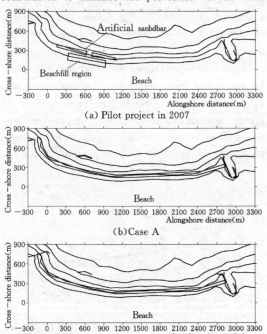

(a) Pilot project in 2007

(b)Case A

(c)Case B

(d)Case D

Figure 3　Schematic diagrams for each case

beach. In the experiments, all the above mentioned breakwaters are submerged. However in the original designing schemes, these low-crest breakwaters are designed to be above the water level. Of course, breakwaters above water level could provide much protection to the filled beach than the submerged breakwaters, however, the latter are less intrusive on the coastal landscape and this property is extremely desirable for tourism dominant beach, and if these submerged breakwaters are built into the shapes of artificial reef, they will add a new habitat to sandy foreshores.

2.1　Scale factors

Kamphuis described lots of dimensionless parameters that should be invariant between the model and prototype to ensure the correct reproduction of coastal sediment transport in a laboratory[7]. However, it is not feasible to make all the dimensionless parameters invariant, as suggested by Work and Rogers[7]. Thus, the dominant parameter must be judged carefully and given the priority of invariance. In our experiments, similitude of bed forms and similitude of beach accretion and erosion are judged as the dominant parameters, based on field conditions and laboratory conditions[8-10]. All scales are summarized in Table 1. Note that, (1) the geometric distortion ratio is (150/36) =4.2, this distortion is necessary as the same scale will yield water depth so small that the physics of the problem would be severely affected and making sufficiently accurate measurement extremely difficult; (2) the hydrodynamic time scale and morphological time scale is the same (=6) due to the adoption of the natural sand in the experiments.

Table 1　Scales used in designing laboratory model

Items	Horizontal scale	Vertical scale	Falling velocity	Morphological time scale & Hydrodynamic time scale
Scales	150	36	1.44	6

2.2　Data collection

Time series of wave elevations, beach profile elevation and shoreline positions are collected in the experiments. Wave elevations are recorded by resistance type wave gages. All the wave gauges are sampled simultaneously at a sampling frequency of 20Hz, mainly aiming to investigate the effects of submerged breakwaters on reducing wave energy. Nearshore currents are also recorded by using ADV (Acoustic Doppler Velocimeters), which makes us clearly know the patterns of the nearshore currents under the influence of submerged breakwaters and artificial sandbars. The shoreline position is defined as the distance from the fixed base line (dashed line marked on the concrete work platform, see Figure 1) to the front of beach berm, and this distance is measured by a meter ruler with the accuracy within 0.1cm. A total of 21 cross-shore sections, numbered from 0 to 20, are surveyed. The seabed elevations along these cross-shore sections are collected using a higher-accuracy bathymetry measurement system—URI-Ⅲ, which consisting of survey rail, seabed elevation detector and terminal PC controller.

A snapshot of URI-Ⅲ seabed elevation detector is shown in Figure 4, which is mounted on a 5m, shore-normal rail (also see Figure 1). One end of the rail rests on an iron frame with four wheels scrolling on the concrete work platform and the other end is supported by a movable iron frame in the deep water. Thus the measurement system could be freely moved to specific cross-shore positions during survey. The main procedure of survey follows, (1) move the survey rail to the specific cross-shore positions; (2) setting parameters on terminal PC; (3) seabed elevations are detected by the probe; (4) these collected data are transmitted via wire to terminal PC, on which there is a user-friendly

interface and the real time change of beach profiles can be plotted on the screen; (5) move the survey rail to the next position and repeat the above steps.

For each case in the experiments, the filled beach is first subjected to the storm waves, and the erosive shoreline position and elevations of beach profiles are measured. Then smaller waves are used in a suitable duration as recovery forces to make the beach reach the equilibrium state and the corresponding surveys including measuring shoreline positions and beach profile elevations are conducted.

Though wave elevations and neashore currents are also recorded in the present investigation, the corresponding results will be published separately in another paper and will not be presented here due to the limitation of the paper scope.

2.3 Experiment validation

As mentioned above, a pilot project has been completed previously on the investigated prototype beach. And the beach profiles have been surveyed at five different cross-shore sections (approximately covering the range between $x = 300$m and $x = 900$m) for a period of time, which provides valuable data to validate the designing method and efficiency of the present laboratory model. The surveyed data in laboratory along two cross-shore sections are illustrated in Figure 5, compared with the field data. These two

Figure 4　A snapshot of URI-Ⅲ

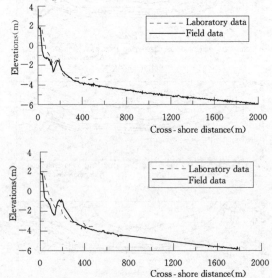

Figure 5　Comparisons between laboratory data and field data at two cross-shore sections

cross-shore sections are approximately located at the one-third and two-third of the sandbar in the along-shore direction. It is seen in the figure that laboratory data and field data are generally in good agreements and hence the validation and efficiency of the present laboratory model is confirmed.

3. Results and discussions

3. 1　Shoreline response to waves

3. 1. 1　*Shoreline response to storm waves*

Under storm waves, the beach tends to be eroded and the sand on the upper part of the beach will be moved to the lower part of the beach to form the storm sandbars and thus the coastal line is retreat. The shoreline positions for different cases under 2-year waves are shown in Figure 6. It is seen that the entire beach is eroded in Case A and, this erosion phenomenon in Case B is slightly decreased while in Case D the only 24 percent beach is eroded and the parts under the protection of offshore breakwaters even be accreted.

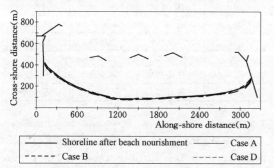

Figure 6　Shoreline positions for Cases A, B and D after 2-year waves

To make things clear, the statistics including the maximum (or minimum) erosion (or accretion) value and the average width of beach berm are listed in Table 2. It is seen in the table that, under storm waves, i. e., 1-year, 2-year and 5-year waves, the average berm width for all cases is negative, denoting erosion of the beach. The decrease of the average erosion value denotes, Case D has the optimum protection effect with minimum erosion width. Case A is the poorest with the maximum erosion width while the performance of Case B falls in between Case A and Case D. This may be explained as follows, the sandbar spreading along the entire beach in Case B leads to

earlier wave breaking and thus reduce the energy received by beach. However, the effect of sandbar is only visible under 1-year and 2-year wave conditions and negligible under great waves, 5-year wave. The three offshore breakwaters in Case D effectively protect the beach from being eroded under all storm wave conditions. This demonstrates the better stability property of Case D in maintaining the filled sands on the beach.

Table 2　Erosion/accretion width of shoreline after storm waves

Waves	Case	Maximum (m)	Minimum (m)	Average (m)
1-year	A	5. 1	−10. 7	−5. 66
	B	7. 3	−10	−3. 27
	D	8. 2	−3. 5	−0. 89
2-year	A	−4	−15	−10. 68
	B	7. 6	−14. 3	−5. 69
	D	9. 5	−10. 5	−2. 83
5-year	A	−7. 5	−31. 4	−19. 99
	B	−4. 4	−31. 1	−17. 67
	D	1. 5	−25. 7	−10. 84

3. 1. 2　*Shoreline response to recovery waves*

The erosive beaches will be recovered to some extent under long time impact of small waves. Figure 7 presents the recovery shoreline of erosive beaches under 2-year storm waves for different cases, and the corresponding statistics are listed in Table 3. It clearly shows, Case D has the optimum recovery ability with maximum recovery width, Case A is the poorest with the minimum recovery width while the performance of Case B falls in between Case A and Case D.

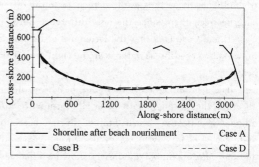

Figure 7　Shoreline positions for Cases A, B and D after recovery waves

Table 3 Statistics of shoreline position after storm waves

Waves	Case	Maximum (m)/ position (m)	Minimum (m)/ position (m)
1-year	A	12.5/1200	−6.7/375
	B	13.1/1050	−5.9/2400
	D	16.0/1500	0.8/2400
2-year	A	12.3/975	−10.6/375
	B	14.6/1050	−9.4/2505
	D	17.5/1500	−6.2/2400
5-year	A	6.0/1500	−23.0/2750
	B	10.2/1000	−16.0/2475
	D	20.5/1050	−14.2/2400

3.2 Beach profile erosion and recovery

It is observed in the laboratory that the beach profiles shifted between two typical beaches, namely, winter profile and summer profile. By analyzing experimental data, three characteristics of beach profiles are summarized as, (1) winter profiles tend to occur under storm waves and the location of the sandbars moves to the deep water with the increase of the characteristic wave height; (2) for summer profile, these sandbars are flatted by recovery waves and beach profiles commonly exhibit concave shape and relatively mild bed slopes.

It is shown in section 3.1 that Case D has the optimum protection effect and recovery ability on shoreline restoration. So the following analysis on change of beach profiles focuses on this case.

3.2.1 *Erosion of beach profile under storm waves*

The erosion/accretion volume per width of beach profiles is calculated to the water depth−2.5m (prototype size). And these values for Case B are subtracted from those for Case to demonstrate the difference between them since the only difference between Case B and Case D is that the latter has a third additional submerged breakwater near the right end of the beach.

It is seen in the table that 65% beach profiles in all are in the accretion state under all storm wave conditions. Specifically, beach profiles in the accretion state are 71% for 1-year wave, 65% for 2-year wave and 59% for 5-year wave, clearly showing a decrease trend with the increase of storm wave height. This is reasonable as greater storm waves lead to heavier ero-

sion. It is interesting to note that profiles P6, P10, P13, and P17 are always in the accretion state under different storm waves. These four profiles are exactly or approximately located at the shelter region of submerged breakwaters. The above analyses demonstrate the protection effect of submerged breakwaters to the nourished beach.

Table 4 Relative erosion/accretion volume per width for Case D after storm waves (m³/m)

No.	1-year wave	2-year wave	5-year wave
P1	17.047	3.085	−4.375
P2	34.920	−30.308	−15.236
P3	−18.621	17.054	3.237
P4	39.526	−20.070	12.398
P5	−37.522	19.173	19.194
P6	0.596	26.115	28.868
P7	28.662	−8.602	1.076
P8	24.491	−1.264	−8.815
P9	−14.933	6.254	9.280
P10	31.571	18.681	2.463
P11	9.095	12.90	−45.21
P12	−10.308	49.024	−17.191
P13	13.551	5.981	29.612
P14	−32.184	−24.827	42.461
P15	5.952	−9.127	−55.154
P16	43.903	34.332	−4.286
P17	41.109	50.925	5.405

3.2.2 *Beach profile recovery under small waves*

The erosion/accretion volume per width of beach profile for Case D is listed in Table 5. The average value is 1.11m³/m for 1-year wave, 0.44m³/m for 2-year wave and 2.08m³/m for five-year wave. Case D clearly exhibits beach recovery ability under all storm waves.

Table 5 Relative erosion/accretion volume per width after recovery waves (m³/m)

Wave conditions	values	
1-year wave	maximum	20.34
	minimum	−16.92
	average	1.11
2-year wave	maximum	26.78
	minimum	−27.61
	average	0.44
5-yearwave	maximum	15.92
	minimum	−15.01
	average	2.08

4. Conclusions

A laboratory experiment was conducted to test three designing schemes for beach nourishment project. The change of shoreline and beach profiles under different wave conditions is surveyed to investigate the effects of different protection disciplines on beach fill and protection. The main conclusions are drawn as follows.

(1) The artificial sandbar leads to the earlier wave breaking and hence reduces the wave energy received by the filled beach. The protection effect is visible under relatively smaller storm waves (1-year and 2-year wave) and negligible under great storm waves (5-year wave).

(2) The submerged breakwater in deep water indeed provides significant protection to the filled beach and enhances the stability of the nourished beach.

(3) Of all investigated three designing schemes, Case D is the optimum one, and has most significant protection effect to the filled beach and the best beach recovery property, thus it is recommended for the practical use.

(4) Considering the scale effect and complexity of conducting such an experiment on sands transportation, the present investigation generally fulfill its aim. However further investigation is still needed.

Acknowledgment

The authors would like to thank the financial support from National Development and Reform Commission (NDRC), People's Republic of China.

References

[1] Pan Y., Kuang C. P., Yang Y. X., et. al. Study on beach nourishment scheme of west beach in Beidaihe. Port and Waterway Engineering, 2008 (7): 23-28. (In Chinese)

[2] Yang Y. X., Zhang J. B., Qiu R. F., et. al. Feasibility study of the beach nourishment project in Beidaihe. Research Report, 2009, Qinghuangdao Mineral Resouce and Hydrogeological Brigade. (In Chinese)

[3] Hamm L., Capobianco M., Dette H. H., et. al. A summary of European experience with shore nourishment. Coastal Engineering, 2002, 47: 237-264.

[4] Ji X. M., Zhang Y. Z., Zhang D. K. The development of beach nourishment: a review, Marine Geology Letters, 2006, 22: 21-25.

[5] Zhang J. B., Yang Y. X., Hao W. H.. Longshore current simulation and study for shore nourishment and projection project in Beideaihe. Advances in Marine Science, 2009, 27 (3): 324-331. (In Chinese)

[6] Zhang Y., Liu S. G., Kuang C. P., et al. Numerical study on flow field in Nourishment area on west beach of Beidaihe. 2008, 7: 7-11. (In Chinese)

[7] Work P. A., Rogers W. E.. Laboratory study of beach nourishment behavior. Journal of Waterway, Port, Coastal, and Ocean Engineering, 1998, 124: 229-237.

[8] Hughs S. A.. Physical models and laboratory techniques in coastal engineering, World Scientific, Singapore, 1993: 568.

[9] Kennedy J. F.. The mechanics of dunes and antidunes in erodible-bed channels. Journal of Fluid Mech., 1963, 16: 521-550.

[10] Zou Z. L., Fang K. Z., Li G. W., et. al. Laboratory investigation of beach nourishment project in Beidaihe. Research Report, 2009, Dalian University of Technology. (In Chinese)

Experimental Study on Decoupling Algorithm of Automatic Control Canal

Cui Wei　Chen Wenxue　Guo Xiaochen　Wang Qi　Mu Xiangpeng

Department of Hydraulics, China Institute of Water Resources and HydropowerResearch, Beijing 100038

Abstract: Coupling effects among pools are the key reason to induce persistent water fluctuation and decrease response and recovery characteristic of automatic control canal. Combined with theoretical analysis and physical experiment method, revised Decoupler I is studied herein. The algorithm decouples in upstream direction by means of feedforward control and in downstream direction by appending a flow controller. The experiment is carried out in a 317m long, 0.6m wide, 5-pool test canal in China IWHR. The canal is operated using constant downstream depth operation method and Proportional-Integral (PI) decentralized control scheme. Different decoupler coefficients ranging from 0 to 1.0 are tested. Experimental results are proved to be consistent with simulation ones. Revised Decoupler I functions well both in upstream direction and downstream direction, on condition that proper decoupler coefficients are selected.

Key words: open channel; decoupling algorithm; physical model test; decoupling coefficients

1. Introduction

Canal is the main way to transfer water to irrigation districts or urban areas in the world nowadays. More and more canals are automatically controlled to improve water transfer efficiency and reduce water losses. Check gates are the main regulation structures and they divide canal into pools in series. The operation of one check gate affects pools both upstream and downstream. When several gates operate simultaneously, their influences will overlap and pools are coupled to each other. Research shows that coupling effects are the key reason to induce persistent water fluctuation and decrease response and recovery characteristics. So effective decoupling algorithm is fundamental for canal control system.

Though some decoupling algorithms are developed in the past decades, such as Decoupler I, Decoupler II and revised Decoupler I, they are mostly tested in simulation and seldom verified in physical model test. For canal, simulation model are not the same as physical ones for some hypotheses or approximations used in mathematic model. What's more, disturbances in field are hard to simulate precisely and completely. So it is significatant to research canal decoupling algorithms in physical model test. In this paper, the revised Decoupler I algorithm is tested in a 317m long canal physical model loca-

ted in IWHR (Institute of Water Resources and Hydropower Research) in China. Its decoupling effects will be verified and the influence of decoupling coefficients on system performance will be analyzed.

2. Coupling effects among constant downstream depth operation pools

Many canals operate with constant downstream depth method[1] (as shown in Figure 1), including the largest water transfer project in China, i. e., the Middle Route of South-to-North Water Transfer Project. For canals using this operation method, canal pool water surface pivot point is located at the downstream end and the depth remains constant while the water surface slope varies. Major turnouts usually are located near the downstream end of the canal pools. This allows turnouts to be designed for a maxi-

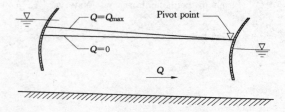

Figure 1　Constant downstream depth
method of operation

mum and relatively constant depth in the canal, and also prevents problems in water delivery to users caused by low or fluctuating water depths.

For demand-oriented operation canal, pool storage must change oppositely to the natural tendency. As shown in Figure 2, when a decrease in pool outflow occurs, the tendency is for pool storage to increase. To pivot the water surface about the downstream end, however, pool volume must decrease. The same problem exists with an outflow increase. The inverse tendency causes longer delay time and complicated gate regulations in upstream direction. In addition, turnout flow variation also induces downstream water level fluctuation and gate adjustments. Obviously, for pools in series with multiple turnouts, gate controllers effects are coupled.

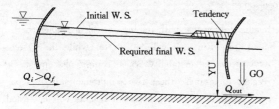

Figure 2　Downstream flow change with the constant downstream depth method of operation

Coupling effects cause canal control algorithms fine tuned for individual pools become less effective or even unsteady when adopted to pools in series. They complicate the design of controllers and the global system must be taken into account. As experiments described in Schuurmans 1992 have shown[2], the gains of water level controllers must be smaller in order to avoid instability. Due to this, performance degrades as compared to a situation in which the controller would not couple. For reasons above, it pays to reduce the coupling as much as possible.

3. Decoupling algorithms of revised Decoupler I

The disturbing effects of control signals of neighbour controllers can be reduced by decouplers, which measure the disturbing control input to determine a control action. For canal using constant downstream water depth method of operation, the decoupler increases the variation of gates in upstream direction to reduce the disturbing effect from control error downstream. Figure 3 presents Decoupler I suggested

by Schuurmans[3]. The blocks labeled "K_{DI}" represent the transfer function for Decoupler I. A fraction of a gate's movement is passed to all of the upstream gates. The decoupler acts as a feedforward controller that anticipates disturbing control actions. Figure 4 presents Decoupler II suggested by Schuurmans. It attempts to prevent the control loop disturbances from moving in the downstream direction by passing a fraction of a gate's movements to the next gate downstream.

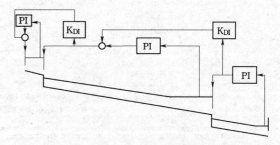

Figure 3　Block diagram for downstream proportional-integral controller with Decoupler I on first three pools of ASCE Test Canal 1

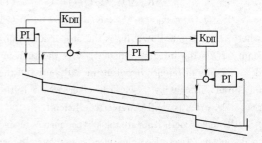

Figure 4　Block diagram for downstream proportional-integral controller with Decoupler II on first three pools of ASCE Test Canal 1

Because Decoupler II is so complicated to implement, in recent years, that more researchers have adopted revised Decoupler I for decoupling in both upstream direction and downstream direction. Revised Decoupler I takes the same form as Decoupler I, but its output is not gate opening but flow rate, as depicted in Figure 5. A flow controller is appended to the original decouper. Simulation research results[4] have shown that decoupling effects can be greatly improved both in upstream direction and downstream direction. That's because the revised Decoupler I takes the form of cascade feedback control in automatic control theory, which provides timely adjustment of the manipulated variable thereby decreasing the error for disturbances affecting the process.

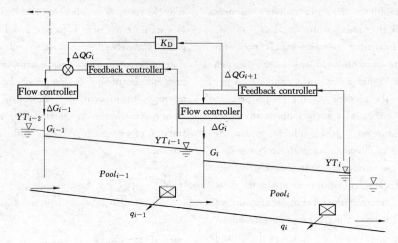

Figure 5   Scheme of feedback controller with revised
Decoupler I for pools in series

The revised Decoupler I will be tested in the experiment. The control law for the five-pool experimental canal can be expressed as

$$\Delta Q_i(k) = K_{P_i} \Delta e_i(k) + K_{I_i} e_i(k-1)$$
$$+ K_{Di} \Delta Q_{i+1}(k) \quad (i=1-4) \quad (1)$$
$$\Delta Q_5(k) = K_{P_5} \Delta e_5(k) + K_{I_5} e_5(k-1) \quad (2)$$

where Q is the gate flow; K_P is the proportional constant; K_I is the integral constant; k is the time index; e is the difference between the set point and the measured output (i. e. , water-level error); i is the pool index; K_D is the Decoupler coefficient.

Decoupler coefficients are the key parameters related to decoupling effects. Recommended values by different researcher are mostly within the range of 0-1. Simulation results[4] show that with the increase of decoupler coefficient, response time will be shorter while the system tends towards instability. So a compromise must be reached to optimize these two requirements. In this paper, the influenc of decoupler coefficients will be tested in physical model experiment.

4. General description of the experimental facility

The experimental facility is located in Daxing District, Beijing City, which is a part of IWHR Laboratory of Hydraulics. It's composed by a control room, an automatic canal, a traditional canal and a reservoir, as shown in Figure 6 and Figure 7.

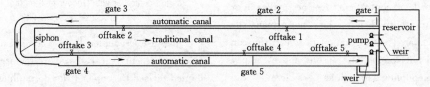

Figure 6   Schematics of experimental facility

Control room is located by the side of the canals. The gate controller and server PC are installed in it. In the PC, a SCADA application developed by Labview is installed to ensure the automatic canal control and supervision, as shown in Figure 8. Gates may be controlled by the control room for all the considered variables are available through a modbus connection protocol.

Automatic canal is a lined rectangle canal with 0. 6m cross section bottom width, and 1. 0m deep. With a length of 317m 0. 001 longitudinal bottom slope, it is designed for delivering water at a rate of 0. 143m³/s. The design maximizes water level variations for a certain flow perturbation, in order to detect and record the smallest flow perturbations.

The canal is equipped with 5 rectangular sluice

Figure 7   Experimental facility

Figure 8   SCADA system of the canal

gates (G1, G2, G3, G4, G5), operated by servo-motors. The gate-opening rate should not exceed 1 cm/s. These gates divide the canal into 5 pools with lengths of respectively 40m、70m、53.8m (including a siphon 9.5m long)、100m and 51.23m. A triangular weir is located at the downstream end of the canal. The discharge can be known very accurately by measuring the head water. There are also 5 offtakes, one for each gate. Each of them was designed as an

orifice in the canal wall, equipped with an additional steel pipe where an turbine flowmeter is installed as well as a butterfly valve. The outflows go directly to the traditional canal, located at a lower level. Three water level sensors were installed for each pool, located respectively at the beginning, middle and end of the pool.

The gates are controlled and monitored by a local programmable logic controller, after appropriate programming. Each PLC receives water depths information from the sensors. These PLC's are linked to the master PLC at the control room with a modbus network.

The traditional canal is a lined return canal, located paralleled to the automatic. It's 1.5m wide and 150m long. Water storage of it is about 140m³ and water is pumped to upper reservoir by 3 pumps. The constant level reservoir is located at the head of the automatic canal. The objective of this element is twofold: to dissipate the flow energy and to provide the canal with enough and virtually unlimited water.

5. Experimental test of revised Decoupler Ⅰ

The schematic longitudinal view of automatic canal is shown in Figure 9. A decentralized PI controller with revised Decoupler Ⅰ [as shown in Figure 5 and described in Formula (1) and Formula (2)] is tested to control the canal. Manning coefficient and gate discharge coefficients are tuned using measured water level data. Controller coefficients (K_P、K_I) are tuned with ATV[5] method by the means of simulation. PI controller parameters are shown in Table 1. Initial operation conditions of the experimental canal are shown in Table 2. A set point of 620 mm for all gates are defined. The sample time is 1 second and time interval between actuator starts is 10 seconds.

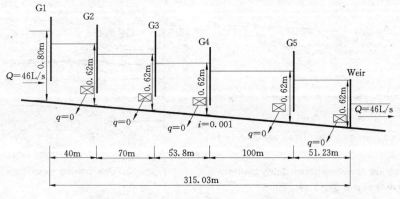

Figure 9   Schematic longitudinal view of automatic canal

Table 1 PI controller parameters

Gate No.	K_P	K_I
1	0.1950	0.0093
2	0.1980	0.0053
3	0.2010	0.0085
4	0.2100	0.0038
5	0.2070	0.0074

Table 2 Initial operation condition of the experimental canal

Pool No.	1	2	3	4	5
Water depth set point upstream of the gate (m)	0.62	0.62	0.62	0.62	0.62
Flow rate of the check gate (m³/s)	0.046	0.046	0.046	0.046	0.046
Flow rate of the turnout (m³/s)	0	0	0	0	0

The test time is 2400s. At time $t = 885$ (s), a sudden demand of 12 L/s for offtake 4 occurs, as shown in Figure 10. For comparison, different decouping coefficients ranging from 0 to 1 are tested in four experiments respectively. The response of water depth variations immediately upstream of Gate 2 are shown in Figure 11.

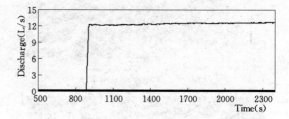

Figure 10   Offtake flow schedule of automatic canal

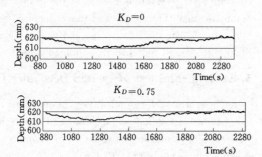

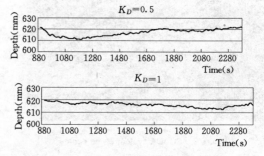

Figure 11   Water depth variations immediately upstream of Gate 2 to different decouping coefficients

Water levels begin dropping after the offtake change. Evidently, the speed at which water levels return to the desired setpoint are faster for $K_D = 0.5$ and $K_D = 0.75$, and former speed is relatively faster. However, when $K_D = 1$, water level have't returned to the set point in 2400s. Refering to simulation research conclusions before[4], Decouplers I may be aggressive when K_D exceeds certain value and

oscillations will happen with long recovery time. Therefore, experimental results are consistent with simulation ones. Proper K_D is in the range of 0.5 and 0.75 for present test case.

Figure 12 and Figure 13 show the decoupling effects of flow controller in downstream direction. The variations of water depth and gate opening downstream of Gate 5 are rather small. Therefore the

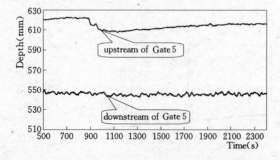

Figure 12   Water depth variations immediately upstream and downstream of Gate 5 ($K_D = 0.5$)

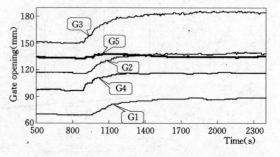

Figure 13   Gate opening variations ($K_D = 0.5$)

decoupling effects of flow controller are significant.

6. Conclusions

An experiment is carried out in a 317m long 5-pool canal to study the decoupling effects of revised Decouper I . Experimental results prove to be consistent with simulation ones. (1) The revised Decoupler I functions well both in upstream direction downstream direction, on condition that proper decoupler coefficients are selected. (2) Decoupling coefficients have great influence on control system performance. With the increase of K_D within the range of 0 to 1, system will be aggressive and recovery time will be short. However, when K_D exceeds some value, which is about 0.75 in the experiment, oscillations happens, recovery time become long, and overall performance decreases. (3) Flow controller shows distinct decoupling effects in downstream direction. The influence of disturbance from upstream is greatly reduced and less gate regulation is required.

Acknowledgment

The work is financially supported by the National Key Technologies R & D Program of China during the 11th Five-Year Plan Period (2006BAB04A12) and was also supported by the National Natural Science Foundation of China (Grant No. 50909104).

References

[1] Buyalski C P, Falvey H T. Rogers D S. *Canal systems automation manual Volume* 1. Denver Colo: US Bureau of Reclamation, 1991.

[2] J. Schuurmans. Controller design for a regional downstream controlled canal. The Netherlands, Delft University of Technol, 1992, A6681.

[3] Brian T. Wahlin, Albert J. Clemmens. Performance of Historic Downstream Canal Control Algorithms on ASCE Test Canal 1. *Journal of Irrigation and Drainage Engineering.* 2002, 128 (6): 372-375.

[4] Cui W., Chen W., Guo X. Research on decoupling of constant downstream depth operation of canal. *Journal of Irrigation and Drainage.* 2009. (in Chinese)

[5] X. Litrico, P.-O. Malaterre, J.-P. Baume. Automatic tuning of PI controllers for an irrigation canal pool. *Journal of Irrigation and Drainage Engineering.* 2007, 133 (1): 27-30.

Design of Spillway Bucket for Kuhrang Ⅲ Dam by Using a Physical Model

Reza Roshan[1]   AbdolReza Karaminejad[1]   Hamed Sarkardeh[2]

Ebrahim Jabbari[2]   Amir Reza Zarrati[3]

[1] *Hydraulic Structures Division, Water Research Institute (WRI), Tehran, Iran*
[2] *School of Civil Engineering, Iran University of Science and Technology, Tehran, Iran*
[3] *Department of Civil and Environmental Engineering, Amirkabir University of Technology, Tehran, Iran*

Abstract: Hydraulic structures are used to control the flow of water in dam developments. Since flow pattern in hydraulic structures is very complex, optimum design is usually found by hydraulic model studies. Kuhrang Ⅲ Dam is constructed on the Birgan River in the province of Charmahal-O-Bakhtiari, Iran, to support and regulate downstream water demand. The dam is a double arch concrete type, 88 meters high from the river bed. In the present work the 1 : 30 physical model of Kuhrang Ⅲ Dam was used to design the spillway bucket and improve its hydraulic performance. Performance of the bucket was examined with more than seven possible flood discharges including critical conditions of very low and very high flows. Primary experiments were focused on the jet trajectory of the bucket and its aeration. In order to prevent the contact between dam body and released water from the bucket in lower flood discharges, two alternatives were examined for the bucket lip. It was concluded that by installing a continuous wedge shape wall below the bucket, low flow jets will be well separated from the bucket lip. Moreover, scour hole at the jet impact was studied and the effect of bucket splitters on reduction of scour depth was investigated.

Key words: bucket; jet trajectory; scour hole; physical model; kuhrang Ⅲ dam; splitter

1. Introduction

In the last decade, many developments have made both in theoretical and numerical modeling of hydraulic structures. A number of softwares have been developed for simulating flows with complex boundaries. However, many aspects in design of hydraulic structures still remain unsolved owing to their complex behavior. For example air entrainment to the flow or scouring phenomenon can not easily be modeled by numerical models. To solve these design problems in details, physical modeling seems to be necessary and is still a usual practice in design procedure of hydraulic structures.

Spillways are among the most important structures in a dam project. Spillways release excess or flood water in a controlled or uncontrolled manner to ensure the safety of the dam. In arch dams it is desirable to construct the spillway on the dam crest and deliver the spillway discharge directly to the river without additional streambed protection works. In this design a deflector bucket which acts as an energy dissipater at the base of an ogee spillway throw the jet further away from the dam body. Flip bucket energy dissipaters are often used in association with high overflow dams to reduce the project cost when spray from the jet can be tolerated and the erosion by the plunging jet can be controlled. Most of the energy is dissipated when the jet plunges into the tail water. The extent of the scour hole is based on judgment of stable slope of material surrounding the deepest hole. A physical model is normally used where topography is complex and where scour can endanger project structures. Some times, a number of splitters are used over the bucket to reduce the scouring depth of the jet at the tail water (Coleman, 2004).

In this study by using a comprehensive physical model, the hydraulic performance of the overflow spillway and the bucket of Kuhrang Ⅲ Dam were studied to choose the best design alternative. In order to reduce the depth of scour hole, splitters were used over the bucket. Moreover, the bucket was modified to prevent contact between the dam body and released water in lower flood discharges.

2. Experimental Setup

To supply water demand and regulate water for irrigation purposes, Kuhrang Ⅲ Dam was constructed on the Birgan River in province of Charmahal-O-Bakhtiari, Iran. The dam is a double arch concrete type, 88 meters high from the river bed. The flood control system consisted of an overflow spillway and a bucket. The effective length of spillway was 58m separated by three middle piles with thickness of 2.95m each. Two middle bays were 15m and two side bays were 14m wide and side bays' crests were at a higher elevation (Figure 1).

Figure 1　Face view of the primary model of
Kuhrang Ⅲ Dam spillway

As a part of design, a 1 : 30 comprehensive scale model was constructed. To include the effect of reservoir geometry on spillways inlet, the whole dam crest and a part of the reservoir length were constructed in the model. A sharp crested rectangular spillway installed downstream of the model was used for discharge measurement. A point gauge with 0.1 mm accuracy was used to measure elevation of the water surface.

Experiments were conducted for seven possible design discharges of 37, 95, 340, 435, 800, 1295 and 2330m³/s. Elevation of the spillway crest in the middle and side bays were 2233 and 2235 meters above sea level (masl) respectively. Two conditions in spillway operation needed especial attention i) separation of issuing jet from the dam body at low and ii) scour depth at high flood discharges.

3. Model Results

It was observed that at low flows the issuing flow from the spillway bucket touches the dam body. Owing to negative consequences of this contact, design of a separator structure below the bucket was proposed. For this purpose, two alternatives were considered i) a groove below the bucket (Figure 2) and ii) installing a continuous wedge shape wall bel-

low the bucket (Figure 3).

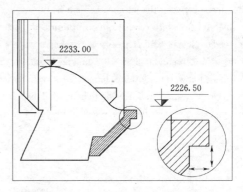

Figure 2　The groove bellow the bucket

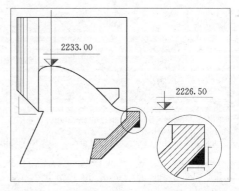

Figure 3　The wedge shape wall bellows the bucket

For evaluating the performance of the two separator alternatives, model was operated at few low discharges. Experiments were conducted for long time to ensure that the issuing jet did not contact the dam body (Figure 4).

Figure 4　Comparison between two alternatives

Observations showed that both alternatives separated the jet at low discharges from the dam body. However the performance of the wedge shape wall was better with more space between the jet and the dam body.

More tests were carried out to optimize the size of the wedge shape wall. Three wedge shape wall sizes (0.30m×0.30m and 0.60m×0.60m and 1.0m×1.0m) were tested at various low discharges. Observations indicated that a 1.0m×1.0m wedge shape wall had better performance and was therefore selected (Figure 5).

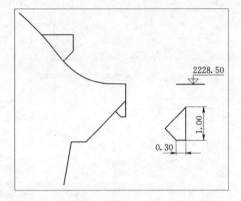

Figure 5   The final alternative

Buckets are the most common structures to dissipate flow energy downstream of a dam. Moreover, if the scour hole depth downstream of the dam body was not acceptable, splitters may be used to aerate the flow and dissipate its energy before the impact zone. Optimizing the splitter dimensions is done in the present physical model studies (Wei et al., 2004).

To select the best dimensions and location for splitters, more than 10 alternatives were tested (Figure 6) and their hydraulic performance was observed in the model.

Splitters performance was selected based on better aeration at high discharges. Different splitter alternatives are listed in Table 1.

To choose the best alternative, all designs were examined in the four bays. This means that by changing the dimensions and location of the splitters, the best performance was found. It should be mentioned that bays No.1 and No.4 had different elevation in comparison with bays No.2 and No.3 (Figure 1). Therefore the designed splitters for them were different in size and position. By testing all possible alternatives, it was concluded that performance of the first alternative was good for bays No.1 and No.4 and also the alternative No.8 was good for bays No.2 and No.3. In these alternatives the jet was fully aerated

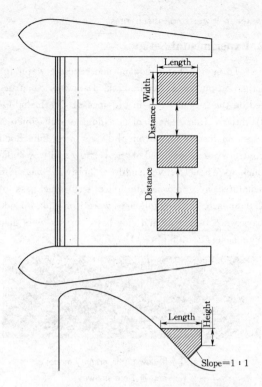

Figure 6   Dimensions of splitters alternatives

and the performance of the splitters was acceptable. The final alternative is shown in Figure 7, Figure 8 and Figure 9.

Table 1   Designed and tested splitters in the model

Alternative	Splitter Dimensions (m)			Distance (m)
	Length	Width	Height	
No.1	2.0	1.5	1.0	1.6
No.2	2.5	1.5	1.0	1.6
No.3	3.0	1.5	1.0	1.8
No.4	3.5	1.5	2.7	1.8
No.5	3.8	1.5	3.0	1.8
No.6	4.0	1.5	3.0	1.8
No.7	3.3	2.5	3.0	2.5
No.8	3.3	2.5	3.0	2.0

To examine effect of the selected splitter alternative on scour hole, its performance was compared with spillway without splitter in 1000 (800m³/s) years flood. Measurements showed that with splitters the scour hole depth reduced from 18m to 14m but the position of maximum scour depth was about 20m closer to the dam body with splitters (Figure 10).

For final assessments, the jet lengths with the final

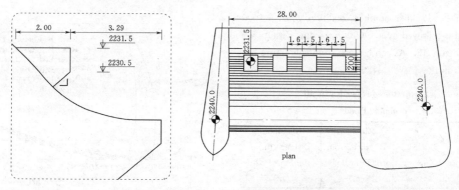

Figure 7   The final alternative for bays No. 1 and No. 4

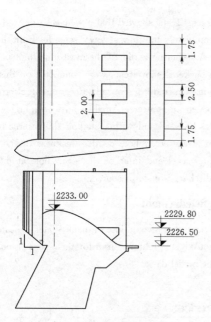

Figure 8   The final alternative for
bays No. 2 and No. 3

Figure 9   Fully aerated jet in the final
alternative in the model

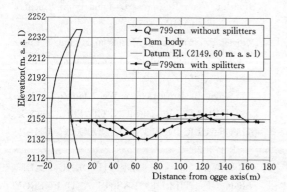

Figure 10   Effect of presence of splitters on scouring

splitter design were measured at seven different floods
with return periods of 2, 5, 50, 100, 1000, 10000

years and PMF.

　　Table 2 shows the measured minimum and maximum boundaries of jet at different flow discharges. These
figures show the thickness of the jet at tail water elevation. As can be seen from this table jet lower and upper
surfaces cover a range of 30 m in different discharges.

　　Scour depth with the final splitter design was also
measured at 3 different flood discharges and results are
presented in Table 3 and plotted in Figure 11. The limit
of the scour hole is also given in Table 3. The prototype
stone size was assumed 0.25m based on rock crack distances reported by geologists. Tests were carried out for 2
hours. It was decided in the next stage of experiments to
compare the scour hole shape and bed topography in

10000 year flood and in another condition when a 1000 years flood is followed by a 10000 years one. Results are shown in Figure 12. As can be seen from this figure bed topography in these two cases is similar.

Table 2   Jet trajectory properties in all probable floods

Flood Return Period (year)	Discharge (m³/s)	Minimum Distance From the Dam Body (m)	Maximum Distance From the Dam Body (m)
2	37	26.0	36.0
5	96	36.0	43.0
50	340	35.0	47.0
100	435	36.5	53.0
1000	800	36.0	56.0
10000	1295	37.0	56.5
PMF	2330	33.5	46.5

Table 3   Results of scouring tests

Discharge (m³/s)	Depth of Scouring (m)	Distance from Dam Body (m)	Starting point of Scouring from Dam Body (m)
800	14	45	18
1300	18	54	24
2330	21	45	7.5

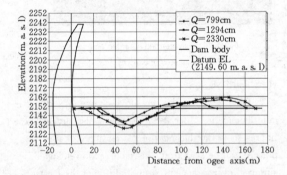

Figure 11   Comparison between scouring of three discharges

4. Summary and Conclusions

In the present work, using a physical model, flow condition at the overflow spillway of Kuhrang Ⅲ Dam was studies and optimized. Tests were carried out to design a separator structure bellow the bucket to prevent a contact between issuing jet at the dam body at low flow

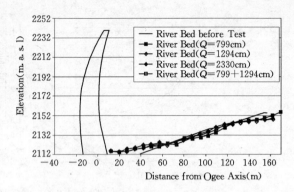

Figure 12   Effect of flood durability on scouring

discharges. Tests showed that a wedge shape wall separates completely the flow at low discharges from the dam body. As scouring is one of the most important parameters in bucket design, tests were conducted on this phenomenon. To reduce scouring 10 different alternatives were tested for splitters. An alternative with higher aeration was selected and with installing this alternative reduction of the scour hole depth, thickness of the jet at tail water level and limit of the scour hole at different flood discharges were measured.

Acknowledgement

The authors would like to thank Water Research Institute (WRI), Tehran, Iran for their permission in using the model data.

References

[1]   Kobus, H. Hydraulic Modeling. German Association for Water Resources & Land Improvements, Bulletin 7, IAHR, 1980.

[2]   Novak P., Cabelka J. Models in Hydraulic Engineering - Pitman Advanced Publishing Program, London, 1981.

[3]   H. Wayne Coleman, C. Y. Wei, James E. Lindell. HYDRAULIC DESIGN OF SPILLWAYS. HYDRAULIC DESIGN HANDBOOK, CHAPTER 17, 2004.

[4]   C. Y. Wei, James E. Lindell. HYDRAULIC DESIGN OF STILLING BASINS AND ENERGY DISSIPATORS. HYDRAULIC DESIGN HANDBOOK, CHAPTER 18, 2004.

Design of Aerator for a Rectangular Bottom Outlet by Physical Model

Khodadad Safavi[1]   AbdolReza Karaminejad[1]

Reza Roshan[1]   Amir Reza Zarrati[2]

[1] *Hydraulic Structure Division, Water Research Institute (WRI), Tehran, Iran*

[2] *Department of Civil and Environmental Engineering, Amirkabir University of Technology, Tehran, Iran*

Abstract: In high speed flows, such as flow in bottom outlets any change in the flow boundary may result in high negative pressures and cavitation attack. In the present work, the bottom outlet of a large buttress dam was studied in a physical model. This bottom outlet had a rectangular section and was damaged by cavitation attack every year after operation. By conducting many experiments, it was revealed that a very mild expansion along the right wall was the reason for high negative pressures and cavitation attack. Installation of an aeration system right upstream of the expansion was proposed for elimination of cavitation damage. Aeration system included a ramp on the conduit bed and right wall. Since the right wall tended away from the flow along the mild expansion, enough space for aeration was provided for the wall ramp. Results of experiments showed that a completely aerated jet formed on the conduit aerator. Moreover measurements showed that owing to aeration, negative pressures downstream of the wall expansion were decreased to an acceptable level. The results of present work are useful in cavitation studies and also design of aerators in bottom outlets.

Key words: bottom outlet; cavitation; physical model; aerator

1. Introduction

Bottom outlets are used for emergency drawdown of the reservoirs of dams, regulating the reservoir water level in impounding and sometimes for sediment flushing[1]. In bottom outlets, a high speed water jet issues from the gate which is located in the conduit. High speed flow in the bottom outlet drags and entrains a lot of air. If air-demand of the flow is not supplied, pressure reduction and vacuum occurs downstream of the gate. High negative pressures downstream of the gate can cause pressure fluctuations and increase the danger of cavitation and vibration and therefore should be avoided. When the flow velocity near the outlet boundaries is high, the cavitation danger on flow boundaries increases. It is usual practice to prevent cavitation damage by flow aeration. Peterka[2] and Russell and Sheehan[3] showed that 5% to 8% of air concentration near the spillway invert prevents cavitation damage on concrete surfaces. A step is designed after the outlet gate to form a jet and facilitates aeration of the high speed flow is-

suing from the gate. If along the conduit the danger of cavitation persists, aerator should be designed along the outlet too. Principal types of aerators usually consist of deflectors, grooves and/or combination of those[4]. The purpose of the deflector is to lift the flow from the boundary so that air can be entrained underneath the flow surface. Such aeration systems have been installed at number of bottom outlets[5]. Installed aerators in bottom outlets entrain air from the open space above the flow. Cavitation occurrence is predicted by calculating Cavitation index σ defined as:

$$\sigma = \frac{P_o - P_v}{1/2 \rho V_o^2} \qquad (1)$$

where P_o and P_v are reference pressure and vapor pressure respectively, ρ is the water density and V_o is the flow velocity. In fact, σ defines the ratio of decrease of pressure through vapor pressure to potential of pressure reduction by kinetic energy of flow. On basis of this definition, when the value of cavitation index is less than a critical value, cavitation occurs. The critical value of cavitation index is different in various surface conditions and is a function of surface irregularities. Major damages on tunnels have been reported[6] at cavitation indices less than 0. 2.

Sometimes flow separation due to change in flow boundary direction such an expansion, may cause severe reduction of local pressure and consequently increase in potential of cavitation damage. In higher flow velocities, large pressure drop is expected due to any change in flow direction.

In the present work, by using a physical model of a bottom outlet, the hydraulic condition of flow was analyzed, and then the probability of cavitation occurrence in the bottom outlet tunnel was investigated. On the other hand influence of outlet geometry on reduction of local pressures and increase in cavitation danger were investigated. By use of model results an aeration system was designed for the bottom outlet.

2. Description of the bottom outlet damages

To supply water demand for irrigation a buttress-gravity dam was designed and constructed with 106 m height from the river bed. The dam crest length was 425 m and the dam crest elevation was 277 m above sea level (m. a. s. l). Five bottom outlets with rectangular cross section were constructed through the dam body. Figure 1 shows the bottom outlet No. 3. This outlet is designed for 142m³/s discharge at the maximum reservoir water surface elevation of 272 (m. a. s. l). After many years of operation volume of sediment settled in the reservoir has increased to a level that endangered operation of the dam outlets. And therefore sediment flashing operation was started in the last 7 years. For flushing every year after irrigation

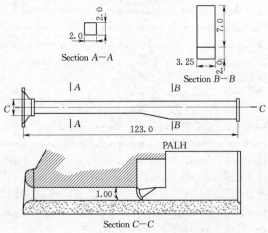

Figure 1   Layout of bottom outlet 3 (Dim. in meter)

period the bottom outlets were opened and reservoir was completely emptied and as a result a large amount of sediments was removed to the downstream of the dam. Upon inspection, several areas of damages were discovered along the bottom outlet No. 3 (Figure 2). Major damages occurred on the right wall of the outlet. Damages were repaired each year and occurred again the next year after the flushing operation. There was a very mild expansion at this area to increase the outlet width from 2 m to 3. 2 m by an expansion with an angle of 2 degrees. Since with drawdown the reservoir, large values of sediment passed through the bottom outlet tunnel, it was thought that abrasion is the reason for damages. There was also an idea that the damages may be a result of cavitation attack. Owing to the importance of this bottom outlet for downstream irrigation, it was necessary to find the causes of damages and eliminate the damages in future. Therefore physical model study of this outlet was conducted to find the problem and its solution.

Figure 2   Photo of damages in the bottom outlet

3. Physical model studies

A 1 : 15 scale Froude-based hydraulic model of bottom outlet No. 3 was constructed in Water Research Institute Lab in Tehran. This model was used to verify operating conditions as well as test modifications aimed for preventing future damages. The model included a steel cylindrical reservoir with 5 m diameter and 10 m height, of bottom outlet inlet and pressurized portion of the outlet upstream of the gate, radial service gate and the rectangular tunnel downstream from the gate. To see the flow pattern along the outlet, the model was made from Perspex.

A sharp crested rectangular weir installed downstream of the model was used for discharge measurement. A sensitive inclined manometer was used to measure negative pressures. Flow velocities were measured by a pitot tube.

The scale of model was chosen to minimize scale effects for studying aeration and air entrainment. Model Reynolds Numbers at the gate location ranged from 9.0×10^5 to 2×10^6. Prior researches[7-8] have shown that to model free surface flows with air entrainment, the flows in the model need to be fully turbulent, $Re_m > 10^5$.

4. Experimental results

All experiments conducted in two reservoir water elevations of 260 and 272 (m. a. s. l) and 3 gate openings 20%, 60% and 100%. To evaluate possibility for occurrence of cavitation through the tunnel, flow velocities and piezometric pressures at different sections of tunnel after the gate were measured. Cavitation index at various points were then calculated from Equation (1). These analysis showed that σ is less than 0.2 at various points and velocities were more than 30 m/s to 40 m/s near the gate. Regarding the results, occurrence of cavitation was expected in about 70 m length of the tunnel downstream the gate. Inspection report also indicated that about 50 m of outlet downstream of the gate was damaged and intense damages were occurred on the section just after of tunnel expansion. Regarding the results, to avoid the cavitation attack, design of an aerator was necessary.

Hydraulic gradient line at bottom of the conduit, 30 cm from the right wall was also plotted along the conduit for the 3 gate openings (Figure 3). This figure shows that at about 10m distance from the tunnel expansion, pressures are negative in all flow conditions.

R. W. L. =272(m. a. s. l), H. G. L at Bottom of Tunnel near the Right Wall

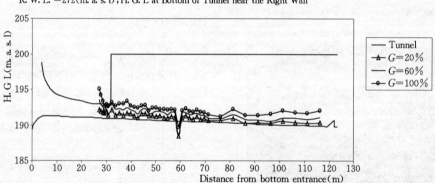

Figure 3　Hydraulic gradient line along the conduit

The reason for negative pressures can be the 2 degrees expansion in the right wall. Maximum negative pressure was 2.15 m of water column at reservoir level of 272 (m. a. s. l) and 20% gate opening.

5. Design of Aerator System

Due to cavitation damage aeration system was proposed. The flow aeration system includes a ramp, slot and an offset or combination of those. In the present work since the danger zone was downstream of the right wall expansion it was decided to install a ramp on the right wall and invert of the outlet right upstream of the expansion (Figure 4). No groove was necessary on the right wall as the expansion itself provided enough space below the separated jet from the right wall.

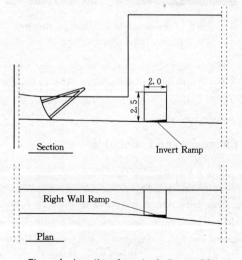

Figure 4　Location of aerator in the conduit

The geometry of ramp includes its angle and which are necessary to be designed. The height of ramp at the bed should be so that flow disturbance and splashing is minimum. By studying the existing projects, it became evident that angle of ramp changes from 5 to 7 degrees. Therefore by studying different alternatives in the model, finally an aeration system including a ramp on the bottom as well as the right wall with an angle of 5° and length of 2m were designed. Experiments showed that increasing in the angle of the ramps could lead excessive splash.

It was also shown that a clear jet was formed on the ramps 30 long (Figure 5). The milky air-water mixture showed that all zones in danger of cavitation were aerated. Also experiments showed that jet with minimum disturbance was formed above the ramps. After installing aerators, negative pressures along the expansion reduced to a maximum of 40 cm of water column and pressure fluctuations were eliminated.

Figure 5   Formation of fully aerated jet after the
aerator in the physical model

The relative depth of flow in the outlet was less than 25%, therefore with enough open space above the flow the tunnel outlet could supply the air demand of the aerator system[9].

6. Summary and Conclusion

During the recent years, one of the bottom outlets of a buttress-gravity dam was damaged each year after operation. The problem was investigated in a hydraulic model and analysis by measuring pressures and flow velocities showed that cavitation is the cause of the damage. It was also revealed that a zone of negative pressure exists along a very mild expansion on the right side

wall of the outlet. Aeration system was therefore designed and tested in the model. This system included a ramp on the right wall upstream of the expansion together with a ramp on the tunnel bed. Size of the ramps was finalized by the model for minimum disturbance and splash in the outlet. Measurements showed that after installation of the aeration system absolute value of negative pressures were reduced considerably.

Acknowledgements

The authors would like to thank Water Research Institute (WRI), Tehran, Iran for their permission in using the model data. All experiments were conducted in the water research institute in Tehran, Iran, affiliated to the Ministry of Energy, their all collaboration is acknowledged.

References

[1]   Vischer D. L. , Hager W. H. . *Dam Hydraulics*, Wiley, Chichester, 1997: 190-213.

[2]   A. J. Peterka. The Effect of Entrained Air on Cavitation Pitting. *IAHR – ASCE Joint Conference*, Minnesota, Minneapolis, USA, 1953: 507-518.

[3]   S. O. Russell, G. J. Sheehan. Effect of Entrained Air on Cavitation Damaged, *Canacaian Journal of Civil Engineers*, 1974, 1: 97-107.

[4]   D. Vischer, P. Volkart, A. Siegenthaler. Hydraulic Modelling of Air Slots in Open Chute Spillways. *International Conference on the Hydraulic Modelling of Civil Engineering Structures*, BHRA, Fluid Engineering, Coventry, England, 1982: 239-252.

[5]   K. W. Frizell. Hydraulic Model Studies of Aeration Enhancements at the Folsom Dam Outlet Works: Reducing Cavitation Damage Potential. Water Resources Research Laboratory, www. usbr. gov/pmts/hydraulics _ lab/wfrizell/ folsom, 1988.

[6]   Falvey H. T. . *Cavitation in Chute and Spillways*, Engineering Monograph No. 42, Denver, USA, 1990.

[7]   L . Toombes, H. Chanson. Free-Surface Aera-

tion and Momentum Exchange at a Bottom Outlet. *Journal of Hydraulic Research*, 2007, 45 (1): 100-110.

[8] Wood I. R. . *Air Entrainment in Free-Surface Flows*, *Hydraulic Structures Design Manual 4*, ASCE, Balkema, Rotterdam, Netherlands, 1991.

[9] K. Safavi, A. R. Zarrati, J. Attari. Experimental Study of Air Demand in High Head Gated Tunnels. *Journal of Water Management*, 2008, 161 (WM2): 105-111.

The Application of Wavelet Analysis Technology in the Signal Processing of Flow-Induced Vibration of Gate Model Test

Zhao Jianping   Yan Genhua

State Key Laboratory of Hydrology-Water Resources and Hydraulic Engineering,
Nanjing Hydraulic Research Institute, Nanjing, 210029

Abstract: Through the application of the Wavelet Analysis Technology, the paper mainly analyzes the gate vibration model test signal, achieves the real helpful signal signature. It provides a high efficient and reliable processing mode for flow-induced gate vibration test, and expands the application in signal locating of gate strong vibration. Starting from the model test, it provides the advantages and disadvantages that wavelet technology be used in signal processing during gate operation. On the whole, wavelet analysis technology is an effective method that can be applied in flow-induced gate vibration signal processing, which can be used in various time-frequency characteristic analysis about flow-induced gate vibration signals.

Key words: wavelet analysis; flow-induced vibration; signal processing; gate

1. Introduction

The gate of the canal is designed to open or close face to face along the circular orbit in the plane; the total width reached 90m, it is a novel design, the arrangement indication as shown in Figure 1. The gate makes overdraft adjustment by way of the floating box water filling and drainage which fixed inside the gate, use hoist dragging move in plane orbit.

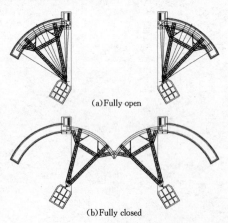

(a) Fully open

(b) Fully closed

Figure 1   Schematic diagram of the gate position

The research on vibration similar model test of this project finds that: under the condition of the gate partial opening, if the water level of the floating box inside the gate be controlled improperly, it will occur strong vibration along the vertical line; in the gate operation, the dynamic change of the upper and lower water level and flow may cause the operating mode of strong vibration; in this case, the identification of real and effective signal as well as the extraction of the signal signature has become the key of the test analysis.

2. Data processing requirements

Random signal analysis of gate vibration is generally about the analysis of time-domain statistic characteristic and frequency-domain statistic characteristic, the common-used time-domain statistic method often do following statistic analysis: (1) probability distribution function; (2) probability density function; (3) mean value; (4) mean square value; (5) variance and so on. Frequency-domain analysis, also known as spectrum analysis, is the transform processing that based on Fourier transformation, the main statistical parameters of the random vibration frequency-domain characteristics is power spectral density function as well as frequency response function and coherence function that derived from power spectral density function.

Currently in gate opening and closing process

test, when the gate occur strong vibration, the vibration process has an obvious non-stable features, the vibratory response signals appear transient signals and stacking unknown noise signals, thus it is not only to analyze the components of steady signals, but also to pinpoint the possible location of the strong vibration signals. It increases the difficulty of randomized time-domain process analysis, for these signals signatures may represent the normal or strong vibration status that occur in gate actual operation, while the true signals may be submerged in the noise signals. Based on the frequency spectrum of traditional Fourier transformation, it need to fully expand the entire time-domain information of signals in frequency-domain, no more time-domain information, which is appropriate for the steady signal monitoring, but no reflection of the signal frequency components' changing situation over time. Time and frequency are two important physical quantities in describing signals, the time-domain and frequency-domain of signals are linked tightly. In this case, the paper intends to do the vibration response signal analysis by means of wavelet analysis, mainly for the frequency-domain analysis and measurement as well as noise reduction and signal locating, etc. through the Wavelet analysis of vibration response signal to carry out such an analysis, the main frequency domain, noise and signal measurement and positioning.

3. The fundamental theory of wavelet analysis

Wavelet, namely the wave in small area, is a kind of special wave with limited length and the mean value of Zero.

Given $\psi(t)$ is a square-integrable function, which means $\psi(t) \in L^2(R)$, if its Fourier transformation $\psi(\omega)$ matches the condition:

$$C_\psi = \int_{-\infty}^{+\infty} \frac{|\hat{\psi}(\omega)|^2}{|\omega|} d\omega < \infty \qquad (1)$$

then call $\psi(t)$ as basic wavelet or mother wavelet function.

Definition: expand the function $f(t)$ of any space $L^2(R)$ under wavelet, which is called continuous wavelet transform (CWT) of the function $f(t)$, here is the expression:

$$(WT_f)(a,\tau) = \langle f(t), \psi_{a,\tau}(t) \rangle$$
$$= |a|^{-\frac{1}{2}} \int_R f(t) \overline{\psi\left(\frac{t-\tau}{a}\right)} dt \qquad (2)$$

Signal $f(t)$ is discrete series, a, τ must be discretized, too, which is called discrete wavelet transform.

For the a, τ of $\psi_{a,\tau}(t) = |a|^{-\frac{1}{2}} \psi\left(\frac{t-\tau}{a}\right)$ is usually doing following discretization:

(1) Do power series discretization to measure, namely make $a = a_0^m$, $a_0 > 0$, $m \in Z$, then the corresponding wavelet function is:

$$a_0^{\frac{-j}{2}} \psi[a_0^{-j}(t-\tau)], j = 0,1,2,\cdots \qquad (3)$$

(2) Discretize the displacement, basically make uniform discrete value selection to τ, and then the corresponding wavelet function is:

$$a_0^{\frac{-j}{2}} \psi[a_0^{-j}(t-ka_0^j\tau_0)] = a_0^{\frac{-j}{2}} \psi[a_0^{-j}t - k\tau_0]$$

was marked as: $\psi_{a_0^j, k\tau_0}(t)$, Discrete wavelet transform is defined as:

$$WT_f(a_0^j, k\tau_0) = \int f(t) \overline{\psi_{a_0^j, k\tau_0}(t)} dt$$

$$j = 0,1,2,\cdots, k \in Z \qquad (4)$$

In engineering application, it is often used as approximate rectangle method numerical integration:

$$WT_f(a, kT_s) = \frac{\Delta T}{\sqrt{a}} \sum_n f(nT_s) \psi\left[\frac{(n-k)T_s}{a}\right] \qquad (5)$$

$\Delta T = T_s$ is sampling interval.

If set $a_0 = 2$, $\tau_0 = 0$, this wavelet is called dyadic wavelet, which is correspond to the continuous wavelet making quantization at measure a, the translation parameter is still in continuous variation without be discretized.

Generally speaking, the commonly referred wavelet analysis has "the nature of self-adaption" and "the nature of mathematical microscope", namely can use different measure to make multi-resolution analysis to the signals, the quality factor under different resolution is constant, which has the ability of highlighting the local feature of the signals in time-domain and frequency-domain.

4. The application of wavelet analysis in frequency domain analysis

Select a section of vibration signal in non-stationary gate operation, the original signal as shown in Figure 2, directly make decomposition treatment to the entire process signal through wavelet transform, the signal multi-scale wavelet decomposition coefficient diagram as shown in Figure 3. Each level of signals decomposed by the signal multi-scale wavelet and

then the signals reconstruct respectively as shown in Figure 4.

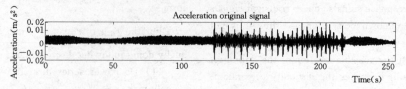

Figure 2　The vibration acceleration signal to be analyzed

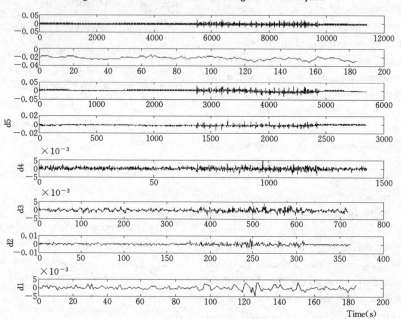

Figure 3　The multi-scale wavelet decomposition coefficient of the
typical vibration acceleration signals

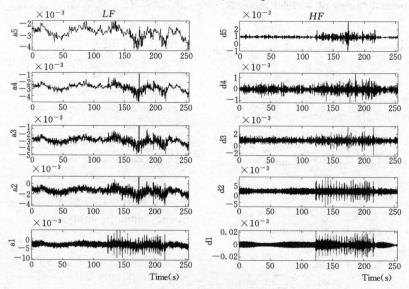

Figure 4　The wavelet reconstructed coefficient of the
typical vibration acceleration signals

The figure shows: in the analysis of the gate non-stationary vibration signal analysis, it takes the signal processing method of wavelet decomposition and reconstruction will accurately extract the effective signal components, reveals the frequency-domain and time-domain characteristics, obtains the gate vibration characteristics with time, which plays an important role in discovering regularity property, improving the accuracy of monitoring and test prediction.

When the gate is in opening and closing process, the submerged effect of noise signal on effective signal is quite apparent, especially when it appears strong vibration, the noise may make it fail in locating position. Usually the source of the signal noise is varied which need accurate separation of noise or exclusion of unwanted information; the signal noise reduction often requires: the denoised signal should have similar smoothness to the original signal, the estimate of

variance between the denoised signal and the original signal should be the minimum at the worst. The method of wavelet analysis used for noise reduction process: select a wavelet, decompose the wavelet, select the threshold value, and reconstruct the denoised signal. When the noise reduced in the denoised time-domain process diagram, it makes higher demands of the noise identification, only with rich experience and more reliable actual data can really achieve "sift the true from the false". Currently there are many methods of threshold value selection, and after adopted the wavelet decomposition, it estimates the signal noise intensity by the hierarchical detail coefficients as the threshold value and use global threshold value. The noise reduction of the signal in typical gate running process is based on this method, and the first group to the third group noise reduction process as shown in Figure 5.

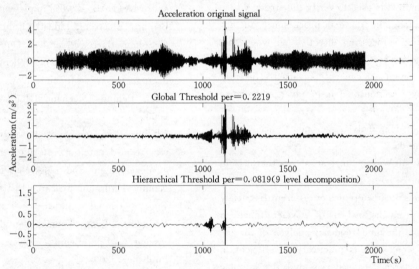

Figure 5　The wavelet decomposed coefficients and the denoised time-domain process diagram

The presence of various strong vibration signals and more accurate locating can be clearly seen from the analyzed signals. If treat the gate strong vibration state as a fault, it can be recognized as the transient signals carrying the important information of fault; in differentiate fault information, it often be compared with the steady state, or the known fault condition. By observing the denoised vibration acceleration time-domain process diagram of Figure 5, and combining with field observation find that: the de-

noised time-domain process remain the transient signals, while the process which have no strong vibration don't retain more transient signals. The maximum point of the transformed wavelet which caused by normal noise is declining rapidly, thus come out the maximum point of the transformed wavelet which caused by fault. Therefore, wavelet analysis can denoise the non-stationary signals conveniently and efficiently, detect transient signals in low noise-signal ratio signals, and extract characteristic parameter.

5. Summary

In engineering application, on the premise of the gate running in multi-group hydraulic parameter changing with time and difficult to predict the vibration status accurately, use wavelet analysis, effectively reveal the varying characteristic with time of the gate body vibrating, which can get the gate vibrating characteristic under some time-varying parameter from time-domain and frequency-domain simultaneously, and discover regular conclusion to conduct the gate running. When the gate occur strong vibration, the vibration process has obvious non-stationary characteristic, treat by the conventional methods will lead to errors.

The currently adopted analysis method provides powerful help to the identification of strong vibration operating mode in the engineering practical implementation; since now it is used to verify the strong vibration operating mode in post-data processing, but not used in real-time processing directly, so when this method is used for real-time prediction in actual prototype measurement, it can only be used efficiently after collecting a section of data in advance for noise processing and effective signal extraction, and integrating use of data analysis methods to accumulate certain running data.

The study on Wavelet analysis shows that it is an effective means of non-stationary signal research, the transient signals' characteristics of different frequency band that get from the analysis of wavelet transform are important to the time frequency analysis of the abrupt impulse signals, it is practical and efficient in the noise interference cancellation and the impulse signals' locating, which achieves a satisfying result in model test.

References

[1] Zhang Defeng. MATLAB wavelet analysis [M]. Beijing Mechanical Industry Publisher, 2009.

[2] Yan Genhua, Zhao Jianping. Research about flow induced vibration test technology on gate which in ZhongLou Flood Control Project in Beijing-Hangzhou Canal-Changzhou Urban change line region [R]. Nanjing: Nanjing Hydraulic Research Institute, 2007.

[3] Cheng Lizhi, Wu Hongxia, Luoyong. Wavelet Theory and Applications (Science) [M]. Beijing: Science Publisher, 2004.

[4] Ge Zhexue, Sha Wei. Wavelet analysis theory and MATLAB R2007 implementation [M]. Beijing: Electronic Industry Pubisher, 2007.

[5] C. Sidn ey Burrus Ramesh A. Gopinath, Haitao Guo, Translator: Cheng Zhengxing. Introduction to Wavelet and Wavelet Transform [M]. Beijing: Mechanical Industry Publisher, 2008.

[6] Li Yun, Yan Genhua et al. Dam hydraulics key long-term dynamic monitoring sensor technology development and research [R]. Nanjing: Nanjing Hydraulic Research Institute, 2009.

[7] Puec (United States), Translator: Liu Yongqing, Fang Yanmei. Digital signal processing [M]. Beijing: Electronic Industry Publisher, 2007.

Investigation of an Artificial Island on Soft Ground Improved by SCP by Means of Centrifuge Modeling Test

Gu Xingwen   Xu Guangming   Mi Zhankuan

State Key Laboratory of Hydrology-Water Resources and Hydraulic Engineering,
Nanjing Hydraulic Research Institute, Nanjing, 210029

Abstract: A centrifuge model test was performed to study the behavior of an artificial island on soft ground improved by SCP using lattice type steel plate piles. The test included 2 phases, the fist concerned behavior of artificial island from its formation till excavation and the second studied the long-term deformation behavior of artificial island after its completion. PIV (Particle Image Velocimetry) method was used on flight to analyze the soil movement on the cross-section of the model.

Key words: centrifuge model test; artificial island; PIV; SCP

1. Introduction

The improvement efficiency of composite foundation, especially the stability and long term deformation of the composite foundation, has been a hot issue and one of the difficulties in the geotechnical engineering field. In this paper, with regard to the cellular steel sheet pile scheme for the soft foundation improvement of an artificial island by use of the sand compaction pile (SCP) method, a centrifugal model test was carried out to reveal the deformation characteristics and stability of the natural soft foundation during the construction and operation periods. The lateral displacement of the cellular steel sheet piles is measured to validate the feasibility of design schemes for the cofferdam structure after the excavation of the foundation pit, which will provide technical support for the optimum structural design of the artificial islands.

A typical section was selected to carry out the test. The main construction sequences include the important time nodes of formation of SCP composite foundation, the insertion of the cellular steel sheet piles, dewatering and excavation of the foundation as well as the final backfill. The observed lateral displacement of the cellular steel sheet piles is first simulated when the foundation pit is excavated. Simultaneously, the total settlement and deformation characteristics of the sloped embankment with riprap are observed during the construction period. The post-construction settlement deformation

of the tunnel foundation and the embankment after the project completion is simulated.

2. Test facilities

The centrifugal model test was carried out by use of large geotechnical centrifuge 400gt (Figure 1) in Nanjing Hydraulic Research Institute in China, whose largest radius is 5.5m, the largest centrifugal acceleration is 200g, the largest corresponding load is 2 tons, and the volume (the product the largest centrifugal acceleration multiplied by the largest corresponding load) is up to 400gt.

Figure 1   NHRI 400gt large geotechnical centrifuge

3. Model design

The determination of model length scale is the

first step for the model design. There are 2 factors for the model length scale: one is the dimension of the model box for the model, which is dependent on the geotechnical centrifuges and the box conditions; the other is the prototype range simulated by the model. The simulated range by the model should be large enough, and the standards to be determined are as follows: if instability failure occurs in the artificial island, the existence of model box boundaries should not stop the instability failure phenomenon from producing and developing freely. Therefore, the centrifugal model tests employ the plane strain model box with a net dimension of 1200mm (length) ×400mm (width) ×800mm (height). Owing to the plane

symmetry of the artificial islands, the area of one side along the central line is selected as the simulating range. Accordingly, along the width direction only 120m is simulated. Along the depth direction, the bottom elevation to be simulated is −65m and the top elevation of the cofferdam is +8m, thus, the total vertical height to be simulated is 73m.

After a comprehensive consideration according to the above conditions, the selected similarity scale for the model length is 100, that is, $n = 100$.

Figure 2 and Figure 3 are layout for model with excavated tunnel pit and with the tunnel set up respectively. They show different phases of the model test, stated later.

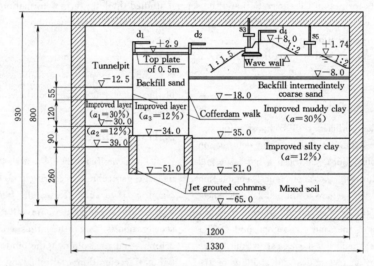

Figure 2　Model with excavated tunnel pit（unit: m for elevations, mm for others）

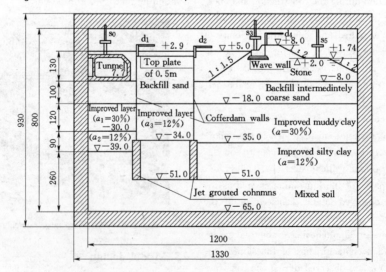

Figure 3　Model with the tunnel set up（unit: m for elevations, mm for others）

4. Model measurements

The most important measurement items in the model test include the settlements and horizontal displacements of the foundation of the artificial islands during various stages by use of the SCP method as well as the lateral displacements of the cellular steel sheet piles. Simultaneously, the most advanced PIV (Particle Image Velocimetry) technology is also employed.

In view of the specialty of measuring data in high gravitational fields by use of transducers, the settlement of island surface and the lateral displacement of cellular steel sheet piles of the artificial islands are measured by non-contact laser displacement gauges which can bear high centrifugal accelerations and exhibit good anti-interference property. They are special inductors developed by Wenglor Co., Ltd. of Germany for centrifugal models. The laser transducers employed in the model tests are YP11MGVL80 and YP05MGVL80. The measurement ranges are 50mm and 10mm, and the precisions are 20mm and 2mm, respectively. As shown in Figure 2 and Figure 3, 6 laser displacement transducers are arranged on the model surface for the artificial islands. They are respectively for measuring the lateral displacements (d_1, d_2 and d_4) of the cellular steel sheet piles and the wave walls and the settlements (s_3, s_5 and s_0) of the embankment surface and the tunnel.

The displacements of model island profiles are measured by the PIV technology, which is the most advanced technology for analyzing images. A PIV displacement field image measuring system is newly equipped. The ES11000 CCD camera is 11000000 pixels, the resolution is 4000×2672, and the sampling frequency is up to 5 frames/s. In the test, layout of measurements by the PIV technology is shown in Figure 4. Through the correlation analysis of 2 neighboring images, the displacement of soil elements and displacement vector fields of the whole section are gained.

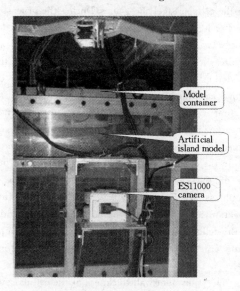

Figure 4   Layout of measurements
by use of PIV technology

5. Model preparation

The preparation of model foundation is based on the properties of various prototype soil strata (Table 1) and model similarity scales. Different preparation methods are required with regard to different kinds of soil layers. As for the non-cohesive sandy soil layer, its dry density or relative density should strictly agree with that of the prototype; as for the cohesive soft soil layer, its in-situ undrained shear strength is controlled to agree with that of the prototype, and the stage preloading drained consolidation method should be employed for the preparation of model soils; as for

Table 1   **Physical and mechanical properties of main soil strata**

Name of soil stratum	Thickness of soil stratum t(m)	Moisture content w(%)	Unit weight (kN/m³)	In-situ strength s_u(kPa)	Compression index C_c	Permeability coe. k(cm/s)
Stone cushion	16 (+8－−8)		20			3.0×10^{-4}
Backfill medium/coarse sand layer	10 (−8－−18)		14.4*			5.0×10^{-3}
Improved muddy clay ($A_s=30\%$)	17 (−18－−35)	60	16.9	27 (46)*		
Improved silty clay ($A_s=12\%$)	16 (−35－−51)	46	17.2	58	0.27	
Mixed layer	14 (−51－−65)	33	18.3		0.22	

Notes   1. * stands for the dry density.

2. Value in bracket being increased in-situ strength through SCP improvement.

the composite foundation with sand compaction piles, its equivalent strength and compressibility should agree with those of the prototype.

In the present model tests, the key structures are the cellular steel sheet piles. During the process of dewatering and excavation of the foundation pit, they mainly bear the differences of earth pressures and water pressures between the inner and outer sides, that is, action of bending moment. Therefore, it is designed according to the similarity principles of the flexural rigidity. If the model preparation selects non-prototype materials, the wall thickness should be designed based on the transformation of relevant module. In the bending plane, that is, in the vertical profile of the cellular steel sheet pile cofferdam, it should satisfy $E_p I_p = n^3 E_m I_m$, and along the width direction, that is, along the circumference direction of the cellular steel sheet pile cofferdam, the components are prepared according to the model length scale, i. e. , $b_p = b_m * n$.

The equivalent wall thickness of the cellular steel sheet piles for the model is calculated according to the following equation:

$$\delta_m = \frac{\delta_p}{n} \sqrt[3]{\frac{E_p}{E_m}} \qquad (1)$$

where δ is the wall thickness of the cellular steel sheet piles; E is the elastic modulus. Through calculation, the model cellular steel sheet piles for the final design are made of aluminum alloy plates 5mm in thickness, 369mm in length and 390mm in width, which is 10mm smaller than that of the model box. The height along the bending direction is 220mm, and the net height is 210mm. The net spacing between division plates is 210mm.

The water level at the outer side of the artificial islands to be simulated in the model tests corresponds to the elevation of the prototype +1.74m.

6. Test procedures

To simulate the two working conditions: island formation after the foundation treatment and excavation elevation of the foundation pit of $-12.5m$, steady operation of 4-5 years after the project completion. The test procedures are as follows.

(1) Preparation of the mixed layer with the elevation of $-51 - -65m$ for model natural foundation using remolded soils.

(2) Simulation of preparation silty clay improvement layer after SCP treatment with the elevation of $-35 - -51m$ (replacement rate of SCP=12%).

(3) Simulation of preparation muddy clay improvement layer after SCP treatment with the elevation of $-18 - -25m$ (replacement rate of SCP =30%).

(4) Simulation of preparation of medium/coarse sand layer for replacement with the elevation of $-8 - -18m$.

(5) Insertion of model cellular steel sheet piles.

(6) Simulation of preparation of island embankment with stone blocks and backfill of medium/coarse sand behind the cellular steel sheet piles.

(7) Simulation of excavation of the foundation pit, the excavation surface corresponds to elevation of $-12.5m$.

Now, the centrifuge starts and accelerates to the design acceleration of 100g by stepless adjustment and stops after 1 hour.

(8) Simulation of backfill of the foundation pit and arrangement of the tunnel, medium/coarse sand is employed to fill the spacing between the tunnel and steel sheet piles and the upper part of the tunnel till the design elevation (+5.0m).

(9) Owing to the occurrence of settlement on the island embankment surface at the first time of operation, crushed stone is employed to place the island embankment till the original design elevation (+8.0m).

The centrifuge starts again and directly accelerates to the design acceleration of 100g and stops after 4 hours so as to simulate the operation of 4-5 years after the projection completion.

7. Deformation behavior of artificial island from its formation till excavation

Shown in Figure 2, the model with excavated tunnel pit is moved onto the platform of centrifuge and brought into flight of 100g in steps. Each step is 20g. On arrival of 100g, one hour is taken for model of soil to be stressed under self-weight. The surface settlement and lateral movement of structure are measured and given on prototype scale in the plot of displacement versus time in Figure 5, in which lateral movement pointing to tunnel pit is accounted to positive, and downward vertical, settlement, denotes positive.

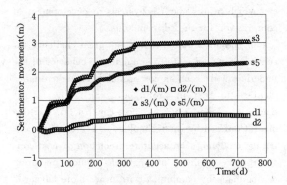

Figure 5   Displacement versus time
(Tunnel pit excavated to El. −12.5m)

It can be seen from the plot that the lateral movement of two walls of cofferdam structure at their top, d1 and d2, are of positive sign and of identical value. Therefore, the cofferdam structure of cellular steel sheet piles moves toward tunnel pit due to excavation. The lateral movement of cofferdam structure at the top continues to develop with a decreasing rate after being excavated to El. −12.5m but that the value of lateral movement soon reached a stable one, indicating that the tunnel pit and the cofferdam structure is of good stability during excavation.

Following the plots of Figure 5, two walls of the cofferdam structure of cellular steel sheet piles move toward the tunnel pit by 0.46m at top. And according to analysis result of PIV measurement (see Figure 6), the cofferdam structure of cellular steel sheet piles move toward the tunnel pit by 0.30m at its bottom. As a result, a difference of about 0.16m of lateral movement is formed during excavation.

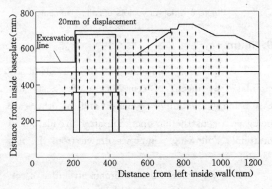

Figure 6   Map of vectors of soil movement

The plots of Figure 5 shows that a great amount of settlement of the artificial island occurs from for-

mation of embankment till excavation of tunnel pit, mainly resulted from consolidation of the underlain compressible layers. The settlement develops with time at a decreasing rate and approaches to a constant value, showing that the sloped embankment with riprap is stable at the end of construction.

Following the plots of Figure 5, the total settlement of the embankment from its formation till excavation is of about 2.31-3.06m, which are listed in Table 2 later.

Figure 6 shows that the soils enclosed by cellular steel sheet piles and their adjacent soils move toward tunnel pit while the soils outside embankment moves a little toward seaside during their settlement. It is found that on the whole, the direction of displacement experience by artificial island is downward, indicating no possibility of sliding at the end of construction.

8. Post-construction deformation behavior of artificial island after its completion

As shown in Figure 3, the model with the tunnel set up is moved onto the platform of centrifuge once again and brought into flight of 100g to be stressed under self-weight. This second flight lasted four hours, simulating a term of about 4.6 years of prototype time after the completion of artificial island. The change of the surface settlement of embankment and tunnel and the lateral movement of cofferdam structure are depicted in the plots of displacement versus time in Figure 7. It can be noticed that the sign of readings is negative of lateral movement of two walls of cofferdam structure, indicating that the upper wall parts moves away from tunnel pit because of the presence of tunnel structure and backfilled sand.

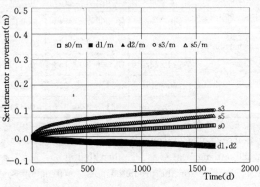

Figure 7   Displacement versus time after the tunnel is set up

It can be seen that the magnitude of lateral movement of the upper walls develops gently at a reducing rate with prototype time. During 4.6 years, the movement amounts to 0.062m at d1 and 0.044m at d2, respectively.

It can be also seen from the plots in Figure 7 that the magnitude of settlement of the embankment and the tunnel develops gradually at a reducing rate with prototype time. During 4.6 years, the settlement amounts to 0.103m at s3, 0.081m at s5, and 0.044 m at s0, respectively.

Since it is hard to simulate post-construction such as 50 years even 120 years of operation by running centrifuge without a stop, the prediction of post-construction settlement cannot be made reliably. But it can be anticipated that the settlements will come to their steady values as long as the artificial island is put into use long enough. If the power curves are used to fit the plots of settlement versus time obtained, it can be roughly estimated in a extrapolation method that the embankment of artificial island and the tunnel structure will have 0.38-0.43m and 0.18m of post-construction settlement in 120 years, respectively (Table 2).

Table 2　Total settlement at the end of construction and predicted settlement of post construction

Gauging point		s0 (m)	s3 (m)	s5 (m)
Total settlement at the end of construction		—	3.06	2.31
Predicted settlement of post construction	2 years	0.027	0.078	0.045
	4 years	0.039	0.097	0.073
	6 years	0.045	0.117	0.086
	8 years	0.051	0.131	0.101
	10 years	0.057	0.143	0.113
	12 years	0.062	0.153	0.125
	14 years	0.066	0.163	0.136
	50 years	0.119	0.268	0.271
	120 years	0.178	0.377	0.434

It is noted in Table 2 that the magnitude of the total settlement at s5 is less than that at s3, and that it is the case for the predicted post-construction settlement at s3 over a period of less than 46 years. This is attributed to the boundary effect of model containing er, which brings about a hystesis of development of compression for the adjacent soils.

It should be noted that it is a great challenge for centrifuge model tests to simulate the behavior of the underlain soils such as compressibility characteristics identical to that in prototype, which governs the magnitude and rate of post-construction settlement. Confined by the technology of geotechnical centrifuge modeling, an accurate prediction of post-construction settlement of post-construction operation can be made by combining with other more advanced approach.

As shown in Figure 8, a map of vectors of soil movement on the cross-section is acquired by means of PIV measurement technology for the second stage of centrifuge flight. It can be seen that on the whole, the direction of displacement experience by artificial island is downward during 4.6 years of post-construction. It is noticed that the soils adjacent to the cofferdam structure still move slightly toward tunnel pit while the soils outside embankment moves a little toward seaside during their settlement.

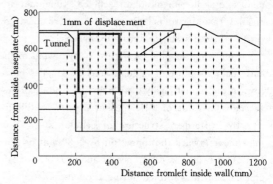

Figure 8　Map of vectors of soil movement

9. Summary

Two working conditions of artificial island are simulated by means of centrifuge model test, one case being excavation of tunnel pit and the other case being post-construction operation after the placement of tunnel. Following points are drawn from the test results.

(1) Two walls of the cofferdam structure of cellular steel sheet piles move toward the tunnel pit by 0.46m at the top and 0.30m at the bottom due to the excavation, so the difference of lateral movement of the cofferdam structure is about 0.16m between the

top and the bottom. It is shown that the tunnel pit and the cofferdam structure are stable during excavation since the displacement developed can come to a steady value.

(2) A lot of settlement occurs for the sloped embankment of artificial island from its formation till excavation. The part of artificial island adjacent to the cellular steel sheet piles moves laterally toward tunnel pit and the outside part of artificial island moves laterally toward seaside when they settle. The total settlement of the embankment from its formation till excavation is of about 2. 31-3. 06m.

(3) The upper part of two walls of the cofferdam structure of cellular steel sheet piles move away from tunnel pit by 0. 044-0. 062m after placement of tunnel structure and backfilling of middle coarse sand. New settlement of 0. 08-0. 10m occurs for the sloped embankment of artificial island during the 4. 6 years of operation modeled. The displacement of embankment and underlain soil layers is composed of settlement and movement of small magnitude, whose direction depends on the position of the part of embankment.

(4) From the development tendency of settlement curve, settlement develops with time at a reducing rate. Confined by capability of geotechnical centrifuge modeling, an accurate prediction of post-construction settlement can be made by combining with other physical simulation measures and analytical methods with more precision. it is estimated by the power fitting curve that the embankment of artificial island will have a post-construction settlement of 0. 38-0. 43m in 120 years, and the tunnel will experience 0. 18 m of settlement in 120 years.

Study on Application of Interior Energy Dissipator in Filling and Emptying System of Navigation Locks

Li Jun[1,2]   Zhang Ruikai[1,2]   Xuan Guoxiang[1,2]   Huang Yue[1]

[1] *State Key Laboratory of Hydrology-Water Resources and Hydraulic Engineering,*
Nanjing Hydraulic Research Institute, Nanjing, 210029
[2] *Key Laboratory of Navigation Structures Construction Technology,*
Ministry of Transport, PRC, Nanjing, 210029

Abstract: Aiming at the key technical problems of high head single-lift navigation locks, the application of interior energy dissipator (IED) in filling and empting system (FES) of navigation locks is put forward and the basic layout principles are indicated. On the basis of analyzing the dissipating characteristics of various typical IEDs, the drop-shaft-type IED is adopted, and the specific form of the FES with drop-shaft-type IED and the methods to determine its main controlling dimensions are studied. The layout of this new FES is been put up for the Yinpan navigation lock on Wujiang River which has the third highest water head among the single-lift navigation locks around the world, and the hydraulic characteristics of this new FES are studied in detail by the experiments, its advantages and disadvantages compared with the common manifold FES of the same lock are also analyzed.

Key words: Interior energy dissipater; Navigation lock; Filling and emptying system; Drop shaft

1. Introduction

The filling and emptying system (FES) of high head navigation locks is commonly composed of intake, culvert, outlet and energy dissipator. In the early stage of water delivery, most of the head is undertaken by the valve, and after the valve is wholly opened, most of the energy through the FES is dissipated by the outlet culverts and the energy dissipating cover plates (or open ditch) in the lock chamber. So with the head increasing, the working condition of the valve during the opening course will become worse, and this will make it much more difficult to solve the induced cavitation problem. At the same time, the energy entering into the chamber is also increasing, and the resultant dissipating problems in chamber will become more prominent. Thus it is clear that the key technology for high and superhigh head lock is to provide a well working condition for valve and solve the dissipating problems in chamber.

In recent years, energy dissipators laid inside the spillway tunnel are studied and developed, and have been already applied in some hydraulic pro-

jects. If the above interior dissipating concept were adopted in the water delivery course, the key technology of single-lift lock having over 40m head might be better solved, because whether laying the interior energy dissipator (IED) in front of the valve to dissipate the flow energy or laying it behind the valve to reduce the head of the valve, will all obviously improve the valve's working condition and consumedly reduce the energy entering into the chamber.

2. Analysis on the applicability of IED in FES of navigation locks

2.1 Main types of IED

There are four main types of common IED: (1) throat-type, such as orifice plate and plug; (2) rotary-type; (3) shaft-type; (4) combined-type.

2.2 Analysis on the applicability of each IED in FES of navigation locks

At home and abroad, the measures to suppress cavitation of high head lock valve can be usually divided into two kinds: active and passive protection measures. The former kind aims at no cavitation existing, such as decreasing the working head of the valve (dividing the water head into several steps), optimizing the form of valve stretch culvert (top-abrupt-expansion, bottom-abrupt-expansion or three-dimension-expansion), quick opening the valve, increasing the submerging depth of valve stretch culvert and so on. The other kind allows the existence of cavitation, but tries to weaken the impact force when cavitation bubble collapsing so as to protect valve and culvert against cavitation erosion and afford a fit and steady working condition for valve.

The main measures to improve the vessel's berthing condition in the lock chamber are: optimizing the layout of FES, adopting better energy dissipating measures or decreasing the discharge by optimizing the valve's operation mode (partial opening, intermittent opening).

According to the above engineering measures and the dissipating characteristics of each interior energy dissipator, we can know that:

(1) Throat-type. The orifice plate type has a simple layout and a better energy dissipation effect, and it is advantageous to the berthing condition in chamber. However, because the best aperture ratio of the orifice plate is about 0. 7, the valve section becomes the control section when its opening is between 0. 2 and 0. 7, during this period valve and culvert are not protected by the orifice plate and they are easy to suffer cavitation erosion. The plug type also has a better energy dissipation effect but a smaller aperture, so the valve can be well protected during its opening. However, the plug section becomes the control section after the valve is wholly opened, the flowing capacity of the FES will decrease, and as a result, the filling time will increase. Moreover, the above two IEDs all have the problem of its autologous cavitation, and they are only fit for the situation with high downstream stage.

(2) Rotary-type. The energy dissipation effect of the rotary-type IED is also very well and the flow energy entering into the chamber can be consumedly reduced. Because energy dissipation needs aeration, complicated water-air two phase flow will come into being in the flow passage, if do not adopt effective measures, the flow regime in chamber will become worse and the vessels berthing in the chamber will be in danger. Moreover, the energy dissipation principle of rotary-type is on the basis of forming steady vortex, and

this needs to build rotary flow generator and also needs a certain discharge and time. However, the discharge during water delivering is much smaller compared with the value in spillway tunnel, but the filling time needs to be as short as possible, so the condition for rotary energy dissipation could not be easily satisfied.

(3) Shaft-type. The shaft-type IED has simple structure type, the energy dissipation can be easily realized and also have a better effect. If put it after the valve, the working head of the valve could be decreased and its working condition could also be improved. However, the aeration problem should also be taken into account.

Above all, it indicates that, the orifice plate type is helpless to solve the cavitation problem, the flowing capacity of plug type is restricted by its own figure, the premise to realize rotary energy dissipation can not be easily satisfied. However, the shaft-type might be applied in FES if the aeration problem could be well solved.

3. Analysis on the layout of the FES with IED

3. 1 General principles of layout

Three general principles must be followed when designing, there are: (1) Guarantee the valve's operating safety. (2) Dissipate most of the energy before water entering into the chamber, improving the berthing condition and even simplify the layout of FES. (3) Try to shorten the filling time. On the basis of these principles and the characteristics of each IED, the drop-shaft-type is selected and the sketch of the FES with drop-shaft-type IED is shown in Figure 1.

3. 2 Analysis on specific layout

3. 2. 1 *Layout of overflow well*

It mainly includes determining its dimension, top elevation of the overflow weir and layout of the energy dissipating grating, etc. The top elevation of the overflow weir is the most important parameter because it directly influences the dividing of the water head which is mainly limited by the filling time under the low upstream stage and the flow regime in the drop shaft. Synthetically considering the factors such as the amplitude of the upstream and downstream stage, the weir top elevation could be selected between the lowest upstream navigable stage and the highest downstream navigable stage, and should be as close as possible of the latter one, on the one

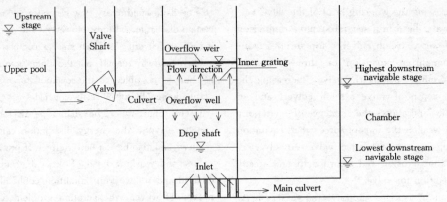

Figure 1　Sketch of the FES with drop-shaft IED

hand, it can improve the valve's working condition, and the filling time can be shortened at the same time. The well's dimension and its inner grating's layout should make the overflow as smooth as possible.

3.2.2　Layout of valve stretch culvert

Section area and elevation of culvert ceiling are the main controlling parameters here. The section area is limited by the filling time under different stage combinations, and it can be calculated using the following equation:

$$\omega = \frac{2C(h_u - h_d)}{\mu T \sqrt{2g(h_u - h_y)}} \qquad (1)$$

where ω is the section area (m^2), C is the water area (m^2), μ is the discharge coefficient when the valve is wholly opened and can be selected between 0.5 and 0.7 when initially calculating, T is the filling time and it's often between 12 and 15 minutes, g is the acceleration of gravity and is nearly 9.81m/s², h_u, h_d, h_y represents the upstream stage (highest and lowest), downstream stage (lowest) and the top elevation of the overflow weir respectively.

The elevation of the culvert ceiling should be lower than the top of the weir to guarantee the steady aeration during valve opening, and it needs not to be so low that will increase the investment of the project, usually, the elevation could be 3-5m lower than the overflow weir's top.

3.2.3　Layout of main culvert

Because of the setup of the IED, the resistance coefficient of FES becomes larger, as a result the section area of the main culvert should be larger (1 – 2time) than the valve stretch culvert. The elevation of the main culvert should consider the submerged depth of the culvert inlet in the shaft and the downstream outlet.

3.2.4　Layout of drop shaft

The cross-sectional area is the most important controlling dimension here. According to the principles of layout, the overflow well should not be submerged to guarantee a steady aeration condition during the period when the valve opening between 0.2 and 0.7 in which cavitation will easily appear, the controlling opening could select 0.8, namely, the water level in the shaft will not reach the top of the weir until the valve's opening reaches 0.8. However, the cross-sectional area should not too large to lengthen the filling time, for this, the controlling opening could select 1, in other words, when the valve has been wholly opened, the water level in the shaft should not lower than the top of the weir. When calculating, the following assumptions are adopted: (1) water entering into the drop shaft do not enter into the chamber during valve opening, (2) the discharge coefficient linearly varies, and according to principle of water balance, the cross-sectional area interval of the drop shaft is as follows:

$$0.32\beta\chi \leqslant \Omega_0 \leqslant 0.50\beta\chi \qquad (2)$$

$$\chi = \frac{\mu_1 \omega_1 t_v \sqrt{2gH_1}}{H_2} \qquad (3)$$

where Ω_0 is the cross-sectional area, μ_1 is the discharge coefficient of the culvert upstream the shaft, t_v is the valve's opening time, H_1 and H_2 is the divided water heads by the overflow weir, H_1 is the valve's working head, and H_2 is the initial dropping depth, β is the correction coefficient, it is used to correct the influence of considering no water entering into the chamber, and it is about 0.5.

4. Engineering case study

4.1 General introduction of Yinpan Lock

Yinpan hydro-junction is located in Wulong country, Chongqing City, and it is the 11th cascade plant built on Wujiang River. The navigation lock is one of the main structures of the pivot and it is mainly composed of upper approach, lock and lower approach, its length, width, sill depth and maximum lift are 120m, 12m, 3m and 36.46m respectively, the highest and lowest upstream stage are 215.00m and 211.50m respectively, and the highest and lowest downstream stage are 193.42m and 178.54m which means that the maximum range of downstream stage reaches 14.88m.

4.2 Layout of the FES with drop-shaft-type IED for Yinpan Lock

According to the analysis in Section 3.2, the specific layout of the FES with drop-shaft-type IED for Yinpan is shown as follows:

(1) The final determined elevation of the overflow weir top is 194.10m.

(2) For Yinpan lock, $C = 1704m^2$, $h_u = 215.00m$ and 211.50m, $h_d = 178.54m$, $h_y = 194.10m$, $T = 15min$, $\mu = 0.5$-0.7, so $\omega = 9.34$-13.64m^2 ($h_u = 215.00m$) and $\omega = 9.65$-13.51m^2 ($h_u = 211.50m$), from which we can know that because the upstream stage range is not very large, the difference of the result is insignificant, considering other factors, the final determined section area of the valve stretch culvert (two culverts) is 11.44m^2, the width and height of it are 2.2m and 2.6m respectively.

(3) The type of valve stretch culvert is the simplest layout with flat ceiling and flat bottom, and its top elevation is 190.10m, 4m lower than the overflow weir top. The upstream inlet adopts the type with two front vertical orifices for each side, and the dimension of each orifice is 4.0m × 3.3m (width × height). The bottom elevation of the inlet is 187.50m and its minimum submerged depth is 20.9m which has already met the design code.

(4) The final determined section area of the main culvert (two culverts) is 14.52m^2, 1.27 times larger than valve stretch culvert, the width and height of it are 2.2m and 3.3m respectively, and the bottom elevation is 168.40m.

(5) The drop shaft is located outside the culvert,

and its bottom elevation is also 168.40m. In order to determine the cross-sectional area, the discharge coefficient μ_1 is estimated to be 0.8 and the valve's opening time adopts 2min, besides there are $H_1 = 20.9m$ and $H_2 = 15.56m$, so we can easily obtain that the range of the cross-sectional area is between 229m^2 and 357m^2, and the final determined area is 275m^2, the length and width of each shaft are 20m and 6.875m respectively.

The detailed layout of the drop-shaft-type IED and the FES with the IED for Yinpan Lock are shown in Figure 2 and Figure 3 respectively.

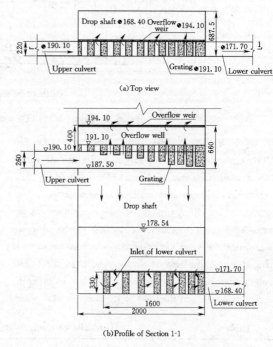

(a) Top view

(b) Profile of Section 1-1

Figure 2   Layout of the drop-shaft IED

5. Model test study

According to the above layout, a physical model (scale 1 : 25) has been built for the hydraulic test study. The water head adopted in the test reached the ultimate value 38.46m (corresponding combined stage 217.00-178.54m) under the allowable condition of the laboratory, and different working conditions combined with four shaft cross-sectional areas and three valve opening times are adopted in the test.

5.1 Hydraulic characteristics of the filling course

The hydraulic characteristics obtained from the tests

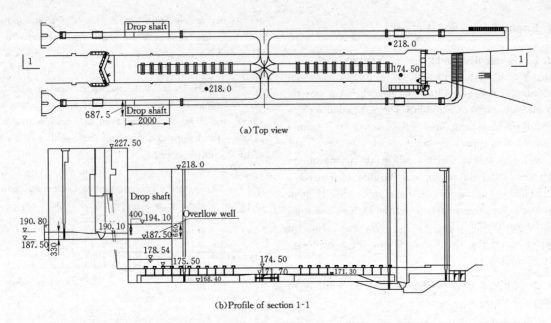

(a) Top view

(b) Profile of section 1-1

Figure 3　Layout of the FES with drop-shaft IED for Yinpan Lock

are shown in Table 1, and it indicates that: (1) The filling times under three operation modes are all within 15min until the cross-sectional area exceeds 275m². (2) When the cross-sectional area decreases from 275m² to 200m², the filling time under each mode does not decrease obviously, moreover, according to the observation during the valve's opening course, because there is little space for the aerated air escaping from the falling water, there are a certain amount of bubbles entering into the chamber from the inlet of the main culvert which might worsen the berthing condition. (3) Considering both the filling time and the scale effect of aeration, the cross-sectional area of the shaft should be chosen between 250m² and 300m², so the layout principles introduced above are proved to be reasonable. (4) The final recommended cross-sectional area is 275m², and the valve's opening time is 2min. The stage hydrograph of the chamber and shaft and the discharge hydrograph under the above working condition are shown in Figure 4 and Figure 5 respectively.

Table 1　　　　　　　　　　　　　　Hydraulic characteristics during the filling course

Ω_0^* (m²)	Opening mode	t_v (min)	T (min)	Q_{1max} (m³/s)	Q_{2max} (m³/s)	H_{0max} (m)	U_{zmax} (m/min)	U_{xmax} (m/min)
350		3.0	15.25	182.78	138.76	13.85	4.89	11.90
		2.5	14.92	188.08	140.16	14.18	4.94	13.06
		2.0	14.58	192.41	142.69	14.24	5.02	14.18
275	double valves continuously opening	3.0	14.92	175.04	142.31	14.44	5.01	14.86
		2.5	14.67	180.18	144.54	14.79	5.09	16.06
		2.0	14.33	186.78	146.98	15.05	5.18	18.51
200		3.0	14.83	169.10	145.62	15.20	5.13	17.07
		2.5	14.50	172.85	148.44	15.68	5.23	19.44
		2.0	14.25	177.64	150.68	15.86	5.30	25.46

Notes　Ω_0 is the total shaft cross-sectional area, t_v is the valve's opening time, T is the filling time, Q_{1max} and Q_{2max} are the maximum discharge in the culvert upstream and downstream the shaft, H_{0max} is the maximum stage difference between the shaft and the chamber, U_{zmax} and U_{xmax} are the largest rising velocities of the stage in the chamber and the shaft respectively.

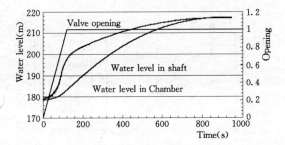

Figure 4　Water level hydrograph

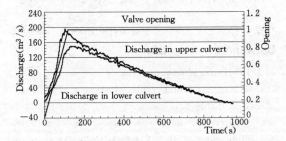

Figure 5　Discharge hydrograph

5.2　Pressure characteristics of the culvert

The pressure characteristics of typical survey point on the culvert are shown in Figure 6. It indicates that when double valves operating, the pressure of the typical survey point gradually descends with the

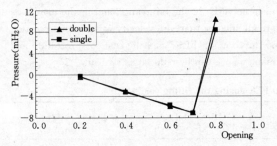

Figure 6　Pressure characteristics

opening increasing from 0 to 0.7, and reaches the lowest valve −7.02m at the opening of 0.7. However, when single valve operating, the pressure characteristic of the same survey point has good agreement with the result under double valves operating and its lowest valve is about −7.12m. The pressure has a speedy rise after down to the lowest valve, after the opening of 0.8, the pressure has already become positive. According to the pressure characteristics, we can easily know that, because of the setup of the shaft, the valve's working condition under double

valves or single valve operating are basically coincident, as a result, the valve's working condition under single valve operating will be greatly improved. Moreover, when the opening between 0 and 0.7, the pressure of the culvert ceiling keeps negative pressure, so the condition of aerating at ceiling could be guaranteed, if further combining the measure of aerating at sealing sill, the filling valve's cavitation problem will be well solved.

5.3　Discharge coefficient of the FES

According to the steady flow test results, the discharge coefficients of the culvert upstream and downstream the shaft under double valves operating are 0.812 and 0.609 respectively, however, the corresponding valves under single valve operating are 0.814 and 0.711. It indicates that because of the setup of the shaft, the discharge coefficient of the culvert upstream the shaft keep basically coincident no matter when double valve operating or single one operating.

5.4　Energy dissipation rate of the IED

The total energy and the energy of typical sections can be calculated by the hydrographs of the chamber stage, discharge and the pressure of the shaft's inlet and outlet, and the ratio of the difference between the shaft's inlet and outlet energy and the total energy represents energy dissipation rate of the IED. The hydrographs of above typical energy and the energy dissipation rate are shown in Figure 7 and Figure 8 respectively.

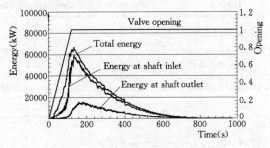

Figure 7　Energy hydrograph

From the hydrographs we can know that most of the energy has been dissipated in the range of the IED, the maximum energy dissipation rate under double valve operating is about 77%, and the average rate is about 62% during the middle period of the filling course. At the end of the filling course the energy

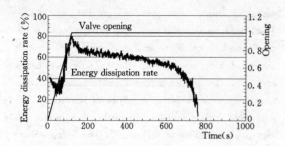

Figure 8   Energy dissipation rate hydrograph

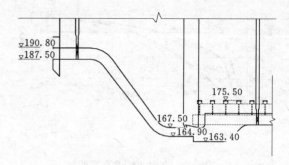

Figure 9   Layout of valve stretch culvert in common FES

of each section becomes even, so the rate becomes gradually decreasing.

5. 5   Flow regime in the drop shaft and chamber

The flow regime in the drop shaft is quite complicated before the overflow well is submerged because of the impact jet flow and its inducing aerated water. But during the entire filling course, the flow regime in the chamber is quite smooth and there are no air bubbles escaping from culvert.

6. Comparison between the FES with drop-shaft-type IED and the common FES

6. 1   Layout of common FES for Yinpan Lock

The valve culvert stretch and the upstream inlet of the common FES (manifold FES) for Yinpan Lock are connected by a goose neck pipe, and the valve culvert stretch adopts the abrupt-expansion type (shown in Figure 9). The other layout is the same as the FES with drop-shaft-type IED.

6. 2   Comparison of the experimental results between the two FESs

6. 2. 1   *Hydraulic characteristics*

The experimental results of above two FESs are shown in Table 2. It indicates that, under the same initial water head and the same valve operation mode, the filling time of the IED type is longer than the common type because of the larger amount of filling water for the IED type. However, the maximum discharges in the culvert upstream and downstream the drop shaft are all smaller than the common type, the maximum discharge in the culvert downstream the drop shaft is even smaller than the common type under the slow operation mode. So, the velocities in the main culvert, in the bifurcation and in the manifolds are all smaller than the velocity at the same position in the common FES, and the cost for protection and maintenance can be reduced as a result.

Table 2   **Hydraulic characteristics of the two FES during the filling course**

Type of FES	Valve's operation mode	t_v (min)	T (min)	Q_{1max} (m³/s)	Q_{2max} (m³/s)	U_{zmax} (m/min)
FES with IED	Double	2. 0	14. 33	186. 78	146. 98	5. 18
Common FES (manifold FES)	Double	2. 0	10. 92	201. 60		7. 05
		6. 0	13. 53	165. 36		5. 78

6. 2. 2   *Valve's working condition*

In order to reduce cavitation, the integrative measure composed of large buried depth (11.05m) for valve, optimizing valve culvert stretch type and aeration has been adopted in the common FES, this measure is quite complicated and can not well solve the cavitation problem when only one valve working. However, in the FES with drop-shaft-type IED the cavitation problem can be easily solved by a-

dopting simpler valve culvert stretch type assisted with aeration, especially under only one valve working. At the same time, the valve structure and its related boom structure and headstock gear design can be all simplified because of the reduced working head and improved working condition.

6. 2. 3   *Berthing condition in the chamber*

Because of the drop shaft, the filling course in the FES with IED has been divided into two steps, and during the course that the water entering into the chamber from the shaft, the water head increases first and then decreases, as a result the maximum growth rate of the discharge in the main culvert does not appear in the initial filling period and it appears when the overflow well has been just submerged, the water head at this moment has been much more smaller, so the growth rate of discharge is also much more smaller than it in the common FES. According to the definition of the wave force during the filling course, under the condition of the same vessel dimension, berthing position and layout of culvert in the chamber, the wave force in the chamber is only proportional to the discharge growth rate, so it indicates that adopting the FES with IED can consumedly reduce the wave force, as a result the berthing condition in the chamber which mainly influenced by the wave force will also be well improved.

6.2.4 *Energy*

The energy in the FES has been evidently reduced when adopting the drop-shaft-type shaft, the energy of typical section has been reduced from 34402kW and 18047kW to 13844kW and 11396kW under single and double valve working respectively and the decreasing amplitude are 60% and 37% respectively. It indicates that the dissipating effect of the FES with IED is quite remarkable, and the energy entering into the chamber has been consumedly reduced.

6.3 Advantages and disadvantages of the FES with drop-shaft-type IED

According to the above analysis, we can know that the advantages of the new FES are:

(1) Provides the valve steady and well working condition.

(2) Can simplify the valve structure and decrease the design difficulty of its controlling system.

(3) Consumedly improved the valve working condition when one valve working.

(4) Evidently dissipates the flow energy entering into the chamber, and improves the berthing condition in the chamber.

(5) Consumedly decreases the wave force in the chamber, and further improves the berthing condition in the chamber.

And its main disadvantages are:

(1) The filling time is a little longer than the common FES.

(2) The flow regime in drop shaft during valve opening is complicated, if adopts inadequate measures, the air bubbles will entering into the chamber and influence the berthing condition.

Although there are some disadvantages, compared with the common FES, the developing potential of the FES with drop-shaft-type IED is still quite huge. Because the existing techniques can well solve the valve problems under 30m water head, if these techniques could be used in the FES with IED, the applied range of the new FES can be widely expanded.

7. Conclusions

In recent years, in order to promote the economy of the western regions, China has been developing the western inland navigation energetically, and large quantities of inland navigation facilities have been or will be constructed. So, the selection, layout and hydraulic problems of high head and super high head navigation lock suitable for the mountain rivers are all becoming the focus of the future navigation hydraulics.

Directing against the characteristics of western rivers in China such as large stage range, high water head of the navigation structure, etc, this paper has firstly studied the application of the IED in the FES of high head navigation lock. The results indicate that, the FES with drop-shaft-type IED can well solve the technical difficulties brought by the characteristics of western rivers, and it is a new choice for the design of high head navigation lock on western rivers in China.

References

[1] Xu Qinghua, Zhang Ruikai. Fundamental Study on Application of Navigation Structures [M]. China Water Power Press, 1999.

[2] Zheng Linping, Li Yuejun. Study and Application Development of Interior Energy Dissipater in Hydraulic Tunnel [J]. Water Power, 2006 (09): 81-83.

[3] Hydraulic experiment report of the double-line three-step double-valve plan under 150m water level for the TGP lock [R]. Nanjing Hydrau-

lic Research Institute, 1987.

[4] Yang Lixin. Summary of high head lock's construction and research in Soviet Union [J]. Port & Waterway Engineering, 1986 (2): 21-25.

[5] Zhou Huaxing. Rational layout for the drop-shaft short-culvert filling and emptying system [J]. Journal of Waterway and Harbor, 1980 (4): 38-43.

[6] Li Jun, Zhang Ruikai, Xuan Guoxiang. Application of interior energy dissipater in the filling and emptying system of navigation locks [J]. HYDRO-SCIENCE AND ENGINEER-ING, 2007 (3): 40-46.

[7] Li Jun. Study on application of interior energy dissipater in filling and emptying system of navigation locks [D]. Master thesis of Nanjing Hydraulic Research Institute, 2007.

[8] Wang Zuogao. Navigation lock design [M]. China WaterPower Press, 1992.

[9] Design Code for Filling and Emptying System of Shiplocks (JTJ306—2001) [S]. China Communications Press, 2001.

[10] Analysis of the filling and emptying system and its hydraulic experiment study of Yinpan Lock on Wujiang river [R]. Nanjing Hydraulic Research Institute, 2006.

Study on Vibration-Reducing Design and Vibration Monitoring Applicable to the Gate Impacting by Tidal Surge

Chen Fazhan   Yan Genhua

State Key Laboratory of Hydrology-Water Resources and Hydraulic Engineering, Nanjing Hydraulic Research Institute, Nanjing, 210029

Abstract: On the basis of analysis on the cause of formation of tidal surge and the characteristics of impact load, this essay presents the types and patterns of impact load, and proposes truss-like gate structure applicable to this kind of stress. According to the research results of tidal surge simulation technology and introduction of application of tidal surge, we prove the correctness of simulation experiment and design form through prototype observation and measurement.

Key words: tidal surge; Truss-like gate structure; impact load; experimental verification; vibration monitoring

1. Introduction

With the continuous improvement of people's living standards, environmental improvement and ecological beautification are gradually paid attention to during river and sea treatment. Riverine and coastal development is more and more popular day by day, beautiful and fantastic beaches blend into people's life. The river improvements along river and sea as well as turning the harm into a benefit become key emphasis in work. Prevent ocean tide and storm tide from doing harm to inland. Besides building up embankment, it is of social benefit and economic benefit to construct gate system structure, guide diffluence and store water resources for controlling waterlogging, optimization of water resource utilization, improving land and water transportation, improving river network environment, optimization of investment situation and other aspects. The energy of tidal surge and storm surge is huge with vast destructiveness of impact. It is urgent to design a type of gate applicable to this load action and safe operation.

2. Causes and Characteristics of Tidal Surge

The tide water from open seas enters into river mouth narrow and shallow and great waves surge and accumulate tidal surge. The world-famous tidal surge is Qiantang River spring tide in China, which is a tidal surge in Hangzhou Bay (river mouth section of Qiantang River). As Hangzhou Bay is a big trumpet mouth that is wide outside and narrow inside. The width of marine outfall is 100km, and gradually reduces to 20km after entering into inlandwater. The river surface is only 3km wide during falling tide. When rising tide, plenty of sea water is swallowed in the middle of the river. When it pushes inward, due to the riverway suddenly getting narrow and tide water surging and accumulating to form upsurge. Counting vast sandbank under water, the average depth of water of river bed reduces from 20m to 2-3m rapidly, and there forms a "threshold". The tide water getting inside is blocked; the back rise catches the front rise and forms upstanding "water wall". The tidal head reaches to 3.5m, and tidal range reaches to 8.9m. It is the most famous tidal surge around the world. This essay conducts overall research through Qiantang River Cao'ejiang gate as supporting project and investigates the gate system structure applicable to this kind of tidal surge. Cao'ejiang tide-resistance and drainage sluice gate is located at the intersection of Cao'ejiang and Qiantang River. The river mouth is wide, and the engineering project is designed with tide gate and plug dam combined. The overall trim width of orifice of tide-resistance and drainage sluice gate is 560m with 28orifices, and the design flow is about 11340m^3/s with large scale. The gate type of working gate is sealing water plane slide gate, and the elevation of parapet is 4.5m, the size of orifice is 20.0m×5.0m.

It is known from the specific characteristics of the strong tidal surge in Qiantang River that the stress characteristics of Cao'ejiang gate are different from that in other projects. Besides of meeting the requirements for convention sluicing, it needs to stand the acting force of tidal surge in Qiantang River. Under combined action of dynamic load of flow and tidal surge, the strained condition of gate is complicated. In fact, this acting force of tidal surge has impacting characteristics to make the dynamic action the structure bearing bigger than dynamic force of drainage. And the energy and the directions of acting force working on gate or other buildings have the randomness themselves as the direction, height, speed and impacting position of tidal surge. In the meantime, the form and energy of tidal surge in Cao'ejiang vary from hour to hour along with time, hydrology and climatic conditions. The tidal surge impacting gate body varies along with the impacting position and the directions of tidal surge, and there's obvious difference among the impacting forces on every part. And some of the greatest impact wave comes at the first impact wave, and some appears at the following flow reverberation impact. The impact load is larger in some regions of flow reverberation with lower frequency, and then becomes weak gradually. According to the observed data, tidal surge of Qiantang River proceeds at flow speed from 4m/s to 7m/s, the max flow speed reaches to 10m/s. The max impacting pressure measured in Cao'ejiang ranges from 47kPa to 95kPa (changing along with the directions of tidal surge).

3. Designs of Gate

At present, the gate structures adopted are plane gate, radial gate, miter gate, arch gate, spherical gate, cylinder-shaped gate and so on. As to the natural characteristics of strong acting force of tidal surge and wide river mouth, we use low content alloy structural steel that is with high intensity and splendid impact toughness to make truss-like system gate structure with long span and large aspect ratio, which is more applicable to physical conformation of wide river mouth and impact load (Figure 1-Figure 2). The specific strong tidal surge in Qiantang River

induces strong surge impact load on working gate, and its dynamic load action is controllability vibration source inducing severe vibration on gate. Once the characteristics of acting force of dynamic force load and the natural characteristics of gate structure produce adverse combination effect, it will cause harmful damage to gate. In a sense, whether the flow-excited vibration problems of tide-resistance and drainage sluice gate are properly resolved becomes a controlled factor of long-term safe operation of Cao'ejiang gate.

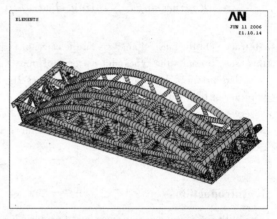

Figure 1　The mathematical model of gate structure

Figure 2　Project examples of gate structure

This experiment and research take physical model test as principal, analysis and calculation as auxiliary. In order to increase precision of experimental results, we take the following measures: take the four orifices of the big gate to build hydraulic model with large scale, and the geometric scale is 1 : 25. We should minish the effect of model reduced scale as far as possible to build normality hydro-dynamic force model according to gravity law. As this sluice gate must retain water (tide) in both directions, in order to simulate the water flow in the infall and outlet of sluice gate more correctly, making the scale of upper and lower reaches of sluice gate model enlarged to keep the similarity of the characteristics of Cao'ejiang effluent and outer river (Qiantang River) tidal surge of this gate. And we do static and dynamies analysis and calculation of gate structure through three-dimensional finite element numerical value calculating models, conducting aggregate analysis on the test and calculating results to bring forward the schemes for optimized design of gate structure.

Through contrast between experimental modes and analysis on finite element, the trial value of vibratory fundamental frequency with the gate being fully closed ranges from 5.0Hz to 7.0Hz, and the calculated value ranges from 5.8Hz to 8.2Hz of fundamental frequency at the first step, and the results are basically the same. The analyzed results of transient response simulation when the gate is under the action of tidal surge show that when the gate is fully closed with 90kN/m² of amplitude of tidal surge action, the max stress of gate is 620.0MPa, which is at the end of the circular tube of φ530mm of truss-like structure. The stress on crunodes offside of arc tube of truss-like pipe system is 194.0MPa. The max displacement of gate is 22.4mm, and there's stress centralization on gate structure. As to the max stress appearing on the junction panel at the end of the circular tube of truss-like structure and the joint of rectangular pipe xb1 and square pipe xb2, the max stress reduces to 156MPa through optimization of appropriately adding facing bars, strengthening ribbed slab or sloping plate and the thickness as well as increasing structures with little steel.

4. Tests on contrast between tidal surge and actual measurement

In order to simulate the impacting load of tidal surge, in this test we use single-orifice gate passageway presenting the tidal surge characteristics, making water in wide expanse reservoir rise to a certain water level difference. And we use rapid gate to open gate and drain off, doing contrast of the flow speed, tidal height, form of tidal head and actual measurement of model tidal surge through adjusting the water head of model reservoir to make tidal surge simulation into formation (tidal surge simulation and field picture of tidal surge are shown on Figure 3 and Figure 4).

Figure 3   Simulation of tidal surge

Figure 4   Field picture of tidal surge

On this basis, we do parameter experiment of impact load, impact acceleration, impact stress of flow and son on to gate, gate pier and bottom plate of sluice gate. The test results show that the impacting pressure working on gate body comes to the biggest when tidal surge impacts on the gate at that instant, and then there's a decreasing course of vibrating at low frequency. The impacting pressure of tidal surge acting on gate in positive direction varies from 28.38kPa to 82.24kPa, and in side long direction ranges from 39.85kPa to 114.51kPa, and in parallel direction ranges from 11.25kPa to 72.98kPa. There's a big difference of the impacting force on each part the acting force tidal surge impacting on gate from the field measured value along with the impacting position and the directions of tidal surge, and some of the biggest impact waves appear in the first impact wave, and some appear in the following flow reverberation impacting. And the impact load is bigger in a certain region of flow reverberation with lower frequency and then becomes weak gradually. The max flow impacting force tested is 53.78kPa, and the characteristics of impact load of tidal surge in model experiment and prototype test are shown on Figure 5.

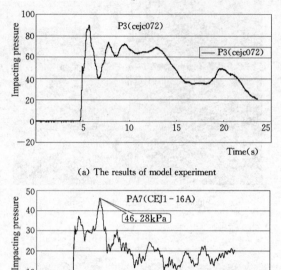

(a) The results of model experiment

(b) The results of field test

Figure 5   Impacting pressure of gate tidal surge in model experiment and prototype test

The max impacting vibration acceleration of working gate model comes at the instant of tidal head impacting, and then there's damped oscillation soon after. When the tidal surge impacts in positive direction, the max vibration acceleration in fair current direction of gate body midspan ranges from 8. 65m/s² to 11. 49m/s², in landscape direction is 9. 04m/s², in vertical direction is 6. 39m/s². When the tidal surge impacts in sidelong direction, the max vibration acceleration in fair current direction of gate structure ranges from 5. 859m/s² to 13. 89m/s². As the tidal surge impacts in vertical direction, the max vibration acceleration in fair current direction ranges from 4. 274m/s² to 11. 78m/s², and in landscape direction ranges from 5. 74m/s² to 5. 97m/s². As the tidal surge impacts in parallel direction, the max vibration acceleration in fair current direction ranges from 4. 62m/s² to 7. 02m/s², and in vertical direction is 2. 86m/s² to 4. 156m/s², and in landscape direction ranges from 2. 15 m/s² to 2. 76m/s². The measured results show that the vibration acceleration magnitude of gate structure increases along with impacting force of tidal surge getting strong, converse-

ly, the impacting force gets weak, and the vibration acceleration of gate structure decreases along with it. The vibration acceleration changes along with the form of tidal surge load. Under vibratory load action, the vibration acceleration of gate structure changes in an oscillatory way. The max impact load response of vibration acceleration measured is 6. 30 m/s². The vibratory characteristics of tidal surge impacting in model experiment and prototype test are shown on Figure 6.

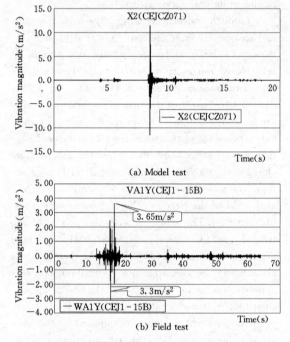

(a) Model test

(b) Field test

Figure 6   The vibratory characteristics of tidal surge impacting in model experiment and prototype test

The max impacting stress of working gate model comes at the instant of tidal head impacting, and then there's damped oscillation soon after. When the tidal surge impacts in positive direction, the larger stress of gate body appear at both sides and the dynamic stress on the midspan is comparatively small. The larger stress measured is 70. 45MPa, which is at the end of arc circular tube under gate. The max impacting stress tidal surge impacting on gate structure in sidelong direction ranges from 59. 84MPa to 150. 63MPa. The max impacting stress in parallel direction ranges from 46. 32MPa to 54. 2MPa. The dynamic stress of gate structure tested changes along with the impacting force of tidal surge. The stronger the impacting force of tidal surge is, the larger the stress of gate structure will be, conversely, the weaker the impacting force of tidal surge is, the smaller the stress of gate structure will be. The stress also changes along with the form of tidal load. Under vibratory load, the stress of gate structure changes in an oscillatory way. The max structural stress measured is 147.67MPa. And the stress characteristics of tidal surge impacting in model experiment and prototype test are shown on Figure 7.

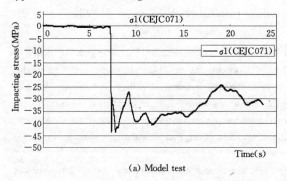

(a) Model test

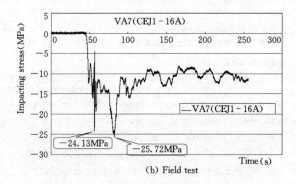

(b) Field test

Figure 7　The stress characteristics of tidal surge impacting in model experiment and prototype test

5. Conclusions

Cao'ejiang gate is the biggest tide-resistance and drainage sluice gate in present China and its long span truss-like gate structure is the first time to develop and research in China with creative structural design. Through simulation of flow-excited vibration and field observation of gate under typical orifice tidal surge load action and normal drainage, we acquire a fairly complete dynamic response of hydro-dynamic load and gate structure since the gate runs, which offers scientific basis for running safety of sluice

gate. From dynamic response data of tide-resistance working condition, the max vibration response of gate comes at the instant of tidal surge impacting, and then the vibration magnitude diminishes quickly. Under the present observing working condition, the gate structure meets the design requirements for structural rigidity and intensity with tide-resistance, and it runs safely. The prototype observation results prove that the gate structure possesses the characteristics of normal operation performance, convenient operation, smooth running and being safe and reliable. It offers good examples for design and application of long span specific type of gate.

The model experiment proves the creative design of long span truss-like structure in this engineering project a success, especially possesses many advantages in river mouth water resistance project with tidal surge action. The prototype and model data of tidal surge load and dynamic response characteristics have strong correspondence, and both of them have the same results. That indicates it is correct of simulation and test methods for flow-excited vibration modes design and tidal surge, and the research results can be spread and adopted.

References

[1] Specification for Design of Steel Gate in Hydraulic and Hydroelectric Engineering (DL/T 5013—95). Water Resources and Electric Power Press, June, 1995.

[2] Yan Genhua. Flow-excited Gate Vibration and Dynamic Optimized Design. Hydro-science and Engineering Scientific Research, January, 1999.

[3] Research Report of Gate Flow-excited Vibration Experiment in Cao'ejiang Gate Key Project. Nanjing Hydraulic Research Institute, September, 2006.

[4] Experiment and Research on Prototype Observation of Flow-excited Vibration in Cao'ejiang Gate Project Tide-resistance and Drainage Gate. Nanjing Hydraulic Research Institute, December, 2009.

Study on the Key Technologies of Diversion and Drainage Sluice Gate

Chen Fazhan   Yan Genhua   Hu Qulie

State Key Laboratory of Hydrology-Water Resources and Hydraulic Engineering, Nanjing Hydraulic Research Institute, Nanjing, 210029

Abstract: On the basis of the characteristics of inflow in side long direction, reasonable arrangement of diversion sluice gate in sidelong direction assuming responsibility of diversion and drainage are made in the research. Through experiments of energy dissipation and erosion control, the arrangement of utility-type energy dissipation threshold with less investment and engineering quantities are brought forward. In addition, we conduct research on the effect the arrangement of the back plate behind gate having on vibration-reducing of gate. According to contrast between prototype and model experiment results, this essay makes a summary about the basic design theory and application prospect of this kind of gate.

Key words: sluice gate; diversion and drainage; energy dissipation and erosion control; vibration-reducing measure; experimental verification

1. Preface

With the social economic developing at a fast speed, water resources development and utilization, water environment improvement and ecological beautification are gradually paid attention to. Urban construction transforms from developing relying on waters to diversion, water treatment and water creation. Developing and using riverine and coastal water resources in a reasonable way is critically important to advance regional development. In order to fully exploit and use riverine and coastal water resources, promote what is beneficial and abolish what is harmful, we build gate system to guide diffluence, store water resources besides of constructing embankment, which has social and economic benefit for controlling waterlogging, optimizing and improving water resources utilization, improving land and water transportation, river net environment, optimizing investment environment and son on. It is urgent to develop and research new type of diversion and drainage sluice gate that is highly efficient and safe in order to lessen project investment as far as possible and play the max economic benefit to ensure project safety.

2. Flow characteristics and engineering requirements

The axis of flow of sluice gate of branches diversion and drainage projects connects with the axes of embankment with 90°, and the characteristics of diversion and drainage are inflow in side direction. And the flow direction angle θ changes from 90° to 0°, this angle is determined by the changing amplitude of flow speed of mainstem and diversion flow quantity as well as water level of sluice gate. Therefore, θ changes along with the position, water level and flow quantity at the position of flow field, and varies in a small amplitude along with time. The project measures usually adopted are to enlarge inflow water-carrying section of sluice gate, reduce the flow speed of entrance flow, adjust the flow direction in front of sluice gate orifice to being parallel with longitudinal axis of gate orifice, well-distributed in lateral direction and small flow state difference through arc flarewall and the following long approach channel, in order to make the flow go through the gate evenly and smoothly. Generally, large and middle scale diversion and drainage projects are located in towns and regions with developed economy, and not only should the project arrangement meet regional diversion and drainage, but also take the requirements of environmental beautification, ecological coordination into consideration, which meets double requirements of practicability and sightseeing. Relying on two diversion and drainage key projects of southwest sluice gate and Lubao sluice gate on Beijiang dam in Guangdong Province, this essay introduces arrangement, energy dissipation and erosion control, gate optimized design and vibration-reducing measures of this kind of projects, and sums up experience to makes it popular (the arrangement of projects relying on is shown on Figure 1).

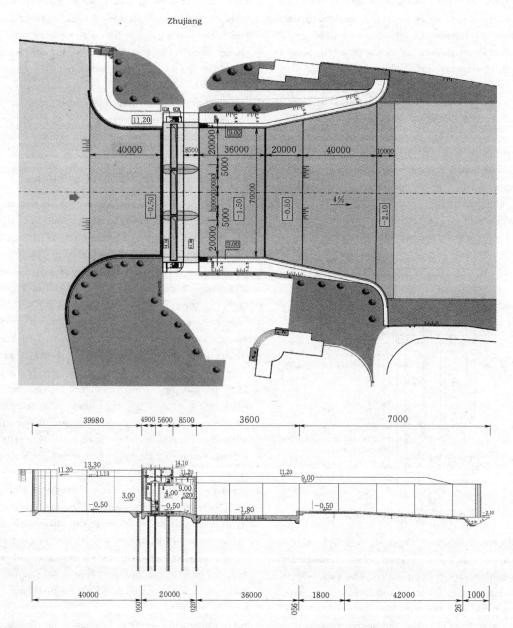

Figure 1   The plane arrangement of Southwest sluice gate

3. Research on the key technical problems of Projects relying on

As an important embankment in Guangdong Province, Beijiang Dam North River is significant protective screen for protecting developed areas like Guangzhou, Foshan and Pearl River Delta, and the overall investment of this project achieves 2.545 billion RMB. Southwest sluice gate and Lubao sluice gate are the main outlet structures in Beijiang dam, and they are important components of flood control system of lower reaches in Beijiang dam. The hinge of this outlet structure is mainly used for flood diversion, and takes diversion and improving local water environment into consideration. Southwest sluice gate is plane vertical lift gate with long span, large high-wide scale (5.7 : 1), which is the largest hydraulic ratchet down-the-hole plane gate home and abroad. The hydraulics and hydro-dynamic load action during gate running are complicated, especially the flow state under gate is complicated, which is harmful to gate safety. The flow-excited vibration problems during gate running are highlighted.

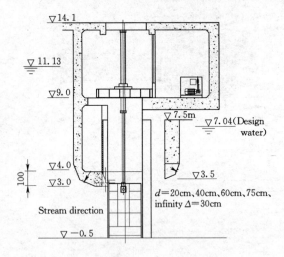

Figure 2 Sectional view of gate

Lubao sluice gate needs to run with partially opened to regulate flow magnitude, and the flow state behind gate is complicated. The gate vibration problems can't be ignored. And Lubao sluice gate is in sandbank river bed with loosening and soft sandiness, large gate drainage and severe flow scouring problems. Under flood discharge working condition, there's slight submerging or critical effluent behind gate, the flow has great dynamic effect on gate, and the gate flow-excited vibration problems stick out.

When we design flow magnitude of flood diversion, the Froude number of flow under gate is $Fr \approx 1.6 - 2.0$, and there's undulance water jump combining flow state in lower reaches. The efficiency of flow energy dissipation is low, and sandiness river bed under gate, small flow speed of anti-impact. The stability problem of anti-impact sticks out. Southwest and Lubao sluice gate are the only two large-scale discharge constructions in Beijiang dam, and whether the sluice gate structure is safe directly affects the safety of the whole Beijiang dam. Smooth construction and safely running of these sluice gates symbolize that the experimental and theoretical level of plane vertical lift gate with large high-wide scale achieve to a new high.

4. Research on measures of solutions

For the key technologies problems that the sluice gate system facing, we propose the following measures on basis of experiment on energy dissipation of lower reaches flow through contrast of systematical research: ① approach channel straight section with a certain length in front of gate; ② the tail of gate pier is modified to semiellipse curve; ③ improve the combination form of sluice chamber and sidewall of absorption basin; ④ set grid tail bucket with 2m of the height; ⑤ extend dumped rocks protecting foundation by 20m behind anti-scour channel. The research results show that with the above comprehensive measures, the energy dissipation problem of flow with low Froude and low speed are properly resolved, acquiring the results of well-distribution of flow, straight direction of flow, smooth and steady water surface, fully energy dissipating and even distribution of flow speed. The max bed velocity of river way in lower reaches of anti-scour channel decreases from 2.0m/s to 1.4m/s, and basically meets requirement for flow speed of anti-scour. The grid tail bucket adopted by the project plays an important part in adjusting distribution of vertical velocity, reducing bed velocity of river way, advancing energy dissipation of flow, improving combining fluid state under gate and weakening vibration force of gate, at the meantime, it meets diversion requirements of lower reaches with low water level. On basis of analysis technology of experimental modes and analyzing and calculating net dynamic force, taking the capacity of hoist into account, we do the following modifications to gate structure on premise of keeping hemline angles unchanged: ① increase the width of main beam of gate by 50cm; ② add portiforium close board on the panel behind gate to minish water jump effect behind gate. The research results show that with the effect of aqueous medium considered: ① the fundamental frequency of original scheme is about 5.4Hz; ② the fundamental frequency of modification scheme is about 6.4Hz, increasing by 18%. The flow-excited vibration of gate experiment uses wholly similar hydro-elastic model developed to satisfy the requirements for simulation of flow action, structural action and hydro-elastic vibration. The model material is developed by poly-component special materials of heavy metal powder and polymer materials, and we do test and selection to these developed materials. The results indicate that hydro-elastic materials for research meet the requirements of density $\rho m = \rho p$ and elastic model of structure $Er = Lr$. The physical dimension meets Lr, and the models we make satisfy the requirements for complete simulation. On this basis, as the changing amplitude of lower reaches water level of this kind of sluice gate is big, the effluent situation is complicated and severe running condition of gate, the research suggests that back plates should be added behind gate in a certain distance (Figure 2) to separate the gate from floe, in order to reduce flow-excited vibration of gate. The experiment results show when flood diversion is 1100m³/s, the vibration control and reducing of gate have a good effect after adding back plate behind gate. As the water level of upper reaches is 11.13m, the space between back plates is 75cm, the max vibration displacement in horizontal and vertical direction decreases by 28% and 12%, respectively. As the water level of upper reaches is 10.5m, the max vibration displacement in horizontal and vertical direction decreases by 37% and 30%, respectively. As the water level of upper reaches is 10.0m, the max vibration displacement in horizontal and vertical direction decreases by 42% and 24%, respectively (Figure 3).

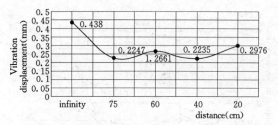

Figure 3　The effect different spaces of back plates behind gate having on vibration-reducing

5. Observation of engineering prototype

The field monitoring results show that the vibration acceleration of gate changing along with water level is drawn on Figure 4, and the max RMS value of vibration acceleration is less than 0.05m/s², the RMS value of gate vibration displacement is less than 0.2mm. The test results of vibration displacement are shown on Figure 5. It is in the scope of vibration at little amplitude, which makes sure of project safety in an effective way. Through field monitoring we concludes that the flow direction angle of diversion in side direction in Beijiang is $\theta = 20° \sim 30°$. The type, position of happening, vacillating route and submerging courses and modes of swirl in front of gate orifice are basically similar, and swirl in prototype is obvious and strong. The flow state at the absorption basin section in gate lower reaches is similar, and both model and prototype possess characteristics of smooth and steady flow, not greatly waving of water surface, well-distributing of flow section. The variability of erosion topography in lower reaches river way is similar. Under the same discharge working condition, the topography of lower reaches river bed is as the same steady. The drainage capability is basically the same, and both meet the project requirements.

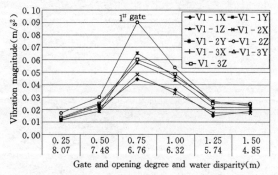

Figure 4　The changing relationship between gate vibration acceleration and water level

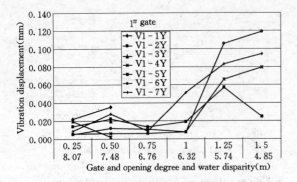

Figure 5　The changing relationship of
gate vibration displacement

6. Conclusions

The working gate of Southwest sluice gate is of plane gate structure with low water head, high flow magnitude, long span and large aspect ratio (up to 5. 7 : 1), and the running condition and fluid state under gate possess obvious characteristics. We combine numerical models, physical models and prototype observation to conduct systematical and intensive research on characteristics of hydraulic power and structural static dynamic force, characteristics of flow-excited vibration response and safety and reliability of this new kind of gate, which offers sufficient theory evidence and technology support for successful design, construction and maintaining of the first plane gate with large aspect ratio in China. The tail bucket of absorption basin in Lubao sluice gate uses grid tail bucket developed to improve the fluid state of effluent and reduce flow-excited vibratory force of gate, motivating flow dissipation of lower reaches, making flow distribution even, decreasing bed velocity of section and strengthening anti-impact stability of river bed, which is a creative measure for resolve flow energy-dissipation in low Froude number. The back plates set behind gate and reasonable space between back plates proposed reduce the vibration magnitude of gate at large amplitude. The whole hydro-elastic models are developed through hydro-elastic vibration theory and research technology. And we conduct full research on the characteristics of flow-excited vibration of this new kind of gate structure under different running condition, which offers scientific evidence for design, construction and running of this kind of sluice gate.

References

[1]　Specification for Design of Steel Gate in Hydraulic and Hydroelectric Engineering (DL/T 5013—95). Water Resources and Electric Power Press, June, 1995.

[2]　Experiments and Researches on Hydraulic Models of Lubao Sluice Gate (four orifices) in Guangdong Province Beijiang Dam. Nanjing Hydraulic Research Institute, May, 2006.

Design of Large Span Gobe Gate Structure and Study on Anti-Resonance Measure

Chen Fazhan   Yan Genhua   Hu Qulié

State Key Laboratory of Hydrology-Water Resources and Hydraulic Engineering,
Nanjing Hydraulic Research Institute, Nanjing, 210029

Abstract: It takes both the design ideas of urban flood control and landscape environment into account in large span gobe gate, this essay represents the basic design principle and the implementing condition of large span gobe gate. Through applying of perfect hydro-elastic simulation technology as well as thorough study on gate hydro-elastic vibration and vibration reduction measure and by field prototype survey, we bring forward operating regulations for appropriately avoiding vibration.

Key words: large-span gobe gate; hydro-elastic simulation technology; vibration-reducing measure; experimental verification

1. Preface

With China economic and social developing at a fast, urban construction progresses with each passing day, the vigor from mountains and water gradually integrates into the urban planning, and ecological landscaping as well as natural harmony are gradually paid more attention to. Whether it is dynamic of city water system directly affects the city beautification, and as the controlling hinge of water system, the water gate is treated as the source of urban vigor. Not only can the water gate regulate water to maintain the city vigor, but also can be suitable for requirements for urban overall planning. Therefore, the design of water gate should not only meet practicability, but also should satisfy functionality, artistry, esthetics and coordination. In that case, there emerges lots of gates with novel, simple, long-span and easily operated. In recent years, during improvement of urban water environment, rail steel radial gate, hydraulic flip gate, underlying gate and space truss system gate are gradually popularized and applied. The remarkable characteristics of this kind of gates are long span, noble shape, conveniently adjusting for water level and strong coordination with surroundings. It needs high requirements for stability of operation. Long span, large gate pocket small gate gobe type of steel gate is a new type that possesses the above features. We conduct research on safety of operation in this kind of gate, which is important for popularizing the utilization.

2. Working Conditions and design concept of the gate

Nanjing External Qinhuai Sanchahe flood gate is located at the entrance to the Changjiang River, which is a key part of environmental improvement for Qinhuai River. The main function of Sanchahe flood gate is to store water in non-flood season, to raise the water level of Qinhuai River, to use overflow on top of gate to make the river water flow smoothly, and form landscape falls, to maintain the city landscape water system, to open the gate to assure urban flood control and drainage during flood season. It has double-orifice gobe gate in this program, 40.0m of the width of single-hole gate, and the gate height can be adjusted in the range from 5.15m to 6.65m. The gate is in semi-circular three-hinge arch structure, the radius of arch axis of 22.0m (Figure 1). The arch springing on both sides of semi-circular gate's gate leaf of the gobe is supported on abutment pier and central pier after connecting through butt hinges. The gate retains water or overflow op top under horizontal condition, rising and falling of the dodge gate leaf on the gate can adjust the side water level of upstream (5.50-6.5m), which satisfies requirements for city landscape. On the top of the gate sets pedestrian passageway that can commute traffic on both sides and is convenient for people leisure travel. When open, the gate rotates upward taking cardinal axis as center, stops and locks in as moving to 60°. The river course is in flood passage and overflowing, the upturning gate combines with the water reflection, as a pair of beautiful glasses, showing the beauty of the city, so it is named as gobe gate (Figure 2). The gate sculptures wide river mouth water system, which is significant for maintaining the city water system arterie.

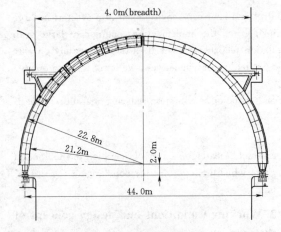

Figure 1　The gate structure

Figure 2　The majestic appearance of gobe gate

3. Research on Dynamic Safety Measures

Gobe gate is a new type with arc shape, and there's no application precedent in China. The span of this gate gets to 40.0m with thin structure. And the stress on it under each working condition is complicated, particularly overflow on gate's top. The flow centers with changeful fluid state when the gate is open or partially opened. The flow exerts complex dynamic force on the gate, which brings the gate vibration more serious. As to the unique characteristics of the gate and the flow conditions, in this experiment we conduct research hydrodynamic load characteristics of gate structure by the way of whole hydrodynamic model combining with local amplification cross-section modals. And we do researches on static force and dynamic force characteristics of gate structure through large-scale model and calculations of three-dimensional finite element numerical values so that we have study on flow-excited vibration characteristics of gate under hydro-dynamic force by whole hydro-elastic simulation models. According to general analysis on research results of gate hydro-dynamic load acting characteristics, structural dynamic characteristics and hydro-elastic vibration characteristics, the existing problems are found out and improvement and optimization measures are brought forward to ensure safe operation.

3.1　Hydraulics test

When the sluice gate is under the condition of being opened to discharge and scour and fill, the flow under gate experiences submerged discharge, critical effluent and free discharge. When it is in submerged discharge, discharge flow goes centripetally along gate width direction, and the out-gate flow disturbs each other at the hemline of downstream. When it is in critical submerged discharge, the rotating water body behind gate beats the gate body, which has exciting force load on gate. When it is in free discharge, the discharge flow presents unstable water jump or full-face far-forth driving type of water jump, and forms critical water jump at the local range of downstream hemline close to gate's trunnion. When the water gate is under flood discharge condition, due to the high water level of downstream, with most opening degrees the effluent under gate is submerged, and the water surface of downstream is stable on the whole. Once the sluice gate discharges, the pulsating pressure working on different parts of the gate changes along with the opening degree of the gate. Except that as the opening angle of gate is smaller (Eg: 0.5°) the hemline of downstream is affected by unstable dynamic action, with other opening degrees most areas under gate are in free flow, and the downstream hemline of gate body is not affected by flow action. When it runs in flood discharge, the pulsation magnitude acting on gate body will increase along with the opening degree getting big. The vibration magnitude with small opening degree is relatively smaller, and the RMS value of the max pulsating pressure on gate downstream hemline is 0.28m Water Column. The frequency distribution of main energy working on upstream and downstream hemline of gate body concentrates in the range from 0Hz to 5.0Hz, and the dominant frequency ranges from 1.0Hz to 1.5Hz. The frequency domain of pulsating energy on upstream hemline is comparatively large. The research indicates that the pulsating load working on upstream and downstream hemline of gate is the main source inducing gate vibration.

3.2 Modal Analysis and Finite Element Calculation of Gate Structure

The first-step fundamental frequency of gate is 1.2Hz, flexural deformation vibration on gate panel along with arch ring. The second-step frequency is 3.0Hz, flexural vibration deformation in radial direction on the central part of panel. The third-step vibration frequency of gate structure in radial direction is 7.0Hz; in lateral direction is 5.6Hz. The frequency above the third-step reflects high-step crankle vibration deformation. The hydro-dynamic experiment of gate structure shows that the RMS values of the max flow pulsating pressure acting on the side wall of sluice gate and gate body are 0.38m Water Column and 0.29m Water Column, respectively. The frequency distribution of pulsating energy ranges from 0Hz to 5.0Hz, and the dominant frequency ranges from 1.0 to 1.5Hz. The high energy region of flow pulsating pressure is coincident with the gate fundamental frequency, which is easy to induce low-frequency vibration of gate structure.

3.3 Hydro-elastic Vibration Experiment

When the gate structure is in discharging and scour-and-fill with upstream water level of 5.5m, downstream water level of 3.0m and opening degree of 0.5, there happens huge amplitude resonance vibration. The tested results of vibratory acceleration changing along with downstream water level are drawn on Figure 3. The vibration dominant frequency is 4.9Hz (model values), on prototype about 1.1Hz. At that time, the hydro-dynamic load excites the first-step fundamental frequency of gate structure,

resulting in low-frequency large-amplitude oscillation. The analysis results of frequency spectral are drawn on Figure 4. The vibration magnitude of gate in radial direction is the biggest, followed by that in lateral direction, in vertical direction the smallest. The max vibration RMS value in radial, lateral and vertical direction are 18.23m/s^2、8.50m/s^2 and 2.6m/s^2, respectively. The max RMS value of vibration displacement in radial direction reaches to 162.9mm, in lateral direction 135.5mm, and in vertical direction 63.5mm. The max RMS value of vibration stress is 15.9MPa, which acts on upper main crossbeam in the middle part of gate, bending stress in lateral direction. The RMS value of vibration stress in lateral direction of bottom main crossbeam under the hemline is smaller than the hydro stress on upper main crossbeam, and the max RMS value of hydro stress is 14.2MPa. When the water level of downstream is 2.77m, although there's a decrease in the vibration magnitude of gate and a relief in severe vibration, the gate is still in resonance region, the absolute value of vibration displacement is still a big number. As the water level of downstream gets high, the gate starts to enter into stable zone. The vibration stress with other water level of downstream is less than 1.8MPa. When the opening degree of gate is bigger than 1.0, the gate will have skipped resonance zone, and the vibration magnitude tends to decrease. Therefore, it needs to be paid high attention to and we must take the necessary measures to solve these problems in an appropriate way. On the premise that the gate's body size can't be adjusted in a great move, the most effective way is to adjust the operational regulations to avoid gate's severe vibration region. After avoiding resonance zone, the total stress magnitude inducing vibration on the big gate and the little gate is not large when the big gate is open to conduct scour-and-fill. The RMS values of dynamic stress on all test points

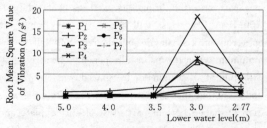

Figure 3　The changing relationship between vibration acceleration of gate and downstream water level

are within 1.0MPa as the little gate is in overflow condition. Consequently, the dynamic stress on the big gate structure when the little gate is in overflow condition shouldn't be viewed as controllability factor.

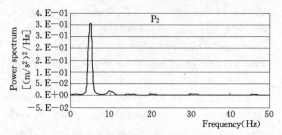

Figure 4   Frequency spectrum of strong vibration of gate

4. On-site monitor and analysis

In order to more fully grasp the operating characteristics of water gate structure after completion of construction to evaluate the operating safety in a scientific way and draw up reasonable operating regulations for gate's safe running, this prototype observation sticks to the principles of comprehensive coverage, plenty of testing points and complete working conditions. The observation of static displacement of gate structure shows that the max value of the dead weight flexivity of 1# gate (left bank) is 5.8cm, 4.0m smaller than that before optimization and amendment, and 69% lesser of flexivity. The stress distribution on gate under dead weight Gate under the dead weight indicates that the max stress on lifting lug is on the root part, and the max stress value is 60MPa from the test (the changing relationship shown on Figure 5). The stress on stiffening plate of lifting lug portiforium is 20MPa. The max stress value of the hinge part is −26MPa from measurement, the compression stress. The max stress on flanged beam's flange at the crosspoint of the gate pier's upper part and the bridge as well as web plate is 100MPa. The stress will get to the max when the gate runs in a small opening degree, and fall to the minimum when it runs in full-opened. The stress distribution of all sections of the gate downstream panel indicates that the max stress is in the gate cross-section U-shaped bottom part of steel spring plate and the central part of V-shaped weak joint cover, about 58MPa. The stress on other parts is less than this

value, and the max stress value exists with small opening degree, which falls to the minimum with the gate fully opened. Overall, the dead weight stress of gate measured is not a big value, which satisfies requirements for engineering safe operation. The max stress of gate occurs at the gate-opening starting phase, and the stress value of structure decreases gradually with the increase of gate opening degree, the stress of some measuring points presents the changing characteristics from pulling stress at gate-opening time to compression stress during gate-opening. The stress at the moment of gate-opening is 35.0MPa pulling stress. The stress on other sections of gate is within 20.00MPa. The stress on gate's trunnion reaches to a peak value at the moment of gate-opening, and the max stress value is 16.0MPa compression stress. Then, after experiencing unexpectedly lessening, the stress gradually increases along with the increase of opening degree. The max stress value comes at with the gate fully opened, 18.0MPa. The max stress on lifting lug of gate is 68.0MPa. The max stress on the bridge on both sides of gate middle section central pier is 100.00MPa. The max stress on bridge at downstream of right side of middle pier is 67MPa. The stress on other parts is less than 70.0MPa. The test results of flow pulsating pressure show that the max flow pulsating pressure of gate central section appears with gate opening degree of 1.03, which is at the hemline behind gate. The RMS value of pulsating pressure is 18.5m Water Column, and there appears before-the-gate water jump and pulsating load in a low frequency (Figure 6). The RMS value of pulsating pressure at hemline before gate is 10.0cm Water Column, and the pulsating pressure with other opening degree is significantly reduces. That means under the combination condition of this water level and opening degree, the gate will undergo flow pulsating pressure load, relatively large. That is to say, it will cause strong vibration of gate structure under this condition. The pulsating pressure of flow is in low frequency, and its main energy concentrates in low frequency zone within 2.0Hz. The model observation results of gate structure indicate that the gate overall first-step fundamental frequency is 1.27Hz, the second-step frequency is 2.1Hz, the third-step frequency is 2.26Hz, the fourth-step frequency is 5.56Hz, and model frequency at other high step is 10.0Hz. The model damping of gate structure varies in the range from 0.32% to 5.38%. The test results of vibration acceleration of gate show that the vibration magnitude of gate has an increase along with the working water head getting high. Especially when the gate opens at about 1.0°, there is before-the-gate water jump of downstream, which excites the gate structure and causes strong vibration of gate body in a low frequency. The vibration characteristic is doing movement in low frequency and vast amplitude. The phenomenon is identical with the model testing results. Although the vibration time-domain course with other opening degree is in low frequency, not in vast amplitude. From the spectral density curve of vibration power of gate, the vibration acceleration of gate in frequency domain shows the characteristics of vibrating in low frequency, and the dominant frequency of vibration focuses on low frequency zone ranging from 0Hz to 3.0Hz. This is identical with the distribution of energy of flow pulsating pressure in frequency zone. The test results of vibration stress of gate show that under the condition of downstream water level of 4.26m and opening degree of 1.03°, the vibration stress of each parts shows the characteristic of simple harmonic vibration, which is relevant to before-the-gate water jump behind gate, pulsating pressure working on gate body with the above opening degree coming to the max. And it is identical with the changing characteristics of gate vibration acceleration. Once the opening degree is bigger than 2.0°, the vibration stress will varies randomly. At that time, the pattern of before-the-gate water jump behind gate changes, the factor exciting water gate resonance vibration is gone. The larger stress on gate body measured is at the bridge site on both sides of middle pier of gate, and the max RMS value of dynamic stress is 13.28MPa, the RMS value of stress on gate panel is within 4.0MPa, the stress on other parts is not in a big value. The max stress value comes with opening degree of 1.03°, and the stress with other opening degrees significantly decreases (Figure 7). The energy distribution of vibration stress in frequency domain possesses distinct characteristics of being in low frequency, and its main energy still concentrates in low frequency within 3.0Hz.

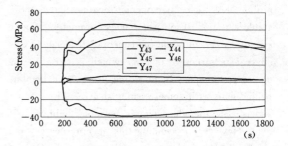

Figure 5　The changing relationship of stress on gate

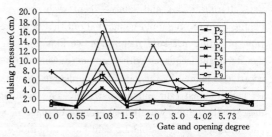

Figure 6　The changing relationship between the RMS value of flow pulsating pressure and opening degree

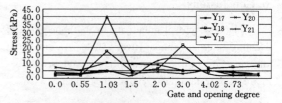

Figure 7　The changing relationship between the RMS value of dynamic stress of gate and opening degree

5. Operation scheduling mode

When the gate runs with being partially opened, the max flow pulsating pressure working on gate comes with the opening degree of about 1.03°, at the hemline behind gate. At that time, there appears strong vortices excitation on the flow at hemline behind gate, which is over four times bigger than that under the working condition of high water level of upstream and downstream and small drop height. That means under the combination condition of this water level and opening degree, the gate will undergo flow pulsating pressure load, relatively large. It is easy to cause strong vibration of gate structure. The pulsating pressure with other opening degrees significantly falls; the main frequency of flow pulsating pressure is 1.0Hz, which is identical with the fundamental frequency of gate structure. The flow-excited vibratory response of gate shows harmonic vibration in vast amplitude, but with other opening degrees shows more stable vibration. From the modes parameters of gate structure, the vibration fundamental frequency of this sluice gate structure is low, which is close to the dominant frequency of pulsating pressure of hemline behind gate with some opening degrees (Eg: 1.0°) and is easy to induce simple harmonic vibration of gate. However, if avoiding this opening degree, the gate will run stably with partially opened. On the whole, it's easy and convenient to operate this sluice gate. It runs stably and safely except with some opening degrees. It should be pointed out that both the flow pulsating pressure magnitude working on gate and dominant frequency with main energy concentrated are determined by the gate opening degree and water level of upstream and downstream. The opening degree with strong vibration appearing changes in a certain range, so it should be paid close attention to during operation to avoid them.

6. Conclusions

The new type of gobe gate in the sluice gate of Nanjing external Qinhuai River Sanchahe is the result of joint effort of design, scientific research, construction and operation management department. We ac-

quire significant breakthrough and achievements in structural design, function of use and safe operation. This successful construction and safe operation of the sluice gate mark that the level of design, scientific research, construction and management of sluice gate achieves ne high in China. The model test technology adopted in this essay offers scientific evidence and technical support for long term, safe and reliable operation of sluice gate structure, and offers references for construction of the same kind of sluice gate.

References

[1] Specification for Design of Steel Gate in Hydraulic and Hydroelectric Engineering (DL/T 5013—95), Water Resources and Electric Power Press, June, 1995.

[2] Yan Genhua. Flow-induced gate vibration and dynamic optimization design. Scientific Research of Water Conservancy and Water Transportation, January, 1999.

[3] Experiments and Researches on hydro-elasticity vibration of Nanjing Sanchahe gate. Nanjing Hydraulic Research Institute, January, 2005.

Study of Function Extended Application of Large-Size Plain Emergency Gate

Chen Fazhan   Yan Genhua   Hu Qulie

State Key Laboratory of Hydrology-Water Resources and Hydraulic Engineering,
Nanjing Hydraulic Research Institute, Nanjing, 210029

Abstract: Based on the study about operation management of large size diversion tunnel design, from practical operating management as well as reducing project investment, the feasibility and safely applied scope of large size plain emergency with gate partial opening release water were studied in this paper. Through analysis on typical engineering application examples, openness scope and rational operating rules of emergence gate of diversion tunnel in certain project was given in the paper.

Key words: large size plain emergency gate; partial opening; discharging; safety verification

1. Interview

As high dam construction technology is constantly improved, construction period shortened, to safely get over flood season assured, water delivery of ecological environment and during the early period water for industry and life from downstream in construction time, the utilization ratio of diversion tunnel will constantly increase, in order to reduce the engineering construction cost and increase availability in an effective way, rebuilding diversion tunnel into discharging tunnel, emptying tunnel and desiting tunnel is becoming more popular and technology level is also rising day after day. To greatly explore the utilization ratio of diversion tunnel in the early period, especially in the middle construction period. A great many researches have been conducted to greatly explore the utilization ratio of diversion tunnel in the early period, especially in the middle construction period, to save investment for engineering and guarantee the safety of this engineering project. The extended application scope and technical feasibility are elaborated through this engineering example.

The diversion tunnel in Shenxigou Hydropower is targeted for research to study the feasibility of function extension of large-size diversion tunnel in this paper, which is located on the border of Dadu River in Hanyuan and Ganluo County of Sichuan Province. It is the 18th stairstep power station in Dadu River main stream planning with the main function of power generation. It has 4 installations, 660MW of total capacity, and 32.35 billion kW · h of annual energy output. The power plant discharge structure is formed of two discharging tunnels and a three-hole sluice gate. 1 # and 2 # discharging tunnel are used for the construction diversion at the first place during construction, and then are rebuilt into discharging tunnels. Besides during flood discharge, it also will be used for sand washing and desilting when the reservoir silting gets serious. To properly resolve the setting off period of gate-closing and temporary water-getting of downstream problems in Pubugou hydroelectric power station, we will adopt the plan of cofferdam in Shenxigou for water retaining to supply water to the downstream to satisfy its normal water requirements for production and life. The plan of cofferdam in Shenxigou for leakage control need to make use of 2 # discharging tunnel entrance emergency gate (also as diversion tunnel closure gate) to retain water, and it is partially open to make the discharge under control, thus, we can make sure of the need of downstream water supply. If the 2nd diversion tunnel in Pubugou power station doesn't make gate closing one time success, in order to ensure the safety of Shenxigou pit, we should take contingency plan of opening 1 # diversion tunnel emergency gate by flowing water. As the running mode of the emergency gate changes, the running water head is high when it is partially opened, and the gate hole size is large, especially the four joints under 1 # diversion tunnel emergency gate below weld into a whole and the two above weld into a whole, which forms a special structure. In case of an emergency, if the dynamic water elevates the gate leaves of the two joints above to discharge, the gate leaves of the two joints above and below gate can't meet the specified size requirements of hemline structure in the regulations, therefore, we need to carry out experiments and researches on the hydraulics and flow-induced vibration of the entrance emergency gate model, aiming to determine the gate opening degree and range of applications and to ensure safe operation.

2. Research on Dynamic Characteristics with Emergency Gate partially opened

We conduct a series of experiments and researches aiming at the large size of gate orifice of 2 # diversion tunnel, the emergency gate can only be closed by flowing water and can't be opened by flow water, the tail gate partially opening should satisfy water need flow rate of downstream. The research results show that when it meets the downstream water delivery requirements, the water level of downstream is low and the gate opening is small, in the tunnel behind the emergency gate it is free flow. The maximum wind speed of venthole is 15.65m/s, which meets the standard. In tunnel the headroom above the water is large, and we set three ventholes at the top parts behind the gate with sufficient air entrainment, smooth air vent and small difference of air pressure between inside and outside the tunnel, which does not affect the dynamic pressure distribution in tunnel, especially the gate slot section. The running water head under discharge control is low, water flowing not fast, so it's with small chance that there's cavitation on the downstream tunnel at the gate slot section of emergency gate and scouring on both the two banks of downstream at entrance. The tested characteristics of pulsating pressure of gate structure changing show that the pulsation magnitude of dynamic pressure is small with the gate opening degree above $e = 2.1$m. And as the opening degree becomes bigger and the discharge flow rate increases, the flow pulsating pressure magnitude enhances in a significant way (Figure 1). Through the experiments and researches it comes to conclusion: ① We adopt back sealing in the engineering project entrance emergence gate, the gate hemline is set in the front of the middle part,

the space in the back gate slot is small. Under the working condition of $H_{top} = 658.0$m and the gate opening degree of $e = 1.5 \sim 3.0$m, there's no whirlpool in gate slot, the flow in the gate groove angle goes in the middle with "gathering" phenomenon. ② When the gate opening degree is $e = 1.5 \sim 2.0$m, after contraction, the flow simulation of effluent on both sides of the gate rebounds, centers and stirs up. The slope section and water flow are divorced to form a lateral cavity with thin bottom aquifer layer. The thinness of aquifer layer it is easy to appear disturbance and aerification with thin aquifer layer, which improves the anti-cavitation performance of gate slot under running condition of small opening degree. ③ The water level of the reservoir falls to 650.0m, and the gate opening degree is $e = 2.5$m, the slope segment sometimes flows, and sometimes voids, which is the limit state in lateral cavity. The flow aerification is smaller than that with the gate opening degree of $e < 2.0$m. With the gate opening degree of $e = 3.0$m the water flows along the slope section of the gate slot with limpid water and zero aerification. ④ According to the flow state at gate slot section, it is suggested that the water level of the reservoir be $H_{up} = 658.0$m, the opening degree of emergency gate be $e \leqslant 2.0$m; when $H_{up} < 650.0$m, gate opening operation can be increased to 2.5m; when $H_{up} < 636.0$m, the gate opening degree allowed will be $e \leqslant 3.0$m. In order to reduce the cavitation and crosion under adverse working conditions, it should be avoided of running with high water level of the reservoir when the emergency gate is partially opened to discharge. Since it is more favorable of high water level with small opening degree and low water level with large opening degree to resist cavitation, therefore, it would be appropriate to make that as a guideline for gate open operation. 1 # diversion tunnel is treated as a contingency reserve installation. The flow velocity at this orifice is smaller with weaker dynamic action. The gate structure is small and light, so it's convenient to open and close the gate. However, the bottom edge of the gate does not fit design specifications; circumfluence of the weak border is inclined to induce elastic vibration on gate structure and flow cavitation and crosion at the gate slot section. When the gate is with small opening degree, the effluent tongue gathers to the central part, there forms lateral cavity at both sides in a certain range, the upper air can be transported to the low part of the water tongue. However, the gate is opened with big opening degree till fully opened, the lateral cavity of water tongue disappears, and it is difficult to supply air. The water level in cavity increases so that the cavity disappears, accompanying which the uplifting pressure, dynamic pressure and friction force acting on low joint gate will change. And it would be possible that the low joint gate fails to be stable. Therefore, for the gate safely running, the gate should run with the opening degree of $e < 4.0$m, and the opening and closing action should be slow instead of fast.

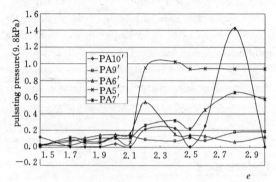

Figure 1　The changing relationship between pulsating pressure of diversion tunnel emergency gate and opening degree

3. Research on Gate vibration and stability of gate running

The experimental results of perfect hydroelastic vibration of 2 # diversion tunnel emergency gate point out that the vibratory magnitude of gate increases along with the water level of the reservoir getting high. There exists the trend of increasing along with the opening degree getting bigger (Figure 2). During the effluent with the gate partially opened, the vibratory accelerations of emergency inspection gate in three directions are in the relationship of $V_x > V_z > V_y$, that is the vibration magnitude in horizontal direction the maximum, followed by that in vertical and in lateral direction the minimum. The max RMS values of vibratory acceleration of gate structure under each working condition are 0.400m/s² in direct flow direction (X); 0.151m/s² in lateral direction (Y); 0.184m/s² in vertical direction (Z), respectively. The vibration magnitude is comparatively smaller with the opening degree of 1.5m. The RMS value in direct flow direction is 0.22m/s², which accounts for

50% of the vibration magnitude with 2.6m of opening degree. Consequently, the gate under dynamic action is easy to vibrate in horizontal direction, followed by in vertical direction. Comparatively speaking, the vibration magnitude with the gate opening degree in the range from 1.5m to 2.0m is much smaller. Accordingly, on the premise of discharge meeting the requirements, it is more appropriate that the gate opening degree should be small. The vibration magnitude of effluent of 1 # diversion emergency gate possesses the following change law, along with the water level of the reservoir getting high, the vibratory acceleration of gate structure intends to increase, and so does the gate opening degree (1.5 - 3.0m), the vibration magnitude increases progressively. With comprehensive survey of each working condition, vibratory accelerations of Shenxigou's 1 # diversion tunnel emergency inspection gate in three directions are in the relationship of $V_z > V_x > V_y$. From the test, the max RMS values of vibratory acceleration of 1 # diversion desilting tunnel under each running working condition are 0.190 m/s² in direct flow direction (X); 0.165m/s² in lateral direction (Y); 0.346m/s² in vertical direction (Z), respectively. Apparently, the max vibration magnitude of diversion tunnel emergency gate is in vertical direction, followed by that in direct flow direction and in lateral direction the minimum. When the gate opening degree is $e > 3.0$m, as the water tongue of effluent becomes thick, the lateral cavity is plugged off, there's no air supplying source for the bottom cavity below the water tongue, the vacuum of cavity increases, back flow rotation strengthens, which stimulates instability vibration on low joint gate leaf. Since backwater behind gate rises, it makes the total thrust of low joint gate leaf in horizontal direction lessens, the friction force becomes smaller. Under uplifting pressure of water, pulsating pressure of rolling flow and cavity suction, the vibration on the low joint gate leaf gets strong with instability. Once the low joint gate leaf loses stability, it is easy to form the upper and lower over-current. And as time goes on, there's vibration up and down in low frequency and impact on the bottom plate of diversion tunnel, which causes damage on cavity body, meanwhile, with that the vibration magnitude of upper and lower joint gate leaf reinforces. Therefore, it is recommended that the gate should run with the opening degree limited below $e=3.0m$, and the extreme running working condition can't exceed 4.0m.

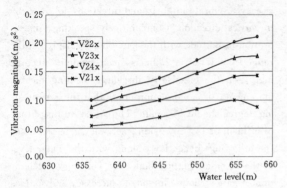

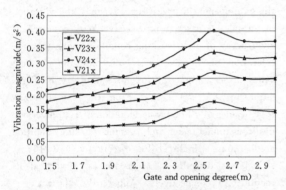

Figure 2　The changing relationship between vibration magnitude of emergency gate and opening degree

4. Conclusions

According to the results we obtain the following conclusion: 2 # diversion tunnel entrance emergency gate should run without high water level of the reservoir as far as possible. In consideration of high water level with small opening degree and low water level with large opening degree being more favorable to resist cavitation on gate slot, it would be appropriate to make that as a guideline for gate open operation.

When 1 # diversion tunnel emergency gate is with small opening degree, the effluent tongue gathers to the central part, there forms lateral cavity at both sides in a certain range, the upper air can be transported to the low part of the water tongue. However, the gate is opened with bigger opening degree, the lateral cavity of water tongue disappears, and it is difficult to supply air. It is suggested that under the condition of meeting discharge requirements, the gate should run with small opening degree as far as possible. With cavitation and crosion on gate slot taken into consideration, it is inappropriate that the emergency inspection gate runs with being partially opened for a long time so as to avoid cavitation and damage on gate bracing structures. The research conclusions chime with the field test result, and the engineering project safely runs, which greatly reduces investment, it can be adopted in similar engineering projects.

References

[1] Specification for Design of Steel Gate in Hydraulic and Hydroelectric Engineering (DL/T 5013—95), Water Resources and Electric Power Press, June, 1995.

[2] Yan Genhua. Flow-induced gate vibration and dynamic optimization design. Scientific Research of Water Conservancy and Water Transportation, January, 1999.

[3] Experiments and Researches on Hydraulics Modals of diversion tunnel entrance emergency inspection plane gate in Dadu River Shenxigou Hydropower Station. Nanjing Hydraulic Research Institute, June, 2009.

Study of Flow-induced Vibration and Operating Rules of Vertical Spindle Rotary Gate

Chen Fazhan   Yan Genhua   Hu Qulie

State Key Laboratory of Hydrology-Water Resources and Hydraulic Engineering,
Nanjing Hydraulic Research Institute, Nanjing, 210029

Abstract: Based on analysis on design idea and applied scope of vertical spindle rotary gate, we conduct static dynamic analysis on gate structure through analysis technology of experimental models and three dimensional finite element mathematical models. This helps us to acquire the vibration modes parameters and distribution characteristics of stress displacement of gate structure. On basis of this we do hydro-elastic vibration simulation experiment and acquire the vibratory response parameters of gate. According to the dynamic characteristics and flow-induced vibration response data of structure, we raise anti-vibration optimization plan and improvement measures. Reasonable way of gate running and operation is put forward through different way of running simulation experiment.

Key words: vertical spindle rotary gate; dynamic characteristics; vibration response; vibration-reducing measure; experimental verification

1. Preface

With China economic and social developing at a fast and people's living standard constantly improving, environment renovation and ecological beautification are gradually paid more attention to. The function of sluice gate develops from water storage and drainage flow control to taking urban beautification and art scene into consideration. The design of sluice gate is expanded from practicability to functionality, artistry, esthetics and coordination. The design form of gate develops to novelty, terseness and long span. Plane gate, radial gate, miter gate, arch gate, spherical gate and cylindrical gate and other gate structure that is commonly used have been in a period of innovation and breakthrough. In recent years, up-welling steel arch gate, railed steel radial gate and space truss framing system structural gate are popularized and applied progressively for improving city water environment, and the outstanding feature of this kind of gates is long span, novel formation, convenient for adjusting water level, and excellent coordination with surroundings. The vertical spindle rotary gate is a new system of gate that integrates all the above features, doing systematic study on this gate system offers technical support for popularization and application of this kind of gate.

Based on revolving gate, The relevant experts in China Anhui Survey and Design Institute of Water Conservancy and Hydropower acquire inspiration to put two arch gate structures upright in side direction together by connection of structural elements to form a new type of gate structure that is applicable to retain water in both directions as well as being inspected without blocking the flow. The picture of vertical spindle rotary gate is showed on Figure 1.

Figure 1   Illustration of model of gate structure

This kind of gate has been used in floodgate of Tangxi River in Hefei City, Anhui Province. The sluice gate is with single orifice, clear width of gate orifice is 30m, height of gate is 9.4m, and radius of turn is 25.0m. The ally arm is in truss type structure, and the sluice-pump station is a roundness frame structure with diameter of 55m, two floors. And the second floor can spin. There're 12 little gates in gate leaves of each side of the vertical spindle rotary gate, the size of orifice (width * height) is 1.5 * 3.0m. The overall architectural composition and ecological wetland of upper reaches as well as roads around the lake of lower reaches enhance each other's beauty, which forms distributing center for tourism, sightseeing and leisure. The design drawing of this engineering is shown on Figure 2

Figure 2　The design drawing of sluice gate

2. Analysis on finite element of gate structure

The analysis on vibration modes of the vertical spindle rotary gate structure is conducted through adopting three-dimensional finite element method, and shell element SHELL63 and beam element BEAM188 are used in the calculation to do joint modeling of structure. The shell element is used to do part modeling of gate leaf, and real constants represent plates with different thickness. The beam element is used to do modeling of ally arm bars, defining attributes of different section representing main post and chord member. The number of the whole model unit is 78,256, and the total number of degrees of freedom is 383,136. The calculation of boundary conditions takes two situations into account: (1) simply support of trunnion, gate leaf being leaf; (2) simply support of trunnion, mudsill of gate leaf being constrained in vertical motion freedom direction. The results of analysis on vibration modes characteristics indicate that different boundary conditions and constraint conditions of gate structure have influence on vibration characteristics. Additionally, fluid exerts

certain influence on fluid-structure interaction vibration of structure. The vibration modes frequency distribution at the first three steps under free situation of gate structure is 0.8360Hz, 1.093Hz and 1.153Hz, respectively, lateral-torsional vibration deformation on gate body, the main bar of ally arm and connecting structure of tension rod. When the simple support of trunnion, mudsill of gate leaf being constrained by chain wheel into consideration, the natural frequency of vibration on gate structure greatly increases. The vibration modes frequency distribution at the first three steps under free situation of gate structure is 3.624Hz, 6.126Hz and 6.726Hz, respectively, still lateral-torsional vibration deformation on gate body, the bar of main ally arm and connecting tension rod independent or combination. As fluid-structure interaction taken into consideration, the model vibration frequencies at the first three steps are 2.540Hz, 3.426Hz and 4.023Hz, respectively. The finite element models and vibration modes are shown on Figure 3 and Figure 4.

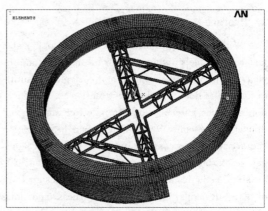

Figure 3　Finite element mathematic model of gate structure

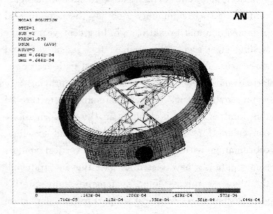

Figure 4　The vibration modes of gate structure

3 Physical model test

The hydraulic model is designed according to gravity law, scale of 1 : 20. The water flows unsymmetrically during gate opening, there's rotating flow in sluice chamber with complicated flow pattern. The flow has a special dynamic force action on sluice gate structure, especially the main parts of the gate body, ally arm and upright post of trunnion are impacted by dynamic load when the gate is fully opened. The vibration problems of gate are highlighted. According to proving of hydraulics experiment and optimization, we get the following conclusions: (1) the drainage capability of sluice gate is determined by checking flood discharge working condition and the big gate being fully opened, the drainage capability meets the design requirements; (2) the four crest outlets of spillway in the optimized installation can automatically adjust the water level of upper and lower reaches as well as discharge. And it is convenient for running and management of sluice gate, making the design of sluice gate more reasonable; (3) when the big gate is opened to discharge flood, the flow in the river way of upper and lower reaches is smooth, the flow in sluice chamber is smooth, too. Although the dynamic load action has action on the ally arm, upright post and other structures emerging in water, increasing the flow's drag force, it has small influence on the whole gate structure; (4) the structural arrangement after optimization further improves safety and reliability.

The hydro-elasticity experiment of gate structure shows that when under the working condition with the big gate fully opened, the intersection angle between gate's main ally arm and flow direction is big, blocking water flowing and bearing strong flow impacting force, so the vibration magnitude of the main ally arm of upper reaches is bigger. And the vibration magnitude on main ally arm of lower reaches and pull rod of ally arm in transverse direction is comparatively small. The test results show that the vibration acceleration gets high along with the increasing of discharging flow. The tested max RMS value of acceleration with the big gate fully opened is 0.211m/s^2. The vibration with the little gate being closed and the big gate being partially opened is much stronger than that with the little gate being partially opened and the big gate being opened. Under the same discharge, the max RMS value of vibration acceleration with little gate being fully opened and big gate being partially opened is 0.424m/s^2; the max RMS value of vibration acceleration with little gate being fully closed and big gate being partially opened is 0.733m/s^2 (Figure 5). Under the same water level, the vibration magnitude on all the parts of gate changes along with opening degree, which is relevant with the pulsating load magnitude of flow impacting on gate structure. Generally speaking, the vibration magnitude increases along with the increasing of opening degree. The max vibration accelerations of gate structure under each water level difference are: 0.720m/s^2 with 4m of water level difference, 0.675m/s^2 with 3m of water level difference, 0.495m/s^2 with 1m of water level difference. Under the working condition of big gate being closed and little gate being partially opened, as the tie rod of the gate's ally arm is in vertical direction with flow direction, and the vibration acceleration magnitude improves along with the water-blocking force. The impact from water-blocking flow of main ally arm is striking. The experiment indicates that the vibration magnitude of gate increases along with the opening degree increasing, the max RMS value of vibration acceleration tested is 0.272m/s^2. In upper reaches there's the same changing discipline with little gate being partially opened. Under the same drainage, the vibration acceleration magnitude of gate with little gate of upper reaches being opened is larger than that with little gate of lower reaches being opened (Figure 6). The vibration acceleration of gate with little gate in upper reaches being opened in symmetric interval way changes along with the opening position. As to the gate is evenly opened, the same drainage flow rate is $100 \text{m}^3/\text{s}$. the little gate is evenly opened out of phase, the vibration acceleration of gate structure is bigger. The max RMS value of gate vibration acceleration with little gate in upper reaches being opened in symmetric interval way is 1.053m/s^2. From the test we find that the gate vibration with drainage of sluice gate is on ally arm, and the vibration on big gate body is weak, the vibration magnitude accounts for 20%-30% of the vibration magnitude on ally arm. The measurement results of vibration displacement and stress are in accordance with the changing discipline of vibration acceleration.

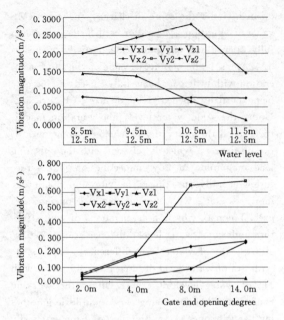

Figure 5　The vibration acceleration of gate changing
with the opening way of big and little gate

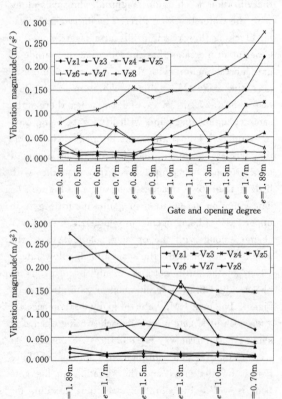

Figure 6　Vibration acceleration changing with the opening
height of upper and lower reaches little gate

4　Running and operation mode of gate proposed

The analysis on hydraulic Characteristics of the sluice gate show that the big gate and little gate are of use and attend to their duties. According to this, the basic mode of running of big and little gate is determined: (1) The two big gates in front and back are used to drain off floodwaters, so their basic running mode is being fully opened to drain without being impeded. During the gate being opened, the water level difference of upper and lower reaches should be small instead of large. To ensure the stability of the gate structural having stress on and safe running, the working water head should be limited. when the discharge flow is $150m^3/s < Q < 240m^3/s$, it is more favorable of the little gate in upper and lower reaches being fully opened and turning the big gate to weaken the gate vibration and for energy dissipating as well as erosion protection of lower reaches. (2) During nonflood season, the little gates both in front and back control drainage. The combinations of opening degree of the little gates in front and back are as follows: the max flow rate under the given water level working condition, it should be the little gates both in front and back being fully opened. The discharge rate is $150m^3/s > Q > 100m^3/s$, it should be the twelve little gates in front being fully opened and the twelve little gates in back being partially opened in an even way. The flow rate is controlled by the opening degree of little gate in back. The discharge rate is $50m^3/s < Q < 100m^3/s$, it should be the little gates in both ways being partially opened, but the opening degree of little gates in front bigger than that in back. The discharge rate is $Q < 50m^3/s$, the little gates in both front and back should be partially opened in an even way, and other opening modes are without limits.

5. Conclusions

When the big gate is opened to discharge flood, the flow in the river way of upper and lower reaches is smooth, the flow in sluice chamber is smooth, too. Although the dynamic load action has action on the ally arm, upright post and other structures emerging in water, increasing the flow's drag force, it has small influence on the whole gate structure. The little gate is opened to drain off floodwaters, although the flow in sluice chamber is in a mess with increasing turbulence, if controlling the floe speed in a proper way, it will not affect the stability of gate structure. The big gate covers the little gate and they run in a way of cooperating each other, which creates favorable condition for safe running. The design theory is reasonable. The gate bears the water head action in two-way and integrates tourism, sightseeing, urban planning and water conservancy into one. Its form is the very first proposed both at home and abroad, which is used for the first time with creativity. The optimized and amended arrangement set four spillover orifices to automatically adjust water level of upper reaches and discharge flow rate, making the running and management of sluice gate more convenient and the design of gate more reasonable. The main frequency of flow pulsating pressure of gate structure under different working conditions ranges from 1.0Hz to 3.0Hz. If simple support of trunnion, mudsill of gate leaf being constrained by chain wheel taken into consideration, the vibration modes frequencies at the first three steps of gate structure are 3.624Hz, 6.126Hz and 6.726Hz, respectively. Therefore, the fundamental frequency of gate structure is slightly higher than the dominant frequency zone in flow high-energy region. There's small possibility of resonance vibration on gate structure, but the running condition should be paid close attention to formulate rational operating regulations, avoiding higher flow pulsating

pressure frequency energy zone appears at the low-frequency zone of gate's ally arm, then causing strong vibration on structure. According to analysis on the gate vibration and energy dissipating as well as erosion protection of lower reaches, it is recommended that the little gate in upper reaches is fully opened or with large opening degree, the little gate in lower reaches is partially opened when little gates is partially opened to drain off floodwaters. That is to say, regulate and control based on inflow with small water level difference, effluent with large water level difference. Tourism and sightseeing have high requirements for water environment, so using the new type of gate structure is an ideal choice.

References

[1] Experiments and Researches on Hydraulics of Key Project of Floodgate in Tangxi River in Hefei City and Hydro-elastic Vibration Model of Steel Gate. Nanjing Hydraulic Research Institute, June, 2010.

[2] Specification for Design of Steel Gate in Hydraulic and Hydroelectric Engineering (DL/T 5013—95), Water Resources and Electric Power Press, June, 1995.

Research on Wave Forces Interference Action Main Girders of Twin-Deck Bridges

Zhang Xiantang[1,2]   Chen Airong[2]   Wang Chen[1]

[1] *Department of Bridge Engineering, Tongji University Shanghai, 200092*
[2] *Shandong Provincial Key Laboratory of Civil Engineering Disaster Prevention and Mitigation (Shandong University of Science and Technology), Qingdao, 266510*

Abstract: When a bridge crossing a waterway or in ocean is partially or entirely submerged during a flood event or ocean wave, its deck may be subjected to significant wave force. According to the time-varying wave forces acting on the bridges for various submergences, wave force interference actions between twin decks bridges were investigated preliminarily based on section physical model tests. The load distribution acting on bridge were calculated and analyzed. The experimental results have been analyzed, the Froude number of main girder and geometrical parameters of the bridge are discussed and compared against relevant literature. Due to the presence of a free surface, force coefficients can be either larser (by more than a factor of 2) or lower than the corresponding values of the unbounded domain. The experimental drag coefficients are then compared with the results obtained by the momentum equation. The experiment and research aimed at the optimization of the structure of the bridge, the distance between the two bridges and the plan layout of the shore connection structure of the bridge were proposed. The experiment and research have not only provided the designers with the actual wave loads on the bridge structure, but also discussed how the wave force acting on two bridges should be estimated, thus proposing a calculation method that can be referred to during designing.

Key words: bridge engineering; twin decks; wave force; interference action

1. Introduction

When a bridge crossing a river or sea is partially or entirely submerged during a flood event or huge wave, its main girder may be subjected to significant wave forces. Together with the information of the probability of occurrence of flooding or huge wave, the proper estimate of load exerted by the flow on the girder is relevant for designing or evaluating its vulnerability. When considering the specific problem of river bridge load, the presence of the channel walls and a water-free surface can cause some deviations from the behavior typical of infinite domains [1]. The effects of the river walls and floor can be neglected in some practical situations for large river cross sections and elevation of the bridge girder. However, a combination of such geometry with large submergences of the bridge girder, mimicking an unbounded domain, is of limited practical interest. For the design and vulnerability assessment of bridges, the most interesting situation occurs when the presence of a free surface

prevents one from directly applying models of fluid-structure interactions derived for unbounded domains. Interactions between free surface flows and man-made structures are mainly studied with reference to circular cylinders [2-4] and are mainly focused on the analysis of flow field distortion. The main influence of the shape of the structure renders these results usable only for a limited number of general considerations. The force on the beams of a four-and three-beam bridge was investigated [5], for a totally and partially submerged bridge. The structure was positioned in a laboratory flume at an elevation such that the effects of the channel floor were negligible. The force acting on the bridge was calculated by measuring the pressure distributions on the girder located at the center of the flume, within a two-dimensional scheme, and disregarding the contribution of the shear stresses along the bridge surfaces. A model was built to reproduce a full-scale bridge and the results were up-scaled to the prototype by Froude similarity, assuming that the process is valid with respect to the Reynolds number, R, within the range 1×10^4 $<R>5 \times 10^6$. Denson[6] presented an experimental study of the drag, lift, and momentum coefficients for partial and total submergence of three different types of girders decks. He studied the dependence of the force coefficients on a bridge Froude number. The drag and lift coefficients are evaluated using the parameters s and l, l is the total bridge length in the flow direction and s is the total bridge thickness. Respectively, as characteristic lengths, while l^2 was used for analyzing the momentum coefficient. Although an extensive series of data are presented, no interpretation of the physical meaning of the evidenced dependencies is offered. Both literatures[5-6] assumed the parameters to be independent of R. The global drag acting on a girder bridge model was directly measured by means of a dynamometer[7]. The following effects are analyzed: (1) the number and the wheelbase of the girders; (2) the elevation of the bridge; and (3) the angle between the flow and the bridge axis on time-averaged hydrodynamic loads for the partially submerged bridge. They introduced the following functional relationship between the local hydrodynamic load coefficient and the controlling parameters[7]:

$$C_D' = \frac{1}{n} \frac{F(x)/h}{\rho v_m^2 /2} = f\left(\frac{v_m}{\sqrt{gh}}, \frac{d}{h-h_b}, \frac{d-h_b}{c}, \frac{a}{c}, \alpha, n\right)$$

$$(1)$$

where, n is number of girders; $F(x)$ is time-averaged local hydrodynamic load per unit length at (transverse) location x on the bridge; v_m is depth averaged velocity of flow; c is wheelbase of girders; a is width of the horizontal girder plates; D is drag component of average force on bridge deck; g is gravitational constant; d is water depth; h_b is elevation of bridge deck from channel floor; R is Reynolds number; α is angle between flow direction and bridge axis. Dimensionless quantities in parenthesis include: (1) the local Froude number, which is computed as the Froude number of the channel flow; (2) a measurement of the obstruction, expressed by the submergence of the bridge; and (3) the shape factor of the girders. After measuring the total dynamic load on the bridge, literatures[5-6] calculated C_D' as:

$$C_D' = D\left[nh \frac{1}{\sin\alpha} \frac{\rho V^2}{2} \int_0^b \left(\frac{v_m}{V}\right)^2 dx\right]^{-1} \quad (2)$$

Data were then interpreted by a Karman-Prandtl-type equation for rough-wall resistance as a function of the submergence of the bridge while keeping the remaining parameters as constants during the experiments.

The Federal Highway Administration (FHWA) (1995) proposed the following equation for submerged or partially submerged bridge superstructure[8]:

$$F_D = C_D^* \rho H \frac{V^2}{2} \quad (3)$$

where, F_D is time-averaged drag force per unit length of bridge; the water density, ρ is assumed equal to 1000kg/m^3; H represents the flow area per unit length obstructed by the deck (m); V is mean flow velocity (m/s); and C_D^* is drag coefficient per unit length. FHWA suggested a constant value within the range $2 \leqslant C_D^* \leqslant 2.2$.

We present and discuss the results of an experimental study of the interaction between a free surface flow and twin bridge girders. The latter is modeled as a practical cross section, with piers in the sea, since a simple geometry is more suitable for the basic understanding of the mechanisms governing the phenomenon. The practical engineering is the coastal bridge of one bank protection engineering of the coastal road.

The bank protection work of the coastal road stretches for tens of kilometers, it locates in tropical storm area, and the sea conditions is adverse and complex. The design standard is high and the project scale and investment is large. The road is made up of coastal bridge, land road, non maneuvering revetment driveway and a full line of landscape engineering. Construction requires that the coastal bridge should be coastal highway two-way bridge and the two bridges are not far from each other. Different from bridges on land, the bridge will be under great wave lift and impact because of the low bridge elevation. The research is aimed at finding an economic, reasonable, safe and reliable bridge elevation, and providing the wave load on the bridge for structure design.

Based on model test and considering different bridge elevation, elevation of sea bed, different spacing between the two bridges and the influence of the revetment, the author determined the wave force on bridge girder and pile under designed wave action. The author verified the rationality of the bridge design and made suggestions on engineering measures for the purpose of safety and economic rationality of construction investment. Besides, the author researched the functional rule and calculation method of wave force on the bridge, proofed the reliability of the model test results.

2. Test and research content

2.1　Test section

(1) Bridge elevation are $+5.5$m, $+6.0$m, $+6.5$m, $+7.0$m, elevation of sea bed are -4.0m and -10.0m. The distance from the revetment outside to the centerline of the two bridges is 19.5m[9], the distance between the two bridges is 2.0m. Elevation of the revetment front line is $+5.5$m, bottom elevation is -2.0m, the shore protection is a curved wave wall, the cross section of the bridge is shown in Figure 1. The scale of the section and measuring points are shown in Figure 2.

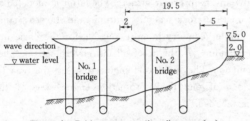

Figure 1　Bridge cross-section diagram (m)

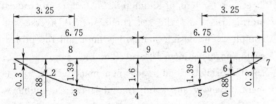

Figure 2　Section of bridge superstructure indicating example locations of pressure transducers imbedded in the model (m)

(2) Bridge elevation is $+5.5$m; elevation of sea bed is -4.0m. The distance from the revetment outside to the centerline of the two bridges is 18.75m; the distance between the two bridges is 0.5m. Elevation of the revetment front line $+5.5$m, bottom elevation is -2.0m, and the shore protection is a curved wave wall.

(3) Bridge elevation is $+5.5$m, elevation of sea bed is -10.0m, the distance between the two bridges is 2.0m. Make Contrast test in four situation: the distance between the revetment and the bridge is 0m or 1m, far from the revetment or there is no revet-

ment[9-10].

2.2　Wave conditions

Extreme high levels: $+4.33$m (return period: 50a, the elevation of Yellow Sea in 1956.)

The design high water level: $+3.16$m

The designed wave parameters are shown in Table 1.

Table 1　The designed wave parameters (return period, 50a)

Water level	$H_{1\%}$ (m)	$H_{4\%}$ (m)	$H_{5\%}$ (m)	$H_{13\%}$ (m)	H_m (m)	T (s)	L (m)
Extreme high levels $+4.33$m	3.58	3.06	2.96	2.50	1.60	5.6	45.72
	5.12	4.45	4.32	3.70	2.46	9.7	97.2
The design high water level $+3.16$m	3.44	2.94	2.85	2.42	1.57	5.6	44.82
	5.04	4.36	4.24	3.64	2.40	9.7	93.34

Considering different bridge elevation, elevation of sea bed, different spacing between the two bridges and the influence of the revetment, we can determine the wave force on bridge girder and pile under designed wave action and make observation of the wave level of the bridge and the revetment. In the research, we mainly used irregular wave, moreover the regular wave and irregular wave are compared, and JONSWAP spectrum is used in the imitation of irregular wave[9].

3. Analysis of research results

3.1　Influence of different bridge elevation on wave force

Research results of the maximum horizontal force F_{Hmax}, maximum vertical force F_{Vmax}, and horizontal force of simple pile are shown in Table 2. In Table 2, bottom elevation is -4.0m, -10.0m, the distance between the two bridges is 2.0m, the distance between the revetment outside and the inside bridge is 5.0m. It can be seen from Table 2 and Table 3 that there is no obvious regularity because of the comprehensive influence of external factors. Taking wave force and the landscape effect into consideration, the bridge elevation ($+5.5$m) is the best choice. The relation curve of the bridge elevation and F_{Vmax} is shown in Figure 3.

Table 2   The total wave force of seabed elevation (−10.0m) with revetment

DE (m)	DL (m)	WD	WF	The total wave force (kN/m)			
				F_{Hmax}	F_{SV}	F_{Vmax}	F_{SH}
+5.5	+4.33	S	$H_{1\%}=3.58m$ $T=5.6s$	58.9	350.5	504.7	47.2
+6.0				119.7	500.6	557.2	52.2
+6.5				92.2	399.0	643.7	74.2
+7.0				70.9	408.7	542.2	61.5
+5.5	+3.16		$H_{1\%}=3.44m$ $T=5.6s$	93.3	358.2	637.4	12.4
+6.0				55.7	228.4	719.2	51.8
+6.5				43.9	485.8	620.9	−3.9
+7.0				21.3	87.3	610.7	−14.7
+5.5	+5.5	SE	$H_{1\%}=5.12m$ $T=9.7s$			438.6	59.6
+5.5	+5.5		$H_{1\%}=5.12m$ $T=9.7s$			721.9	41.7

Table 3   The total wave force of seabed elevation (−4.0m) with revetment

DE (m)	DL (m)	WD	WF	The total wave force (kN/m)			
				F_{Hmax}	F_{SV}	F_{Vmax}	F_{SH}
+5.5	+4.33	S	$H_{1\%}=3.58m$ $T=5.6s$	88.6	350.5	597.1	2.0
+6.0				126.8	542.9	665.1	32.6
+6.5				78.7	599.0	599.0	78.7
+7.0				65.1	504.7	546.4	0.86
+5.5	+3.16		$H_{1\%}=3.44m$ $T=5.6s$	69.5	496.3	710.4	−29.6
+6.0				69.0	569.1	686.9	−15.4
+6.5				55.7	451.3	639.0	−31.3
+7.0				50.8	595.5	595.5	50.8
+5.5	+5.5	SE	$H_{1\%}=5.12m$ $T=9.7s$			478.1	85.3
+5.5	+5.5		$H_{1\%}=5.12m$ $T=9.7s$			749.7	76.6

Notes   1. In Table 2: DE is deck elevation; DL is design water level; WD is wave direction; WF is wave factors; F_{Hmax} is the total maximum horizontal force; F_{SV} is the simultaneous total vertical force; F_{Vmax} is the maximum total vertical force; F_{SH} is the simultaneous total horizontal force.

2. Distance between the bridge and bank protection is 5.0m (average case), distance between the two bridges is 2.0 m. Comparison of results under different seabed elevation. S represents test results, SE represents calculated results.

3. No. 1 bridge is outside, No. 2 bridge is inside.

3.2   Influence of different distance between the two bridges on wave force

The bottom elevation is −4.0m, the bridge elevation is +5.5m and the distance between the two bridges is 2.0m and 0.5m. Results of the comparative trial of wave force is shown in Table 4. In Table 4, F_V represents the resultant force of wave uplift pressure on the bridge and water pressure on the bridge girder.

Table 4   Comparison of wave force test results of different distances between the two bridges

				5.5	5.5	5.5	5.5	5.5
DE (m)				5.5	5.5	5.5	5.5	5.5
DL (m)				3.16	3.16	4.33	4.33	4.33
WH (m)				$H_{13\%}$ 2.42	$H_{13\%}$ 3.64	$H_{13\%}$ 2.50	$H_{13\%}$ 3.70	$H_{1\%}$ 5.12
T (s)				5.6	9.7	5.6	9.7	9.7
2.0m	No. 1 bridge	F_H (kN/m)	$F_{1/3}$	32.2	43.9	21.5	41.0	53.4
			$F_{1/100}$	58.4	72.8	46.7	96.9	56.2
		F_V (kN/m)	$F_{1/3}$	221	198.1	150.5	189.0	162.9
			$F_{1/100}$	371.5	432.0	290.9	365	172.8
1.5m	No. 1 bridge	F_H (kN/m)	$F_{1/3}$	22.9	45.6	18.8	33.9	47.1
			$F_{1/100}$	51.6	81.1	34.0	74.6	51.6
		F_V (kN/m)	$F_{1/3}$	211.9	244.5	152.0	249.8	164.6
			$F_{1/100}$	311.5	430.8	240.0	438.5	171.0

Notes   1. Test condition: bottom elevation -4.0m, distance between the bridge and bank protection is 5.0m.

2. F_H is the total horizontal force, F_V is the total vertical force, WH is the wave height.

Observing from Table 4: When the bridge elevation is $+5.5$m and the two working conditions have the same water level and wave action, the test results of wave force are nearly the same. But when the distance between the two bridges gets smaller, there is little space for wave energy to release, and the water is squeezed, the wave delta and fountain between the two bridges get serious; as a result, it is not good for vehicle.

3.3   Influence of different distance between the revetment and the bridge on wave force

The bottom elevation is -10.0m, the bridge elevation is $+5.5$m and the distance between the two bridges is 0.15m, 1.0m and 5.0m. Results of comparative trial of wave force are shown in Table 5. Observing from Table 5: Total vertical force decreases with the increasing distance from the bridge and bank protection under the same water level and wave force. But when the distance between the two bridges gets smaller, the role of water acts after the bridge and bank protection, and there is not enough space for the release of wave energy, leading to the role of the wave forces on bridge girders significantly increased. Therefore, the distance between the proposed bridge and bank protection should not be smaller than 1.0m.

Table 5   Comparison of wave force test results of different distances between the two bridges

				5.5	5.5	5.5
DE (m)				5.5	5.5	5.5
DL (m)				3.16	3.16	4.33
WH (m)				$H_{13\%}$ 2.42	$H_{13\%}$ 3.64	$H_{13\%}$ 2.50
T (s)				5.6	9.7	5.6
2.0m	No. 2 bridge	F_H (kN/m)	$F_{1/3}$	4.9	30.7	19.5
			$F_{1/100}$	10.0	78.5	38.0
		F_V (kN/m)	$F_{1/3}$	292.3	642.9	576.0
			$F_{1/100}$	529.0	1258.7	620.0

1.5m	No. 2 bridge	F_H (kN/m)	$F_{1/3}$	6.6	47.8	18.6
			$F_{1/100}$	19.7	93.6	41.9
		F_V (kN/m)	$F_{1/3}$	94.0	517.2	196.3
			$F_{1/100}$	197.6	545.9	315.3
1.5m	No. 2 bridge	F_H (kN/m)	$F_{1/3}$	12.3	53.4	17.0
			$F_{1/100}$	28.3	97.7	28.5
		F_V (kN/m)	$F_{1/3}$	57.2	255.4	81.2
			$F_{1/100}$	119.0	431.6	138.0

Notes 1. Test condition: bottom elevation−10.0m, distance between the bridge and bank protection is 2.0m.

2. F_H is the total horizontal force, F_V is the total vertical force, WH is the wave height.

3.4 Wave force without revetment

When the bridge is far away from the revetment and the effect of dissipating waves is good, it is workable that there is no bank protection. The theoretical calculation of the wave force is done on the basic of that there is no bank protection. In the model test, the distance between the two bridges is 2.0m, the bottom elevation is −4.0m and there is no bank protection a contrast test is done for the purpose of the accuracy of the calculation and test. Test results show that the F_{Hmax} and F_{Vmax} of No. 1 bridge are both bigger than that of No. 2 in the same deck height, the same water level and the corresponding wave force. The regularity is the same as the test of total wave force.

3.5 Comparison of design scheme and selection of the measured wave loads

After experimentally optimized design, the distance between the two bridges is 2.0m, the bridge elevation is +5.5m, the distance between the bridge and the revetment is 5.0m. As the revetment is constructed behind the bridge, at present the design departments choose the test results in Table 2 and Table 3 as a design basis.

4. Calculation and analysis of wave forces on the bridge and analysis

4.1 The wave lift forces on the bridge

In practical engineering design, "Port Terminal Structural Design Manual" is used. The pressure of wave uplift forces acting on the bottom of the bridge girder can be calculated in the following formula[11]:

$$p = \beta\gamma(\eta - h_1) \qquad (4)$$

where: γ is the gravity of water, kN/m^3; η is the highness of waveform above still water surface, m; h_1 is the heights from still water surface to the bottom of the bridge, m; β is a coefficient of pressure.

When $\lambda < 10m$, then $\beta = 1.5$; if the panel width is a bit greater or bank slope is connected with the panel, $\beta = 2.0$. λ is the length of the panel along the wave propagation direction, m.

$$\eta = \frac{H}{2}\cos\frac{2\pi x}{L} + \frac{\pi H^2}{4L}\left(1 + \frac{3}{2\sinh^2\frac{2\pi d}{L}}\right)\coth\frac{2\pi d}{L}\cos\frac{4\pi d}{L} \qquad (5)$$

Where, x is the horizontal distance from wave crest, m; d is depth of water, m; L is wave length, m; H is the highness of incident wave, m.

The lateral pressure on the bridge panel includes hydrostatic pressure and hydrodynamic pressure. There is no hydrostatic pressure in the wave surface, but on the still water surface[10]:

$$p_s = \gamma\eta \qquad (6)$$

Hydrodynamic pressure is:

$$p_d = 1.7\frac{\gamma}{2g}u^2 \qquad (7)$$

Where, u is orbital motion of water particles of the horizontal velocity.

Orbital motion of water particles of the horizontal velocity on the wave crest is[11]:

$$u = \frac{\pi d}{T}\coth\frac{2\pi d}{L} = \frac{H}{2}\sqrt{\frac{2\pi g}{L}\coth\frac{2\pi d}{L}} \qquad (8)$$

If there is a certain height from the crest to the bridge girder, then the wave will be broken on the bridge girder and produces downward hydrodynamic pressure, When the certain height is y_0, broken water will be at a distance x_B from the bridge girder.

$$x_B = \frac{U\sqrt{2gy_0}}{g} \qquad (9)$$

Where, $U = 0.75C + u$, U is velocity of water particles when wave crest is broken; C is the speed of wave propagation, m/s.

The maximum downward hydrodynamic pressure p_{max} at x_B is:

$$p_{max} = 1.7\frac{\gamma}{2g}\left[U^2 + \left(\frac{gx_B}{U}\right)^2\right]\cos(90° - \alpha) \quad (10)$$

Included angle between broken water and the deck is α.

$$\tan\alpha = gx_B/U^2 \qquad (11)$$

Pressure distribution can be approximately taken as the isosceles triangle graph, that means the pressure will be 0 at the deck edge and $2x_B$, but p_{max} at x_B.

4.2 Comparison between theoretical calculation with the measured results under the condition of no revetment

Wave force theory is calculated for single-bridge with no bank protection behind. Table 6 shows the theoretical calculation comparison with the measured results when the bridge elevation is $+5.5$m. Data in Table 6 indicate that under the conditions in the absence of bank revetment, the total vertical force of the test point and the calculation results agree well. It verifies the accuracy of test results under the conditions in the absence of bank revetment.

Table 6   Theoretical calculation comparison with measured wave force (No. 1 bridge)

WD	DL (m)	WF	Vertical floating force (kN/m)			Horizontal force (kN/m)	
			C	T		C	T
			N	N	W	N	N
S	+4.33	$H_{1\%} = 3.58$m $T = 5.6$s	360.29	385.4	597.1	41.06	70.9
	+3.16	$H_{1\%} = 3.44$m $T = 5.6$s	473.97	483.1	710.4	21.62	68.9
SE	+4.33	$H_{1\%} = 5.12$m $T = 9.7$s	478.10	419.8	573.6	85.34	65.4
	+3.16	$H_{1\%} = 5.04$m $T = 9.7$s	749.66	620.0	770.0	76.56	139.4
SE	+4.33	$H_{1\%} = 3.75$m $T = 9.7$s	501.45		467.8	43.21	
	+3.16	$H_{1\%} = 3.7$m $T = 9.7$s	712.07			57.36	

Notes   1. C is calculate results; T is test results.

2. N is no bank protection; W is with bank protection.

3. The remaining symbols as described above.

5. Conclusions

(1) The results show that the bridge girder form is beneficial to alleviate the wave action, and it is feasible that deck elevation is $+5.5$m. It not only reduced the engineering investment and took into account the landscape effect, but also avoided the visual pollution of the tourists. Besides, it has created a precedent for the construction of a low bridge on the sea.

(2) The bridge should be away from the bank protection, and avoid enclosed type, otherwise wave energy will not have enough space to release, and wave forces the bridge suffered will be too much, the structure is very unfavorable, both of them are not good for the structure. The distance between the proposed bridge and bank protection should not be too small to make wave energy leakage, if necessary, structures for eliminating waves can be used, or set

the pressure relief holes and other measures to alleviate the force of waves on the bridge.

(3) Distance between the two bridges should not be too small, we recommend it to be 2.0m, which is appropriate to alleviate the effects of waves on the bridge is appropriate. If the distances between the two bridges are smaller or close to 0, release space for the wave energy is too small, water is squeezed and there will be a clear fountain phenomenon between the two bridges not conducive to vehicular traffic.

(4) Test results and theoretical calculation results agree very well under the conditions in the absence of bank revetment, and it verifies the accuracy and reliability of test results. In this paper, methods of calculating wave forces on the bridge are available for estimating wave forces on the bridge with no bank protection, and it has accumulated much information to improve and perfect the existing formula and supplement the relevant specifications.

(5) Experimentally optimized design are: distance between the two bridges is 2.0m, deck height is +5.5m, and the distance between bridge and bank protection is 5.0m.

Acknowledgment

This research is part of the ongoing Project of Shandong Province Higher Educational Science and Technology Program. Authors wish to thank Shandong Provincial Education Department.

References

[1] Stefano Malavasi, Alberto Guadagnini. Hydrodynamic Loading on River Bridges. *Journal of Hydraulic Engineering*, 2003 (37): 854-861.

[2] Arntsen O. Disturbances, lift, and drag forces due to translation of a horizontal circular cylinder in stratified water. *Exp. Fluids*, 1996, 21 (5): 387-400.

[3] Hoyt J. W., Sellin R. H. J. A comparison of tracer and PIV results in visualizing water flow around a cylinder close to the free surface. *Exp. Fluids*, 2000, 28 (3): 261-265.

[4] Zhu Q., Lin J. C., Unal M. F., Rockwell D. Motion of a cylinder adjacent to a free-surface: flow patterns and loading. *Exp. Fluids*, 2000, 28 (6): 559-575.

[5] Tainsh J. Investigation of forces on submerged bridge beams. *Rep. No.* 108, *Dept. of Public Works*, New South Wales Univ., Sydney, Australia, 1965: 1-25.

[6] Denson K. H. Steady-state drag, lift, and rolling moment coefficients for inundated inland bridges. *Rep. No. MSHD-RD-82-077*, *reproduced by National Technical Information Service*, Springfield, Virg., 1982: 1-23.

[7] Naudascher E., Medlarz H. J. Hydrodynamic loading and backwater effect of partially submerged bridges. *J. Hydraul. Res.*, 1983, 21 (3): 213-232.

[8] Federal Highway Administration (FHWA). *Stream stability at highway structures*. Rep. No. FHWA-HI-96-032, FHWA, Washington, D. C., 1995: 56-59.

[9] Tong Desheng, Xie Shanwen, Pang Tiezheng, Yang Xiaobin. Experiment and Research in Wave Forces Acting upon Bridge between Naval Berth and Yanwu Road in Phase III of Huandao Highway in Xiamen. *China Harbour Engineering*, China, 2003 (6): 41-44.

[10] Ministry of Transport of the PRC. *Wave model test regulation*, JTJ/T 234—2001, China Communications Press, Beijing, 2002: 7-19.

[11] The First Harbour Engineering Investigation and Design Institute of the Ministry of Communications. *Handbook for design of sea Harbours*. China Communications Press, Beijing, 2001.

Similarity Law of Physical Model on Air Entrainment to Alleviate Cavitations

Zheng Shuangling   Li Zhongyi   Ma Jiming

State Key Laboratory of Hydro-science and Engineering,
Tsinghua University, Beijing, 100084

Abstract: There are water flow and air flow in air-entrained hydraulic model. But the similarity of air flow is often ignored in the normal-pressure model, which makes the large difference of the aeration discharge in model-prototype comparison test. Therefore, the subatmospheric experiment is conducted to make the air flow fit to the similarity law, but it's very difficult to observe the test. For the similarity of discharge and concentration of entrained-air, the model experiment can ignore the similarity of aeration hole in normal pressure, and the aeration concentration can be calculated by the formula in the paper. In the cavitation model test with aeration device the cavitation and aeration discharge can be fitted to the similarity law by changing the area of air hole. And the aeration concentration is almost similar in subpressure zone where it is easy to cavitate. To satisfy similar conditions of air-entrained flow, it's insignificant to enlarge the model scale alone.

Key words: similarity law; aeration; cavitation; scale effect; model experiment

1. Introduction

In the high-head hydraulic structures, aeration devices, such as bucket and drop sill, are used to prevent cavitation damage by entraining air in bottom flow. The physical model without aeration device can satisfy the similar conditions of flow, i. e., the distribution of pressure and velocity, if the geometry and gravity fit to the similarity, and Reynolds number is in drag square zone. Since the 1960s, aeration devices have been used widely in hydraulic structures. Therefore, there are two kinds of flow, water flow and air flow, which interact and effect on each other. Although there are many studies in the literature of air entrainment to alleviate cavitations, few researches are taken into account the difference of air discharge of prototype and model. At the present aerated model experiment is still conducted in the atmospheric air pressure.

The unit discharge of air-carrying of aerated jet flow is expressed by the following empirical equation[1]:

$$Q_a = KV_0L \qquad (1)$$

In which V_0 is the velocity of flow, L is the cavity length, and K is a nondimensional experimental coefficient. The cavity length, L is difficult to be measured because of the low velocity of air. The values of K obtained in laboratory works are 0.01-0.042[1]. In the spillway of Guri

hydraulic engineering, the K value presented by prototype is 0.073[1], which is much bigger than the experimental values. Maybe the dissimilarity of model results in scale effect. How to avoid the scale effect? Some people suggested enlarging the model scale, which are very costly and still insufficient to overcome the effect of scale. It's an important research task for the high-speed flow by simulation technique of aeration to alleviate cavitations.

2. Effect of cavity subpressure and ambient pressure on cavity length

2.1   Effect of cavity subpressure on cavity length

The jet flow of unit length is analyzed as shown in Figure 1. The direct of out-jet flow is x-axis, and

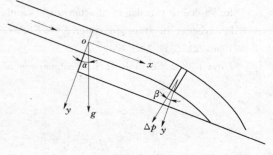

Figure 1   Sketch of jet flow with aeration device

the y-axis refers to the vertical line to x-axis. α is the angle between gravity and y-axis, and β is the angle between acting direction of subpressure and y-axis.

Based on Figure 1, the equilibrium equation can reduce as follows:

$$\frac{d^2 y}{dt^2} = g\cos\alpha + \frac{\Delta p}{b\rho}\cos\beta \qquad (2)$$

$$\frac{d^2 x}{dt^2} = g\sin\alpha - \frac{\Delta p}{b\rho}\sin\beta \qquad (3)$$

In which, b is the jet width, ρ is the density of water, and Δp, the cavity subpressure, is the difference between the atmospheric air pressure and lower slot cavity air pressure.

For the aeration slot (bucket), $\beta = 0 \rightarrow \beta$ in the cavity, and the variety is very low, i.e., $\beta \approx 0$, so the solution of the equation above is as follows:

$$x = V_0 \sqrt{\frac{2y}{g\cos\alpha + \dfrac{\Delta p}{\rho b}} + \frac{yg\sin\alpha}{g\cos\alpha + \dfrac{\Delta p}{\rho b}}} \qquad (4)$$

Equation (4), the trace equation of centroid of jet flow, is used in both prototype and model experiment. The cavity length is the straight distance between the points of intersection of the centroid trace and trench floor. The space under the jet hemline is cavity. The hemline often breaks up without steady boundary because of the turbulence aeration. Therefore, the cavity length is often calculated by the jet intersection point (where the pressure is the largest). The equation above indicates that greater the cavity subpressure, lower the trajectory distance and the capacity of air carrying. There is still no research about the effect of subpressure on length of cavity.

2. 2　Relation between subpressure and cavity length

The discharge of entrained-air, Q_a, can be calculated in terms of the cavity subpressure, as follows:

$$Q_a = \mu A \sqrt{\frac{2\Delta p}{\rho_a}} \qquad (5)$$

where, A is the control area of aeration hole; and μ is discharge coefficient. Q_a in Equation (1) is the air carrying of flow, and Q_a in Equation (5) is the air demand of aeration pipe. The pressure can reach a balance if air carrying discharge is equal to the air demand discharge. So by Equation (1) and Equation (5) the cavity length, L, is expressed as follows:

$$L = \frac{\mu A}{K V_0} \sqrt{\frac{2\Delta p}{\rho_a}} \qquad (6)$$

L and Δp can be obtained by simultaneous Equation (4) and Equation (6).

2. 3　Effect of ambient pressure on cavity subpressure

For air-entrained model experiment with two kinds of different density of air, by Bernoulli's Equation their relation of pressure of cavities is as follows:

$$\frac{\Delta p_m}{\Delta p_a} = \frac{\rho_m v_m^2}{\rho_a v_a^2} \qquad (7)$$

where, ρ is air density in the model, and Δp is cavity subpressure. Subscript a and m indicate the model values of atmospheric air pressure and similar ambient pressure, respectively. L_r is the model scale. By gas equation of state:

$$\frac{\rho_m}{\rho_a} = \frac{1}{L_r} \qquad (8)$$

$$\frac{\Delta p_m}{\Delta p_a} = \frac{v_m^2}{L_r v_a^2} \qquad (9)$$

By Equation (5), for the similarity of cavity subpressure in atmospheric pressure model, namely $\Delta p_a = \Delta p_m$, discharge of air demand of aeration pipe can be expressed as follows:

$$Q_a = Q_m / \sqrt{L_r} \qquad (10)$$

In atmospheric pressure model, the air-supplying discharge is not enough to keep the supply-demand balance, which will make the subpressure increase, and cavity length and air-carrying decrease. At last the pressure will reach a new balance. Now the discharge of aeration is as follows:

$$Q_m / \sqrt{L_r} < Q_a < Q_m \qquad (11)$$

For the similarity of discharge of aeration in normal-pressure model, i.e., $\Delta Q_a = \Delta Q_m$, the pressure of cavity should be enlarged as follows by the Equation (5):

$$\Delta p_a = L_r \Delta p_m \qquad (12)$$

But when the cavity subpressure increases, the cavity length and the air-carrying will decrease, thus the air-supplying is larger than the air-carrying. At last the pressure will reach a new balance, as expected.

$$\Delta p_m < \Delta p_a < L_r \Delta p_m \qquad (13)$$

Figure 2 is the comparison of jet trajectory distances in atmospheric pressure and subpressure in the middle holes of three gorges project [2]. From Figure 2, the cavity length in subpressure is much larger than that in atmospheric pressure. It also shows that air discharge in subpressure is also much larger than that in atmospheric pressure. Therefore, Figure 2 explains that the obvious difference of aeration of prototype and model is the result that the ambient pressure isn't reduced by gravity similarity law. This shows that it's ineffective to enlarge the model scale only.

If Δp_m and ρ_m ($=\rho_a/L_r$) replace Δp and ρ_a of Equation (6), respectively, the new simultaneous Equation (4) and Equation (6) can be solved to obtain the similar subpressure of model, i. e., Δp_m. The wind velocity of prototype can be calculated by the pressure of model. If the wind velocity is more than 100m/s, aeration area should be increased. Otherwise, if it is lower than 60m/s, the area should be decreased. By the design code, reasonable suggestion is proposed.

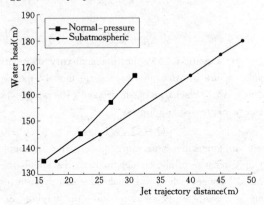

Figure 2　Comparison of jet trajectory distances in different pressure in model test of three gorges project

2.4　Correction of aeration discharge in normal-pressure model

The aerated flow is the result of the interaction of water and air. The main research parameters of scale model should be aeration concentration, pressure, velocity distribution and reasonable size of aeration hole. And it should satisfy the main similarity law of water and air flow. For the water flow, it's important to satisfy the similarity of geometry and gravity, which is also significant to air flow. But the size and shape of air hole depend on the test. The boundary of cavity of the flow is kinematic, and to satisfy the similarity of flow boundary and gravity, it must fit to the similarity law of the boundary pressure. So the ambient pressure of model should reduce by the scale of gravity similarity. And the model test

should be made in the leakproof depression tank, but it is impossible because the test equipments in water probably conduce to the flow corrosion. The similarity of aeration discharge is the precondition of similarity of air concentration. Thus, for the similarity of aeration discharge and flow conditions, the effective area of air hole, $\mu_m A_m$, can be increased $\sqrt{L_r}$ times in normal pressure. Now although the size of air hole isn't similar, the main similarity laws are satisfied at the result of the similar air discharge. At last, the aeration concentration in the water should be corrected further.

3. Similarity and correction of aeration concentration

By the corrected method above, the main parameters such as air discharge, cavity pressure, jet trajectory distance, pressure distribution and so on, are similar, but the aeration concentration is not similar. The aeration concentration, percent of air unit discharge, means the ratio of air discharge to air-water mixture in unit discharge, which is a nondimensional coefficient and can be used directly in prototype. Although the scale effect has been studied, it is mainly about the problem such as the water viscosity and surface tension. The aeration flow is two-phase, in which water is incompressible and air is compressible. The discharge of air bubble in moving changes by the pressure. Because the air pressure in model is not reduced by the pressure similarity law, the air bubble discharge and air concentration are not similar with those of prototype. When the air bubble pressure is equal to p_a (the atmospheric pressure) and p, its volume are $V(0)$ and $V(p)$, respectively, which ignore the precipitation and dissolution of air in water. By gas equation of state,

$$V(0)=\frac{p_a+p}{p_a}V(p)=(1+y)V(p) \qquad (14)$$

where, $y=p/p_a$. If $p=0$, aeration concentration is,

$$C(0)=\frac{V(0)}{V_w+V(0)} \qquad (15)$$

$$\frac{V(0)}{V_w}=\frac{C(0)}{1-C(0)} \qquad (16)$$

where, V_w is the volume of water in water-air mixture. When the pressure of mixture is p, the aeration concentration is that,

$$C(p)=\frac{V(p)}{V_w+V(p)} \qquad (17)$$

$$\frac{V(p)}{V_w}=\frac{C(p)}{1-C(p)} \tag{18}$$

By Equation (14), Equation (16) and Equation (18), aeration concentration at pressure p is as follows:

$$C(p)=\frac{C(0)}{1+y[1-C(0)]} \tag{19}$$

The equation above is a universal formula that calculates aeration concentration changing with pressure. The subscript m and p means model and prototype, respectively. Namely,

$$C_m(p)=\frac{C_m(0)}{1+y_m[1-C_m(0)]} \tag{20}$$

or,

$$C_m(0)=\frac{C_m(p)(1+y_m)}{1+y_m C_m(p)} \tag{21}$$

$$C_p(p)=\frac{C_p(0)}{1+y_p[1-C_p(0)]} \tag{22}$$

In normal-pressure model $y_m \neq y_p$. Just when $y_m = 0$, $y_p = 0$, the aeration is similar, and the aeration concentration of model is equal to that of prototype, i. e. , $C_m(0)=C_p(0)$. The pressure in sensor is very low, $y_m \approx 0$, so Equation (22) can be expressed as follows,

$$C_p(p)=\frac{C_m(0)}{1+y_p[1-C_m(0)]} \tag{23}$$

Table 1 shows the aeration concentration comparison of normal-pressure model ($L_r = 30$) and prototype for the same air-water mixture under different pressure. From Table 1, the aeration concentration in model is different with that in prototype because of the dissimilar air pressure. So the aeration concentration in model can't be used directly in prototype, which should be calculated and conversed by the pressure attained in model and prototype.

Table 1 Comparison of aeration density

$p(kPa)$	$C_m(p)$	$C_m(0)$	$C_p(p)$
50	20	20. 26	14. 48
	10	10. 15	7. 01
	5	5. 06	3. 45
100	20	20. 53	11. 44
	10	10. 3	5. 43
	5	5. 16	2. 65
150	20	20. 79	9. 5
	10	10. 45	4. 46
	5	5. 24	2. 16

4. Similarity of cavitation model

Since the 1960s, aeration device are widely used to prevent corrosion and alleviate cavitation damage of structures in hydraulic engineering. The subatmospheric hydraulic model can present the effect of aeration. However, the test technique can't advance to put forward more reasonable cavitation similarity law. In the cavitation model test without aeration device, the pressure should reduce by the principle of equal cavitation value. If it has aerator device, the model should satisfy the similarity of air discharge at the same time. The ambient pressure p_m is equal to p_a/L_r, and to satisfy the cavitation similarity, the pressure p_{mk} should be that [2],

$$p_{mk}=p_a\left(\frac{1}{L_r}+\frac{L_r-1}{L_r}\times\frac{p_v}{p_a}\right) \tag{24}$$

where, p_v is the saturation vapor pressure of water, and ρ_{mk} is the air density, which is expressed as follows.

$$\rho_{mk}=\rho_a\left(\frac{1}{L_r}+\frac{L_r-1}{L_r}\times\frac{p_v}{p_a}\right) \tag{25}$$

Eq. (24) shows that the vacuum degree by cavity similarity is different with that by ambient pressure similarity, and $p_{mk}>p_m$. If the vacuum degree is similar by ambient pressure similarity, the air discharge is similar, but the cavitation is not similar and less than that of prototype. Cavitation occurs in prototype but not in model. Under the condition of similar vacuum degree by cavitation similarity, the cavitation occurs at the result of less air discharge. So it should satisfy the cavitation and aeration similarity for cavitation model test. Thus, by the method above, the area of air hole can be adjusted to make the air discharge and subpressure similar in the pressure equal to p_{mk}. $\mu_{mk} A_{mk}$ refers to the effective area of air hole adjusted, and the aeration discharge in cavitation similarity is,

$$Q_{mk}=\mu_{mk}A_{mk}\sqrt{\frac{2\Delta p_m}{\rho_{mk}}} \tag{26}$$

The aeration discharge of model in pressure similarity is,

$$Q_m=\mu_m A_m\sqrt{\frac{2\Delta p_m}{\rho_m}} \tag{27}$$

where, $\rho_m=\rho_a/L_r$, by Equation (26) and Equation (27),

$$\frac{Q_{mk}}{Q_m}=\frac{\mu_{mk}A_{mk}}{\mu_m A_m}\times\sqrt{\frac{\rho_m}{\rho_{mk}}}=\frac{\mu_{mk}A_{mk}}{\mu_m A_m\sqrt{1+(L_r-1)p_v/p_a}} \tag{28}$$

If $Q_{mk} = Q_m$, to satisfy the similarity of the cavitation and aeration, the area of aeration hole is,

$$A_{mk} = \sqrt{1+(L_r-1)p_v/p_a} \mu_m A_m / \mu_{mk} \qquad (29)$$

Although the air discharge fit to similarity law by adjusting the size of air hole in conditions of cavitation similarity, the aeration concentration is dissimilar, yet. The ambient pressure in prototype is larger than that in model, so the aeration concentration is dissimilar except that at the place of zero pressure $[C_m(0) = C_p(0)]$. If the aeration concentration is similar at the low pressure where it is easy to cavitate, the cavitation value of model is equal to that of prototype.

5. Conclusion

(1) In hydraulic out-let structures with aerator device there are water flow and air flow, which interact and effect on each other. And the physical model should satisfy the similarity of geometry and gravity.

(2) In the normal-pressure model, air flow doesn't satisfy the gravity similarity, and cavity length doesn't satisfy the geometrical similarity, yet. For the aeration similarity, it can enlarge the area of air hole $\sqrt{L_r}$ times, which is the precondition of the similarity of aeration concentration.

(3) If the ambient pressure fit to the similarity law, the discharge and concentration of the aeration are satisfy the similar conditions, too. In the normal-pressure model, the aeration concentration is dissimilar except that at the place of zero pressure, which can be obtained by correcting the value from the model test according to the equations.

(4) In the cavitation model, the air discharge is dissimilar, and the area of air hole can be enlarged by

Equation (29). The aeration concentration is similar at the subpressure zone, where the cavitatioin is easy to occur.

References

[1] Pan S B, Shao Y Y. Design and application of aeration equipment to alleviate cavitation [A]. Hydrochina zhongnan engineering corporation. Special subject of design code for spillway [C]. Beijing: China WaterPower Press. 1990. (in Chinese)

[2] Nanjing Hydraulic Research Institute. China Institute of Water Resources and Hydropower Research. Hydraulic model test (2nd Edit) [M]. 1984. (in Chinese)

[3] LI Z Y. Study on aeration to alleviate cavitation of model experiment in normal and atmospheric pressure of TGP's deep hole [R]. 1997. (in Chinese)

[4] Yang Y S. Research progress of aeration to alleviate cavitation of high-head outlet structure [A]. A volume of papers about outlet engineering and high speed [C]. Chengdu: Chengdu University of Science and Technology Press, 1994. (in Chinese)

[5] A. Marcano, N. Castillejo. Model Prototype Comparsion of Aeration Devices of Guri Dam Spillway [A]. Symposium on Scale Effects in Modelling Hydraulic Structures Esslingen Neckar [C]. Germany, September 3-6, 1984.

Application of Series Model Extrapolating Method in Research of Local Scour Around the Closure

Chunhua Wang   Yufang Han   Zhichang Chen

State Key Laboratory of Hydrology-Water Resources and Hydraulic Engineering,
Nanjing Hydraulic Research Institute, Nanjing, 210029

Abstract: The normal series model extension method aims to simulate the local scour around the buildings, and it is widely applied in the research of changes in local topography around the hydraulic structures such as spur dikes, bridge piers, and pile foundation. Based on the simulation of the whole scour morphology around the closure and the north dike in the closure process of the Qingcaosha Reservoir Project, the present study discusses the application of normal series model extrapolating method in the research of local scour in such a large-scale project. The results show that the method can be used to simulate the scope and depth of the local scour around the closure and to put forward effective protection measures for project design and construction.

Key words: the normal series model; the model extension method; scale; local scour; scouring range

1. Introduction

The local scour around the construction usually depends on its surrounding and complicated vortex system and the circulation structure. This flow phenomenon is closely related to the vertical flow structure. For the simulation of the three-dimensional water-sand characteristics correctly[1], it is difficult to choose the model scale and the model sand. Abundant experience has been accumulated in the use of the normal series model extension method in the local scour around the buildings, and the method is often used in the local scour induced by bridge pier, caisson, spur and dams projects[2]. Early in the 1960s, Yuqing Sha and Soviet Zrelov studied siltation in reservoir and local scour downstream buildings respectively, and obtained satisfactory results[3]. Dechun Jin employed the series model test with light sand in the research of local scour depth around a bridge caisson, and discussed the theoretical basis of the series model test[2]. Hao Lu adopted prototype sand, non-prototype natural sand and plastic light sand to study the local scour and protection tests of many bridge piers, caissons, cofferdam and the construction platform legs[4]. Xiaodong Zhao studied the scour and protection about the spur dikes using prototype sand and light model sand[5]. Xijun Han used prototype sand in the local scour tests of the trestle pile foundation. Practice shows that satisfactory results can be obtained in the research of the local scour around the buildings with

the series model test method.

In the closure process of the Qingcaosha Reservoir Project, local scour is caused around the closure and the outside of the north dike because of complex flow structure during flood and ebb. Therefore, the difficulty is the choice of a rational model scale in the study of local scour around the closure by use of the series model test method.

2. The normal series model extension method

The geometric scale of the normal movable-bed model is generally large because of the restriction of the engineering scale and the testing ground. At the same time, the model tests should satisfy the similarity of sediment suspension and starting, which requires the model sand has small density and small particle size. However, fine model sand can bring some problems such as flocculation and adhesive force, which makes it difficult to meet the similarity of sand starting. Thus, the model sand with larger density and larger particle size is required, and then a larger model is made, resulting in large flow velocity in the model. In fact, it is difficult to meet these conditions simultaneously, and then difficult to choose the model sand and the scale for the design of a workable normal model which is strictly in compliance with the similarity law. The series model test is the method for the solution of such problems. By this method, test results from a series of models with different scales can be extrapolated and then the deviation caused by non-similarity of sediment movement can be reduced.

The geometric scale of the normal model λ_{H_0} is designed according to the similar conditions. When the model fully meets the similar conditions, $\lambda_{hs} = \lambda_H = \lambda_{H_0}$ (λ_{hs} is the depth scale of scour or siltation, λ_H is the geometric scale of the normal model); when the model deviates the similar conditions, $\lambda_{hs} \neq \lambda_H \neq \lambda_{H_0}$. λ_{hs} deviating λ_H is caused by λ_H deviating λ_{H_0}. The relation can be given as follows:

$$\frac{\lambda_{hs}}{\lambda_H} = \left(\frac{\lambda_{H_0}}{\lambda_H}\right)^{\alpha}$$

or

$$h_{sp} = h_{sm}\lambda_L \left(\frac{\lambda_{L_0}}{\lambda_L}\right)^{\alpha} \qquad (1)$$

where, h_{sp} is the scour depth in the prototype, and h_{sm} is the scour depth in the model. The results can be drawn a line in the log-log coordinates where h_{sm} is the longitudinal axis and λ_H is the transverse axis. When $\lambda_H = \lambda_{H_0}$, the corresponding scour depth in the model h_{sp} is the depth in the prototype.

According to λ_H and h_{sp} of two models with two different scales, the slope of the extending line can be directly obtained as follows:

$$\alpha = \left[\lg(h_{sm_2}\lambda_{H_2}) - \lg(h_{sm_1}\lambda_{H_1})\right]/(\lg\lambda_{H_2} - \lg\lambda_{H_1}) \quad (2)$$

Then the scour depth in the prototype can be calculated with Eq. (1).

In the series of model tests, only two different model scales are needed to obtain scour depth or siltation depth in prototype. More models are needed just for the correction of the errors and more reasonable results in the movable-bed model tests.

In the series of model extension tests with prototype sand, it is equivalent to extending to the prototype when $\lambda_{H_0} = 1$. This method is simple and easy to obtain satisfactory results. When model sand is adopted ($\lambda_{H_0} \neq 1$), a scale $\lambda_H = \lambda_d\lambda_{\gamma_s - \gamma}^{3/2}$ will be obtained according to similar conditions of sediment movement in the normal model. Then, the similar model test results can be obtained by extending the model results, and the prototype results can be calculated.

3. Application of the series model extrapolating method in research of local scour around the closure

3. 1 Situation of the closure

The Qingcaosha Reservoir, situated in the water areas of the Central Sand, the Qingcaosha and the North Kobuchi which lie respectively to the north and the west sides of the Changxing Island lying off the upper reaches of the south and the north channels of the Yangtze Estuary, has a total area of 66.26km². In the project, there are two closures, one at the north dyke and the other at the east dyke. In the present study, the experiment is focused on the local scour around the east dyke in the protective period with the north dyke having been closed. The closure in the east dyke is 900m wide and is a compound section. The elevation is -4m for the 800m width in the middle part, and 0m for the two 50m sides.

3. 2 Design of the model

The difficulty of the normal series model for simulation of the local scour near the closure of the Qingcaosha Reservoir is that the project scale is large and the local scour is caused by many factors, such as the change of flow structure near the building caused by the closure, the change of the north channel flow caused by the whole reservoir engineering and the closure tide properties different from the north channel caused by the reservoir tidal prism. So, for reasonably simulation of the local scour around the closure, the selection of scale is the key to the design of series model.

In order to design the reasonable series normal models to study the local scouring problems around the closure, all factors for the selection of the model scale should be considered, such as the similarity of flow, sediment movement and the model sand erosion morphology. In the model tests, in order to satisfy the similarity of sediment starting, wood flour is often selected as the light model sand to calculate the eligible scale λ_{H_0} of the normal model, and then other scales about the series normal model are determined.

In the local scouring tests, the model sand $d_{50} = 0.67$mm, $\gamma_s = 1.15$t/m³, and the prototype sand $d_{50} = 0.15$mm, $\gamma_s = 2.65$t/m³. For cohesionless sediment, the incipient velocity should satisfy $v_c = k\left(\frac{\gamma_s - \gamma}{\lambda}gD\right)^{1/2}\left(\frac{H}{D}\right)^{1/6}$. Comprehensive considering the similar conditions of the flow and sediment movement, we can obtain the eligible scale of the light sand normal model $\lambda_{H_0} = \lambda_D\lambda_{(\gamma_s - \gamma)/\gamma}^{3/2}$, that is $\lambda_{H_0} = 8.17$.

For reasonable simulation of the local scour morphology, the scour or siltation of model sand should be similar to the prototype. Therefore, the scale of the series model of the light sand is $50 \leqslant \lambda_H \leqslant 400$. According to the dimension of the model testing ground with a length of 70m and a width of 30m, two geometric scales of the normal models are selected as $\lambda_H = 150$ and $\lambda_H = 300$ respectively. The layout of the two scales of the models are shown in Figure 1. The thick line shows the model range for 1 : 300, while the thinner one for 1 : 150. Part of the river bed of the north channel near the reservoir is simulated in the local model with $\lambda = 300$. According to the situation of the model ground, the region of the Qingcaosha Reservoir is divided into two parts for simulation: the main part is connected with the supplement reservoir through a pipeline (as shown in Figure 2), which can effectively simulate the whole region of the reservoir. The hydrological verified data of the $\lambda = 150$ model is provided by the $\lambda = 300$ model, ensuring the similarity of the whole scour and siltation morphology around the closure.

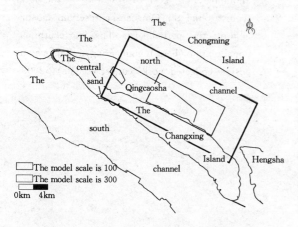

Figure 1　Layout of the two scale normal models

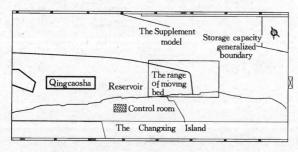

Figure 2　Layout of the normal model with $\lambda = 300$

3.3　Test results

Movable bed scouring tests are carried out in the two scale models after flow verification. According to

the previous introducing method to count the data obtained in the local scouring tests, the average of the extension curve coefficient can be obtained $\alpha = 124.0$, then the extension formula is:

$$h_{sp} = h_{sm} \lambda_H \left(\frac{\lambda_{H_0}}{\lambda_H} \right)^{0.124} \qquad (3)$$

In this way, according to the extension results of the series model tests, we can calculate the data of the local scouring test and then obtain the local scouring scope and depth of the prototype.

Figure 3 shows the largest scouring depth around the inside of the closure obtained form the extension results of the series model tests under a set of experimental conditions. Figure 4 shows the scour depth and morphology of the inside and the outside of the closure. These results have been submitted to the design department for the successful closure in the Qingcaosha Reservoir Project.

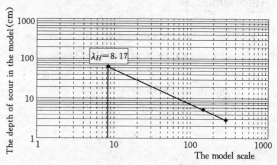

Figure 3　The results of the series model extension tests

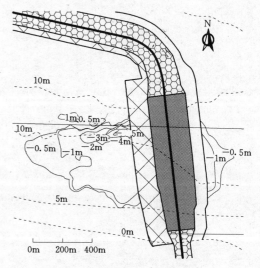

Figure 4　The scour morphology of the inside and outside of the closure

4. Conclusion

The series model extension method has some theoretical basises in the research of local scour around buildings, and it is feasible after a number of engineering practice. In the research of local scour around the Qingcaosha Reservoir closure and the dyke body, the reasonable selection of model scale and test condition not only achieves scouring shape around the closure and in a certain neighboring region, but also satifies the similarity of 3D flow structure for local scour, ensuring that reasonable scour depth can be obtained. The results show that the method can be used to simulate the scope and depth of local scour around the closure and put forward effective protection measures for project design and construction.

References

[1] Yufang Han, Xiaofeng Luo. Model test in research of local scour around the Qingcaosha reservoir closure [R]. Nanjing: Nanjing Hydraulic Research Institute, 2008, 09.

[2] Changhua Li, Dechun Jin. River model test [M]. Beijing: The People Traffic Publishing House, 1981.

[3] Yuqing Sha. Introduction of sediment [M]. Beijing: The Chinese Industry Publishing House, 1965.

[4] Hao Lu. The series model extending method and the choice of its test sand [J]. Sediment research, 1987 (1): 10-18.

[5] Xiaodong Zhao, Lihua Wu, Zhichang Chen. The test research of scour and protection about the head of the spur dike in the Yangtze estuary channel regulation [J]. The fourteenth national hydrodynamics symposium collected works, 2000.

[6] Xijun Han, Zude Cao, Shusen Yang. The series model extrapolating method in research of local scour around the building in the silty seabed [J]. The marine journal, 2007, 1 (1).

[7] Yufang Han, Mingxia Qian, Xuelan Wang, Shoubing Yu. Test in research of the local scour around the Qingcaosha reservoir closure in protective period [J]. The fourteenth national marine (bank) engineering academic symposium collected works (under volume), 2009.

Distorted Model Scale and Appling of Unsteady Flow in Conduit System of Hydropower Station

Zhang Shaochun

Hydrochina Kunming Engineering Corporation, *Kunming*, 650033

Abstract: This paper has derived the distorted model scale of unsteady flow in conduit system of hydropower station and applied it to the model test study of the waterway system of a hydropower station. The distorted model scale has proved correct by prototype observation of the hydropower station.

Key words: model scale; distorted model; unsteady flow; hydropower station

1. Introduction

The calculation of unsteady flow in diversion of station is well-developed, however, it is still necessary to conduct physical model test in case that the shape of the surge tank is complicated or special patterns of flow such as water spray and air inlet should be known.

In the study of unsteady flow model test of the waterway system which consists of a power tunnel, a surge tank and penstocks of a hydropower station, the power tunnel and the surge chamber are often considered as a system and the penstocks as another system. When the surge in the surge tank and the water hammer of penstocks are simulated at the same time on the same physical model, because both power tunnel and penstocks are very long, if no distorted model is used, the length requirement can be satisfied but the cross section of the power tunnel or the penstock should be too small. If the cross section requirement of the power tunnel or the penstock is satisfied, either the power tunnel or the penstock would be too long, which can not be implemented on the model. What's more, the roughness of the model can not be infinitively small, it is also needed to adopt the distorted model, so that the initial water level of the surge tank could be kept in line with that of the prototype. That's why the distorted model scale problem, i. e. correlations of proportions of various parts of the power tunnel and the penstock, was put forward.

2. Model scale

To simulate surging process in a surge tank and wa-

ter pressure in penstocks, the model scale must satisfy basic equation of unsteady flow in penstocks and variation equation of water level in the surge tank and corresponding boundary condition at the same time.

2.1 Equation of water flow movement in power tunnel and model scale

Equation of water flow movement in power tunnel[1]:

$$H_T = -\frac{L_T}{g}\frac{\mathrm{d}V_T}{\mathrm{d}t} - h_\Sigma \qquad (1)$$

In which, H_T is water level difference between reservoir and surge tank; L_T is length of power tunnel; V_T is flow velocity in power tunnel; h_Σ is total water head lose in power tunnel; the subscript T stands for Tunnel.

Derived from Eq. (1):

$$\lambda_{H_T} = \lambda_{L_T}\lambda_{V_T}/\lambda_t = \lambda_{h_\Sigma}$$

That is:

$$\frac{\lambda_{H_T}}{\lambda_{L_T}} = \frac{\lambda_{V_T}}{\lambda_t} \qquad (2)$$

$$\lambda_{h_\Sigma} = \lambda_{H_T} \qquad (3)$$

The distorted model scale of power tunnel has 5 parameters (λ_{H_T}, λ_{L_T}, λ_t, λ_{V_T}, λ_{h_Σ}) and 2 relative equations.

2.2 Equation of water level variation in surge tank and model scale

The water level variation in surge tank is determined according to the following equation.

$$As\frac{\mathrm{d}Zs}{\mathrm{d}t} = Q_s \qquad (4)$$

In which, A_s is cross sectional area of surge tank; Z_s is water level of surge tank; Q_s is discharge of flowing into surge tank.

Derived from Eq. (4):

$$\lambda_{A_s} \frac{\lambda_{Z_s}}{\lambda_t} = \lambda_{Q_s}$$

Considering $\lambda_{A_s} = \lambda_{D_s}^2$, D_s is section line dimensions of surge tank, the above formula is changed to

$$\lambda_{D_s}^2 \frac{\lambda_{Z_s}}{\lambda_t} = \lambda_{Q_s} \qquad (5)$$

Distorted model scale of surge tank has 4 parameters (λ_{D_s}, λ_{Z_s}, λ_t, λ_{Q_s}), and 1 relative equation.

2.3 Equation of water hammer in penstock and model scale

The movement of water flow in penstock is decided by the following equation[1]:

$$g \frac{\partial H_p}{\partial x} + V_p \frac{\partial V_p}{\partial x} + \frac{\partial V_p}{\partial t} + \lambda_p \frac{|V_p| V_p}{2 D_p} = 0 \qquad (6)$$

$$V_p \frac{\partial H_p}{\partial x} + \frac{\partial H_p}{\partial t} - V_p \sin\alpha + \frac{a^2}{g} \frac{\partial V_p}{\partial x} = 0 \qquad (7)$$

In which, H_p, V_p are piezometric tube water head and flow velocity in penstock respectively; a is wave speed of penstock water hammer; D_p is diameter of penstock; x is length of penstock; λ_p is the Darce-Weisbach coefficient.

Eq. (6), Eq. (7) yields

$$\frac{\lambda_{H_p}}{\lambda_x} = \frac{\lambda_{V_p}^2}{\lambda_x} = \frac{\lambda_{V_p}}{\lambda_t} = \lambda_{\lambda p} \frac{\lambda_{V_p}^2}{\lambda_{D_p}} \qquad (8)$$

$$\lambda_{V_p} \frac{\lambda_{H_p}}{\lambda_x} = \frac{\lambda_{H_p}}{\lambda_t} = \lambda_{V_p} \lambda_{\sin\alpha} = \frac{\lambda_a^2 \lambda_{V_p}}{\lambda_x} \qquad (9)$$

Eq. (8) yields

$$\frac{\lambda_{H_p}}{\lambda_x} = \frac{\lambda_{V_p}^2}{\lambda_x} \Rightarrow \lambda_{H_p} = \lambda_{V_p}^2 \qquad (10)$$

$$\frac{\lambda_{V_p}^2}{\lambda_x} = \frac{\lambda_{V_p}}{\lambda_t} \Rightarrow \lambda_{V_p} = \frac{\lambda_x}{\lambda_t} \qquad (11)$$

$$\frac{\lambda_{V_p}^2}{\lambda_x} = \lambda_{\lambda p} \frac{\lambda_{V_p}^2}{\lambda_{D_p}} \Rightarrow \lambda_{\lambda p} = \frac{\lambda_{D_p}}{\lambda_x} \qquad (12)$$

Eq. (9) yields:

$$\lambda_{V_p} \frac{\lambda_{H_p}}{\lambda_x} = \frac{\lambda_{H_p}}{\lambda_t} \Rightarrow \lambda_{V_p} = \frac{\lambda_x}{\lambda_t} \qquad (13)$$

$$\lambda_{V_p} \frac{\lambda_{H_p}}{\lambda_x} = \lambda_{V_p} \lambda_{\sin\alpha} \Rightarrow \frac{\lambda_{H_p}}{\lambda_x} = \lambda_{\sin\alpha} \qquad (14)$$

$$\lambda_{V_p} \frac{\lambda_{H_p}}{\lambda_x} = \frac{\lambda_a^2 \lambda_{V_p}}{\lambda_x} \Rightarrow \lambda_a^2 = \lambda_{H_p} \qquad (15)$$

Out of Eq. (10)-Eq. (15), Eq. (11) and Eq. (13) are the same, hence only 5 equations are independent.

Since $\sin\alpha = \Delta Z_p / \Delta x$, piezometer head $H_p = Z_p$

$+ p/\gamma$, so

$$\lambda_{\sin\alpha} = \frac{\lambda_{Z_p}}{\lambda_x} = \frac{\lambda_{H_p}}{\lambda_x}$$

That is Eq. (14) is satisfied automatically.

Therefore there are only Eq. (10), Eq. (11), Eq. (12), Eq. (15) 4 independent scale equations and λ_{H_p}, λ_{V_p}, λ_x, λ_t, λ_{D_p}, λ_a, λ_{λ_p} 7 scale parameters for the penstock.

2.4 Scale relationship of power tunnel, surge tank and penstock combined model test

The scale relationships that should be satisfied when a power tunnel, a surge tank and a penstock are tested separately have been derived from the above equations. When a power tunnel, a surge tank and a penstock are simulated at the same time on a model, i. e, the combined test, the respective scale relationship of the power tunnel, the surge tank and the penstock shall be satisfied but also the scale relationship of their jointly combined test shall be satisfied.

2.4.1 Time scale

Time scale λ_t shall be the same for the power tunnel, the surge tank and the penstock when the combined test is carried out.

2.4.2 Hydraulic condition in joints

Hydraulic parameter in joints must fit continuity condition and pressure is same, that is

$$Q_T = Q_s + Q_p \qquad (16)$$

$$H_T = H_p = Z_s \qquad (17)$$

In above equations, subscript T denote tunnel, subscript S denote surge tank, subscript p denote penstock.

Eq. (16) yields

$$\lambda_{V_T} \lambda_{D_T}^2 = \lambda_{Q_s} \qquad (18)$$

$$\lambda_{V_p} \lambda_{D_p}^2 = \lambda_{Q_s} \qquad (19)$$

Eq. (17) yields

$$\lambda_{H_T} = \lambda_{Z_s} \qquad (20)$$

$$\lambda_{H_p} = \lambda_{Z_s} \qquad (21)$$

Hence when a combined test is carried out, the scale relationships Eq. (18), Eq. (19), Eq. (20), Eq. (21) shall be fitted in addition to their respective relationships.

The scale relationships that shall be satisfied for the combined test can be obtained by synthesizing the scale relationships of Eq. (21) - Eq. (24).

$$\frac{\lambda_{H_T}}{\lambda_{L_T}} = \frac{\lambda_{V_T}}{\lambda_t} \qquad (22)$$

$$\lambda_{h_{\Sigma}} = \lambda_{H_T} \qquad (23)$$

$$\lambda_{D_s}^2 \frac{\lambda_{Z_s}}{\lambda_t} = \lambda_{Q_s} \qquad (24)$$

$$\lambda_{H_p} = \lambda v_p^2 \qquad (25)$$

$$\lambda v_p = \frac{\lambda_x}{\lambda_t} \qquad (26)$$

$$\lambda_{\lambda_p} = \frac{\lambda_{D_p}}{\lambda_x} \qquad (27)$$

$$\lambda_a^2 = \lambda_{H_p} \qquad (28)$$

$$\lambda v_T \lambda_{D_T}^2 = \lambda_{Q_s} \qquad (29)$$

$$\lambda v_p \lambda_{D_p}^2 = \lambda_{Q_s} \qquad (30)$$

$$\lambda_{H_T} = \lambda_{Z_s} \qquad (31)$$

$$\lambda_{H_p} = \lambda_{Z_s} \qquad (32)$$

The above 11 equations are independent from each other, and there are altogether 15 scales, of which 4 can be selected at random as the basic scales. In general 4 scales are selected from the geometric scales (λ_{L_T}, λ_{Z_s}, λ_{D_s}, λ_{D_T}, λ_{D_p}, λ_x).

3. Applied example

The waterway system of a certain hydropower station is composed of a power tunnel with a diameter of 5.05m and a length of 5, 207.617m, a penstock with a diameter of 3.6m and a length of 347.800m and an upper differential surge tank with a raiser diameter of 5.8m. A physical model was designed to simulate the changing process of hydraulic parameters of water flow in the power tunnel, the penstock and the surge tank when the operational condition of a water turbine changes.

In the test, in order to keep the geometrical shape of the surge tank undistorted, make

$$\lambda_{Z_s} = \lambda_{D_s} \qquad (33)$$

In order to keep reflection and transmission of pressure wave at the surge tank unchanged[2] in prototype and model, make

$$\lambda_{D_s} = \lambda_{D_p} \qquad (34)$$

Now there are 13 scale equations, hence only has 2 basic scale.

The length scale $\lambda_x = 41$ of the penstock was selected according to the test field condition. In order to reduce work load of making power tunnel model power tunnel use the pipe that can be purchased, taking $\lambda_{D_T} = 36.04$ as the basic scale, and other scales were determined by Eq. (22)-Eq. (34). For details, see Table 1.

Table 1   **Model Scale**

	Scale	Formula	Scale ratio	Characteristic Parameter to Prototype	Characteristic Parameter to Model	Remarks
Basic model scale	λ_x		41	347.8m	8.48m	Length of penstock
	λ_{D_T}		36.04	5.55m	0.154m	Diameter of tunnel
Power Tunnel	λ_{L_T}	$\lambda_{D_T}^2/\lambda_x$	31.68	5207.617	164.38m	Length of tunnel
	λv_T	$\lambda_x^{5/2}/\lambda_{D_T}^2$	8.287			
	λ_{Q_T}	$\lambda_x^{5/2}$	10763.65	46.2m³/s	0.00429m³/s	Total discharge passing through turbine
	λ_{n_T}	$\lambda_{D_T}^{5/3}/\lambda_x^{3/2}$	1.498			
	λ_{H_T}	λ_x	41			
Penstock	λ_{D_p}	λ_x	41	3.6m	0.0878m	Diameter of penstock
	λ_{x_p}	λ_x	41	347.801m	8.48m	Length of penstock
	λ_{H_p}	λ_x	41			
	λv_p	$\lambda_x^{1/2}$	6.403			
	λ_{Q_p}	$\lambda_x^{5/2}$	10763.65			
	λ_a	$\lambda_x^{1/2}$	6.403	1000m/s	156.17m/s	Water hammer Wave speed
	λ_{n_p}	$\lambda_x^{1/6}$	1.857	0.012	0.00646	
Surge tank	λ_{Z_s}	λ_x	41	50.688m	1.236m	Height of surge tank
	λ_{D_s}	λ_x	41	5.8m	14.15m	Diameter of surge tank
	λ_t	$\lambda_x^{1/2}$	6.403	7s	1.093s	Shut-down time of unit
	λ_{L_s}	λ_x	41			

It can be seen from the above table that the surge tank and the penstock in the model are both normal and only the power tunnel is distorted. The model was made for test and study according to the scales indicated in Table 1. The model test results have been adopted for the prototype design.

4. Comparison between model test and prototype survey

After completion of the hydropower station construction prototype observation on surging of both the surge tank and the water hammer pressure was made.

Table 2 gives the comparison between the prototype observation results and the model test results[2]. Because the geometrical dimensions of the power tunnel and the surge tank and their initial conditions after their completion often differ from the model dimensions and conditions in the test, the model test results were modified. For the modification, the numerical method[3] was used to calculate the surging process corresponding to both prototype observation and model test conditions, which was modified by the difference between the two.

It can be seen from Table 2 that:

The highest surge of the upper chamber in the model test under load rejection is little higher than the prototype and maximum difference is 0.19m, while maximum difference of the highest surge of the riser between the prototype and the model test is 1.6m. When load is increased, the lowest surge in the inclined shaft in the model test is 1.51m lower than the prototype.

The law of variation of both the highest surge and the lowest surge with operating condition in the model test is consistent with the prototype.

The time when either the highest surge or the lowest surge occurred in the model test basically coincided with the prototype. The fluctuation period of water mass in the model test is identical to the prototype.

Table 2   Compare of results of prototype servey with model test about surge in surge tank

Item / Operating condition		Full Load Rejection of Single Unit	Full Load Rejection of 2 Units	Full Load Rejection of 3 Units	Full Load for Units 2 and 3, No Load to Full Load for Unit 1
Reservoir water level (m)	prototype	2260.2	2260.14	2259.4	2253
	model	2260.3	2260.3	2260.3	2250
Initial water level (m)	prototype	2260	2256.92	2252.53	2249.65
	model	2259.66	2257.63	2254.48	2245.24
Highest surge in upper chamber (m)	prototype	2261.52 (241s)	2261.03 (536s)	2260.34 (603s)	—
	model	2261.48 (237s)	2261.22 (367s)	2261.23 (601s)	—
	After model modification	2261.63	2261.12	2260.41	—
Highest surge in raiser (m)	prototype	2262.41 (37s)	2263.85 (46s)	2266.51 (33s)	—
	model	2263.16 (42s)	2265.58 (38s)	2266.0 (38s)	—
	After correct in model	2262.75	2265.45	2265.76	—
Lowest surge in inclined shaft (m)	prototype	—	—	—	2238.35 (68s)
	model	—	—	—	2231.19 (56s)
	After correct in model	—	—	—	2236.84

Table 3 gives the comparison of Water Hammer pressure when three generating units underwent full load rejection at the same time. It shows that the time that maximum water hammer pressure occurred in the prototype observation is different from that occurred in the model test, and the maximum pressure in the model test occurred at the end of guide vane closing, while that of the prototype occurred at the beginning of guide vane closing. The maximum pressure rise in the model test is 19.5% higher than the prototype. This difference is due to that the discharge character of turbine needle valve simulated on the model is different from the real turbine. Attenuation of water hammer wave of the prototype is faster than that in the model test.

Table 3   Comparison of Water Hammer Pressure between Model Test and Prototype Observation under Simultaneous Full Load Rejection of 3 Generating Units

Item	Model test	Prototype survey
Reservoir water level (m)	2260.3	2259.4
Initial Discharge (m³/s)	46.2	41.09
Closing time of guide vane (s)	7.616	7.616
Water pressure before load rejection (kPa)	1718.11	1659.18
Water hammer pressure rise (kPa)	438.7	367.19
Time of max pressure occurred (s)	7.49	2.336
Speed rise ratio (%)	—	40.45
Time of max speed occurred (s)	—	5.804
Character of water hammer	Limit water hammer	Limit water hammer

Note   The model test results refer to the statistical results of many groups of tests.

5. Conclusions

The distorted scale ratio derived in the paper proved correct by both model test and prototype observation.

It is feasible to simulate water hammer and surge at the same time on the same model by using a distorted model.

The model test can well simulate surge process in a surge tank.

When a water hammer test is carried out by using a needle valve to simulate discharge characteristics of a Francis turbine, the water hammer pressure in the model test is significantly different from the prototype. And the maximum deviation of water hammer pressure rise is as high as 19.5% in the above-mentioned case.

References

[1] Wang Shuren. Constructure of hydropower station. Tsinghua University Press, July, 1984.

[2] Quyendeisan (Japanese). Water Hammer and Pressure Fluctuation. China Electric Industrial Press, February, 1981.

[3] Zhang Shaochun. Test and Study on Surge of Surge Tank in Luo Shiwen Hydropower Station. Yunnan Hydropower, Third, 1999.

[4] Zhang Shaochun. Model Test and Numerical Calculation of Surge of Surge Tank. Thesis Compilation, The Second Youth Science Annual Symposium of Yunnan Province, Yunnan Provincial Science and Technology Publishing House, September, 1996.

The Research on the Impact of the Variability of Sluice Gate on the Discharge Capacity

Zhu Chao[1]   Zhang Congjiao[2]   Wu Caiping[1]   Li Yuanfa[1]

[1] *Yellow River Institute of Hydraulic Research*, *YRCC*, *Zhengzhou*, 450003
[2] *North China University of Water Resources and Electric Power*, *Zhengzhou*, 450011

Abstract: In the drastic change rate movable-bed model test, it has been concerned about how to ensure similar discharge of the flood diversion gate. Six models with variable rates 1, 3, 5, 8, 10 and 13.3 separately have been chosen to compare their different discharges and to analyze the impact of the change rate on the discharge capacity. The results showed that: when their water heads were same, There were no obvious rules between the flow coefficients of the five distorted models and the variation rates, but when compared to the flow coefficient of the normal model, the former were on the higher side; when their water heads were different, the distorted models' flow rate increased with the variation rate, and when the rate of change came to 5, the flow increasing value tended to stable and close to 20%.

Key words: distorted model; similarity; discharge capacity; flow coefficient capacity

1. Introduction

Physical model test is an important method of studying and solving hydraulics problems in hydraulic engineering. Physical models include normal and distorted models; the normal models are mainly used for hydraulic model tests, while the river models are almost all distorted models. In theory, the distorted model does not fully meet the requirements of similarity theory: it may have different degrees of errors in flow field and sediment movement. Therefore, it has become a long-term debate among those people who work with physical model tests and those scientific and technical personnel to find out the application scope of distorted model and the variation rate limit.

In movable-bed model test, the model design generally selects large variation rate. It has been concerned about how to ensure the discharge similarity of the flood diversion sluice in these large variation rate models, some use water measurement equipment for the calibration test, others determine the rough discharge capacity by measuring the flow rate, but the results are not satisfactory.

After the outlet structure is distorted, the boundary conditions and flow conditions are changed, it is difficult to judge the impact of the change rate on the discharge capacity just from the theoretical analysis, and the only way to solve this problem is through a series of model tests.

2. Experimental design

Referring to a flood-diversion sluice in Shandong reaches of the Yellow River, the models and the test process were designed, and six models in which the variability were 1, 3, 5, 8, 10 and 13.3 were selected. Among them, the horizontal scales of these five distorted models were 180, 300, 480, 600 and 800 respectively, and the vertical scales were all 60, the corresponding strain rates were 3, 5, 8, 10 and 13.3. The horizontal and the vertical scale of the normal model, whose variation rate was 1, were 100. Test the impact of the variation rate on the discharge capacity was tested by comparing the discharge capacity of each model.

The entrance of the flood diversion sluice was a flat sluice gate. The gate contained 7 holes, and the clear width of each hole was 20m. The bottom of the chamber gate connected with the top of the surface cover of the upstream entrance, the back of the sluice gate connected with the prime stilling pool by setting up steep slopes. The flood diversion sluice was shown in Figure 1.

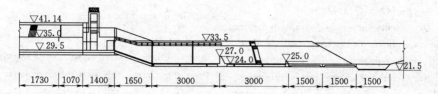

Figure 1　The longitudinal section profile of the diversion
gate（elevation：m，their sizes：cm）

3. The discharge capacity and flow coefficient features of different variation rate models

The test was held in a tank of 2.0m × 8.0m, and the width of the approach channel in front of each gate was 2.0m, and the leading edge gate's location of each model did not change. The water gauge was set 150m ahead from the sluice gate（prototype）. There was also movable measuring pin in front of the gate. It was used to observe the water surface drop curve of the entering water flow. The relationship of the measured weir head and the flow coefficient are shown in Figure 2. Among them, the flow coefficient was calculated by the formula：

$$m = Q/B \sqrt{2g}H_0^{3/2} \tag{1}$$

Where：Q represents the measured discharge capacity of the model；B the over-current net width；g acceleration due to gravity，and H the head of the weir including the running flow velocity head.

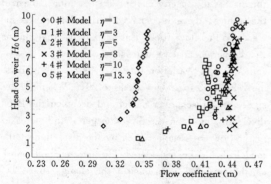

Figure 2　The relationship between the head of the weir
and the flow coefficient（the side contraction
coefficient not included）

The results shows that measuring points groups of the five distorted models are disperse, and there are no apparent laws between the flow rate coefficient and the variable size, but compared to the normal model flow coefficient, the flow rate coefficient of the distorted models are on the higher side when the head

on weirs are the same.

Analysis shows that in the normal model, with the increase of the head on weirs, the flow regime transits from the flow state of the open channel to the broad-crested weir, and then to the practical weir. This phenomenon also existed in the distorted models. Meanwhile, when the head of the weir is fixed, with the variation rates increased, the chamber length would be shorten and the water regime would transit from the water state of the broad-crested weir to the practical weir even to the thin plate weir. The results shows that when the hydraulic head are the same, the discharge capacity of the distorted models is significantly greater than the normal model, and the flow coefficient tend to similar to the practical weir's and even the thin weir's flow coefficient, that is，$m = 0.44 - 0.47$[2-4].

4. The comparison between the distorted models with different variable rates and the normal model

To further illustrate the variation trend of the discharge flow of the distorted model, Figure 3 showed the relationship between the change of rate and the discharge capacity deviation values by using the measured data according to the different head of the weir.

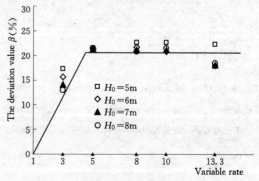

Figure 3　The relationship curve of the model variability
and the deviation values of the flow volume

$$\beta = (Q_D - Q_N)/Q_N \qquad (2)$$

Where, Q_D presents the discharge capacity of the distorted model; Q_N shows the discharge capacity of the normal model. It can be seen that, under different head on weirs, the discharge capacity of the variability increase with the variability value, when the variable rate arrives at 5, the flow deviation value stabilizes around 20%.

Therefore, in the river model production, according to model variability and the deviation values shown in Figure 3, we can increase the over-current width of the diversion gate appropriately to revise the discharge volume.

5. The spread and application of the test results

The main element which affects the discharge capacity of the flat gate (or the weir) is the shape ratio $\frac{\delta}{H_0}$ of the gate (or the Weir) (δ is the thickness of the weir crest). According to the shape ratio $\frac{\delta}{H_0}$, the weir types can be divided into broad crested weir, practical weir and thin weir. The six models of the test also could be seen as serials models of different flat sluice gates which have same pocket floor height but different lock chamber length and gate width. When $\frac{\delta}{H_0}$ values are different, the over-current flow patterns are different. From the measured dropdown curve of the from gate, we can see the obviously different water regime.

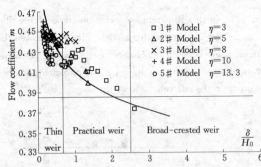

Figure 4　The relationships of the shape ratio $\frac{\delta}{H_0}$ and the flow coefficient m (the side contraction coefficient not included)

From the actual measurement data of the six serials models, the relationships of the shape ratio $\frac{\delta}{H_0}$ and the flow coefficient $\frac{\delta}{H_0}$ (not included the side contraction coefficient), is shown in Figure 4. Ac-

cording to the different $\frac{\delta}{H_0}$ to classify the Weir-type, there are three weir-type areas (thin weir, practical weir and broad-crested weir) in the test data. As seen from the chart, the five different weir-type series model, the measured point groups of the model could link up effectively, and had a good trend: with the increase of the shape ratio $\frac{\delta}{H_0}$, the flow coefficient decrease, when the ratio $\frac{H_0}{\delta} > 1.49$, that is $\frac{\delta}{H_0} < 0.76$, the weir type belongs to thin weir, and the flow coefficients m no longer increases, closing the flow coefficient of the steep slope and the hydraulic drop, that is 0.44-0.47. Based on the test data, empirical formula could be written as follows:

$$m = 0.41 - 0.08 \lg\left(\frac{\delta}{H_0}\right) \qquad (3)$$

The flow coefficient of the three weir-type areas could be calculated by using Formula (3), and the calculation error does not exceed ±4%.

6. The main conclusions

As the distortion of the model, the chamber length was cut short, the slope behind the gate was steepened, and the location of model's critical depth, with the model variability increasing, moved from the sluice gate to the upstream of the gate, that is, the discharge capacity control cross-section moved to the front of the gate, and with the change rate increased, the flow coefficient increased, that means the discharge capacity increase.

When the variation rate exceeds 5, the increasing value tend to stable, close to 20%. When we designed long reach river model and designed the flood diversion sluice and the diversion gate, the increases of the flow coefficient caused by the distortion of the model should be considered. The over-current width can be amended according to the change rules. Amendment can refer to Figure 3. After the amendment, the model can automatically satisfy the over-current capability as the normal model, do which avoid the calibration work needing use the gate to control the discharge volume as before.

For the flood diversion sluice and the diversion gate in the long reach distorted river model, the flow coefficient m (which does not include the side contraction) can be calculated according to the actual $\frac{\delta}{H_0}$

of the distorted scale model and using the empirical formula $m = 0.41 - 0.08 \lg \left(\dfrac{\delta}{H_0} \right)$. Then the side contraction coefficient of the abutment wall can be determined by observing the abutment wall contraction condition. After that, the actual width of the sluice hole of the model is to be calculated in accordance with the design flow requirements of the flood diversion sluice gates and the diversion gate. The width of the gate hole calculated by the described method above can maintain the same over-current capacity as the normal model.

References

[1] Yao Wenyi. Design Theories of Yellow River Physical Model and Their Applications, Nanjing, Hohai University doctoral dissertation, 2005.

[2] Pan Qingshen, Yang Guolu, Fu Renshou. The research on the Sediment problem of the Three Gorges Project, China Waterpower Press, 1999.

[3] Japan Society of Civil Engineering. Hydraulic formulary. Chinese Railroad Press, 1977.

[4] Hydraulic and Electric Engineering institute of Wuhan University. Hydraulic calculation manual. Beijing: China Waterpower Press, 2006.

[5] Xia Yuchang, Zhang Liming. Hydraulics prototype observation and model test. Beijing: China Electric Power Press, 1999.

Hydraulics Mechanism and Application of a Swirling Device in Morning Glory Shaft Spillway

Dong Xinglin   Yang Kailin   Guo Xinlei   Guo Yongxin   Wang Tao   Fu Hui

Department of Hydraulics, China Institute of Water Resources and Hydropower Research, Beijing, 100038

Abstract: According to swirling flow mechanism, a novel shaft spillway which consists of submersible and spiral-flow-generated piers and flaring weir was developed. It has shown that the stable and swirling flow with air core in the vertical shaft was generated under a variety of weir water head, thus the negative pressure was eliminated and also the cavitation was prevented. Besides, the energy dissipation ratio of the shaft spillway was increased significantly. The flow mechanism of this novel piers is: when the water head above the weir is low, the water will flow to the vertical shaft along the piers wall, and the stable and swirling flow with air core is generated, when the depth of water been significantly raised over the top, the inflow angle will be automatically adjust under the inertial forces which leading to the larger discharge capacity, and meanwhile, the water will be synchronously swirling under the drag force of bottom flow. The design theory and the research results have been firstly applied in the shaft spillway of Qingyuan pumped storage power station in Guangdong province.

Key words: swirling flow; morning glory shaft spillway; submersible and spiral-flow-generated piers

1. Introduction

The morning glory shaft spillway was commonly built from the 1940s and 1950s in America[1]. Without setting a gate, the normal water level is the height of the weir crest, this type of spillway are usually adopted in emergency spillway as well as the pump storage power station in unattended operation, especially appropriate for tasks like the west routes of water diversion project in sparsely populated western regions. The maximum discharge capacity was up to 3000m^3/s when it built in relatively open terrain.

There are much weak points in the tradition morning glory shaft spillway. Firstly, the discharge capacity was affected when the vortexes at intakes occurs, even worse, the structure was vibrated causing by the unstable vortex rope, sometimes the air blowback was produced such as the spillway of Owyhee dam in Oregon of the United States[2]. Secondly, the negative pressure was generated between the spill weir and shaft; especially right where they join and when it increases to the vapor pressure, the cavitations are likely to occur. For reasons given above the application of shaft spillway in the design of high head

and large discharge may be limited. And there are a few shaft spillways built in China such as Panzhihua and Dashimen spillways.

In order to prevent the vortex at intakes, the traditional way is to set the diversion piers. And there are also many control measures for preventing cavitation of shaft, such as setting sudden enlargement aerator under the weirs or ringlike aerator[3-4], and forming the bottom of shaft into contraction to increase the wall pressure, and setting aerator between the shaft and the corner of tunnel (e. g., the Charvak spillway[5]). The cavitation may be occurred due to the finite length of protection of aeration. If setting the aeration equipments in the shaft, the ventilation pipe systems were needed and it directly results in the complicate structure and construction.

2. Swirling device in morning glory shaft spillway

To overcome the weaknesses above, a novel shaft spillway was developed. It mainly consists of submersible and spiral-flow-generated piers, flaring (morning glory) weir, vertical shaft and pressure dissipator in the tunnel. According to swirling flow and aeration mechanism, several piers around the flaring weir were symmetrically set, thus the stable and swirling flow with air core in the vertical shaft was generated under a variety of weir water head. For this reason, the negative pressure was eliminated and the cavitation was prevented. Besides, the energy dissipation ratio of the shaft spillway was increased significantly. Although the diameter of the shaft was enlarged, it is also small compared with the diameter of traditional shaft spillway with aerator measurements. The discharge of Panzhihua and Dashimen spillways are 80m³/s and 152m³/s, with the height of vertical shaft are 80m and 48m, respectively. The diameter of these two shaft was 5m, but if the novel shaft spillway adopted, the two values were be decreased to 3.8m and 4.8m.

Without setting a gate, the flood will be overflown when the water level exceeds the height of weir crest. In order to make the swirling flow under the lower water head, the submersible piers must be arranged at a small angle to the tangent direction of flaring weir, the preferable angle is less than 15°. The more number of piers and the fewer angles it sets the stronger rotation of flow and the lower discharge capacity, and vice versa. Therefore, if you want produce the preferable swirling flow in the shaft under the lower water head with the same number of piers, the angle must be small, and it will not meet the requirements of maximum design flow. To resolve this conflict, a swirling device which called the submersible and spiral-flow-generated piers was developed; it can produce effective swirling under the lower water head and large discharge capacity under the high water head at the same time.

The flow mechanism of this submersible piers is: when the water head above the weir is low, the water will flow to the vertical shaft along the piers wall, and the stable and swirling flow with air core is generated, when the depth of water been significantly raised over the top, the inflow angle will be automatically adjust under the inertial forces which leading to the larger discharge capacity, and meanwhile, the water will be synchronously swirling under the drag force of bottom flow. Such a simple structure of the piers with a small size will be easier to popularization and application. A general and fundamental structure of swirling device in morning glory shaft spillway is given in Figure 1.

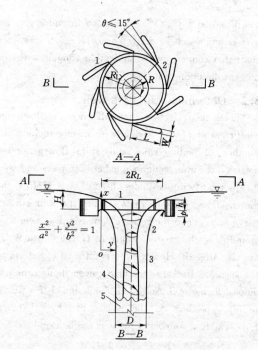

Figure 1   Structure and flow pattern
1—submersible and spiral-flow-generated piers;
2—weir; 3—shaft; 4—swirling flow with air
core; 5—annular hydraulic jump

3. Design of the spillway

3.1   Determination of the diameter

The diameter of the shaft is the control section which determines discharge capacity of the spillway, and other sizes of the weir are associated with it. So first of all, its value need be determined. Based on the previous research experience, the stable and swirling flow with air core as well as the maximum discharge capacity will be achieved as long as the shaft satisfies the virtual number of Fr which can be expressed as

$$Fr=Q/\sqrt{gD^5}\approx1 \qquad (1)$$

where, Q is the maximum discharge of spillway; g is the acceleration due to gravity, and D is the diameter of the shaft.

According to Eq. (1), the specific solution for the diameter of shaft under the maximum discharge capacity is

$$D=\left(\frac{Q^2}{g}\right)^{1/5} \qquad (2)$$

And the above formula has been validated by the physical model tests of Shapai, Wawushan, Xiluodu, Xiaowan and Qingyuan spillways[6-10]. Jain[9] has proposed Eq. (2) according to the minimum air core radius which be derived as $ra/R=0.5$ in vortex shaft that connecting the derivation conduit with the tangent direction of it.

3. 2　Determination of the weir

The major factors that influence the discharge coefficient m of the flaring weir mainly include the number of submersible and spiral-flow-generated piers n, the intersection angle between piers and weir θ, the relative dimensionless head of weir H/R_L, the dimensionless piers height h/H and the weir surface curve form. Then the relative dimensionless height of weir P/H, the relative length L/H and width W/H are also affect it. However, it is difficult to determine the accurate discharge coefficient even though the traditional flaring weir. And so far the design of this weir is mainly depending on the results of physical model tests. The flow formula of the weir can be expressed as

$$Q=m2\pi R_L\sqrt{2g}H^{3/2} \qquad (3)$$

Form Eq. (3), by transforming of $(H/R_L)^{3/2}$ to the equation and it is easy to obtain the radius of the weir in the following

$$R_L=\left[\frac{1}{m2\sqrt{2}\pi(H/R_L)^{3/2}}\right]^{2/5}\left(\frac{Q}{\sqrt{g}}\right)^{2/5}=KD \quad (4)$$

In which the coefficient K can be derived as

$$K=\frac{0.417}{m^{0.4}(H/R_L)^{0.6}} \qquad (5)$$

where, H is the maximum water head above the weir and m is the discharge coefficient.

Notice Eq. (4) indicates that radius of flaring weir R_L can be expressed by shaft diameter D. For easy design, other factors which influence the discharge coefficient need be unified specified as follows: $n=8$, $\theta=10°$, $P/H\approx0.3$, $h/H\approx0.7$, $L/R_L\approx1$ and $W/H\approx0.4$. Since no negative pressure occurs in the weir surface, 1/4 elliptic curve with $b=D/2$ and $a=(2.2-2.5)$ b will be adopted instead of the traditional complex curve. The R_L could be calculated by Eq. (2), Eq. (4) and Eq. (5) as long as the known relationship between H/R_L and m.

4. Application in Qingyuan pumped storage power station

Qingyuan pumped storage power station, with a total installed capacity of 1280MW, is one of the important electric supply projects in Guangdong prov-

ince. It mainly consists of higher and lower reservoirs and the underground powerhouse. The clay core rockfill dam was designed in the lower reservoir and its height is 75.9m. The normal water level is equal to the height of the weir crest, and the morning glory shaft spillway was used to discharge flood flow. The weir head H is 4.75m and the maximum discharge Q is 534.2m³/s. A diversion tunnel with the height of 8m and width of 6m was connected with the vertical shaft. There are catchments of baffle blocks and self-aerated energy dissipator established in the tunnel.

4. 1　Basic dimensions

(1) The diameter of the vertical shaft was selected according to Eq. (2) which gives $D=8m$ for a maximum design discharge of 534.2m³/s.

(2) Dimensions of the weir: in the present study $R_L=8m$ was used, and $n=8$, $\theta=10°$, $p=1.5m$ were also suggested according to section 3.2. The submersible piers with the length are 8m and width are 1.5m and heights are 3m were designed. The curve with $b=4m$ and $a=2.5$, $b=10m$ will be adopted. The shaft structure size and the configuration of Qingyuan pumped storage power station was shown in Figure 2.

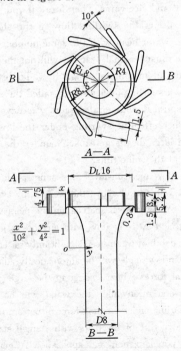

Figure 2　Dimension of Qingyuan spillway (m)

4.2 Results of model test

In the first model test, by setting the height of submersible piers equal to 3m, the discharge of the flow under flood frequency 0.02% was 603.6m³/s which is 13% bigger than the design value. It indicated the height is too low and should be increased to decrease the flow discharge. So in the present model, it is increased to 3.7m with the measured discharge decreased to 521.77m³/s and this is closer to the design value.

The relationship between relative dimensionless head of weir H/R_L and the discharge or the coefficient of flow was studied, as shown in Figure. 3. The dash-dotted line represents the top height of the submersible and spiral-flow-generated piers. The character of the relationship shows that: when the water level is higher than this height, the flow will be linear growth along H/R_L which called submerged flow shown in A area, and the coefficient of flow is nearly constant (0.223). Compared with the A area, the flow discharge and the coefficient will be goes up and goes down along H/R_L respectively in the B area which called the free flow area.

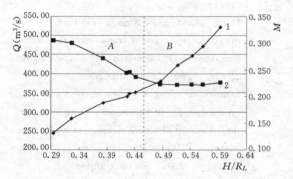

Figure 3   Relationship between H/R_L and
discharge or coefficient

A—submerged flow area; B—free flow area;
1—Relationship between H/R_L and discharge Q;
2—Relationship between H/R_L and coefficient m

Figure 4 and Figure 5 give the flow pattern of the inlet of weir under the flood frequency 5% and 0.02% respectively. These photos and the results have shown that the stable and swirling flow with air core in the vertical shaft were generated under a variety of weir water head which from $H/R_L=0.1$ to $H/R_L=0.6$. Besides, the negative pressure was eliminated and the cavitation was prevented expect in the

Figure 4   Flow pattern under the flood frequency 5%

Figure 5   Flow pattern under the flood frequency 0.02%

condition that the flood frequency is 20%. And the maximum negative pressure head was less than 1m which located up the circular hydraulic jump of the shaft. It indicates that the centrifugal force of swirling flow has played an important role in pressure increasing and cavitation preventing.

5. Design spread of the novel spillway

From the relationship curve in Figure 3 we can see that the maximum discharge in the design of the spillway will be reach to 652.6m³/s if $H/R_L=0.48$ and $H=4.8$m were used in Eq. (3). That is to say, the relationship between H/R_L and m is feasible only for the shaft design within the range $Q \leqslant 650$m³/s. Since the initial design of Qingyuan spillways is based on the traditional method and the head H and radius of weir have been given in advance, the formula Eq. (4) to calculate R_L was not used. The following example describes the systemic design of novel shaft spillway.

For example, the given flow discharge being 300m³/s, find the diameter of the shaft and the dimension of weir. Firstly, we can calculate the diameter of the shaft by Eq. (2), which gives $D=$

6. 2m. For the small excavated volume, H/R_L may be choosing for a little big and $H/R_L = 0.6$ was used in the present design.

Then consider $m = 0.223$ in Figure 3, Eq. (5) becomes

$$K = \frac{0.417}{m^{0.4}(H/R_L)^{0.6}} = \frac{0.417}{0.223^{0.4} \times 0.6^{0.6}} = 1.033$$

From Eq. (4)

$$R_L = KD = 1.033 \times 6.2 = 6.4 \ (\text{m})$$

By adding $H/R_L = 0.6$

$$H = 0.6 \times R_L = 3.84 \ (\text{m})$$

The discharge should be checked as

$$Q = 0.223 \times 2\pi \times 6.4 \times \sqrt{2 \times 9.81} \times 3.84^{1.5}$$
$$= 298.9(\text{m}^3/\text{s})$$

and it meets the requirements of the design flow.

6. Conclusion

The swirling device which called the submersible and spiral-flow-generated piers in morning glory shaft spillway is a novel shaft spillway that it is totally different from the traditional design. The main issues and achievements are as follows.

(1) The design theory of this novel spillway has been proposed. (2) The swirling flow difficulty under the low water head above the weir was settled and the inflow angle will be automatically adjust under the inertial forces which leading to the larger discharge capacity. (3) The piers are not affected by terrain influence, the stable and swirling flow with air core in the vertical shaft was generated under a variety of weir water head, thus the negative pressure was eliminated and the cavitation was prevented. Besides, the energy dissipation ratio of the shaft spillway was increased significantly.

References

[1] W. E. Wagner. Morning-glory shaft spillway Determination of pressure controlled profiles. *Proc. ASCE.* 1954, 80 (432).

[2] H. T. Falvey. *The Aerated Flow in Hydraulic Structures.* Water Resources and Electric Power Press. (in Chinese)

[3] Wang N S, Dong X L. Experimental Study on Flood Discharge Shaft of Guaziping in Panzhihua, China Institute of Water Resources and Hydropower Research, 1987, 8. (in Chinese)

[4] Sun S K, Jiang H, et al. Hydraulic Model Test of the Shaft Spillway of Dashimen Reservoir. China Institute of Water Resources and Hydropower Research, 2008, 1. (in Chinese)

[5] Chen Z H, Han L, et al. Hydraulic Calculation Handbook of Diversion Structure. Water Resources and Hydropower Engineering Design & Research Institute, North China Institute of Water Conservancy and Hydroelectric Power, 1993. (in Chinese)

[6] Dong X L, Gao J Z, Zhong Y J, et al. Design and Study on Supercritical-flow Vortex Shaft Spillway. *Water Power*, 1996 (1). (in Chinese)

[7] Dong X L, Guo J, Xiao B Y. Design Principle of High Head and Large Discharge Vortex Drop Spillway. *Journal of Hydraulic Engineering*, 2001 (11). (in Chinese)

[8] Dong X L, Yang K L, et al. Test and Study on the Rebuild of Diversion Tunnel into Gyrating Current Flood-Releasing Tunnel. China Institute of Water Resources and Hydropower Research, 2005, 8. (in Chinese)

[9] S. C. Jain. Tangential Vortex-Inlet. *Journal of Hydraulic Engineering*, 1984, 110 (12).

[10] Dong X L, Yang K L, et al. Experimental Study on Shaft Spillway of Qingyuan Reservoir in Guangdong Province. China Institute of Water Resources and Hydropower Research, 2009, 12. (in Chinese)

The Sedimentation Effects of Dezful Regular Dam on the Power Generation of Dez Dam

Samad Emamgholizadeh

Department of Water and Soil, Shahrood University of Technology, Iran

Abstract: Dezful regular dam was constructed at 23km downstream of Dez Dam. This dam with height of 20m regulates the released water from Dez Dam among 24 to 48 hours. According to the hydrography after 30 years operation, deposited sediment reduced its storage volume more than 40%. In this research the effect of increasing trend of sedimentation of this dam investigated on the hour power generation of Dez Dam. The result show that the decreasing of the storage volume of the Dezful regulatory dam, in addition to the decrease of the available storage capacity in the regulation of needed water for downstream consumption, it has considerable effect in the decreasing of energy production at the peak time of energy consumption.

Key words: sedimentation; Dezful regulatory dam; energy production; Dez power plant; desilting

1. Introduction

Reservoir sedimentation is one of the important problems in the area of hydraulic engineering. As more reservoirs are constructed across the world, the issue of reservoir sedimentation is of primary concern. During the processes of reservoirs design, research and forecast of future sedimentation should be undertaken to meet the need of the design, otherwise, various negative effects caused by sedimentation may occur after the completion of construction[1]. The problems aroused by reservoir sedimentation include reduction of storage capacity, affects on navigation, increasing inundation and downstream channel erosion. There are no available solutions to these problems so far. Reservoir sedimentation estimates often depend on the analysis and calculation of the amount of the sediment deposited in the reservoir and the input from the adjacent river section. Such information can be used as a rough guideline for reservoir operation[1].

Dams in order to different goals are built, for example, may be needed to provide agricultural water, drinking water, controlling flood, and also for energy production. Currently the thermoelectric and hydroelectric power plants have the most of the share of the power supply compared with other methods of energy in the world. At the future years, due to some of problems such as technological problems, environ-mental aspects, resources restrictions and etc. the trend of electricity production in thermal plants (with fossil fuel or nuclear), would counted to problems. As the name of the hydroelectric power plant is states, for this type of energy, it is need to the construction of dams on the river. Also with construction of dams on rivers, in addition to the production of energy it can be to control, store and regulate flood water for downstream consumptions. Iran with regard to its many rivers, has potential for producing hydroelectric energy about 50 billion kilowatts per hour in the year that can supply 60 percent of the needed electricity of the country. The Khoozestan province with important rivers such as Karoon, Dez, Jarahi, Maroun and Karkhe (about 40 percent of the country's water potential), is one of the most important provinces of Iran. Therefore it has the potential of using these resources. According to the studies which done in the Karoon, Dez and Karkhe basins, they have potential of energy production of 30, 9 and 6 billion kilowatts per hour in a year. Use of hydroelectric power plants has many advantages to other power plants, which it can be mentioned to its coefficient of high readiness, the ability and confidence over an effective role in controlling their frequency and the resistance network. When the energy consumptions have a lot of fluctuations and changes the hydropower plants with high storage capacity are high flexible percentage in coordination with electricity consumption. Furthermore, the fuel for energy production in these power plants is water; therefore these power plants will not create environmental pollution, consequently, these power plants as one of the plants with renewable energy sources. In addition to the mentioned advantage, these types of power plants have other advantage rather than other power plants which can be mentioned to the low depreciation and low cost of the utilization and maintenance.

2. Power market rules regarding hydropower plants

In recent two decades, electricity industry in most countries from the governmental and non-competitive structures changed to the private sector and competitive structures. Similarity, in Iran's Electricity industry, electricity market was launched based on the method of buying and selling electricity in the national network from October 2003. The aforementioned instructions are as follows[2].

(1) Energy production of the hydroelectric power plants has been calculated with regard to the function coefficients of 0.25, 1 and 2.5 for peak power plants, middle and base, respectively.

(2) Separation of energy production has been done in 24 hours of day and night with production priority in the peak, normal, and low hours.

(3) 4, 12 and 8 hours have been elected for peak, normal, and low hours in duration 24 hours. The relevant coefficients for total days in a week were 2.5, 1 and 0.25 respectively. Also for Thursday and Friday that coefficients were equivalent to 1.5, 0.7 and 0.25, respectively. These coefficients multiplied in the readiness base rate.

(4) Coefficient related to the hot months (May, June, July and August) equal to 1.2 and for other months of the year equivalent to 0.9 in which the coefficients multiplied in the readiness base rate.

Any revenues for social projects such as employment, reduce environmental pollution, etc. not considered.

3. The study Area

Dez River is located at the Khozestan Province of Iran. This river after the Karoon River is the biggest river of the Khozestan Province. This river has high potential for using the water resources for producing electricity. Only Dez Dam has been constructed on this

river. This dam is a double curvature concrete arch dam with a height of 203 meters and initial storage capacity of the reservoir of $3315 \times 106m^3$ which operated in the year 1963. It has for 47 years fulfilled a very important role in the areas of power generation (520MW), water supply (125 thousand hectares of irrigation) and flood control. The Lake area behind the dam is 63 square kilometers and the reservoir capacity and its useful capacity is 3400 and 2500 million cubic meters, respectively. Dez power plant with eight 65-megawatt Unit (in sum 520 megawatts) produce annual 2310 million kilowatts hour energy, and meanwhile it regulates the frequency nationwide electricity country. In order to adjust and control the water which needed for Dezful Irrigation Network, the Dezful Regulatory Dam with height of 20 meters constructed at 23km downstream of Dez dam (Figure 1). Table 1 show the technical specification of the Dezful regular dam.

Figure 1　Dezful regular dam

Table 1　Technical specification of the Dezful regular dam

Type of dam	Concrete
Height	20m
Crest Length	136m
Reservoir Area	330hec
Reservoir Capacity	13.7Mm³
Crest Elevation	130m
Maximum level of operation	135.2m
Maximum of passing flood discharge	6000m³/s

The basin area of Dez regulatory dam is 17813 square kilometers and the volume of the lake in its maximum elevation is 13.7 million cubic meters. Because of sedimentation in this dam after operation until now, the volume of its useful decreased. According to the hydrography after 30 years operation, deposited sediment reduced its storage volume more than 40%. In other word, the trend of sedimentation in this dam is equal to 1.2%.

With respect to the sedimentation in the present and future conditions in the Dezul regulatory dam and also with respect to the new instructions market of buying and selling energy, for answering the following questions, it is needed to do the present study.

(1) How many hours can be operating Dez power plant in the peak, normal and low mode?

(2) Does Dezful Regulatory Dam have ability to regulate released water from Dez power plant (according to the most optimum production) for downstream?

4. The results and the discussion

In order to achieve the desired goals of this research, it is necessary to consider the conditions governing the issue, such as the released water from Dez power plant with the probability of various kinds, the storage volume of the Dezful regulatory dam and its sedimentation trend and the hydraulic conditions of Dez River downstream of Dez dam in a system and they have been modeled. Therefore different alternative production options considered for study (Table 2).

Table 2   The study options of this research

Statues of power plant	Studied months	Studied years	The number of options which investigated
8 units with 65MW capacity	June July August September	2011 2021 2031	48

4.1 The outflow hydrograph from Dez power plant for the most optimal situation

For preparation of the outflow hydrograph from Dez power plant, the statistics of outflow discharge during the past 40 years is used and with the use of statistical relations, outflow discharge of the power plant with percent probability of the 25, 50, 75 and 90 percent are calculated (Table 3). The mentioned discharges can be released with any form from the dam. For example the outflow discharge from the

power plant can be a fixed discharge in the duration of 24 hours. But as it was mentioned before, according to law of the buying and selling electricity, the duration of 24 hours divided to three durations:

(1) Peak load time (4 hours).

(2) Normal load time (12 hours).

(3) Low load time (8 hours).

Table 3   Outflow discharge from Dez dam during the 40 past years according to the different probability

Month	Options	Probability frequency (%)	Discharge (m³/s)
June	1	25	320.22
	2	50	240.75
	3	75	186.36
	4	90	98.8
July	1	25	248.22
	2	50	211.8
	3	75	181.8
	4	90	145.6
August	1	25	234.7
	2	50	214.9
	3	75	197.3
	4	90	147.8
September	1	25	234.5
	2	50	220.8
	3	75	193.08
	4	90	173.88

The price of buying and selling energy at the low, normal and peak time is different together (which this difference is to ten times).

Therefore the most benefit will come when the outflow discharge from the power plant in accordance with priority on the above. So the discharge which has been calculated with the probability of that, distributed in the last 24 hours, according to the priority of low, normal and peak times. Also in order to considering the effects of environmental down of the dam, in the preparation of the hydrograph for all options, a constant discharge of a single unit (almost equal to discharge 73 cubic meters per second) is considered which must be released in the downstream river.

By considering the mentioned issues, the outflow hydrograph from Dez power plant are prepared which a sample of it is indicated in Figure 2.

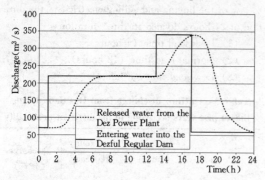

Figure 2　The output hydrograph from Dez Dam and
entrance hydrograph to Dezful regular dam for
June month with probability of 75%

4.2　Run of HEC-RAS

To know that the output hydrograph from Dez Dam in entering the Dezful regular dam what form will be have, HEC-RAS software was used[4].

For running the model the information needed by the model such as cross sections from Dez Dam to Dezful Regular Dam (25km), the relationship between discharge and water level at the downstream and Manning coefficient introduced to the model. After calibration of the model, prepared hydrograph, which mentioned in the last section, introduced to the model.

After running model, the entrance hydrograph to Dezful Regulatory Dam in every option was get for study months such as June, July, August and September with probability of 25, 50, 75 and 90 percentages. For example Figure 2 shows the calculated output hydrograph for months of June with probability of 75%.

4.3　Investigation of the ability of Dezful regular dam in the adjusting of the entering hydrograph (released water from Dez power plant)

The maximum and minimum water level of the Dezful regular dam is 135.5m and 129m respectively. With respect to the sedimentation in the Dezful Regular Dam, the main questions is that the Dezful Regulatory Dam has ability to regulate released water from Dez power plant according to the most optimum production for downstream consumptions. To answer

this question, it first need to know about the relation between reservoir's volume and elevation at the present and future conditions. For this purpose for prediction sedimentation in the Dezful dam, the BRISTARS software was used. With using BRISTARS software the relation between elevation and volume at the Dezul Dam in the preset and future years of 2010, 2021 and 2031 predicted[3] (Figure 3).

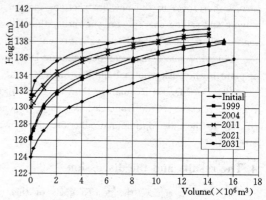

Figure 3　Elevation and volume relationship at
the Dezful regular dam

The ability of the Dezful regulatory dam at all options was studied with regard to knowing the following issues.

(1) The relation between the volume and elevation of the Dezful regular dam in the studied years.

(2) Entering hydrograph to the Dezful regular Dam.

(3) Pumped water from Sabily Pump station at studies months.

(4) Constant released water from Dezful regular dam for downstream consumptions.

With respect to the sedimentation of the Dezful regular dam after operation until now its storage volume decreased annually with trend of 1.2%. The investigations of this study with 48 options show that at the futures years (2011, 2021 and 2031) the average volume of sedimentation would be 100, 580 and 900 thousand cubic meters. Therefore with regard to the sedimentation and therefore decreasing of useful volume the Dezful regular dam it would not be able to regulate water needed for making down in some options. As the function of this regular dam is the adjusting of the released water from Dez power plant with regard to the it's most optimum production, with respect to the sedimentation and also with respect to the new instructions market, the results of this study show in some options the Dezful regular dam cannot adjust the released water from Dez power plant. Therefore it is necessary to carry out dredging, flushing or other methods for removing sediment at the Dezful reservoir.

References

[1] K . D. Fu, D. M. He, X. X. Lu. Sedimentation in the Manwan reservoir in the Upper Mekong and its downstream impacts. *Quaternary International Journal*, 2008, 186: 91-99.

[2] Tavanir Company of IRAN. Instructions for determining interest rates and conditions for buying and selling electricity. 2003.

[3] Sadeghi, M. *Investigation of the trend of sedimentation of the Dezful Regular Dam with BRI- SRARS model*, MS thesis, Shahid Chamran University, 2004.

[4] *Hydraulic Reference Manual of HEC-RAS software* U. S Army Corps of Engineering, Institute for water Resource, Hydraulic Engineering Center, 2008.

[5] J. S. Lai, H. W. Shen. Flushing sediment through reservoirs. *Journal of Hydraulic Research*, 1996, 34 (2): 237-255.

A Laboratory Study of Dynamic Responses of a Moored Rectangular Floating Breakwater to Regular Waves

Zhenhua Huang[1]   Wenbin Zhang[2]

[1] *School of Civil and Environmental Engineering, Nanyang Technological University, Singapore, 639798*

[2] *DHI-NTU Centre, Nanyang Technological University, Singapore, 639798*

Abstract: Floating breakwaters allow water exchange between harbors and their surrounding coastal waters, and are good low cost alternatives for marinas and recreational harbors. This paper presents a new set of experimental data for wave transmission coefficients and response amplitude operators (heave, roll, and sway) for a 2-dimensional, rectangular floating-breakwater. An inertia measurement unit (IMU) was used to obtain accurate measurements of the breakwater responses to regular waves. The results are compared with some existing results for similar breakwater configurations.

Key words: floating breakwaters; response amplitude operators; wave transmissions; barges

1. Introduction

With the development of large number of small marinas and recreational harbors, many new types of breakwaters have been proposed, including caisson breakwaters, slotted breakwaters, and vertical wall breakwaters. However, in deep water environments, bottom-sitting breakwaters may lead to more expensive construction costs. Floating breakwaters can be considered as an efficient approach to protecting marinas and harbors from wave attacks[1].

Compared to conventional bottom-sitting breakwaters, floating breakwaters have several unique features: (1) floating breakwaters might be ideal in those places where foundations are poor or water is deep, thus it is not possible to build bottom supported breakwaters; (2) floating breakwaters have the ecological advantages of allowing water exchange beneath the structure. In contract, the bottom supported breakwaters would prohibit the natural water exchange between oceans and harbors; (3) floating breakwaters have a low profile, especially in the areas with high tide ranges; and (4) floating breakwaters could be installed in a short time and are easy to transport, which could make it reusable in other marinas or harbors.

Previous research has shown that B/L, where B is the width of the breakwater and L is the wave length, is a parameter controlling the characteristics of wave transmission through floating rectangular[1]. The mooring line configurations are found to have insignificant effects on the motion responses of moored floating breakwaters[2]. The scale effects on transmission coefficients and forces in mooring lines are significant only in the neighborhood of $B/L = 0.5$, with no noticeable scale effects being found for waves that are very long compared to the breakwater breath[3].

Previous experimental results also showed that currents might affect the responses of a cage floating breakwater under regular waves[4]. Due to Doppler's effect, the wave length can be increased and wave height can be decreased by wave-following currents. The transmission coefficients of long waves can be slightly increased by a wave-following current for long waves. No significant effects of currents on the transmission coefficients of short waves[4].

The draft of a model is controlled by the weight of a floating breakwater: the draft can be increased with an increase of the model weight. The increase of draft may yield a decrease of resonance frequencies and the transmission coefficient will decrease with the increase of draft (see, for example[5], for theoretical results developed for a freely floating rectangular breakwater).

Stability and motion responses characteristics of

a floating breakwater are significantly affected by the stiffness of its mooring system; higher stiffness of mooring chain may result in a lower transmission coefficient and lower RAOs[6].

The responses of a pair of moored rectangular floating waters were studied theoretically and experimentally by several authors (for theoretical studies see [7-8]; for experimental studies see [8-9]). The diffraction-radiation of rectangular floating breakwaters in directional waves was studied numerically by several authors for multiple structures[10] and theoretically for a single structure[5].

Several papers published in the past several years have examined other aspects of moored floating breakwaters. For a floating breakwater installed in front of a seawall, it was found that the distance between the breakwater and the seawall played an important role[11]. Recent large scale wave basin experiments indicated the layout of a floating breakwater was another factor affecting the performance of the floating breakwater[12]. Recently, a comparison of transmission coefficients was done among single-box, double-box and board-net floating breakwaters recommend the use of board-net breakwater aquaculture engineering in deep water[9].

In this paper, we present another set of experimental data for a floating rectangular breakwater, which is similar to the one studied by Sannasiraj, et al[2]. In addition to the response amplitude operators (RAOs), the transmission coefficients were measured as well. The forces in mooring lines are not included in the present study. The results reported here are derived from one of our on-going projects on the reduction of RAOs and transmission coefficients by using new types of dampers. Only regular waves are discussed in this paper. The experimental setup and our floating breakwater model are described in section 2. The data analysis methods are explained in section 3. The results and discussion are given in section 4, and a short summary is provided in section 5.

2. The breakwater model and experimental setup

The experiments were conducted in a wave flume located in the hydraulics laboratory of the School of Civil and Environmental Engineering, Nanyang Technological University (NTU), Singapore. The flume was 45m long, 1.55m wide. The design working water depth was 1.2m. The flume was constructed of concrete with three glass windows for observations.

A piston-type wave generator was installed on one end of the flume and a wave absorber was installed on the other end to reduce the wave reflection. Wave generator was equipped with an Active Wave Absorption Control System (AWACS) to generate specified incoming waves and absorb possible waves reflected from models and the wave absorbing beach. The wave generator system was designed and fabricated by Danish Hydraulic Institute (DHI).

The breakwater model was designed to be 1.420m long, 0.75m wide and 0.410m high. For the convenience of fabrication and transportation, the model was designed as two identical parts connected by 6 stainless bolts. Each part was divided into 9 identical compartments for adjusting draft and fine-tuning the moment of inertia. The floating breakwater body was made of acrylic plate of a design thickness 10mm. The design density of the acrylic plate is 1190kg/m³. After receiving of the model, the finished breakwater model was labeled into 48 segments, the thickness of each segment was measured, and the total mass was calculated using the measured thicknesses. It was found the thicknesses of the acrylic plates varied from 8.6mm to 10.2mm. The two parts of the model were weighted by a scale and the measured weights were regarded as the true weights in this study. The error between the calculated mass and the measured mass was 0.61% for the part 1 and 0.08% for the Part 2.

The draft of the breakwater model was adjusted by adding iron plates and acrylic plates inside each of the 18 compartments. The designed thickness of the iron plate was 10mm with a design mass of 3.95kg. The designed thickness of the acrylic plate was also 10mm with a design mass of 0.6kg. Each of the plates was measured after receiving them from the contractor. The maximum and minimum masses of iron plates were 3.96kg and 3.92kg, respectively; the maximum and minimum masses of acrylic plates were 0.62kg and 0.53kg, respectively. The relative large weight variation of acrylic plates came from the noticeable variation of the plate thickness. Detailed measurement of the weight for each plate was necessary for accurate estimates of the model weight and moment of inertia.

The mooring system comprised 2 components:

Figure 1   The anchor with mooring line and the ball-bearing structure

mooring lines and anchors. Stainless steel chains were used as mooring cables. The length of each mooring cable was 3.05m and the line density was 0.155kg · m. Concrete cubes were used as anchor (see Figure 1), each having a average dimension of 0.15m × 0.15m × 0.15m and an average mass of 2.265kg. The two anchors were placed on the bottom 2.80m up-wave of the model and the other two 2.80m down-wave of the model (see Figure 2).

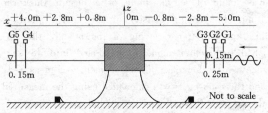

Figure 2   Locations of the 5 wave gauges

Figure 3   A view of the breakwater model in the wave flume

In order to reduce the risk of any damage to the model and the two walls of the wave flume and maintain a two dimensional motion, ball bearing structures were designed and fixed at the two sides of the floating breakwater. When the floating breakwater responds to the regular waves, the can ball bearings prevent the floating structure from colliding with the two side walls of the wave flume. The use of the ball

bearing structures resulted in gaps on the two sides; each gap was about 0.05m between the model and the tank wall. A view of the breakwater model installed in the wave flume is shown in Figure 3.

The responses of the floating breakwater model to waves were recorded by an Inertia Measurement Unit (IMU), which measured the accelerations of three translational motions and angular velocities of the three rotational motions. The IMU was mounted at one corner on the top plate of the model. To prevent the possible damage due to over-topping waters, the IMU was covered with a thin plastic wrapper during the experiments (see Figure 4). The IMU sampling frequency used in this study was 200Hz.

Figure 4   A view of the IMU mounting

The surface elevations were measured by 5 resistance-type wave gauges (Wallingford, UK): three were placed in front of the model to separate the incident waves from reflected and radiated waves; two were placed after the model to measure the transmitted waves (see Figure 2). The distances between the three wave gauges in front of the model were chosen based on the requirement of two-point wave separation method[13]. Two calibrations were done every testing day: one before the tests and another after the tests.

In all experiments, the water depth was fixed at 0.9m and the target wave height was fixed at 0.04m. Wave period varied from 0.8s to 1.8s. The draft of the floating breakwater was fixed at 0.217m for all experiments. All signals were recorded on a computer through a data acquisition system (National Instruments). Each test condition was repeated three times. The results presented in the rest of the paper are the average of three repeated tests.

3. Data Analysis

A two-point method[13] was adopted in the present study to determine the reflection coefficients. The error in this method was believed to be less than 10%. The code used to separate waves in this study was the one we used to study wave interaction with a slotted barrier in the present of a current[14]. The current velocity in this study is zero. Examples of surface elevations recorded by wave gauges G1, G2 and G3 are shown in Figures 5-7.

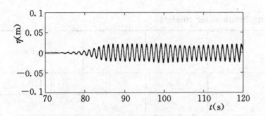

Figure 5　An example of surface elevation recorded by G1

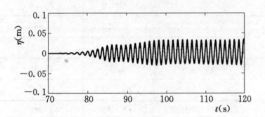

Figure 6　An example of surface elevation recorded by G2

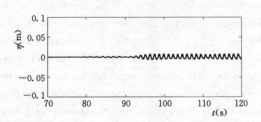

Figure 7　An example of surface elevation recorded by G4

To avoid the effects of the waves reflected back from the wave absorbing beach at the end of the wave flume, the first three waves immediately after the wave front were selected to perform data fitting for separation of waves and calculation of the amplitude of the transmitted waves. For regular waves, a recorded surface elevation can be fitted to

$$\eta(t) = \sum_{n=1}^{N} a_n \sin(n\omega t) + \sum_{n=1}^{N} b_n \cos(n\omega t) \quad (1)$$

where ω is the angular frequency and a_n, b_n are fitting parameters to be found. The amplitude A_n and the phase angle φ_n of the n-th harmonic wave component are determined, respectively, by

$$A_n = \sqrt{a_n^2 + b_n^2}, \quad \tan(\varphi_n) = \frac{b_n}{a_n}$$

In this study, $N = 3$ was used in all data fittings.

The motion responses recorded by IMU were affected by random noises. Before performing the data fitting, a low pass FFT filter with a cutoff frequency of 5 Hz was first used to remove the random noises. Figure 8 shows an example of the recorded roll response and Figure 9 shows the same roll response after possessing the data using the FFT filter.

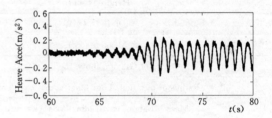

Figure 8　An example of recorded roll response before FFT filtering

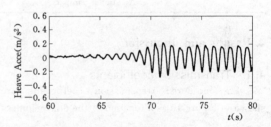

Figure 9　The record roll response after FFT filtering

IMU measured the accelerations of the three translational motions and angular velocities of the three rotational motions. To calculate the response amplitude operators (RAO), the recorded signals were converted to displacements according to the conversion formula given in Table 1, where simple harmonic motions were assumed and the effects of higher harmonics were ignored in this study when performing the signal conversions.

The results derived from this study were compared with those for other two models: Model 1[15]

Table 1 Conversion formula for RAO calculations

	Record parameter (unit)	Displacement (unit)	Conversion function
Roll	$\dot{\theta}$ Angular velocity (deg/s)	θ Angle (deg)	$\dot{\theta}=\omega\theta$
Sway	$\ddot{x}$ Acceleration (m/s²)	x (m)	$\ddot{x}=\omega^2 x$
Heave	$\ddot{z}$ Acceleration (m/s²)	z (m)	$\ddot{z}=\omega^2 z$

and Model 2 (The c-2 configuration in [2]). Model 1 was equipped with oscillating water column (OWC) units as energy absorbers. Our model is called Model 3. Some key specifications of the three models are listed in Table 2 for later discussion. The roll natural frequency and the heave frequency for the Model 3 happen to be very close to each other and both were estimated from the measured RAO curves.

Table 2 Key specifications of floating breakwater models

Items \ Models	Model 1	Model 2	Model 3
Length L (m)	1.800	3.780	1.420
Breath B (m)	0.990	0.400	0.750
Height H (m)	0.650	—	0.410
Draft D (m)	0.450	0.100	0.217
Mass m (kg)	779.0	150.5	230.6
Center of Gravity above the base (m)	0.299	0.026	0.086
Radius of gyration about longitudinal axis (m)	0.322	—	—
Mass MI about longitudinal axis (kg/m²)	—	5.33	15.89
Roll natural frequency (Hz)	0.567	0.592	0.768
Heave natural frequency (Hz)	0.481	1.012	0.768
Additional dampers	OWC	—	—

4. Results and discussion

4.1 Transmission coefficients

The measured transmission coefficients are shown in Figure 10 as a function of the dimensionless parameter $\omega^2 B/2g$, where B is the breath of the breakwater, ω the angular frequency of regular waves, and g the gravitational acceleration. Denote the wavelength by L, $\omega^2 B/2g = B/L$ for deep-water waves. Our data show a trend similar to that of the theoretical prediction reported in Sannasiraj S. A. et al. (1998): there is a minimum transmission coefficient at certain wave frequency and the breakwater is virtually transparent for very long waves. For very short waves, the transmission coefficient is not zero due to the radiated waves generated by the motion of the breakwater. The trend of our results agrees with the other published experimental results[1,6] and numerical results[6]. Note that the experimental data of Model 2 do not show a minimum transmission coefficient as their theory predicts.

4.2 Roll RAO

The measured roll RAOs are shown in Figure 11 as a function of the dimensionless parameter T_n/T,

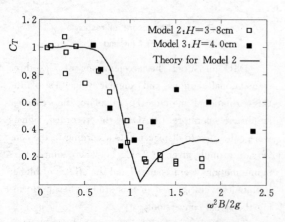

Figure 10 Measured transmission coefficients for Model 2 and Model 3

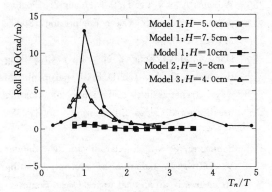

Figure 11   Measured roll RAOs for the
models in Table 3

where T_n is the roll natural period and T is the wave period. For all three models, T_n were estimated from the measured RAO curves.

Our data show a trend similar to the data for Model 2, but of a smaller maximum RAO at the natural period. The RAOs of Model 1 are significantly smaller than those for Model 1 and Model 2, showing that the addition OWC units as dampers can greatly reduce the roll response to regular waves.

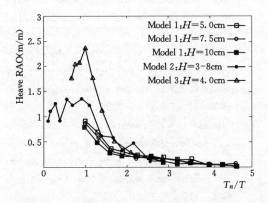

Figure 12   Measured heave RAOs for the models in Table 3

4.3   Heave RAO

The measured heave RAOs are shown in Figure 12 as a function of the dimensionless parameter T_n/T. For Model 1, as the measured RAOs do not show peaks, the longest wave period was used as T_n when preparing Figure 12. Again, our data show a trend similar to that of Model 2. The heave RAOs of Model 3 in general are larger than those of Model 2, even though the Model 2 is almost two times heavier than Model 3. Around $T_n/T=1$, Model 1 has the smallest RAO among the three models, showing again that

the OWC units are good dampers for heave motion.

4.4   Sway RAO

The measured sway RAOs are shown in Figure 13 as a function of the dimensionless parameter T_n/T. As buoyancy does not provide restoring force for sway motion, the natural sway period is strongly depended on the restoring force provided by the mooring lines.

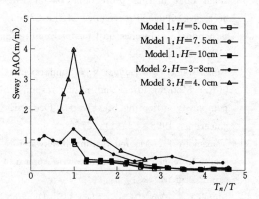

Figure 13   Measured sway RAOs for the
models in Table 3

Again, for Model 1, as the measured sway RAOs do not show peaks, the longest wave period was used as T_n when preparing Figure 13. Model 3 performs poorly in terms of sway responses to regular waves, especially near the sway natural period. This may be due to the larger draft of the Model 3, and light mooring lines used in our experiments. Similar to the heave RAOs, the use of OWC units can be a good measure for sway control.

5. Summary

A series of experiments were carried out to measure the wave transmission through a moored floating breakwater and the characteristics of the RAOs for various regular wave conditions. The trends of measured transmission coefficient and RAOs for our breakwater model agree with other reported measurements. The results provide a benchmark for our further study of damper optimization for RAO control of moored floating breakwaters.

Acknowledgement

The first author would like to thank Nanayang Technological University, Singapore, for providing the financial support through the project SUG-3/

07. The second author would like to thank Mr. Zhenzhi Deng and Mr. Fang He, both are PhD students at NTU, for their help in the experiments.

References

[1] Christian C. D.. Floating breakwaters for small boat marina protection. Coastal Engineering (2000), Proceedings of the 27th International Conference on Coastal Engineering (ICCE 2000), 2000.

[2] Sannasiraj S. A. , Sundar V. , Sundaravadivelu R.. Mooring forces and motion response of pontoon-type floating breakwaters. Ocean Engineering, 1998, 25 (1): 27-48.

[3] Yamamoto T.. Moored floating breakwater response to regular and irregular waves. Applied Ocean Research, 1981, 3: 114.

[4] Murali K. , Amer S. S. , Mani J. S.. Dynamics of Cage Floating Breakwater. Journal of Offshore Mechanics and Arctic Engineering, 2005, 127: 331-339.

[5] Zheng Y. H. , Shen Y. M. , You Y. G. , Wu, B. J. , Jie, D. S.. On the radiation and diffraction of water waves by a rectangular structure with a sidewall. Ocean Engineering, 2004, 31: 2087-2104.

[6] Yamamoto T. , Yoshida A. , Ijima T.. Dynamics of Elastically Moored Floating Objects. Appl. Ocean. Res. , 1980, 2: 85-92.

[7] Williams A. N.. Dual Floating Breakwaters. Ocean Engineering, 1993, 20 (3): 215-232.

[8] Williams A. N. , Lee H. S. , Huang Z.. Floating pontoon breakwaters. Ocean Engineering, 2000, 27: 221-240.

[9] Dong G. H. , Zheng Y. N. , Li Y. C. , Teng B. , Guan C. T. , Lin D. F.. Experiments on wave transmission coefficients of floating breakwaters. Ocean Engineering, 2008, 35: 931-938.

[10] Sannasiraj S. A. , Sundaradivelu R. , Sundar V.. Diffraction-radiation of multiple floating structures in directional waves. Ocean Engineering, 2000, 28: 201-234.

[11] Elchahal G. , Younes R. , Lafon P.. The effects of reflection coefficient of the harbour sidewall on the performance of floating breakwaters. Ocean Engineering, 2008, 35: 1102-1112.

[12] Martinelli L. , Ruol P. , Zanuttigh B.. Wave basin experiments on floating breakwaters with different layouts. Applied Ocean Research, 2008, 30: 199-207.

[13] Goda Y.. Random Seas and Design of Maritime Structures, second ed. World Scientific, Singapore, 2000.

[14] Huang Z.. An experimental study of wave scattering by a vertical slotted barrier in the presence of a current. Ocean Engineering, 2007, 34 (5-6): 717-723.

[15] Rapaka E. V. , Natarajan R. , Neelamani S.. Experimental investigation on the dynamic response of a moored wave energy device under regular sea waves. Ocean Engineering, 2004, 31: 725-743.

Study on the Influence of Groundwater Depth on Air Temperature in Land Surface Processes

Fu Zhimin[1]   Xiang Yan[2]

[1] *College of Hydrology and Water Resources, Hohai University, Nanjing, 210098*
[2] *Department of Dam Safety Management, Nanjing Hydraulic Research Institute, Nanjing, 210029*

Abstract: Based on the data on the hydrological tests of typical soils in North China, this thesis probes into the relationship between groundwater environment and air temperature. Firstly, the relationships between the effect of groundwater depth, air temperature, antecedent rain and time were identified in the method for data mining based on the measured data from a certain field experiment station in North China, and a time sequence model of ground temperature was established according to the results of data mining. And according to the catastrophe theory, a groundwater exploitation depth catastrophe model of the ground temperature was established. The thesis also explores into the criteria for determining the alteration or changes of groundwater exploitation depth and has worked out the cause and effect relationship between excess exploitation of groundwater and the catastrophe of near-surface ground temperature. Based on the high correlation between near-surface ground temperature and near-surface air temperature, the thesis comes to the conclusions that the lowering of groundwater level results in increase in near-surface air temperature, and there is a catastrophe point between the dynamic groundwater level and the air temperature, that is, the groundwater is no longer capable of regulating the air temperature when the groundwater level is lower than the catastrophe point. It is a beneficial supplement to the research of the evolution law of the groundwater environment and climate, which is very important for developing the study of groundwater.

Key words: groundwater; air temperature; data mining; catastrophe theory

1. Introduction

In recent years, land-atmosphere interaction has been one of the focuses of global climatic and hydrological studies. While as an important factor in the land-surface hydrological processes, groundwater is generally neglected or simplified in modeling. Land surface models currently used for global and regional climate studies include IAP94, VIC, BAT, LSM and SIB, all of which only consider the variations in the moisture content in the near-surface soil layer (the bottom boundary of which is taken as the constant moisture content or for loss in gravity discharge), neglecting the possible changes in groundwater depth[1]. This results in inaccuracy of the calculation based on the models[2]. Therefore, it is particularly important to verify the relationship between groundwater level and air temperature. It has been generally acknowledged that there is certain correlation between groundwater, soil moisture and ground tempera-

ture[3], but few studies have been conducted to explore into the impact of the groundwater depth on ground temperature and even air temperature. the change of groundwater level may influence the temperature of groundwater and climate[4], and the year-to-year temperature of groundwater and the air temperature may change slowly, with concealment, hysteretic and correlation on environmental ecology, so it may cause more serious damage, but with slow timeliness. Determination of the relationship between groundwater level and air temperature will facilitate comprehensive determination of the hazards of excess exploitation of groundwater and provide better guidance for the remediation of groundwater environment.

2. Determination of the factors influencing groundwater in land surface processes through data mining

The measured data from field experiment stations for a long period of time and the data mining technologies have great significance in analyzing the various data on the factors influencing air temperature, verifying the correlation between air temperature and groundwater, and proving the significant influence of these factors on groundwater in land surface processes.

2.1 About the testing station

The testing station is situated in Northern China Plain, with the climate of the semi-humid in warm temperature zone. The testing station distributed 10 groups of round measuring cylinders with the surface square of 1m², keeping the groundwater beneath the ground 0.5m, 1.0m, 1.5m, 2.5m and 3.5m respectively. In each measuring cylinder, the homogenous sandy loam, with the typical characteristics of Northern China Plain, is filled. Periodically it observes the air temperature, wind speed and direction, pressure, humidity, sunlight and other weather data, as well as the ground temperature, water content of soil, soil suction and total soil water potential at the different depth inside the measuring cylinder. The testing station collects the data measured from 1991 to 1996. See Figure 1 for location of the proving ground.

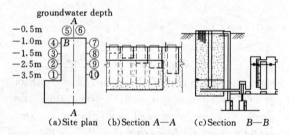

Figure 1   Location of the proving ground

2.2 The correlation between air temperature and ground temperature

Since the one that affects the human beings mostly is the near-surface air temperature, and the ground temperature has close relation with the temperature of near-surface. A temperature bulb capable of measuring the temperature of the groundwater with depth of 350cm was used to observe the variation trends of the groundwater at different depths, and the results are shown in Figure 2. As can be seen from the figure, ground temperature not only changes with the variations of the atmospheric temperature, but also rises with the increase in depth. The maximum

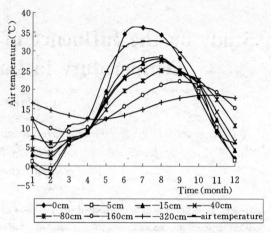

Figure 2   Variety temperature of the different point

temperature at a certain depth in a year indicates obvious hysterics effect of changing with the variations of the depth, that is, the surface temperature is relatively identical to the atmospheric temperature and reaches the peak during the period from June to August; but with the increase of depth, the maximum temperature occurs during the period from July to September in case of 5cm in depth, and from October to November in case of 320cm. In summary, there is certain hysteresis effect between the ground temperature and the depth. Therefore, only changes of the near—surface ground temperature is relatively consistent with the variations of the air temperature. It graphs the hydrograph for the temperature (away 1.5m from the ground), temperature of ground beneath 10cm and temperature of ground beneath 40cm measured by the Testing Station during the year from 1996 to 1997 (See Figure 3). After calculating, the related coefficient between temperature and near-surface ground temperature is more than 0.948. Therefore it uses the near-surface ground temperature to substitute the temperature for analyzing the influence of groundwater depth on temperature.

2.3 Detection of influence factors of near-surface ground temperature

Decision Tree Algorithm, which is widely used in data mining, is a method of inductive classification and can be designed to have good scalability. This algorithm can be combined with a very-large data base to process various types of related data (continuous data, discrete data, and Boolean data). The DecisionTree Algorithm is adopted in this thesis to explore

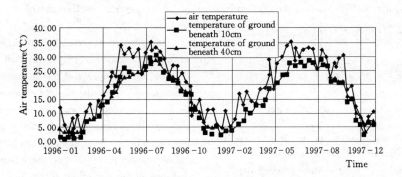

Figure 3   Process lines of the gas temperature and the ground
temperature near the surface

into the effect of groundwater depth on shallow ground temperature[5]. By using the ground temperature on the land surface at the testing station as the target attributes, using the evaporation capacity, groundwater recharge, amount of precipitation, groundwater evaporation capacity, wind speed of land surface, air pressure, sunlight, relative humidity, groundwater depth as the input attribute and using the data mining technology to make correlation analysis, it can detect the influence extent of all influence factors on land surface temperature. See Figure 4 for results.

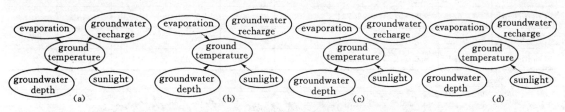

Figure 4   The causes mining of soil temperature near the ground

The influence caused by penetration recharge may disappear when the strength of anticipated link is 30% [see Figure 4 (b)], the influence caused by the groundwater evaporation may disappear when the strength of anticipated link is 62.5% [see Figure 4 (c)], and the influence caused by the groundwater depth may disappear when the strength of anticipated link is 87.5% [see Figure 4 (d)]. From which, it can be seen that the one to affect the ground temperature of near-surface land mostly is the sunlight; following by the groundwater depth, evaporation capacity of groundwater and penetration recharge; therefore the influence caused by underground depth on near-surface ground temperature can not be ignored. Based on the high correlation between the ground temperature and air temperature, it can be seen that the dynamic groundwater level is another factor to influence the air temperature in land surface process, in which without considering the influence of the dynamic groundwater is unreasonable.

3. Catastrophe model of groundwater exploitation depth influencing ground temperature

3.1   Time sequence model

It is discovered from the mining results that ground temperature is mainly influenced by such factors as solar radiation, groundwater depth, recharge from infiltration, etc., therefore, the statistical model of surface temperature is deducted as follows:

$$T = T_H + T_T + T_P + T_\theta \qquad (1)$$

where: T is the value calculated according to the statistical model of surface temperature; T_H, T_T, T_P and T_θ are respectively groundwater influence coefficient, temperature influence coefficient, antecedent rain influence coefficient and time effect coefficient.

Taking into consideration of the influence by the initial measured value of surface temperature, a time series statistical model to obtain the measured value of surface temperature was obtained through putting the various influence coefficients into the equations:

$$T = a_0 + \sum_{i=1}^{4} a_i (H^i - H_0^i) + \sum_{i=1}^{2} \left[\left(b_{1i} \sin \frac{2\pi it}{365} - \sin \frac{2\pi it_0}{365} \right) \right.$$

$$\left. + b_{2i} \left(\cos \frac{2\pi it}{365} - \cos \frac{2\pi it_0}{365} \right) \right]$$

$$+ \sum_{i=1}^{7} c_i P_i + d_1 (\theta - \theta_0) + d_2 (\ln\theta - \ln\theta_0) \qquad (2)$$

where: a_0 is a constant term; H_0 is the groundwater depth on the initial day of the data series; t_0 is the accumulated number of days of the modeling data series from the initial day to the day of initial survey; θ_0 is $t_0/100$; a_i is the regression coefficient of groundwater influencing factors; b_{1i} and b_{2i} are regression coefficients of temperature influencing factors; c_i is the coefficient of regression; and d_1 and d_2

are the regression coefficients of the time-effect factors.

In order to analyze the time-effect variation of ground temperature, the method of simulating the data on measured ground temperature at the test site was adopted. And the method of stepwise regression analysis was adopted, and these factors were put into the regression equation one by one in a descending order based on the level of significance of their influence on surface temperature, and these factors were excluded one by one according to the level of influence of these factors to obtain the coefficient of regression of the time series statistical model, see Table 1. And see Figure 5 for the measured values of the survey point, the fitted value, and the residual error line.

Table 1   **The regressive coefficients of the statistics model of land surface temperature**

Factors	a_0	a_1	a_2	a_3	a_4	b_{11}	b_{21}	b_{12}	b_{22}
Coefficient	3.502	886.601	−96.731	0.00	0.334	−6.223	−17.330	1.433	−1.730
Factors	c_1	c_2	c_3	c_4	c_5	c_6	c_7	d_1	d_2
Coefficient	−0.194	−0.222	−0.233	−0.339	−0.306	−0.304	−0.195	1.194	−0.546

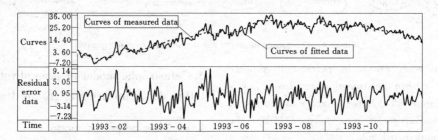

Figure 5   Curves of soil water measured data, fitted data and residual error data

The coefficient of total correlation of this statistical model is 0.966, and the residual standard deviation is 2.78, which demonstrates the high accuracy of the statistical model and that the ground temperature of the survey area can be predicted based on the relevant factors.

3.2   Catastrophe model

In order to study the critical depth of groundwater extraction depth, the corresponding statistics model established above has to be used. And then study the catastrophe criteria based on the catastrophe theory. It uses the Grey theory to process the data series in this paper. Supposing that the irreversible deformation series is $S^{(0)}$, it can generate 1-AGO

accordingly; from which, it can obtain $S^{(1)}$. For the $S^{(1)}$, it can establish GM (1, 1) model normally. However the result of which is in an index form, which dissatisfies with the requirements of the analysis based on the catastrophe theory. Therefore, it uses Taylor formula to unfold. Considering the influence caused by sufficient accuracy, it adopts 5 cutoffs of Taylor expand form.

$$S^{(1)}(x) = a_0 + a_1 x + a_2 x^2 + a_3 x^3 + a_4 x^4 + a_5 x^5 \qquad (3)$$

By making primary mathematical manipulation after calculating the reciprocal for Formula (3) said above and omitting the constant term (the constant term will never change the nature of V), i.e.:

$$V = \left(\begin{matrix}+\\-\end{matrix}\right)\frac{1}{4}Z_s^4 + \frac{1}{2}uZ^2 vZ \qquad (4)$$

where: $u = \dfrac{-6a_4 + 10a_3 a_5}{5a_5 \sqrt{|5a_5|}}$, $v = \dfrac{8a_4^3 - 30a_3 a_4 a_5 + 50a_2 a_5^2}{25a_5^2 \sqrt[4]{20|a_4|}}$

Based on the above basic theory, it conducts the calculation on finding the reciprocal of potential function, and making $\partial V/\partial Z = 0$, so:

$Z^3 + uZ + v = 0$ (regular cusp catastrophe)

Or $-Z^3 + uZ + v = 0$ (antithesis cusp catastrophe) (5)

The critical point set determined by Formula (5) is the profile of equilibrium (M); this is called also the catastrophe manifold. The graph of M in (Z, u, v) space is a spectrum curve with certain wrinkles, consisting of three parts on the top, in the middle and at the bottom; where the parts on the top and at the bottom are stable, the one in the middle is unstable. Even though (u, v) changes in any forms, the phase point (Z, u, v) may change evenly on the top part (or bottom part) only, and leap the middle part when reaching to the wrinkle edge of the part. Therefore all points at the vertical tangent line on the profile of equilibrium shall constitute the catastrophe point set of status (i. e. : set of singularities) S, with the equation as follows:

$\dfrac{\partial^2 v}{\partial Z^2} = 3Z^2 + u$ (regular cusp catastrophe)

Or

$\dfrac{\partial^2 v}{\partial Z^2} = -3Z^2 + u$ (antithesis cusp catastrophe) (6)

The projection of the set of singularities on the plane to control the variables (u, v) may constitute the bifurcate point set (B); this is the collection of all points that may make the variables of status leap; by removing Z through the calculation of Formula (5) or Formula (6), it obtains the equation of B as follows, i. e. :

$$\Delta = 4u^3 + 27v^2 = 0 \qquad (7)$$

From the analysis above, it can be seen that the extraction depth of groundwater is mainly subject to the following criteria, i. e. :

If $\Delta = 4u^3 + 27v^2 > 0$, the groundwater produces constant influence on the ground temperature; and the groundwater is extracted reasonably.

If $\Delta = 4u^3 + 27v^2 < 0$, the groundwater produces unstable influence on the ground temperature, and the groundwater is overdrawn.

If $\Delta = 0$, it may reckon that the groundwater depth is the critical point to make ground temperature

rise; if less than the groundwater depth, the groundwater may produce zero influence almost on temperature of ground surface, and the groundwater may have no way to adjust the tropical island effect; so the groundwater will have no way to influence the tropical island effect. In this paper, it calls the current groundwater depth as the critical depth for tropical island effect of groundwater, and the corresponding ground temperature status is called the critical surface of tropical island of groundwater; See Figure 6.

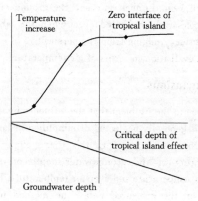

Figure 6　Sketch map of the critical height by the heat island

A typical North China area near the test station was selected for analog computation, and the surface temperature influence coefficient of groundwater was generated based on the above statistical model and was expressed as $S^{(1)}(x)$. And then Equation (3) was used for the multiple regression analysis, the results of which are as follows:

$$S^{(1)}(x) = -3.6477 + 4.4193x - 1.3977x^2$$
$$+ 0.2567x^3 - 0.0224x^4 + 0.0007x^5$$
$$(8)$$

$$V = \frac{dS^{(1)}(x)}{dx} = 4.4193 - 2.7954x$$
$$+ 0.7700x^2 - 0.0894x^3 + 0.0036x^4$$
$$(9)$$

The following was obtained through transformation:

$$V = \frac{1}{4}Z^4 + \frac{1}{2} \times (-0.9188)Z^2 + (-0.3394)Z$$
$$(10)$$

As is known from Equation (6), the catastrophe model of surface temperature influence coefficients based on groundwater depth is a canonical catastrophe model, in which $u = -0.9188$ and $v = -0.3394$, therefore, the discriminate is:

$$\Delta = 4u^3 + 27v^2 = 0.0074 \qquad (11)$$

As is known from $\Delta > 0$, the groundwater exploitation depth of the selected area was stable, but Δ was close to zero and in the critical state. It is known from the calculation that as for the similar soil conditions in this typical North China area, the range of groundwater depth is from 6m to 8m in case $\Delta = 0$; therefore, groundwater no longer regulates the air temperature in case of groundwater depth outside the range of 6-8m. In such as case, the groundwater has been excessively exploited even when the land does not subside, judging from the perspective of environmental evaluation in terms of air temperature.

4. Conclusions

Due to the problem that the influence by groundwater is generally neglected or simplified in land surface processes at present, this thesis explores into the impact of the groundwater depth on shallow ground temperature and even air temperature. Firstly, based on the measured data, the method for data mining was used to prove that groundwater depth is one of the main factors influencing ground temperature and verify the high correlation between near-surface ground temperature and air temperature, reaching the conclusion that groundwater depth has great impact on air temperature and is indispensable in establishing models of land surface processes; whereas, the factor of groundwater is generally neglected in the studies of land surface processes, which will surely result in big errors. And then the catastrophe model of groundwater exploitation depth which influences ground temperature was studied, and reached the conclusion that groundwater level does not always influence the ground temperature, wherein, when the value of groundwater level reaches a certain point, groundwater no longer influences the ground temperature, hence the concept of excess exploitation critical depth as to the influence of groundwater on ground temperature.

Through analog computation of the data of a typical area, this thesis obtains the upper and lower thresholds of the groundwater depth of 6m$<H<$8m influencing air temperature, that is, groundwater no longer regulates air temperature when the groundwater level is lower than 8m. And it is only through studies of the criteria for determining the alteration or changes of groundwater exploitation depth to determine the range of groundwater depth within which the groundwater has influence on air temperature that the groundwater factors in land surface processes can be more specifically defined. Therefore, this study is a beneficial supplement to the studies on the relationship between groundwater environment and climatic evolution and has important significance to the studies of groundwater development.

References

[1] Jing Jihong, Jing Lei, Sun Jichao, Wang Shan. The groundwater resources distribution and its exploitation and utilization in Huang-Huai-Hai plain, http: //www. cgs. gov. cn/zt _more/34/zhaiyao/html/05/52 4. htm, 2005.

[2] Xie Zhenghui, Liang Xu, Zeng Qingcun. A parameteri- zation of groundwater table in a land surface model and it s applications. Chinese Journal of A tmospheric Sciences (in Chinese), 2004 , 28 (3): 374-384.

[3] Sun Xu jin, Zhou Ge, Sun Xiao ping, Zhang Jin bing. The Study of Ground Temperature Change after Groundwater Exceeding Extraction. Journal of North China Institute of Water Conservancy and Hydroelectric Power, 2003 , 24 (1): 37-39.

[4] Lln Xueyu, Fang Yanna, Liao Zisheng, Chen Hongyan. Impact of Global Warming and Human Activities on Groundwater Temperature. Journal of Beijing Normal University (Natural Science), 2009, 45 (5): 452-457.

[5] Babovic, Vladan. Data mining and knowledge discovery in sediment transport. Computer- Aided Civil and Infrastructure Engineering. Blackwell Pub Inc, 2000, 155: 383-389.

Investigation of Critical Submergence Depth in Horizontal Intake of Power Plant for Hydro-Electric Dams

H. Golmohammadi[1]   M. R. Jalili Ghazizadeh[2]

R. Roshan[3]   A. Ahmadi[4]

[1]*MSc. Student, Shahrood University of Technology, Shahrood, Iran*
[2]*Assistant Professor, Power and Water University of Technology, Tehran, Iran*
[3]*Academic Staff, Water Research Institute, Tehran, Iran*
[4]*Assistant Professor, Shahrood University of Technology, Shahrood, Iran*

Abstract: Formation of vortex at the intake of a power plant may cause some problems for the proper operation of the power plant. This phenomenon could make air and floating materials to enter the power plant's conduit, causing damages to its equipments. Providing an enough submergence depth for the intake prevent the vortex formation. But due to some other considerations such as the risk of entering sediments to the power plant, designers are forced to choose higher level for the intake installation. Therefore, determination of critical submergence depth plays an important role on designing a power plant. In recent decades, many research with the aim of determining a Critical (minimum) Submergence Depth (CSD) led to some empirical relationships for the prediction of CSD. A comparison between the results of these formulas show a wide range of discrepancies in predicted CSD, confirming the need for further studies for developing a more comprehensive equation. In this study the CSD data for number of large dams, derived from a series of large-scale hydraulic model studies were collected and based on these data a new empirical relationship for predicting CSD was introduced. As the data were collected from a wide range of experimental studies, it is believed that the new formula has a good integrity and accuracy.

Key words: hydraulic structures; power plant; horizontal intake; vortex; submergence depth

1. Introduction

Vortex is normally formed at the intake of power plants by a rotational flow of water, when the rotational force exceeds the force necessary for the movement of water above the intake. The geometrical specification of a horizontal intake is shown in Figure 1, in which d is the intake's diameter; h, called submergence depth, is the distance between the intake's centre line to water surface and v is the flow velocity in the intake conduit. Critical Submergence Depth (CSD) is the minimum depth, h_c, required for the formation of a vortex with air core at the intake.

Vortex forms when the water depth above the intake is less than the critical submergence depth. It can cause various problems such as increasing head loss, creating noises and reducing the efficiency of turbines.

It may also causes damages to the mechanical equipments of power plant. To prevent the formation of vortex and its negative consequences, the existence of good enough water depth at the intake is always recommended.

2. Vortex types

Vortices are categorized based on different criteria. In 1981, Hecker studied the vortices and according to vortex power, he categorized them in six types[1]. Vortex type 1, can only create a swirling flow on water surface. Vortex type 2, makes a pit on water surface in addition to swirling flow. Vortex type 3, generates a swirling flow with a cone from water surface to the intake. Vortex type 4, has the same shape as type 3,

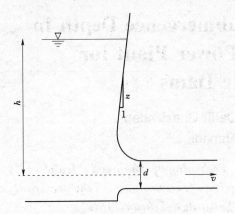

Figure 1　Specifications of a horizontal intake

with the ability to draw floating particles to the intake entrance. Vortex type 5, sucks the air in addition to floating particles to the intake. Finally in vortex type 6, a swirling flow with an air-entraining core is formed. Schematic pictures of these vortices are shown in Figure 2[1].

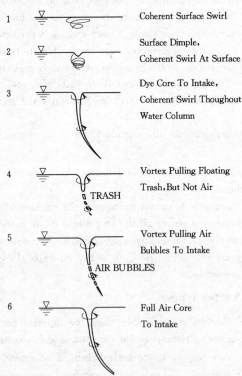

1		Coherent Surface Swirl
2		Surface Dimple, Coherent Swirl At Surface
3		Dye Core To Intake, Coherent Swirl Thoughout Water Column
4	TRASH	Vortex Pulling Floating Trash, But Not Air
5	AIR BUBBLES	Vortex Pulling Air Bubbles To Intake
6		Full Air Core To Intake

Figure 2　Various types of vortices based on their
power categorized by Hecker[1]

Vortices, unable to entrain air to the flow, generate only head loss and are not hazardous for the

power plant equipments, but the air entraining ones are dangerous and their generation should be avoided[2]. Providing an enough submergence depth and using anti-vortex generation equipments are some of important ways to prevent the vortices at the intake. By controlling the minimum operation level and providing CSD at the intake, it is possible to avoid formation of vortex and to prevent any damages to power plant.

3. Determination of Critical Submergence Depth (CSD)

In recent decades, critical submergence depth hag been investigated by many researchers. As a very complex phenomenon, vortex formation may depend on various parameters including flow velocity, diameter of tunnel to the power plant, the level of swirling of the flow, intake's upstream wall slope and the geometry of the intake. To predict the formation of vortex, researchers have examined various parameters and they have proposed different relationships commonly based on Froude number, tunnel diameter and the flow velocity in the tunnel. Some of these formulas for horizontal intakes are listed in Table1.

Table 1　　Different formulas for
horizontal intakes

Equation number	Equation	Ref.
(1)	$h_c/d = 3.95Fr^{0.5} - 0.5$	Amphlett[3]
(2)	$h_c/d = 2.3Fr$	Gordon[4]
(3)	$h_c/d = 2Fr + 0.5$	Knauss[1]
(4)	$h_c/d = 2.25Fr + 1.5$	Novak[5]
(5)	$h_c/d = 2.3Fr + 1$	Sinijer[6]
(6)	$h_c/d = 2 \ (1/z)^{0.008} Fr^{0.334}$	Sarkardeh[7]
(7)	$h_c/d = 4.4v^{0.54}/d^{0.73}$	Nagarka[8]
(8)	$h_c/d = 1.474v^{0.48}/d^{0.24}$	Rohan[9]

In the above relationships, Fr is the Froude number $[v/(gd)^{0.5}]$, v is the flow velocity and d is the diameter of intake tunnel and z is the slope of upstream approaching wall (Figure 1).

The variation of submergence depth with Froude number, using relationship 1-6, is shown in Figure 3. It should be mentioned that in Equation(6), z assumed to be equal 1.

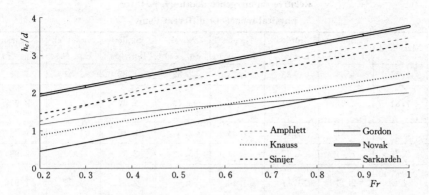

Figure 3　Variation of CSD to tunnel diameter (h_c/d) versus
Froude number (Fr) for relationships 1-6

Figure 3 shows a considerable difference between CSD predicted by relationships 1-6. As it can be seen, Novak Equation (4) provides the most and Gordon Equation (2), the least CSD. The wide range of discrepancies between the results warrents more studies for determining a proper CSD, when designing power plant.

4. Experimental results

In an attempt to examine the accuracy of empirical equations mentioned in Section 3 for predicting CSD, the data from large-scale physical model study of power plants designed for 7 dams in Iran are collected in Table 2.

Table 2　　　　**Specifications of physical models of dams in Iran**

Dam Name	Scale model	Intake's discharge (m³/s)	Tunnel's diameter (m)	Fr	Relative CSD (h_c/d)	Refrence no.
Gotvand olia	1 : 25	400	11. 5	0. 36	1. 77	[10]
Shahid Abasspour 1	1 : 18	185	6. 4	0. 73	2. 81	[11]
Karoon 3	1 : 33	700	12. 6	0. 51	2. 38	[12]
Siyah bisheh	1 : 20	130	5. 7	0. 68	2. 72	[13]
Masjed Soleiman	1 : 66. 6	375	10	0. 48	2. 2	[14]
Seymareh	1 : 50	460	11	0. 47	2. 29	[15]
Shahid Abasspour 2	1 : 30	375	9	0. 63	2. 82	[16]

5. Comparison between experimental and empirical results

Using empirical Equation (1) -Equation (8), critical submergence depth was calculated for the models mentioned in Section 4. The results are shown in Table 3.

The results in Table 3 show that relative CSD for the physical models varies considerably between the empirical relationships and those of experiments which confirm the need for a more accurate and comprehensible equation for evaluating relative CSD.

Table 3   **Relative submergence depth (h_c/d) for**
physical models of different dams

Dam Name	Measured	Amphlett [Eq. (1)]	Gordon [Eq. (2)]	Knauss [Eq. (3)]	Novak [Eq. (4)]	Sinijer [Eq. (5)]	Sarkardeh [Eq. (6)]	Nagarka [Eq. (7)]	Rohan [Eq. (8)]
Gotvand olia	1.77	1.88	0.83	1.23	2.32	1.83	1.41	1.53	1.57
Shahid Abasspour 1	2.81	2.87	1.67	1.95	3.13	2.67	1.77	2.92	2.19
Karoon 3	2.38	2.31	1.16	1.51	2.64	2.16	1.57	1.76	1.84
Siyah bisheh	2.72	2.76	1.57	1.86	3.03	2.57	1.73	2.98	2.12
Masjed Soleiman	2.2	2.24	1.11	1.46	2.59	2.11	1.55	1.91	1.8
Seymareh	2.29	2.2	1.07	1.43	2.55	2.07	1.53	1.79	1.77
Shahid Abasspour 2	2.82	2.63	1.44	1.76	2.91	2.44	1.69	2.31	2.04

6. Introducing a new relationship for relative CSD

As it can be shown in Table 3 the CSD for a studied dams predicted by empirical relationships 1-8 varied considerably. It is believed that this could be due to the fact that these empirical equations were established using the results of a limited number of physical models. It is argued here that using the results of several large-scale hydraulic models of large dams, Table 1, would lead to introducing a reliable and accurate equation for calculating relative CSD. Figure 4 shows the experimental data for 7 hydraulic models of dams. It can be seen that a logarithmic equation can correlate them adequately. This new relationship is in the form of:

$$h_c/d = 1.522 L_n(Fr) + 3.376 \qquad (9)$$

The regression coefficient of the curve in Figure. 4 is 0.942, showing a good correlation between the experimental results.

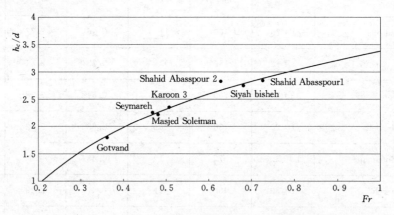

Figure 4   Variation of h_c/d versus Froude number for experimental results

7. Conclusion

(1) Critical submergence depth calculated by existing empirical equations showed a wide range of variation between them.

(2) Using the results of 7 large-scale physical models of large dams in Iran a new relationship for relative CSD was introduced. It is believed that this relationship could accurately predict CSD for horizontal intakes.

(3) The wide range of difference between the results from empirical equations, suggest that critical submergence depth, CSD, may be dependent to other parameters in addition to Froude number, confirming

the need for further studies on this phenomenon.

Acknowledgment

The support and co-operation of Water Research Institute affiliated to ministry of energy in this research is greatly acknowledged.

References

[1] Knauss J. Swirling Flow Problems at Intakes. *IAHR Hydraulic Structures Manual*, Balkema, Rotterdam, 1987.

[2] Anwar H. O., Weller, J. A., Amphlett, M. B.. Similarity of free-vortex at horizontal intake. *J. of Hydraulic Research*, 1978, 16, (2): 95-105.

[3] Amphlett, M. B.. Air-entraining vortices at a Horizontal Intake, *HRS Wallingford*, U. K., Report No. OD/7, 1976.

[4] Gordon J. L.. Vortices at intake structures. *Water Power*, 1970 (4): 137-138.

[5] Novak P, A. I. B Moffat, C. Nalluri, R. Narayanan, *Master eBook*, 1990, *ISBN*10: 0-203-24651-9.

[6] Sinniger, Richard Relié. Constructions hydraulioues. Presses polytechniques et universitaires romandes, 1993.

[7] Sarkardeh H.. Influence of reservoir vortex on flow regime in the intake tunnel of power plant. Amir Kabir University of Technology, Tehran, Iran, 2008.

[8] Nagarka P. K.. Submergence criteria for Hydro-Electric Intakes, *ASCE Journal of Hydraulic Divn*, 1988.

[9] Rohan K.. Conditions of similarity for a drain vortex. *Simposia CWPRS*. Poona, 1965, 2.

[10] Water Research Institute of Ministry of Power. Hydraulic model of intake of power plant-Gotvand Olia Dam. 2009.

[11] Water Research Centre of Ministry of Power Hydraulic model of intake of power plant-Shahid Abasspour 1 Dam. 1991.

[12] Water Research Centre of Ministry of Power Hydraulic model of intake of power plant-Karoon 3 Dam. 1998.

[13] Water Research Institute of Ministry of Power Hydraulic model of intake of power plant-Siyah bisheh Dam. 2007.

[14] Water Research Centre of Ministry of Power Hydraulic model of intake of power plant-Masjed Soleiman (Godar lander) Dam. 1995.

[15] Water Research Centre of Ministry of Power. Hydraulic model of intake of power plant-Seymareh Dam. 1999.

[16] Water Research Centre of Ministry of Power. Hydraulic model of intake of Power plant-Shahid Abasspour 2 Dam. 1994.

River Bed and Bank Stability Using Convex, Concave and Linear Bed Sills

Alireza Keshavarzi   James Ball

School of Civil & Environmental Engineering, F.E.I.T., UTS,
Broadway, NSW, 2007, Australia

Abstract: In this study, to examine the effect of a bed sill on river bed scouring, five types of bed sills with nine different configuration patterns were tested in a laboratory flume. To find the optimum sill configuration, the scouring pattern and the depth of the scour hole at the bed was observed downstream of the sill during the experiment. The depth of the scour hole was measured during and at the end of an experiment. It was found that concave and convex configuration patterns with a circular shape and $D/W=1.2$ was the most effective sill configuration for protection of both the bed and bank of a river. In particular, the convex pattern was found to be very effective for protection of the center of the river and a concave pattern was very effective for protection of the river bank. The results from Particle Image velocimetry (PIV laser) study showed that at the center of the bed sill, the bed sill pushes the wake vortex rings further downstream.

Key words: bed sill; convex circular; concave circular; PIV; bed scouring

1. Introduction

Scouring of the bed and bank of a river is a major source of environmental degradation problem and it is still a major concern for environmental hydraulic engineers. The river bed scouring is very complicated and it is not completely yet understood. The sediment particle that is scoured from the bed and banks of the river creates a major problem in natural and artificial channels, while the deposition of sediment particles produces problems for riverine habitat.

A number of studies for example; Hassan and Narayanan (1985) and Habib et al. (1994) have been undertaken to understand the bed scouring downstream of a linear sill in a hydraulic jump. The results of these studies showed that there was maximum scour downstream of the sill. In addition, Hoffmans and Pilarczyk (1995) measured the maximum scour downstream of large hydraulic structures in rivers. They found different phases of scour under clear water conditions. In the studies by Breusers (1966), Volkart et al. (1973), Borman and Julien (1991), Van der Meulen and Vinji (1995), Hoffmans and Verheig (1997), Marion and Gaudio (1998), Gaudio et al. (2000), Gaudio and Marion (2003), Lenzi et al. (2002) and Ben Meftah and Mossa

(2004), bed scouring was tested for a bed sill constructed perpendicular to the flow direction under different flow conditions. In almost all previous studies, it was found that the bed sill at a right angle to the flow direction produced large uniform bed scouring downstream of the sill. However, no previous work examined a circular bed sill under different arrangement patterns has been reported in the literature. Therefore, in this study, five different types of bed sills with nine installation arrangements have been tested in an experimental study, as a result, a new bed sill pattern is suggested to prevent scouring from the bed and banks of a river.

2. Material and methods

An experimental study was performed in a non-recirculating experimental flume with mobile bed. The experimental flume consists of 15.8m long, 0.5m width and 0.4m height. The bed of the flume was covered with 100mm sand particle with approximately uniform size of D_{50} of equal to 0.63mm. The flow condition for the data set 1 is presented in Table 1.

The data set 2 is measured in an experimental flume 15m long, 0.7m wide and 0.6m depth with glass walls and bed. The flow condition for the experiments of data set 2 is presented in Table 2. The bed

Table 1 Flow condition in the experimental tests series 1 (Data Set 1)

Test No.	Model Test	Flow depth (cm)	Flow discharge (L/s)	r/w
1	M1 (CVX)	8.8	14.75	0.5
2	M1 (CCV)	8.8	14.75	0.5
3	M2 (CVX)	9	17.1	0.27
4	M2 (CCV)	8.8	14.75	0.27
5	M3-30 (L.D)	8.8	14.75	—
6	M3-30 (L.U)	8.8	14.75	—
7	M4-60 (L.D)	9.8	20.3	—
8	M4-60 (L.U)	9.5	19.2	—
9	M5-90	9.2	17.94	—

of the flume was covered with 120mm sand particles with approximately uniform size of D_{50} of equal to 0.63mm. Nine different forms of sill were installed at the bed of the flume and the sills were tested under similar flow condition. The sill patterns were shown in Figure 1. The experimental tests were performed from 6 to 24 hours. Sandy surface meter was used to measure the bed scouring in a grid. A downstream gate was used to control the water surface inside the flume. A pre calibrated V-Notch is used to measure the flow rate at the downstream of the experimental flume. Further experimental studies were performed at the University of Technology Sydney, using PIV to find the flow structure in front of and behind a convex circular bed sill with $D/w = 1.2$. These experimental studies were carried out in a rectangular recirculating flume with length of 6m, width of 0.25m

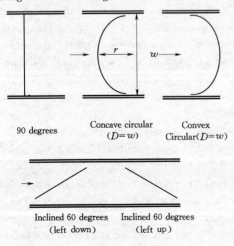

90 degrees   Concave circular ($D=w$)   Convex Circular($D=w$)

Inclined 60 degrees (left down)   Inclined 60 degrees (left up)

Figure 1   Patterns of 5 bed sills used in this study

and depth of 0.3m. Figure 2 shows the setup of experimental study with PIV laser (Particle Image velocimetry). The flow conditions in these experimental tests were presented in Table 3.

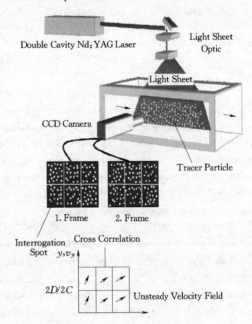

Figure 2   Experimental setup of PIV

Table 2 Flow condition in the experimental tests series 2 (Data Set 2)

Test No.	Model Test	Flow depth (cm)	Flow discharge (L/s)	r/w
1	M1 (CVX)	9.8	18.8	0.5
2	M1 (CCV)	6.6	15.87	0.5
3	M2 (CVX)	8.8	18.8	0.27
4	M2 (CCV)	8	16	0.27
5	M3-30 (L.D)	9.3	18.2	—
6	M3-30 (L.U)	—	—	—
7	M4-60 (L.D)	7.5	17.7	—
8	M4-60 (L.U)	—	—	—
9	M5-90	7.2	14.4	—

Table 3 Flow condition in the experimental tests series 3 (Data Set 3)

Test No.	Model Test	Flow depth (cm)	Flow discharge (L/s)	D/w
1	Convex	22.5	2.67	1.2
2	Convex	22.5	2.67	1.2
3	Convex	22.5	2.67	1.2
4	Convex	19.6	2.67	1.2

3. Result and discussion

To find the most important parameters for bed scouring, a dimensional analysis was done for circular and inclined sills. For circular bed sills, it was found that the following parameters are very effective for scouring:

$$Y_s = f(g, \nu, \rho_w, \rho_s, q, q_s, W, d_{50}, r) \qquad (1)$$

and for inclined sills:

$$Y_s = f(g, \nu, \rho_w, \rho_s, q, q_s, W, d_{50}, \theta) \qquad (2)$$

In which Y_s is the maximum scouring in equilibrium condition, g is the acceleration, ρ_w is the flow density, ρ_s is the particle density, q is the flow discharge per unit width, q_s is the sediment discharge per unit width, d_{50} is the median particle size, W is the width of channel, r is the arch distance of the circular sill and θ is the angle of inclined sill (Figure 1).

The results of the experimental tests are compared in Figure 3 to Figure 4. From the results it was found that the pattern of the bed scouring and maximum scouring occurred in different pattern and location downstream of bed sills. Also maximum scouring hole occurred at different location for different bed sill models. For the convex circular models [Tests M1 (CVX) and M2(CVX)] with $r/w = 0.5$ and $r/w = 0.27$, respectively, the scouring hole occurred in the center of the flume, whereas for the concave models [M1(CCV) and M2(CCV) tests], maximum scouring occurred at the bank or side walls of flume. For inclined sill models (M3-30 and M4-60 tests), bed scouring occurred at a location upstream side of bed

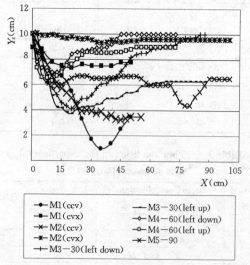

Figure 3   Comparison of all models for data set 1

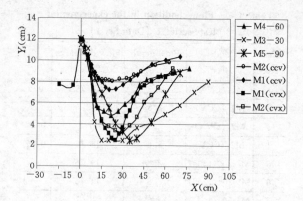

Figure 4   Comparison of all models for data set 2

sill. For 90 degree sill model (M5-90) which is a usual sill model, bed scouring occurred uniformly at all locations downstream of bed sill. Also from comparison of bed scouring in different bed sill patterns, it was found that the maximum scouring depth occurred at the 90 degree model. From the comparison of bed scouring, it was found that in 90 degree sill model, the scour hole depth is uniform along the length of bed sill and no difference was found in scouring hole along the width of channel, whereas, bed scouring for other models are very different at the center and side of channel. Therefore, it was found that every sill pattern has a specific application and it is appropriate for special protection.

In general, from the results it was found that the bed scouring in convex circular (CVX) is much less than that of concave circular (CCV) model. From the comparison of circular models, it was found that in convex pattern with $r/w = 0.27$ (Test M2), the bed scouring is less than that of convex pattern with $r/w = 0.5$ (Test M1). The profile of bed scouring for concave and convex circular models is shown in Figure 4 and they are compared with the right angle bed sill. It is shown that two circular bed sill models produced less bed scouring than 90 degree model.

From the results of inclined sill presented for 30 degree and 60 degree models (Tests M3-30 and M4-60) in Set 1 of the experimental test, it was found that for both models, there was no significant difference for the left up and left down patterns. The only difference was that in left up pattern, the scouring occurred in the left side of the channel, whereas in left down the scouring occurred in the right side of the channel. Due to this reason, only the left down pattern was tested in experimental study. This is important to select which side of bank protection is needed. It was found that 60 degree model produced less scouring depth when compared with the 30 and 90 degree models. No significant difference was found for 30 and 90 degrees models. It is shown that the convex circular model with $r/w = 0.35$ produced minimum scouring when compared to all models.

It was concluded that without consideration of scour location, the circular convex model with $r/w = 0.35$ produced least bed scouring at the downstream. In a case where scour location is very important, circular convex or circular concave can be selected. For example to prevent scouring downstream of a bridge pier, the circular concave with $r/w = 0.35$, and to prevent scouring at the river bank, the circular convex with $r/w = 0.23$ are suggested to prevent bed scouring. Also to prevent bed scouring at river bends, the inclined sill with 60 degree can be used.

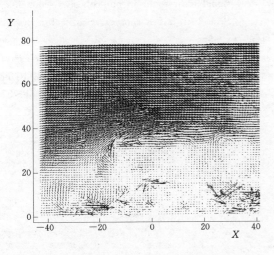

Figure 5   Flow structure in front and behind of convex bed sill (Test 5) (X and Y are in mm)

The results from PIV study also confirmed the above results in which at the center of the bed sill, the flow diverges to the river banks and the bed sill pushes the wake vortex rings further downstream. Figure 5 and 6 show the flow structure in font of and at behind of convex bed sill.

4. Conclusions

In this study, the effect of shape and sill patterns on the bed scouring at the downstream of the

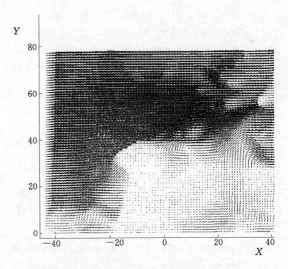

Figure 6   Flow structure in front and behind of convex bed sill (Test 6) (X and Y are in mm)

sill was investigated. Five different shapes with nine patterns including two circular sills with $D=W$ and $D=1.2W$, in concave and convex patterns, two inclined sills with 30 and 60 degrees in two patterns and a sill with 90 degree normal to flow direction were used in the experimental study. The measurement of bed scouring at the downstream of the sills showed that in convex pattern, the minimum bed scouring occurs at the center of channel, whereas in concave pattern, the minimum bed scouring occurred at the bank. In general it was found that the $D=1.2W$ is the most effective sill pattern and shape to prevent scouring from the center and river banks. For the inclined sills the scouring occurred at the location where initially flow attached the sill. It was found that the 60 degree inclined sill produced minimum scouring at the bed. The 90 degree bed sill produced uniform maximum downstream bed scouring along the sill. The results from PIV study showed that at the center of the bed sill, the bed sill pushes the wake vortex rings further downstream.

References

[1]   Arlinghaus R, Engelhardt C, Sukhodolov A, Wolter C. Fish recruitment in a canal with intensive navigation: Implications for ecosystem management. *J. Fish Biol*, 2002, 61: 1386-1402.

[2] Armstrong JD, Kemp PS, Kennedy GJA, Ladle M, Milner NJ. Habitat requirements of Atlantic salmon and brown trout in rivers and streams. *Fisheries Research*, 2003, 62 (6): 143-170.

[3] Ben Meftah M. , Mossa M.. Experimental study of the scour hole downstream of bed sills. RiverFlow, 23-25 June, 2004. Napoli, Italy. Edited by M. Greco. , A. Carravetta. , &. Della Morte. 2004, 2: 585-592.

[4] Bormann N. E. , Julien P. Y.. Scour downstream of grade-control structures. J. of Hyd. Eng. , 1991, 117 (5): 579-594.

[5] Breusers H. N. C.. Conformity and time scale in two dimensional local scour. Proc. Symposium on model and prototype conformity, 1-8. Hydraulic Research laboratory, Poona, 1966.

[6] Gaudio R. , Marion A. , Bovolin, V.. Morphological effects of bed sills in degrading rivers. J. Hyd. Res. , 2000, 38 (2): 89-96.

[7] Gaudio R. , Marion A.. Time evolution of scouring downstream of bed sills. J. Hyd. Res. , 2003, 41 (3): 271-284.

[8] Habib E. , Mossa M. , Petrillo A.. Scour downstream of hydraulic jump. Modeling, Testing &. Monitoring for Hydro Power Plants Conference, Budapest, Hungary, 1994: 561-602.

[9] Hassan N. M. K. N. , Narayanan R.. Local scour downstream of an apron. J. of Hyd. Eng. , ASCE, 1985, 111 (11): 1371-1385.

[10] Hoffman G. J. C. M. , Pilarczyk K. W.. Local scour downstream of hydraulic structures. J. of Hyd. Eng. , ASCE, 1995 121 (4): 326-340.

[11] Hoffmans G. J. C. M. , Verheij H. J.. Scour manual. A. A. Balkema, Rotterdam, Brookfield, 1997.

[12] Lenzi M. A. , Marion A. , Comiti F. , Gaudio R.. Local scouring in low and high gradient streams at bed sills. J. of Hyd. Res. , IAHR, 2002, 40 (6): 731-739.

[13] Marion A. , Gaudio R. V.. Scour down stream of bed sills. HR Wallingford Research Report TR7, Wallingford, United Kingdom, 1998.

[14] Van der Meulen T. , Vinji J. J.. Three-dimensional local scour in non-cohesive sediments. Proc. 16th IAHR congress. Sao Paulo, Brazil, 1995.

[15] Volkart P. , Tshopp J. , Bisaz E.. The effect of sills on river bed. Proc. of Int. Symp. on River Mechanics, IAHR, 1, Bangkok, 1973: 167-178.

Research and Application of the Flow-Dividing Characteristics of Intake-Outlets in Upper and Lower Reservoirs in Xiangshuijian Pumped Storage Power Station

Zhang Landing

State Key Laboratory of Hydrology-Water Resources and Hydraulic Engineering,
Nanjing Hydraulic Research Institute, Nanjing, 210029

Abstract: Based on the hydraulic model tests, the "unit velocity method" was proposed and applied to analyze the flow-dividing characteristics of intake-outlet in upper and lower reservoirs of Xiangshuijian pumped storage power station. The proposed method has the advantages of clear physical meaning and exact quantification. By applying the "unit velocity method", the spacing between the splitter walls at the entrance of diffuser was adjusted to equal spacing in order to fit the bend effect of the tailrace conduit in the intake-outlet of lower reservoir. This arrangement is an improvement compared with the conventional unequal spacing usually adopted for the splitter wall configuration design of side-type intake-outlet at home and abroad. During model tests the proposed method can be used to adjust quickly the splitter wall spacing of side-type intake-outlets to find a satisfactory configuration with small hydraulic loss and uniform distribution of flow velocity.

Key words: unit velocity; intake-outlet; splitter wall head spacing; hydraulic model tests

1. Introduction

The side-type of intake-outlet is widely employed in a lot of pumped storage power stations in China such as Ming Tombs, Tianhuangping, Guangzhou, Yixing, Langyashan, Tai'an pumped storage power stations etc. The configurations of side-type intake-outlets abide usually by the scheme given in Lu Youmei et al. (1992) Granted by former State Power Corporation, the author has systematically studied hydraulic characteristics of the side-type intake-outlets, based on the Langyashan pumped storage power station. The study involved side-type intake-outlets with three-channel type (two splitter walls) and four-channel type (three splitter walls). The study has revealed that the hydraulic characters of side-type intake-outlets are affected by many complicated factors. Presently the configuration of side-type intake-outlets employed in engineering design is only a basic scheme corresponded with the common laboratory condition. However, this basic scheme could not meet the conditions of different projects, because of their individual layouts such as boundary conditions at upstream and downstream, asymmetry or irregular outer boundary of intake-outlets, the length and form of section of conduit, etc. Furthermore some hydraulic structures may be constructed near intake-outlets, for example, the emergency gate shaft, the frameworks of trash racks etc. All these factors would lead the flow pattern always to differ from the common laboratory condition, and would probably deteriorate hydraulic characters of intake-outlets. The harmful flow pattern might even endanger their operation safety.

The problems mentioned above have been partly solved in Zhang Landing (2003). Some effective suggestions proposed in Zhang Landing (2003) have been employed in Langyashan station, which involve the configuration of emergency gate shaft near by intake-outlet, the slop of headrace conduits, the joint between conduit and upper intake-outlet, the joint between tailrace channel and lower intake-outlet, as well as the configuration of the diffuser, etc. Those suggestions have been verified again by later model tests of side-type intake-outlets of upper and lower reservoirs in Xiangshuijian station. Based on the latter,

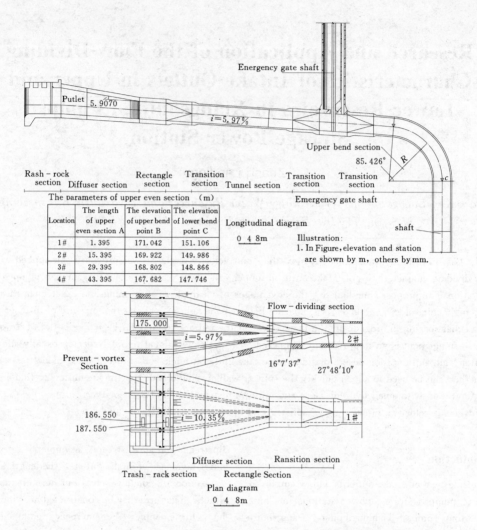

Figure 1   Diagram of upper intake-outlets of Xiangshuijian Power Station

Longitudinal diagram

0 4 8m

Illustration:
1. In Figure, elevation and station
   are shown by m, others by mm.

The parameters of upper even section (m)

Location	The length of upper even section A	The elevation of upper bend point B	The elevation of lower bend point C
1#	1.395	171.042	151.106
2#	15.395	169.922	149.986
3#	29.395	168.802	148.866
4#	43.395	167.682	147.746

Plan diagram

0 4 8m

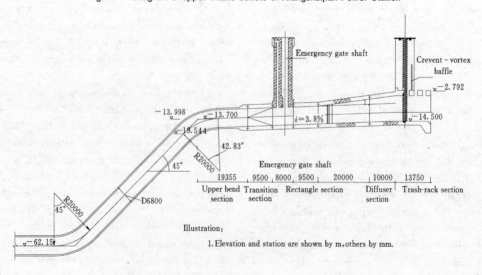

Figure 2   Diagram of lower intake-outlets of Xiangshuijian Power Station

Illustration:
1. Elevation and station are shown by m, others by mm.

author has proposed further suggestions aiming at that how to overcome the bend effect of upper bend sections in headrace and tailrace conduits linked respectively with the upper or lower intake-outlets, how to avoid the funnel vortex forming in the space between the frameworks of trash racks at lower reservoir, and how to improve the inflow field in front of intake-outlet at upper reservoir etc. New suggestions have been applied in design of Xiangshuijian station. One of them is to apply "the unit-velocity method" to analyze the flow-dividing characteristics of intake-outlets in upper and lower reservoirs.

To sum up, the hydraulic characteristics of intake-outlet depend on numerous factors. For any project there are some key factors which are usually studied at first. The spacing between the splitter walls at the entrance of diffuser section (called "the splitter wall head spacing" for short hereafter) would be adjusted after key factors settled down. But it is also important because it would make flow-dividing more uniform and get less water head losses of intake-outlets. The control operation mode for side-type intake-outlets is outflow mode, because the flow diffuses in both horizontal and vertical planes, where flow-separating and unequal flow-dividing may occur easily. Meantime the local head loss in intake-outlets is larger.

2. Layout of the Splitter Wall Head Spacing

Presently in design of side-type intake-outlets the layout of unequal splitter wall spacing is generally employed at home and abroad. The spacing proportions are 0.35B, 0.3B, 0.35B for three-channel type (two splitter walls) and 0.28B, 0.22B, 0.22B, 0.28B for four-channel type (three splitter walls) respectively, in which B is total width of the entrance of diffuser section. However, such an experiential spacing proportion was not always proper for any individual intake-outlets. In Xiangshuijian station the side-type intake-outlets with four-channels (three-splitter-walls) are accepted in both upper and lower reservoirs. Both upper and lower intake-outlets have similar diffuser sections, but the relative locations of reservoirs, intake-outlets, gate shafts and upstream

conduit bending sections are different. It is verified by model tests that the unequal splitter wall head spacing would lead worse flow-dividing characteristics and larger water head losses in the lower intake-outlets than in the upper intake-outlets.

The diagram of upper intake-outlets of Xiangshuijian power station is shown in Figure 1, the gate shafts and upper conduit bending sections are far from the intake-outlets. The diagram of lower intake-outlets of the station is shown in Figure 2, the gate shafts and upper conduit bending sections are very near to the intake-outlets.

3. Research on the Flow-dividing Characteristics of Intake-outlets in Upper and Lower Reservoirs Based on the Model Test

The research of the flow-dividing characteristics of intake-outlets of the Xiangshuijian project focused on pumping mode in upper reservoir and generating mode in lower reservoir. This is because in outflow mode the water flow diffuses with larger water head losses. So the outflow mode becomes always the control operation mode to evaluate the flow-dividing characteristics of intake-outlets.

The layout of upper intake-outlets is shown in Figure 1. The gate shafts are far from intake-outlets and linked four upper bent sections with unequal-length conduits.

Model test have shown that the dividing-flow characteristics in upper intake-outlets are fine enough. The differences of divided discharges between any adjacent channels are less than 10% at any reservoir water levels and operation discharges. There are stable flow pattern and uniform velocity distribution at the diffusers of intake-outlets. The average coefficient of head loss is 0.27 in outflow mode, which is relatively low among the built pumped storage power stations in China. Parts of test results are shown in Table 1.

In the table, the maximum coefficient of un-uniform is $2.73 < 3.0$. But the index "un-uniform of flow-dividing", which is a typical index of flow-dividing characteristics, is much more than 10% and to be modified. It was supposed that the unequal splitter wall head spacing might lead the flow-dividing not uniform.

Table 1　The velocity distribution in upper intake-outlets （the chosen scheme， pumping mode；discharge：555. 93m³/s； water-lever of reservoir：el. 190. 1m）

Intake-outlet （1#）												
Channel No.	1			2			3			4		
Vertical No.	1	2	3	1	2	3	1	2	3	1	2	3
Splitter wall head spacing	1.8			1.4			1.4			1.8		
Mean-velocity at vertical	0.65	1.02	0.86	0.74	1.06	0.73	0.92	0.95	0.64	0.81	0.94	0.50
Mean-velocity at diffuser exit	0.84			0.84			0.84			0.75		
Coefficient of un-uniform	1.64			2.21			1.82			1.66		
Un-uniform of flow-dividing	0.09											

Intake-outlet （2#）												
Channel No.	5			6			7			8		
Vertical No.	1	2	3	1	2	3	1	2	3	1	2	3
Splitter wall head spacing	1.8			1.4			1.4			1.8		
Mean-velocity at vertical	0.58	0.94	0.90	0.70	1.06	0.81	1.00	1.00	0.65	1.09	0.98	0.53
Mean-velocity at diffuser exit	0.81			0.86			0.88			0.87		
Coefficient of un-uniform	1.65			2.32			2.08			1.96		
Un-uniform of flow-dividing	0.05											

Note　In the table，the velocity is in m/s， the splitter wall head spacing is in m. The coefficient of un-uniform equals to the maximum point velocity at diffuser exit plane/the mean-velocity at diffuser exit plane. Un-uniform of flow-dividing is the maximum difference of velocities between adjacent channels and so on.

　　The configuration of intake-outlets in lower reservoir is similar to that of the upper intake/outlets， and is also side-type with four-channels （three splitter walls）， shown in Figure 2. Firstly， the unequal splitter wall head spacing was employed by the conventional layout. In the table 2， the model test results of lower 1# and 3# intake-outlets are given， with the unequal spacing. And the width of the diffuser entrance of 1# intake-outlet is different from that of 3#.

Table 2　　　　The Velocity Distribution in Lower Intake-outletswith Unequal Splitter Wall Head Spacing（in generating mode； discharge：639. 26m³/s， water-level of reservoir：EL. 2. 95m）　　（m/s）

Intake-outlet No.	1#				3#			
Splitter wall head spacing	1.8	1.4	1.4	1.8	1.74	1.36	1.36	1.74
Mean-velocity at diffuser exit	1.01	0.72	0.78	1.09	1.11	0.65	0.73	0.92
Mean-velocity at diffuser entrance	3.44	3.16	3.42	3.72	3.92	2.93	3.30	3.25
Unit-velocity	1.91	2.26	2.44	2.07	2.25	2.16	2.42	1.87
Coefficient of un-uniform	1.61	2.73	2.54	1.70	2.10	2.25	1.74	1.73
Un-uniform of flow-dividing	0.31				0.46			

4. Analysis and computation of flow-dividing characteristics of lower intake-outlets

　　The configurations of both upper and lower intake-outlets are similar. The layout of conventional unequal splitter wall head spacing may be employed for the both. But the model test showed that the effects of flow-dividing were very different. The effect is fine in the upper one and is not good in the lower one. This is because the upper bend section of tailrace conduit is near the gate shaft which itself abuts on the lower intake-outlets. The bend effect of the upper bend section leads the mean velocities at both side verticals at gate shaft to nearly equal or larger than that at central vertical. [4] Consequently the flow-dividing characteristics of splitter walls in lower intake-outlet are affected by the upper bend section. For this reason, author has studied the results of model test in detail, and proposed the "unit-velocity method" based on the "total-flow analysis method"[6] to determine the splitter wall head spacing. A favorable result has been obtained.

Main steps are listed as follows.

(1) Based on continuity equation: $Q_{div} = Q_{channel}$;

The total width of the entrance of diffuser section should be divided into four parts. The discharge of every part Q_{div} equals to $Q_{channel}$, the discharge of corresponding channel.

(2) Then the area of every part S_{div} is divided by Q_{div}, to get a mean velocity $V_{div} = Q_{div}/S_{div}$, $S_{div} = bh$, b is splitter wall head spacing of the part, h is the height of channel (located at entrance of diffuser section).

(3) If the flow-dividing characteristics of diffuser were un-uniform, we would expect to modify flow-dividing characteristics better by adjusting the head spacing. We meet with a difficult problem here. It is always known only the mean velocity of dividing section but not known the velocity distribution at the dividing section. How to adjust the divided head spacing to realize modification of flow-dividing characteristics? By analysis and computation repeatedly, author

proposed the "unit velocity method" as follows.

"Unit velocity" $V_b = \overline{V}_{div}/b$

(4) To adjust the splitter wall head spacing:

Let a new head spacing $b' = b \pm \Delta$; Δ is a tiny space, plus to (or minus from) the original head spacing.

(5) If new head spacing $b' = b - \Delta$, then $\overline{V}'_{div} = V_b b'$, where V_b is the original "unit velocity" of the channel.

If $b' = b + \Delta$, then $\overline{V}'_{div} = V_b \cdot b + V_{b\Delta} \cdot \Delta$, where V_b is the original one, $V_{b\Delta}$ is "unit velocity" at adjacent channel.

(6) Thus the adjusted discharge of the channel will be $Q_{div} = \overline{V}'_{div} S'_{div}$, $S'_{div} = b'h$, where the meaning of every item is the same as in Step (2).

Because Q_{div} equals $Q_{channel}$, a new discharge of the channel after adjusting the head spacing can be obtained. Thus we can find appropriate head spacing only by simple computation. Therefore the "unit velocity" becomes a representative parameter reflecting the flow-dividing characteristics of each channel at entrance of diffuser section. We can decide the appropriate dividing-discharge only by adjusting the tiny space.

Base on the results of model test listed in Table 2, the flow-dividing characteristics with equal splitter wall head spacing were forecasted by applying the "unit velocity method". The computation results are listed in Table 3.

Table 3   The Computation Results of Flow-dividing in Lower Intake-outlets
with Equal Splitter Wall Head Spacing
(in generating mode; discharge: 639. 26m³/s, water-level of reservoir: EL. 2. 95m)   (m/s)

Intake-outlet No.	1#				3#			
Splitter wall head spacing	1. 6	1. 6	1. 6	1. 6	1. 55	1. 55	1. 55	1. 55
Mean-velocity at diffuser exit	0. 80	0. 92	1. 00	0. 86	0. 88	0. 85	0. 92	0. 73
Mean-velocity at diffuser entrance	3. 06	3. 54	3. 83	3. 30	3. 49	3. 36	3. 65	2. 89
Unit-velocity	1. 91	2. 26	2. 44	2. 07	2. 25	2. 16	2. 42	1. 87
Un-uniform of flow-dividing	0. 14				0. 19			

It is shown in Table 3 that how to apply the "unit velocity method" to forecast the flow-dividing characteristics with equal splitter wall head spacing. The computation results showed that the flow-dividing characteristics would be improved. And the 1# was a little better than 3#. It implied that the un-uniform of flow-dividing may be also caused by other reasons. For example, the total width of the entrance of diffuser section was too short. So the "unit velocity method" could be also used to make a "diagnosis".

Because the subcritical flow occurs in intake-outlets, the upstream and downstream flow-boundary condition may affect the flow inside intake-outlets. The distribution of discharge in the entrance of diffuser section would be formed under special upstream and downstream boundary condition. Based on the results of "diagnosis", the total width of the entrance of diffuser section has been finally selected as 6. 4m for total four intake-outlets in the lower reservoir.

5. The verifying test results of flow-dividing characteristics of lower intake-outlets under the equal splitter wall head spacing

Based on forecast results (Table 3) by the "unit-velocity method", the equal splitter wall head spacing has arranged in intake-outlets of lower reservoir. And the total width of the entrance of diffuser section of the 3 # intake-outlet has been widened to the same with the 1 #. Parts of model test results of lower intake-outlets (the chosen scheme) are shown in Table 4.

The model tests indicate that the flow-dividing characteristics of the intake-outlets with equal splitter wall head spacing have been improved and much better than those with unequal spacing (see Table 2). Applying "unit-velocity method", the discharge of flow-dividing in each channel can be inferred according to a known mean velocity at diffuser exit, based on equation of continuity. And then the mean velocity at diffuser entrance is obtained.

Table 4   **The Verifying Test Results of Flow-dividing in Lower Intake-outlets with Equal Splitter Wall Head Spacing (the chosen scheme)**

(in generating mode; discharge: 627. 92m³/s, water-level of reservoir: EL. 12. 60m)   **(m/s)**

Intake-outlet No.	1 #				3 #			
Splitter wall head spacing	1. 6	1. 6	1. 6	1. 6	1. 6	1. 6	1. 6	1. 6
Mean-velocity at diffuser exit	0. 79	1. 03	0. 99	0. 87	0. 91	0. 91	0. 98	0. 82
Mean-velocity at diffuser entrance	3. 03	3. 95	3. 80	3. 34	3. 49	3. 49	3. 76	3. 15
Unit-velocity	1. 89	2. 47	2. 37	2. 09	2. 18	2. 18	2. 35	1. 97
Coefficient of un-uniform	1. 50	1. 82	1. 94	1. 63	1. 46	2. 12	1. 79	1. 55
Un-uniform of flow-dividing	0. 24				0. 16			

The mean velocity at entrance of each channel reflects objectively the hydraulic characteristics at special upstream and downstream boundary conditions. So "unit-velocity method" can truly reflect the hydraulic characteristics at diffuser entrance. Applying "unit-velocity method" to determine the splitter wall head spacing has the advantages of clear physical meaning and exact quantification.

6. Research on arrangement of splitter wall heads

Research on arrangement of splitter wall heads has also carried out in the model tests of intake-outlets of upper and lower reservoirs in Xiangshuijian station. Generally there are three arranging patterns for side-type intake-outlets.

Pattern 1: The head of three splitter walls are arranged abreast at the entrance line of diffuser.

Pattern 2: The head of middle splitter wall

moved properly backward, while the heads of both side splitter walls are located at the entrance line of diffuser.

Pattern 3: The heads of both side splitter walls moved properly backward, while the head of middle splitter wall is located at the entrance line of diffuser.

Presently all the three arranging patterns of splitter wall heads have been employed in different pumped storage power stations in China. The author has conducted preliminary study to the three patterns in Zhang Landing (2008a, 2008b). In Langyashan station, two pump-turbine sets link to one tailrace conduit and operation discharge is large. So the size of lower intake-outlets is relatively big. The Pattern 1 arrangement of splitter wall heads is employed. The coefficients of water head loss in lower intake-outlet are 0. 27 in outflow mode and 0. 19 in inflow mode. In

Pattern 1 the boundary of flow-dividing is clear. If the entrance of diffuser section is wide enough, Pattern 1 will be suitable. In Xiangshuijian station, one conduit connects with only one pump-turbine set and the sizes of both upper and lower intake-outlets are less than those in Langyashan station, the entrance of diffuser section is relatively narrow. If Pattern 1 were employed, the flow would be congested probably at narrow entrance, and water head loss would be increased. If Pattern 2 were employed, the problems mentioned above would be mitigated. In the middle part of entrance section flow is relative steady and uniform. If the head of middle splitter wall moved properly backward, the uniformity of flow-dividing would not be affected, and the water head loss would be decreased. During model tests of upper and lower intake-outlets of Xiangshuijian station, contrast model tests about Pattern 1 and Pattern 2 have been carried out. The model tests showed that in outflow mode the head loss coefficients in lower intake-outlet were 0.47 with Pattern 1 and 0.39 with Pattern 2 under the same discharge and the same water level. Finally the Pattern 2 has become the chosen scheme of lower intake-outlets applied in prototype engineering, but with equal splitter wall head spacing. The average head loss coefficients of four intake-outlets are 0.32 in outflow mode and 0.10 in inflow mode, under any discharge and water level of design. The Pattern 2 has also been employed in upper intake-outlets, but with unequal splitter wall head spacing. The average head loss coefficients of four intake-outlets are 0.27 in outflow mode and 0.13 in inflow mode, under any discharge and water level of design. It should notice that the head loss coefficients of intake-outlets mentioned above did not include the loss of trash rack.

Pattern 3 may relate to the uniformity of flow-dividing in side-channels and the possibility of head loss reduction by moving two side splitter wall heads backward. So it will be necessary to study further.

7. Concluding remarks

In the paper author proposed the "unit-velocity method" to adjust the splitter wall head spacing based on the "total-flow analysis"[6]. The "total-flow analysis" is a kind of quantifying analysis method based on mean-velocity of outlet or cross section of hydro-structures. This method is widely applied to hydro-design and hydro-research. It has been employed for hundreds of years and will be still employed in the future.

The side-type intake-outlet is a kind of pressure conduit. The velocity distribution in model test can only be measured near the exit section. Based on the data of point velocity the tester can compute in sequence the mean-velocity in exit section and discharge in each channel, then to infer the flow-dividing characteristics of the diffuser. It is difficult to observe directly the velocity field at the entrance of diffuser section or at splitter wall heads.

By traditional model test method, in order to find a satisfactory flow-dividing scheme, the splitter wall head spacing in the model must be adjusted and tested repeatedly. So the diffuser roof and splitter walls have to be dismounted and re-mounted again and again. The model adjustment will be a complicated and time-consuming work. Therefore author expects to look for a proper method, with which a satisfactory flow-dividing scheme could be decided rapidly. And this method should have clear physical meaning and exact quantification. By repeatedly calculating and analysis, the "unit-velocity method" was proposed. It can meet above mentioned requirement, and has been verified by model tests. This method can be also applied to side-type intake-outlets of three-channel type (two splitter walls).

The comparative tests about side-type intake-outlets with four-channel (three splitter walls) have been carried out successively in the models of Langyashan pumped storage power station and Xiangshuijian pumped storage power station. The water head loss coefficients about different arrangement of splitter walls were shown in the paper. The model scales are 1:30 in Langyashan station, 1:40 and 1:50 in lower reservoir and upper reservoir of Xiangshuijian station respectively. The research results have been applied in two stations mentioned above. Author expects the "unit-velocity method" could also be employed in other similar projects.

Acknowledgement

By the end of the paper, the author would give the honest thankfulness to Mr. Xiao Gongyuan, the chief engineer in charge of designing the lower and upper reservoirs of Xiangshuijian pumped storage power station. He has given valuable suggests and has encouraged the author to overcome difficulties in the research.

References

[1] Lu Youmei, Pan Jiazeng, editors in chief. Pumped-storage Power Station [M]. Peking. Hydro-power Publisher, 1992. (in Chinese).

[2] Zhang Landing. Hydraulic Research about Side-type Intake-outlets [R]. Nanjing: Nanjing Hydraulic Research Institute, 2003. (in Chinese).

[3] Zhang Landing. Hydraulic Model Test on Verified Configuration of Upper Intake-outlets in Xiangshuijian Pumped-storage Power Station [R]. Nanjing:
Nanjing Hydraulic Research Institute, 2009. (in Chinese).

[4] Zhang Landing. Hydraulic Model Test about Lower Intake-outlets in Xiang Shuijian Pumped-storage Power Station (Volume I) [R]. Nanjing: Nanjing Hydraulic Research Institute, 2008a. (in Chinese).

[5] Zhang Landing. Hydraulic Model Test about Lower Intake — outlets in Xiang Shuijian Pumped-storage Power Station (Volume II) [R]. Nanjing: Nanjing Hydraulic Research Institute, 2008b. (in Chinese).

[6] Zhang Changgao. Hydro-dynamics [M]. Peking. High Education Publisher, 1993. (in Chinese).

[7] Hua Shaozeng, Yang Xuening. Translator and Editor, Handbook of Fluid Resistance in Practice [M]. Peking. National Defence Industry Press, 1985. (in Chinese).

Laboratory Experiment for the River Bank Sedimentation

Md. Lutfor Rahman[1]   B. C. Basak[2]   Md. Showkat Osman[3]   Md. Altaf Hossain[4]

[1] *Chief Scientific Officer, River Research Institute, Faridpur and Ph. D. Student, Bangladesh*
[2] *Professor & Chairman, Department of Civil Engineering, Stamford University Bangladesh*
[3] *Professor & Head, Department of Civil Engineering, DUET, Gazipur, Bangladesh*
[4] *Director General, River Research Institute, Faridpur-7800, Bangladesh*

Abstract: An experimental investigation was conducted with the construction of Bandals structure in the mobile bed at the laboratory river channel of the River Research Institute, Faridpur, Bangladesh. A series of test runs were conducted with the controlled condition by placing bandals at the river bank of the channel. It was found that water flow diverted towards the main river due to bandal structures resulting maximum velocity accumulated at the main channel whereas comparatively less velocity appeared near the river bank where bandals structures are placed resulting sediment deposition.

Key words: investigation; bandals; structures; velocity; sediment; deposition

1. Introduction

River bank erosion and channel shifting are recurrent problems in Bangladesh that usually occur during the monsoon (more specifically, during rising stage and recession stage) when huge sediment load is generated by means of bank erosion and bed changes. Conventionally, spurs, groins, revetments or combination of them are used in order to manage sediment load thus generated and mitigate river erosion and related problems. Spurs, groins or revetment-like structures are too expensive to adapt along the longer reaches of the large-scale alluvial rivers in Bangladesh. Therefore, it is important to develop alternative low cost approaches that can be adaptive within local socio-economic and environmental condition.

In Bangladesh, over the years, channel width is increasing and depth is decreasing because of unfavorable geographic location and discharge control by the countries in the upstream reaches that lead to unexpected erosion-siltation processes along the major rivers. It is very difficult and even impossible to maintain in-stream flow requirement that is very important for the maintenance of river ecology and aquatic habitat necessary for the healthy life cycle of plants and animals. Rivers are loosing their navigability and waterways are severely obstructed during the dry season. On the other hand, conveyance capacity of rivers is reducing and is insufficient for safe and expeditious passage of floodwater and sediment discharge during the monsoon. As a result, country had experienced severe flood disasters during the past such as in 1988 and 1998. However, the situation seems to get more severe gradually as compared with the past events.

Floodplains and riverbanks are developed from recent deposits consisting mostly silt and fine sand that are highly susceptible to erosion. As a result, the river channels often shifts within wide range of river belt. To prevent the river erosion groins, spurs, revetments, porcupines, sand bags, boulders etc. are applied. Some of these methods (groins, revetments) are very expensive considering the large river dimensions and corresponding limited financial strength of Bangladesh. On the other hand, porcupines, sand bags, boulders are being used from experience of local people against river bank erosion and none of these methods have been proved to be effective for the protection of river erosion in long-term basis.

Bandals are one of the local structures developed in the Indian sub-continent that obstruct flow near the water surface and allow it to pass near the riverbed. These are made of naturally available materials such as bamboo, wood etc. and regarded as inexpensive method over conventional structures and mostly applied for the improvement of navigational channels during the low flow season. But application of Bandalling for riverbank protection is not yet practiced in Bangladesh. In recent past some field tests along the Jamuna was executed for the bank protection using Bandalling (FAP 21/22, 2001). At the laboratory scale, the preliminary idea on the possibility of use of Bandalling for sediment management (erosion and siltation) was discussed (Rahman et al., 2003, 2004).

2. Objectives of the Study

The main objective of the laboatory experimental study are being carried out at River Research Institute, Bangladesh in order to investigate the sediment deposition and the flow field around bandals that are to be conducted under different scenarios. The specific research objectives are as follows:

(1) to investigate the river bank sedimentation pattern;

(2) to understand the characteristics of the flow field around the bandal when placed in the different scenarios;

(3) to get idea about the navigational phenomenon in the river channel.

3. Methodology

To achieve the objectives of the study, an experimental set are taken in re-circulated straight flume that is 22m long, 2m wide and 1.5m deep at River Research Institute. The flume setup was in live-bed condition. The effect of a series of bandals structures are examined in terms of the arrangement of spacing & with certain angle of the flow direction in the flume river channel. The laboratory experimental set-up with the circulated water supply system is shown below in Figure 1.

Figure 1   Experimental set-up for the river channel with the water supply system

The bandals structures are placed with laboratory river bank i. e. with the water flow direction at 40 degrees. The different parameters of the experimental river channel are summarized in the Table 1 is shown in below.

Table 1   Parameters used in the river channel (Navigation & Bank protection options)

Flow angle θ (°)	Bandals length (m)	Spacing bet^n. bandals (m)
40	0.66	1.32

4. Data collection & analysis

From the experimental laboratory as in Figure 2 which is in the running condition, data both for the bathymetric & hydraulic are collected.

Figure 2   River channel during the test run in case of river navigation options

The bed level data before the test runs as well as after the test run were collected. These bed level data are taken in every 20cm interval along the cross-section of the 20m length river channel. There are 40 cross-sections along the 20m long river channel at an interval 0.50m.

5. Result & discussion

From Figure 3 and Figure 4, it is clear that the river channel bank is deposited where as main river central channel is deepen to improve the navigational channel development. Although there was an idea from Figure 2 and Figure 5 so that the river channel may deep near the central line and at the same time the sedimentation near the river bank. From Figure 6 and Figure 7, it is clear that the river channel bank is deposited where as the main river channel away from the river bank is deepen to improve the navigational channel development. Although there was an idea from Figure 8 and Figure 9 so that the river channel may deep away from the river bank & at the same time the sedimentation near the river bank.

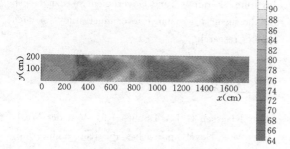

Figure 3   River channel bed contour map
after the test run navigation option

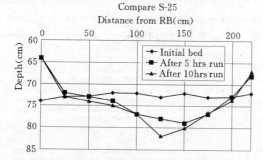

Figure 4   River Bank sedimentation due to effect of
Bandalling for the initial & final bed conditions

Figure 5   River channel during the dry bed condition
after the test run for the navigation option

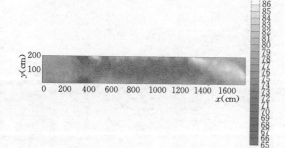

Figure 6   River channel bed contour map after the
test run for bank protection option

X-section 20, Run-08, Spacing 66cm, & Angle-40, B. P

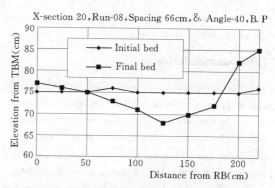

Figure 7   River bank sedimentation after the test run
for bank protection option

Figure 8   River channel during the test run in case
of river bank protection options

Figure 9   River channel during the dry bed condition
after the test run for bank protection option

6. Conclusion

In conclusion, we can say that bandals are capable for flow diversion towards the main channel leading to deep navigational channel formation. On the other hand, flow velocities are reduced near the bank lines that ensure sediment deposition. If the bandal structure functions optimistically, the river can get sufficient time for its adjustment and new main channel and bankline development.

7. Recommendation

Pilot projects are recommended in the field are very important to execute before applying such method for the formation/restoration of navigational channels in alluvial river. The out put of the present research for the stabilization of river course can solve the problems of Bangladesh that is more or less inherent due to its complex geographical location at the lower riparian of the catchments

Acknowledgement

The authors are grateful to the River Research Institute, Faridpur, and Government of Bangladesh for the financial assistance to conduct study for such type of research.

References

[1] Klaassen G. J. , Douben K. , Van der Waal, M. . Novel approaches in river engineering, River Flow, 2002: 27-43.

[2] Rahman M. M. , Nakagawa H. , Ishigaki T. , Khaleduzzaman, ATM. . Channel stabilization using bandalling, Annuals of Disaster Prevention Research Institute, Kyoto University, 2003 (46B): 613-619.

Eliminating Vortices at the Proposed Folsom Dam Auxiliary Spillway

Steven L. Barfuss[1]   William J. Rahmeyer[1]   Nathan C. Cox[2]

[1]*College of Engineering, Utah State University, Logan, Utah   84322, USA*
[2]*US Army Corps of Engineers, Sacramento, California 95814, USA*

Abstract: Folsom Dam is on the American River about 20 miles northeast of Sacramento, California. The dam was designed and built by the U. S. Army Corps of Engineers (USACE) and transferred to Reclamation for operation and maintenance in 1956. The dam is a concrete gravity structure 340 ft high and impounds a reservoir of a little more than one million acre-ft. An auxiliary spillway has been proposed that will enhance flood reduction capabilities of the dam. The auxiliary spillway and gate structure will be located in a side channel near the left abutment of the existing spillway. A physical model study of the proposed auxiliary spillway structure at Utah State University revealed that the original proposed design was susceptible to strong surface vortices that were deemed unacceptable. After several unsuccessful attempts to reduce the strength of the vortices, a successful configuration was eventually determined by inclining the upstream head wall (vertical face wall) across the top of the gate structure into the flow at an incline of 20 degrees. This paper focuses on the design changes made at the spillway gate structure to reduce surface vortices resulting from the high velocity approach flow in the approach channel to the auxiliary spillway. Additionally, this paper discusses the potential vortex problems that are associated with submerged radial gate structures located in reservoir approach channels with higher approach velocities in comparison to those seen with typical submerged gates in a reservoir. The iterative process used during the model study to find an acceptable solution and to implement the working model configuration into the final USACE design is also reviewed.

Key words: physical model Study; vortices; submerged radial gates; spillway capacity; flood passage

1. Project description

Folsom Dam is on the American River about 20 miles northeast of Sacramento, California. The dam was designed and built by the U. S. Army Corps of Engineers (USACE) and transferred to The Bureau of Reclamation (USBR) for operation and maintenance in 1956. The dam is a concrete gravity structure 340 ft high and impounds a reservoir of a little more than one million acre-ft. The dam features two tiers of four outlets each, controlled by 5- by 9-ft slide gates. The outlets consist of rectangular conduits of formed concrete passing through the dam and exiting on the face of the service spillway. The proposed auxiliary spillway will allow passage of the Probable Maximum Flood (PMF) without encroaching into freeboard. It will also enhance flood reduction capabilities of the dam, making it possible to limit releases to 160, 000 cfs (downstream levee capacity) during passage of a flood event that will occur on the average

once in 200 years.

The proposed auxiliary spillway will be located in a side channel near the left abutment of Folsom Dam. The design of the auxiliary spillway was patterned after the design of the submerged tainter gate spillway at Oroville Dam, located near Oroville, California on the Feather River. The auxiliary spillway control structure will contain six 23- by 34-ft submerged tainter gates (also known as top-seal radial gates) that can discharge, in concert with the existing main dam sluices, at least 160,000 cfs discharge at pool elevation 418. 0 (existing dam crest, based on NGVD 1929 datum). The maximum discharge capacity through the auxiliary spillway will be about 320, 000 cfs at maximum pool elevation 481. 0. A chute will extend at a 2% slope from the auxiliary spillway control structure to an energy dissipater that terminates about 1500 feet downstream of the main dam.

To insure satisfactory hydraulic performance of the auxiliary spillway, a physical model study at

Utah State University was conducted (UWRL 2008). The model was a 1 : 30 Froude scale model that included: approximately 2500 feet of the upstream reservoir approach; the gated control structure consisting of gate intake, piers, gates, trunnions, and discharge conduits; and 800 feet of the downstream concrete-lined chute. The model was a fixed bed model constructed of cemented gravel for the approach and Plexiglas and steel for the gated control structure and outlet channel.

2. Project objectives

The objectives of the physical hydraulic modeling included evaluating the original auxiliary spillway design and conducting tests as needed to improve the structure's hydraulic performance while minimizing construction costs. It was necessary that the prototype auxiliary spillway be able to safely release the anticipated design flows without any restrictions due to the formation of damaging vortices, potential cavitation damage, damaging gate vibrations, debris and rock passing through the structure or flow instabilities in the chute downstream of the control structure. This paper focuses primarily on the surface vortices that were occurring as a result of the high velocity approach flow in the preliminary design and the iterative effort necessary to eliminate them.

3. Physical model

3.1 Model construction

The physical model of the proposed auxiliary spillway was constructed at a model scale of 1 : 30 at the Utah Water Research Laboratory (UWRL) in Logan, Utah. The scale of 1 : 30 was selected so as to adequately model hydraulic phenomena such as the formation of vortices and negative pressures, and provide for accurate discharge ratings. The scale of 1 : 30 also allowed the model to include 2500 ft of the upstream reservoir approach and was able to contain the maximum reservoir pool elevation in the five foot high walls of the model head box. The footprint of the physical model on the laboratory floor was approximately 55 ft × 100 ft. Figure 1 shows the approach channel on the reservoir side of the modeled auxiliary spillway. Figure 2 is a sectional drawing of the original submerged tainter gate design for the proposed auxiliary spillway at Folsom Dam.

The model included reservoir topography up to

Figure 1　Photo of model reservoir and head box

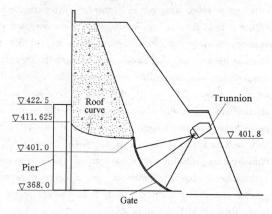

Figure 2　Gate and control structure as originally proposed

2500 feet upstream of the control structure, and 1690 feet across, from elevation 350 to 481 (NGVD 1929 datum). The gate structure included the gate intakes, six moveable top-sealed radial gates, gate piers and trunnions, and six discharge conduits. Approximately 800 feet of the downstream concrete-lined chute was also modeled. The topography was modeled using plywood templates of cross sections, which were then filled in with gravel and capped with mortar. The piers, gates and roof curves were all constructed using clear Plexiglas and Lexan to aid in visualization of the flow and any vortices that might form. The model was built according to drawings provided by the USACE.

Water was supplied to the model from a reservoir near the UWRL. A 48-inch pipe carried flow rates of up to 64cfs to the model. A 55-ft long diffuser system evenly distributed large flows to the reservoir in the model.

3.2 Vortex modeling

It is difficult to simultaneously satisfy similitude of all forces when modeling free surface vortices. Inertial, viscous, and surface tension forces all have an effect on vortices described by the Reynolds and Weber numbers. Hecker notes (Hecker 1981) that inertial and gravitational forces are reduced similarly in Froude-scaled models but prediction errors are inherent due to the impossibility of reducing all pertinent forces by the same factor. Viscous and surface tension forces cannot simultaneously be reduced as much. Therefore, both qualitative and quantitative vorticity data was collected during the model study. Although small, intermittent vortices were both expected and considered acceptable during the passage of large flood events for design, large, fully aerated, and stable vortices were always deemed unacceptable. Therefore, when unacceptable vortices were observed during a model run, the structural configuration and the hydraulic condition in which the vortices were observed were carefully noted. Steps were then taken to modify wetted components of the structure to either reduce the intensity or eliminate the vortices entirely.

The goal was to be able to release the required flows without restrictions due to the formation of damaging vortices, potential cavitation damage, damaging gate vibrations, or flow instabilities in the downstream chute. Often, the presence of vortices produces vibration problems, flow instabilities, and localized negative pressure regions. To ensure acceptable operation:

(1) Visual observations were used to determine acceptable vortex action. During visits to the lab, the principal investigators and engineers from the USACE and USBR made visual observations and judgments as to what degree of intensity of vortices was acceptable.

(2) Piezometer taps in the intake roof curves, invert, and walls were used to determine pressures. The USACE would accept moderate negative pressures as long as they were significantly above the-20 feet of water threshold stated in EM-1110-2-1602 (USACE, 1980).

(3) Pressure taps on the faces of the gates were used to measure dynamic pressure fluctuations and determine if any unacceptable conditions existed.

(4) Visual observations were made and water surface elevations were measured in the gate conduit and the downstream chute to identify flow instabilities.

(5) The location, intensity, type and duration of vortices occurring in the model were noted and as-signed a problem index value. This numerical problem index was then used to compare one test configuration against another to note relative improvement.

4. Modeling configurations

4.1 Original design

The original design of the control structure was generally patterned after a spillway at Oroville Dam that also utilizes large submerged tainter gates with about 100 feet of head at maximum pool. The following photos show the original structure design (looking upstream through the acrylic vertical face wall of the model). Significant vortex action was observed at the gate structure at most of the higher reservoir pools. The fully aerated vortices were the strongest on the far left and far right corners of the structure (Figure 3 and Figure 4). The presences of the vortices caused uneven and bulked flow, which caused the flow though the outside conduits (Figure 1 and Figure 6) to hit the gate trunnions (Figure 5 and Figure 6). Figure 7 shows flow hitting a gate trunnion. The model vortices were also strong enough to induce vibrations that could be felt when standing on the model.

Figure 3　Photo of strong surface vortex upstream of gate 6#

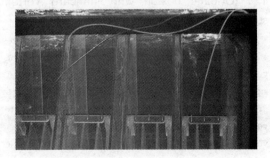

Figure 4　Photo of strong surface vortex upstream of gate 1#

Figure 5　Photo of flow bulking and trunnion
impact within conduit 6#

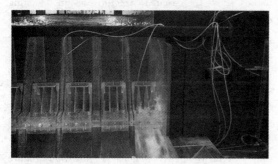

Figure 6　Photo of flow bulking and trunnion
impact within conduit 1#

Figure 7　Photo of trunnion impact within conduit 1#
(flow moving right to left)

4.2　Original design with standard vortex suppressors

A vortex suppressor was constructed and installed on the original design of the structure. The vortex suppressor design was patterned after the design at Theodore Roosevelt Dam. This vortex suppressor (Figure 8 and Figure 9) did not eliminate vortex action. Several other variations of this first vortex suppressor were also installed and tested in the model with limited success. Under some flow conditions the modeled vortex suppressors even enhanced the strength and persistence of the vortices. Even with

the vortex suppressor installed, vortices were able to form and pass behind the suppressor and go through the structure.

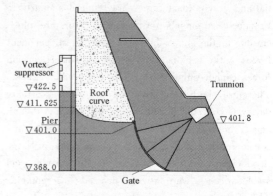

Figure 8　Vortex suppressor installed on
original design structure

Figure 9　Vortex suppressor installed on
original design structure

Figure 10　Photo of vortex occurring with vortex
suppressor installed on original design

4.3　Pier and approach walls extended vertically

Because the original piers and approach walls only extended up to elevation 422.5 ft, significant cross-flow

416

Figure 11 Photo of vortex occurring with vortex
suppressor installed on original design

would occur when the pier and walls were submerged, which increased the intensity of the vortices at the structure. To help reduce the amount of cross-flow, the initial configuration was modified by raising the piers and approach walls up to the full height of the structure (elevation 481). The original curvature of the approach walls was maintained up to the top of the walls. Figure 12 shows the vertical extensions in the model.

Figure 12 Vertical extensions to the model
piers and approach walls

Only limited improvement was noted in the intensity of the vortices that occurred in the model with the pier and approach wall extensions. It was concluded at this time that the majority of the vortex strength was induced by the high approach velocities in the reservoir approach channel and the sudden downward draw of water to the submerged conduits. The cross-flow added to the strength, but did not initiate the formation of the vortices. It was also noted that the pressures on the roof curve of the conduit were reaching vapor pressure. Since this was not acceptable, the USACE determined that a major change was necessary to the structure itself. It was concluded that the structural design change should

not only increase conduit pressures, but hopefully reduce vortex intensity.

4.4 Modified roof curve

The initial design of the roof curve within the gate structure was modified to have a milder curvature and extend further upstream. The roof curves included bulkhead slots that will be constructed in the prototype. The configuration shown in Figure 13 through Figure 15 was tested with limited reduction in vortex action but significant improvement in negative pressures along the roof of the gate structure.

Figure 13 Trial modified roof curve

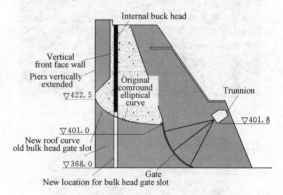

Figure 14 Profile drawing of modified roof curve

4.5 Approach walls modified

With the modified roof curve installed and the extreme negative pressures on the original roof curves (−39 ft) increased to an acceptable level, the approach walls were redesigned with a longer and milder curvature in plan view. The top of the curved wall was extended to elevation 465 with a sloping embankment on top in an attempt to reduce the construction costs for the wall (Figure 16 and Figure 17). Several variations of the approach wall (length, curvature and height) shown in Figure 16 and Figure 17 were tested in an effort to

Figure 15　Photograph of completed modified
roof curve in the model

smooth out flow patterns approaching the structure and
reduce vortex intensity (see Figure 18).

Figure 16　Milder and longer approach walls

Figure 17　Milder and longer approach walls
(strong vortices still present)

Figure 18　Approach wall extension trial
to suppress vortices

When the pool elevation was above 465 severe
vortices would form in the dead space above the approach walls on both sides. This led to the decision
that the approach walls had to go all of the way up to
elevation 481. Although the flow patterns approaching the structure were improved the strong surface
vortices near the face of the structure continued to be
a problem.

4. 6　Piers lengthened

In an attempt to reduce the strength of the rotational and cross-flow patterns near the face of the
structure, the piers were extended into the reservoir
(Figure 19). Although some reduction in the strength
of the vortices was noted, this modification did not
reduce or eliminate the vortices to an acceptable level
of operation. As before, it was again noted that the
reservoir approach channel velocities in conjunction
with the sudden down-flow component near the face
of the structure was setting up the vortices. A modification was necessary that would smoothly transition
the flow from the surface (vertically) to the submerged tainter gate openings.

Figure 19　Trial pier extension to suppress vortices

4.7 Face wall modified

Using a piece of plywood (Figure 20), several options were tried to eliminate the vortices by placing it across the front of the piers and varying the degree of incline into the flow. A face wall with zero incline was tried first, and then an inclined wall of 10, 20 and then 30 degrees were tried. It was immediately apparent that the inclined wall was very effective in eliminating the vortices. With the 10 degree wall in place, a few unacceptable vortices were observed and therefore, this configuration was not adopted. Because a hanging wall is not cost friendly, the 30 degree wall was immediately eliminated for design. Consequently, the 20 degree wall was accepted as an acceptable design as it removed all but a few insignificant, non-damaging type vortices from the face of the structure.

Figure 20   Trial inclined face wall installed in the model to suppress vortices

To structurally support the 20 degree sloping face wall, the piers were lengthened into the reservoir and raised to support the hanging structure. Figure 21 shows the trial wooden piers and sloping face wall in the model. Figure 22 shows the final design of the 20 degree inclined front face wall, the extended piers, the approach walls raised to elevation 481 and the approach walls lengthened. Figure 23 illustrates a sectional drawing of this particular configuration.

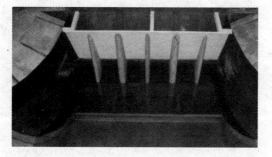

Figure 21   Temporary modified piers and face wall

Figure 22   Modified piers and face wall

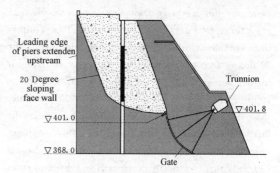

Figure 23   Profile of the modified piers and face wall

5. Summary of results

For vortex suppression, the final design configuration included: (1) modifying the shape and height of the concrete approach walls to reduce flow separation and vortex action at the gate structure; (2) the addition of a 20 degree incline or face wall (into the flow) to the upstream face of the control structure to accelerate and streamline and turn the surface flow down into the gate inlets; (3) extending the gate piers upstream to the leading edge of the 20 degree inclined face wall to further reduce vortex action from the down turn of approach flow and the contraction of flow from the channel sides; and (4) significantly increasing the length and height of the elliptical roof curves to eliminate excessive negative pressures along the roof curve.

The final design also included the widening of the approach channel and the addition of a rock spur dike to further improve the approach flow conditions. These modifications are not discussed in this paper.

419

6. Conclusions

The Folsom auxiliary spillway has a reservoir approach channel that produces relatively high approach velocities to the structure. Prototype structures that have a similar design to the original Folsom Dam auxiliary spillway design are normally not used when an approach channel is part of the design. Essentially, high velocities approaching the structure that have to rapidly change directions to enter the submerged conduits set up strong vortices.

A number of vortex-suppressing configurations were tested in the model, yet while the configurations that were tested appeared to work well at one or more specific reservoir pool elevations and discharges, these configurations were found to be entirely unacceptable at other conditions. The addition of the 20 degree face wall was by far the major contributor to the success of the model study in regards to vortex suppression. This single modification to the structure essentially eliminated the shear boundary near the face of the structure where the surface flow approaching the structure in the reservoir approach channel suddenly turns vertically downward and dives to enter the submerged conduit openings. It was determined that this diving flow in shear with a dead zone near the structure wall was the primary cause of the vortices forming during the study.

In summary, the 20 degree sloping face wall in conjunction with the curved approach walls created a venturi-effect. This enabled the flow streamlines along the water surface and the side walls to smoothly accelerate into the structure thereby eliminating any dead zones where vortex action and formation normally occur. By smoothly accelerating the flow from the surface into the gate openings, not only were the separation and stagnation zones near the face of the structure eliminated, but the negative pressures on the roof curve and near the bulkhead gate slot locations were also reduced. Additionally, the sloping wall reduced flow turbulence and pressure disturbances and improved the discharge water surface so that the trunnions were no longer impacted by the flow. The inclined face wall allowed for a larger range of acceptable flow conditions, including; gate operation, gate openings, reservoir pool elevations, and discharges. The use of the inclined face wall also helped remediate approach flow problems and the effects of surface and wave effects. One possible negative effect of the inclined face wall is the possible deflection of floating debris into the gate passages, but this was not studied.

Authors

Steven L. Barfuss is a research assistant professor at Utah State University. He has over 23 years of experience performing physical model studies at the Utah Water Research Laboratory in Logan, Utah. His modeling expertise includes: dams, spillways, outlet works, energy dissipation basins, pumping plants, intakes, outfall diffusers, and fish/debris screens.

William J. Rahmeyer is the Department Head of Civil and Environmental Engineering at Utah State University. He has been active in physical modeling of hydraulic structures since 1976.

Nathan C. Cox is a Hydraulic Engineer with the U. S. Army Corps of Engineers, Sacramento District. He has been with the Corps for over 5 years cumulatively. He has a M. S. degree in civil engineering from Utah State University.

References

[1] U. S. Army Corps of Engineers (USACE), October 1980. Hydraulic Design of Reservoir Outlet Works, EM 1110-2-1602.

[2] Utah State Water Research Water Research Laboratory (UWRL), October 2008. Model Study of the Folsom Dam auxiliary Spillway, CESPK-ED-DH Contract W91238-05-D-0009, Task Order 0007, Report No. 680, prepared for Ayres Associates, Inc. and the U. S. Army Corps of Engineers, Sacramento District.

[3] Hecker, Model-Prototype Comparison of Free Surface Vortices, Journal of the Hydraulics Division, ASCE, 1981, 107 (HY10).

Application of Thermal Infrared Imager to Measure Temperature in Cooling Water Models

Tong Zhongshan[1,2]   Wang Yong[1]   Xu Shikai[1]   Ruan Shiping[1]

[1]*State Key Laboratory of Hydrology-Water Resources and Hydraulic Engineering,
Nanjing hydraulic research institute, Nanjing, 210029*
[2]*Key laboratory of navigation structures, Nanjing, 210029*

Abstract: The working principle, components and development of thermal infrared imager was expounded, and the application problems were analyzed from a technical perspective. A comparative investigation of non-contact infrared temperature measurement by means of thermal infrared imager and traditional contact measuring temperature in a coastal power station model was employed.

Key words: thermal infrared imager; infrared measuring temperature; application study; cooling water; model test

1. Introduction

There are three methods in the studying of cooling water models, namely, prototype observation, mathematical model and physical model test, and the last one is the main method, in which temperature measurement is particularly important. Temperature measurement includes contact and non-contact temperature measurement, the last one, with the advantage of fast response time, non-contact, safety and long life, the non-contact thermal infrared imager technology developed rapidly, the performance improved continuously, function grown, the growing species, also expanded the scope of application in resent 20 years. Typically, as a non-contact measurement method, thermal infrared imager is used in industry and field-work when great accuracy is not needed, and it is difficult to get high measuring precision in the model test of cooling water with it.

2. Thermal infrared imager

2.1 Principle

Infrared radiation is one of the most common electromagnetic radiations in the world, which produced by the thermal infrared energy come from the rule-less movements of molecular and atomic. The more intense movement, the more radiation energy, and vice versa. Using the infrared detector, radiation power signals can be transformed to electrical signals which can be processed by electronic system and display to screen by means of thermal images.

Figure 1   Thermal imaging system

2.2   Thermo Vision A40M infrared thermal imaging system

Thermo Vision A40M is a high-precision infrared thermal imaging system, it has a resolution of 0.08℃ at the ambient temperature of − 40℃ to 500℃. It can produce noise-free high-resolution infrared thermography (320 × 240 pixels), and each image contains more than 76800 temperature measurement points with the refresh rate of 50-60Hz. It does contribute positively to the measuring of temperature field and recording of the motion process of cooling water. Figure 1 is the picture of thermal imaging system.

3. Application of thermal infrared imager in cooling water models

3.1　Project overview

The studied coastal power plant has units of (2 ×125) MW and (2×300) MW in the first and the second phase respectively, and the discharge of cooling water is 11. 2m³/s and 22. 14m³/s. The proposed unit is (2 × 660) MW and the discharge of cooling water is 42m³/s, the temperature rise is 8. 90℃.

3.2　Model design

The simulation scope is 10km× 8km in prototype, horizontal scale of the model is $\lambda_l=300$, vertical scale is $\lambda_h=120$, and the distortion rate is $\eta=2.5$.

3.3　Test Instruments

Contact temperature sensors and thermal infrared imager were used simultaneously to measure the surface temperature field near the cooling water outfall and the test results were compared.

3.4　Test results

Figure 2 is part of infrared image produce by Thermo Vision A40M during the spring tide. Limited by the built and proposed coal berths, cooling water discharge with a temperature rise of 8. 9℃ had a flow direction of northwest. Because the coal berths blocked the free diffusion, the influence of cooling water to the intake was weakened. With the rise, ebb flow during the reciprocating motion, the thermal discharge flowed in the direction of northwest and east, but its influence to proposed intake appeared limited.

Take Point A (at the front intake of the first and the second phase), Point B (at the front intake of proposed phase), and Point C (at the midpoint of the two breakwaters) as feature points, the test results of temperature sensors and thermal infrared imager was listed in Table 1 and a comparison can be drawn.

The measurement shows that the measured results by using the Thermo Vision A40M system agree with those by temperature sensors.

Table 1　　Test results from temperature sensors and thermal infrared imager

Feature points	Test Instruments	Spring tide rapid rise	Spring tide rapid drop	Neap tide rapid rise	Neap tide rapid drop
A	A40M	0. 6	0. 9	0. 2	0. 6
A	Temperature sensors	0. 8	0. 4	0. 4	0. 4
B	A40M	0. 2	0. 2	0. 2	0. 3
B	Temperature sensors	0. 2	0. 2	0. 2	0. 4
C	A40M	4. 2	4. 7	4. 3	4. 6
C	Temperature sensors	4. 3	4. 8	4. 3	4. 7

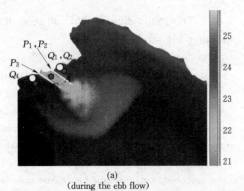
(a)
(during the ebb flow)
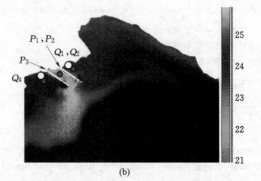
(b)

P_1, P_2, P_3 —the outlet of the first, second and the third phase;
Q_1, Q_2, Q_4 —the intake of the first, second and the fourth phase.

Figure 2　Infrared image of the spring tide

4. Conclusions

By using of the thermal infrared imager in the physical model test, we arrive at the following conclusions:

(1) Thermal infrared imager can reflect the diffusion process of the cooling water in the tide directly and quickly, and this will make up the insufficient of temperature sensors.

(2) Using calibration of ambient temperature and actual temperature, we can establish the relationship between thermal infrared imager and temperature sensors.

Reference

Tong Zhongshan, Xu Shikai, Wang Yong. *Experimental study on the second phase of Huaneng Weihai Power Plant*. Nanjing Hydraulic Research Institute, Nanjing, 2006.

Study on Big Scale Ratio (1:10) Hydraulic Model Test on Energy Dissipater of Bypass for Water Supply for a Reservoir in Iran

Lin Jinsong   Liu Hansheng   Lv Hongxing   An Mengxiong   Ding Limin

College of Water Resources and Architectural Engineering,
Northwest A&F University, Yangling, 712100

Abstract: The variable frequency and constant pressure water supplywas applied in a big scale ratio (1:10) hydraulic model test on energy dissipater of bypass for water supply for a reservoir in Iran. The model simulated headrace tunnel, needle valve, energy dissipation box, tailrace surge chamber and tailrace tunnel, observed and measured such hydraulic parameters of energy dissipater as discharge, velocity, flow pattern, pressure and flow profiles at various reservoir water levels and openings of the needles valve, analyzed the dissipater style, cavitations and erosion for the flow of high velocity, studied the flow patterns in the tailrace surge chamber. The results shown that hydraulic characteristics of energy dissipation box are favorable, which meet the demand of this project. All the characteristics of turbulent pressure are conventional, which are normal probability distribution and have favorable time correlativity. The preponderant frequencies are low, and the standard deviations of turbulent pressure are small that meets the demand of the project. Meanwhile a vertical dissipation grid is used in the surge chamber to improve energy dissipation efficiency, a large number air bubbles was avoided to enter the tailrace tunnels, the water surface in surge chamber was stabilized effectively and demand of the project was satisfied.

Key words: reservoir; big scale ratio; model test; bypass for water supply; variable frequency and constant pressure water supply

1. Introduction

Taleghan dam & HPP project is one of most important water control projects in Iran, which consists of rock-fill dam with clay core wall, overflow spillway on the left bank, bottom outlet, tunnel for water supply and underground powerhouse on the left side in the reservoir. The dam is 103m high. Installation capacity of power station is (2×8.9) MW, which runs according to water demand of irrigation and municipal use. The turbine tail water and bypass pipe are designed to supply water of 15m³/s, from which the flow converges into tailrace surge chamber then enter the tailrace tunnel for water supply.

Main hydraulic structures for water supply and power in the project are headrace tunnel, pressure steel pipe for water delivery, bypass pipe for water supply, underground powerhouse transportation vertical well, tailrace surgechamber and tailrace tunnel.

The reservoir has normal level of 1780.0m,

dead level of 1742.0m, where there is maximum drawdown of 38m. The water level is up to 1784.5m with 10000-year flood. The normal tail water level of power station is 1710.5m.

For hydraulic model, scale ratio is most important. Generally speaking, the model with larger scale ratio is better when the field and equipment in laboratory are allowed. Large scale ratio is very important for raising test quality. According to experiment of Bureau of Reclamation in U.S.A which canbe taken as guide for selecting scale radio[1], outlet works having gates and valves are constructed to scale ratios ranging from 1:10 to 1:30, and gates and valves should be at least 75mm diameter for satisfactory study of an outlet works energy dissipater.

The model scale was chosen by 1:10 because the diameter of bypass pipe is only 1.8m and the discharge is 15m³/s, the water head in model is high upto 8.0m, and there is maximum drawdown of 4.25m for model reservoir level. The minimum diameter of pressure pipe in model is 160 mm. this scale ratio accords with experience of bureau of Reclamation, which is reasonable for this project.

In order to solve the problem, the variable frequency and constant pressure water supply was applied in a big scale ratio (1 : 10) hydraulic model test on energy dissipater of bypass for water supply.

2. Test purpose and test content

2. 1　Test purpose

Hydraulic model test is an effective method for energy dissipater problem, which are difficult to be solved by hydraulic calculation. Through this hydraulic model test, the validity and rationality of dissipater are evaluated on hydraulic, and secure and economical layout for bypass pipe is recommended. The main problems that this test is necessary to solve are as follows:

(1) The form of energy dissipater is determined for water transmission system. Hydraulic characteristics of flow pattern, velocity and pressure are given around outlet of needle valve in all the flow condition.

(2) The layout and configuration of dissipation structure for water supply system are studied and optimized. It is demanded that the flow is stable in the surge chamber in all the flow conditions. The velocity of flow entering surge chamber is controlled under 3-5m/s. According to the demand of tailrace gate design, the orifice above the end still is not larger than 3.7m×2.8m (width×height)

(3) Through this hydraulic model test, the measures to protect surge chamber are given.

2. 2　Test content

(1) Measure such hydraulic parameters of energy dissipater as discharge, velocity, flow pattern pressure and flow profiles at various reservoir water levels and openings of the needle valve.

(2) Analyze and improve the dissipater, and provide the measures of control of capitation and erosion for the flow of high velocity.

(3) Analyze the end sill of energy dissipation box, and suggest proper configuration for the end still.

(4) Observe the flow patterns in the tailrace surge chamber, and suggest improving measures.

(5) Testify the effectiveness of the dissipater, and suggest the proper configuration of the dissipater

and measures of erosion control for tailrace surge chamber.

3. Hydraulic model and test method

3. 1　Similarity principle

Hydraulic model is a physic model which simulates the flow of water control project. The flow in model is in the similitude of flow in prototype. The geometric similarity dynamic similarity and cinematic are required.

For this and this project, the flow is caused by gravity. The principle of gravity similarity (Froude law) is used to design the model. If Froude number between model and prototype is the same, dynamic similarity and cinematic similarity are achieved when geometry and roughness are similar between model and prototype, and Reynolds number in model high enough.

The principle of gravity similarity can be expressed as follows:

$$Fr_r = \frac{v_r}{\sqrt{g_r L_r}} = 1 \qquad (1)$$

where Fr_r is the ratio of Froude number; v_r is the velocity ratio; g_r is the ratio of the acceleration due to gravity; L_r is the geometric ratio. Since the accelerations due to the gravity between model and prototype are same, i. e. , $g_r = 1$, the last formula turns into:

$$v_r = \sqrt{g_r L_r} = \sqrt{L_r} \qquad (2)$$

According to this formula, the following ratio are obtained:

Length ratio　　$L_r = \dfrac{L_p}{L_m} = 10$

Area ratio　　　$A_r = L_r^2 = 100$

Velocity ratio　$v_r = L_r^{1/2} = 3.162$

Time ratio　　　$t_r = L_r^{1/2} = 3.162$

Discharge ratio　$Q_r = L_r^{5/2} = 316.2$

Roughness similarity requirement

$$n_r = L_r^{1/6} = 1.468$$

Where n_r is the roughness ratio.

The roughness coefficient of concrete in prototype is about 0.013. According to last formula, the roughness coefficient in model should be 0.0088. This model of organic glasses which has roughness coefficient of 0.0085, roughness similarity is satisfied.

In addition, similarity of flow pattern is required. For the model of qualitative similarity, essence flow pattern must be similar, i. e. the flow pattern in model must be turbulence. This model is designed completely according to quantitative similarity of flow pattern. The flow pattern in model should be high enough, which should meets[2]:

$$Re_{min} = \frac{q_{min}}{v} \geqslant 10000 \qquad (3)$$

where Re is the minimum Reynolds number in model; q is the minimum discharge per unit width in model; v is the cinematic viscosity of water, which is about $1.2 \times 10^{-6} \, m^2/s$.

For this model, the scale ratio is 1 : 10, calculation with last formula gets $Re_{min} = 38300$, which ensures that the flow in model is in the region of resistance square. This also shows that the scale ratio of this model is appropriate.

3.2　Test instruments and accuracy control

3.2.1　*Transducer for steady pressure*

The variable frequency and constant pressure water supply was applied in the hydraulic model test, pump supplies directly water to model pressure pipe. In order to ensure the water head of model is steady, transducer for steady pressure, which is manufactured by Northwest Hydro Technical Research Institute in China, is adopted to control the pump. In all the test conditions, the pressure fluctuation is ($\pm 20.4 \times 9.8$) pa (± 1.5mm Hg column).

3.2.2　*Instrument for turbulent pressure*

CYGB Type of turbulent pressure sensor is used, the error of which is 0.21%. SG-60 Type of date collection system process is used in the electronic analog signals, then computer analyses the date.

3.2.3　*Pressure meter of mercury with U shape*

The water head in this model is high up to 8.0m, pressure meter of mercury with U shape is set up in front of the needle valve, which is used to measure the water pressure. The computation method for the pressure is: upstream reservoir level minus head loss between inlet and valve leaves total head in front of needle valve, which minus velocity head leaves pressure head. The pressure meter of mercury measures the pressure head, the accuracy of which is

± 0.5mm Hg column.

3.2.4　*Weir for measuring discharge*

Downstream rectangle weir is set up to measure the model discharge, which has the width of 60.10 cm, and the height of 38.20cm, calculating with T. Rehbock formula, the error of discharge is $\pm 1\%$ within the range of $p/2 \geqslant h \geqslant 3$cm ($p$ is height of weir, h is the head over the top of weir). Model discharge is controlled by opening or closing the valve to make the discharge to be 18m³/s, 15m³/s or 8m³/s. The corresponding model heads over the top of weir are shown inTable 1.

Table 1　Model head over the top of weir

Discharge in prototype (m³/s)	Discharge in mode (L/s)	Model head over the Top of weir (cm)
18	56.9	13.59
15	47.4	12.07
8	25.3	8.00

3.2.5　*Other instrument*

Water surface profile and water level of tailrace surge chamber are measured with measuring needle, the error of which is ± 0.1mm, piton cylinder with the accuracy of which is ± 0.5mm H_2O column. Theodolite and level gauge are employed to control model layout. The plane error is within $\pm$ (0.5-1.0) mm, and the elevation error is± 0.2mm.

Model scope is from upstream section to downstream section. The upstream section at headrace tunnel is 30m in front of the needle valve; the downstream section at tailrace tunnel is 25 m behind the surge chamber. The model includes headrace tunnel, needle valve, energy dissipation box tailrace surge chamber and tailrace tunnel it is required that the model flow is in the similitude of actual flow. The water level signs for reservoir and tailrace surge chamber are set. Energy dissipation area may be taken as fixed-bed in this test.

4. Test of original design scheme

4.1　Configuration of original design scheme

Configuration of original design scheme is shown in Figure 1. Actual operating conditions in the test are listed in Table 2.

Table 2 Actual operating conditions in the test

No.	Reservoir level (m)	Discharge (m³/s)	Water level of tailrace surgechamber (m)
Condition 1	1780. 0	15	1710. 5
Condition 2	1780. 0	15	Free efflux[2]
Condition 3	1780. 0	8	1710. 5
Condition 4	1780. 0	8	Free efflux
Condition 5	1765. 0	15	1710. 5
Condition 6	1765. 0	15	Free efflux[2]
Condition 7	1765. 0	8	1710. 5
Condition 8	1765. 0	8	Free efflux
Condition 9	1742. 0	15	1710. 5
Condition 10	1742. 0	15	Free efflux[2]
Condition 11	1742. 0	8	1710. 5
Condition 12	1742. 0	8	Free efflux
Condition 13	1784. 5	15	1710. 5
Condition 14	1784. 5	15	Free efflux[2]
Condition 15	1784. 5	18	Free efflux

Note Free efflux[2] (during the period of the filling tailrace tunnel with water).

The characteristics are, in order to make flow smoothly, short straight pipe is set up behind the needle valve. Sudden expansion of sideway is used for energy dissipation. High end sill is adopted in the tail of energy dissipation box for forming submerged jet at outlet of pressure pipe. The low sill is set to protect the chamber floor turns 90⁰ and flows to the tailrace tunnel. In order to form submerged orifice for inlet of tailrace tunnel, the floor of tailrace tunnel has lowered 2. 4m.

4. 2 Actual operating condition in the test

In order to know more hydraulic characteristics of tailrace surge chamber based on demanded operating conditions, 3 operating conditions (Condition 12-Condition 14) are added in the text for free efflux of tailrace tunnel. In additions, a supplement operation condition (Condition 15) is also added to study fully the energy dissipation box, which has the reservoir level of 1784. 50m, and has discharge of 18m³/s.

4. 3 Flow pattern

4. 3. 1 *Jet outlet of pressure pipe*

The pressure flow behind the needle valve enter the energy box in the form of submerge jet, which dissipate the water energy. The flow pattern of jet is satisfactory because of showing by the aubergine water which is added in experiment, the water jets out forwards, not jets towards the sidewalls. This is propitious to the stabilization of the sidewalls.

4. 3. 2 *Flow pattern in energy dissipation box*

In all operating conditions, the flow in energy dissipation box is stable, which is pressure flow in the fore part, and is weir flow (with free-surface) in the back part. The flow is clear and transparent if there is not tracer, and the fluxion is not detects by eyeballing. The surface wave height in front of the high end sill is shown in Table 3.

Table 3 Water level and water wave height in front of the high end sill

No.	Water level (m)	Water height (cm)
Condition 1	1713. 20	2. 0
Condition 2	1713. 20	2. 0
Condition 3	1712. 70	1. 8
Condition 4	1712. 70	1. 8
Condition 5	1713. 20	2. 0
Condition 6	1713. 20	2. 0
Condition 7	1712. 70	1. 8
Condition 8	1712. 70	1. 7
Condition 9	1713. 20	2. 0
Condition 10	1713. 20	2. 0
Condition 11	1712. 70	1. 6
Condition 12	1712. 70	1. 0
Condition 13	1713. 20	2. 0
Condition 14	1713. 20	1. 0
Condition 15	1713. 25	8. 0

From operating Condition 1 to Condition 14, the wave height is not over 2cm, even for Condition 15 (discharge is 18m³/s), the wave height is not more than 8cm. The flow satisfies the project demand. It is get from Table 3 that the water level in front of the high end sill mainly lays on discharge. This result is reasonable. Analyze as follows:

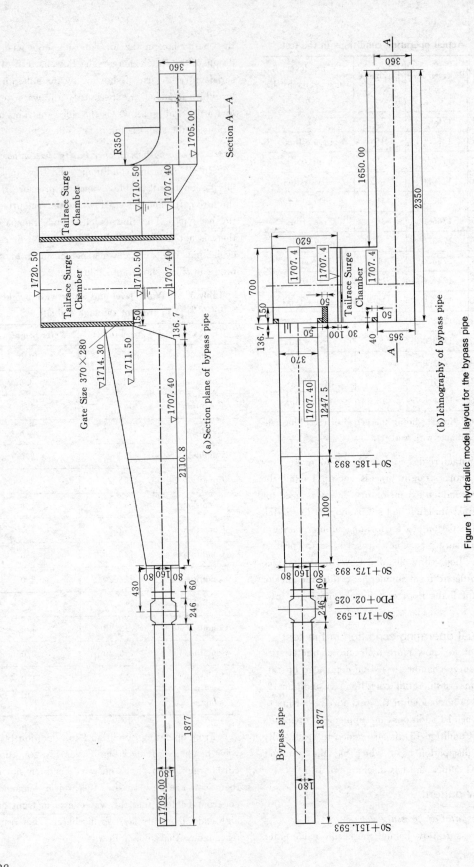

(a) Section plane of bypass pipe

Section A—A

(b) Ichnography of bypass pipe

Figure 1　Hydraulic model layout for the bypass pipe

The discharge formula of weir flow for free efflux is

$$Q = m\varepsilon B \sqrt{2g}H^{3/2} \qquad (4)$$

where, Q is discharge; m is discharge coefficient; ε is contraction coefficient; B is width of weir; g is acceleration due to gravity; H is total water head over the top of weir.

From this formula, the relation between Q and H is monotone function. This means that one discharge has one water head, and one water level in front of the water.

4.3.3 *Flow pattern in the tailrace surge chamber*

Flow pattern in the tailrace surge chamber lies on two factor: (1) discharge; (2) water level in the surge chamber. There are two discharges for this project, $15m^3/s$ and $8m^3/s$. Water levels in the surge chamber also have two conditions: (1) condition of normal supplying water, water level in the surge chamber is 1710.50m; (2) condition during the period of filling tailrace tunnel with water. Tailrace tunnel is free efflux.

The hydraulic characteristics in the surge chamber are as following: in the form of weir flow, the flow drops into tailrace surge chamber, which hits the tail water, and has some impact on floor. Then the flow turns 90°, after a short distance, it turns 90° again, and enters tailrace tunnel.

The flow patterns are observed in detail, the characteristics of which are as follows:

(1) When discharges are same, and water levels in surge chamber are same, flow patterns in surge chamber are also same for different reservoir level. This is a rational result. The reason is that the weir flows in the back part of energy dissipation box mainly lies on discharge.

(2) When discharge is $15m^3/s$, the drop flow over the high end sill makes a quantity of air enter water. Since the short distance, air bubble can not get out of water, and a great deal of air bubble enters the tailrace tunnel. Generally speaking, onefold water is better for tailrace tunnel. So many air bubbles in the water make against the run of tailrace tunnel.

When discharge is $8m^3/s$, less air enters water, and the air bubbles get out of water. Almost no air bubble goes into the tailrace tunnel. This flow is satisfactory.

(3) During the period of filling tailrace tunnel with water, the discharge is $15m^3/s$, and the tailrace tunnel is channel flow. It is also supercritical flow. The depth of water is small, not over the third of tunnel. Although there are strong shock waves, it is impossible to come out transition flow between free flow and pressure flow. Because the condition of the period of filling tailrace tunnel with water is temporary, It is not necessary to demand smooth flow in the tunnel. This flow pattern satisfies the need of this project.

(4) The flow over the high end sill is drop, which makes the surge chamber fluctuate like water wave. The wave height is shown in Table 4.

Table 4   Wave height at the sidewall opposite the inlet of tailrace tunnel

Discharge (m^3/s)	Wave height (cm)	Wave level of tailrace surge chamber (m)
15	35.0	1710.5
8	19.0	1710.5
15	30.0	During the period of filling tailrace tunnel with water

When the discharge is 15 m^3/s, the maximum height of the wave is 35 cm, which is on the high side for the tail water of turbine, and has some disadvantageous effect on run of the turbines. Engineering measure should be taken to control the wave.

(5) The flow patterns over the high end sill are: The nape of drop flow disperses along the top of the sill, and the nape of the right side has dropped the outside of the low end sill on the floor of the surge chamber which is not a good flow pattern. It is necessary to set upa sidewall on the right side on the top of the high end sill, which leads all the flow into the inside of the low end sill.

4.4   Pressure distribution

The pressure distributions of energy dissipation box for the 15 operating conditions have been measured. Pressure characteristics on the two main parts are measured: (1) the energy dissipation box, pressure characteristics on sudden expansion of sidewall are focused on; (2) the drop low over the high end sill. The pressure impacting on the floor of the surge chamber is measured. The values of pressure are satisfying, and distribution characteristics are preferable.

In normal water supply conditions, the pressure changes gently, and the pressure distributions on the floor of surge chamber mainly lies on discharge. The drop flow over the high end sill has little effect on the pressure distributions on the floor. The reason is that the water depth in surge chamber is deep, which cushion greatly the impact of drop flow. So the pressure changes gently.

During the period of filling tailrace tunnel with water, the water depth insurge chamber is shallow, which has weak cushion function. So the drop has effect on pressure on the floor of surge chamber. In this condition, pressure distributions still lie on discharge, which is 15m^3/s and is a fixed value. The pressure distributions are almost same among these conditions. Because of the impact of drop flow, these are a peak in the figure of pressure distribution. The peak value is among (3.70 $\times$ 9.8) - (3.90 $\times$ 9.8) kPa (Table 5), which meets the demand of this project.

Table 5   Maximum pressure on the floor of surge chamber

No.	Maximum pressure (9.8kPa)
Condition 2	3.70
Condition 6	2.80
Condition 10	3.77
Condition 14	3.90

In addition, the pressure distribution on the floor of surge chamber is measured when the discharge is 8 m^3/s and the tailrace tunnel is free efflux. The test results are for reference.

4.5 Velocity distribution

Velocity distribution at outlet of pressure pipe is a very important hydraulic characteristic for the bypass pipe, which has nothing to do with water level in surge chamber. Based on the combinations are measured, which are even, has a good form and are prepositional todischarge. The velocity value of supplying 15m^3/s is almost twice of the velocity value of 8m^3/s. The velocity decreased rapidly when the flow goes into energy dissipation box. This is the very low, not over 2.5m/s. Capitation does not happen in energy dissipation box.

4.6 Turbulent pressure

This test shows that the pressure fluctuation is different along the energy dissipation box. There are three turbulent pressure sensors—one is on the bottom, another is on the top and the third is on the sidewall. All the representative values are listed in table 6. All the characteristics of turbulent pressure are conventional. Turbulent pressure test results show they are normal probability distribution, and its time correlativity is favorable. All preponderant frequencies are low and are in the range among 0-4.5Hz. All the standard deviations of turbulent pressure are small, less than 0.3kPa, which meets the demand of this project.

Table 6   Representative values of turbulent pressure

No.	Reservoir level (m)	Discharge (m^3/s)	Position	Average Pressure (kPa)	Standard Deviations (kPa)	Main Frequency (Hz)	Preponderant Frequency (Hz)
Condition 1	1780	15	Bottom	5.07	0.20	0.23	0-3.40
			Sidewall	3.43	0.20	0.07	0-2.80
			Top	1.41	0.15	0.20	0-0.80
Condition 3	1780	8	Bottom	4.99	0.07	0.10	0-2.70
			Sidewall	3.33	0.07	0.02	0-1.35
			Top	1.34	0.07	0.10	0-0.30
Condition 5	1765	15	Bottom	5.05	0.18	2.04	0-3.30
			Sidewall	3.42	0.19	0.12	0-2.40
			Top	1.38	0.13	0.31	0-1.05

No.	Reservoir level (m)	Discharge (m³/s)	Position	Average Pressure (kPa)	Standard Deviations (kPa)	Main Frequency (Hz)	Preponderant Frequency (Hz)
Condition 7	1765	8	Bottom	5.01	0.07	0.01	0-1.90
			Sidewall	3.33	0.07	0.09	0-2.20
			Top	1.36	0.06	0.23	0-1.00
Condition 9	1742	15	Bottom	5.04	0.19	0.17	0-4.10
			Sidewall	3.42	0.19	0.40	0-2.10
			Top	1.40	0.13	0.18	0-1.45
Condition 11	1742	8	Bottom	5.03	0.06	1.32	0-2.70
			Sidewall	3.31	0.06	0.11	0-1.45
			Top	1.33	0.05	0.07	0-0.50
Condition 13	1784.5	15	Bottom	5.07	0.19	0.18	0-3.60
			Sidewall	3.44	0.20	0.37	0-3.00
			Top	1.42	0.15	0.11	0-0.60
Condition 15	1784.5	18	Bottom	4.85	0.26	2.49	0-3.30
			Sidewall	3.35	0.27	0.48	0-4.45
			Top	1.28	0.20	0.37	0-1.60

4.7 Condition of filling the energy dissipation box with water

From the safe run of the bypass pipe, the energy dissipation box should be filled with water before it is put into use. The method is to open the needle valve a little, after the box is full of water, the needle valve is turned on to the opening required. Three small opening are tested, which are 1.6%, 3.3% and 6.6% in normal reservoir level. The test shows that the smaller the opening of the needle valve is 1.6%, it is open flow at the outlet of pressure pipe. And the flow in the box is most stable in the 3 openings of needle valve. When the opening is 3.3% or 6.6%, the flow in the pipe behind the needle valve is annular. The reason is that the smaller the opening is, the more loss of water head. The lower the velocity is, and the better the pattern is.

5. Test of energy dissipation grid

5.1 Necessity of energy dissipation grid

Energy dissipation contradicts the tail water of power plant. They are usually separated from concrete wall, or they are laid in different places so as to avoid its interference. For this project, bypass pipe and tail water of power plant use same chamber. Energy dissipa-

tion grid can solve the problems of power plant use some surge chamber. Energy dissipation grid can solve the problems of water surface fluctuation and the pattern of flow entering tailrace tunnel.

5.2 Engineering example of energy dissipation grid

Energy dissipation grid is a special dissipater, which is mainly applied in small or medium project. There is this type of dissipaters in the northeast of China, northwest of China and north China. For example, energy dissipation grid is adopted in Fugui Drop and Qinjia Drop[3] in Heilongjiang province, china. For common small drop, there are design methods and plenty of experiences. Energy dissipation grid in dissipation chamber, the total Fuliutan Ship Lock[4] adopted double energy dissipation grid in dissipation chamber, the total water head of which is 13.08m. In order to dissipate more energy, Hongjiang Water Power Station[5] introduced a certain amount of energy dissipation beams in inverse outlet dissipater by means of hydraulic model test. There beams are energy dissipation grid from the view of dissipation mechanism, KaqunWater Station sets up energy dissipation grid along the chute to reduce flow velocity[6].

5.3 Test on energy dissipation grid

The vertical energy dissipation grid, suggested by this test, is adopted to stabilize the water surface and improve the pattern of flow entering tailrace tunnel. A horizontal energy dissipation grid is tested firstly. The effect is not good. The flow pattern in the surge chamber does not change obviously.

Five vertical energy dissipation grids are tested, the test shown that if the hole of the grid is too small, the upstream water level is raised, forming critical flow in the drop in the high end sill. Sometimes it is submerge flow, sometimes it is free efflux. This flow does not meet the demand of the project. If the hole of grids is too large, the function for tail water of power plant is not obvious. By comparing these vertical energy dissipation grids, the size of the hole is suggested to be 35cm × 35cm, Figure 2.

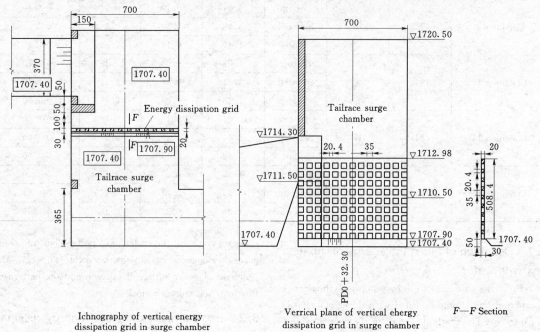

Ichnography of vertical energy dissipation grid in surge chamber

Verrical plane of vertical ehergy dissipation grid in surge chamber

F—F Section

Figure 2   Layout of vertical energy dissipation grid in surge chamber
(size of grid hole: 35cm×35cm; length unit: cm; elevation: m)

Applying this size of grid hole, the drop in the high end sill is free flow, the water surface is stabilize effectively, and a large number of air bubbles are avoided to enter tailrace tunnel. The wave heights are listed in Table 7, which shows that energy dissipation grid can reduce 37% of wave height. it is obvious that the water surface with the grid is much stable.

Table 7   Comparison of wave height between with the grid and without grid

Discharge (m³/s)	Wave height with grid (cm)	Wave height of No grid (cm)	The value reduced (%)
15	35	22	37.10
8	19	12	36.80

Note   The water level in surge chamber is 1710.5; The size of grid hole is 35cm×35cm.

The pressure distribution with the grid is reasonable, because of the effect of the grid; a water cushion is formed in the downstream of the high end sill during the period of filling the tailrace tunnel with water.

6. Conclusion

This test adopts large scale ratio (1：10), related hydraulic parameters are measured hydraulic problem of the bypass pipe are discussed. Conclusions obtained are as follows.

(1) The variable frequency and constant pressure water supply was applied in a big scale ratio (1：10) hydraulic model test on energy dissipater of bypass for water supply for a reservoir, the result shown that the test is reasonable and successful. It is effective method to solve the high water level and lit-

tle discharge hydraulic experimentmodel.

(2) By measuring pressure distribution, turbulentpressure, velocity distribution, flow pattern and water level, it is shown that hydraulic characteristics of energy dissipation box are favorable, which meet the demand of this project. As to original scheme, the fluctuation of water surface in surge chamber is on the high side, and a large number of air bubbles enter the tailrace tunnel when the discharge is $10m^3/s$.

(3) All the characteristics of turbulent pressure are conventional, which are normal probability distribution and have favorable time correlativity. The preponderant frequencies are low, and the standard devotions of turbulent pressure are small, less than 0.3 kPa, which meets the demand of this project.

(4) After setup of the sidewall on the right side on the top of the high end sill, the sidewall leads all the flow into the inside of the low end sill in all the operating conditions, which arrives at the design requirement.

(5) It is an effective method to set up energy dissipation grid in order to improve the flow pattern. Through tests of 5 configurations of energy dissipation grid, the hole size of the grid is suggested 35cm × 35cm. The energy dissipation grid avoids a large number air bubbles entering the tailrace tunnel, stabilizes effectively the water surface in surge chamber, and meets the demand for the flow pattern in the high end sill.

Acknowledgements

This paper is supported by Northwest A & F University, Returning from abroad fund (01140507) and Specific research fund (A213020503).

References

[1] U. S. Department of the Interior Water and Power Resources Service. Hydraulic Laboratory Techniques [C]. United State Government Printing Office, Denver, Colorado.

[2] Zuo Dongqi. Theory and Method for Hydraulics Model test [C]. Press of water Conservancy and Hydroelectricity (in Chinese).

[3] Wang Wenkai, Ling Chuanwei. Design of Energy Dissipation Box [J]. Water resource and hydropower Engineering, 1984 (1): 12-18, (in Chinese).

[4] Zou Bingsheng, etc. Application of Double Energy Dissipation Grid in the Water Supply System in the Ship Lock with Middle Water Head and Short Conduit, Port & waterway Engineering [J]. 2002 (8): 53-57, (in Chinese).

[5] Li Danmei, Yin penwei. Application of Flowing through the inverse opening Energy Dissipater to Water Conveyance System of Ship-lock in Hongjing Peoject. Hongshui River. [J]. 2002 21 (3): 35-38 (in Chinese).

[6] Qiu Xiuyun, Hou Jie etc. A new measure of eliminating the torrent shock wave in steep bend channel, Water Power [J]. 1998 (11): 19-22, (in Chinese).

Application of FL-NH Six-Component Test System for Motion of Ship Model

Peng Li   Jie Jin

State Key Laboratory of Hydrology-Water Resources and Hydraulic Engineering,
Nanjing Hydraulic Research Institute, Nanjing, 210029

Abstract: In this thesis, the understanding and analysis of VR devices for real-time follow-up of the motion is summarized; the developed FL-NH measurement system for motion of ship model can be applied in the physical model test on moored floating body in the wind wave and current tank in the harbor engineering. This system can be used in all weather; the non-contact single multi-point transmission sensor (9g) synchronously receives the six-degree-of-freedom quantity of motion of the measured floating body. The dynamic collection of data is real-time and accurate and has good repeatability; the measured process data of the test on six-component motion of the single ship is repeated under the 3D software. The whole system has a wide prospect for expansion and application.

Key words: VR; motion of floating body; non-contact measurement; 3D demonstration

1. Introduction

In the physical model test for ocean engineering, the simulated wind wave and current need to be controlled in the wind wave and current tank in the harbor engineering; under the action of steady wind or non-steady wind, it is required to research the dynamic response of single/multiple ships which are moored at the structures of the wharf's harbor engineering, or anchored beside the chain of floating body as well as its similar problems. The stress motion of the mooring structure has six degrees of freedom along the three-axis direction in the space rectangular coordinates (Descartes). For the motion of the ship model mooring and berthing alongside, it refers to the displacement of center of gravity of the mass point in the center of its rigid body after the ship model is loaded with different weights in the harbor basin as well as the swaying size of the swinging angle.

In 2007, Harbor and River engineering Department of the Nanjing Hydraulic Research Institute introduced the FL real-time spatial VR motion tracking and positioning system, researched and developed the NH application software, and applied such software in the all-weather test which aims to measure the berthing stability motion of the ship model at the wharf under the action of wave and water current in a non-contact way, so as to rapidly, accurately and reliably collect the data in an efficient and labor-saving manner with good repeatability, and obtain the efficient and labor-saving actual effect.

The instrument operators analyzed the technical conditions for FL hardware interface and transmitting/receiving sensor, etc. and their application environment, calibrated the six parameters by analogy as well as the reference values calibrated according to the measurement criteria (CTC), compared the tradi-tional six-component instruments (like mechanical contact type, photoelectric capture type and gyro ac-celerometer, etc.) which have been used at home and abroad, found out the regular factors of the FL technical conditions and their advantages and charac-teristics, reformulated the new standard for technol-ogy application of FL device in the six-degree-of-free-dom motion test of the mooring structure (e. g. space limitation of metallic materials after the floating body is loaded with the weight; permissive conditions for peripheral ambient electromagnetic field; and meas-uring range coverage of the transmitting/receiving sensor, etc.), found out some problems not stated and interpreted in the original manual of FL device, explored the methods and approached and further ex-panded the application field of FL-NH system.

In order to accurately fix the T4 transmission source of the FL hardware and save the total time of

434

labor and test during the modeling of different model engineering topography schemes in the harbor basin and the arrangement of harbor engineering buildings, we designed the spatially three-axis pendent and hung 3SRL servo dragging system in the test hall, respectively realized the double-direction remote control or wired control in three directions of X Y and Z, namely left and right, front and rear and up and down and satisfied the accurate positioning any of spatially pendent location and height of T4 transmission source within $24\,m^2/s$ range of the spatially pendent X-Y plane. 3SRL system has been debugged and installed in the irregular wave generator test hall of our wind wave and current tank (the harbor basin is 60m long and 17.5m wide). After the combined use of NH-FL and 3SRL, the integral model test for berthing stability of the ship models of two bulk carriers weighed respectively 100,000 tons and 250,000 tons in the dock engineering has been finished.

2. Brief introduction to hardware of the FL device

FL, by taking lower-frequency magnetic conversion as technical core and taking the DSP digital signal processing method, creates the industry standard for tracking of motion state. FL can effectively eliminate the sound wave caused by signal blockage and interference and the transmission signal distortion caused by the laser equipment source. FL has no obstruction to vision of transmission source T4 and spatial receiving sensor R2 in its application. It is a positioning and tracking system with maximum accuracy at present. FL shows good efficiency in capturing and measuring the motion data of the nonmetal targets; when the receiving sensor R2 is attached to the rigid body of a moving target and then is stressed, the additional mechanical inertia and lagged effect in the synchronous follow-up process may be neglected. The deadweight of the microminiaturized R2 is only 9g, and the three-axis signals are received by the same mass point. (Additionally, the deadweight of the R1 receiving sensor is 2.8g.)

Placing rigid But box in the closed bodies of T4 and R2 is an advanced technology. Sensors in this box can ensure a harsh working environment (to reduce the influence of integrated interference factors like

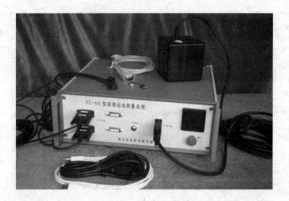

Figure 1　The FL device

strong magnetism, high humidity, high temperature and high pressure, etc.), so as to ensure the accurate and reliable transmission of the transmitted/received signals. The transmitting/receiving source also can realize one transmitting/ multiple "receiving". The basic configuration of FL is one transmitting and receiving (1-4); and after it is expanded, it can reach one transmitting and 24-receiving; a variety of models of transmission sources (T4/T2/T1/LR) under different measuring range coverage are available for selective purchasing. The above four kinds of transmission sources can be matched with the same FL interface device. (see Figure 1) FL fully shows a non-contact state because of T4 and R2 in the six-component measurement on the motion of the moored floating body, so as to ensure the authenticity and completeness of wave current water area and water field in the testing harbor basin.

3. Performance indexes

3.1　Accuracy

(1) Displacement in three-axis direction: longitudinal shift X, transverse shift Y and heaving $Z \leqslant$ 0.08mm (RMS).

(2) Three-axis rotation angle (0 – 360°): transverse swaying $x°$, longitudinal swaying $y°$ and gyration $z°$, $\leqslant 0.2°$ (RMS).

(3) RMS (root-mean-square) refers to the root mean square value and effective value.

3.2　Resolution

(1) Displacements X, Y and Z are 0.005mm.

(2) The rotation angles $x°$, $y°$ and $z°$ are 0.025°.

3.3 The decimal 4-digit valid figure of PC screen

It is displayed as — "✕✕.✕✕" [e. g. 22. 19 (cm); 12. 66°].

3.4 Sampling frequency and refresh rate

(1) Single R2 sensor is selected with a sampling frequency of 120 Hz.

(2) In a set of FL hardware configuration, the refresh rate is 120 Hz/for one R2 and the quantity of sensors is equal to the quantity of channels. (e. g. , the rate of two R2 is 60 Hz)

(3) The default Baud rate of the FL hardware configuration is 115. 2K.

(4) When clicking to select single or successive data acquisition, the sampling delay is 4 ms.

3.5 Communication between port and PC machine

(1) Common RS-232 serial port is connected to the COM1 socket of the PC machine.

(2) The USB port can be selected.

(3) Either of the communication modes, RS-232 and USB, can be selected. The RS-232 communication mode is suggested.

4. Application software

NH software, based on the original executive software of FL, is a kind of application software with a Chinese interface which is subject to secondary development for the purpose of six-component test measurement on motion of the moored floating body. The operating system of the selected PC is required to be WIN-XP/2000/98 (or WIN7). Click twice the "FL" folder on the desktop, so as to enter into the operating procedures.

The common display format of data on the display screen is ASC Ⅱ or the selected binary code.

When the harbor basin is at a still water state, before the wave generator is started, the function is provided for setting the initial test data into "zero". Starting from the "zero condition" static ship model, after the wave generator is running, click to begin measurement (In the duration of the data acquisition, click once again to stop measurement at any time). If resetting the data into "zero" once again in the same group of repeated test (in view of the actual state of the water surface undulation and calmness in the harbor basin), begin the secondary measurement, which can provide convenience for the later test data processing and reduce the trifles of repeatedly calibrating the initial "zero" value.

The test data can be imported to EXCEL or CDA for analysis processing.

3D software can simulate the demonstration of real-time six-component motion process of the single ship in a dynamic way on the screen or the projector after the test data are loaded into the document.

5. Several descriptions and precautions

In use, the distance between the geometric center of T4 and the R2 sensor should be 3 times less than the distance between the metal materials or electromagnetic producing sources (like motor) within a spatial solid scope; or else part of the measured data will distorted, but has good repeatability.

After the weighted and ballasted ship model including galvanized iron blocks is laid down, the maximum level of the iron blocks should not exceed the ground level of R2; the nearest linear distance between the iron blocks and the four horizontal sides of R2 should be greater than 30 cm; the linear distance between the ballasting iron blocks under the plane where R2 is located and R2 should be larger than the linear distance between T4 and R2.

When R2 exceeds the radius of T4 measuring range coverage, the items 1-4 among the collected data are slightly deviant. Through analysis, after the transmission signal exceeds the covered measuring range, attenuation causes the data distortion; but the data have good repeatability. Through contrast calibration, the effective value within the measuring range is increased by 1%-3%. It is suggested to select the LR spherical transmission source.

In the saline water (semi-saturated, with a concentration of 4%) which contains colored (single-colored) slush and different proportions of model water mixed with sand (such as sawdust, bakelite powder, pulverized coal, moldable polyester powder and fine glass bead powder, etc.), we carried out a lot of fundamental work for different contrast tests in the application of FL device, and achieved the quite satisfactory breakthroughs, which will not be described in detail.

T4 and R2 are transmitted/ received for measurement mainly in the air medium, but also can be transmitted/received in the gas-liquid medium and in the water. Relevant methods will not be stated one by one.

As the application of FL device in the long glass tank or internal wave tank is limited by the tank's metal framework and other similar conditions, we are seeking the pilot plant test methods and striving for breakthroughs.

6. Application instances

The six-component measurement system for the motion of ships has been successfully applied in the ship model test for the 100,000-tonnage steel wharf and the 250,000-tonnage ore wharf in the Port of Fangcheng, (see Figure 2) and the test result provides a very good reference for the design organization's design optimization. If compared with the past six-component measurement systems for motion of the ship model, this system is less affected by the external conditions, like light ray, magnetic field, temperature and humidity. The higher the accuracy of measuring result is, the better the repeatability is. It also greatly improves the efficiency of ship model test in the test room and has a very strong expandability. With the continuous use and renewal of the equipment, it will have a wide application prospect.

Figure 2   Application of the system in the 250,000-tonnage ore wharf in the Port of Fangchengang

7. Exploration of application field and prospect forecast

(1) Six-component measurement of the moored floating body. Motion path and range of single/multiple ships with the tide at the anchorage ground.

(2) Semi-floating or completely immersed motion measurement of the submarine, measurement of return navigation path and working state of the submarine under the water.

(3) Interaction between the ship waves incurred when the two ships are navigating in the same direction/reverse direction, influence of ship waves on the materials of armor blocks and research on dragging stability.

(4) Six-component measurement on the swaying or dragging process of ocean engineering ships, working platforms, navigational lights and buoys.

(5) Influence of wave current motion on dike collapse process and path of the parapet and revetment materials.

(6) Six-component measurement on the impact process of hydraulic structures in the harbor engineering, like multi-pile foundation and bridge pier at the wharf, by the wave current and on the swaying process of cable-stayed bridge by the wind with fixed or unfixed length; based on such specialties like hydraulics, converting the effect on the stress on objects from the numerical value and promoting the application of related software.

(7) Six-component measurement on the swaying process of the mooring ropes.

(8) Six-component measurement on engineering structures in the wind tunnel.

(9) Synchronous measurement on respectively six-component motion value of multi-point targets on the deck.

(10) Hoisting stability process of alongside replenishment operation of the two ships which are navigating; analysis on stability of mooring operation.

References

[1] Qihua Zuo. *Water Wave Simulation*. Ocean Press , Beijng, 2006.

[2] Dianpu Li. *Ship Movement and Simulation*. National defense industry Press, Bejing, 2008.

[3] *FK Manual*. PS Corporation, 2005.

[4] 3*SRL Manual*. YT Corporation, 2008.

[5] Peng Li, Yiren Zhou. Ship model test of Fangchenggang Port. *Nanjing Hydraulic Research Instituter*, Nanjing, 2009.

Study on Lock Part-Diverging Filling and Emptying System

Liu Benqin   Zong Muwei   Xu Xinmin

State Key Laboratory of Hydrology-Water Resources and Hydraulic Engineering,
Nanjing Hydraulic Research Institute, Nanjing, 210029

Abstract: A new lock part-diverging filling and emptying system is put forward based on analyses on end filling and emptying system and diverging filling and emptying system. Its diverging section is arranged in upstream lock head and partial lock chamber. It is applicable for ship locks whose distinguishing coefficient for types of filling and emptying system is 2.5 to 3.5. The layouts and hydraulic characteristics of two schemes are introduced in this article. The layout of its energy dissipation should be designed according to demands of diverging filling and emptying system. And the length of its diverging section in chamber should be designed based on initial wave force. Physical modeling studies indicate that its hydraulic characteristics during lock filling are between end filling system and diverging filling system. And this type of lock filling and emptying system can reduce the wave force coefficient, raise the initial valve opening speed, shorten the filling and emptying times and has no need for mitigative segment like diverging filling and emptying system except the merits of cost saving and convenient construction similar to end filling and emptying system. It is a fine filling and emptying system type suitable for middle head ship locks.

Key words: ship lock; filling and emptying system; part-diverging; hydraulic characteristic; physical modeling test

1. Introduction

There are mainly two types of lock filling and emptying systems. One is end filling and emptying system and the other is diverging filling and emptying system[1]. The former's obvious advantage is cost saving and convenient construction. It is widely used in ship locks with water head under 10 meters[2], but it has greater wave forces on ships mooring in lock chamber because of unsteady wave movement from upstream and stronger current forces. Therefore, the valve opening speed should be reduced in order to meet the ship mooring conditions and thus may result in longer filling time. And it's difficult to come up to designed demands for those ship locks with high head. While diverging filling and emptying system has complex construction and the costs are larger than end filling system, although it has better water pattern and mooring conditions in lock chamber. Meanwhile, the layout of diverging filling and emptying system usually needs certain terrain and structure conditions. So it is necessary to develop a new filling and emptying system suitable for middle head ship locks to overcome the defects of end filling and emptying system in wave forces, at the same time, its advantages are still remained.

Through more and more hydraulic physical modeling studies[3-5], a new part-diverging filling and emptying system has been developed. The diverging part of this type of filling and emptying system is arranged in upstream lock head and partial lock chamber. During lock filling, water flows into lock chamber through the diverging part near upstream lock head. It does not need mitigative segment. And its hydraulic characteristics are between end and diverging filling and emptying system. The design demand of this type of lock filling and emptying system is that the energy dissipation of its diverging part in chamber should fulfiu ship's mooring conditions.

This article mainly introduces the scheme selection, layout and hydraulic characteristics of lock part-diverging filling and emptying system.

2. Scheme selection and layout

Scheme selection and layout of lock part-diverging filling and emptying system should be completed according to demands of diverging filling and emptying system. It is known to us all that the diverging filling and emptying system can be divided into lock wall longitudinal culvert and lock bottom longitudinal culvert according to the arrangement of the main culvert. Usually the lock wall longitudinal culvert system is more popular for those middle head ship locks in order to obtain high hydraulic characters. So setting the main culvert of the part-diverging filling and emptying system in lock wall is more reasonable, especially for gravity lock wall structure.

2.1 Scheme selection

There are two layout schemes for lock wall longitudinal culvert. One is lock wall longitudinal culvert side holes part-diverging filling and emptying system. Water flows into lock chamber through a series of side holes on each side lock wall culvert in this scheme. The distance between the two adjacent side holes should be about one quarter of the width of lock chamber, and the size of the hole also should be under certain conditions. So the length of its diverging part will be longer.

Another option is connecting several transverse branch culverts under the chamber floor to the lock wall main culvert. Thus the water flows into the branch culverts along the main culvert, and then into lock chamber through the branch culvert's side holes. In the latter layout scheme, the holes are arranged on the transverse branch culverts and are wide distribution, so it can shorten the length of the lock wall main culverts. At the same time, water from upstream is dissipated within the lock bottom and then flows into the chamber. Therefore, better energy dissipation can be obtained in this lock part-diverging filling and emptying system scheme. But its disadvantage is that the flow is not so dispersive as the former scheme along the longitudinal chamber and its wave force coefficient is greater[3]. In addition, the arrangement of the branch culverts will reduce the lock wall main culvert elevation and complicate the floor structure.

2.2 Layout of part-diverging filling system

Physical modeling tests are completed with some ship lock as a support project in order to fully understand the hydraulic characteristics of the two types of part-diverging filling and emptying system. The support ship lock's effective scales are 180 meters long, 23 meters wide and its minimum water depth are 3.5 meters. The lock's filling time is 10 minutes of design

with 10 meters maximum head. According to "Design Code for Filling and Emptying System of Shiplocks (JTJ306-2001)"[1], the culvert's section scales are calculated to be 3.6 meters wide and 3.0 meters high near the filling and emptying valves.

For lock wall longitudinal culvert part-diverging filling and emptying system with side holes, the height of the main culvert in lock wall is increased to be 3.4 meters and its width is still 3.6 meters. So the section areas are 24.48 square meters and the ratio to areas of filling valve is 1.13. After detailed hydraulic calculations, 20 side holes are arranged on each side lock wall longitudinal culvert. And each hole's section areas are 0.62 square meters. The ratio of the total side holes' section areas to valve areas is 1.15. The layout characteristics of this scheme are as Table 1 and Figure 1.

Table 1   Layout characteristics of filling system for scheme with side holes

Position	Areas (m²)	Ratio to areas of valve
Water inlet	68.20	3.16
Culvert near valves	21.60	1.00
Culvert in lock wall	24.48	1.13
Side holes	24.80	1.15

For lock wall longitudinal culvert part-diverging filling and emptying system with transverse branch culvert, the culvert's section dimensions and areas near filling valve are the same as side holes scheme. It is 3.6 meters wide and 3.0 meters high and the total areas are 21.6 square meters. With work amount and effects of water dispersion considered altogether, one transverse branch culvert is arranged in lock chamber. The width of the transverse branch culvert is 4.7 meters and the height is still 3.0 meters. On each side of the transverse branch culvert, 18 side holes are arranged and the total section areas are 27.0 square meters. Water which flows out of the branch culvert is dissipated through double open ditches. The width of each ditch is 3.5 meters. Thus the length of the diverging part near upstream lock head is 23.2 meters along longitudinal lock chamber, 18.2 meters of which are in chamber and the other 5.0 meters are in upstream lock head. The filling system's layout characteristics of this scheme are as Table 2 and Figure 2.

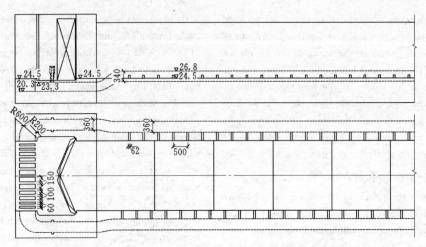

Figure 1 Layout of filling system for scheme with side holes

Table 2 Layout characteristics of filling system for scheme with transverse branch culvert

Position	Areas (m²)	Ratio to areas of valve
Water inlet	68. 20	3. 16
Culvert near valves	21. 60	1. 00
Branch culvert	28. 20	1. 31
Holes on branch culvert	27. 00	1. 25

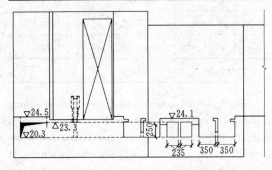

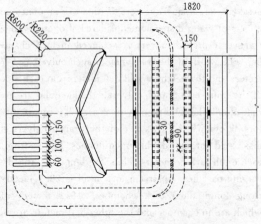

Figure 2 Layout of filling system for scheme with
transverse branch culvert

3. Hydraulic characteristics during lock filling

Resistance coefficient and discharge coefficient are the basic hydraulic characteristics. They can be calculated according to pressure values measured in model under steady flow pattern. The continuous changes of water flow rate and water head are measured during lock filling. And lock over-filling can also be obtained.

3. 1 Basic hydraulic characteristics

Measuring points are arranged on the culvert. Pressure of different part of the culvert and water level in chamber and water inlet are measured in physical modeling test through piezometric tubes under steady flow pattern. Then the resistance coefficient and discharge coefficient are calculated. The calculation results are as table 3. From the table we can see that the discharge coefficient of side holes scheme and transverse branch culvert scheme is 0. 65 and 0. 77 respectively. Lock over-filling is measured during lock filling. The values are 0. 16 meters and 0. 11 meters, and it can fulfill the criterion demand in both schemes. According to the measured lock filling times under different valve opening modes, we can obtain the relationship between them presented in Table 3. In the table, T and t_v refers to lock filling times and valve opening times respectively.

Based on before-mentioned layouts of two schemes, the ratio of branch culvert section areas to filling valve areas in latter scheme is bigger than that of lock wall culvert section areas to filling valve areas in former scheme. And it is similar for side holes. Therefore, the part-diverging filling system's discharge coefficient is bigger in transverse branch culvert scheme than in side holes scheme, which is proved by physical modeling test results in Table 3.

Table 3 Basic hydraulic characteristics

Schemes	Side holes scheme	Transverse branch culvert scheme
Resistance coefficient	2. 367	1. 687
Discharge coefficient	0. 65	0. 77
Lock over-filling (m)	0. 16	0. 11
$T = f(t_v)$	$T = 0.425\,t_v + 431$	$T = 0.455\,t_v + 363$

3. 2 Hydraulic characteristics during lock filling

For lock wall longitudinal culvert part-diverging filling and emptying system scheme with side holes, according to water pattern and hawser forces in chamber measured in physical modeling test during lock filling, a 0. 3-meter-high sill is arranged in front of the side holes in order to meliorate water conditions in chamber, and it is 1. 0 meter away from the holes. At the same time, valve opening time is determined to be 360 seconds according to the test results of initial wave forces in physical modeling test. Thus it is secure for ships mooring in chamber during lock filling.

The varieties of water flow rate and water head during lock filling are as Figure 3. The maximum value of measured water flow rate in Physical modeling Test is 132m³/s presented in Table 4. And other maximum hydraulic characteristics during lock filling are also in Table 4. From the table we can see that filling times are 585 seconds when valve opening times are 360 seconds, which is less than designed filling times (600 seconds).

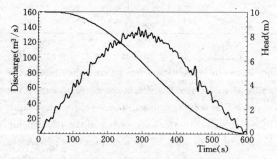

Figure 3 Hydraulic characteristic curves during lock filling for side holes scheme

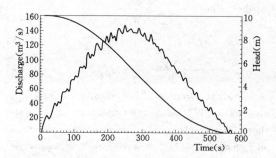

Figure 4 Hydraulic characteristic curves during lock filling for transverse branch culvert scheme

Table 4 Maximum hydraulic characteristics during lock filling for two schemes

Schemes	Side holes scheme	Transverse branch culvert scheme
Valve opening times (s)	360	360
Filling times (s)	585	527
Maximum flow rate (m³/s)	132	144
Maximum energy (kW)	8361	8700

For lock wall longitudinal culvert part-diverging filling and emptying system scheme with transverse branch culvert, the maximum value of water flow rate is bigger because of the bigger discharge coefficient when the valve opening times are same in both schemes. The maximum water flow rate is 144 m³/s if valve opening times are 360 seconds. And the filling times are 527 seconds. The above-mentioned hydraulic characteristics are coincident to criterion and design.

4. Analyses on wave force and energy dissipation

4. 1 Analyses on wave force

Lock filling is unsteady flow, and change of water flow rate will cause significant long-wave movement. The gradient of the long wave is proportional to change ratio of water flow rate and inverse to velocity of the wave, in addition, it is relevant to the type of lock filling and emptying system. Furthermore, the change ratio of water flow rate depends on water head and change ratio of opened valve area. Wave velocity depends on water depth. Usually at the beginning of the valve opening with uniform speed, the gradient in chamber is maximum because of the little initial water depth and maximum head. And in this time the wave forces on the ships mooring in chamber can be calculated as follows[3]:

$$P = W i_0 g \qquad (1)$$

$$i_0 = D \left(\frac{\mathrm{d}Q}{\mathrm{d}t}\right)_0 / [g(\omega_c - \chi)] \qquad (2)$$

$$\left(\frac{\mathrm{d}Q}{\mathrm{d}t}\right)_0 = \frac{k_r \omega}{t_v} \sqrt{2gH} \qquad (3)$$

In above formulas, "P" means wave forces (unit: kN). "W" means the number of tons of water that a ship displaces. "i_0" means the initial water surface gradient. "D" is wave force coefficient, relevant to filling type, length of chamber, scale of ship and mooring position, and it can be calculated according to "Design Code for Filling and Emptying System of Shiplocks". "$\left(\frac{\mathrm{d}Q}{\mathrm{d}t}\right)_0$" means change ratio of water flow rate at the beginning when filling valves are opened. "ω_c" refers to initial cross-sectional area of chamber under the water. "ω" refers to valve area. "χ" refers to cross-sectional area of ship under the water. "H" refers to lock head. "t_v" refers to valve opening times. "k_r" is a coefficient relevant to valve type, its value is 0.725 for flat valve.

Values of wave force coefficient of the support ship lock for different filling and emptying system have been given in Table 5. We can see from the table that filling and emptying system type has obvious affection to wave force coefficient. The minimum wave force coefficient is not more than 0.1 for great ships if diverging filling and emptying system is applied. And the maximum value is not more than 0.3. For part-diverging filling and emptying system, the part of lock chamber in which diverging water flows out of the culvert during lock filling is short relatively and the outflow is more in the front of the chamber. The characteristics of the wave force are a little similar to end filling system, but it can reduce the value of wave force coefficient. The value of the wave force coefficient is 0.196 to 0.996 according to length of diverging section if part-diverging filling and emptying system is applied. While it reaches 1.235 with ending filling system applied. Therefore, it is important to select suitable filling and emptying system type for ship locks with higher head and hydraulic characteristics in order to meet the demands of filling and emptying times and ship berthing conditions.

Table 5   Values of wave force coefficient for different types of filling and emptying system

Type of filling system		Value of wave force coefficient
End filling		1.235 (Calculation)
Part-diverging	Side holes scheme	0.196 (Physical modeling test)
	Transverse branch culvert scheme	0.996 (Physical modeling test)
Diverging	Type one (simply diverging)	0.1-0.3 (for great ships), 0.5-0.65 (for small ships) (From Design Code)
	Type two (relatively complicated)	0.1-0.3 (for great ships), about 0 (for small ships) (From Design Code)
	Type three (complicated)	$\leqslant$0.1 (for great ships), about 0 (for small ships) (From Design Code)

According to the maximum longitudinal hawser forces in chamber during lock filling measured in physical modeling test, the wave force coefficient can be calculated as the following formula[1]. The meanings of symbols in this formula are the same as abovementioned.

$$D = \frac{t_v P_L}{k_r \omega W} \frac{(\omega_c - \chi)}{\sqrt{2gH}} \qquad (4)$$

Physical modeling studies on part-diverging filling and emptying system prove that it may reduce the value of wave force coefficient compared to end filling system. The reduction degree is relevant to the length of diverging part in lock chamber and the relationship is showed in Figure 5[3-5]. In the figure, "k" refers to ratio of the length of diverging part to the length of the whole lock chamber, and "j" refers to ratio of

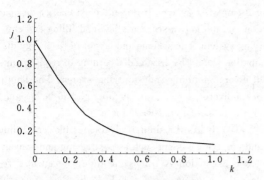

Figure 5   Relationship between *j* and *k*

Table 6   Hawser forces in chamber during lock filling

Valve opening times (s)	Longitudinal hawser force (kN)	Transverse hawser force ahead (kN)	Transverse hawser force behind (kN)
240	22. 23	8. 32	7. 86
300	17. 77	8. 26	8. 40
360	14. 82	6. 57	8. 62
420	12. 68	8. 15	7. 02
480	11. 11	5. 65	6. 83

wave force coefficient of part-diverging filling and emptying system to end filling and emptying system.

For transverse branch culvert scheme, because its diverging part is in upstream lock head and lock chamber near upstream lock head. The total length of its diverging part is 23. 2 meters along longitudinal lock chamber although, 18. 2 meters of which are in chamber and the other 5. 0 meters are in upstream lock head. Therefore, only that diverging part which is in lock chamber can reduce the wave force coefficient. The part which is in upstream lock head can be considered as end filling system and its wave force coefficient is the same as that of end filling system. The average value weighted according to the length proportion of the two parts is the total wave force coefficient of part-diverging filling and emptying system with transverse branch culvert.

The value of the coefficient (D) can be calculated according to hawser forces measured in physical modeling test by above-mentioned formula. The values of hawser forces under different valve opening modes are presented in Table 6. Thus the average value of wave force coefficient of part-diverging filling and emptying system is calculated to be 0. 996. And the coefficient is 1. 235 for end filling system. Therefore, we can obtain the wave force coefficient of the diverging part which is in lock chamber. And the value may be used to deduce the relationship of "*j*" and "*k*".

4. 2   Analyses on energy dissipation

During lock filling, huge energy conversion and dissipation will produce. Potential energies convert into kinetic energies, and then they are exhausted in culverts, dissipater, water in chamber and dynamic actions on ships. For end filling system, the remaining kinetic energies are mostly absorbed by hawsers and presented as hawser forces of tows. While for diverging filling and emptying system, water flows out of the culverts from many side holes. And they are dissipated by water diffusion or cover board, ditches and other dissipater, then act on ships through water body. Therefore, it is usually easier to solve the problem of energy dissipation for diverging filling and emptying system than eng filling and emptying system.

Part-diverging filling and emptying system has no energy dissipation and mitigative segment like end filling system. So the modes of its energy dissipation, especially its transverse hawser forces of tows have the same characteristics as those of diverging filling and emptying system. But longitudinal water flow still exists. The scheme selection and layout of energy dissipation in part-diverging filling and emptying system should be the same as diverging filling and emptying system. In this article, the above-mentioned principle is followed to determine the scheme and layout of energy dissipaters for two types of part-diverging filling and emptying system. In layout of lock wall longitudinal culvert part-diverging system with transverse branch culvert, double open ditches are used for energy dissipation. Physical modeling test proves that the energy dissipation effects are satisfied. And in layout of lock wall longitudinal culvert part-diverging system with side holes, sills are arranged in front of the side holes in order to meet the mooring conditions in chamber during lock filling, at the same time, slowing valve opening speed or varied valve opening speed is usable if necessary.

5. Application of part-diverging filling and emptying system

According to "Design Code for Filling and Emptying System of Shiplocks"[1], there is a formula $m = T/\sqrt{H}$ for choosing the type of lock filling and emptying system. End filling and emptying system is preferable if $m > 3.5$, otherwise diverging filling and emptying system is more suitable if $m < 2.5$. When the coefficient is between 2.5 and 3.5, technical and economical evaluation should be carried out or by reference to similar projects. So we can think that part-diverging filling and emptying system can be used in lock projects whose coefficient (m) are between 2.5 and 3.5 firstly. Second, comprehensive technical and economical comparison should be made with other probable types of filling and emptying system in hydraulic performance (including filling and emptying times, water conditions, mooring conditions in lock chamber and valve operation conditions), construction performance (including cost, construction period and complexity) and running performance (including management and maintenance).

For the support ship lock project: (1) Its coefficient is 3.16, which is between 2.5 and 3.5; (2) It is difficult to meet the designed hydraulic demands for its higher head, shorter filling and emptying time if end filling and emptying system is selected. So it is suitable to use lock wall longitudinal culvert part-diverging filling and emptying system with side holes or transverse branch culvert by further comparison.

6. Conclusion

Types, layout schemes and hydraulic characteristics of part-diverging filling and emptying system are introduced in this article. Furthermore, certain analyses and discussions are put forward. Conclusions are summarized as below through physical modeling studies.

(1) Lock part-diverging filling and emptying system is suitable for ship locks whose distinguishing coefficient for types of filling and emptying system is between 2.5 and 3.5. Comprehensive technical and economical comparison should be made with other type of filling and emptying system in hydraulic performance, construction performance and running performance before it is applied in some ship project.

(2) Its diverging section is arranged in upstream lock head and partial lock chamber. And its hydraulic characteristics are between end and diverging filling and emptying system. It doesn't need energy dissipation and mitigative segments like end filling and emptying system. The scheme and layout of its energy dissipation should be arranged according to requirements of diverging filling and emptying system. The length of its diverging section in chamber should be designed on the basis of initial wave force.

(3) It has the similar advantages like end filling and emptying system such as cost saving and convenient construction. At the same time, it can also reduce the initial wave forces in lock chamber, raise the initial valve opening speed and shorten lock filling times. So it is a fine filling and emptying system type suitable for middle head ship locks.

References

[1] JTJ 306—2001. *Design Code for Filling and Emptying System of Shiplocks*, China Communications Press, Beijing, China, 2001.

[2] Liu Benqin, Zong Muwei, Xuan Guoxiang. Study on lock's concentrated filling and emptying system with no need of mitigative segment. *Port & Waterway engineering*, Editorial department of Port & Waterway engineering, Beijing, China, Nov. 2008: 139-143.

[3] Zong Muwei, Liu Benqin, Xu Xinmin. *Hydraulic research report of part-diverging filling and emptying system for middle and low head ship locks*, Nanjing Hydraulic Research Institute, Nanjing, China, 2005.

[4] Xuan Guoxiang, Liu Benqin. *Hydraulic model experiment study of the Longzhouyuan ship lock's filling and emptying system on the water transfer project from Yangtze River to Hanjiang River*, Nanjing Hydraulic Research Institute, Nanjing, China, 2006.

[5] Li Jun, Xuan Guoxiang, Huang Yue. Hydraulic model tests on filling and emptying system of the shiplock of Shihutang navigation project on Ganjiang River. *Hydro-science and Engineering*, Editorial department of Hydro-science and Engineering, Nanjing, China, Mar. 2009: 17-21.

The Visualization of Open Channel Flow with Submerged Flexible Vegetation by PIV

Wu Fusheng   Jiang Shuhai   Zhou Jie   Wang Yong   Ruan Shiping

State Key Laboratory of Hydrology-Water Resources and Hydraulic Engineering,
Nanjing Hydraulic Research Institute, Nanjing, 210098

Abstract: The PVC material is used to simulate the flexible vegetation to conduct the research in glass flume. Two-dimensional high-frame rate PIV is used to measure the flow field with submerged flexible vegetation. And then time average velocity field, vorticity field and turbulent entropy are obtained. The investigation gives a new approach to further reveal the ecological hydraulics research with submerged flexible vegetation.

Key words: submerged flexible vegetation; high-frame rate PIV; turbulent entropy; velocity field; vorticity field

1. Introduction

Vegetation growing in a river is one of the basic elements of river and wetland system which constitutes the habitats of animals in the water. The ecosystem function of river largely depends on water, distribution process of energy and the cycle of water and material accompanying this process.

When water flows through flexible vegetation, vegetation bends and sways along the flow. Vegetation in rivers not only affects the capacity of discharge, but plays an important role in the ecological restoration. The flow field with vegetation is complicated and can be easily influenced. Therefore, it is of great significance to conduct the research about flow field with vegetation to provide the basis results of rivers for the management. At present, research of this area in home and international is still in stage of development. The flexible vegetation is simulated by specific materials in the flume. The river flow characteristics of submerged vegetation section was researched and measured by two-dimensional high frame rate Mini-PIV (sampling frequency of approximately 200Hz), for studying its velocity field, vorticity field distribution, turbulence quasi-entropy and characteristics of turbulent dissipation.

2. The velocity field and vorticity field of open channel flow with submerged flexible vegetation

2.1 Equipments and experimental conditions

As has been stated, the experiment is processed in the glass flume. The elevation and measuring system diagram are shown in the table 1. The vegetation is simulated by flexible PVC pieces which are 50mm high, 5mm wide, and 0.2mm thick, the elastic modulus and density of the PVC materials are 2000MPa and (1.4×10^3) kg/m^3 respectively which is used to simulated and determined the nature of similar choked flow characteristics and turbulent characteristics of flexible vegetation distorted by channel flow flooding. In the flume the length of vegetation is 4m, the plane layout of flexible vegetation is shown in Figure 2. The water depth of the vegetation reference section start section is about 11.5-11.7cm when the experiment is processed, and three different discharge of flow are carried out; there are five different vegetation densities, the flexible vegetation densities are 0, 175, 225, 350, 475 sapling/m^2 when the experiment is processed. In the experiment Reynolds number is about $(1.33 - 1.68) \times 10^4$, vegetation flow is turbulent and fully developed. The Froude number is 0.143-0.178, less than 1, which is slowly variable flow. The 1.5W semiconductor laser, which belongs to Mini-PIV system and SM-SEMI1500 series laser is used. Its measuring range is 6cm×6cm, sampling frequency is about in the 200Hz which satisfies measure request of vegetation flow turbulent characteristics. The Mini-PIV system sampling frequency is 170-200Hz, that is used at this performance, which can be used to research turbulent pulsating quantity and turbulent dissipation characteristics of vegetation flow. The experimental condition is shown in Table 1.

445

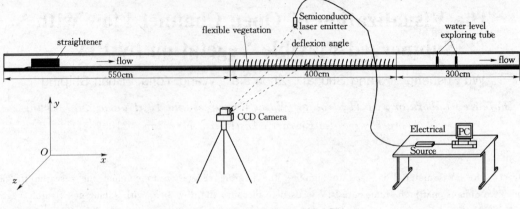

Figure 1　Flume with vegetation and Mini-PIV system

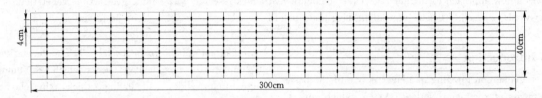

Figure 2　Layout diagram of flexible vegetation

Table 1　　Experimental condition

Vegetation density (IP/m²)	Q (m³/s)	h (cm)	Re	Fr
175	0.0088	11.7	16845	0.176
225	0.0088	11.7	16845	0.176
350	0.0084	11.5	16086	0.172
475	0.0070	11.5	13340	0.143

In order to make PIV measure vegetation flow velocity implement effectively in experiment, bottom and vegetation are sprayed by black Matte painting so as to try to weaken the impact of the laser sheet light. Glass flume side wall must be completely clear. Before the experiment degreased cotton must be dipped in alcohol to wipe the glass clean. In order to improve the measurement effect, appropriate tracer particles were added. Tracer particles with the hollow glass balls, is 2-3μm in size, and is about 1.02 kg/m³, close to the proportion of water.

Sampling frequency of Mini-PIV is about 170-200Hz. Each time, 2000 samples were collected for calculation to get the time averaged velocity field. Meanwhile, the velocity processed the standard deviation analysis. Velocity standard deviation is defined as: the sample flow rate of each value and the sample mean square difference between the arithmetic average of the square root. The calculation formula is:

$$Std(V) = \sqrt{\frac{1}{n}\left[(V_1 - \overline{V})^2 + (V_2 - \overline{V})^2 + \cdots + (V_n - \overline{V})^2\right]}$$

(1)

Velocity standard deviation is a measure volume of fluctuations of a velocity sample. The more standard deviation in the sample velocity, the greater fluctuation of the sample data.

From instantaneous flow velocity field measured by Mini-PIV, the pulse velocity of various point were obtained. According to the definition of vorticity, the instantaneous vorticity field can be acquired from the instantaneous velocity field which can evaluate the turbulence entropy.

2.2　Experiment results of time averaged velocity field and vorticity field

The instantaneous velocity field can be measured by Mini-PIV, so that the time-averaged velocity field and vorticity field are obtained. The flow field of the vegetation canopy at 13.5cm from the flume wall at a vertical cut was measured during the experiment, which was compared with that under non-vegetation situation. The contour maps of the averaged-time velocity and the averaged-time vorticity field which are at the vertical aspect are given as follows.

446

2.2.1 *The distribution of the time-averaged velocity field and time-averaged vorticity field without vegetation on the vertical aspect*

The distribution of the time-averaged velocity field and the time-averaged vorticity field on the vertical aspect is shown in Figure 3 and Figure 4.

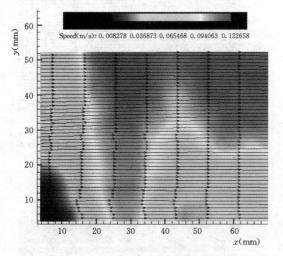

Figure 3　The distribution of the time-averaged velocity of non-vegetation on the vertical aspect

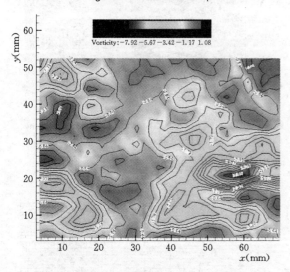

Figure 4　The distribution of the time-averaged vorticity of non-vegetation on the vertical aspect

2.2.2 *The distribution of the time-averaged velocity field and the time-averaged vorticity field when the canopy of flexible vegetation was on the vertical aspect*

（1）The density of flexible vegetation is 175 sapling/m².

The distribution of the time-averaged velocity field and the time-averaged vorticity field of flexible vegetation on the vertical aspect is shown in Figure 5 and Figure 6.

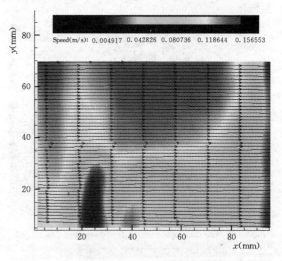

Figure 5　The distribution of the time-averaged velocity of flexible vegetation on the vertical aspect

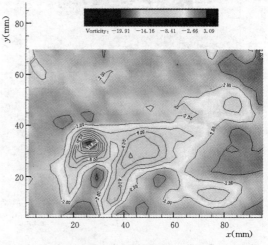

Figure 6　The distribution of the time-averaged vorticity of flexible vegetation on the vertical aspect

（2）The density of flexible vegetation is 225 sapling/m².

The distribution of the averaged-time velocity field and the averaged-time vorticity field of flexible vegetation on the vertical aspect is shown in Figure 7 and Figure 8.

（3）The density of flexible vegetation is 350 sapling/m².

The distribution of the time-averaged velocity field and the time-averaged vorticity field of flexible vegetation on the vertical aspect is shown in Figure 9

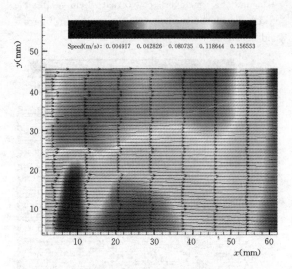

Figure 7　The distribution of the time-averaged velocity of flexible vegetation on the vertical aspect

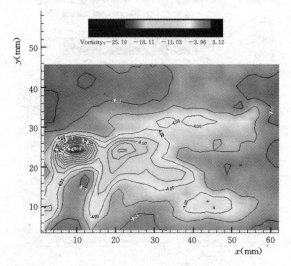

Figure 8　The distribution of the time-averaged vorticity of flexible vegetation on the vertical aspect

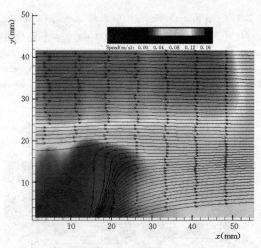

Figure 9　The distribution of the time-averaged velocity of flexible vegetation on the vertical aspect

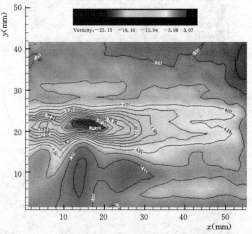

Figure 10　The distribution of the time-averaged vorticity of flexible vegetation on the vertical aspect

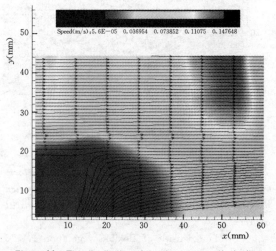

Figure 11　The distribution of the time-averagedvelocity of flexible vegetation on the vertical aspect

and Figure 10.

(4) The density of flexible vegetation is 475 sapling/m².

The distribution of the averaged-time velocity field and the averaged-time vorticity field of flexible vegetation on the vertical aspect is shown in Figure 11 and Figure 12.

From the experiment results above, the vegetation has a certain impact on the time-averaged velocity field and vorticity field. The distribution regulation of the time-averaged velocity at the vegetation canopy

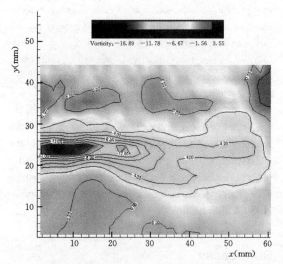

Figure 12   The distribution of the time-averaged vorticity of flexible vegetation on the vertical aspect

changed obviously, under the plant canopy the velocity of flow significantly decrease, which caused by plants, at the plant canopy the velocity gradient got larger. Therefore, the intensity of reversed time-averaged vorticity significantly increased.

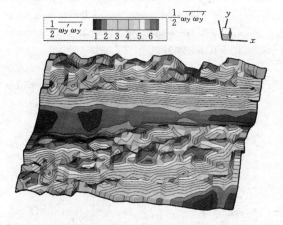

Figure 13   The distribution of turbulence entropy on the horizontal plane under water surface about 6cm

3. The characteristics of the turbulent entropy of open channel flow with submerged flexible vegetation

The turbulent entropy on the horizontal plane of flow with submerged flexible vegetation is analyzed as follows. Meanwhile, non-plant section at the upper about 1m was measured, compared with non-plant conditions to reveal characteristics of the turbulent dissipation with submerged flexible vegetation.

3.1   The distribution of non-vegetation turbulence entropy $\frac{1}{2}\overline{\omega_y'\omega_y'}$ at the horizontal plane under water surface about 6cm

The distribution of non-vegetation turbulence entropy on the horizontal plane under water surface about 6cm was shown in Figure 13. It can be seen that, due to the impact of fluid viscosity or the impact of the background disturbances caused by the experimental system, the intensity of magnitude decreased, to a value of about (1-6) s^{-2}.

3.2   The distribution of turbulent entropy $\frac{1}{2}\overline{\omega_y'\omega_y'}$ on the horizontal plane above the vegetation canopy

Figures 14-17 present the turbulent entropy distributions on the horizontal plane above the vegetation

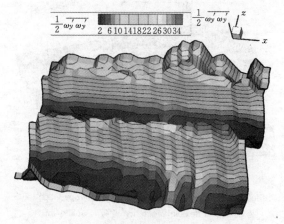

Figure 14   The distribution of turbulent entropy when the density of flexible vegetation is 175 sapling/m²

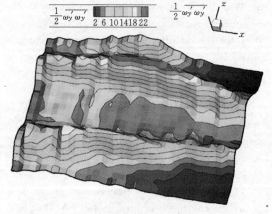

Figure 15   The distribution of turbulent entropy when the density of flexible vegetation is 225 sapling/m²

449

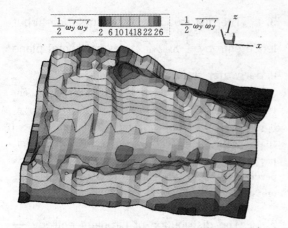

Figure 16　The distribution of turbulent entropy when the density of flexible vegetation is 350 sapling/m²

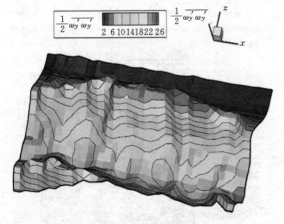

Figure 17　The distribution of turbulent entropy when the density of flexible vegetation is 475 sapling/m²

canopy. Clearly, the magnitude of turbulent entropy is about five times more than that without vegetation. With different density of flexible vegetation, the magnitude scope of turbulent entropy changes very little, but the distributed shape of its field changes apparently.

4. Conclusions

(1) Two-dimensional high-frame rate PIV is used to measure the instantaneous velocity field with submerged flexible vegetation, and then the instantaneous vorticity field, time-averaged velocity field and vorticity field are obtained.

(2) Vegetation has a certain impact on the time-average velocity field and vorticity field. At the plant canopy the distribution law of time-average velocity changes significantly; under the plant canopy the velocity of flow significantly decreased due to vegetation; the velocity gradient is larger at the plant canopy. Therefore, the intensity of reversed time-averaged vorticity increased significantly.

(3) The magnitude of turbulent entropy on the horizontal plane of the canopy of submerged flexible vegetation is about five times more than that without vegetation. With different density of flexible plant the magnitudes scope of turbulent entropy field changes very little, but the shape of its field distribution changes significantly.

References

[1] Wu Fusheng, Wang Wenye, Jiang Shuhai. The research and development of hydrodynamics in open channel flow with vegetation [J]. Advance in water science, 2007, 18 (3): 456-461.

[2] Nezu I., Onitsuka K.. Turbulent structures in partly vegetated open-channel flows with LDA and PIV measurements. Journal of Hydraulic Research, 2001, 39 (6): 629-642.

[3] Shi Zhong, Li Yanhong. Research and prograss of the average velocity distribution of plants [J], Journal of Shanghai Jiaotong University, 2003 (8): 1254-1260.

[4] Nehal Laounia, Yan Zhongmin, Xia Jihong. Study of the flow through non-submerged vegetation, Journal of Hydrodynamics Ser. B, 2005, 17 (4): 498-502.

[5] Yang Kejun, Can Shuyou, Knight Donald W. Flow patterns in compound channels with vegetated floodplains [J]. Journal of Hydraulic Engineering, 2007, 133 (2): 148-159.

[6] Wu Fusheng. Characteristics of flow resistance in open channels with non-submerged rigid vegetation [J]. Journal of Hydrodynamics Ser. B, 2008, 20 (2): 239-245.

[7] Tong Binggang, Yin Xieyuan, Zhu Keqin. Vortex Movement Theory, University of Science and Technology of China Press, 2009.

Test and Research on Flow-Excited Vibration of Working Gate of Upper Stream Pumping Unit in Qingcaosha Reservoir

Yan Genhua   Chen Fazhan   Hu Qulie

State Key Laboratory of Hydrology-Water Resources and Hydraulic Engineering,
Nanjing Hydraulic Research Institute, Nanjing, 210029

Abstract: Through using hydro-elasticity similar model we had this research on flow-excited vibration of working gate of upper stream pumping unit in Qingcaosha reservoir, elaborating on the similar principle and mode-making method of gate vibration test. According to test on flow-excited vibration of gate with two working conditions of water diversion and drainage under different combination of upper and downstream water conditions, we got the hydrodynamic pressure load statistical characteristics, pulsating pressure coefficient as well as structural vibration acceleration, dynamic displacement and stress parameters working on gate body. We had discussed about the effect the draining grating before the gate and lateral opening boundary conditions exert on fluid state and gate vibration, offering scientific proofs for making operation regulations and safety evaluation.

Key words: upper stream pumping unit; working gate; hydro—elasticity vibration; hydrodynamic load; dynamic response

1. Introduction

The engineering of water-transmission pumping unit in Qingcaosha reservoir is located in the entrance of Changjiang River, which is the main intake structure of division works. Its major function is to make flow divert automatically to reservoir and to lead freshwater in the period of non-salt tide as well as to drain beforehand with emergences (earthquake). Five-orifice gate of underflow aperture type is set in water-taking sluice, and the trim width of each orifice is 14.0m, the height of ground sill −1.50m, the level of parapet 3.5m, the height of orifice 5.0m. The working gate is of underflow aperture type of flat fixed roller steel, which is operated by inversion—type hydraulic hoist.

According to the requirement of locks function, it is needed of water—taking sluice to retain water in two directions. It is needed of frequent hydraulic actuation and partially opening to adjust flow capacity when water is diverted. The tail water is high enough to form submerged effluent. The gate and its hemline are affected by the striking of circumfluence tail water, which is easily to arouse vibration, besides; there is wind wave in the entrance of Changjiang River with complicated flow situations. Therefore, starting test and research on flow-ex-

cited vibration of working gate is to make sure it is important for safe operation of gate and water diversion project. Essentially, the flow-excited vibration of gate is in the category of hydro-elasticity vibration. The hydraulic load exerts excitement on gate body as one of vibration sources, but there appears dynamic response on gate structure after fluid coupling. And the reciprocity between this fluid and gate structure constitutes all dynamic characteristics of gate hydro-elasticity vibration. In order to actually invert the operation conditions of sluice, it has to be researched and developed of completely similar model of gate hydro-elasticity vibration in order to acquire the dynamic effect hydraulic load exerting on gate structure under all kinds of working conditions, the response characteristics of gate vibration, including vibration acceleration, dynamic displacement, dynamic stress and such vibration parameters to offer a reliable basis for dynamic safety design of gate structure.

2. The design of hydro-elasticity model and test macro

2.1 The relationship between hydrodynamic load and gate vibration

The hydrodynamic load is external factor inducing structural vibration, and the internal factor affecting vibration intensity is the dynamic response of structure itself. It is critical to solve the engineering problems by having research on features of both sides and their relationship and adopting relative measures. The dynamic response spectrum density of gate structure under the effect of hydrodynamic load is:

$$S_x(\omega) = \sum_{r=1}^{N} \sum_{s=1}^{N} H_{xP_r}^*(\omega) H_{xP_s}(\omega) S_{P_r P_s}(\omega) \quad (1)$$

In this formula, $S_x(\omega)$ is structural response spectrum under the effect of hydrodynamic load; $H_{xP_r}^*(\omega)$ or H_{xP_s} is the transfer function between structure's points discussed and excitement points P_r (or P_s), reflecting the structural characteristics; $S_{P_r P_s}(\omega)$ is the cross spectral density between input force r and s, expressing the nature of forces. Therefore, once the gate structural body is determined, vibratory output will be settled under the effect of certain external load. Therefore, making the gate vibration model has to satisfy the corresponding similarity criterion in order to make test results invert to the engineering prototype.

2.2 The design and manufacture of hydro-elasticity gate model

There appears vibration on gate structure under the effect of hydrodynamic action, which could be expressed by discrete matrix differential equation as follows:

$$[M]\{\ddot{X}\} + [C]\{\dot{X}\} + [K]\{X\} = \{F(t)\} \quad (2)$$

In this formula, $[M]$, $[C]$ and $[K]$ respectively are the structure quality, damping and rigidity matrix under $N \times N$ step; $\{\ddot{X}\}$, $\{\dot{X}\}$ and $\{X\}$ respectively are acceleration, speed and displacement array as structure makes movements at discrete nodes; $\{F(t)\}$ is vector of outside force.

According to equation on structure's movement, the test of hydro-elasticity gate model has to satisfy both load (including dynamic load) response and dynamic response being similar, and also has to meet both hydraulic conditions and structural dynamics conditions being similar. The hydro-elasticity gate model should be content with geometric scale, water flow movement and structural dynamic (the similarity of the mass density, elasticity modulus, Poisson's ratio and such parameters) being similar and it is deduced that the scale of parameters is as follows:

geometric scale $L_r = 25$, the scale of the mass density $\rho_r = 1$, the scale of elasticity modulus $E_r = L_r$, the scale of Poisson's ratio $\mu_r = 1$, the scale of damping $C_r = L^{2.5}$.

The input pump working gate in the upper stream of Qingcaosha reservoir is made by using steel splice, whose basic physical mechanics indexes are: volume weight $7.85 \times 10^4 \, \text{N/m}^3$, elasticity modulus $8.4 \times 10^3 \, \text{MPa}$, Poisson's ratio 0.3. Currently, in the market there's no proximate matter that can simultaneously satisfy all the above parameters, so special hydro-elastic material that is specifically developed in our institute is used in this experiment, which is developed through multi-component specialty material by heavy metal powder, polymer materials and so on as well as by using special moulds to make proximate matter for this project. The experiment tests show that the preferred hydro-elastic materials basically meet the demands of materials density $\rho_m = \rho_p'$ and structure's elastic modulus $E_r = L_r$.

2.3 The arrangement of test points and measurement analysis system

In order to get the flow-excited vibratory characteristics of working gate during the operation, 5 three-dimensional vibration test points are set in purpose-made hydro-elasticity gate model (Figure 1 is about arrangement of test points). Through KD5018 charge integrator amplifier we get the characteristics of vibration acceleration and vibration displacement, and at each test points we measure the vibration acceleration and vibration displacement of gate in downstream direction (X), vertical direction of flow (Y) and plumb direction (Z). The vibration data should process by adopting random vibration theory and the spectral analysis method, and then we get the spectrum characteristics and digital features during the process of gate vibration to reveal the energy distribution of vibration frequency domain and vibration magnitude.

We synchronously process the measurement of dynamic pressure and vibration magnitude, of which we focus on the former one. The test points include gate's panel, cantilever diaphragm and flange girth as well as web longeron and flange girth and so on (Figure 2). After technology treatment on gate structure surface and through putting on bended strain gage stress and strain measurement is processed. By stress and strain amplification measuring system we process the measurement of dynamic strain capacity at each part during the gate's operation, and then through special software analyzing and processing random data we do the statistical computation. Using special miniature pulse pressure sensor we process the measurement of flow pulsating pressure, and the arrangement of test points is shown in Figure 3. According to Strain measurement analysis system we process the data analysis and treatment on pressure pulse signal. The flow of measurement and analysis on random data is shown on Figure 4.

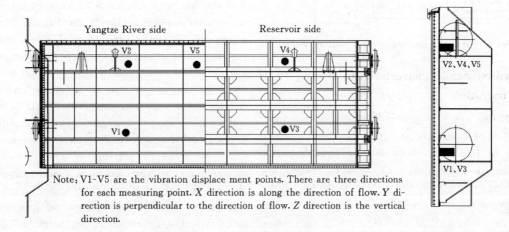

Note: V1-V5 are the vibration displace ment points. There are three directions for each measuring point. X direction is along the direction of flow. Y direction is perpendicular to the direction of flow. Z direction is the vertical direction.

Figure 1　The arrangement of test points of vibration and dynamic displacement on working gate

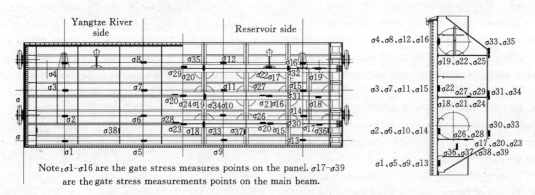

Note: σ1-σ16 are the gate stress measures points on the panel. σ17-σ39 are the gate stress measurements points on the main beam.

Figure 2　The arrangement of test points of stress on working gate

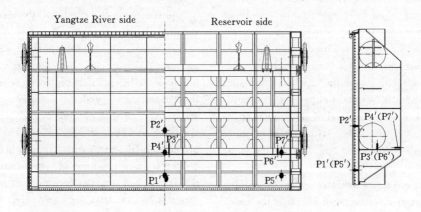

Figure 3　The arrangement of test points of pulsating pressure on working gate

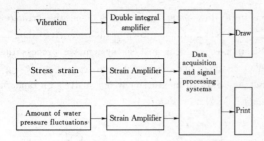

Figure 4　The flow of measurement and analysis system

3. Hydrodynamic characteristics during the gate operation

During gate operation with sound water-sealing two parts of acting force constitute the effect flow exerting on gate body: time-averaged hydrodynamic pressure, and flow pulsating pressure. Therefore, the above two parts of acting force form the total load on gate body:

$$P = P_0 + P'(t) \tag{3}$$

In this formula, P_0 is time-averaged hydrodynamic pressure; $P'(t)$ is pulsating pressure.

Time-averaged hydrodynamic pressure reflects static action, which can be used to check on the strength of structure. The flow pulsating pressure load is the main source inducing structural vibration.

In order to study up on the flow pulsating pressure load working on gate structure, we set 7 test points of pulsating pressure on gate body, among which there are 3 on the surface of upper stream, and 4 on the surface of lower stream and the beam behind gate (Figure 3). The test is conducted in high tide and ebb tide of Changjiang River. In non-saline period the water level of Changjiang River is 5.03m (once two years), and the water level of reservoir −1.50-2.00m. The gate is partially opened, the discharge per unit width is $q = 3$-$10\text{m}^3/(\text{s} \cdot \text{m})$; The water level of Changjiang River is at 3.35m of mean high tide, and that of reservoir is 2.00-3.00m. with the gate being partially opened, when the discharge per unit width is $q = 3$-$13\text{m}^3/(\text{s} \cdot \text{m})$ we open the gate to divert water. When the water level of reservoir is at the range from 2.00m to 7.00m, with 0.92m of water level of Changjiang River and $5\text{m}^3/\text{s}$ of the discharge per unit width we process.

The experiment data indicates under the working condition of water-diverting with ebb tide the maximum RMS value of pulsating pressure measured is 0.287m Water Column, located at the edge of gate's under-part. Under the working condition of water-diverting with high tide the maximum RMS value of pulsating pressure measured is 0.303m Water Column, also located at the edge of gate's under-part. Under the working condition of water drainage with ebb tide the maximum RMS value of pulsating pressure measured is 0.141m Water Column, located at the main beam of the edge of gate's under-part. According to spectral curve, the high energy region of current pulse pressure power spectral density on the working gate of upper stream pump sluice concentrate at low frequency zone from 0Hz to 5.0Hz, and the dominant frequency varies at the range from 1.0Hz to 2.5Hz, which is the same as the measuring results of pulsating pressure at sluices' boundary. Under the working condition of water-diverting with ebb tide the curve about the time domain process and spectral density of flow pulsating pressure at gate's typical test points is drawn on the following figures (Figure 5 and Figure 6).

If considering pressure fluctuations as a random smoothly-ergodic process, probability density will be distributed normally. The amplitude of 99.7% of appearing odds on the max random pulsating pressure can be evaluated by 3 times of pulsation RMS:

$$P_{\max99.7\%} = 3\sqrt{\overline{p'^2}} \tag{4}$$

If define C_1 as pulse pressure coefficient:

$$C_1 = \sqrt{\overline{p'^2}} / \left(\frac{1}{2}\rho u^2 \right) \tag{5}$$

Then

$$\sqrt{\overline{p'^2}} \approx C_1 \left(\frac{1}{2}\rho u^2 \right) \tag{6}$$

In this formula, $\sqrt{\overline{p'^2}}$ is RMS value of flow pulsation; u is average flow velocity.

According to the experiment results on gate's pulsating pressure, we get the larger gate's pulse pressure coefficients C_1 with free flow and submerged flow beneath gate under the working conditions of water-diverting with ebb tide or high tide respectively are 0.141-0.186 and 0.228-0.67. The gate's pulse pressure coefficient C_1 with free flow beneath gate under the working condition of water drainage with ebb tide is 0.111-0.202. The max value of pulsating pressure can be calculated by using three times of pulse RMS as the design value of pulsating load on gate.

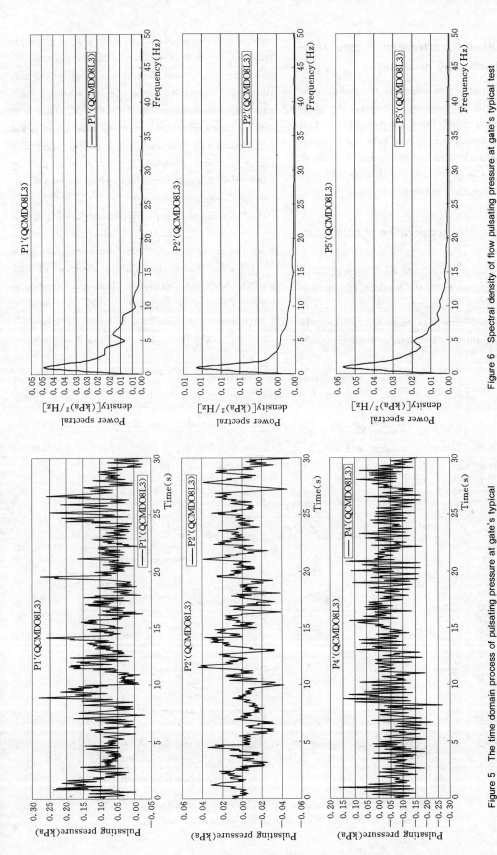

Figure 6 Spectral density of flow pulsating pressure at gate's typical test points under the working condition of water-diverting with ebb tide

Figure 5 The time domain process of pulsating pressure at gate's typical test points under the working condition of water-diverting with ebb tide

4. The vibration characteristics of gate structure with sluice water-diverting

According to the gate arrangement and operating characteristics, it has been researched on the effect hydraulic characteristics of lock basin exerting on gate vibration after drain grating being removed from the gate, and it was taken into consideration of the flow-excited vibration on gate when flow changes with high and ebb tide under the working condition of water-diverting.

4.1 The vibratory acceleration of gate structure

The changing relationship of RMS value of gate's vibratory acceleration under the working condition of drain grating being removed from the front side of gate is drawn on Figure 7. In order to compare, the evolution of vibratory acceleration on each test point

under the working condition of setting drain grating is drawn on Figure 8. It is drawn on Figure 9 of time domain curve and power spectral density of typical gate vibration. The relation among the gate vibration magnitude in three directions is $V_z > V_y > V_x$. That is to say, the gate vibration in vertical direction is larger than that in side direction, which is also larger than that in horizontal direction. Without drain grating the RMS value of larger vibration acceleration on 3# orifice in downstream direction is 0.035-0.052m/s²; the RMS value of larger vibration acceleration in side direction is 0.05-0.058m/s², which is located at 4.91m of upper stream, 2.0m of lower stream and q = 12.83m²/s under the working condition of high tide. The RMS value of larger vibration acceleration in vertical direction is 0.16m/s², which is located at 4.91m of upper stream, 2.0m of lower stream and q= 12.83m²/s under the working condition of high tide.

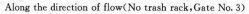

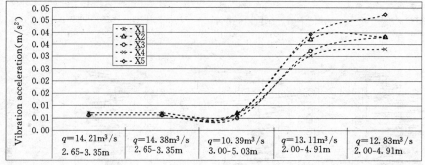

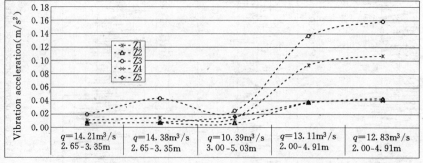

Figure 7 The changing relationship of gate's vibration acceleration
adding rectifier pier without drain grating before water gate

456

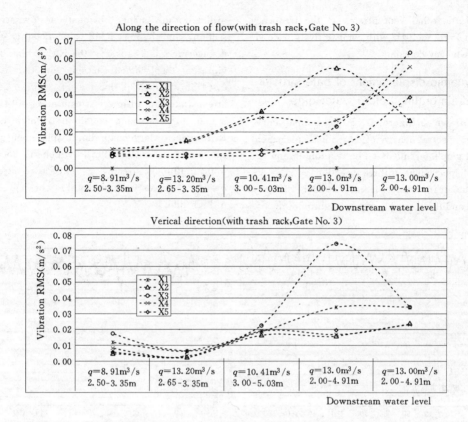

Figure 8   The changing relationship of gate's vibration acceleration with drain grating before water gate

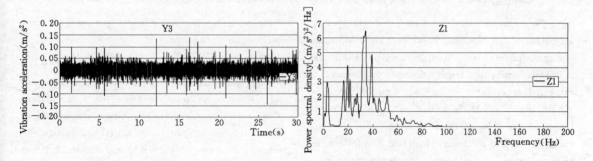

Figure 9   Gate's vibration spectral density on typical test points

The RMS value of larger vibration acceleration on 5# lateral opening of left side in downstream direction is 0.05-0.057m/s²; the RMS value of larger vibration acceleration in side direction is 0.08-0.09 m/s², which is located at 4.91m of upper stream, 2.0m of lower stream and $q=12.83m^2/(s \cdot m)$ under the working condition of ebb tide. The RMS value of larger vibration acceleration in vertical direction is 0.16m/s², which is located at 4.91m of upper stream, 2.0m of lower stream and $q=13.11 m^2/(s \cdot m)$ under the working condition of high tide.

The experiment results indicate it is not that influential on vibration of gate structure with or without drain grating, but the boundary conditions have certain effect on the gate vibration. Comparing 3# middle-orifice gate with 5# lateral opening, the vibration magnitude of 2-hole gate in horizontal and vertical direction is basically the same, and the vibration magnitude on 5# gate in side direction is 50% larger than that on 3# gate. That is because there appears lateral force on lateral-opening flow. The main energy of vibratory acceleration achieves to a peak value at

5Hz, and the other concentrates at the frequency band from 15Hz to 40Hz with comparatively wide vibratory frequency domain.

4. 2 Dynamic displacement of gate structure without drain grating before water gate

It is drawn in Figure 10 of time domain curve and power spectral density of dynamic displacement on typical gate's test points. The relation among the dynamic displacement in three directions is $V_z > V_y > V_x$. The test results indicate under different working conditions the vibration magnitude on 3 #

orifice doesn't change obviously in downstream direction, and the max RMS value of larger vibration displacement is 0. 15mm; the RMS value of larger vibration displacement in side direction is 0. 32-0. 44mm, which is located at 3. 35m of upper stream, 2. 65m of lower stream and $q = 14. 38 m^2 /$ s under the working condition of high tide. The RMS value of larger vibration displacement in vertical direction is 0. 80mm, which is located at 5. 03m of upper stream, 3. 0m of lower stream and $q = 10. 39 m^2 / (s \cdot m)$ under the working condition of high tide.

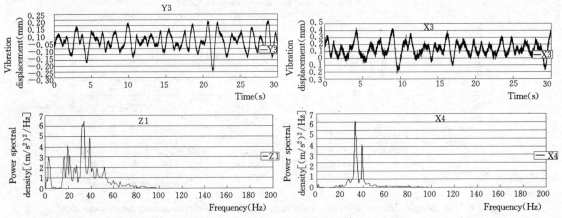

Figure 10   Time domain curve and power spectral density of vibration displacement on typical gate's test points

The vibration displacement on 5 # lateral-opening gate is larger than that on 3 # orifice, The RMS value of larger vibration displacement in downstream direction is 0. 14-0. 16mm; the larger vibration displacement in side direction comparatively increases (the RMS value is 0. 62-1. 1mm), which is located at 5. 03m of upper stream, 3. 0m of lower stream and $q = 10. 39 m^2 / (s \cdot m)$ under the working condition of high tide. The RMS value of larger vibration displacement in vertical direction is 1. 1mm, which is located at 5. 03m of upper stream, 3. 0m of lower stream and $q = 10. 39 m^2 / (s \cdot m)$ under the working condition of high tide.

In general, affected by boundary conditions, flow turbulence of lateral opening is larger than that of mid-orifice gate, and the vibration magnitude is also larger than mid-orifice gate. It is obvious of vibration magnitude in side direction. The vibration displacement on 5 # orifice gate in side direction is 1. 5 times larger than that on 3 # orifice. The main energy of vibration displacement concentrates at the frequency band from 15 Hz to

40 Hz with comparatively wide frequency domain distribution of energy.

4. 3 The dynamic stress on gate structure

The research on dynamic stress on gate structure and the vibration test are processed synchronously. We have research on the changing of dynamic force of critical stressed parts in horizontal and vertical direction. It is drawn on Figure 11 and Figure 12 of the changing relationship of the RMS value of dynamic stress on 3 # gate with or without drain grating under typical working condition. Time domain curve and power spectral density of dynamic force on typical gate's test points under working condition of water-diverting with ebb tide. RMS value of the max dynamic stress measured is 0. 718MPa. The distribution of dynamic stress's energy at frequency domain mainly concentrates at low frequency zone less than 2. 5Hz. On the whole, the vibration stress on gate structure under hydrodynamic action is not strong, which is not the controlling factor of strength of structure.

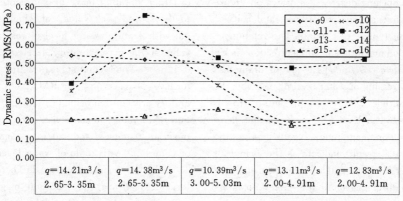

Figure 11　The changing relationship of dynamic Stress on 3# gate without drain grating

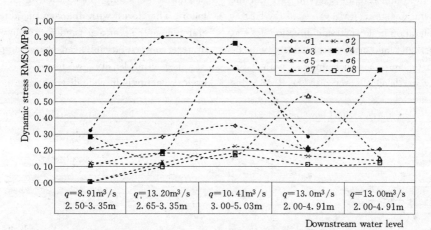

Figure 12　The changing relationship of dynamic Stress on 3# gate with drain grating

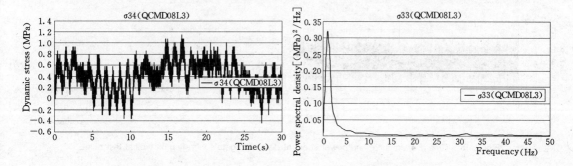

Figure 13　Time domain curve and power spectral density of dynamic force on
typical gate's test points under working condition of water-diverting with ebb tide

5. The vibration characteristics under the working condition of water drainage

The vibration experiment of gate structure under working condition of drainage with ebb tide is launched with 0.92m of water level of Changjiang River, 2.00-7.00m of water level of reservoir, gate being partially opened and $q=5.0\text{m}^3/(\text{s}\cdot\text{m})$ of discharge per unit width. It is drawn on Figure 14 about the relation the RMS value of vibration acceleration changing as the opening degree. It is drawn on Figure 15 of time domain curve and power spectral density of gate vibration on typical test points. The relation among the vibration magnitude in three directions is $Vz > Vx > Vy$. With $q = 5.0\text{m}^3/(\text{s} \cdot \text{m})$ of drain discharge as the increasing of water level of reservoir the vibration magnitude of gate presents growth trend. When the water level of reservoir achieves to 7.0m, the vibration magnitude of gate gets to the max value. The test results indicate as the change of reservoir's water level the gate's vibration magnitude at mid-orifice is different from that at lateral opening, among which the gate vibration at lateral opening (5♯) in downstream is much larger than that at mid-orifice. When the water level of reservoir varies at the range from 2.0m to 4.0m, RMS values of the largest vibratory acceleration in downstream direction (X) at lateral opening and mid-orifice respectively will be 0.23m/s² and 0.05m/s², which appears with 3.5m and 2.0m of water level of reservoir. The RMS value

of larger vibration in vertical direction (Z) achieves to the second peak value, of which the first comes out with 2.5m of water level of reservoir, the RMS value of the max vibratory acceleration being about 0.12m/s². The second appears with 3.4-4.0m of water level of reservoir, the RMS value of the max vibratory acceleration being 0.135m/s². The max vibration RMS values in horizontal direction (Y) respectively are 0.06m/s² and 0.07m/s², and the peak value comes out with 2.5m of water level of reservoir.

The test results show as the increasing of water level of reservoir the vibration magnitude of gate presents growth trend. When the water level of reservoir increases from 4.0m to 7.0m, the vibration magnitude of gate increases to 3 times than before. Therefore, with drainage the gate should operate under the working condition of low water level of reservoir as possibly can. Research on vibratory power spectral density, we found that the gate's vibratory energy is in wide distribution.

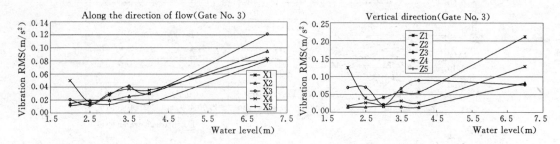

Figure 14   The relationship RMS value of 3♯ gate's vibration changing as water level of reservoir

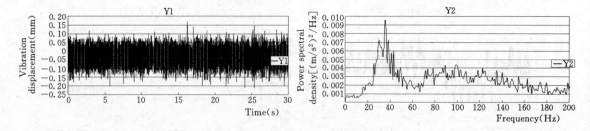

Figure 15   Time domain curve and power spectral density of gate's vibration on typical test points

6. Evaluation on the safety of gate operation

There appears gate vibration under dynamic water due to hydro-dynamic load, and the vibration magnitude and hazard rating are closely related to the magnitude of hydro-dynamic load, energy distribution and gate's natural vibration characteristics. The resonance vibration is the most harmful. The experi-

ment results point out that the high energy region of power spectral density of flow pulsating pressure working on gate mainly concentrates at the low frequency zone from 0Hz to 5.0Hz, and the main frequency varies at the range from 1.0Hz to 2.5Hz. The analysis results on dynamic characteristics of gate structure show that the first step vibration frequency of the combination system of gate structure and hoist stem in horizontal direction is 0.75Hz, and the fundamental vibration frequency of gate's bending deformation with flow mineral extraction is 7.6Hz. Apparently, the whole vibration frequency of gate structure is on the low side, which needs to adopt effective lateral restraint to control and avoid overlarge transverse vibration or resonance vibration. Under the condition of gate structure being well restrained in side direction, there is no resonance vibration on gate structure itself due to the effect of hydrodynamic load.

The results of vibration response on gate structure under different working conditions indicate that with different working conditions of water-diverting the gate vibration magnitude is the largest in vertical direction, and then in side direction and horizontal direction. Because of the effect of boundary condition, the vibration magnitude of lateral-opening gate is larger than that of mid-orifice gate. The RMS value of larger vibratory acceleration of 5# lateral opening at left side in downstream direction is 0.057m/s², 0.09m/s² in side direction, 0.16m/s² in vertical direction. The main energy of vibratory acceleration gets to a peak value at about 5Hz and the other concentrate at the frequency band from 15Hz to 40Hz with wider vibratory frequency domain. On the whole, there is no harmful vibration on gate structure.

7. Conclusions

Through hydro-elastic vibration model we research on the hydrodynamic load on gate structure and acquire the flow pulsating pressure working on gate, the acceleration, the dynamic displacement, the dynamic stress of gate's vibration and such structural parameters under different working conditions of water-diverting and drainage. And then we get the conclusion as follows.

(1) The test results point out the larger gate's pulse pressure coefficients C_1 with free flow and submerged flow beneath gate under the working conditions of water-diverting with ebb tide or high tide respectively are 0.141-0.186 and 0.228-0.67. The gate's pulse pressure coefficient C_1 with free flow beneath gate under the working condition of water drainage with ebb tide is 0.111-0.202. The max value of pulsating pressure can be calculated by using three times of pulse RMS as the design value of pulsating load on gate.

(2) If it is effectively handled of the horizontal restraint of gate structure, there will be no violent vibration and resonance vibration on working gate.

(3) As the increasing of water level of reservoir the vibration magnitude of gate presents growth trend. Therefore, with drainage the gate should operate under the working condition of low water level of reservoir as possibly can.

(4) It has no obvious influence on gate's vibration magnitude setting or removing drain grating in front of gate. Affected by boundary conditions, the vibration magnitude of lateral-opening gate is larger than that of mid-orifice gate. The vibration displacement on 5# orifice gate in side direction is 1.5 times larger than that on 3# orifice. Therefore, the flow boundary condition is one of the important factors affecting gate's vibration.

References

[1] Yan Genhua. Hydrodynamic Load and Gate Vibration, Hydro-science and Engineering, Feb, 2001.

[2] Yan Genhua, Chen Fazhan, Zhao Jianping. Experiment and Research on Gate's Hydro-elastic Vibration in Nanjing Sancha Estuaries, Journal of Vibration Engineering, Sept., 2005, Supplement, 18 (5): 123-127.

[3] Yan Genhua. Research Process of Flow-excited Vibration of Hydro-gate, Hydro-science and Engineering, Jan, 2006, 107: 66-73.

Characteristics Research of Vortexesat Dissymmetrical Boundary Intake

Dang Yuanyuan   Han Changhai   Yang Yu

State Key Laboratory of Hydrology-Water Resources and Hydraulic Engineering,
Nanjing Hydraulic Research Institute, Nanjing, 210029

Abstract: A vortex test device was set up for physical modeling test of relationship between critical submerged depth and Fr both at symmetrical and dissymmetrical boundary intakes. It is obtained from comparison that dissymmetrical boundary intake was easier to induce vortex than symmetrical boundary intake. PIV was used to measure flow field and vorticity field at intakes. Under the same condition, the max. value of vorticity at dissymmetrical boundary intake is smaller than that at symmetrical one, but Nr at dissymmetrical boundary intake is bigger than that at symmetrical one.

Key words: dissymmetrical boundary; vortex; submerged depth; vorticity; circulation

For a long time, inducement of vortexes at hydraulic intakes, which has been an important subject of hydromechanics and engineering application, is an extremely complex flow phenomenon[1]. However, due to its complexity, research of vortexes has not yet made a breakthrough. There are more studies on practical measures while less studies on vortex theory. There are more experiments for specific projects while less studies on vortex regularity. Therefore, it is of great significance to seek effective research means to reveal vortex inducement regularities and motion characteristics and to analyze factors affecting vortex motion for promoting vortex research. It was focused on the research of inducement and characteristics of vortexes at symmetrical and dissymmetrical boundary intakes in this paper. Especially PIV[2] was used to measure flow field and vorticity field at symmetrical and dissymmetrical intakes so as to get some new ideas and findings.

1. Boundary impact on inducement of vortexes

Boundary conditions include factors like intake shape, size, structure and layout, nearby terrain and inflow conditions. When an intake body is unreasonable and its layout is asymmetric, circulation will be generated at the intake, which can induce vortexes[3].

In the study of vortexes at intakes, it was thought that under certain boundary condition,

i. e. when the submerged depth met $H/D \approx 2.0$ (H was orifice head, D was orifice height or diameter), the vortex intensity was the highest, resulted from a stronger water turbulence and a high flow velocity, whereas the deviation from this condition would neither generate strong water turbulence and high water velocity, nor produce strong water vortexes. But vortexes are no more limited by the submerged depth, i. e. $H/D \approx 2.0$, as boundary condition changes. Deng Shuyuan[4] concluded that the relative submerged depth of some vortexes was not only of large range but also of high values. Taken the Baozhusi Hydropower Station as an example, the relative submerged depth of the bottom outlet is 2.8-6.9, and even reaches to 7 when boundary is asymmetric. It can be seen that the intake boundary has crucial effect on the formation of vortexes.

2. Test device and method

2.1 Test device

The test device[3,5] was an open water tank of 120cm long, 80cm wide and 30cm high. A mobile gate divided the tank into a steady inflow area and a test area, as shown in Figure 1. One inlet valve and a group of parallel impact flow meters were fixed upstream, and one outlet valve was fixed downstream to regulate discharge.

Steady inflow area: Inflow port was close to the

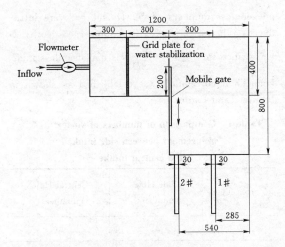

Figure 1　Test device

sink bottom. The grid plate for water stabilization which was 40cm wide was installed in the tank, 30cm downstream away from the inflow port. Its lower half was a baffle of 20cm high to amortize the jet of inflow and to reduce flow fluctuation. Its upper half was formed by the grid plates with each of 1cm wide, 10cm high and an interval distance of 1cm to ensure smooth and uniform flow.

Mobile gate was 40cm wide and 30cm high. It could move up and down in the chute to achieve different openings between 0-40cm for the flow control in vortex area.

Area of initial conditions: After water flowed through the mobile gate, different initial conditions were produced for vortex formation because of different flow conditions.

To study vortex regularity in different working conditions, square holes of 3cm×3cm, 4cm×4cm and 5cm×5cm respectively were reserved in the middle part and near edge of the sink lateral sides, 3cm away from the bottom. And vortex formation was researched successively under the conditions of near edge and central part.

2.2　Test method

Observation was used for studies on vortexes, which was to observe vortex occurrence regularities by the control of water level and discharge. Meanwhile, PIV flow field measurement system was applied to measure and analyze the flow field around vortexes.

PIV was a relatively advanced measurement technique, which could get transient flow field mainly through the particle imaging technology of its hardware and related algorithms of its software.

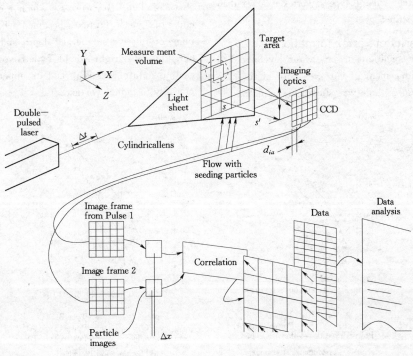

Figure 2　PIV working principle diagram

Its basic principle[6] is as following. A large number of tracer particles (less than 10 microns) are casted in a flow field and move with the flow field. A laser beam is extended to sheet beams by composite lens to illuminate the flow field. A digital camera is used to shoot particle images in the flow field. A quantitative velocity distribution within a cross-section of the flow field is obtained by the cross-correlation calculation of two successive particle images. After further processing, the parameter distribution of flow field properties, such as vorticity, streamline, equal velocity lines, etc., can be obtained. The principle is showed in Figure 2.

3. Test results and analysis

3.1 Vortex classification

From a practical engineering point of view, according to the literature on the classification of vortex development process, combined with the vortex impact on normal operation of intakes, vortexes observed in tests are classified into the following three categories[3], Type A, almost no vortex, Type B, non-through suction vortex and Type C, through suction vortex. It is stipulated in this paper that the grade of Type A vortex is lower than that of Type B vortex, and the grade of Type B vortex is lower than that of Type C vortex.

3.2 Regularity of vortex formation

Under the conditions of same water level and discharge, the inducement of vortexes both at the side hole and the central hole was shown in Table 1. Dimensionless numbers of S_{σ}/D and Fr were introduced, where, S_{σ} was critical submerged depth; D was orifice side; $Fr = \dfrac{V}{\sqrt{SD}}$ was dimensionless numbers. Their correlation was established as shown in Figure 3 and Figure 4.

Table 1　Comparison of numbers of vortex inducement between side intake and central intake

Vortex Type	Side Hole (numbers)	Central Hole (numbers)
A	21	33
B	24	18
C	15	9

From Table 1, the grade of vortexes induced at the side hole was mostly higher than that induced at the central hole. The side hole was easier to induce vortexes than the central hole, and vortex intensity of the side hole was stronger than that of the central hole. From Figure 3 and Figure 4, test points of Type A and Type C vortexes formed an envelope curve of Type B vortexes, which meaned that the curve of Type B vortexes becomed a boundary in the plane of relative submerged depth and Froude number. The lower right side of Type B vortex curve was easy to induce vortexes, and its upper left side was not easy to induce vortexes.

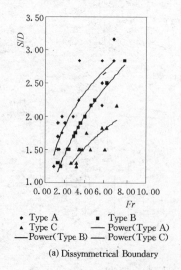

(a) Dissymmetrical Boundary

- ◆ Type A　　■ Type B
- ▲ Type C　　— Power(Type A)
- —Power(Type B)　— Power(Type C)

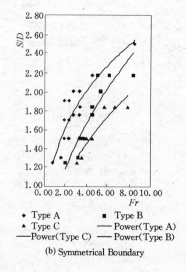

(b) Symmetrical Boundary

- ◆ Type A　　■ Type B
- ▲ Type C　　— Power(Type B)
- —Power(Type C)　— Power(Type A)

Figure 3　Vortex inducement regularity at mobile gate opening of 20cm

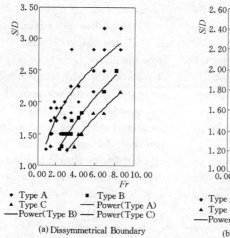 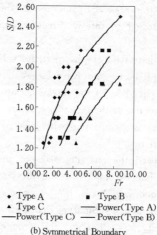

(a) Dissymmetrical Boundary   (b) Symmetrical Boundary

Figure 4   Vortex inducement regularity at mobile gate opening of 40cm

Correlation between critical submerged depth and Froude number when Type B vortexes were induced at dissymmetrical boundary intake:

$S_{cr}/D = 1.172Fr^{0.3864}$ (Opening of mobile gate was 20cm)

$S_{cr}/D = 1.099Fr^{0.3453}$ (Opening of mobile gate was 40cm)

Correlation between critical submerged depth and Froude number when Type B vortexes were induced at symmetrical boundary intake:

$S_{cr}/D = 1.0842Fr^{0.3519}$ (Opening of mobile gate was 20cm)

$S_{cr}/D = 0.9771Fr^{0.3834}$ (Opening of mobile gate was 40cm)

Therefore, a conclusion can be reached by a quantitative analysis, i.e. at the same submerged depth, dissymmetrical boundary intake is easier to induce vortexes than symmetrical boundary intake.

3.3 Flow field and vorticity field

Take a intake with the diameter of $3cm \times 3cm$ for PIV flow velocity measurement. At the submerged depth of 0.055m, the measured flow field and vorticity field were shown in Table 2.

Talbe 2   Comparison of flow field and vorticity field between side intake and central intake

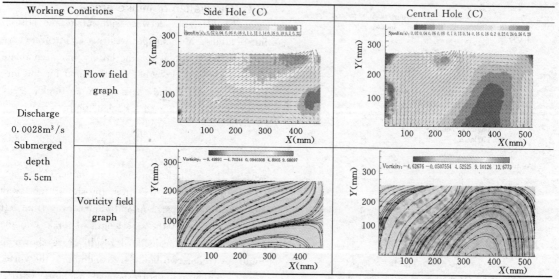

Working Conditions		Side Hole (C)	Central Hole (C)
Discharge 0.0028m³/s Submerged depth 5.5cm	Flow field graph		
	Vorticity field graph		

From flow field graph by PIV, the closer it was   to intake, the higher velocity was. The velocity vector mainly pointed to the intake. Streamline near the boundary curved according to inflow direction and boundary type.

According to vorticity field graph, vorticity value of the whole flow field was not very large, but vorticity was much bigger nearby the intake because of the inducement of vortice or the generation of rotating flow. This situation was consistent with local rotational flow characteristics.

Experiments showed the characteristics of the max. vorticity point under symmetrical and dissymmetrical boundary conditions. With the same discharge, the smaller the intake submerged depth was, the greater the vorticity was, that was the stronger the vortex intensity was. At the same submerged depth, the smaller the intake discharge was, the smaller the vorticity was. Location of max. vorticity was random.

Table 3 showed the max. vorticity values of symmetrical and dissymmetrical boundaries under the same conditions. It could be seen under the same type of vortexes that the max. vorticity value of dissymmetrical boundary was smaller than that of symmetrical boundary. That was to say, dissymmetrical boundary intake was easier to induce vortexes than symmetrical boundary intake, but the max. vorticity value at dissymmetrical boundary intake was smaller than that symmetrical boundary intake. Even when Type C vortexes were induced at dissymmetrical boundary intake and Type B vortexes were induced at symmetrical boundary intake, the vorticity value at dissymmetrical boundary intake was smaller than that at symmetrical boundary intake.

Table 3 Comparison of max. vorticities between symmetrical and dissymmetrical boundary intakes

Working Conditions	Dissymmetrical Boundary			Symmetrical Boundary		
	Max. Vorticity (rad/s)	Location of Max. Vorticity (mm)	Vortex Type	Max. Vorticity (rad/s)	Location of Max. Vorticity (mm)	Vortex Type
Discharge 3.3L/s Submerged depth 6.5cm	7.54	219	B	12.74	210	B
Discharge 2.8L/s Submerged depth 6.5m	5.48	223	B	8.46	163	B
Discharge 2.8L/s Submerged depth 5.5m	8.63	85	C	13.19	216	C
Discharge 2.8L/s Submerged depth 4.5m	8.92	58	C	10.15	202	C
Discharge 1.7L/s Submerged depth 4.5m	7.91	111	C	13.52	242	B

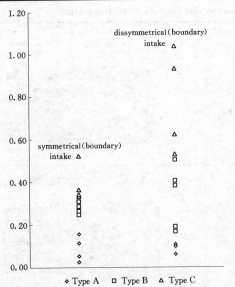

Figure 5   Comparison of circulation between symmetrical and dissymmetrical boundary intakes

3.4   Circulation of intake area

The difference between side intake and bottom intake studied in this paper was that an integral curve could be easily found for the latter, but a complete closed integral curve could not be found for the former. In order to study circulation at intake area, Stokes formula was used for calculation:

$$\Gamma = \iint\limits_{S} \Omega \, dx dy \qquad (1)$$

where, Ω is a single point vorticity value. The selected area was a square area from the site 45 mm away from the intake upstream and 45mm away from left side (1.5 times of the side length of intake) to the intake center. The calculated results were shown in Figure 5. Under two boundary conditions, the vortexes of a higher grade corresponded to a larger vorticity value. The vorticity value at dissymmetrical boundary intake was larger than that at symmetrical boundary intake through the comparison under two different boundary conditions. In fact, the vorticity value at intake had a great impact on vortex inducement. If only the intake size D, inflow Froude number F_r and initial circulation constant N_r were taken into account, $\frac{S_c}{D} = f(F_r, N_r)$ was obtained. Because of the limit of test group times, this relationship could not be given in this paper. But according to PIV measurement and calculation, a close relationship between inducement and properties of vortexes and flow circulation was confirmed.

4. Conclusion

Vortexes at dissymmetrical boundary intake are proposed in this paper. Both observation and PIV measurement system are used to study the vortex problem at dissymmetrical and symmetrical boundary intakes. Flow field and vorticity field at dissymmetrical and symmetrical boundary intakes are measured, and different characteristics of vortexes at dissymmetrical and symmetrical boundary intakes are analyzed. Not only the impact of boundary conditions on the inducement of vortexes, but also the characteristics of flow filed and vorticity field which induce vortexes are explored. A conclusion that dissymmetrical boundary intake is easier to induce vortexes than symmetrical boundary intake is drawn in this paper. Under the same condition, the max. vorticity value of dissymmetrical boundary intake is smaller than that of symmetrical boundary intake, but the circulation of dissymmetrical boundary intake is larger

than that of symmetrical boundary intake. These innovative research findings can provide inspirations and references for future studies.

References

[1] He Xuemin, Ru Shuxun. Study on Funnel Vortex Flow Field of Discharge Structure Intake [J]. Journal of Sichuan Water Resources, 1991, 1: 7-8.

[2] Tang Hongwu, Xu Xirong, Zhang Zhijun. PIV Technology and Its Application in Vortex Flow Field of VerticalIntake [J]. ChineseJournal of Hydrodynamics, 1999, 14 (1): 128-134.

[3] Chen Zongna, Wu Chao, Chen Yunliang. Experiment StudyofVertical Vortex at Hydraulic Intakes [J]. Journal of Southwest University for Nationalitics, 2006, 32 (4): 799-804.

[4] DengShuyuan. Probeof Formation and Countermeasures of Vortexat Discharge Structure Intake [J]. Hydro-Science and Engineering, 1986, 4: 51-63.

[5] Ye Mao, Wu Chao, Chen Yunliang. Physical and Numerical Modeling of Vertical Vortex [J]. Journal of Hydroelectric Engineering, 2007, 26 (1): 33-36.

[6] AdrianRJ. Particle-imaging Techniques for Experimental Fluid Mechanics [J]. Annual Review Fluid Mechanics, 1991 (23): 261-304.

Discussion on Improving Measures for Navigation Flow Condition at the Entrance Area of the Approach Channel and Connecting Section in Low Head Dams

Teng Juan   Zhao Jianjun   Li Fang   Cheng Lu

State Key Laboratory of Hydrology-Water Resources and Hydraulic Engineering,
Nanjing Hydraulic Research Institute, Nanjing, 210029

Abstract: As to the ships, entrance areas and the connecting sections are accident-prone locations, so whether the ship is safe or not when it comes into the approach channel is not only a great concern to the shipping sector, but also a difficulty for navigation structure design of the hydropower project. Thus far, in order to improve the navigation flow condition at the entrance area of ship-lock approach channel and connecting section, various measures have been proposed by many scholars from different countries. On the basis of physical model test and self-propelled ship model test, this paper has an investigation about the influences on the flow condition, which brought by the structure of guiding wall, riverside shape, routing selection and the tail water of the hydropower station. According to the problems in the original design, we put forward many optimization measures, which not only creates a favorable navigation flow condition, but also provides a technical basis for the navigable structure design.

Key words: lock; approach channel; entrance area; connecting section; navigation flow conditions; riverside shape; optimization measures; self-propelled ship model

When the hydraulic junction is established, we should set up some navigation structures like locks or ship shifts to ensure the safety of navigation. But after that, due to the flood discharge and the power station tailrace, oblique flow and backflow will be generated at the entrance areas of upstream and downstream approach channels, which make the ships (fleets) produce lateral drifting and rotation. In severe cases, the ships (fleets) will be out of control and maritime affairs will happen, which have a bad effect on navigation. According to overall model test of Laokou hydro-junction in the Yujiang river, this article analyzes some problems existing in the entrance areas of approach channels and connecting sections for the junction established in curved channel, in which time the lock don't discharge, so we propose many engineering measures to improve the flow conditions and achieve an effective result.

1. Project overview of Laokou Junction in Yujiang River

Laokou Junction in Yujiang River of Guangxi Province, which is based on navigation, flood control and mainly in power generation, is the seventh step in cascades planning for multipurpose utilization. Meantime, it's a comprehensive hub for creating conditions to improve the water environment of Nanning. After established, it can improve the ranking of waterway to Grade III and help to achieve full range of shipping for 1000t ships (or ship teams), which make the construction of main channel of sea exiting line in Southwest Water Transport finished.

Laokou Junction in Yujiang river is located at the anti-S in the downstream section of the convergence point between Zuojiang river and Youjiang river. The Layout from left to right: joint earth dam at the left bank, lock, gate chamber, sluice, hydropower station, joint concrete gravity dam at the right bank with the height of 92.35m. A total of 19 holes are arranged in discharging sluice dam, and the clear width of every hole is 16m; weir type is broad crested weir, with the elevation of 61.5m; orifice size (width×height) is 16m×14m; use plane sluice gate to retain water. The navigation structure is located at the left of riverside. Entrance area in upstream approach channel is locates in concave back, while another one in downstream approach channel is locates in convex bank. This Layout is adverse solve navigation flow conditions[1].

In "lock general design specification" (JTJ 305—2001), the maximum flow rate limits for entrance area in approach channel are provided. As to 1-4 lock, vertical velocity is not greater than 2m/s, transverse velocity is not greater than 0.3m/s, backflow rate is less than 0.4m/s; while on the 5-7 level lock, vertical velocity is not greater than 1.5m/s, transverse velocity is not greater than 0.25m/s, backflow rate is less than 0.4m/s[2].

2. Model Design and Construction

At the smallest navigable discharge (243m³/s), the depth of prototype channel is about 3m and average velocity in section is 0.60m/s. Considering the model Reynolds number requirement, model scale ratio should not be smaller than the scale 1 : 100. Taking into account comprehensively from many factors, we select the model geometric scale as 1 : 100. Considering the impact on navigation flow condition brought by upstream and downstream boundary, the upstream boundary is located at the place which is 1km away from upstream bend and the downstream boundary is located at the place which is 1km away from downstream bend. The length of model is equivalent to the total length of about 7km in prototype. Simulation range of model is shown in Figure 1.

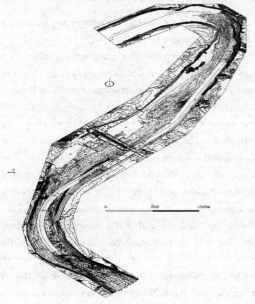

Figure 1　Physical model

The ship tonnage through the lock is (2×1000) t, with designed represent fleet are 2 barge fleet and 1 pusher barge—a designed fleet size is 160.0×10.6 ×2.0 (length × width × draft), and a 1000t self-propelled ship model, of which the scale is 67.5× 10.8×2.0 (length×width×draft).

3. Navigation flow conditions under the original design

Because the entrance area in upstream approach channel is located at concave bank of the lower bend, lateral pressure gradient exist in the vertical direction and transverse gradient of the water surface produces lateral pressure difference affected by the river topography, all of which lead to the emergence of transverse circulation in curved section[3].

3.1　Entrance areas in upstream approach channel and connection section

At all levels of navigation flow rate ($Q \leqslant$ 13900m³/s), the flow pattern at upstream entrance area is good, but there is weaker backflow zone in the approach channel and entrance area. According to the measured data, at the entrance area, the lateral velocity and vertical velocity can basically meet the requirements except that there is lateral velocity of local point is greater than 0.3m due to the impact of tortuous flow in the head of diversion dike. In the connection section, which is about 600m away from the entrance area, horizontal velocity > 0.3m/s, vertical velocity > 2.0m/s for part of the point due to the effect of upstream bend flow and mutation of local topography.

3.2　Entrance areas in downstream approach channel and connection section

When the junction discharges, the mainstream tends to move to right bank affected by the curved channel flow, meanwhile due to the effect brought by backflow close to the back, ski-jump of separation levee in approach channel, tail water of hydropower station and convex hank topography, the flow pattern at the entrance area in the downstream approach channel is complex and flow condition is very bad. At the largest navigation flow rate ($Q = 13900$m³/s), vertical velocity at the right of navigation waterway center line at the entrance area is up to over 2.5m/s (maximum velocity is 3.01m/s), while the maximum velocity is up to 2.73m/s and the most velocity is over 2. 0m/s in connecting section, where the ship is very difficult to get into the lock.

4. Improving measures on navigation flow at the entrance areas in upstream and downstream approach channels and connection sections

4. 1 Entrance area in upstream approach channel (Guiding wall modification)

Opening holes[4] in the head of guiding wall (dike) and the wall itself is the main measures to improve flow conditions at the upstream entrance area. According to the engineering practice in open-type guide wall, while the open-type guide wall in the United States is longer (about between 70m and 200m), and the orifices are spaced uniformly and have a uniform height. We choose the orifice with uniform space and height.

4. 1. 1 *Revised scheme* I (*open holes at the bottom of guiding wall ＋ backfill the slope of dyke*)

To improve the bad flow patterns such as tortuous flow in the head of guide wall in the approach channel, we modify the outer section in guiding wall. By opening holes at the bottom of guiding dyke, we make part of flow move into the approach channel from the entrance area, and during the time, some water discharge from the approach channel all the way through the bottom holes to weaken the impact brought by tortuous flow in the head of dyke. At the same time, we can smooth the flow boundary and reduce the transverse velocity by backfill the nearby slope of dyke.

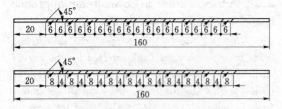

Figure 2　Opening scheme in the guide wall
in upstream approach channel

Layout of opening holes in guide wall is shown in Figure 2. Open holes from the height of 68. 5m to 70. 5m. IA program modification: opening width is 6m; IB program modification: opening width is 8m.

To compare IA with IB from a different hole size and flow, something can be found. As the opening area of modified scheme IA is smaller than the scheme IB, the discharge introduced is slightly smaller and the flow state is bad, too; modified scheme IB increased the opening area by 1/3 than scheme IA, the discharge introduced is slightly larger and the phenomenon of flowing around dike decreases significantly due to the flow velocity near guide wall entering into approach channel gets larger and intensity of backflow increases slightly, but it is agreeable.

Something can be seen from lateral and transverse velocity distribution at the entrance area of the downstream approach channel under the revised scheme: after opening the guide wall, the flow pattern near the entrance area improves and water flows more smoothly; meanwhile, from the numerical point of view, we can find that the stability of water flow also improves in the course of doing experiments, and lateral with vertical velocity flow conditions also meet the requirements of navigation flow conditions at all levels of navigable discharge.

4. 2 Adjustment of flow conditions in the upstream connecting section

Just like the entrance area, indexes used to measure navigable flow conditions in connecting sections are vertical and horizontal velocity, backflow velocity, and a hydraulic parameter called wave which cannot be ignored in addition[5].

4. 2. 1 *Revised scheme* II (*a submerged dam located in upstream connecting section*)

At the vicinity of connecting section which is about 600m away from entrance area, due to curved bend flow and local topography, lateral velocity ＞ 0. 3m/s (locally up to 0. 6m/s), and vertical water ＞2. 0m/s (locally up to 2. 86m/s), which is very difficult for the ship to pass. In the experiment, we try to build a submerged dam of which the top elevation is 68. 5m at the concave bank in order to improve navigation flow condition. In the test, we measures various velocities at the lowest navigable level ($Q=$ 6300m³/s), but we find that a submerged dam plays a limited role in improving navigation flow condition from the measured velocity distribution. Through the analysis of its causes, we see that affected by the river region it's very difficult to change the phenomenon that bend flow tend to move to concave bank in a relatively short distance artificially.

4.2.2 Scheme change Ⅲ (local topographical excavation)

During excavating topography locally, we should consider to smooth shoreline to prevent the mutations of flow conditions caused by mutations in topography. So we adopt two schemes to amend underwater prominent point: scheme Ⅲ A is to excavate from axis of waterway to the center line of the river; while scheme Ⅲ B is to excavate along the shore line to the center line of the river. From the comparative tests, we draw a conclusion that the two schemes have a certain effect on reducing lateral velocity at high water level, but we also need to take further measures at low water level ($Q=6300\text{m}^3/\text{s}$).

4.2.3 Scheme change Ⅳ (local topographical excavation + backfill bank slope)

From the results of revised scheme Ⅲ, we can see that navigation flow conditions can't meet requirements by excavating solely, because the phenomenon of water flow singularity caused by local terrain is not entirely improved. On the basis of revised scheme Ⅲ B, the scheme Ⅳ takes steps to smooth shoreline by partly backfilling the prominent point on the upstream and downstream side of bank, thus flow condition will be improved.

We can be see that from the measured longitudinal and lateral velocity distribution, modified scheme Ⅳ can significantly reduce the transverse velocity in connecting section, and also reduce the recirculation zone and longitudinal velocity. Meanwhile, vertical velocity is also slightly lower.

To the channel graded Ⅳ[6], vertical, horizontal velocity are less than 2.7m/s and 0.45m/s respectively in connecting section when the connection section is arranged on the side of main channel; the proposed limiting value of vertical velocity is 2.60m/s and the horizontal flow rate is 0.40m/s, when the connecting section is located on the sides of non-main channel. As to the revised scheme Ⅳ ($Q=8690\text{m}^3/\text{s}$), flow conditions in connecting section are improved significantly: the vertical velocity is below 2.3m/s, and the horizontal velocity is less than greater than 0.35m/s, except individual points, of which the horizontal velocity is over 0.3m/s. While $Q=13900\text{m}^3/\text{s}$, the vertical velocity is below 2.6m/s, and the horizontal velocity is less than greater than

0.45m/s in upstream connecting section.

4.3 Entrance areas in downstream approach channel and connection section

4.3.1 Revised scheme Ⅰ (adjust the route)

From the downstream routes point of view, the angle between axis of waterway at entrance area and the centerline of main flow in river channel is larger and the route is in the mainstream zones, so the horizontal velocity and vertical velocity are larger. For this junction, we decide to reduce water flow velocity by shifting the axis of route to the shore side to increase curvature radius (increase R from 1000m to 1100m). However, the zone at the left of centerline for waterway overlaps with the backflow zone close to banks, which makes the water flow side to side significantly.

4.3.2 Revised scheme Ⅱ (backfill the backflow zone close to bank)

Due to the effect of downstream bend flow, there is strong backflow near shore at convex bank, which make the cross section area compressed and increases the vertical velocity of route, meanwhile, transverse velocity is also larger for the existing of backflow, thus we decide to backfill the backflow zone close to bank. The intensity of backflow is decreased, but the backflow at entrance area caused by the effect of flow deflecting from guide wall still exist in this scheme.

4.3.3 Revised scheme Ⅲ (set diversion piers)

Let's set some diversion piers at the head of guide dike to weaken the impact of oblique flow and reduce angle between the oblique flow and transverse flow velocity component; at the same time, we can decompose the big backflow into a large number of small one to narrow the scope of backflow and to reduce the backflow velocity, due to the gap between the piers is relatively small. Therefore, we change successive guiding wall to eight piers.

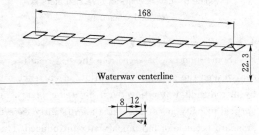

Figure 3 Layout chart for diversion piers in revised scheme Ⅲ

When Q is 8659m³/s, the vertical and horizontal velocity at entrance area can basically meet the requirements of navigation flow condition, but the transverse velocity is generally greater than 0.3m/s at the section which is about 50-100m away from entrance area and the intensity of backflow at entrance area is still large at the maximum navigable discharge.

4.3.4 *Revised scheme* Ⅳ

Based on the revised scheme Ⅲ, we adjust the angle of layout of the piers and it's external angle (adjust the angle between the centerline of piers and channel centerline to 8°) to increase the introduction of discharge and reduce the intensity of backflow at entrance area.

When Q is 8690m³/s, the vertical and horizontal velocity at entrance area can basically meet the requirements, but the transverse velocity at right of route at entrance area is generally greater than 2.0m/s and lateral velocity seldom exceeds standard at the maximum navigable discharge.

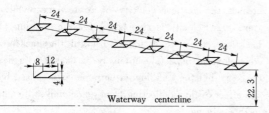

Figure 4　Layout chart for
diversion piers in revised scheme Ⅳ

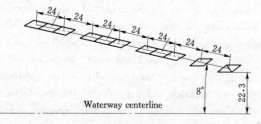

Figure 5　Layout chart for diversion piers in
revised scheme Ⅳ

4.3.5 *Revised scheme* Ⅴ

Something can be seen from the test results, we can weaken the strength of backflow at entrance area by introducing water flow with diversion piers, but vertical velocity at entrance area significantly exceeds standard because of too much water introduced. Thus we plug up individual drainage holes to reduce the in-

troduction of flow for decreasing vertical velocity.

When Q is 8690m³/s, the vertical and horizontal velocity at entrance area can basically meet the requirements, but the vertical and horizontal velocity can basically meet the requirements except certain points at the maximum navigable discharge. Meantime, backflow off bank will form behind the diversion piers because distribution of drainage flow discharge is more concentrated, and the flow pattern is worse, too.

4.3.6 *Revised scheme* Ⅵ

Just like scheme Ⅲ, the scheme Ⅵ is described below: center distance between piers is 15m; the expanding angle of the axis is 8°; angle between piers and the centerline is 20°; meanwhile, we plug up the upper part of holes to form a submerged hole so as to improve flow pattern; height of opening holes is 15m; an extended guide wall with a length of 25m located at the end of the piers.

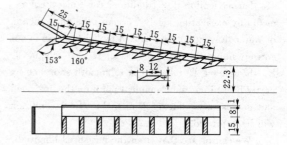

Figure 6　Layout chart for diversion
piers in revised scheme

When Q is 8690m³/s, the vertical and horizontal velocity at entrance area can basically meet the requirements, so is the working conditions at the maximum navigable discharge ($Q=13900$m³/s). Meantime, vertical velocity is less than 2.5m/s in connecting section and the horizontal velocity is less than 0.45m/s, which can meet the navigation flow conditions. We choose the revised scheme Ⅵ finally according to the results of hydraulic model test and self-propelled ship model test.

5. Conclusion

We can get the following conclusions from the research on navigation flow condition at the entrance area of upstream and downstream approach channel and connecting section for Laokou hydrojunction.

(1) From velocity distribution and flow pattern

for junction, we can see that bend flow has effect on the upstream velocity distribution: the mainstream tends to move to the side of concave bank and the flow before sluice is a bit smoother; the left velocity distribution before sluice was slightly larger than the right; the bend flow has effect on discharging flow, so mainstream tends to move to the side of at the entrance area of downstream approach channel and will become uniform in downstream narrow sections. By adjusting model test, the discharging flow from junction does not affect the navigation flow conditions at entrance areas.

(2) By adjusting topography partly at the entrance area in upstream channel and opening holes to introduce water flow at the bottom of guiding wall upstream channel, we can improve the upstream navigation flow conditions and make the distribution of longitudinal, transverse and backflow velocity fully satisfy the requirements of navigation flow conditions at all levels of navigable discharge.

(3) By selecting the upstream excavation scheme on local terrain, optimizing the upstream backfill scheme on local coastline, and setting a submerged dam in upstream regions, we can improve the navigation flow conditions in upstream connecting section, and make the flow condition satisfy the requirements at the maximum navigable discharge.

(4) By adjusting and optimizing the type of divergent section in guiding wall at downstream approach channel, adjusting and optimizing the axis of route and the boundary condition of excavation at convex bank, we can make the distribution of longitudinal, transverse and backflow velocity fully satisfy the requirements of navigation flow conditions at all levels of navigable discharge, however, the vertical velocity of individual points exceed the standard slightly at the right of route, which don't affect the operation of the ships.

(5) When designing the lock at curved reach, we should not only analysis the navigation flow condition at the entrance area, but also consider the impact on the navigable ships caused by the bend flow in connecting sections. At present, the limits about flow conditions on the connecting section are not given in "Lock design specification", we should undertake certain researches as soon as possible.

(6) To some place where the navigation flow conditions can't meet the needs, it's necessary to adopt real ship model tests besides hydraulic model test and the self-ship model tests.

References

[1] Huang Lunchao, Li Wei, Xiao Zheng. Influence of Different Riverside Shape on Flow Condition of Low-head Hydraulic Lock's Outlet Area [J]. Port & Waterway engineering, 2006 (04): 53-57.

[2] Lock Design Specification (JTJ 305—2001 [S]).

[3] Dai Yuting. Analysis on Status of the basic choracterics of Bend Flow. Communication Science and Technology in Hunan, 2008, (01): 127-130.

[4] Zhou Huaxing, Zheng Baoyou. An Approach to the Limiting Value for the Flow Conditions at the Entrance Area of Lock Approach channel [J]. Port & Waterway engineering, 2002 (02): 81-86.

[5] Li Yibing, Jiang Shiqun, Li Fuping. On Standard of Flow Conditions for Navigation in Transitional Reach Outside Entrance [J]. Journal of Waterway and Harbor, 2004 (12): 179-185.

[6] Gong Decheng, Bai Chenliang, Yang Shisheng. Discussions on Navigation Flow Conditions at the Entrance Areas and the Connecting sections of Navigable Structure [J]. China Water Transport, 2007, 07 (1): 31-32.

[7] Huang Bishan, Zhang Xujin, Shu Ronglong, etc. Study on the Flow Conditions at the Entrance Area of ship lock Approach Channel in Xinzheng Hydro-power Station [J]. Academic Journal of Chongqing Jiaotong University, 2003, 22 (3): 120-124.

[8] Han Changhai, Wang Fuwen. Study on the Measures about Improving Flow Conditions at the Downstream Entrance Area of ship lock Approach Channel in Feilaixia River [J]. Port & Waterway engineering, 1996, 12: 26-29.

[9] Li Yan, Zheng Baoyou, Liu Qingjiang. Effect of Submerged Opening of Guiding Dike

on Flow Condition for Navigation of Approach Channel [J] . China Harbor Engineering, 2007 (10): 12-16.

[10] Li Yan, Meng Xiangwei, Li Hejin. Effect of Arrangement of Upstream Approach Channel of Flow Conditions of the Three Gorge Project [J] . Journal of Waterway and Harbor, 2002, 23 (4): 281-287.

[11] Wu Xueru. Research on Navigation Flow Condition of Hydro-junction [J] . Port & Waterway engineering, 2006 (9): 52-55.

[12] Zhou Daixin. Flow Condition Improvement Countermeasures in the Outlet Area of Approach Channel for Water Project Ship Gate [J] . Pearl River, 1997 (5): 29-31.

[13] Chen Guifu, Zhang Xiaoming, Wang Zhaobin, etc. Influence of Open-type Guide WOEKS on Flow Condition for Navigation of Ship lock [J] . Port & Waterway engineering, 2004 (9): 56-58.

[14] Meng Xiangwei, Li Jinhe, Li Yan, etc. Influence of Training Wall with Holes on Navigable Condition [J] . Ioural of Waterway and Harbor, 1998 (2): 17-24.

[15] Analysis on General Hydraulic Model Test Results in Laokou Junction in Yujiang River of Guangxi Province at Preliminary Design Stage-Selection of Dam line [R] .

Exploratory Experimental Study of Rocky Ramp Type Nature-Like Fishway

Felipe Justo Breton[1]   Abul Basar M. Baki[2]   David Z. Zhu[3]
N. Rajaratnam[4]   Christos Katopodis[5]

[1]*Visiting Student, Department of Civil and Environmental Engineering, University of Alberta, Canada*
[2]*Graduate Student, Department of Civil and Environmental Engineering, University of Alberta, Canada*
[3]*Professor, Department of Civil and Environmental Engineering, University of Alberta, Canada*
[4]*Professor Emeritus, Department of Civil and Environmental Engineering, University of Alberta, Canada*
[5]*Katopodis Ecohydraulics Ltd.; formerly of the Freshwater Institute, 501University Crescent, Winnipeg, Manitoba, Canada*

Abstract: Fish passage facilities are often required to provide a pathway for fish and other aquatic species to reach their preferred habitats in the presence of barriers like dams and other hydraulics structures. In the present study, a ramp type fishway on a 5% slope is investigated, with particular attention on the reduction of flow velocity through the obstruction and the preferential flow path along the fishway. The experimental work was conducted in a flume; its width, height, and length were 0.91m, 0.61m, and 8.89m, respectively. The experimental setup with the rocky ramp type design consisted of a staggered arrangement of isolated boulders 15cm diameter on the channel bed and spaced at a distance of 38cm in both longitudinal and lateral directions. Experiments were performed for three different discharges and results are analysed and presented. Experimental results indicate that the effect of boulders on the water surface profile is fairly consistent throughout the flume length. In this rocky ramp nature-like fishway, incoming flow velocity can be reduced more than 55%, which is a significant reduction and may enable many fish species to move upstream. The velocity contours indicate that fish may rest behind the boulders and navigate upstream following any of the potential flow paths.

Key words: ramp type fishway; water depth; velocity; experiment; hydraulics

1. Introduction

Ecological preservation has increased worldwide in the past several decades. In-stream constructions, such as dams, weirs and under the road culverts have modified the essential role of rivers as a longitudinal corridor for different species. Both the engineering and science communities have expended large efforts to restore the "river continuum" concept and fish migration as outlined by Vannote et al[1]. Once it was realized that fish migration had been considerably reduced, fishways started to be constructed. Several designs were developed to generate flow regimes suitable for fish to swim upstream, such as pool, slot, denil, eel and culvert fishways, or fish locks and fish lifts[2-3]. Concrete, steel or wood were common materials chosen for these designs.

In recent years, a more holistic ecosystem approach[3] has led to the re-discovery of "stream simulation" or nature-like fishway designs, in which target fish are not confined to commercially important species only. In accordance with ecological design principles, a fishway should be able to accommodate all species living in a water body[4-5]. Nature-like fishways are amenable to be used by many species of fish as they create more diverse flow conditions within a cross section than those constructed with conventional materials and flat surfaces. The construction costs of these fishways could be more economical if the construction materials are locally available[6-7]. The technology and construction techniques required for low barriers are relatively simple, with limited engineering expertise required[8]. Moreover, fishways constructed with naturally occurring materials may be more easily modifiable and more acceptable for both the public and regulatory agencies.

Mainly, two kinds of nature-like fishway designs can be distinguished: pool and riffle type and rocky ramp type [9]. Some guidelines have been developed based on field experience. For instance, US-BR[10] provides practical guidelines for design and construction of fish ramps. DVWK[2] included a chapter with useful guidelines for nature-like fishways based on German experience, and provided some examples of field applications. Parasiewicz et al. [11] reviewed the design basis of nature-like bypass channels based on Australian experience.

The use of rocks aims to both mimic natural conditions and to satisfy fishway requirements. A certain rock arrangement increases the channel roughness so that it is possible to obtain lower velocities. It also provides some rest areas where fish can rest during their migrations, especially within the wake created just downstream of a rock. Those effects have been widely studied. For example, in order to understand this habitat environment, Shamloo[12] studied the flow around a hemisphere and its generated wake is well described in Shamloo et al[13]. Archarya et al[9]. provided some recommendations on how to space isolated roughness elements based on the experimental results. More recently, Sadeque et al[14]. investigated the flow around cylinders. In the present study, natural rocks (boulders) are used in a 5% slope flume for three different flow conditions. The purpose of this study is to investigate the reduction of flow velocity due to rock obstructions and to identify the preferential flow paths for fish swimming from downstream to upstream in a particular zone of the rocky ramp nature-like fishways.

2. Experimental arrangement

The experimental work was conducted in a flume located in the Hydraulics Lab at the University of Alberta. The flume was connected to a head tank, and the discharge was measured using a magnetic flow meter installed on the supply line. The width, height, and length of the flume were 0.91m, 0.61m, 8.89m, respectively (Figure 1). The flume was adjusted to have a longitudinal slope of 5%.

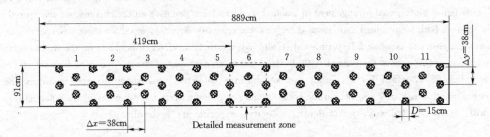

Figure 1　Schematic arrangement of the rocky ramp type nature-like fishway

The rocky ramp type design consists of a staggered arrangement of isolated boulders that were placed throughout the length of the flume (Figure 1 and Figure 2). A total of fifty-eight boulders, 15cm inequivalent diameter, were used in this experiment. These boulders were chosen as round as possible. The boulder diameter in all directions ranged between 10cm to 20cm. Isolated boulders were arranged in 23 rows alternating between two and three boulders in each one, as shown in Figure 1. The centre to centre distance between two boulders along longitudinal (Δx) and lateral (Δy) directions was about 38 cm. Boulders were glued at the ramp bottom with silicon. The measurements were taken in zone 6 indicated in Figure 1, at a distance of 4.19m from the upstream end of the flume, so that neither upstream nor downstream conditions could affect the flow between isolated boulders and flow is fully developed in this zone. This position was defined after the preliminary tests. Three sets of experiments were conducted for three different flow rates, 25L/s, 45L/s, and 60L/s, while keeping the same channel slope of 5%. The experimental conditions are summarized in Table 1.

Figure 2   Detailed arrangement of the rocky ramp type fishway in the flume

Table 1   Experimental conditions

Experiment	Channel slope (%)	Discharge (L/s)	Measured average flow depth (cm)	Flow observations
1	5	25	6. 7	Unsubmerged
2	5	45	10. 2	Overflow
3	5	60	11. 6	Submerged

An aluminium frame was used to support the Acoustic Doppler Velocimeter (ADV) to avoid the vibration effect of the flume (Figure 2) . The support frame provided a Cartesian coordinate reference system where x, y, z are defined as the longitudinal (stream-wise), lateral (cross-stream), and vertical directions, respectively.

3. Methodology

3. 1   Experimental Procedure

The experiments were conducted under three flow conditions: unsubmerged, overflow and submerged. The unsubmerged flow condition was used to calculate the mean flow velocity following the DVWK[15] guidelines. In this condition, flow depth was insufficient to use the ADV for velocity measurement; a Pitot tube of external diameter of 3mm was used to measure the point velocity in the x direction (u) . A total of 73 points were chosen within the particular zone for measuring the u velocity at a depth of 2. 5cm from the bottom. This is approximately 38% of the water column height considering water depth at all 73 points. In the case for of the 25L/s flow rate,

the point velocity at 2. 5cm water column height may represent the mean velocity in the longitudinal direction. An ADV (SonTek) was used to measure the instantaneous 3D velocities for the overflow and submerged conditions. The standard sampling volume for the ADV is a cylinder with a diameter of 6mm and height of 9mm. The distance of the sampling volume is 5. 7cm from the acoustic transmitter[16]. Data were collected at 50Hz frequency with a sampling duration of 2 minutes. A total of 123 points were chosen for the three velocity components (u, v, w) in the three directions (x, y, z), at a depth of 3. 0cm from the bottom which were approximately 29 and 26 % of the water column height for 45 and 60L/s respectively. It was not possible to measure the point velocity at the 40% water column height due to the insufficient water depth for the ADV considering all 123 points. As a result, for the 45 and 60L/s flow rates, the point velocity at 3. 0cm water column height may represent lower than the mean velocity in longitudinal direction.

At each point, the fluctuating water level was measured twice with the help of a point-gauge; once at the lower stage and again at the higher stage. For the longitudinal water surface profiles a total of 36 point gauge measurements were taken for each flow along the centre line of the entire flume. For each zone, a total three point gauge measurements were taken at positions 1 (downstream of the boulder), 2 (in between two boulders) and 3 (upstream of the boulder) as shown in Figure 6 (a) . A Pitot tube was also used to measure the point velocity in the x direction at an interval of 1cm from the channel bottom to develop the vertical velocity profiles at 1, 2 and 3 positions.

3. 2   Mean flow velocity calculation

In the unsubmerged condition, the mean flow velocity within the particular zone was calculated according to the recommended method for hydraulic design calculations of running waters, which have been compiled in the DVWK-Guidelines[15] " Hydraulic calculations of running waters" . The calculation of mean flow velocity in open channels is based upon the Darcy-Weisbach flow formula:

$$V_m = \frac{\sqrt{8gR_hS_0}}{\lambda_m} \qquad (1)$$

where, V_m means flow velocity in m/s; R_h means hydraulic radius; and S_0 means slope of the flume/channel. The total resistance coefficient λ_{tot} in Equation 1 results from the superposition of the bottom roughness λ_0 and the resistance of the perturbation boulders λ_s:

$$\lambda_{tot} = \frac{\lambda_s + \lambda_0\,(1 - \varepsilon_0)}{1 - \varepsilon_v} \qquad (2)$$

where, λ_0 means bottom roughness and can be expressed as,

$$\frac{1}{\sqrt{\lambda_0}} = -2\log\frac{k_s/R_h}{14.84}$$

k_s means equivalent sand roughness co-efficient, in which the equivalent sand roughness diameter k_s is replaced in the calculation by the average rock diameter d_s, in the case of a rockfill bottom.

$$\varepsilon_v = \frac{\sum V_s}{V_{tot}} = \frac{immersed\ vol.\ of\ boulders}{total\ volume}$$

$$\varepsilon_0 = \frac{\sum A_{0,s}}{A_{0,tot}} = \frac{surface\,area\ of\ boulders}{total\ basal\ area} \quad and$$

λ_s means resistance co-efficient due to boulder = $4C_w\dfrac{\sum A_s}{A_{0,tot}}$.

For boulder, C_w has been considered as 1.5, A_s = wetted surface area of individual boulder and $A_{0,tot}$ = unobstructed flow cross section (without boulders).

In the overflow and submerged conditions, the DVWK-Guidelines[15] are not applicable. In this case the mean velocity within the particular zone was estimated from averaging the 123 points absolute velocities from the ADV measurements. The efficiency was estimated from the following equation.

$$difficiency = \frac{V_1 - V_2}{V_1} \cdot 100 \qquad (3)$$

where, V_1 is the "unregulated" velocity and V_2 is the "regulated" velocity with the rocky ramp in the flume.

3.3　Construction of flow paths

In natural channels, fish usually follow the low velocity path to ascend. A velocity contour is a line connecting points of equal velocities and the contour interval of a contour map is the difference in velocities between successive contour lines. To identify the low velocity path in a particular zone, a velocity contour map was constructed for three flow conditions. For the unsubmerged flow condition, a total of 73 points were used; only the u component was used to plot the velocity contour with in a particular zone. But for the overflow and submerged flow conditions, a total of 123 points were used; the u, v and w components were used to plot the velocity contours within that zone. Surfer-8.0 software was used to generate the velocity contour map.

4. Results and discussions

4.1　Flow characteristics

The rating curves based on "regulated" with boulders and "unregulated" conditions were generated from the test flows in the flume (Figure 3). At the low flow rate of approximately 15L/s and the high flow rate of approximately 60L/s, the water depth with boulders is almost twice and more than three times higher than the depth without boulders, respectively.

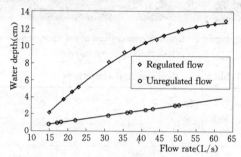

Figure 3　The rating curves under "regulated" and "unregulated" flow conditions

The longitudinal water surface profiles along the centre line of the flume were generated for the three flow rates as shown in Figure 4. According to Figure 6 (a), the water depth at positions 1, 2 and 3 at any zone should be lower, higher and in between, respectively. Figure 4 also represented the same result for three flow rates as we expected. The average water depth at positions 1, 2 and 3 for 25 (45/60) L/s were 5.9 (9.5/11.0) cm, 7.4 (11.0/12.7) cm and 6.8 (10.2/11.2) cm respectively within the entire boulders ramp. The drops in water level from position 3 to position 1 were 0.9cm, 0.7cm and 0.2cm for the 25, 45 and 60L/s flows, respectively. On the other hand, for these three flow conditions, the jumps of water level from position 1 to 2 were respectively 1.5, 1.5 and 1.7cm. So the drops of water level decreased with increased flow and jumps of water level increased with increased flow. The results indicate that the water depth was not constant at all locations and it varied depending on the location of the boulders. For the three flow rates, similar patterns of longitudinal water surface profiles were found over the entire flume. Therefore it can be said that the effect of boulders on the water surface profiles was almost consistent throughout the flume length.

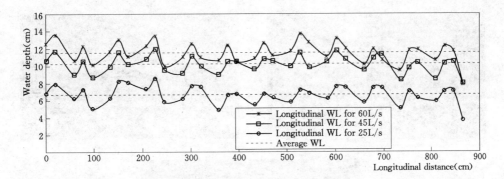

Figure 4   Longitudinal water surface profiles along the centre line of the flume

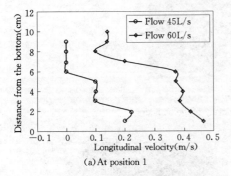

(a) At position 1

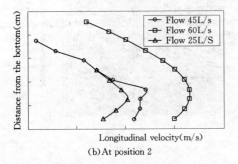

(b) At position 2

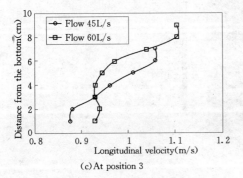

(c) At position 3

Figure 5   The vertical velocity profiles at different positions for different flow rates

Figure 5 represents the vertical velocity profiles at positions 1, 2 and 3 in a particular zone for different flow rates. This actually characterizes the downstream, side (left/right), and upstream areas of the boulders respectively i. e. all around the boulders. From these figures, it is apparent that the vertical velocity distribution at positions 1 and 3 was not as smooth as at position 2 due to higher turbulence; it also didn't follow the general vertical velocity distribution as occurred in the uniform flow. At position 1, the longitudinal velocity distribution along the water depth was inverse, i. e. velocity decreased with increasing water depth, because of the spherical shape of boulders [Figure 5 (a)]. At position 2, the vertical velocity distribution along the water depth became more regular as the flow rate increased [Figure 5 (b)]. Finally at position 3, the velocity increased with increasing water depth due to the overflowing water [Figure 5 (c)].

The major observations of the flow visualization tests were detected by injecting dye at a point along the centre line as shown in Figure 6. The dye spreads in the longitudinal and lateral directions from a very regular to an irregular pattern with increasing flow rate from low to high. The lateral extent of the dye decreases with increasing flow as well as from unsubmerged to submerged flow conditions.

4. 2   Velocity reduction

The design of a fish pass, the effectiveness of which is closely linked to the water velocities and flow patterns, should take into account the behavior of the target species[17]. Thus the water velocities within the fishpass must be compatible with their swimming capacity and behavior. A large water level difference between pools, excessive aeration or turbulence, large eddies or low flow velocities can act as a barrier for

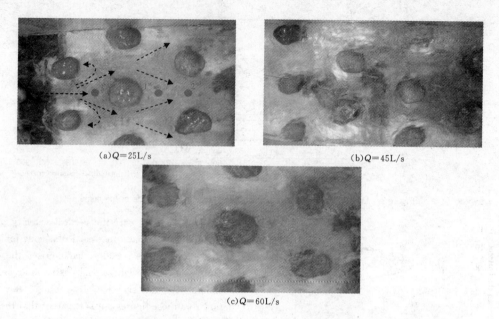

(a)$Q=25$L/s   (b)$Q=45$L/s

(c)$Q=60$L/s

Figure 6   Flow visualization in the vicinity of the boulders

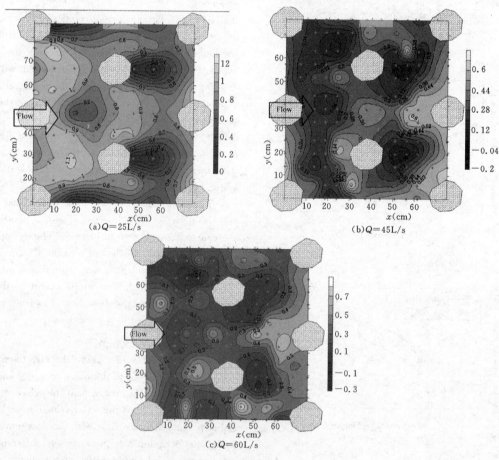

(a)$Q=25$L/s   (b)$Q=45$L/s

(c)$Q=60$L/s

Figure 7   Contours for the longitudinal （u） velocity distribution within the particular zone
for different flow rates

fish. According to DVWK[2], the maximum permissible flow velocity in fishways is 2m/s. Therefore, the main objective to construct any fishpass is to reduce flow velocity to within the permissible limit. Experimental results indicate that the approaching velocities were decreased from 1.11 to 0.48m/s for 25L/s, from 1.63 to 0.51m/s for 45L/s and from 1.98 to 0.54m/s for 60L/s flows between the "unregulated" and "regulated" flume conditions, respec-

tively. So a significant velocity reduction has occurred between the unregulated and regulated mean approaching velocities within the particular zone which may be defined as efficiency (Equation 3). The efficiency is 57, 69 and 73% for the 25, 45 and 60L/s flows, respectively (Table 2). It indicates that the efficiency increases with increasing flow rate and this is a significant reduction of flow for many fish species.

Table 2   **Experimental results of mean velocity**

Set no.	Flow (L/s)	Unregulated cross sectional velocity (m/s)	Source	Regulated mean velocity (m/s)	Source	Efficiency (%)
1	25	1.11	Manning's equation*	0.48	DVWK[14] guidelines	57
2	45	1.63	Manning's equation*	0.51	Absolute velocity from ADV	69
3	60	1.98	Manning's equation*	0.54	Absolute velocity from ADV	73

Note * Manning's roughness coefficient for smooth steel bottom = 0.012.

4.3 Velocity contours

From the velocity contours, a significant variation of the u velocity within the particular zone were found for three flow conditions due to boulders placement at the channel bottom. The velocity contour lines were more regular for the low flow condition compared to medium or high flow conditions. For the 25L/s flow, the contour lines were more consistent in between, upstream and downstream of the boulders [Figure 7 (a)], but for 45 and 60L/s, the contours lines were inconsistent in those locations within the particular zone [Figure 7 (b) and Figure 7 (c)]. The maximum and minimum velocity for high ($Q = 60$L/s), medium (45L/s) and low (25L/s) flow conditions were 0.8 and -0.3m/s, 0.68 and -0.2m/s, and 1.2 and 0m/s respectively within the zone. Here the longitudinal maximum velocity decreased with increasing flow rate, although hypothetically this is not correct. The probable strong reason was velocity measurement accuracy at 29% and 26% water column height for 45 and 60L/s flow rates respectively (described in the experimental procedure).

It was also noticed that the velocity was very high between two boulders compared to the velocity just behind the boulders. In between two boulders there were several potential flow paths for different velocities. So fish can take rest behind the boulders and can navigate up stream following any of the potential flow paths. The experimental results from this section indicate that by placing multiple obstructions at the channel bottom some low velocity zones can be generated for fish to use when traveling in upstream direction.

Figures 8 and 9 show the lateral (v) and vertical (w) velocity contours respectively for 45 and 60L/s flow conditions. The distributions of lateral velocity contours within the zone were almost consistent for both flow conditions. Usually the higher lateral velocity was found at the upstream of the boulders due to the diversion of flow and lower velocity in between the boulders (Figure 8). The vertical (w) velocity contours was more consistent for 45L/s compared to 60L/s (Figure 9). In general higher vertical velocity due to circulation of flow downstream of the boulders; the vertical velocity contours for 45 and 60L/s represented similar result.

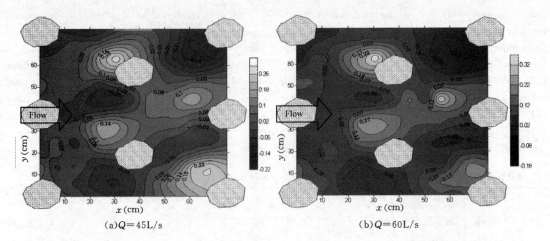

(a)$Q=45$L/s   (b)$Q=60$L/s

Figure 8   Contours for the lateral （v）velocity distribution within the particular zone for different flow rates

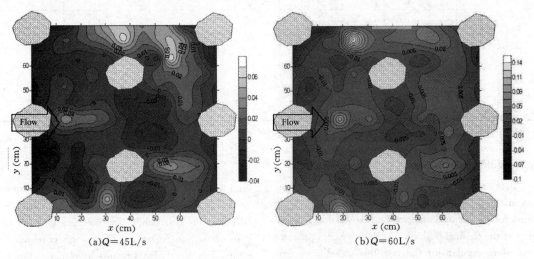

(a)$Q=45$L/s   (b)$Q=60$L/s

Figure 9   Contours for the vertical （w）velocity distribution within the particular zone for different flow rates

5. Conclusions

In this study, the reduction of flow velocity due to rock obstructions was investigated. Preferential flow paths were identified for fish to swim in a particular zone of the rocky ramp nature-like fishways. Several conclusions have been drawn from this study, including: (1) at the low flow rate of approximately 15L/s and the high flow rate of approximately 60L/s, the water depth with boulders was almost two times and more than three times higher than the depth without boulders, respectively; (2) the average water depths for 25, 45, and 60L/s were 6.7, 10.2, and 11.6cm respectively within the entire boulder ramp; (3) the drops in water level from positions 3 to 1 decreased with increased flow and

jumps of water level from positions 1 to 2 increased with increased flow; (4) for the three flow conditions, similar patterns of water surface profiles were found over the entire flume. It can be said that the effect of boulders on the water surface profiles was almost consistent throughout the length of the flume; (5) the vertical velocity distribution along the water depth at the positions 1 and 3 was not as smooth and regular as at position 2 due to more turbulence. It didn't follow the general vertical velocity distribution profiles that occurred in the uniform flow; (6) a significant velocity reduction had occurred between the unregulated and regulated mean approaching velocities within the particular zone which may be defined as efficiency: these were estimated as 57, 69 and 73% for 25, 45 and 60L/s, respectively; (7) the maximum and minimum longitudinal velocity for high ($Q =$ 60L/s), medium (45L/s) and low (25L/s) flow conditions were 0.8 and $-$ 0.3m/s, 0.68 and $-$0.2m/s, and 1.2 and 0m/s respectively within the particular zone; (8) the contours indicated that by placing multiple obstructions at the channel bottom some low velocity zones can be generated for fish to use when traveling in the upstream direction; (9) the distributions of lateral velocity contours within a particular zone were almost consistent for both flow conditions; and (10) the vertical velocity contours were more consistent for the 45L/s compared to the 60L/s flow condition.

References

[1] R. L. Vannote, G. W. Minshall, K. W. Cummins, J. R. Sedell, and C. E. Cushing. The River Continuum Concept. *Canadian Journal of Fisheries and Aquatic Sciences*, 1980, 37: 130-137.

[2] DVWK. *Fish Passes - Design*, *Dimensions and Monitoring*. FAO: Rome, Italy, 2002.

[3] C. Katopodis. Developing a Toolkit for Fish Passage, Ecological Flow Management and Fish Habitat Works. IAHR J. Hydraulic Research, 2005, 43 (5): 451-467.

[4] C. Katopodis. Riverine Fish Habitat: Mitigation and Restoration Approaches. Proc. ECO-SET95 (International Conference on Ecological System Enhancement Technology for Aquatic Environments), Tokyo, Oct. 29-Nov. 3, 1995, 31: 356-360.

[5] T. Tsujimoto, N. Horikawa. Methodology of Ecological Fishway Design. Proceedings, Congress of the International Association of Hydraulic Research, IAHR Environmental and Coastal Hydraulics: Protecting the Aquatic Habitat, San Francisco, CA, August 10-15, 1997, B: 1378-1383.

[6] J. H. Harris, G. Thorncraft, P. Wem. Evaluation of Rock-ramp Fishways in Australia. Proceedings, ⅩⅩⅧ IAHR congress-Graz 1999, Hydraulic Engineering for Sustainable Water Resources Management at the Turn of the Millennium, Vienna, Austria, August 22-27, 1999: 331-347.

[7] P. Wang and C. Katopodis. Fishway Studies for the Lower Churchill River Water-Level Enhancement Project in Manitoba. Proceedings, 3rd International Symposium on Ecohydraulics, Salt Lake City, UT, July 13-16, 1999.

[8] T. J. Marsden, G. Thorncraft, K. L. Woods. Reconstruction of Culverts and Causeways to Assist Migrations of Adult and Juvenile Fish Project. NHT Project No. 2012102, Final Project Report. Queensland Department of Primary Industries, Fisheries, 2003.

[9] M. Acharya, J. A. Kells, C. Katopodis. Some Hydraulic Design Aspects of Nature-like Fishways. Proceedings of 2000 Joint Conference on Water Resources Engineering and Water Resources Planning and Management, EWRI, ASCE, Minneapolis, Minnesota, USA, July 30-August 2, 2000: 10.

[10] U. S. B. R.. *Rock Ramp Design Guidelines*. U. S. Department of the Interior Bureau of Reclamation Technical Service Center Denver, Colorado, USA, 2007.

[11] P. Parasiewicz, J. Eberstaller, S. Weiss, S. Schmutz. Conceptual Guidelines for Nature-like Bypass Channels. In: M. Jungwirth, S. Schmutz, and S. Weiss, eds, *Fish Migration and Fish Bypasses*, Oxford: Fishing News Books, Blackwell Science Ltd, 1998: 348-362.

[12] H. Shamloo. Hydraulics of Simple Habitat Structures in Open Channels. Ph. D. Thesis, Department of Civil Engineering, University of Alberta, Edmonton, Canada, 1996: 342.

[13] H. Shamloo, N. Rajaratnam, C. Katopodis. Hydraulics of Simple Habitat Structures. *Journal of Hydraulic Research*, 2001, 39 (4): 351-366.

[14] M. A. F. Sadeque, N. Rajaratnam, M. R. Loewen. Shallow Turbulent Wakes Behind Bed-mounted Cylinders in Open Channels. *Journal of Hydraulic Research*, 2009, 47 (6): 727-743.

[15] DVWK. Ökologische Aspektebei Ausbau und

Unterhaltung von Fließgewässern'. DVWK Merkblatt 204, Bonn, 188 S.

[16] SonTek. ADV Technical Documentation. SonTek, 1997.

[17] M. Larinier, G. Marmulla. Fish Passes: Types, Principles and Geographical Distribution an Overview. Proceedings of the second international symposium on the management of large rivers for fisheries, 11-14th February, 2003, Phnom Penh, Kingdom of Combodiya.

The Rate of Headcut Migration of Cohesive Soil Homogeneous Embankment

Xie Min[1]   Shen Dengle[2]

[1]*State Key Laboratory of Hydrology-Water Resources and Hydraulic Engineering,*
Nanjing Hydraulic Research Institute, Nanjing, 210029
[2]*Chuzhou Survey and Design Institute of Water Conservancy and*
Hydropower, Chuzhou, 239000

Abstract: Through the physical model test of the headcut migration rate and the threshold friction velocity test of cohesive soil, it can be found that the water content has an important influence on the breach process of earth dam. With the same degree of compaction (in this paper, the degree of compaction is close to 100%), the larger the water content is, the lower the headcut migration rate is. After the water content exceeds the optimum moisture content, the speed is very slow. The smaller the degree of compaction is, the faster the headcut moves. In this paper, the relationship between the "threshold friction velocity" U_{*c} and material rate parameters C is established, which can significantly improve predicating accuracy of the headcut migration rate.

Key words: headcut; material rate parameters; threshold friction velocity

1. Introduction

One of the most difficult problems for earth-rock dam breach is the determination of the erosion rate of dam soil under the high-speed unsteady flow conditions. The theory of sediment transport cannot solve this problem adequately, which usually leads to a substantial degree of uncertainty in prediction of the speed of earth-rock dam breach and breach process [1,2]. A large number of tests confirm that headcut is one of the main features of cohesive soil dam breach [3-5]. The headcut migrates from downstream to upstream crest of the embankment gradually, once it reaches to the dam crest the peak flow will appear soon [6]. Evidently, the headcut migration rate is an important parameter for cohesive soil dam breach, which is very important to the prediction and warning of dam-break flood process. Many valuable studies have been carried out to predict the headcut migration rate [7-9]. In 1989, De Ploey combined the flow kinetic energy in the vertical plane of headcut with the rate of headcut migration and presented a formula for calculating headcut migration rate [10]; in 1992, De Ploey analyzed the relationship between earth-rock dam materials coefficient (anti-scouring indicators) and the flow parameters (driving force indicators of erosion), and

established the formula for calculating the headcut migration rate; in 2001, Hanson etc., analyzed the "headcut" stable condition based on soil mechanics, and established an simple calculation model of the headcut migration rate [11]; Overall, prediction of the rate of headcut migration is a very complex problem. Most of previous formulas are semi-theoretical semi-empirical and some parameters of them exist larger uncertainties. In the current, sediment transport theory in the high-speed flow has no substantive breakthrough. Carrying out some research for formula's uncertainty based on the existing formula and improving their theories are of great significance to improve the forecast accuracy of headcut migration rate. In this paper, seven flume experiments for cohesive soil homogeneous earth dam have been conducted and a formula for the headcut migration rate has been presented.

2. Physical model test for headcut migration rate

The main factors to impacting the headcut migration rate can be attributed to two aspects, one is the incoming flow upstream when dam breaching; the other is the anti-erosion ability of dam soil. It is relatively simple to describe the magnitude of dam-break incoming flow. For example, it can be expressed as a function of water head, flow rate, discharge, etc. The anti-erosion ability of dam soil is closely related to water content, degree of compaction, consolidation duration, mineral composition of soil particles, soil particle size grading, and so on. It is impossible to consider all these factors one by one in the test, here the experiments have done mainly considered two main control parameters (degree of compaction, water content) which must be considered in the compaction construction of homogeneous earth dam. Based on these tests an efficient calculation method for headcut migration rate may be sought.

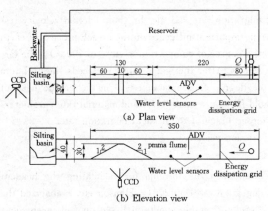

Figure 1 Model layout (cm)

The physical model is as shown in Figure 1. The model dam is 0.3m high, 0.3m wide, and the downstream and upstream dam slope is both 1 : 2 ($V : H$). The dam is divided into 5 filling layers, and each is 0.06m high. Each compacted layer was constructed according to the design degree of compaction and water content. In order to check the actual quality of the filling, random sampling is carried out to measure the degree of compaction and water content in the process of construction. Constant flow is provided by pumping water from the neighboring reservoir. In the front of dam, a Doppler Velocimeter ADV is set up to record the flow velocity process, and two water pressure sensors are set up to gather data through the United States multi-channel acquisition system Wavebook; On the right side of the dam, the top of the dam and the downstream of the dam, high resolution video cameras are respectively set up to record the development process of the breach erosion.

In view of the actual situation of the general homogeneous earth dam, the range of the dam material water content adopted was from 10% to 20% in the tests. The test groups designed for A1-A5 are as shown in Table 1. It is easier to understand the relationship between the degree of compaction and the process of dam breach (the lower degree of compaction is, the more easily the dam breaches), so only four test groups for variable degree of compaction were designed, and for convenience, A3 and A5 acted as the two groups with a larger degree of compaction. The main parameters of variable degree of compaction are shown in Table 2.

Table 1   The main parameters of variable water content

Group	Wet density (g/cm³)	Degree of compaction	Water content (%)
A1	1.89	0.99	10.59
A2	1.96	1.00	12.50
A3	1.93	0.99	13.58
A4	1.96	0.98	15.81
A5	2.03	1.00	17.82

Table 2   The main parameters of variable degree of compaction

Group	Wet density (g/cm³)	Degree of compaction	Water content (%)
A6	1.80	0.89	17.20
A5	2.03	1.00	17.82
A7	1.86	0.88	11.16
A3	1.93	0.99	13.58

Note   A3, A5 are the same with Table 1.

In order to eliminate the influence from other factors on the test results, the same material, the same dam body size and the same flow conditions are adopted in the tests.

3. Results

3.1 The impact of water content and degree of compaction on headcut migration rate

For convenience, the headcut migration rate is definite as $dx/dt = \dfrac{\text{area of dam erosion}}{\text{height} \times \text{time of erosion}}$. According to the tests the relationship between the water content and the rate of headcut migration is shown in Figure 2. When water content was less than the optimal water content (in this paper, the optimal water content of dam soil is 13.96%) the headcut migration rate decreased with the water content increasing. When the water content exceeds the optimal water content, headcut migrates very slowly. As the water content continues to increase, the rate of headcut migrate is almost unchanged (the range of design water content in the experiment is 10%-20%).

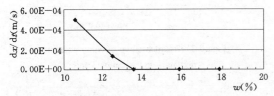

Figure 2　The relationship curve of water content and rate of headcut migration

The impact of degree of compaction on the rate of headcut migration is obvious. According to the results of tests, it can be seen that the changes in degree of compaction speed have a great influence on headcut migration rate. With degree of compaction changing from 88% to 99% the erosion rate reduces 20-50 times as shown in Table 3.

Table 3　The impact of changes in degree of compaction on headcut moving speed

Group	Degree of compaction	Erosion rate (m/s)
A6	0.89	3.59E-04
A5	1.00	1.37E-05
A7	0.88	9.47E-04
A3	0.99	1.78E-05

3.2　Calculation expression for the rate of headcut migration

Temple proposed a calculation model for rate of headcut migration based on an integrated material coefficient C and hydraulic parameters A, which can be written[12]:

$$dx/dt = CA \qquad (1)$$

where, C is integrated material coefficient; A is hydraulic parameter,

$$A = q^{1/3} H^{1/3} \qquad (2)$$

where, q is discharge per unit width; H headcut height.

Hanson[4] found a relatively independent relationship between the water content w and integrated material coefficient C for different body materials, same degree of compaction. The equation of relationship can be expressed as

$$C = 3000w^{-6.5} \qquad (3)$$

In this article for the convenience of application, q is defined as average discharge per unit width in the whole process of dam break, which can be approximately calculated by releasing water W, breach duration T and the average width of final breach B, and the headcut height directly adopted the height of dam. Based on Eq. (1) and Eq. (2), material rate parameters of dam material C for A1-A5 were calculated and listed in Table 4.

Table 4　Results of material rate parameters of dam material C

Water content w (%)	Headcut migration rate (m/s)	Discharge per unit width q (m³/m/s)	Headcut height H (m)	C
A1	5.03E-04	0.02	0.3	2.75E-03
A2	1.42E-04	0.02	0.3	7.75E-04
A3	1.78E-05	0.02	0.3	9.71E-05
A4	2.01E-05	0.02	0.3	1.10E-04
A5	1.37E-05	0.02	0.3	7.49E-05

Using the Hanson's method[11] the empirical relationship between water content w and material rate parameters of dam material C can be established:

$$C = 29265\omega^{-7} \qquad (4)$$

Compared with Eq. (3), the predicted result of Eq. (4) is bigger than that of Eq. (3) when the water content is small; when the water content is great, both of them are closer. The differences are mainly because they had slightly different definition of discharge per unit width and headcut height in the formulas.

The material rate parameter C has a good correlation with water content, however, material rate parameter of dam is the comprehensive performance of the anti-scouring ability of dam body. It is not only interrelated with water content, but also with the degree of compaction (the relative tightness for noncohesive homogeneous dam), consolidation duration, particle size gradation, mineral composition of soil particles. Only establishing the relationship between water content and material rate parameter will greatly limit the application and promotion of the formula. The anti-scouring ability indicators of dam materials are always the focus of dam breach problem for its complexity. In 1993, Xie Renzhi gave the coefficient table of soil after compiling more than 400 dam break data Domestic and Overseas[13]. In 2007, Hanson presented that a certain correlation was existed between water content and material rate parameter [11]. But Xie's coefficient is related with the reservoir storage capacity (not an independent parameter) and Hanson's formula considers too little factors. In 2001, Hanson introduced soil cohesion and internal friction angle to the calculation formula of headcut migration rate, but whether these parameters directly from the soil mechanics can describe anti-scouring ability precisely should be verified. In 2005, Hong Dalin had researched the relationship between the shear strength of soil and soil starting shear stress and found the relationship was relatively scattered[14]. Therefore, the author suggest that calculating material rate parameter should start with theory of sediment transport in dam-break problems, and recommend adopting the threshold friction velocity U_{*c} as an important indicator (in sediment transport problems threshold velocity is often as important parameters but it is associated with water depth).

For establishing the relationship between material rate parameter C and threshold friction velocity U_{*c} five threshold friction velocity tests were carried out (using the same kind of soil, degree of compaction, and different water content). The specific test method refers to the literature[7]. The water content of remoulded soil is slightly different from that of tests above (degree of compaction is pretty much the same), as shown in Table 5. The relationship between the threshold friction velocity and water content is shown in Figure 3. It can be seen when the degree of compaction are the same, with the water content increasing (the range of water content is 12%-20%) the threshold friction velocity U_{*c} increases. This can explain the relationship between water content and headcut migration rate mentioned above.

Table 5 **Threshold friction velocity U_{*c}**
with different water content

Group	Water content (%)	Dry density (kg/m³)	Degree of compaction
B1	12	1690	0.98
B2	14	1690	0.98
B3	16	1690	0.98
B4	18	1690	0.98
B5	20	1690	0.98

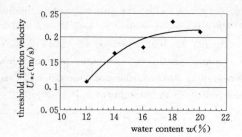

Figure 3　Relation curve of water content and start-up friction velocity

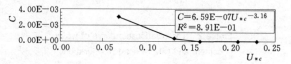

Figure 4　The relation curve between start-up friction velocity U_{*c} and integrated material coefficient C

In order to eliminate the influence of different water content between B1-B5 and A1-A5, the A1-A5 threshold friction velocity U_{*c} corresponding to water content is interpolated using cubic spline interpolation method, and the relation between threshold friction velocity U_{*c} and material rate parameters C is shown in Figure 4.

Based on the characteristics of the data, a empirical formula can be obtained

$$C = 6.59 \times 10^{-7} U_{*c}^{-3.16} \tag{5}$$

Combined with Eq. (5), Eq. (2) and Eq. (1), Eq. (6) can be written as:

$$\begin{cases} A = q^{1/3} H^{1/3} \\ C = 6.59 \times 10^{-7} U_{*c}^{-3.16} \\ \mathrm{d}x/\mathrm{d}y = CA \end{cases} \tag{6}$$

Eq. (6) can predict the headcut migration rate when dam breach. The material rate parameter C of this paper responses the comprehensive anti-erosion capacity of dam soil. It is a comprehensive embodiment of soil viscosity, degree of compaction, characteristics of mineral particles, and so on. Despite the formula in this article is based on the cohesive homogeneous dam, it still has a certain practical value for non-cohesive homogeneous earth dam.

3.3　Example of application

To check the prediction result of Eq. (6) experimental data came from two large-scale field tests in Dawa reservoir, Chuzhou, Anhui Province (carried out by Nanjing Hydraulic Research Institute in 2008) are used[15]. The calculation results are shown in Table 6. Compared with the results of Hanson formula and Eq. (4), the headcut migration rate calculated by Eq. (5) is very close to measured values. Obviously, the calculation formula of material rate parameter based on the characteristic parameters U_{*c} has a great practical significance. It can improve the predic- tion accuracy of the headcut migration rate. For the different characteristics of soil in the natural world it is not satisfactory to predict the erosion characteristics of the material only using the indirect parameters such as water content and degree of compaction.

Table 6   **Comparison of calculation results of prediction formula**

Test date	Water content (%)	Start-up friction velocity (m/s)	Breach duration (min)	Measured erosion rate (m/s)	Erosion rate calculated by Eq. (4)	Erosion rate calculated by Hanson formula	Erosion rate calculated by Eq. (5)
Dec. 8, 2008	19.40	0.07	40	4.85E-03	5.30E-05	2.39E-05	5.50E-03
Oct. 10, 2008	17.6	0.15	75	2.20E-04	5.28E-05	2.27E-05	2.49E-04

4. Conclusion and discussion

This paper presents that water content of cohesive homogeneous earth dam and degree of compaction have a great impact on headcut migration rate through physical model test of headcut migration rate. The calculation of material rate parameter C should consider the effect of threshold friction velocity U_{*c} which is the inherent characteristics parameter of dam soil and is relative to erosion. The results and suggestions are as follows.

(1) The water content affects greatly the process of cohesive earth dam breach. With the same degree of compaction (in this paper, the degree of compaction is close to 100%), the larger the water content is, the slower the headcut migration rate is. After the water content exceeds the optimum moisture content, the speed is very slow.

(2) The smaller the degree of compaction is, the faster the headcut advances.

(3) In dam-break problems, it is proposed to use "threshold friction velocity" U_{*c} to express material rate parameters C.

(4) In this paper, the relationship between "threshold friction velocity" U_{*c} and material rate parameters C has been established. It is proved to have good practicality and can significantly improve the predict accuracy of the headcut migration rate.

References

[1] Mohamed M A, Samuels P G, Morris M W. Improving the Accuracy of Prediction of Breach Formation through Embankment Dams and Flood Embankments, In: Bousmar & Zech (Eds), River Flow 2002, Swets & Zeitlinger, Lisse, the Netherlands, 2002: 663-673.

[2] Faeh R. Numerical Modeling of Breach Erosion of River Embankments [J]. Journal of Hydraulic Engineering, 2007, 133 (9): 1000-1009.

[3] Ralston D C. Mechanics of Embankment Erosion During Overflow [J]. Hydraulic Engineering, Proceedings of the 1987 ASCE National Conference on Hydraulic Engineering, Williamsburg, Virginia, 1987: 733-738.

[4] Temple D M, Hanson G J. Headcut Development in Vegetated Earth Spillways [J]. Applied Engineering in Agriculture, 1994, 10 (5): 677-682.

[5] Temple D M, Moore J S. Headcut Advance Prediction for Earth Spillways [J]. Presented at the 1994 ASAE International Winter Meeting, Paper No. 94-2540, Atlanta, Georgia, December 13-16, 1994.

[6] Hanson G, Cook K, Hahn W, et al. Observed erosion processes during embankment overtopping tests [C]. American Society of Agricultural Engineers Meetings, Paper No. 03-2066, 2003.

[7] Robinson K M, Hanson G J. Large-scale headcut erosion testing [J]. Transactions of the ASAE, 1995, 38 (2): 429-434.

[8]　Robinson K M, Hanson G J. Gully headcut advance [J]. Transactions of the ASAE, 1996, 39 (1): 33-38.

[9]　Römkens M J M, Prasad S N. Hydrologically driven mechanisms of headcut development [J]. International Journal of Sediment Research, 2005, 20 (3): 176-184.

[10]　De Ploey J. A. Model for headcut Retreat in Rills and Gullies, CATENA Supplement 14 [M]. Cremlingen, Germany 1989: 81-86.

[11]　Hanson D G, Temple D. Determination of material rate parameters for headcut migration of compacted earthen materials [J]. Final Report on Coordination and Cooperation with the European Union on Embankment Failure A-nalysis, 2007: 128-138.

[12]　Temple D M, Moore J S. Headcut advance prediction for earth spillways [J]. Transactions of the ASAE, 1997, 40 (3): 557-562.

[13]　Xie Ren-Zhi. Hydraulic of breach [M]. Jinan: Shan Dong Science and Technology Press 1993: 103-135. (in Chinese)

[14]　Hong D L, Miao G B, Den D S, et al. Jump-start of clay soil and its application in engineering [M]. Nanjing: Hohai University Press, 2005. (in Chinese)

[15]　Jianyun Z, Yun L, Guoxiang X, et al. Overtopping breaching of cohesive homogeneous earth dam with different cohesive strength [J]. Science in China Series E: Technological Sciences. Accepted, 2009.

Experimental Study on Flucuating Pressure of Flow around 8 Cylinders

Zhang Luchen   Luo Shaoze

State Key Laboratory of Hydrology-Water Resources and Hydraulic Engineering,
Nanjing Hydraulic Research Institute, Nanjing, 210029

Abstract: Characteristics of flucuating pressure of flow around 8 cylinders in radial arrangement are studied though hydraulic model test. Research shows that flucuating pressure can be regarded as stationary random process. The influence of flow interference between two cylinders in side-by-side is greater than in tandem. Affected by clearance flow and wake flow, valley value appears at medial cylinder. Inclined cylinder deflects separation layer and wake, and superposition causes a fluctuating pressure peak. Fluctuating pressure on cylinders in lateral flow is larger, and it's distribution is more complex. In small space, Strouhal numbers of eight cylinders are about 0.03.

Key words: pile group; flucuating pressure; vortex shedding frequency; Strouhal number

1. Introduction

Flow around cylinder is a classics subject in fluid mechanics. It is a complex phenomenon referring to boundary layer, separation shear layer, wake. In actual project, there are usually cylinder groups, such as smokestacks, aerial cables, pile group of wind turbine, etc. Problem of flow around polycylinder is under the spotlight. M. M. Zdravkovich[1] summarized flucuating of two cylinders within the subcritical Reynolds number regime. Youheng Xu[2], Zhifu Gu[3], etc. studied distribution of aerodynamic force of three cylinders. K. Lam[4] researched force coefficients and Strouhal numbers of four cylinders in cross flow. Although there is much research, understanding of the physical essence of flow around polycylinder remains incomplete.

Polycylinder in radial arrangement is applied widely in offshore structure, foundation design, etc. The study of eight-cylinder in ridial could be helpful to the understanding of the flow physics, of the flow interaction mechanisms and provide reference for the engineering application. This paper based on a offshore wind turbine project, studies characteristics of flucuating pressure of flow around 8 cylinders through hydraulic model test.

2. Experimental approach

The structural type of the foundation of a off-shore wind turbine is shown as Figure. 1. Hydraulic model is designed based on gravity similarity and the scale is 1:34. According to tide level, measure points are arranged on A, B and C level, and on each cylinder eight points are arranged equably. Measure points total 64.

Flucuating pressure is measured using micro pressure sensor and specialized equipment is applied in signal sampling and processing. Test simulates 5 tide levels: 2.55m, − 2.09m, 1.86m, − 1.34m and 0.23m. Experiments are conducted within the subcritical Reynolds number regime. Reynolds number of slab is about 1.8×10^5, and pile is about 2×10^4.

3. Results

Characteristics of tide fluctuating pressure on eight cylindrical pile group fundation are closely related to flow regime, displayed through random distribution characteristics, time-domain amplitude characteristics and frequency-domain energy characteristics.

3.1 Random distribution characteristics

Experiment was carried out with constant water level, constant flow velocity, constant flow direction and hydraulic elements remained essentially the same. So procedure of discharge was constant and flucuating pressure can be regarded as stationary random process. Probability density is near normal distribution, as shown in Figure 2.

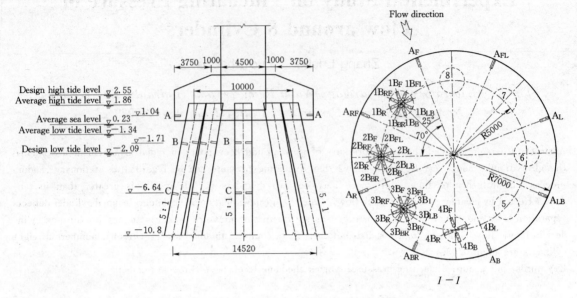

Figure 1 Structual type and measure point layout

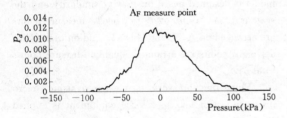

Figure 2 Probability density curve

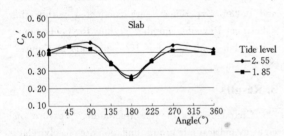

Figure 3 Flucuating pressure coefficient of slab

3. 2 Time-domain amplitude characteristics

Figure 3-Figure 7 show the distribution of fluctuating pressure coefficients in five tide level. Definition of angle: the angle of measure point which is over against flow direction is 0°, increasing clockwise.

Slab is single cylinder, with little interference. The distribution is almost symmetrical: there is a peak both in 90°and 270°, and valley value appears at rear stagnation point which is in wake flow. This result is in accordance with that of Norberg[5] which was got also within the subcritical Reynolds number regime: fluctuating pressure coefficient C'_p gradually increases form front stagnation point and arrives maximum in separation point, and then gradually decreases to rear stagnation point in which it reaches minimum.

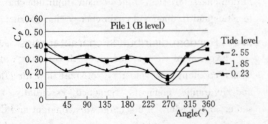

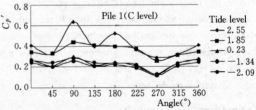

Figure 4 Flucuating pressure coefficient of Pile 1

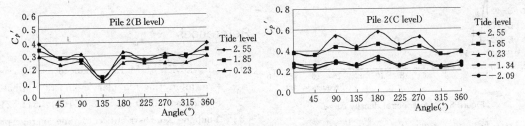

Figure 5　Flucuating pressure coefficient of Pile 2

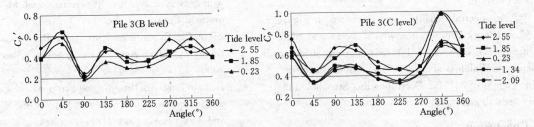

Figure 6　Flucuating pressure coefficient of Pile 3

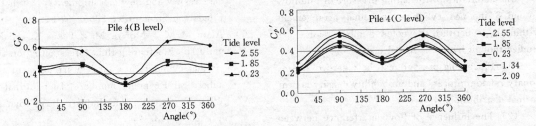

Figure 7　Flucuating pressure coefficient of Pile 4

Pile 1 is at the most upstream. Medial flow is extruded by Pile 8, while lateral flow is restricted by Pile 2. Thus separation point moves downstream to around 180°and valley value appears in 270°. Flucuating pressure coefficient of C level is bigger than B level. That agrees with the results of Alam[6] who reported that when space ratio is between 2. 0-3. 0 (B level is 2. 45 and C level is 2. 89), interference between two cylinders in side-by-side is most significant.

Distribution of C'_p of Pile 2 grossly appears as symmetrical. Affected by staggered Pile 1 and tandem Pile 3, peak and valley appear alternately every other 45°. And there is a valley on down stream surface in 135°, mainly because of the influence of wake flow of Pile 1.

The inference on Pile 3 is greatest. Distribution of C'_p is relatively disorder and difference between B level and C level is comparatively bigger. Mainly affected by Pile 2, there is a peak both in 45°and 315° on B level. Reason is that the separated shear layers

from Pile 2 attach to Pile 3 symmetrically. According to experimental results of Guo Mingmin[7], Alam[6], and Arie[8], this distribution is a typical one of downstream cylinder in two tandem cylinders.

Distribution on C level displays opposite tendency in 45° and 90° compared to B level. And value in 315°is nearly twice than in 45°. This phenomenon indicates that vortex shedding or boundary layer separating diverges and superposition occurs in 315°.

Pile 4 is at the most downstream. Clearance flow and wake flow of upstream cylinders has diffused adequately. It's influence on Pile 4 is little. So distribution of Pile 4 is also symmetrical and is very similar to single cylinder.

3. 3　Frequency-domain energy characteristics

Frequency-domain energy characteristics of hydrodynamic pressure can be presented by auto-spectral density S_{xx}:

$$S_{xx}(f) = 2\int_{-\infty}^{\infty} R_{xx}(\tau)e^{-j2\pi f\tau}\,\mathrm{d}\tau \qquad (1)$$

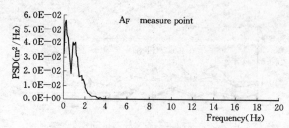

Figure 8   Auto-spectral density curve

$$R_{xx}(\tau) = \lim_{T \to \infty} 1/T \int_0^T x(t)x(t+\tau)\mathrm{d}t \text{ is autocorre-}$$

lation function.

Auto-spectral density curve is shown in Figure 8. Main energy concentrates within 10Hz and main frequency is between 0.1-1.5Hz. Vortex shedding frequency of slab is about 0.19Hz, while that of inclined pile is of 0.25Hz. Corresponding Strouhal numbers are 0.2 and 0.03.

4. Conclusion

Characteristics of flucuating pressure of tide flow around 8 inclined cylinders in radial arrangement within the subcritical Reynolds number regime are studied though hydraulic model test:

(1) Flucuating pressure can be regarded as stationary random process and probability density is near normal distribution.

(2) The influence of flow interference between two cylinders in side-by-side is greater than in tandem. Affected by clearance flow and wake flow, valley value appears at medial cylinder. Inclined cylinder deflects separation layer and wake, and superposition causes a fluctuating pressure peak. Fluctuating pressure on cylinders in lateral flow is larger, and it's distribution is more complex.

(3) Main frequency is between 0.1-1.5Hz. Strouhal number of slab is 0.2 and in small space, Strouhal number of eight cylinders is about 0.03.

References

[1]   Zdravkovich M. M. Flow induced oscillations of two interfering circular cylinders. Journal of Sound and Vibration, 1985 (101): 511-521.

[2]   Youheng Xu. Distribution of time average pressure and aerodynamic force of flow around three cylinders. Aerodynamic Experiment and Measurement Control, 1993, 7 (2): 18-25.

[3]   Zhifu Gu. Experimental study of flow around three cylinders. Acta Aerodynamica Sinica, 2000, 18 (4): 441-446.

[4]   K. Lam Force coefficients and Strouhal numbers of four cylinders in cross flow. Journal of Fluids and Structures, 2003 (18): 305-324.

[5]   Norberg C. Fluctuating lift on a circular cylinder: review and new measurements, Journal of Fluids and Structures, 2003: 57-96.

[6]   Alam M. M., Moriya M., Takai K., Sakamoto H.. Fluctuating fluid forces acting on two circular cylinders in a tandem arrangement at a subcritical Reynolds number [J]. Journal of Wind Engineering and Industrial Aerodynamics, 2003: 139-154.

[7]   Mingmin Guo. Simultaneous measurement of surface pressure of two circular cylinders and characteristic of fluctuating aerodynamic forces. Fudan University, Shanghai, 2004.

[8]   Arie M., Kiya M., Moriya M., Mori, H., Pressure Fluctuations on the surface of two circular cylinders in tandem arrangement, Trans. ASME: Journal of Fluids Engineering 1983: 161-167.

Velocity Profiles and Flow Resistances of Open-Channel Flows over Gravel Beds

C. Zeng   C. W. Li

Department of Civil and Structural Engineering, The Hong Kong Polytechnic University, Hong Kong

Abstract: An experimental study of the velocity profiles and flow resistances of uniform open-channel flows over gravel beds under different bed slopes and flow rates is reported. The measured vertical profiles of the mean streamwise velocity follow the modified log law and the vertical profiles of the turbulence intensity follow the exponential function proposed by Nezu and Rodi. The roughness length scale (k_s) is found to increases with the bed slope (S_0). Using the slope dependent k_s, the Darcy resistance function exhibits a log relationship with the relative submergence, and the roughness parameter $n/k_s^{1/6}$ is roughly a constant.

Key words: laboratory experiments; open channel; gravel bed; velocity profile; flow resistance

1. Introduction

Open channels are used to convey water for land drainage, irrigation, flood protection and navigation improvement. To prevent erosion, gravels and boulders are often deployed onto the beds of these channels. An undesirable effect is that the flow capacities of these channels will be reduced due to the increase in resistances. The engineering design of open channels thus requires the knowledge of the resistance induced by gravels on the flow.

Laboratory experiments of free-surface flows in open channels with rough beds have been carried out by several researchers (e.g., Powell, 1950[1]; Bathurst, 1985[2]; Ferro ed al., 1991[3]; Andreas DITTRICH et al., 1997[4]; Ferro, 2003[5]). Field tests were performed in natural channels by Hey[6], Griffiths[7], Bathust[2] and Smart[8]. These previous studies showed that the flow resistance induced by large roughness depends on the form drag of the roughness elements and the disposition of the elements in the channel. The resistance parameters can be expressed as functions of the Reynolds number (which affects the near-bed boundary layer), the Froude number (which affects the free surface drag), the roughness geometry (which affects the effective roughness concentration), and the channel geometry (which affects the relative roughness area). The nature of the roughness effects can be repre-

sented by the size, shape, and spatial distribution of the roughness elements[9]. According to these results, a bed roughness length scale k_s is often chosen to quantify the roughness effects. For non-uniform bed roughness, k_s is generally assumed directly proportional to a characteristic grain diameter:

$$k_s = C_x d_x \qquad (1)$$

where C_x is a constant corresponding to a characteristic grain diameter d_x, and x denotes the percentage of roughness elements with diameter smaller than d_x. Eq. (1) is empirical and different values of C_x and d_x have been proposed (e.g., Einstein et al., 1952[10]; Leopold et al., 1964[11]; Kamphuis, 1974[12]; Hey, 1979[6]; Bray, 1980[13]; Ferro et al., 1991[3]). Most of the proposed equations use values of x which are greter than 50. For example, Leopold et al.[11], Bray[14] and Hey[6] suggested $k_s = 3.5d_{84}$; Gladki[15] suggested $k_s = 2.5d_{80}$; Millar[16] suggested $k_s = d_{50}$ and Whiting and Dietrich[17] suggested $k_s = 3d_{84}$. However, Bray[13,18] claimed that there was no significant difference among the using of d_{50}, d_{84} or d_{90} as the characteristic grain diameter. The large range of the reported values of k_s in field experiments is not suprising as the bed roughness length scale not only depends on the grain roughness, but also on the size of the bed forms (the aggregation of sediment grains), the bed load (rolling or saltating near-bed sediment layer), the lateral velocity gradients and the varying conditions immediately upstream. Smart[8] concluded that the fixed relation between k_s and d_{90} will be meaningful only when the bed is not mobile, the grain rough roughness is dominant, the upstream conditions are uniform, the bed forms are not apparent, and the lateral velocity gradient is close to zero.

Extensive experimental studies of the mean flows and the turbulence structures of open-channel flows have been performed. Nezu and Rodi [19] suggested that in open-channel flows two different flow regions can be identified: the inner (or near-wall) region and the outer (or near-water-surface) region. In the inner region, the logarithmic velocity distribution is valid. This region extends approximately up a distance of $z/\delta \approx 0.20$, where δ = distance from a reference level to the point of the maximum velocity. For the case of rough-bed, the wall region behaves in the same way as the classical inner layer in boundary-layer flows and obeys the law of the wall with a roughness length scale k_s[20] and a velocity scale (shear velocity) as follows:

$$\frac{u}{u^*} = \frac{1}{\kappa} \ln \frac{z}{k_s} + 8.5 \qquad (2)$$

where κ ($= 0.41$) is the universal von Karman constant; u^* is the bed-shear velocity; and z is the distance measured from a reference level (Figure 1). For fully turbulent flows over smooth boundaries, the zero level of the velocity profile, $z = 0$, should be set at the solid boundary. However, for rough boundary flows some researchers[12,21] hold the view that the actual reference level h_0, called the "hypothetical bed" (i.e., the level where the mean velocity is zero) lies between $z = 0$ and $z = k$, where k is the average height of the roughness elements (Figure 1). In the laboratory it is easy to define h_0, while in the field it is easier to define h_1, the displacement of the zero-velocity plane below the top of the roughness elements (Figure 1). Einstein and El-Samni[22] suggested $h_1 = 0.2k_s$. Bayazit[23] reported that h_1 varies from $0.15k$ to $0.35k$. Jackson[24] found that $h_1 = 0.25k$, with h_1 decreasing with the density of the roughness elements.

The outer region is governed by the maximum velocity u_{max}, and the flow depth D (or the distance from the zero-velocity plane to the plane of the maximum velocity δ), and the velocity distribution can be described by the velocity-defect law:

$$\frac{u_{max} - u}{u^*} = -\frac{1}{\kappa} \ln\left(\frac{z}{\delta}\right) \qquad (3)$$

In this region, Cebeci and Smith[25] and Hinze[26] suggested that the velocity-defect law is also valid for flows over rough surfaces. However, Kirkgöz[27] found Eq. (3) is not accurate enough. Zippe and Graf [28] found that the following Coles's wake law is more accurate.

$$\frac{u_{max} - u}{u^*} = -\frac{1}{\kappa} \ln\left(\frac{z}{\delta}\right) + \frac{\Pi}{\kappa} 2\cos^2\left(\frac{\pi z}{2\delta}\right) \qquad (4)$$

where Π ($= 0.55$) is the Coles's wake strength paramter.

The turbulence intensity of open-channel flow over a smooth bed has been measured by many researchers. Nakagawa and Nezu et al. [29] seems to be the first to measure all the three components of turbulence intensities u'_{irms} ($= \sigma_{u_i} = \sqrt{\overline{u'_i u'_i}}$; $u_i = u$, v, w) in open-channel flows by using dual-sensor hot-film anemometers. According to their studies, the vertical distributions of the turbulence intensities in an open-channel flow follow the semi-empirical formula given by:

$$\frac{u'_{irms}}{u_*} = C_{u_i} e^{(-z/H)} \qquad (5)$$

where C_{u_i} is a constant which is different for different velocity components ($C_u = 2.3$, $C_v = 1.63$ and $C_w = 1.27$). This formula has been verified experimentally by velocity measurements using LDVs over a wide range of Reynolds and Froude numbers. To the author's knowledge, Richardson and McQuivey[30] and Blinco and Partheniades[31] could be the pioneer researchers in studying the influence of the boundary roughness on the turbulence intensities in open-channel flows. By analysing the experimental data of Blinco and Partheniades[31], Bayazit[21] suggested a formula as follows:

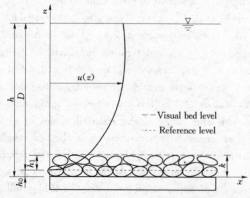

Figure 1   Velocity distribution in rough-boundary flows

$$u'/u^* = A\log(z^+) + B \qquad (6)$$

where $z^+ = u^* z/v$, $v =$ kinematic viscosity of water, $A = -0.39$ and $B = 3.7$. He argued that this formula can be used for the range of $z^+ > 10$, independent of the Reynolds number and the boundary roughness. However, it should be noted that there was a large scatter of the measured data around the regression line determined by Eq. (6). Wang et al. [32] proposed that the distribution of u'/u^* depends greatly on the relative submergence D/k_s, and has nothing to do with the size distribution of the roughness particles. Furthermore, they introduced the following equation which well describes the vertical distribution of u'/u^* for flows over gravel beds with large D/k_s values.

$$u'/u^* = 214\exp(-0.8z/D) \qquad (7)$$

Well-established flow resistance formulas, such as the Darcy-Weisbach, the Manning, and the Chezy equations, have long been used for open-channel flows. The flow resistance is commonly represented by roughness parameters such as the Manning's roughness coefficient (n), the Chezy's resistance factor (C), or the Darcy-Weisbach friction factor (f). These factors are related to the mean flow velocity through the following equations.

$$U = \frac{K_n}{n} R_h^{2/3} S_f^{1/2} \qquad (8)$$

$$U = \sqrt{\frac{8g}{f}} \sqrt{R_h S_f} \qquad (9)$$

$$U = C \sqrt{R_h S_f} \qquad (10)$$

in which $R_h =$ hydraulic radius, $S_f =$ friction slope ($= S_0$, the bed slope in uniform conditions); $g =$ gravitational acceleration; and $K_n = 1$ for U and R_h in SI units, 1.486 for R_h in ft and U in ft/s. From Eq. (8) -Eq. (10), the resistance coefficients can be related by

$$\sqrt{\frac{f}{8}} = \frac{n\sqrt{g}}{R_h^{1/6} K_n} = \frac{\sqrt{g}}{C} = \frac{\sqrt{g R_h S_f}}{U} \qquad (11)$$

Therefore, knowing the value of one resistance coefficient, the values of the other resistance coefficients can be calculated. The friction factor of a gravel-bed river is generally determined by the following logarithmic and power formulas:

$$\left(\frac{8}{f}\right)^{1/2} = C_1 \log\left(\frac{R_h}{d_{xx}}\right) + C_2 \qquad (12)$$

$$\left(\frac{8}{f}\right)^{1/2} = C_3\left(\frac{R_h}{d_{xx}}\right) + C_4 \qquad (13)$$

where $d_{xx} =$ diameter of the bed particles for which xx percent of the particles are finer; and C_1, C_2, C_3, C_4 are empirical constants. Bathurst et al. [2] found that in the large-scale resistance region ($h_m/d_{50} < 2$, h_m is the hydraulic depth) the power formula fits the experimental data best while in the transition region ($2 < h_m/d_{50} < 7.5$) the logarithmic formula is better. Graf[33] subsequently proposed that the friction factor f should depend on the ratio of the hydraulic radius to the size of the bed elements, the Froud number Fr and the shear intensity Ψ. The relationship is given by

$$\sqrt{\frac{8}{f}} = fn\left(\frac{R_h}{k_s}, Fr, \Psi\right) \qquad (14)$$

He assumed R_h/k_s was the most important factor and obtained the following equation by a regression analysis of the experimental data.

$$\sqrt{\frac{8}{f}} = 5.75\log\left(\frac{R_h}{k_s}\right) + A \qquad (15)$$

where $A = 3.25$ when $k_s = d_{50}$, or $A = 4.0$ when $k_s = d_{85}$. The experimental data of Keulegan[34] and Limerinos[35], gave $A = 6.25$ and 1.00 when $k_s = d_{50}$, respectively.

The realtionship between the Manning's n and the relative smoothness R_h/k_s was reported by Sturm[36] as follow:

$$\frac{n}{k_s^{1/6}} = \frac{\dfrac{K_n}{(8g)^{1/2}}\left(\dfrac{R_h}{k_s}\right)^{1/6}}{2.03\log\left(12\dfrac{R_h}{k_s}\right)} \qquad (16)$$

where $K_n = 1.0$ for SI units and 1.49 for English units. The value of $n/k_s^{1/6}$ was found to be roughly constant over a fairly wide range of values of R_h/k_s and therefore is not a function of the flow depth. In the literature different values of $c_n = n/k_s^{1/6}$ were reported. The minimum value is 0.039 in SI units (0.032 in English units), which is close to that reported by Henderson[37] (0.041 in SI units, 0.034 in English units). Several other sources, including Hager[38], reported the value of $c_n = 0.048$ in SI units (0.039 in English units).

This paper is a further contribution to the subject and reports the results of an experimental study of gravel-bed open-channel flows carried out in the Hong

Kong Polytechnic University.

2. Experimental setup

The experiments were conducted in a 12.5 meter long, 0.31 meter wide, and 0.4 meter deep flume which can be tilted with the longitudinal bed slope varying from-0.83% to 2%. The inlet and outlet of the flume are connected to a recirculating system providing a continuous stable flow. The water discharge rate was measured by a built-in flow meter installed in the flow return pipe. The uniform flow condition is obtained by adjusting the tailgate at the outlet of the flume. The flume has transparent glass walls and a steel bottom on which gravels of uniform size were used to roughen the bed surface. Figure 2 shows the set up of the experiments. The thickness of the gravel bed was about 35mm and the median and standard deviation values of the gravels were $d_{50} = (23 \pm 3.2)$ mm. The gravel bed was considered fixed as all these gravels were deployed into 9 plates which were then placed consecutively on the bed of the open channel. In the experiments no motion of the gravels was observed.

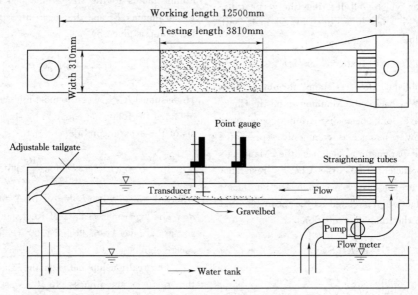

Figure 2　Experimental installation (not to scale)

Other equipment includes point gauges for water surface measurements and an ultra-sound velocity profiler (UVP) for flow velocity and turbulence measurements. Water surface elevations were measured at several points along the channel centreline to ensure uniform flow was maintained. The UVP measured the instantaneous velocity profile along a line. The transducer of the UVP was placed horizontally inside the water body and measured the longitudinal profile of the centerline velocity. The uniform flow condition is reconfirmed by checking the uniformity of the longitudinal velocity profiles. The vertical profiles of the centerline velocity were obtained by measuring the longitudinal velocity profiles at different levels. The vertical measurement interval was 0.5cm. The accuracy of the velocity measurements was checked by additional veloicty measuremnts using an acoustic doppler velocimeter (ADV) for selected flow cases. The two sets of profiles measured by using different instruments are found to be consisent and have high correlation.

The flume experiments were conducted at a range of bed slopes (0.1%–0.4%) for a variety of flows (30–80m³/h). A summary of the experimental conditions is shown in Table 1. In this table Q is flow rate; S_0 is bed slope; D is flow depth measured from the reference level; δ is the distance from the reference level to the point where $u = U_m$ (the maximum velocity); W/D is the aspect ratio; W is the channel width (=0.31m); U is mean (depth averaged) velocity in flume cross section; Fr is Froude number $[=U/\sqrt{(gD)}]$; Re is Reynolds number (= $4DU/\nu$).

Table 1   **Summary of flow parameters**

Run No.	Q (m³/h)	S_0	D (m)	δ (m)	W/D	U (m/s)	U_m (m/s)	Fr	$Re \times 10^5$
1	33.6		0.080	0.068	3.89	0.378	0.456	0.427	1.20
2	44.7		0.095	0.083	3.27	0.423	0.500	0.439	1.60
3	56.1	1/1000	0.110	0.103	2.82	0.458	0.531	0.442	2.01
4	67.9		0.120	0.118	2.59	0.508	0.574	0.469	2.43
5	80.6		0.133	0.113	2.34	0.544	0.587	0.477	2.89
6	33.6		0.075	0.064	4.15	0.404	0.481	0.472	1.20
7	44.7		0.090	0.084	3.46	0.447	0.539	0.477	1.60
8	56.1	1/600	0.101	0.094	3.08	0.500	0.587	0.503	2.01
9	67.9		0.110	0.104	2.83	0.555	0.653	0.535	2.43
10	80.6		0.123	0.114	2.53	0.589	0.663	0.537	2.89
11	33.6		0.072	0.065	4.33	0.420	0.497	0.501	1.20
12	44.7		0.083	0.080	3.75	0.485	0.580	0.538	1.60
13	56.1	1/400	0.095	0.080	3.28	0.531	0.657	0.551	2.01
14	67.9		0.105	0.100	2.96	0.581	0.680	0.574	2.43
15	80.6		0.115	0.110	2.70	0.630	0.727	0.594	2.89
16	33.6		0.071	0.066	4.39	0.426	0.519	0.512	1.20
17	44.7		0.078	0.076	3.99	0.516	0.608	0.591	1.60
18	56.1	1/300	0.089	0.081	3.50	0.567	0.698	0.608	2.01
19	67.9		0.100	0.091	3.11	0.610	0.723	0.617	2.43
20	80.6		0.109	0.101	2.85	0.665	0.762	0.644	2.89
21	33.6		0.068	0.057	4.56	0.443	0.544	0.543	1.20
22	44.7		0.077	0.067	4.03	0.521	0.605	0.600	1.60
23	56.1	1/250	0.087	0.077	3.57	0.578	0.680	0.626	2.01
24	67.9		0.095	0.077	3.27	0.641	0.714	0.664	2.43
25	80.6		0.106	0.097	2.93	0.682	0.778	0.669	2.89

3. Results and discussions

3.1 Determination of shear velocity

Bed-shear velocity u^* is an important parameter in open-channel flows but it is difficult to be measured directly. Generally there are two methods to estimate the parameter. The simpler method is to use the force balance calculation (e.g., Raichlen, 1967[39]; McQuivey et al., 1969[40]; Blinco et al., 1971[31]; Ferro and Baiamonte, 1994[41]):

$$u^* = \sqrt{gR_h S_f} \qquad (17)$$

As the roughness of the glass walls is negligible compared to the roughness of the bed in the present experiments, the hydraulic radius can be set equal to the water depth. Eq. (17) can be rewritten as:

$$u^* = \sqrt{gDS_0} \qquad (18)$$

in which $S_0 =$ bed slope and $D =$ flow depth measured from the reference bed level. The shear velocity evaluated from Eq. (17) or Eq. (18) is a bulk value rather than the local value. Another method to estimate u^* is to fit the measured mean vertical profile of the streamwise velocity against the modified logarithmic law [Eq. (2)]. This method was also adopted by many investigators (e.g., Wang et al., 1993[32]; Chen and Chiew, 2003[42]; Rodríguez and García, 2008[43]). Nezu and Rodi[19] showed that the shear velocities obtained by these two methods have difference less than 5%. In their experiments velocities were measured accurately by a Laser Doppler Anemometer (LDA). However, uncertainty arises when these two methods are applied to open-channel flows over gravel beds, since there are more than one unknown required to be determined: the position of the reference level h_1 (or h_0, see Figure 1) and u^* must be determined simultaneously in the first method, and h_1, k_s and u^* must be determined simultaneously in the second method. To solve the problem, in the present study the reference level is determined by setting $h_1 = 0.20k_s$, which was proposed by Einstein and El-Samni[44] and Bayazit[23].

Using the above empirical relation, h_1, k_s and u^* can be obtained by solving Eq. (2) and Eq. (18) simultaneously. The results show that the values of k_s for the experiments with identical bed slopes were very close. Subsequently, the experiments with the same bed slope were grouped together for analysis and the average value of k_s was determined. The results are listed in Table 2. It can be seen that k_s increases with the bed slope S_0, and the relative submergence (D/k_s) ranged from 2.76 to 36.07.

Table 2   **Parameter estimation results**

Run No.	k_s (m)	D (m)	D/k_s	u^* (cm/s)
$S_0 = 1/1000$				
1		0.080	21.67	2.80
2		0.095	25.74	3.05
3	0.00368 ($=0.16d_{50}$)	0.110	29.82	3.28
4		0.120	32.54	3.43
5		0.133	36.07	3.61
$S_0 = 1/600$				
6		0.075	9.27	3.49
7		0.090	11.13	3.83
8	0.00805 ($=0.35d_{50}$)	0.101	12.50	4.06
9		0.110	13.62	4.23
10		0.123	15.23	4.48
$S_0 = 1/400$				
11		0.072	5.43	4.19
12		0.083	6.26	4.50
13	0.01320 ($=0.57d_{50}$)	0.095	7.17	4.82
14		0.105	7.93	5.07
15		0.115	8.68	5.30
$S_0 = 1/300$				
16		0.071	3.86	4.81
17		0.078	4.24	5.04
18	0.01831 ($=0.80d_{50}$)	0.089	4.84	5.38
19		0.100	5.44	5.71
20		0.109	5.94	5.96
$S_0 = 1/250$				
21		0.068	2.76	5.16
22		0.077	3.13	5.49
23	0.02456 ($=1.07d_{50}$)	0.087	3.54	5.84
24		0.095	3.86	6.10
25		0.106	4.31	6.45

3. 2 Velocity and turbulence intensity distribution

With known values of shear velocities, the dimensionaless streamwise velocity profiles are calculated and plotted in Figure 3. The experimental data fit the modified log law [Eq. (2)] well. The equations [Eq. (3) & Eq. (4)] for the velocity-defect law are plotted in Figure 4. The data fit Eq. (3) reasonably well, but depart from Cole's wake law [Eq. (4)]. The present results indicate that the vertical profiles of the streamwise velocity in open-channel flows with gravel beds follow the velocity-defect law without the wake correction. Similar results are also found in Rodríguez and García[43].

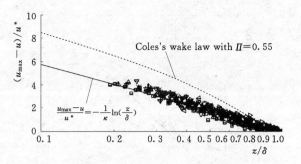

Figure 4   Velocity-defect distributions
[Comparison with Eq. (3) & Eq. (4)]
(See Figure 3 for symbols)

Figure 5 shows the vertical profiles of turbulence intensities u'/u^* for a wide range of relative submergence (D/k_s), together with the empirical equation proposed by Nezu and Rodi[19] and Wang et al. [32]. Although a scattering of the data is observed, the overall agreement between the equation and the data is reasonable. The turbulent intensity profiles for flows over gravel bed and smooth bed thus exhibit not much difference.

The comparison between the streamwise velocities measured by the ADV and UVP for RUN4, RUN9, RUN14, RUN19 and RUN24 are shown in Figure 6. The good agreement indicates that the accuracy of the measuring instruments is acceptable.

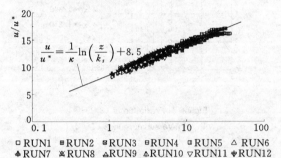

□ RUN1 ⊞ RUN2 ⊠ RUN3 ⊟ RUN4 ⊡ RUN5 △ RUN6
♠ RUN7 ⊠ RUN8 △RUN9 ▲RUN10 ▽RUN11 ▼RUN12
⊠RUN13 ▽RUN14 ▼RUN15 ◁ RUN16 ◀RUN17 ⋈ RUN18
◀RUN19 ◁RUN20 ▷RUN21 ♠RUN22 ⋈ RUN23 ▷RUN24
▷RUN25

Figure 3   Dimensionless streamwise velocity profiles
[Comparison with Eq. (2)]

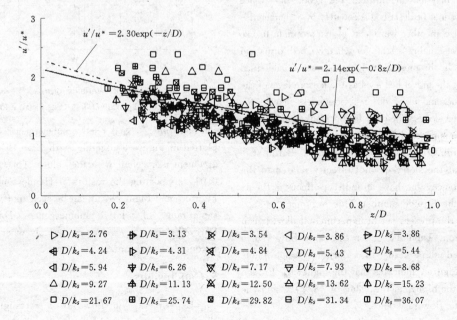

▷ D/k_s=2. 76	⊕ D/k_s=3. 13	⋈ D/k_s=3. 54	◁ D/k_s=3. 86	⊳ D/k_s=3. 86
◀ D/k_s=4. 24	⊳ D/k_s=4. 31	⋈ D/k_s=4. 84	▽ D/k_s=5. 43	◀ D/k_s=5. 44
◁ D/k_s=5. 94	▽ D/k_s=6. 26	⋈ D/k_s=7. 17	▽ D/k_s=7. 93	▼ D/k_s=8. 68
△ D/k_s=9. 27	♠ D/k_s=11. 13	⋈ D/k_s=12. 50	⊿ D/k_s=13. 62	▲ D/k_s=15. 23
□ D/k_s=21. 67	⊞ D/k_s=25. 74	⊠ D/k_s=29. 82	⊟ D/k_s=31. 34	⊡ D/k_s=36. 07

Figure 5   Distributions of turbulence intensities u'/u^*

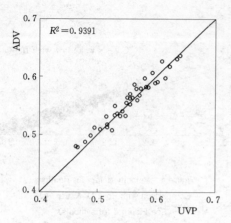

Figure 6　Comparison between streamwise
velocities measured by UVP and ADV

4. Flow resistance analysis

The Darcy friction factor f can be related to the other flow parameters by the following equtaion.

$$\left(\frac{8}{f}\right)^{\frac{1}{2}} = \frac{u}{u^*} = \frac{Q}{A(gDS_0)^{1/2}} \quad (19)$$

The relationship between $(8/f)^{1/2}$ and log (D/k_s) is plotted in Figure 7 together with Eq. (15) ($A = 6.25$). The data fit the equation well and the fitting is better than those obtained by Bathurst[2], Graf[33] and Aberle[45] based on their flume data and field data. Aberle[45] suggested that the discrepancy in the fitting may be due to that the geometric surface structure cannot be described adequately by a characteristic grain size. In other words, it is inappropriate to use one identical roughness scale for whatever bed forms and bed slopes. In the present study, this issue is addressed by defining the gravel-bed level as the bottom level of the open channel and using different roughness lengths for different bed slopes. The good fitting results can be considered as a support of the argument.

Regarding the effect of bed slope, Bathurst[2] argued that the slope was not directly related to the flow resistance, but have an indirect effect via the agency of the Froude number. This argument is supported by the present set of experimental data. It can be seen from Table 1, under the same flow rate a steeper bed slope leads to a larger Froude number. And Figure 7 indicates that the higher the Froude number, the higher is the value of $(8/f)^{1/2}$ and the lower is the flow resistance.

The relationship between the values of the resistance

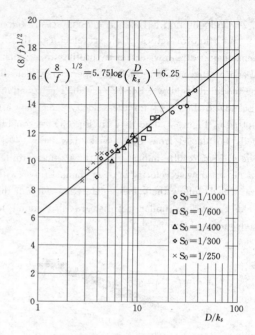

Figure 7　Variation of $(8/f)^{1/2}$
with relative submergence (D/k_s)

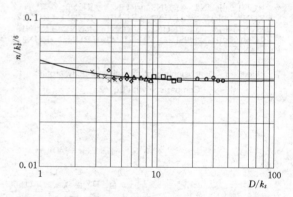

Figure 8　Variation of $n/k_s^{1/6}$
with relative submergence (D/k_s) (See Figure 7 for symbols)

function, $n/k_s^{1/6}$, and relative submergence (D/k_s) is plotted in Figure 8 together with Eq. (16). The agreement is excelent with the values of $n/k_s^{1/6}$ around 0.04, supporting the results of Henderson[37]. From Figure 7 and Figure 8, it can be seen that within the same range of relative submergence (D/k_s), the Darcy-Weisbach's f varies moderately whereas $n/k_s^{1/6}$ is roughly a constant.

5. Conclusions

Experiments on flows in a rectangular open channel with gravel bed have been carried out and the results are consistent with the previously reported results. The bed roughness scale (k_s) is found to increase with the bed slope (S_0). Using the slope dependent k_s, the Darcy resistance function exhibits a log relationship with the relative submergence, and the roughness function $n/k_s^{1/6}$ is roughly a constant.

Acknowledgment

This work is supported by the Research Grant Council of the Hong Kong Special Administrative Region under Grant No. 5221/06E.

References

[1] Powell R. . Resistance to flow in rough channels. *Transactions-American Geophysical Union*, 1950, 31: 572-582.

[2] Bathust J. . Flow resistance estimation in mountain rivers. *Journal of Hydraulic Engineering*, 1985, 111 (4): 625-643.

[3] Ferro V. , Giordano, G. . Experimental study of flow resistance in gravel-bed rivers. *Journal of Hydraulic Engineering*, 1991, 117 (10): 1239-1246.

[4] Andreas DITTRICH, Katinka KOLL. Velocity field and resistance of flow over rough surfaces with large and small relative submergence. International Journal of Sediment Research, 1997, 12 (3): 21-33.

[5] Ferro V. . Flow resistance in gravel-bed channels with large-scale roughness. *Earth Surface Processes and Landforms*, 2003, 28: 1325-1339.

[6] Hey R. . Flow resistance in gravel-bed rivers. *Journal of the Hydraulics Division*, 1979, 105 (4): 365-379.

[7] Griffiths G. . Flow resistance in coarse gravel bed rivers. *Journal of the Hydraulics Division*, 1981, 107: 899-918.

[8] Smart G. . Turbulent velocity profiles and boundary shear in gravel bed rivers. Journal of Hydraulic engineering, 1999, 125 (2): 106-116.

[9] Rouse H. . Critical analysis of open-channel resistance. *Journal of the Hydraulics Division*, 1965, 91 (HY4): 1-25.

[10] Einstein H. A. , Barbarossa N. L. . River channel roughness. *Transactions of the American Society of Civil Engineers*, 1952, 117: 1121-1146.

[11] Leopold L. B. , Wolman M. G. , Miller J. P. . *Fluvial processes in geomorphology*, San Francisco, W. H. Freeman and Co. , 1964.

[12] Kamphuis J. . Determination of sand roughness for fixed beds. *Journal of Hydraulic research*, 1974, 12 (2): 193-203.

[13] Bray D. . Evaluation of effective boundary roughness for gravel-bed rivers. *Canadian Journal of Civil Engineering*, 1980, 7 (2): 392-397.

[14] Bray D. . Estimating average velocity in gravel-bed rivers. *Journal of the Hydraulics Division*, 1979, 105 (9): 1103-1122.

[15] Gladki H. . Resistance to flow in alluvial channels with coarse bed materials. *Journal of Hydraulic Research*, 1979, 17 (2): 121-128.

[16] Millar R. . Grain and form resistance in gravel-bed rivers. *Journal of Hydraulic Research*, 1999, 37 (3): 303-312.

[17] Whiting P. J. , Dietrich W. E. . Boundary shear stress and roughness over mobile alluvial beds. *Journal of Hydraulic Engineering*, 1990, 116 (12): 1495-1511.

[18] Bray D. . Flow resistance in gravel-bed rivers. in R. B. Hey, *Gravel-bed Rivers*, Chichester, England, John Wiley and Sons, 1982: 109-133.

[19] Nezu I. , Rodi W. . Open-channel flow measurements with a laser doppler anemometer. *Journal of Hydraulic Engineering*, 1986, 112 (5): 335-355.

[20] Nezu I. , Nakagawa H. . *Turbulence in Open-Channel Flows*, A. A. Balkema, Netherlands, 1993.

[21] Bayazit M. . Free surface flow in a channel of large relative roughness. *Journal of Hydraulic Research*, 1976, 14 (2): 115-126.

[22] Einstein H. A. , El-Samni E. A. . Hydrodynamic force on a rough wall. *Reviews of Modern*

Physics, 1949, 21 (3): 520-524.

[23] Bayazit M.. Flow resistance and sediment transport mechanics in steep channels. in B. M. Sumer & A. Müller (Ed.), *Mechanics of Sediment Transport, Proceedings of EU-RO-MECH 156 Colloquium*, A. A. Balkema, Netherlands, 1983.

[24] Jackson P. S.. On the displacement height in the logarithmic velocity profile. *Journal of Fluid Mechanics*, 1981, 111: 15-25.

[25] Cebeci T., Smith A. M. O, *Analysis of turbulent boundary layer*, Academic Press, New York, 1974.

[26] Hinze J. O.. *Turbulence.* McGraw-Hill, New York, 1975.

[27] Kirkgöz M. S.. Turbulent velocity profiles for smooth and rough open channel. *Journal of Hydraulic Engineering*, 1989, 115 (11): 1543-1561.

[28] Zippe H. J., Graf W. H.. Turbulent boundary-layer flow over permeable and non-permeable rough surfaces. *Journal of Hydraulic Research*, 1983, 21 (1): 51-65.

[29] Nakagawa H., Nezu I., Ueda H.. Turbulence of open channel flow over smooth and rough beds. *Proceedings, Japan Society of Civil Engrs*, 1975, 241: 155-168.

[30] Richardson E. V., McQuivey R. S.. Measurement of turbulence in water. *Journal of the Hydraulics Division*, 1968, 94 (HY2): 411-430.

[31] Blinco P. H., Partheniades E.. Turbulence characteristics in free surface flows over smooth and rough boundaries. *Journal of Hydraulic Research*, 1971, 9 (1): 43-71.

[32] Wang J., Dong Z., Chen, C., Xia Z.. The effects of bed roughness on the distribution of turbulent intensities in open-channel flow. *Journal of Hydraulic Research*, 1993, 31 (1): 89-98.

[33] Graf W.. Discussion of "Flow resistance estimation in mountain rivers", *Journal of Hydraulic Engineering*, 1987, 113 (6): 819-822.

[34] Keulegan G.. Laws of turbulent flow in open channels. Journal of Research of the National Bureau of Standards, 1938, 21: 707-741.

[35] Limerinos J.. *Determination of the Manning coefficient from measured bed roughness in natural channels*, United States Geological Survey, Washington, D. C., 1970.

[36] Sturm T. W.. *Open Channel Hydraulics*, McGraw-Hill, New York, 2010.

[37] Henderson F. M.. *Open Channel Flow*, The Macmillan Co., New York, 1966.

[38] Hager W. H.. *Wastewater Hydraulics*, Springer Verlag, Berlin, 1999.

[39] Raichlen F.. Some turbulence measurements in water. *Journal of Engineering Mechanics*, 1967, 93 (EM-2): 73-97.

[40] McQuivey R. S., Richardson E. V.. Some turbulence measurement in open-channel flow. *Journal of the Hydraulics Division*, 1969, 95 (HY-1): 209-223.

[41] Ferro V., Baiamonte G.. Flow velocity profiles in gravel-bed rivers. *Journal of Hydraulic Engineering*, 1994, 120 (1): 60-80.

[42] Chen X., Chiew Y. M.. Response of velocity and turbulence to sudden change of bed roughness in open-channel flow. *Journal of Hydraulic Engineering*, 2003, 129 (1): 35-43.

[43] Rodríguez, José F., García Marcelo H.. Laboratory measurements of 3-D flow patterns and turbulence in straight open channel with rough bed. *Journal of Hydraulic Research*, 2008, 46 (4): 454-465.

[44] Einstein H. A., El-Samni E. A.. Hydrodynamic force on a rough wall. *Reviews of Modern Physics*, 1949, 21 (3): 520-524.

[45] Aberle J., Smart G. M.. The influence of roughness structure on flow resistance on steep slopes. *Journal of Hydraulic Research*, 2003, 41 (3): 259-260.

Theory and Practice of Thermo-Hydraulic Modeling

Chen Huiquan

China Institute of Water Resources and Hydropower
Research (IWHR), Beijing, 100048

Abstract: The paper presents a historical resume of the IWHR's study on physical modeling of circulated heated water effluened from thermal/nuclear power plants. General features of thermo-hydraulic simulation of such cooling water circulation and treatment of the related scaling contradictions are briefly described. More complicated modeling with consideration of simulating windy, icing, waving, saltwater intrusting environment of the concerned water region are demonstrated in some detail. Emphasis is put on the understanding of some planning philosophy and its practical application. The methody to approach the problems encountered in modeling is illustrated by the relevant case study.

Key words: physical modeling; cooling water; similarity

1. Introduction

Governed by the second law of thermodynamics, the real efficiency of thermal/nuclear power plants is very low. Majority of the thermal energy generated is dissipated to the surrounding environment through cooling water, resulting temperature increment of the water region. Understanding the thermo-hydraulic behavior of the cooling water circulation in its receiving water body is therefore crucial for the optimization of planning the cooling water system and correct assessment of the thermal environmental impact.

The study on cooling water by hydraulic modeling in China started in late 1950s. [1] For the first time, heated water was used as test medium in the model to simulate the cooling water circulation in a closed boundary water region. Model law with consideration of simulating cooling process at water surface and heat balance was established[2]. Modeling of an existing cooling pond and the relevant field observations were accomplished in the same year[3]. Good agreement of the data obtained from the two brought forth more conviction for its application. Since then, more than 200 thermo-hydraulic models using heated water as test medium and designed accordingly have been performed, mostly in IWHR. Meanwhile, extensive studies have been undertaken in simulation technique, both in theory and in its methodology of practical application. The new exploration includes the modeling of wind impact on water circulation and sim-

ulation of icing process in the heated effluent receiving water region. The paper presents a brief resume of the work with main points on ways to approach the problem encountered in model scaling, and verification of model prediction.

2. General features of thermo-hydraulic modeling

The main difference between the thermo-hydraulic modeling and the conventional hydraulic modeling is that the variation of the water temperature should be studied with water having temperature difference as test medium in the experiment. The additional similar requirements of heat balance and heat exchange at the free surface should be satisfied. The latter is directly related to the meteorological condition above the water surface. The similarity criteria could be summarized as:

Geometric requirement: $l_r = z_r = 1$

Kinematic and dynamic requirement:

$$(Fr)_r = (F_\Delta)_r = (Re)_r = 1$$

Heat balance requirement: $(\Phi/\Phi_0)_r = 1$

Where, Fr, F_Δ, Re are respectively Froude Number, Densimetri Froude Number and Reynolds Number, Φ, Φ_0 are heat dissipated to air through free water surface and heat discharged into the water region; subscript r denotes the ratio between prototype (p) and model (m); l_r is the length ratio, $l_r = l_p/l_m$, z_r is the depth ratio.

Fulfillment of all the above is not possible; three

synthetic parameters were thus induced to relax the situation.

(1) Critical effluent discharge Q_{cr}. It has been found through a series tests with a variety of boundary conditions that the general flow patten of a closed flow region remains nearly the same when the recirculating discharge is greater than the critical discharge. i. e. $Q_{min} > Q_{cr}$ in the model could be used instead of Reynolds' similarity criterion[4].

(2) Natural water temperature T_∞. The temperature somewhat representing the onsite meteorological condition including the strength of solar radiation.

(3) Water surface heat dissipation coefficient K. The parameter was first defined by IWHR in 1954 as the rate of the total heat dissipation per unit free surface area for unit temperature difference through convection, evaporation and radiation. It is now world-wild used.

By using these parameters, the similarity criterion above mentioned could be turned to

$$(F)_r = 1 \qquad (a)$$
$$(F_\Delta)_r = 1 \qquad (b)$$
$$Q_m > Q_{cr} \qquad (c) \qquad (1)$$
$$(Kl^2/Q)_r = 1 \qquad (d)$$
$$(T - T_\infty)_r / (T_1 - T_\infty)_r = 1 \qquad (e)$$

T_1 is the effluent temperature, with $T_1 = T_2 + \Delta T_{1-2}$, T_2 is the intake water temperature and ΔT_{1-2} is a given value governed by the power plant.

Since the length dimension of the water region investigated is usually in higher order than the depth dimension, geometrical distorted model is more often to be adopted accompanied with relevant scale effects.

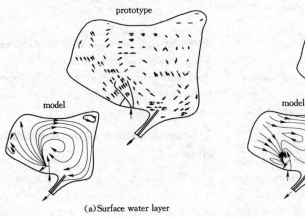

(a) Surface water layer   (b) Bottom water layer

Figure 1   Comparison of flow patterns in Taiyuan Cooling Pond

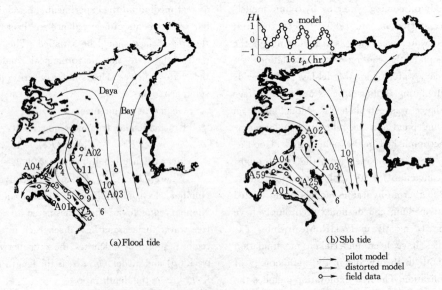

(a) Flood tide   (b) Sbb tide

→ pilot model
•→ distorted model
○→ field data

Figure 2   Field verification of the tidal flow in Daya Bays

A lot of thermo-hydraulic modeling has been conducted and not a few field investigations have been followed. There has been always a good agreement of the general flow patterns obtained from the model and from the field measurement. Figure 1 shows such comparison for Taiyuan Cooling Pond with water surface about 5km² and water depth of 4-5m.

Comparison of the flow figures of Daiya Bay at flood and ebb tidal range was shown in Figure 2. It is the very picture revealing the real tidal flow behavior of the water region nearby the planning Daya Bay Nuclear Power Plant, leading to an ideal scheme of cooling water system obtained from the thermo-hydraulic modeling with good field verification[5]. The cooling water system has been satisfactorily operated for more than 15 years.

For the verification of temperature distributions, due to its high time-dependence, it is a hard task to get the time-dependent data of the surface water temperature at free surface and that of the relevant heat content in the receiving water body. A detailed model and field observation of the Cangxian Cooling Pond was an exceptional example[6].

The pond was only 9,000m² in water surface area with an average water depth of 2.8m. Such small dimensions made it possible for getting a complete set of dynamic synchronous information of the heat content variation in the pond as well as the amount of heat exchange across the water surface by rather simple measuring device. A total of 56 hours of continuous monitoring of about a thousand measuring points were conducted and the 3-D temperature patterns were obtained. The average weather condition of this period was used in the design of the model.

Comparisons of the temperature drop curves of the water surface and the variation of heat content of the pond gained from model and field measurement are shown in Figure 3, Figure 4. The comparisons reveal their pronounced conformity.

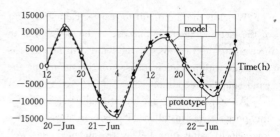

Figure 4　Time variation of heat content in Cangxian Cooling Reservoir

3. Thermo-hydraulic modeling with consideration of simulating windy environment

3.1　Modeling requirement and its implementation

The wind impact on the thermo-hydraulic characteristics of the water region mainly involves in two folds: forming wind-induced water circulation or wind drift and enhancing heat exchange at the water surface. For the former, neglecting the wind inception velocity, velocity of wind drift $w = $ constant $\times V$ which leads to its similarity criterion:

$$w_r = V_r \qquad (2)$$

For the latter, as the coefficient of heat dissipation at free surface is mainly a function of wind velocity, $K = \psi(w)$, the modeling requirement of Eq. (1d) turns to

$$\psi_r(w) = K_r = Q_r / l_r^2 = V_r z_r / l_r$$

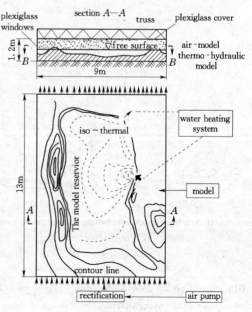

Figure. 5　Experimental Setup

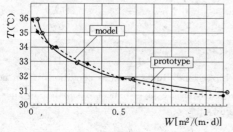

Figure 3　Temperature-drop curve in Cangxian Cooling Reservoir

Which is more difficult to realize. Taking account of the fact that the main purpose of the modeling is to investigate the response of the wind drifting, this is put aside with attention focused on the simulating the wind boundary condition.

An experimental set up specially designed for this purpose was installed. Models were operated in a low-velocity wind tunnel. The test chamber was well fabricated with both sides and ceiling framed with plexiglas plates. All the velocity and temperature measurements for air and water were carried out by arrays of probes inserted through the holes reserved on the ceiling. As demonstrated in Figure 5, it was really an experimental facility for operating a combined model of two, a conventional undistorted air model above the water surface and a conventional distorted thermo-hydraulic model below the water surface. Two models were coupled with conjugation of mass and heat transfer at the air-water interface[7].

3. 2　Field verification

Several model investigations were accomplished in this apparatus with success. One of which was the study of Zhangze Cooling Reservoir. It had been found a significant temperature rise of the intake water during some windy days in summer, leading to a bad operating condition for the power plant. The case was carefully studied with both wind velocity and wind direction simulated.

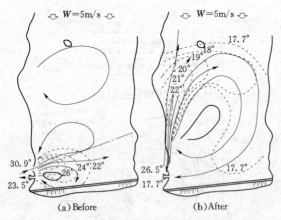

(a) Before　　　　　(b) After

Figure 6　Flowing and temperature pattern before and after the improvement

There came the conclusion that the irregular temperature rise was due to the short circuit of heated effluent induced by unfavorable prevailing wind. Optimization of the intake orientation was made through model test and adopted. The improved cooling water

system has been working well round the year. The fact indirectly reflects the reliability of the modeling. The flow pattern and temperature distribution of the upper layer of the reservoir before and after the improvement are shown in Figure 6.

A direct verification has been performed for the Dou He Cooling Reservoir. Both model-prototype comparisons under no wind and under significant wind condition were made. The model test was conducted in the above-motioned experimental facility with air velocity modulated as Eq. (2). Relevant field data under nearly the same air environment were obtained through thermal scanning from air flight. There was a good qualititive agreement between model and field data with the overall pattern and direction of the thermal plume being rather closely similar. Considering the variation of the time-dependent meteorological and ambient water temperature conditions, a closer comparison of the data seems unwarranted.

4. Thermo-hydraulic modeling with consideration of simulating of icing environment

4. 1　Modeling requirements and its simplification

In the northern China, some cooling water regions are found to be partially frozen in cold weather with the possibility to ice-block the intake. Big difference may occur to the cooling water circulation and to the environmental impact in icy condition. Modeling of cooling water circulation with simulation of the possible icing phenomena have thus been exploited early in 1980s[8].

A general feature of a closed-boundary heated cooling water circulation under icing environment is sketched in Figure 7 with a lot of parameters whose definition self-shown in Figure 7.

If the model meteorological environment could be kept basically the same as that of prototype, in line with the kinematic and dynamic simulation criteria, of Eq. (1), the modeling requirements could be expressed by

$$
\begin{aligned}
z_r &= l_r^{2/3} &\text{(a)} \\
V_r &= z_r^{1/2} &\text{(b)} \\
t_r &= z_r &\text{(c)} \\
(T' - T^*)_r &= z_r &\text{(d)}
\end{aligned}
\qquad (3)
$$

In A_2 zone, under balance condition, from heat conveyed from ice layer = heat dissipated to the air it follows

$$(T' - T^*)_r = \left(\frac{h}{\beta + h}\right)_r (T' - T_0)_r$$

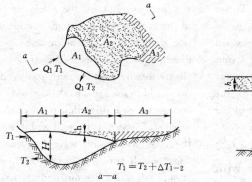

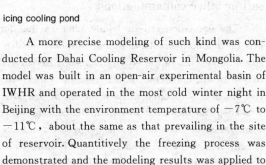

Figure 7   Sketch of an icing cooling pond

where $\beta = k'/K' =$ coefficient of heat transfer in ice layer/heat dissipation coefficient at ice surface.

With $(T'-T_0)_r = 1$, the above formula could finally be deduced to

$$\left(\frac{h}{\beta+h}\right)_r = 1$$

the condition could only be existed in A_1 ice-free zone.

The simple derivation indicated the fact that for simulating the freezing or melting process, a time scale different from that of water flowing should be used.

Hence another synthetic parameter t' has been induced to meet the heat transfer requirements. The relation between t'_r and t_r can be obtained from heat balance in the freezing or melting process,

$$\eta = t'_r/t_r = (\beta+h)_r \qquad (4)$$

where h is thickness of the ice layer. In zone A_1 with no ice, $h_p = h_m = 0$, $\eta = 1$, no two time scale exists. For big ice thickness, $\eta = z_r$ is the another extreme condition. In that case, however, the heat dissipated through the ice cover would share even smaller percentage of the heat load adding to relevant water body through cooling water circulation.

4.2   Practical application

Eq. (3) and Eq. (4) are the simplified modeling requirements when the model is operated under somewhat the similar air temperature environment. Some modeling has been accomplished accordingly. Taking the study of 211 Cooling Reservoir as an example, the model was built in a large warehouse located near the tested reservoir in northeastern China. It was operated with the heating system of the warehouse closed and its windows opened to meet the similarity requirement of its air environment. The tested results have been well used in the planning of the intake-outlet structure.

A more precise modeling of such kind was conducted for Dahai Cooling Reservoir in Mongolia. The model was built in an open-air experimental basin of IWHR and operated in the most cold winter night in Beijing with the environment temperature of $-7℃$ to $-11℃$, about the same as that prevailing in the site of reservoir. Quantitively the freezing process was demonstrated and the modeling results was applied to the optiminization of the cooling-water system[9].

4.3   Approach of the modeling under non-negative air-temperature environment

A simplified approach to predict the ice boundary of the water region with no reproduction of the icing process in the model has been developed[10]. The similarity criteria are basically the same as that for modeling the ordinary cooling water circulation. The only difference is that the attention should be put on the adjustment of the ambient water temperature T_n with

and
$$(T_n-T_0)_r = 1$$
$$(T_1-T_n)_r = 1 \qquad (5)$$

The model thus designed could be operated under room temperature with the water temperature data from the model transferred back to the prototype by

$$(T-T_n)_r = 1 \qquad (6)$$

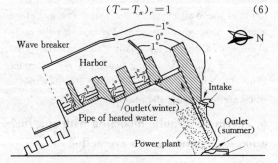

Figure 8   Yingkou Power Plant and its adjacent harbor

The method has been used for Yingkou Power Plant. It was planned to alleviate the freezing problem of the adjacent harbor by inducing in the heated cooling water discharged from the plant, the layout was well studied in a thermo-hydraulic model operated under above criteria. As shown in Figure 8, the tested results proved the expected consequence of the fruitful use of the large amount of waste heat.

5. Thermo-hydraulic modeling with consideration of simulating of ambient water with salt or other contaminations

The simulation criteria is the same as that for general thermo-hydraulic modeling with only the density difference $\Delta\rho$ in the criteria increased $\Delta\rho$ induced by salt or other material contented in the water body concerned. It is required to have the $\Delta\rho$ of the latter simulated and measured free from that due to water temperature difference. The tested results thus obtained reveal truly the behavior of the cooling water circulation with the real features of the movement of the salt or other contaminated water.

Some modeling have been accomplished in this way. A typical case was the study on cooling water circulation of Shajiao Power Plants at Pearl River Mouth. The hydrological and topographical environment there are very complicate. Model study unveiled that the heated effluent flowing below the upper layer of the receiving water region which is covered with fresh water in some tidal period. The phenomena with rather complex temperature as well as salinity distribution along the water depth have been well verified in the field investigation followed.

According to the same guideline, colored Rodamin D with density difference simulated has been used as the test medium for the waste water effluent in the thermo-hydraulic model of cooling water circulation of the Waigaoqiao Power Plant. The model was specially planned and operated for exposing the mutual impact of the water quality between the power plant and the neighboring waste water treatment plant. The prediction of the modeling afforded a scientific base for solving the problem.

6. Thermo-hydraulic modeling with consideration of simulating of wave action

The control factor governing the cooling water circulation in tidal water region is the tide action but not the wave. But for the sea-facing intake structure of the cooling water system, simulation of wave action should sometimes be involved in the thermo-hydraulic modeling. Since geometrically distorted models are usually used for such study, it is problematical to add in the same model the wave simulation with non-geometrical distortion required. A compromise of wave scaling is thus induced with the same z_r and t_r to the wave height and wave frequency but with the wave length scaling distorted. The reliability of this scaling relaxation lies on the fact that wave action is mainly manifested in the wave height and wave frequency.

Such simulation compromise has been applied successfully in the study of cooling water problem arise from Lamma Power Station. The station with installed generated capacity of 3300MW was built on the Lamma Island. The cooling water is drawn from the water region at the corner of the sea wall and the wall of an ash-lagoon. The intake is located underneath the sloped sea wall with very limited submergence during lower tidal level. To investigate the possibility of the air entrance to the intake which is strictly prohibited, wave movement was simulated in the thermo-hydraulic model with distortion ratio $\varepsilon = l_r/z_r = 4$. A notified phenomenon so called "water screen" was discovered. It was found in the period of wave trough of lower tidal level, even though the upper part of the intake has emerged from the water, the water climbing up from the preceding wave draw back downward alone the slope of the see wall, forming a "water screen" before the intake thus effectively prevent the air entrance. The "water screen" physical phenomena were verified separately in an undistorted model separately arranged for wave modeling. This phenomenon was then becoming an important factor for adopting the ideal intake layout.

The intake has been operated with satisfaction since its completion.

Figure 9 is an air-scope of the Lamma Power Station, Hong Kong. The day and night continual infrared scanning to the large intake-outlet water area was carried out on the platform at the top of a chimney 210m above the water level accompanied with a large scale boat survey of 3-dimensional temperature measurement for the same water region. Within the precision range of measuring, no response of wave action to the surface and intake water temperature was found yet.

Figure 9　Lamma Power Station, Hong Kong

7. Thermo-hydraulic modeling with consideration of sedimentation

(1) Study on sedimentation problem are occasionally implemented in the same model for investigating cooling water circulation. The model should usually be designed with additional similitude criteria of $(w/V)_r = 1$ and $(V_{cr}/V)_r = 1$, where w and V_{cr} denote settling velocity and inceptional velocity of the sand particle.

A thermo-hydraulic modeling of such kind for Waigaoqiao Power Plant has been exercised in success for optimizing the intake-outlet layout with good solution of its silting problem at the intake.

In order to make sure to prevent silt setting in front of the intake in alluvial water region, an array of air-nozzles was arranged at the bottom of the intake. The arrangement has been studied in other models specially devised.

(2) Modeling the 3-phases movement. The modeling is characterized by inducing air bubble in the test medium.

As sketched in Figure 10, pressured air injected into the ambient water turns to air bubbles mixing with the surrounding water and silt and forms rising plume accompanied with the relevant entrainment and dispersion. Such complicated physical phenomena could be finally simplified and described by

$$(Fr, w/V, V_{cr}/V, Re_b, F_b, We) = 0 \qquad (7)$$

or　　　$(Fr, w/V, V_{cr}//V, Re_b, Bo, We) = 0$

where:

$$Re_b = \frac{V_o d_s^3}{\nu}$$

$$F_{\Delta b} = \frac{V_o}{\sqrt{\dfrac{\Delta \rho_b}{\rho} g d_o}}$$

$$We_b = \frac{V_o}{\sqrt{\dfrac{\sigma}{\rho d_o}}}$$

Bo is Bond number and defined as the ratio of the buoyancy and tension applied the air bubbles.

$$Bo = \frac{g \Delta \rho_b d_b^3}{\sigma d_b} \sim \left(\frac{We_b}{F_{\Delta b}}\right)^2$$

σ is surface tension, subscript o, b denote bottom nozzle bubble respectively.

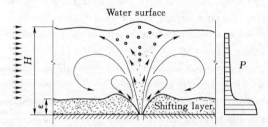

Figure 10　Sketch of Bubble Plume for Releasing Siltation

The simulation requires that the values of each non-dimension term in the above equation should be the same in model and in prototype. It is impossible in practice. However, taking into account the following facts:

(1) the depth and space significantly influenced by the 3-phased plume shares only in a lower order in comparison with that of the whole intake water region;

(2) if there is an environmental flow, the velocity directly effected the behavior of the 3-phased plume is its bottom velocity which also share a lower order in comparison with mean velocity of the coming flow.

The simulation could be approached thereupon by using an undistorted model with length ratio $l_r \approx 1$; taking the bottom velocity of the coming flow to be simulated.

Modeling for the Waigaoqiao Power Plant has been performed in this way with the expected results obtained[11].

8. Conclusions

(1) The basic feature of the thermo-hydraulic modeling is the simulation of heat exchange in the concerned water body and at its free surface. Addition of simulating the nature phenomena such as wind-

drifting, icing ··· in model brings forth much more difficulties in scaling. Such rather complicated modeling has yet been carried out with success is mainly due to (a) correctly focusing the most important feature to be simulated with proper relaxing or neglecting the simulation of others, (b) exploiting some new synthetic parameter and new experimental set up to meet the need of simulation, and (c) making great effort in relevant field investigation.

(2) Emphasis should be put on the field verification. The quality of the modeling lies on the reliability and adoptability to what the model mainly simulated. The effectiveness of the model prediction boils down to the degree of conformity of the model-prototype comparison.

For various kinds of thermo-hydraulic modeling, a variety of model scaling including the geometric-distorted ratio could be chosen. The choice should follow the principle: to have the main concerned feature possibly-best simulated and to make the modeling possibly-feasible in practice under the lab condition provided. The wise solution should be based on the correct evaluation of the related factors as well as the coordination and harmony of these factors.

(3) In line with the rapid development of mathematical modeling, the physical modeling still has its own undisplacible merits and vitality which lie on the fact that not only some complex physical phenomena couldn't be easily described accurately by numerical approach yet; physical modeling itself provides a vast room to be exploited. For planning and designing the intake-outlet system of cooling water circulation, it is favorable to investigate the system by thermo-hydraulic modeling supplemented sometimes with proper mathematical simulation.

References

[1] Chen Huiquan, Yue Juntang. Model Study on Qinghe Cooling Reservoir [R] . IWHR, 1959. (in Chinese)

[2] Chen Huiquan. A Method of Simulation for the Flow in Cooling Reservoir [J]. Proceedings, SCIENTIA Sinica, 1965, 14 (12) .

[3] Chen Huiquan. Modeling of Taiyuan Cooling Reservoir and its Field verification [R]. IWHR, 1959. (in Chinese)

[4] Chen Huiquan, Xu Yulin. Critical Discharge for Flow Simulation [R] . IWHR, 1960. (in Chinese)

[5] Yue Juntang, Lu Yuehui, Lin Youjin. Study on the Planning of Intake-Outlet Works of Daya Bay Nuclear Power Plant [R] . IWHR, Guangdong Hydraulic Research Institute, 1984. (in Chinese)

[6] Chen Huiquan, Yue Juntang. Study on the Cangxian Cooling Resrevoir [C]. Selected Paper of IWHR, no. 3, 1963. (in Chinese)

[7] Chen Huiquan, He Yiying. Theory and Practice of Thermo-hydraulic Modeling with Consideration of Wind Induced Curreny [A]. International Symposium on Hydraulic Reseaarch in Nature and Laboratory [C]. Wuhan, 1992.

[8] Chen Huiquan. Thermo-hydraulic Modeling of Cooling Water Circulation with Icing Simulation [A], Proceedings, IAHR Ice Symposium [C]. Beijing, 1996.

[9] Ji Ping. Model Research of Dahai Cooling Reservoir with Consideration of Freezing Simulation [R]. IWHR, 2003. (in Chinese)

[10] Chen Huiquan. Modeling of Cooling Water Circulation with Icing Simulation under Non-negative Ambient Environment [C] . Selected Paper of IAHR, No. 33, 1990. (in Chinese)

[11] Chen Huiquan. Similarity of the motion of releasing siltation by air-jet [R] . IWHR, 2001. (in Chinese)

Physical Model Test on Cooling Water with PIV and Thermal Infrared Imager System

Wang Yong   Xu Shikai   Tong Zhongshan   Ruan Shiping

State Key Laboratory of Hydrology-Water Resources and Hydraulic Engineering,
Nanjing Hydraulic Research Institute, Nanjing, 210029

Abstract: The impact on flow field and temperature field of water area near discharge outlet usually is forecasted by physical model for studying mixing regularity and diffusing regularity of cooling water which is discharged by heat power plant and nuclear power plant. But people gained little quantitative results because of testing method's limitation. Based on the previous studies, the research in this paper is done on physical model test of cooling water with PIV (Particle Image Velocimetry) and thermal infrared imager system. Based on similarity theory, four variability conditions such as normality, variability two, variability three and variability four are defined. The main work in the paper is summarized as follows: (1) Planar flow field in different variability conditions is measured by PIV; and the data of flow field is compared and analyzed in this article. (2) Surface temperature field is measured by thermal infrared imager system, and the data of surface temperature is compared and analyzed in this article.

Key words: Physical model; cooling water; PIV; Thermal Infrared Imager system

1. Introduction

At present, such as current cooling technique, cooling tower technique and wind cooling technique are major cooling-down method in many heat-engine plant and nuclear power plant. The experimental investigation about some hydraulics problems on cooling water discharging in current cooling technique is studied in some articles. The impact on flow field and temperature field of water area near discharge outlet usually is forecasted by physical model for studying mixing regularity and diffusing regularity of cooling water which is discharged by heat power plant and nuclear power plant. But people gained little quantitative results because of testing method's limitation. Based on the previous studies, the research in this paper is done on physical model test of physical model with PIV (Particle Image Velocimetry) and thermal infrared imager system[1].

2. Laboratory methods and principle

PIV technique is essentially a kind of image analysis techniques. It uses two pulsed lights sources between which there is a very short interval to illuminate the flow field. At the same time by CCD (couple charge of device) the tracer medium (tracking particles) in the flow field are recorded. The tracking particles is hollow glass ball, whose particle diameter is 8-12μm and density is 1. And then with computer to deal with image information of velocity field can be obtained. With PIV tens of thousands of points of the velocity vector can be obtained at the same time. PIV system consists of four parts, including: laser, CCD, synchronization control system, image acquisition and vector calculations. The relationship of each component connected is shown in Figure 1.

The whole PIV system is sequential control by the synchronization control system. Synchronizer ordinally trigger laser pulses in certain time, and through a group of lens pulse beam of light get to the position measured. When the laser illuminating flow field, CCD place into working condition and get the tracking particles distribution of flow field. The images obtained pass through the computer interface board to system memory. The distribution of velocity vector is calculated by software.

The thermal infrared imager system is one kind of precision instrument which use infrared scanning principle to survey object's surface temperature. Any object whose temperature is higher than the absolute zero can send out the infrared radiation. The thermal imaging system uses infrared probe to measure infra-

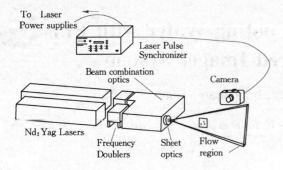

Figure 1　The component of PIV system

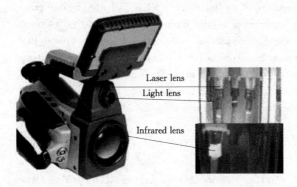

Figure 2　Thermal infrared imager system

red yield of radiation of object, to obtain the object's surface thermal imagery chart. The relations between the object's infrared radiation power with itself temperature is determined by Si Dilao-Boltzmann:

$$E = \varepsilon \delta T^4 \qquad (1)$$

Where E is infrared radiation power launched by objects, ε is emissivity, δ is Boltzmann constant, T is absolute temperature.

The infrared imaging system is shown in Figure 2. Thermal infrared imager system is composed of following part: probe, processor, display, frame record instrument and video record instrument. The probe is used to receive the infrared radiation energy of every target outside part, which is transformed by the signal treatment of the processor, and then the object surface's every part temperature is calculated. Two dimension hot image of the object surface's temperature distribution can be showed from the display.

Temperature range of infrared imaging system used: $-40\,℃$ to $+500\,℃$ (or $-40\,℉$ to $+932\,℉$), accuracy $\pm 2\,℃$. The infrared imaging system can generate high-resolution 14-bit thermal images, and shoot a color image of 640×480 pixels.

3. Test processing and results

In this paper, the coordinate system as follows: x-along the flow direction (ambient flow direction), along the flow direction is positive; y-along the horizontal tank (jet direction); z-vertical, up is positive; discharge outlet center is coordinates origin. Tester system diagram is shown in Figure 3.

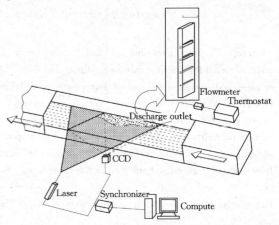

Figure 3　Tester system diagram

Based on similarity theory, four variability conditions such as normality, variability two, variability three and variability four are defined. The prototype corresponding to Model test is a rectangular river with 20 m wide and 4 m deep. The ambient flow velocity is 1.0 m/s. The cross section of the discharge outlet is $50\,\text{cm} \times 30\,\text{cm}$ (width × height). According to the prototype the flow velocity of the discharge outlet is 1.33 m/s. The water temperature of ambient flow is 18 ℃, and the water temperature of discharge is 27 ℃.

Table 1　Experiment parameters[2]

Variability rate	1	2	3	4
λ_l	100	100	100	100
λ_h	100	50	33.33	25
U_a (m/s)	0.10	0.141	0.173	0.20
h (cm)	6.00	12.00	18.00	24.00
R_{ea}	4587	12940	23800	36700
U_0 (m/s)	0.167	0.237	0.290	0.335
d (cm)	0.30	0.60	0.90	1.20
R_{e0}	383	1087	1995	3073

In the table, λ_l is horizontal scale, λ_h is vertical scale, U_a is ambient velocity, U_0 is jet-flow velocity, R_{ea} is ambient Reynolds numbers, R_{e0} is jet-flow Reynolds numbers, h is the depth of water, and d is the height of discharge outlet.

In the test, each test shoot 30 consecutive images with PIV, and then 30 instantaneous velocity vectors are averaged to get the average velocity vector. The average velocity vector of four variable variability rates is shown in Figure 4-Figure 7.

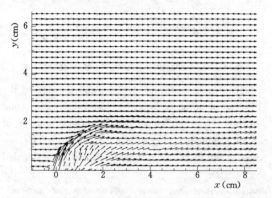

Figure 4   Average velocity vector of normality

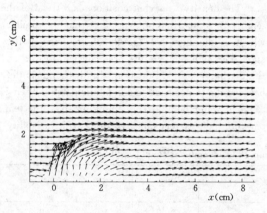

Figure 5   Average velocity vector of variability two

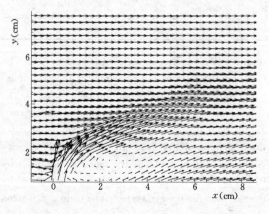

Figure 6   Average velocity vector of variability three

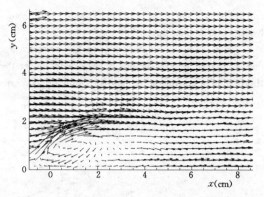

Figure 7   Average velocity vector of variability four

When the initial momentum has greater impact on the jet, ambient flow has little effect on the jet. Then the trajectory of the jet takes on a slight bend under the influence of the ambient flow. With the further spreading of the jet, jet's velocity is decreased. When the jet's velocity and ambient flow velocity is close, the impact of ambient flow to the jets is in a gradual increase. After the curved section, along with the role of ambient flow to further enhance, the jet flow direction is in line with the ambient flow direction.

It is shown in Figure 4-Figure 7, with the variability become greater, the corresponding ambient flow velocity and jet's velocity all increase. As can be seen from the figures, the flow pattern of normality and variability two are similar, but the backflow come forth in the downstream of variability three and variability four.

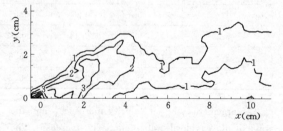

Figure 8   Temperature distribution of normality

The negative velocity turn up in the downstream of variability three and variability four and formed the backflow, which is mainly due to the large velocity of discharge outlet and drain exports have larger initial momentum. With variability rate increasing, the difference between jet velocity and ambient velocity become bigger. According to Bernoulli equation principle, pressure gradient will increase and make water and boundary separate, and swirl come into being, backflow area forming. [3-4]

As is shown in Figure 8-Figure 11, temperature distribution of four variability rates are taken by the thermal infrared imager system. The unit of temperature distribution is ℃.

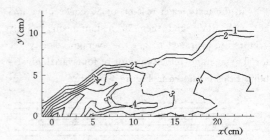

Figure 11   Temperature distribution of variability four

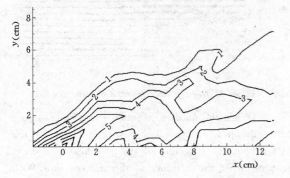

Figure 9   Temperature distribution of variability two

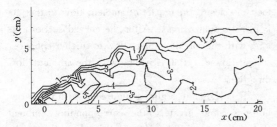

Figure 10   Temperature distribution of variability three

By Figure 8-Figure 11 and Table 2, due to the influence of ambient flow, surface buoyant jet bend and then flow along shore. On the combined action of the longitudinal convection and lateral spreading, hot water in the downstream formed a gradually expanded zonal region. With the variability rate increasing, the distribution of hot water become wider and wider. With the higher diffusion effects, the area of the high temperature zone becomes larger.

The density, relevant positions and trend of the temperature distribution line are roughly similar, which is the inevitable result of the similar flow state. The high temperature area concentrates around the discharge outlet. But after the discharge outlet, because the transportation and dilution ability of environment water body strengthening greatly, the temperature reduces very quickly. Along with variability

Table 2   **Area with temperature rise greater than 18℃ of ambient flow (cm²)**

No.	ΔT 0.5℃	1℃	2℃	3℃	4℃	5℃	6℃
1	94.25	28.96	6.51	2.20	0.57	0.09	0.03
2	132.90	120.14	60.62	16.12	6.45	3.38	0.76
3	172.24	149.72	126.65	49.69	18.23	8.96	2.70
4	216.85	189.81	136.30	65.66	27.07	14.26	4.79

rate enlargement, the jet velocity increases, and the ambient flow velocity also similarly increases. The high temperature zone extends rapidly, and the influence area scope of the same temperature distribution line is bigger. Similarly, the greatest length and the extreme breadth of various temperature distribution line increase[6].

4. Summary and conclusions

The research in this paper is done on flow field and temperature field simulation test of physical mod-

el with PIV and thermal infrared imager system. The measurement method and principles of PIV and thermal infrared Imager system are introduced.

Based on similarity theory, four variability conditions such as normality, variability two, variability three and variability four are defined. The velocity vectors of different variability rate are measured by PIV. The temperature distribution is measured by thermal infrared imager system. The main conclusions are as follows.

(1) With the variability become greater, the

corresponding ambient flow velocity and jet-flow velocity also increase. The flow pattern of normality and variability two are similar, but the backflow come forth in the downstream of variability three and variability four.

(2) The density, relevant positions and trend of the temperature distribution line are roughly similar. Along with variability rate enlargement, the high temperature zone extends more rapidly, and the influence area scope of the same temperature distribution line is bigger. Similarly, the greatest length and the extreme breadth of various temperature distribution line increase.

References

[1] Wygnanski I., Fiedler, H. Some measurements in the self-Preserving jet. J. Fluid Mech., 1969, 38: 577-612.

[2] Yang Haiyan. Study on scale effect of simulation test for temperature field. Hohai University. 1998.

[3] Song T., Graf W. H. Velocity and turbulence distribution in unsteady open-channel flows. J. Hydr. Engrg., ASCE, 1996, 122 (3): 141-154.

[4] JosePh H. W. Lee, Tang H. W. Experiments of a duckbill valve (DBV) jet in cross flow. Proceedings of Hydraulic Engineering for Sustainable Water Resourees Management at Turn of the Millennium, 28th LAHR Congress, 1999, GRAZ.

[5] Kōnig. O, Fiedler H. E. The structure of round turbulent jets in counterflow. A flow visualization study. Advances in turbulence. A. V. Johanson and P. H. Alfredsson, eds., Soringer-Verlag, Berlin, Germany, 1991: 61-66.

[6] Weisgraber T. H., Liepmam D. Turbulent structure during transition to self-similarity in a round jet. Exp. Fluids. 1998, 24: 210-224.

Study on the Internal Flow of Marine Sewage Outfalls with PIV System

Wu Wei

Suzhou University of Science and Technology, 1701 Binhe Road, Suzhou, 215011

Abstract: Saline intrusion into marine sewage outfalls will greatly decrease the efficiency of sewage disposal, especially in some early-constructed structures without special safe-guard measures. In this paper a simplified physical model of the sewage outfall was built in the laboratory to investigate the mechanisms of saline intrusion and purging flow in the system. The PIV (Particle Image Velocimetry) technique was selected to measure the velocity-fields of this internal flow. In order to present the flow characteristics at different inflowing sewage discharges, three series of velocity-fields were measured with PIV. The development of flow patterns at different inflowing sewage discharges and the mechanisms of saline intrusion and purging were analyzed. For an outfall system with certain structure, the inflowing sewage discharge was a decisive factor that affected the internal flow pattern. The internal flow during saline intrusion and purging was three-dimensional, unsteady, density stratified and with intense interfacial entrainment and turbulent mixing. The measured velocity-fields indicated that the interfacial entrainment and turbulent mixing were two main mechanisms to extrude the saline from the system and these two mechanisms would take different proportions due to the different inflowing momentum.

Key words: marin sewage outfalls; internal flow; saline intrusion; PIV system; velocity-field

1. Introduction

Saline intrusion into marine sewage outfalls will greatly decrease the efficiency of sewage disposal, especially in some early-constructed structures without special safe-guard measures. During the periods of low sewage discharges, for example, at the initial operation stage of the outfall or at the abrupt interruption of pumping due to some reasons, seawater would intrude into the outfall due to the density difference of seawater and sewage. Saline wedge would block some of the risers and seawater would circulate in the outfall. Under this condition, sediment would deposit and marine organisms would grow on the inner walls of the outfall pipe, that would result in blockage and decrease the efficiency of the structure [1-2].

For many years researchers had paid more attention to the study of buoyant jet of sewage flow [3-6]. Saline intrusion in long sea outfalls was not investigated until 1970s. The theoretic analysis was limited due to the complex mechanisms of this flow. So for a long time, experimental observation was the main means for the study and engineering design [7-8]. From 1980s, numerical simulation was investigated. In recent years, many researchers have made significant contributions towards this goal [9-12].

For the experiment, of course, spot observation was impractical for the most of the engineering. So in most cases, the laboratory experiment was used to optimize the structural design. In addition, the validation of the numerical model and the discussion of the parameters of the numerical model should resort to the laboratory experiment.

For the laboratory experiment, measuring technique was an important factor besides the similarity principle and scale effect. Due to the limitations of the measuring technique, the early experiments were major at qualitative observation. The physical parameters measured in the experiment were critical discharges for saline intrusion and purging, geometric parameters of saline wedge in the outfall. The early experiment studies achieved some macroscopic results such as the critical conditions of saline intrusion and purging, the effect of structural design [13-21]. However, the internal flow details could not be realized comprehensively. In 2002, the researchers in the University of Belfast measured the internal flow with Laser Doppler Velocimetry (LDV) and set up a two-dimensional numerical model [11-12]. The LDV was used to measure the single-point velocity but not flow field. The multi-points velocities could not be measured synchronously. In order to understand the internal flow field the experiment should be repeated for many times to achieve velocities at different points.

In this paper, the Particle Image Velocimetry (PIV) was used to measure the internal velocity-fields of the outfall. The PIV technique was not interfering with the flow field and could measure flow field synchronously.

2. Experimental study

In order to present the flow characteristics in the outfall at different inflowing sewage discharges, three series of velocity data were measured with PIV. The development of flow patterns at different inflowing sewage discharges and the mechanisms of saline intrusion and purging were analyzed.

2. 1   Design of the experimental system

An experimental system was established in the laboratory. For the experiment a simplified outfall including a main pipe and three risers were made of Perspex with equivalent roughness height of 0. 001mm. The main pipe was 5m long and had a circular cross section with radius of 0. 075m. The risers had a circular cross section with radius of 0. 025m. The distance of adjacent risers was 0. 45m. Each riser had a height of 0. 4m, which was the distance between the center of the main pipe and the exit of the riser.

The receiving saline water was modeled with a flume. For the region size of the receiving water there were three points should be considered. Firstly, the distance of the receiving water surface with the exit of the risers should large enough that the saline intrusion could take place. Secondly, the receiving water surface should keep steady by overflowing. Besides, the range of the receiving water should large enough to be sure that the boundary of the flume would not affect the discharge flow of the risers.

In this experiment, the surface of the receiving water was 0. 275m above the exit of the risers.

The experimental system was shown in the Figure 1.

2. 2   Experimental conditions

The aim of the experiment was to build up pro-

Figure 1   Photo of the experimental system

files of saline wedge formation and saline intrusion within the simplified model outfall. The initial condition during these tests was a fully intruded outfall that the system was full of saline water. And then the sewage discharged into the outfall. Monitoring was carried out continuously from this initial state, through the transient stage until a steady state was achieved. During this process, the development of the flow field within the outfall was measured.

The sewage was modeled with clear water with density of $998kg/m^3$. The receiving saline water was modeled with sodium chloride solution with density of $1021kg/m^3$. The relative density difference between the two fluids was 2. 3%.

The range of flow discharge in the experiment was selected according to the following two points. Firstly, the flow pattern should be kept turbulence. In the experimental flow, the Reynolds Number was 1800. Preliminary tests showed that the internal flow in the outfall would be turbulent conspicuously when the Reynolds Number was 1800 due to stratification of the two fluids. Secondly, the range of the flow discharge should be sure that saline intrusion would occur at the lower discharge and saline purging would occur at the higher discharge. The flow discharges in the experiment were shown in the Table 1.

Table 1   Experimental conditions

Relative density difference	Submergence depth of the riser's exit (m)	Inflowing discharge (m^3/s)
2. 3%	0. 275	$0. 2 \times 10^{-3}$
2. 3%	0. 275	$0. 6 \times 10^{-3}$
2. 3%	0. 275	$0. 72 \times 10^{-3}$

2.3　Measuring system

The velocity-field was measured with PIV made by TSI Corporation. The basic principle of the PIV technique was shown in the Figure 2. The tracking particles were released into the flow water. And then the measured section was illuminated with laser sheet. The particle images of the measured section at different times were photographed by the camera. The velocities were calculated through correlation analysis of the two pieces of particle images.

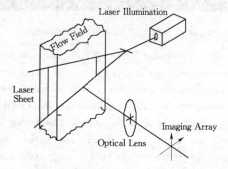

Figure 2　Basic principle of PIV

In this experiment, the laser illumination was double-pulse Nd: YAG system with the maximum laser intensity of 200mJ and the maximum frequency of 15Hz. The image-forming system was 28mm FL D/2.8 Nikon variable focal camera lens. The format of image was 10bit. The effective measured scope of single camera was 350mm × 260mm. The time interval between two images of single camera could be adjust according to the flow velocity of the measured range. The images were dealt with FFT cross-correlation algorithm with inquiry window of 64×64 pixels.

2.4　Points for attention

Firstly, the cross sections of the main pipe and the risers were circular. The particle image would be deformed due to the refraction by different mediums inside and outside of the circular pipe (liquid and air). To eliminate the effect of this refraction, a glass rectangular water tank full of clear water was designed. The measured section of the outfall was put into this water tank and was submerged into the water.

Secondly, the big air bubbles stayed in the outfall would scatter the laser. This would greatly decrease the quality of the particle image. So in the experiment, measures should be taken to exhaust the air bubbles from the measured scope of the outfall pipe.

Thirdly, to avoid the effect of the background light, the experiment should be taken in the darkroom.

3. Experimental results

Measurements of velocities were taken at the centre-line cross section of the outfall. The measured range in the experiment was shown in the Figure 3.

The flow characteristics in the outfall with different inflowing sewage discharges were analyzed. With the discharge of 0.2×10^{-3} m³/s and 0.6×10^{-3} m³/s, the flow patterns in the outfall were similar. Saline intrusion occurred in the outfall. With the discharge of 0.72×10^{-3} m³/s, the flow pattern turned into saline purging.

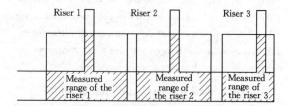

Figure 3　Measured range

Figure 4　Particle image in the experiment

The variation of the flow patterns could be expressed by the flow velocities in the risers. For the convenience of presence, the coordinate system was established as shown in the Figure 5. The measured sections located in the centre-line that was xz plane in the coordinate system as shown in the Figure 5. The coordinate x_0 was the distance from the measuring point to the left wall of the corresponding riser. Variable v was the velocity in the direction z. The positive value of v indicated discharging from the riser and the negative indicated flowing backward into the riser. The coordinate t represented the time of the flow developing from the initial. The velocities of the three risers were expressed by the velocities in the section $1-1$, $2-2$, $3-3$ as shown in Figure 5.

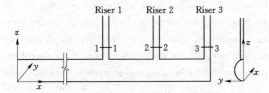

Figure 5　Coordinate system and section-positions

The measured flow fields at the inflowing sewage discharge of 0.6×10^{-3} m³/s and 0.72×10^{-3} m³/s were described in the following.

3.1　Measured flow field($Q_0 = 0.6 \times 10^{-3}$ m³/s)

With the inflowing sewage discharge of 0.6×10^{-3} m³/s, one of the representative flow field measured in the experiment was shown in the Figure 6.

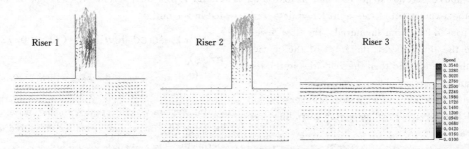

Figure 6　Measured flow field
($Q_0 = 0.6 \times 10^{-3}$ m³/s,　$t = 360$s；unit：m/s)

In the initial stage of the flow, after the flow had persisted 30 seconds in the experiment, the saline was extruded by the inflowing sewage and discharged from the three risers that were indicated by the positive velocities in the three risers. For convenience of analysis, as shown in the Figure 7, the velocities of the three risers were expressed by the velocities in the section $1-1$, $2-2$, $3-3$ as shown in Figure 5. The average velocities in the three risers were closed to each other. The velocities distribution in the outfall pipe indicated that no stratification occurred in the outfall pipe. That is to say, the sewage wedge had not reached the measuring section yet. Under this condition, the discharging of saline from the three risers was due to the extruding by the inflowing sewage.

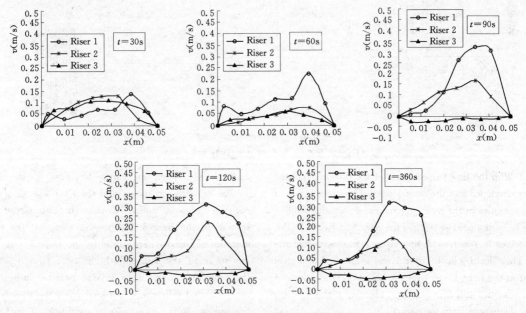

Figure 7　Measured velocities in the risers
($Q_0 = 0.6 \times 10^{-3}$ m³/s)

In the experiment, with the time passed 60s after the flow initiated, the flow developed into the second stage. The front of the inflowing sewage reached the riser 1 and the discharge in the riser 1 began to increase rapidly. In the mean time, the discharge in the riser 2 and riser 3 decreased evidently. This could be seen from the Figure 7.

The above second stage had not lasted long time in the experiment. After a little while, with the time passed 90s from the initial, the flow developed into the third stage, the front of the sewage reached the risers 2 and then the velocities in the riser 2 began to increase. The velocities in the riser 3 continued to decrease and soon flowed inversely that indicated the receiving saline water intruding into the riser 3.

The above flow pattern in the third stage lasted until the end of the experiment. The experimental results indicated that about 300s later the velocities through the three risers tended to remain steady, indicating that the flow pattern tended to be steady. The velocities distribution in the outfall pipe indicated the patterns of the sewage wedge clearly. The stratification of the two fluids was present in the experiment. At the interface of the stratification under the riser 1, the vortex flow due to the entrainment could be seen. The saline fluid in the lower layer of the stratification was entrained by the upper sewage layer and discharged from the riser 1. In the outfall pipe under the riser 3, the lower layer fluid of the stratification flowed backward to the upstream. This indicated that the receiving saline water intruded into the riser 3 and flowed backward to the upstream then discharged from the riser 2 or riser 1. The saline circulated in the outfall.

3. 2　Measured flow field ($Q_0 = 0.72 \times 10^{-3}$ m³/s)

With the sewage discharge flow of 0.72×10^{-3} m³/s, the development of the flow patterns in the outfall system was different with that in the discharge of 0.6×10^{-3} m³/s. The velocities measured in the experiment were shown in the Figure 8.

In the initial stage of the flow, after the flow had persisted 30 seconds in the experiment, the saline was extruded by the inflowing sewage and discharged from the three risers that were indicated by the positive velocities in the three risers. As shown in the Figure 9, the average velocities in the three risers were closed to each other. The velocities distribution in the outfall pipe indicated that no stratification occurred in the outfall pipe.

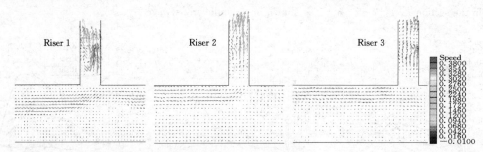

Figure 8　Measured flow field
($Q_0 = 0.72 \times 10^{-3}$ m³/s, $t = 120$s; unit: m/s)

With the time passed 60s after the flow initiated the flow developed into the second stage. The stratification of the velocities in the main pipe indicated that the front of the inflowing sewage reached the riser 1 and riser 2. The velocities in the riser 1 and riser 2 increased evidently. The velocities in the riser 3 was evidently lower than that in the riser 1 and riser 2, however, the values of the velocities still kept positive indicating that the saline still discharged from the riser 3.

The flow pattern in the second stage developed continuously. When the time passed 90s, for the riser 3, the velocities in the part of the riser's section was negative indicating saline intrusion through part of the section of the riser 3. The reason was that the momentum of the discharge flow in the riser 3 was too low to resist the saline. With the time passed 120s, the velocities in the riser 3 were positive indicating that intruded saline was extruded from the riser 3 by the sewage. In the experimental condition, the extruding of the saline from the riser 3 was not intense and the flow pattern in the riser 3 was not steady. The probable reason was that the riser 3 might be in the critical condition of saline intrusion and purging. If the inflowing sewage discharge increased, this extruding flow would be more intensely and saline purging would be evidently.

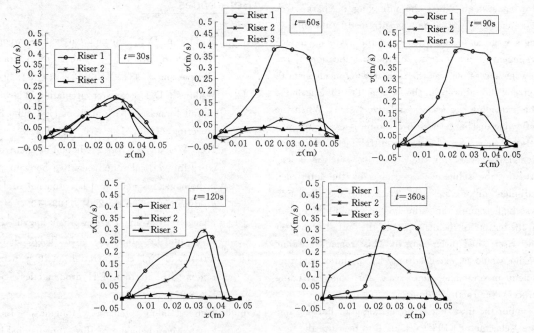

Figure 9   Measured velocities in the risers
($Q_0 = 0.72 \times 10^{-3} \mathrm{m^3/s}$)

At the interface of the stratification under the riser 1, the vortex flow due to the entrainment was more intense than that under the riser 2 and riser 3. The reason for this was that the entrainment was appreciably related to the momentum of the inflowing sewage. When the momentum of the inflowing sewage decreased to a low value, the entrainment was not evident and under this condition, the saline's extrusion from the outfall depended on the interfacial turbulent mixing.

4. Conclusions and discussions

With the technique of PIV, the internal flow fields of the outfall system during saline intrusion and purging in the simplified experimental model were measured. Compared with the previous single-point measuring technique such as LDV, the velocity field could be achieved synchronously and non-interfering. The development of the flow patterns at different inflowing sewage discharges could be clarified by the velocity-fields.

(1) For an outfall system with certain structure, the inflowing sewage discharge was a decisive factor that affected the internal flow pattern in the outfall system. When the inflowing discharge was low, saline circulation would occur in the outfall system. If the practical engineering was under this low discharge working condition, sediment would tend to deposit in the main pipe and result in blockage of the pipe. When the inflowing sewage discharge was high enough, saline circulation in the system could be avoided. So, in the practical engineering, it was very necessary to select a reasonable operation mode for the control of the process of the inflowing discharge. This was particularly important for those early structures without special safeguard measures.

(2) According to the velocity-fields of the internal flow in the outfall system, the flow patterns of saline intrusion and purging developed in several different stages. The development of the flow indicated that the internal flow in the outfall system during saline intrusion and purging was three-dimensional, unsteady, density stratified and with intensely entrainment and interfacial turbulent mixing. These flow characteristics gave some important suggestions for the numerical model.

(3) The measured velocity-fields indicated that the mechanisms to extrude the intruding saline from the system were related to the momentum of the inflowing sewage. When the inflowing discharge was large enough, the saline was entrained by the inflowing sewage and discharged from the system. The entrainment at the interface of the stratification of the sewage and saline was the main driving mechanism to extrude the saline from the system. The time scale for this extruding process was relatively short. When the inflowing discharge was low, some risers in the upstream were purged by the entrainment aforementioned. However, for the other risers in the downstream, the saline was extruded by the interfacial turbulent mixing. Besides, even that the discharge was high enough the saline wedge would still remain in the bottom of the main pipe of the outfall even after the risers had been purged. This remained saline wedge would be extruded only by the interfacial turbulent mixing. The time scale for this extruding process was relatively long.

For the first time, the technique of Particle Image Velocimetry (PIV) was used to measure the flow field of saline intrusion and purging inside the outfall system. The following points should be discussed in the further study.

(1) For the technique of PIV, the performance of the tracking particles following with the fluid was an important factor that affected the accuracy of the measurement. This performance was decided by the density of the tracking particles. The density of the tracking particles was more close to that of the fluid the better performance could be achieved. In this experiment, the tracking particles with the same density were released into the two fluids with different densities. This would have some effect on the accuracy of the measurement. So, the performance of the tracking particles following with the fluid for this density-stratified flow should be discussed in the further study.

(2) In the experiment, the different refractive index of the saline water and clear water would have some effect on the accuracy of the measurement. This should be discussed in the further study too.

Acknowledgment

This research was supported by Laboratory of Engineering Hydraulics of Hohai University. Professor Yan Zhongmin supervised the author patiently.

References

[1] Wilkinson D. L.. Purging of saline wedges from ocean outfalls [J]. Journal of Hydraulic Engineering, 1984, 110 (12): 1915-1829.

[2] Wilkinson D. L. Seawater circulation in sewage outfall tunnels [J]. Journal of Hydraulic Engineering, 1985, 111 (5): 864-858.

[3] Huai Wenxin, Li Wei. Prediction of characteristics for vertical round negative buoyant jets in homogenous ambient [J]. Journal of Hydrodynamics, Ser. B, 2001: 103-110.

[4] Zhang Yan, Wang Daozeng, Fan Jingyu. Experimental investigations on diffusion characteristics of high concentration jet flow in near region [J]. Journal of Hydrodynamics, Ser. B, 2001: 117-121.

[5] Han Huiling, Zhang Hongmin, Liang Sutao. Mean behavior of three dimensional line buoyant jets in cross flows [J]. Journal of Hydrodynamics, Ser. B, 2003: 32-36.

[6] Fan Jingyu, Wang Ddaozeng, Zhang Yan. Three dimensional mean and turbulence characteristics of an impinging density jet in a confined cross flow in near field [J]. Journal of Hydrodynamics, Ser. B, 2004, 16 (6): 737-742.

[7] Mort R. B.. The effect of wave action on long sea outfalls [D]. University of Liverpool, PhD thesis, 1989.

[8] Burrows R., Davies P. A.. Studies of saltwater purging from a model sea diffuser [J]. Water Maritime & Energy, 1996, 118: 77-87.

[9] Guo Zhenren, Sharp J. J.. Numerical Model for Sea Outfall Hydraulics [J]. Journal of Hydraulic Engineering, 1996, 122 (2): 82-89.

[10] Doyle B. M., Mackinnon P. A., Hamill G. A.. Development of a Numerical Model to Simulate Wave Induced Flow Patterns in a Long Sea Outfall [A]. 3rd International Con-

ference on Advances in Fluid Mechanics, Montreal, Canada, 2000: 33-42.

[11] Shannon N. R.. Development and validation of a two-dimensional CFD model of the saline intrusion in a long sea outfall [D]. Queen's University Belfast, PhD thesis, 2000.

[12] Shannon N. R., Mackinnon P. A., Hamill G. A.. Modeling of saline intrusion in a long sea outfall with two risers [A], 1st International Conference on Computational Methods in multiphase Flow, Orlando, USA, 2001: 531-541.

[13] Davies, P. A., Charlton, J. A., Bethune, G. H. M., A laboratory study of primary saline intrusion in a circular pipe, Journal of Hydraulic Research, 1988, 26 (1): 33-47.

[14] Burrows, R., Ali, K. H. M., Davies, P. A., Wose, A. E.. Studies of saltwater purging from a model sea outfall diffuser, Proc. I. C. E., J. Water, Maritime and Energy, 1996, 118: 77-87.

[15] Bethune, G. H. M., M. Sc.. Dissertation, University of Dundee, U. K. 1986.

[16] Burrows, R. Ali, K. H. M., Spence, K., Van Wellen, E.. Experimental observation of salt purging in outfalls with soffit-connected risers, Proceedings Second International Conference on Hydrodynamics, Hong Kong,

1996, 12: 901-906.

[17] Burrows, R., Ali, K. H. M., Spence, K., Chiang, T. T. Experimental observations of salt purging in a model sea outfall with eight soffit-connected risers. Water science & technology, 1998, 38 (10): 269-275.

[18] Burrows, R., Ali, K. H. M., Davies, P. A., Wose, A. E.. Hydraulic performance of outfalls under unsteady flow, Dept. Civil engineering, University of Liverpool, final report on SERC project GR/G/21841.

[19] Wose, A. E., Burrows, R., Ali, K. H. M., Davies, P. A.. Purging studies on a model sea outfall diffuser, in Water Pollution Ⅲ: Modeling, Measuring and Prediction, L. C. Wrobel & P. Latinopoulos (eds.), Computation Mechanics Publications, 277-286.

[20] Charlton, J. A., Davies, P. A., Bethune, G. H. M.. Seawater intrusion and purging in multiport sea outfalls, Proceedings Institution of Civil Engineers, Part Ⅱ, 1987, 83: 263-274.

[21] Burrows, R. Ali, K. H. M., Wose, A. E.. Laboratory studies of saline intrusion, salt wedge formation and sediment deposition in long-sea outfalls, Proceedings of the International Symposium on Environmental hydraulics, Hong Kong, 1991, 12: 269-273.

Model Test for Rectification Method of Pump Intake Channel

Cheng Lu   Zhao Jianjun   Li Fang

State Key Laboratory of Hydrology-Water Resources and Hydraulic Engineering, Nanjing Hydraulic Research Institute, Nanjing, 210029

Abstract: In this research project, a physical model with scale 1 : 10 has been adopted to study the hydraulic characteristics of intake channel in APL Power Station, India. The research emphases include: velocity distribution and flow pattern along intake channel and in suction room, simulation and verification of vortex, tabling a proposal of anti-vortex and improving flow pattern, and so on. Through this research, two kinds of rectifying facilities including dividing columns and beams would be adopted to eliminate bad flow pattern in original scheme. Both two methods have good effect on rectifying flow state and would ensure the safety operation of CW pump, while the beams scheme should be chosen to the recommended scheme.

Key words: intake channel; dividing column; beam; flow pattern; pumping station

1. Introduction

The intake channel of circulating water system is an important part of water supply system in power plant. Whether the design of intake channel of circulating water system is reasonable or not involves the operation safety and economic problems of CW pump.

This article sets the intake channel in India APL Power Station as an example to expound the model test for optimum hydraulic design of front-intake channel of CW pump station.

2. Project overview

In this project of APL Power Station, diversion pipeline and front-intake fore bay have been adopted for water supply of CW pump station. The plan arrangement of CW pump station has been shown in Figure 1.

The cross section size of culvert pipe $B \times H$ is 3.60m $\times$ 3.60m, while the maximum discharge of pipeline is 24.24m³/s. The designed diffusion angle of fore bay is 40°, and the length of fore bay is about 17.0m.

According to the calculation expression for theoretical angle of flow diffusion:

$$\tan(\alpha/2) = 0.204 \times h^{0.5}/v + 0.107 \qquad (1)$$

Where, α is diffusion angle; h is height of pipeline; v is flow velocity out of pipe.

The theoretical diffusion angle designed in fore bay should be about 17°, and the length of fore bay should be about 23.0m. While the maximum discharge of pipe is relatively larger, so a physical model is necessary be built to validate and research flow state in pump station of this project.

3. Model details

According to research purpose and requirement, a physical model with scale 1 : 10 had been adopted, which simulation range includes a part of inlet pipeline, fore bay of pump house, suction rooms of CW pumps, and outlet pipes, etc. The whole model had been made up of organic glass.

3.1 Similarity theory

Basing on the similarity theory of model test, it is impossible to guarantee absolute similarity between prototype and physical model ($\lambda_L \neq 1$). In general case, the hydraulic model test for intake channel of pump station would be designed by Froude similitude principle, and the conditions of Reynolds similarity are released frequently.

Because Reynolds similarity is ignored, while the eddy current influenced by viscous resistance, the simulation of eddy current would be inaccurate. In order to ensure security of engineering, the fluid states of vortices in pump station model would be usually estimated and observed by increasing discharge to 1.5-3 times of the original value (namely the 1.5-3

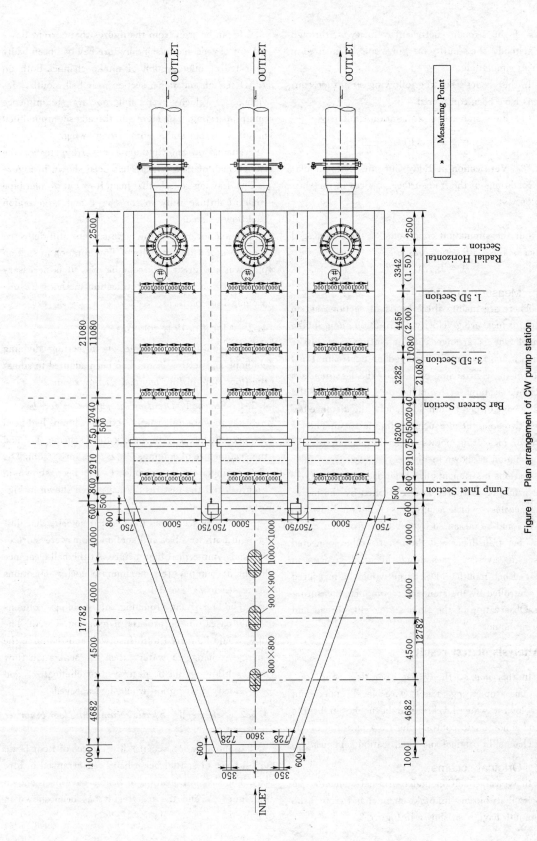

Figure 1 Plan arrangement of CW pump station

times of the Froude coefficient velocity), through this method, the security margin would be ensured in vortices simulation.

In this model test, the following similitude conditions have been considered.

(1) Flow similarity: gravitation similarity

$$\lambda_{Fr} = \frac{Fr_p}{Fr_m} = 1, Fr = \frac{u}{\sqrt{gh}}$$

(2) Verification of Eddy Current. The simulation conditions of the free-surface vortices in the verification tests are:

$$\lambda_v = \lambda_L^{0.2}, \quad \lambda_Q = \lambda_L^{2.2}$$

For the simulation conditions of the submerged vortices, the expressions are:

$$\lambda_v = \lambda_L^0, \quad \lambda_Q = \lambda_L^{2.0}$$

3. 2 Measuring means

There are mainly three kinds of sections set in model for measuring velocity distribution along intake channel. One is arranged at the entrance of intake channel; another one is arranged at upstream 1. 5D far from the central line of pump; the last one is a horizontal section set by radial direction around the leading edge of bell mouth. In model, the distance between two adjacent measuring points is 10cm. The location of measuring points in section layout of CW pump station is shown in Figure 1.

Various instruments had been used in test. Water level would be measured and controlled by the measuring needles, which accuracy is 1/50mm. Flow velocity would be measured by OA-type propeller meter with fiber transducer and Pitot-static tube connected by sensing system (measuring the axial velocity in throat of bell mouth). Discharge would be measured and controlled by the standard rectangular measuring weir. Observation of the fluid state: silk thread and coloring liquid.

4. Analysis of test result

In this paper, the worst operating condition, two pumps operating at low water level (maximum flux, lowest water level), had been chosen be the typical condition to contrast the flow patterns of intake channel in original and modification schemes.

4. 1 Original scheme

In typical condition under original scheme, the velocity distributions in intake channel and around the bell mouth have been shown in Figure 2.

It can be seen from the figures that current flowing out of pipe and passing by fore bay had been badly deflected at inlet section of intake channel. Both on 1. 5D section and radial section near bell mouth, deviations of velocity were still too large to influence pump operating. The eight silk threads set around bell mouth were not stable with strong swing.

Basing on multiple data get from tests, the sketch map of flow pattern has been shown in Figure 3. The drawing reveals that main flow out of inlet pipe couldn't diffuse fully in fore bay, and large-region backflow existed.

In suction room, the main current still deflected strongly; and the intensity of flow rotation in suction pipe was too great to acceptable. So, it is necessary to choose some suitable rectification methods for improving flow patterns.

4. 2 Modification scheme

In modification scheme, both setting dividing columns and setting beams had been adopted to adjust the flow state in fore bay and suction room.

4. 2. 1 Scheme I: Dividing columns set in fore bay

There were four rows, sixteen columns had been arranged in fore bay in modification scheme I, and the section size of columns $L \times B$ is 0. 8m×0. 8m. The plan arrangement of scheme I has been shown in Figure 4 and the test results have been shown in Figure 5.

It can be seen from figures that velocity distributions in both fore bay and suction room were more even and symmetrical than before. At the bell clearance height of pumps, the maximum velocity deviations decreased to 21% and 13%.

The rectifying function of dividing columns mainly bases on the process of gyrating current. The flow would be forced to diffuse well and reinforce the energy exchange of water, then the separation flow and backflow would be destroyed or eliminated, and the velocity distribution would be improved.

4. 2. 2 Scheme II: Spatial beam and floor beam set in fore bay

In modification scheme II, a piece of floor beam and a piece of spatial beam had been arranged in fore bay. The size and location of beams have been shown in Figure 6, and the test result has been shown in Figure 7.

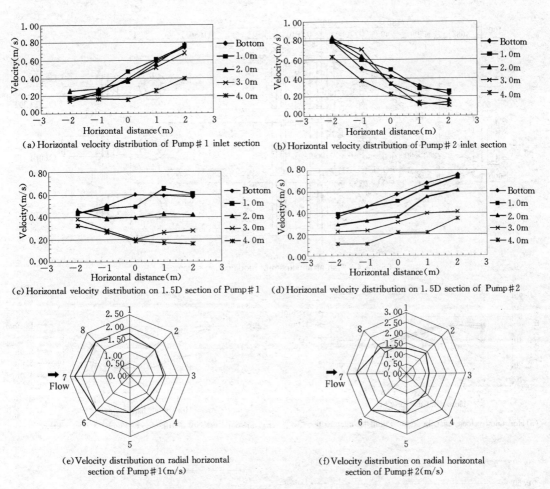

(a) Horizontal velocity distribution of Pump#1 inlet section

(b) Horizontal velocity distribution of Pump#2 inlet section

(c) Horizontal velocity distribution on 1.5D section of Pump#1

(d) Horizontal velocity distribution on 1.5D section of Pump#2

(e) Velocity distribution on radial horizontal
section of Pump#1(m/s)

(f) Velocity distribution on radial horizontal
section of Pump#2(m/s)

Figure 2　Velocity distributions under original program

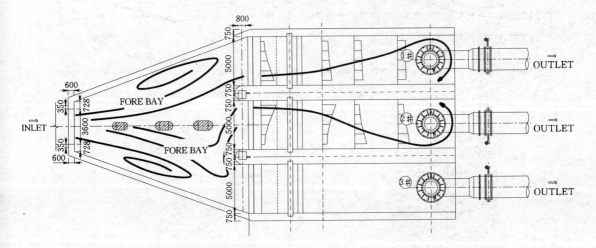

Figure 3　Sketch map of flow pattern in original scheme

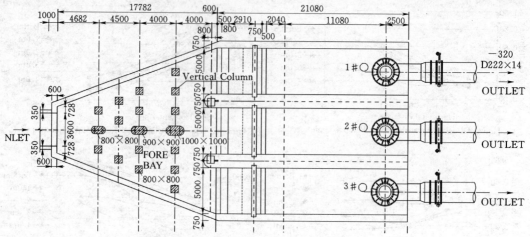

Figure 4   Plan arrangement of modification scheme Ⅰ

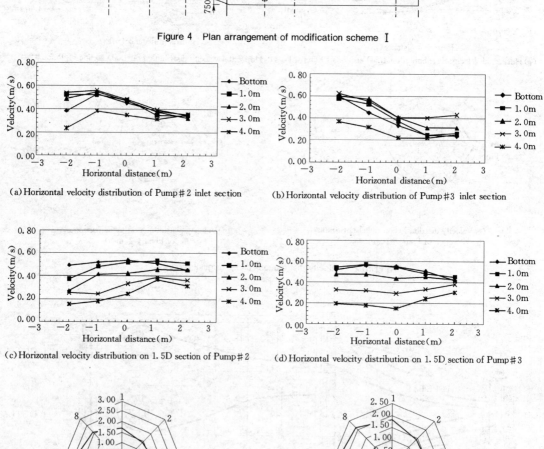

(a) Horizontal velocity distribution of Pump #2 inlet section

(b) Horizontal velocity distribution of Pump #3 inlet section

(c) Horizontal velocity distribution on 1.5D section of Pump #2

(d) Horizontal velocity distribution on 1.5D section of Pump #3

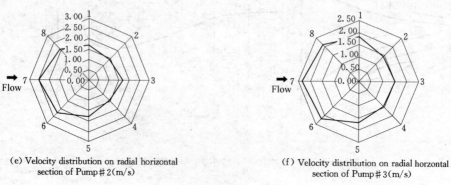

(e) Velocity distribution on radial horizontal
section of Pump #2 (m/s)

(f) Velocity distribution on radial horzontal
section of Pump #3 (m/s)

Figure 5   Velocity distributions in scheme Ⅰ

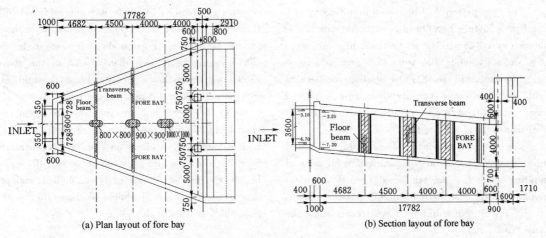

(a) Plan layout of fore bay (b) Section layout of fore bay

Figure 6   Location of beams in modification scheme Ⅱ

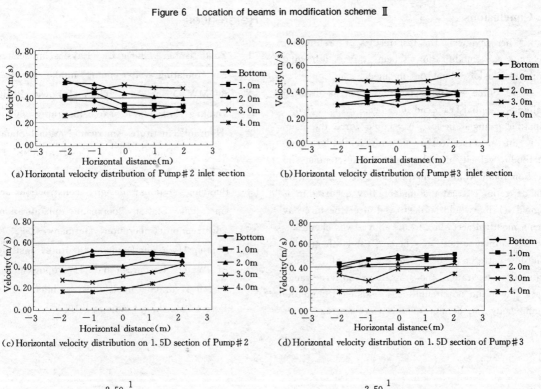

（a）Horizontal velocity distribution of Pump♯2 inlet section

（b）Horizontal velocity distribution of Pump♯3 inlet section

（c）Horizontal velocity distribution on 1. 5D section of Pump♯2

（d）Horizontal velocity distribution on 1. 5D section of Pump♯3

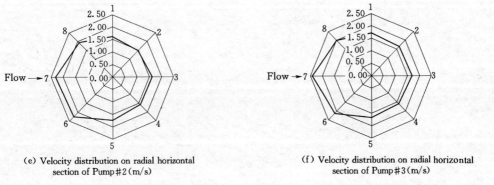

（e）Velocity distribution on radial horizontal
section of Pump♯2（m/s）

（f）Velocity distribution on radial horizontal
section of Pump♯3（m/s）

Figure 7   Velocity distributions in scheme Ⅱ

The flow pattern had been well improved by setting beams in fore bay. On inlet sections, the maximum velocity deviations in vertical direction were 10% and 6%, while they were 73% and 70% in original scheme. The velocity distributions on section 1. 5D and the radial velocity distributions around pump suction flanges were greatly even and symmetrical than original scheme too.

The rectifying function of floor beam and spatial beam mainly depends on the mixing-flow process of gyrating currents, the reinforced mutual turbulent motions redistribute the momentum of water, thereby the velocity distributing downstream would be more symmetrical.

5. Conclusions

When reviewing the test results of preliminary scheme and modification schemes, the following some conclusions should be obtained.

（1） Some rectifying facilities such as dividing columns, spatial beam and floor beam for energy dissipation arranged in fore bay could force the main flow diffuse more early and sufficiently. By these methods, velocity distribution in intake channel and suction room would be more even and symmetrical; the deflecting current and unstable flow occurred in original scheme would be destroyed after the fore bay being modified according to the above two schemes.

（2） In modification scheme Ⅰ and scheme Ⅱ, the flow patterns in suction room were stable, no obvious flow separation, no drop flow, no large-range back flow, and no circumfluence existed around the pump body. Furthermore, no surface vortex or air core vortex greater than grade Ⅱ had been found. Silk threads set around bell mouth were stable without strong swing, the horizontal projection angles of silks swinging were less than 5°. All of the above shows that both scheme Ⅰ and scheme Ⅱ could satisfy the requirement of pump operation.

（3） The rectifying function of floor beam and spatial beam is like the action of dividing columns, both having obvious effect on eliminating separation flow. But the structure of beams in scheme Ⅱ would be easy to construct conveniently and do daily maintenance, so the scheme Ⅱ should be chosen to the recommended scheme in this project.

References

［1］ Zhao Jianjun, Cheng Lu. Physical model test of circulating water pump intake for 4 × 300MW Quang Ninh Thermal Power Plant, Nanjing Hydraulic Research Institute, 2006.

［2］ Hydraulic institute. American National Standard for Pump Intake Design, ANSI/HI 9. 8-1998.

［3］ Pumping stations; design, construction and operation Station pompage; planification, construction et operation, Germany, 1992.

［4］ Nanjing Hydraulic Research Institute, Institute of Water Resource and Hydropower Research. Hydraulic model test, China Water-Power Publishing Company, 1985.

Experimental Investigation of the 2D Flow over Backward Facing Step by PIV

Huang Minghai[1]   Qi Erong[2]

[1] *Department of Hydraulic Engineering, Changjiang River Scientific Research Institute, Wuhan, 430010*

[2] *State Key Laboratory of Water Resources and Hydropower Engineering Science, Wuhan University, Wuhan, 430072*

Abstract: In range of $150 < Re < 6500$, the particle image velocimetry (PIV) was used to measure and visualize the 2D flow over backward facing step. When the flow in this structure were laminar, transitionary and turbulent, the transient velocity distributions were obtained. It showed that the vortices of the representational transient flow field were different when they were in various states of flow. Basing on time-averaged velocity distributions in the backward facing step, which were calculated from the transient velocity, we got relation curve between the length of recirculating flow region and Re, and we discovered that the vortex structure of the transient flow field were also different from that of the time-averaged flow field. In this paper, it was indicated that the flow over the backward facing step possessed simultaneously the characteristic of unsteady flow and steady flow.

Key words: PIV; backward facing step flow; separated flow; length of recirculation region

1. Introduction

Flow separation and reattachment are of great importance in such fields as environmental, aeronautical and mechanical engineering. Among a number of flows with separation and reattachment, the flow over backward-facing step is one of those with the simplest geometries, however, the flow structure is very complex and is the base of many other flows.

Since 1960s, many experimental techniques and theoretical models have been applied to investigate the characteristic of the flow over backward-facing step. In respect of experimental techniques, the flow visualization techniques such as the smoke wire, path line, nydrogen bubble method were used to observing the reattachment, circumfluence and vortices[2-4,11]. And hot-wire method[6], hotting mask wind velocity indicator[7], 1D laser Doppler velocimeter[1,5] were used to measuring time-averaged velocity, tubulence energy statistic and the length of recirculating flow region. In recent years, electrodiffusion technique was implemented on observing the motion of reattachment and recirculation in near wall region[9,10]. The wavelet transform method was used for researc-

hing the pressure characteristic of separated flow near wall. The above studying had obtained fruitful results on the characteristic of the flow over backward-facing step. On the whole, the research about the quantitative and non-intrusive way measurement of vortices was still scarce, especially in 2D flow over backward facing step.

The particle image velocimetry (PIV) is of spatial high-resolution, non-intrusive way and can be applied to measure the instantaneous volume flow rate and investigate the unsteady characteristics. In this paper, we used PIV for investigating the 2D flow over backward facing step.

2. Test facility and measuring technique

2.1   Test facility

Figure 1 shows the diagram of test facility. It is a water self-circulation, and the modelling materials are synthetic glass, 1 cm thick. The cross-section shape of flow channel is rectangular. The high of backward facing step (h_s) is 1cm, the high of upstream channel (h_u) is 1cm, the high of downstream channel (h_d) is 2cm, and the divergenceratio is 2 : 1. The width of flow channel is 36cm. The upstream and downstream length of step are respectively 1500cm and 2500cm, which ensure the flow in both upstream and downstream of step developing adequately.

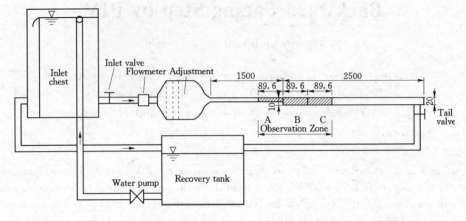

Figure 1　Diagram of test facility（mm）

2. 2　Measuring technique

The fundamental principle of PIV is that: first, measuring the displacement of tracer particles in water, second, basing on the time-of-flight in pulse laser, the velocities approximation of water flow can be got.

The PIV system used in this test is from the United States TSI Inc. and the tracer particles is particulate in water itselft, which's grain diameter is 10-20μm and meet the demand particle's tracking. The PIV system included particle imaging and velocity analyzing, visualizing two subsystems. Detailed instructions about PIV was given the paper of Adrain[12].

2. 3　Measuring method

According to the characteristic of the flow over backward-facing step and the function of PIV system, the observation zone include A, B and C, which are shown in Figure 1. The length of each zone is 8.96 h_s, so the total length of the observation zone is 26.88 h_s.

Reynolds number（Re）is defined that $Re=U_0 \cdot h_s/\nu$, here, U_0 is average velocity of transverse profile, ν is water kinematic viscosity. The flow state was judged by the color water test and the results were that while $Re<498$, $498<Re<1600$ or $Re>1600$, the corresponding flow state were laminar, transitionary and turbulent.

23 tests had been done and Re ranged from 150 to 6500, which include the flow states of laminar flow, transitionary flow and turbulent flow. In each test, 30 images of zone A and C and 40 images of zone B are continuously measured when the frequency is 3.75Hz.

3. Transient flow field analysis

Figure 2 shows the instantaneous streamlines in close time near the step when the flow is separately in three flow states. Comparing two instantaneous streamlines in close time of each flow state, we can see that the streamline in upstream is parallel and steady, the streamline in downstream change differently and appear several vortexs, each vortex experiencing appearance, development and disappear. The related contents of vortex movement near the step for details see the paper of Qi E R[13].

Serial tests showed that there were 5-7, 3-5 and 2-3 vortexs in recirculating flow region, while the water flow were laminar, transitionary and turbulent, there also were one vortex near superior wall while the water flow were laminar and transitionary. Figure 3 shows vortex structures of three representative flow states in recirculating flow region, and the Reynolds numbers were 498, 1029 and 2505. As illustrated in Figure 4, the flow structure of the 2D flow over backward facing step can be divided into seven zones, which were 1-main flow zone, 2-shear layer zone, 3-reattachment zone, 4-back flow zone, 5-corner vortex zone, 6-boundary layer redevelopment zone and 7-superior wall vortex zone. But while the water flow was turbulent, 7-superior wall vortex zone disappeared.

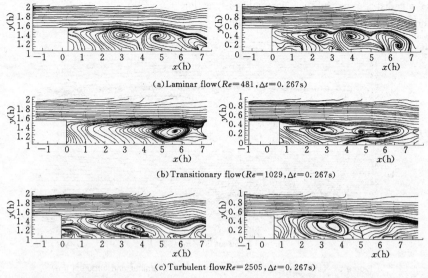

(a) Laminar flow($Re=481$, $\Delta t=0.267$s)

(b) Transitionary flow($Re=1029$, $\Delta t=0.267$s)

(c) Turbulent flow$Re=2505$, $\Delta t=0.267$s)

Figure 2　Comparison between two instantaneous streamline in close time near the step

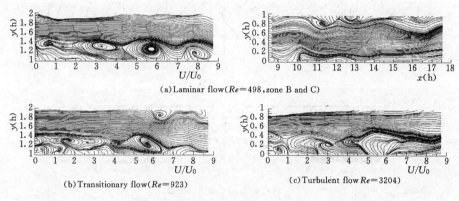

(a) Laminar flow($Re=498$, zone B and C)

(b) Transitionary flow($Re=923$)　　　　(c) Turbulent flow $Re=3204$)

Figure 3　Vortex structure comparison between three flow states in recirculating flow region

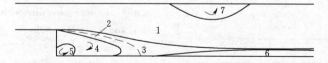

Figure 4　Structure diagram of the 2D flow over backward facing step

1—main flow zone；2—shear layer zone；3—reattachment zone；
4—back flow zone；5—corner vortex zone；6—boundary layer
redevelopment zone；7—superior wall vortex zone

4. Time-averaged flow field analysis

According to the serial tests data of transient flow in laminar and turbulent flow, we got the time-averaged flow streamlines, showed in Figure 5. Comparing the streamlines of the two flow states, we found that the structures were similar but the lengths of recirculating flow region were different,

the length of laminar flow was larger than of turbulent flow.

The time-averaged flow had only one vortex in recirculating flow region, but the transient flow was different from the time-averaged flow.

Figure 6 shows the time-averaged velocity distributions of the laminar and turbulent flow. As showed in this figure, the velocity distributions of laminar flow was parabolic distribution and the velocity distributions of turbulent flow was logarithmic distribution.

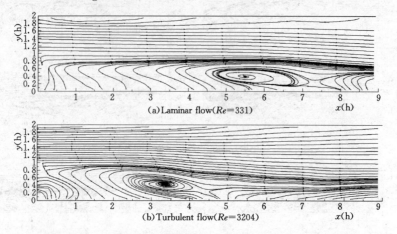

(a) Laminar flow($Re=331$)

(b) Turbulent flow($Re=3204$)

Figure 5   Time-averaged flow streamlines of the laminar and turbulent flow

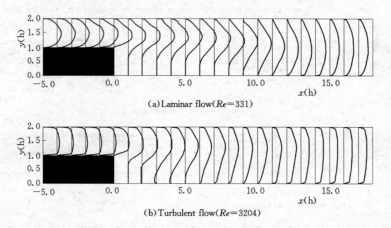

(a) Laminar flow($Re=331$)

(b) Turbulent flow($Re=3204$)

Figure 6   Time-averaged velocity distributions of the laminar and turbulent flow

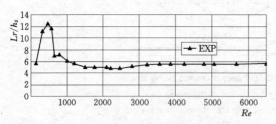

Figure 7   Relation curve between
the length of recirculating flow region and Re

Basing on time-averaged velocity distributions in the backward facing step, which were calculated from the transient velocity, we got relation curve between the length of recirculating flow region and Re, seeing the Figure 7. The length of recirculating flow region (Lr) meet the conditions that $u/y=0$ and $u/xy>0$. The relation curve showed that (a) while Re < 498, Lr increased obviously with the increasing of Re, and when $Re=498$, Lr was $12.2h_s$, (b) while Re was in range of 498—1600, Lr increased with the increasing of Re, inflection point appeared at $Re=716$, (c) while $Re>1600$, Lr increased little with the increasing of Re, while $Re>3500$, Lr was nearly kept on a steady value, which was $5.6 h_s$, and was not associated with Re. These results agreed well with the experiments of Armaly et al[1].

5. Conclusion

Practice showed that PIV was reasonable measuring systems for investigating separated flow like the flow over backward facing step.

The experiment results showed that:

(1) The flow over the backward facing step pos-

sessed simultaneously the characteristic of unsteady flow and steady flow.

(2) The vortices of the representational transient flow field were different when they were in various states of flow.

(3) The vortex structure of the transient flow field were also different from that of the time-averaged flow field.

(4) While $Re > 3500$, Lr was nearly kept on a steady value, which was 5.6 h_s.

References

[1] Armaly B F, Dur st F, Pereira J C F, Schonung B. Experimental and theoretical investigation of backward2f acing step flow [J]. J. Fluid Mech. , 1983, 127: 473-496.

[2] Frederick W R, Kegelman J T. Control of coherent structure of in reattaching laminar and turbulent shear layer [J] . AIAA J. , 1986, 24: 623-629.

[3] Kim J, Kline S, J Johnston J P. Investigation of separate d and reattaching flows of turbulent boundary shear layer In: Flow Over a backward facing step [R] . Rept MD-37, Stanford University, 1978.

[4] Lian Q X. An experimental investigation of the coherent structure s of the flow behind a backward facing step. Acta Mechanica Sinica, 1993, 25 (2): 129-133. (in Chinese)

[5] Wang J J, Lian Q X. An experimental investigation on laminar boundary layer separation over a backward-facing step. Journal of Beijing university of Aeronautics and Astronautics 1993, 19 (1): 52-56. (in Chinese)

[6] Castro I P. Bradshaw P. the turbulence structure of highly curved mixing layer [J]. J. Fluid Mech. 1976, 73: 265-304.

[7] Sinha S N, Gupa A K, Ober ai M M. Laminar separating flow over backsteps and cavities (part 1: backstep s) [J] . AIAA J. , 1981, 12: 1527-1530.

[8] Tihon J, Legrand J, Lege ntilhomme P. Near-wall investigation of backward-facing step flows [J]. Experiments in Fluids, 2001, 31 (5): 484-493.

[9] Lee I, Sung H J. Characteristics of wall pressure fluctuations in separated and reattaching flows over a backward-facing step: Part I. Time-mean statistics and cross-spectral analyses [J]. Experiment s in Fluids, 2001, 30 (3): 262-272.

[10] Lee I, Sung H J. Characteristics of wall pressure fluctuations in separated f lows over a backward-facing step: PartII. Unsteady wavelet analysis [J] . Experiments in Fluids, 2001, 30 (3): 273-282.

[11] Abbott D E, Kline S J. Experimental investigation of subsonic turbulent f low over single and double backward facing steps [J]. ASME J. Basic Engng. , 1962 (84D): 317-325.

[12] Adrain RJ. Pa rticle2imaging Techniques for experimental fluid mechanic s [J] . Annual Review of Fluid Mechanics, 1991, 23: 261-304.

[13] Qi E R, Huang M H, Li W, Li GY, Zhang X, Wu J. Investigation of vortex structure of the 2D backward facing step flow via PIV. Journal of Hydrodynamics, Ser A, 2004, 19 (4): 525-532. (in Chinese)

Studies on Orthogonal Method in Hydraulic Model Test

Chen Huiling   Wu Fusheng

*State Key Laboratory of Hydrology-Water Resources and Hydraulic Engineering,
Nanjing Hydraulic Research Institute, Nanjing, 210029*

Abstract: Hydraulic test is usually a research which includes multi-factor and multi-level, its cost is high and period is long; orthogonal method is suitable for hydraulic test design. Orthogonal method is a way beneficial to comprehensively, scientifically, time-saving and investment-saving designing test, by the example of arranging the test used orthogonal method.

Key words: orthogonal method; model test; test design

1. Introduction

The hydraulic test draws up the outline and technical line according to the requirements of the test. When the test is carried on specifically, it is necessary to design the test. Orthogonal method is a way applied to test design.

Orthogonal method belongs to the category of optimization. Orthogonal method can solve test problems, such as multi-factor, multi-level, multi-index, and the like. It resorts to the ingenious arrangement of orthogonal form, making less times of the test, more resolvedproblems, and more comprehensive results. Orthogonal test can make the effect of numerousfactors on test indicatorclear, and ascertain the relationship between each factor and indicator, how the factor changes with the level, combined with the change range of various factors. Possible scheme of combined parameter can be chosen. The direction of further optimization and thorough test inquiry can be defined by orthogonal experiment.

Because orthogonal form have the characteristic of neat comparability, balanced dispersion, ensuring the statistical foundation of orthogonal test, the large number of combined test picture can be reflected as long as we can accomplish a small group of tests, the way that test is arranged by orthogonal form is calledorthogonal method. Orthogonal design method has a conspicuous advantage over usual testdesign method.

Research on hydraulic test is more often than not optimized contrast under multi-factor and multi-level conditions, and a process of systematically probing into mechanisms, therefore it is suitable for adopting orthogonal method.

As we all know, the cost proportion of test onhydraulic test research is high, and the permissible research period is short, orthogonal method used to design the test is a way beneficial to comprehensively, scientifically, time-saving and investment-saving designing test.

2. The general procedure of orthogonal forms arranging the test

In general, the procedures of orthogonal forms arranging the test are mainly as follows:

(1) Make clear the purpose of the experiment, list the inspected factors and the level of each factor, list the factor bit-level (level) table and set the observed indicator.

(2) Select the orthogonal table, fill in the orthogonal form, this step is usually called the design for the table header.

(3) The experiment is carried out by test combination on the orthogonal form, and the experimentdata is calculated in accordance with requirements of index calculation.

(4) Analyze results, output achievement table and drawing with the help of computer calculation.

(5) After selecting the optimum scheme, doverification test, or do the scattering small network on the basis of the optimum scheme.

3. Orthogonal method adopted in design of hydraulic model tests

Generally, the orthogonal method is applied to a

hydraulic test, it would be best to plan the test procedure at the very start, and identify observed factors, the variation range of each factor besides the grade or level divided into. The number of factors and the density of horizontal classification can be confirmed in line with the importance of the test, the priority of the task, the model area, equipment conditions, process capability, the speed of process and installation, the workload of changing groups and the amount of funds.

3.1 Ascertainment of the factor, level and test indicator

Take the shape optimization of middle hole in an arch dam for example. It is used to explain the selection of factor and the concrete content of test design. The original design of the middle hole which leads to the sluicing tongue hitting the bank slope and attacking the river bed under the dam, makes threats to the safety of dam foundation, therefore the design needs improvements. The location of central and top hole is not easily moved, so it has no choice but to alter the shape of middle hole to satisfy the needs of flow pattern and energy dissipation, meanwhile middle hole should ensure sufficientoverflow dischargeafter optimization. The revised scheme adopts constringent energy dissipator with a drop-step and a small threshold. We can arrived at a conclusion that the effect of energy dissipator bears some relation to the length of constringent section, constringent ratio, the height of drop-step and the height of a small threshold, hence take the length of constringent section, constringent ratio, the height of drop-step and the height of a small threshold as observed factors. Take the workload and time spent on test into consideration, each factor is divided into three levels.

When we specifically analyze and inspect each factor, constringent ratio is considered in the first place, referring to the regular pattern of constringent ratio determined in materialrelevant to slit-type energy dissipation, the flow velocity is generally of a high level in slit-typeenergy dissipator, however, the Fr of flow in middle hole we studied is low, the effect of water flow guidance is inferior to general slit-type, constringent ratio is supposed to be a little larger than that of slit-type, so constringent ratio is set for three levels: 0.647, 0.529, 0.412. The length of con-

stringent section we observe is in the second place, it is confined to stress of dam and construction conditions, for this reason it is notsuitable to be large, initially ranging from 3m to 5m, and is also classified into threegrades: 3m, 4m, 5m. Drop-step of constringentsection is another factor to be considered. If the sill is located in rapid section, it is liable to give birth to low pressure in drop-step section, and negative pressure is detrimental to the safety of middle hole, consequently the height of sill is preliminary determined from 1m to 3m, and three levels classified is as follows: 1m, 2m, 3m. The final factor is small threshold of outlet of constringent section which is set for the sake of efficient energy dissipation, the height of threshold is 0m, 0.5m, 1m. Thus it can be used to determine whether the threshold is set or not and the height of threshold if the threshold is needed.

From the analysis above, we can see that only we master professional knowledge and be familiar with the hydraulic test, can we identify the factor and level. Concrete factor and level is detailed in Table 1.

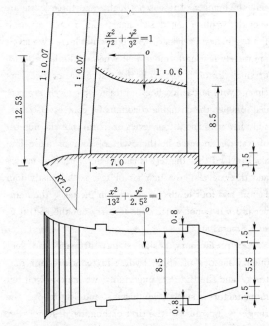

Figure 1 Schematic diagram of mesoporous in arch dam

It is appropriate to select three or four factors and two or three levels in hydraulic test generally. Selecting too many factors and classifying too dense levels makes the number of a set of tests arranged in an orthogonal form high, and the chance of combining secondary factorwith indisposed level is good among numerous factors. As far as combination test itself is concerned, it is unnecessary to do the test. But if it is listed in orthogonal form, we have to do it. The reason is that it is troublesome to calculate and comprehensively analyze experimental result by means of orthogonal method withoutthese experimental data. Based on practical experience and characteristic of hydraulic test, it is appropriate to select three or four factors and classify two or three levels. If the number of parameter is quite high, the test can be carried out at twice. First do the test with few levels, then many levels. Primary and secondary factors are judged by rough test, further the primary factor is divided into several levels for thorough test. It is much more economical and faster to arrange test in this way compared to the arrangement that numerous factors are arranged once and excessive levels are classified.

Table 1

Factors / Level (Grade)	A The height of sake (m)	B Length of contraction (m)	C Contraction ratio of constructional plane	D The depth of falling-sill (m)
1 levels	$A_1 = 0$	$B_1 = 3$	$C_1 = 0.647$	$D_1 = 1.0$
2 levels	$A_2 = 0.5$	$B_2 = 4$	$C_2 = 0.529$	$D_2 = 2.5$
3 levels	$A_3 = 1.0$	$B_3 = 5$	$C_3 = 0.412$	$D_3 = 3.0$

If we select three or four factors to do orthogonal test, choosing three levels is considered the best if conditions permitted. Since factors are divided into three levels, the number of test is not high, and three levels are enough for observing the change trend of each factor. If the test indicator changes with horizontal line, two levels are merely needed to judge that whether the large factor or small factor is better. If the test indicator does not change with horizontal line, at least three levels are needed to judge the changing condition. Of course, it is possibly better to adopt four levels or more, but the number of test can increase in the event of four or more levels. Provided that the test is carried out with four factors and three levels, the number of test is 9. Equally four factors and four levels are adopted in the test, the number is 16. It is not worthwhile to carry out the test in usual circumstances.

In preliminary design stage, primary and secondary factors of some hydraulic tests are not distinct, and the figure is uncertain, we may as well list more factors, select two levels, and figure out the range by indicator of the first orthogonal test in order to judge primary and secondary factor. The secondary factor with minor range is abandoned, and the primary factor with large range is reserved to make the second orthogonal optimization. Thus we can do the second orthogonal test more thoroughly.

A viewpoint deserves to be mentioned, when we select factors, the independence of every factor selected to carry out orthogonal test is remained to be considered, and whether the factors interact each other or not. By interaction we mean that the joint action of several factors has a notable influence on the test index. If the interaction exists among the factors, some special treatments needs taking to design the table header, it amounts to adding the factors giving rise to interaction into considered factors. For instance, the interaction exists between factor A and factor B, described as $A \times B$, in that case A and B respectively occupies a line in orthogonal form, $A \times B$ is supposed to occupy a line and put in interaction line between A and B. The orthogonal form in which interaction is taken into account has ready-made pattern. Hydraulic test stands a little chance of conditions above, and we expect to avoid orthogonal test with interaction to the best of our ability. Owing to the existence of interactive line, the optimizing factor orthogonal form holds can be less, and it brings trouble to the work of test and analysis.

The index of orthogonal test is regarded as the standard weighing the good of the design scheme against the bad. The test indexes of orthogonal test are as follows: pressure, fluctuating magnitude, velocity of flow, vibration, aeration, scouring depth, flow pattern, and so forth. The test index varies correspondingly with the object of problems studied. We require the selected index to be explicit and simple, and reflect the good and bad of test directly, so we are required to acquaint with professional knowledge and test experience. Selecting proper index is favorable to project, and then bring about difficulties to analysis.

In the example, overflow discharge of middle hole must satisfy the design requirements, the water flow should be widen in vertical direction and far away from the bottom of dam in the air, the unit energy of water flow should be tiny after falling into the river channel. There are two measures to identify the test index: if it is the fixed bed test, the indexes weighing the scheme are overflow discharge Q, the jet trajectory distance on outer edge of water tongue L_1, the jet trajectory distance on inner edge L_2, and so on, take Q, L_1 and L_2 as test index. If it is the mobile bed test, take Q, t (the depth of scour hole) and L (the distance from scour hole to the dam bottom) as the test index. Q, L, t mentioned above are ration test indexes, the measured value of every group in orthogonal form is put into form or computer to analyze and calculate. Sometimes a scouring method is set according to the importance of each index, also the importance among the Q, L_1, $L_1 - L_2$, the fraction is used to analyze and calculate. If the index is not ration, we give the emerged results a grade in advance, easy to orthogonal analysis. For example, flow pattern can be disposed in this way. Even if it is ration index, we change the magnitude of index reasonably to facilitate calculation. Adding or subtracting the same value does not affect the results of analysis.

3. 2　The selection of the orthogonal form and the design of table header

When the factors and levels are identified, it is high time to select a proper orthogonal form. The selection of orthogonal form rests with the number of factors and divided levels. Because general users directly apply orthogonal form made by professionally statistical mathematician rather than make it on their own. Selecting orthogonal form works on the principle that the number of studied factors is not more than the number of line of selected form. Since each line holds only one factor. The number of levels is also not more than the number of line permitted in selected orthogonal form. However, the author does not stand for not making the best of capacity of orthogonal form, only when the studied factors and levels are not suitable with the capacity of existing orthogonal form, can these measures be taken. For instance, when it comes to the test with four factors and four

levels, the lines are not enough if we select L_9 (3^4), we have no choice but to select L_{16} (4^5). One line is empty obviously, the capacity of orthogonal form is surplus.

After selecting the orthogonal form, the table header design is in progress. Draw the form, fill the form with factors, classify the levels, and list the levels of each factor of every group of test according to orthogonal form (the experiment condition of every group of test). As the factors and levels in middle hole optimization of arch dam are shown in Table 1, we may as well select orthogonal form L_9 (3^4) shown in Table 2.

Table 2

Column No. \ Test No.	1	2	3	4
1	1	1	1	1
2	1	2	2	2
3	1	3	3	3
4	2	1	2	3
5	2	2	3	1
6	2	3	1	2
7	3	1	3	2
8	3	2	1	3
9	3	3	2	1

The condition of each group is obtained in Table 3, for example, the condition of first group is manifested in the first line, namely, the height of small threshold is 0m, the length of constringent section is 3m, the constringent ratio is 0. 647, and the height of drop-step is 1m. The conditions of other groups are obtained in like manner. The test may be carried out in sequence, or in casual sequence. But when hydraulic test changes groups, the model needs alteration and modification, which brings about heavy workload. So we put the groups with difficult change and heavy workload together to the best of our ability, making less to-and-fro change. In the optimization test of middle hole, the model must be remade when the middle hole shrinks once; while the height of threshold, the height of drop-step and constringent ratio varies easily, the workload is little. We can carry out the test in sequence, B_1, B_2, B_3, etc. (change A, C, D on the identical B condition).

Table 3

Test No.	A Sill hei-ght	B Constringent section length	C Constrin-gent ratio	D Dropstep
1	A_1	B_1	C_1	D_1
2	A_1	B_2	C_2	D_2
3	A_2	B_3	C_3	D_3
4	A_2	B_1	C_2	D_3
5	A_2	B_2	C_1	D_1
6	A_2	B_3	C_3	D_2
7	A_3	B_1	C_3	D_2
8	A_3	B_2	C_1	D_3
9	A_3	B_3	C_2	D_1

If the levels of some individual factors are more than that of others in the test, the mixed orthogonal form is selected. Now and then some factors are implicative of each other; the interaction among the factors is to be taken into account. This book doesn't give unnecessary details on the interaction of orthogonal form and mixed orthogonal test, the readers can refer to relevant documents if they need to understand it in detail.

It is flexible to use orthogonal form. Several kinds of existing orthogonal form are provided to be selected. The orthogonal form is identified according to the concrete requirements of hydraulic test, the characteristics of project and the test condition. As long as we master the fundamental regular pattern and principle for applying orthogonal form, it is not difficult to select a proper orthogonal form.

3. 3 The method of Analysis

When the factors and levels have been defined, orthogonal form selected, the table header designed, data obtained after the test and filled in the table completely according to the stipulation of orthogonal method, choosing proper analytic method is taken into consideration, and the method can analyze obtained information correctly and carefully to arrive at a credible conclusion.

Pros and cons of indexes are easy to judge when the test is carried out with a single index, besides the optimum combination can be obtained through the level number of each factor and range analysis. If when we undertake research into the problems of energy dissipation and erosion control, and the index is nothing but the depth of scour hole, it is easy to use

the depth of scour hole to measure the effect of energy dissipation and erosion control, therefore, the optimum scheme is picked out easily. When we use orthogonal method to arrange the hydraulic test with multi-indexes, synthesized and balanced analysis method (shown in Table 4) is adopted to evaluate the impact of each factor on the indexes. In the example, three indexes are selected: Q (overflow discharge), L_1 (jet trajectory distance), L_2-L_1 (vertically tensile length of water tongue). The most appropriate level corresponding to each indicator is not completely uniform, so we can consider the major factor with large range in the first place, combined with the conditions on constructions, structural stress and quantity of work, then take a moderate scheme.

Analysis Method: Whether the orthogonal trial is single-indicator or multiple-indicator, they are both fit to use the method of directly observing the index. Of course statistical analysis method can be used as well. Statistical analysis consists of range analysis and variance analysis. Each indicator can be counted and analyzed to determine the effects of levels and the function of factors.

The results in the table can be analyzed in the form of drawing to measure up the effect of tow pole's change of various factors on the indicator.

The scheme which is on the optimum shape of constringent energy dissipater in the outlet of the middle hole in a hydropower station, is obtained by analysis and calculation of the orthogonal test.

3. 4 Questions of experimental verification and scattering a small net

(1) Experimental verification. When we use the orthogonal method to arrange hydraulic tests, the conclusion through final analysis is that the level most suitable for each factor forms the optimum scheme. For example, the optimum scheme adopted in optimization test on middle holes in the dam is $A_1 B_3 C_1 D_2$, but the optimal scheme is not necessary the test group contained in the orthogonal form, even though it is sometimes unified, the best set of experimental data shown in orthogonal form does not represent or fully represent all results of the optimum scheme, and test data obtained in orthogonal optimization is far from satisfying our requirements of hydraulic tests. Therefore, in normal circumstances, the optimal scheme selected should undergo systemat-

ical test again, and this is called the final test for scheme in hydraulic test, in which more detailed and more complete information can be measured to be the basis for design parameters.

Table 4 **Analysis table of test result**

No.			A	B	C	D	Test date			
			1	2	3	4	X_i	Q (m³/s)	L_1 (m)	$\Delta l = L_1 - L_2$
1			1 (A_1)	1 (B_1)	1 (C_1)	1 (D_1)	x_1	$Q_1 = 950$	$L_{11} = 69$	$\Delta l_1 = 18$
2			1 (A_1)	2 (B_2)	2 (C_2)	2 (D_2)	x_2	$Q_2 = 960$	$L_{12} = 60$	$\Delta l_2 = 26$
3			1 (A_1)	3 (B_3)	3 (C_3)	3 (D_3)	x_3	$Q_3 = 820$	$L_{13} = 66$	$\Delta l_3 = 23$
4			2 (A_2)	1 (B_1)	2 (C_2)	3 (D_3)	x_4	$Q_4 = 915$	$L_{14} = 65$	$\Delta l_4 = 24$
5			2 (A_2)	2 (B_2)	3 (C_3)	1 (D_1)	x_5	$Q_5 = 690$	$L_{15} = 57$	$\Delta l_5 = 35$
6			2 (A_2)	3 (B_3)	1 (C_1)	2 (D_2)	x_6	$Q_6 = 1100$	$L_{16} = 76$	$\Delta l_6 = 11$
7			3 (A_3)	1 (B_1)	3 (C_3)	2 (D_2)	x_7	$Q_7 = 700$	$L_{17} = 45$	$\Delta l_7 = 39$
8			3 (A_3)	2 (B_2)	1 (C_1)	3 (D_3)	x_8	$Q_8 = 1055$	$L_{18} = 66$	$\Delta l_8 = 18$
9			3 (A_3)	3 (B_3)	2 (C_2)	1 (D_1)	x_9	$Q_9 = 830$	$L_{19} = 68$	$\Delta l_9 = 28$
The analyzation and calculation of test achievements	Analyzing Q	I	$Q_1+Q_2+Q_3=2730$	$Q_1+Q_4+Q_7=2565$	$Q_1+Q_6+Q_8=3105$	$Q_1+Q_5+Q_9=2470$	The sequence of each factor affecting Q is $C \to D \to B \to A$, the optimum combination is $A_1 B_3 C_1 D_2$			
		II	$Q_4+Q_5+Q_6=2705$	$Q_2+Q_5+Q_8=2705$	$Q_2+Q_4+Q_9=2705$	$Q_2+Q_6+Q_7=2760$				
		III	$Q_7+Q_8+Q_9=2585$	$Q_3+Q_6+Q_9=2705$	$Q_3+Q_5+Q_7=2210$	$Q_3+Q_4+Q_8=2790$				
		Range R_1	$2730-2585$ $=145$	$2750-2565$ $=185$	$3105-2210$ $=895$	$2790-2470$ $=320$				
	Analyzing l_1	I	$l_{11}+l_{12}+l_{13}=195$	$l_{11}+l_{14}+l_{17}=179$	$l_{11}+l_{16}+l_{18}=213$	$l_{11}+l_{15}+l_{19}=194$	The sequence of each factor affecting l_1 is $C \to B \to A \to D$, the optimum combination is $A_2 B_3 C_1 D_3$			
		II	$l_{14}+l_{15}+l_{16}=198$	$l_{12}+l_{15}+l_{18}=183$	$l_{12}+l_{14}+l_{19}=191$	$l_{12}+l_{16}+l_{17}=181$				
		III	$l_{17}+l_{18}+l_{19}=179$	$l_{13}+l_{16}+l_{19}=210$	$l_{13}+l_{15}+l_{17}=168$	$l_{13}+l_{14}+l_{18}=197$				
		Range R_2	$198-179$ $=19$	$210-179$ $=31$	$213-168$ $=45$	$197-181$ $=16$				
	Analyzing Δl	I	$\Delta l_1+\Delta l_2+\Delta l_3=67$	$\Delta l_1+\Delta l_4+\Delta l_7=81$	$\Delta l_1+\Delta l_6+\Delta l_8=47$	$\Delta l_1+\Delta l_5+\Delta l_9=81$	The sequence of each factor affecting Δl is $C \to B \to A \to D$, the optimum combination is $A_3 B_1 C_3 D_1$			
		II	$\Delta l_4+\Delta l_5+\Delta l_6=70$	$\Delta l_3+\Delta l_5+\Delta l_8=79$	$\Delta l_2+\Delta l_4+\Delta l_9=78$	$\Delta l_2+\Delta l_6+\Delta l_7=76$				
		III	$\Delta l_7+\Delta l_8+\Delta l_9=85$	$\Delta l_2+\Delta l_6+\Delta l_9=62$	$\Delta l_3+\Delta l_5+\Delta l_7=97$	$\Delta l_3+\Delta l_4+\Delta l_8=65$				
		Rage R_3	$85-67=18$	$81-62=19$	$97-47=50$	$81-65=16$				

Table 5

The living example of arranging the test by orthogonal method

Name of the pilot project	The purposive use orthogonal method	Adoption of orthogonal table	Factor	Level (Grade)	Measurement of Indicators	The method of Analysis	Conclusions
The Guizhou Nanpanjiang Tianshengqiao secondary Hydropower Station of hydraulic model test (1 : 100)	The study of two construction cofferdam removal program in the overall hydraulic model	L_9 (3^4)	A: tidal level B: style of sand sluicing gate apron C: style of removed cofferdam	$A_1=645$, $A_2=640$, $A_3=637$; B_1=flat bottom, $B_2=15°$ inclined oriented angle, $B_3=15°$ directly oriented angle; $C_1=0.2cm$, $C_2=0.52cm$, $C_3=2cm$	I : Maximum scour depth riverbed; II : lowest point elevation of scour hole; III : the distance between guide wall and the scour hole; IV : dig depth of guide wall	Two kinds of analysis method: 1. Rang method, 2. Deviation analysis of variance	1. Factors impact indicators strongest, 2. Tax impacts indicators weakest (The reason is flow pattern has been preferred over), 3. Beneficial for erosion (1980)
The Fujian Mianhuatan Hydropower Station hydraulic model test (1 : 100)	Optimization of energy dissipation spillway dam surface shape in the hinge model	L_4 (2^3)	A: The establishment of flaring pier location B: Bucket trajectory angle C: The wide at the end of sluice gate	A_1: Dam surface A_2: the end of bucket; B_1: bucket angel 34° B_2: bucket angel 0°; C_1: 4.8m, C_2: 6.4m	Horizontal length of jet flow; the shape, depth of scouring pit, Distance to bucket	Rang method	The best program is that flaring gate piers located at the spillway at the end, through test and range analysis; bucket angel 0°, hole width of bucket end 4.8m (1995)
The shape optimization test of discharge bottom hole of Fujian Mianhuatan Hydropower Station (1 : 100)	The design parameter of face slab curve is identified by optimizing the shape of bottom hole inlet, adopts orthogonal method to arrange physics model test and electrical calculation of sluice opening	L_9 (3^4)	A: the long axis of ellipse B: the short axis of ellipse C: the angle between the axis of ellipse and horizontal line D: Terminal plate empty roof slope	$A_1=8m$, $A_2=10m$, $A_3=12m$; $B_1=3.5m$, $B_2=4.0m$, $B_3=5m$; $C_1=0°C_2=10°$, $C_3=20°$; $D_1=1:4$, $D_2=1:6$, $D_3=1:10$	The overflow discharge of bottom hole, the pressure of roof (P_{min}, P_{max}) flow pattern	The result is obtained by model calculations, and then lists the analysis form, the pros and cons of scheme are analyzed by range	The optimum long axis of elliptic in the roof of bottom is 12m, short axis is 5m, the angle between axis and horizontal line is 0, the gradient at the toe is 1 : 40 (year 1995)

(2) The scattering of small network. If the primary factors are not detailed enough and need careful consideration after the first orthogonal test finished, several thin grades should be set around the optimal level of the primary factors, then we select orthogonal table of the second batch of orthogonal test and carry out the second batch of the test. This is called the scattering of small networks or refined test, entering into the excellent stage.

In scattering a small network stage, the principal feature is that general factors are supposed to be fewer than that of the first batch of test, the grade dimension of various factors should also be smaller and its relative accuracy should be higher. Design process of test, for example, selection of orthogonal table, the design of table header, test methods, analysis and calculation, are all the same as the first batch of test.

The above discussions are some main issues of hydraulic test design in which orthogonal method is used. We will run into many problems if we take further research into the design method. For instance, how to determine the credibility of the orthogonal test data, how to choose a mixed orthogonal table when the level of various factors varies, how to carry on statistical analysis, and how to design the table head and analyze the material when interaction exists between the factors. . . As for these problems, the readers can refer to literature on applied probability and statistics to find out the answers.

4. The application of orthogonal test design in hydraulic experiments

Design method of orthogonal test is a kind of method that saves time, effort and funding, and can achieve perfect results if used in hydraulic tests, for instance, the test is economical and the result is both comprehensive and reliable.

Hydraulic test is usually research with multi-factor and multi-level into parameters design. Hydraulic tests used to be done generally by the experience and the knowledge of staffs, optimization process often fixes the condition of partial parameters firstly, then we study the changes of one parameter, the optimized parameters are considered to be the optimal val-

ue. Then we remove the other fixed parameters and repeat optimization above until all parameters are optimized. This optimization process is named after "hill-climbing of a blind ". It is inevitable to omit an excellent program for the test, although it is not all combinations of sub-tests. Even though the experienced researchers do it, the workload is considerable, and the test is wasting time, labors and funding. The efficiency can be promoted by using orthogonal method. Decided by mathematical characteristics of orthogonal method, it reflects the phenomenon comprehensively, so it hardly misses a nice scheme.

In the past few decades, Nanjing Hydraulic Research Institute has adopted the orthogonal test design method in many projects and received pleasing results. The method can be flexibly used in the hydraulic model, such as the shape optimization of spillway, flood discharge tunnel, aeration slot and the step-type dissipater etc. Specific examples are filled in Table 5.

In addition, orthogonal method can also be applied in calculation of numerical model. As we all know, more approximate calculation of parameters is needed in numerical simulation of hydraulic calculation, although the numerical simulation is much faster than the physical model test. However, if all combinations are calculated, the workload is very heavy. If we adopt orthogonal method to arrange the groups of computation and analyze results, it will speed up the numerical calculation and improve the reliability of conclusion. The orthogonal method is used in the calculation of bottom shape in Cotton Beach Hydro-electric Engineering, as is shown in Table 5.

5. Conclusions

(1) Orthogonal method is an excellent method of test design. It is widely applied in agreat many industries at home and abroad, and achieves pleasing effects.

(2) Hydraulic test is usually multi-factor and multi-level, its cost is high and period is long. Therefore orthogonal method is suitable for hydraulic test design.

Operation and Appraisal of Vertical Ship-lift at Shuikou Hydropower Station in Fujian Province

Hu Xiaowen

HYDROCHINA HUADONG Engineering Corporation, *Hangzhou*, 310014

Abstract: The navigation facility at Shuikou Hydropower Station, 2×500t vertical ship-lift, which is of wet full-balanced wire rope-hoisted type with its maximum vertical lift range of 59.00m, is a large-sized one, ranking first among the overseas and domestic ship-lifts of the kind. It is a system engineering involving multi-specialties and multi-disciplines. With the unremitting efforts of the constructors, it was finally completed and successfully put into operation. This article gives a brief presentation with respect to the operation and appraisal of the aforesaid ship-lift.

Key words: Shuikou Hydropower Station; vertical ship-lift; equipment operation; scientific and technical appraisal

The 2×500t wet full-balanced wire rope-hoisted vertical ship-lift at Fujian Shuikou Hydropower Station is a national major technical equipment development project of the Ninth Five-Year Plan and is a system engineering involving multi-specialties and multi-disciplines. It is the largest among the ship-lifts of this kind in China and ranks among the first in the world as well. Years of navigation operation practice indicates that all functions of the ship-lift reach the design requirements, the time for ships passing across the dam is shortened, the navigation capacity is enhanced, the water saved increases power output and the expected effects have been achieved. After the ship-lift was put into operation, it is realized that the two navigation facilities at Shuikou Hydropower Station operate as standby to each other, which improves the navigation reliability and safety, effectively mitigates the bignavigation pressure on the Minjiang River and has great significance for promoting the development of inland water transport in Fujian Province.

Research on the ship-lift technology has been carried out for years in China, and the achievements thereof were obtained in the periods of both the 7th and 8th Five-Year Plans. However, great stride was not made in application of the technology until construction of the ship-lift at Shuikou Hydropower Station. During construction, several new technologies were successfully applied by all the construction parties regarding design, supervision, manufacture, construction, installation, debugging, testing and operation, and an important breakthrough was made in the ship-lift construction sector. The success at Shuikou becomes a milestone in the history of ship-lift construction in China and provides valuable experience and example for construction of the large-sized vertical ship-lift in navigable rivers with high dams, especially for the Three Gorges Project. Therefore, it symbols the new development in navigation technology for rivers with high dams in China and even in the world.

1. Generals of the ship-lift at Shuikou Hydropower Station

Shuikou Hydropower Station is located on the trunk stream of the Minjiang River in Minqing County, Fujian Province. Completion of the reservoir has greatly improved the navigation conditions from Nanping to Fuzhou. In order to make the Minjiang River become a real golden waterway in Fujian, according to the navigation facilities scheme finalized by the National Plan Commission and Fujian Province in March 1987, a single-line 3-flight ship-lock and a single-line vertical ship-lift were arranged on the right bank of the dam so as to meet the scale requirement for an annual freight volume of 4 million tons and log rafting capacity of $2×10^6$-$2.5×10^6$ m³/a. The fleet is composed of a pusher and two barges with total tonnage of $2 \times 500t$, the highest navigation water level is 65.00m upstream and 21.80m downstream ($P = 50\%$), and the lowest navigation water level is 55.00m upstream and 6.00m downstream.

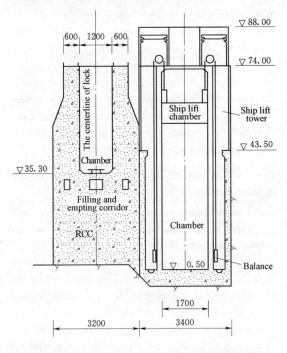

Figure 1　Transverse section of shiplock and ship-lift

The ship-lift at Shuikou Hydropower Station is of wet full-balanced wire rope-hoisted type and the main structures include upper and lower navigation guide walls, upper and lower lock-heads and hoist rooms, bearing tower building, traffic tower building and main machine room. The main electro-mechanical equipment cover main lifting system, boat carrier and equipment, balance system of ship-lift, safety locking device, water retaining gate and hoisters at the upper and lower lock-heads, the boat carrier connection device, electric dragging system, detection system and the computer supervisory control system, etc.

As the large-sized vertical ship-lift is a technically complicated system engineering, in order to ensure its safe and reliable operation, a series of research activities to make technological breakthrough were carried out. The main research results achieved include the following: 1) determining the ship-lift design parameters suitable for operation conditions of the hydropower station; 2) designing the technical measures to ensure the levelness error of the boat carrier under control in the whole operation process; 3) successfully applying for the first time the testing method of the large scale, high simulation integral physical model into systematic test and research on the multi-disciplines of the ship-lift machinery, electric dragging and control, boat carrier balance technology and hydrodynamics, etc. and its success initiates a new testing method for the construction of the large-scale ship-lift; 4) designing the asymmetric tall and slender aseismic structure and construction technical scheme.

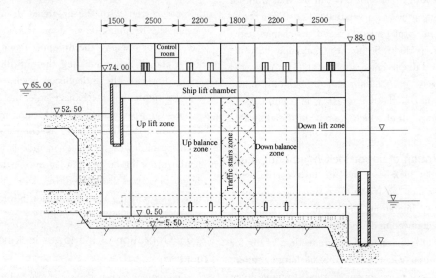

Figure 2　Longitudinal section of the ship-lift at Shuikou Hydropower Station

Construction of the large-scale ship-lift belongs to a bran-new field and the ship-lift project is a system engineering involving various new technologies. With the designer's aborative design, the project owner's normalized management and the constructors' years of unremitting efforts, the ship-lift was smoothly put into navigation. At present, the ship-lift is in good operation as a whole and all technical indicators meet and even exceed the design requirements.

2. Consolidated Indicators for Operation

2. 1 Operation of hydraulic structures

The ship-lift structures are composed of such sections as upstream guide navigation, upper lock-head, upper working gate, tower upper lifting section, tower upper balance, tower traffic ladder, tower lower lifting, lower lock-head (including the lower working gate), and downstream guide navigation. In order to monitor the operation safety of the ship-lift, several observation items regarding deformation, stress, strain, joint measurement and temperature are designed at all positions of the ship-lift, totally there are 142 built-in observation instruments, 6 holes for measuring inclination, 12 points for settlement observation, 2 points for relative displacement monitoring, 10 perpendiculars (20 measuring points), 5 measuring points and 9 components of the seismometer for strong shock. It can be seen from all the monitoring process curves that all positions of the ship-lift are in normal operation and the change regularity appears periodic. So far, the maximum horizontal and vertical displacements of the ship-lift gate wall are respectively -10.69mm and 13.26mm (to the settlement point at the dam crest), maximum opening is 3.99mm, all of which are within the allowable design range.

2. 2 Operation of upper lock-head equipment

The upper lock-head and the boat carrier were connected for more than 280 times and the equipment were in normal operation. The average action time of all upstream connection devices is as follows: 108s for connection, 58s for removal of connection, 130s for filling interstitial water, 126s for discharging interstitial water, 68s for opening the dropping-down door, 88s for closing the dropping-down door, 450s for boat ingoing and 300s for boat outgoing. Mean-

while, due to rise/fall of the upstream water level for 28 times, it was necessary to regulate the water depth of the boat carrier with average time consumption for water depth regulation of 330s.

2. 3 Operation of main lifting equipment

The main lifting equipment was operated for more than 560 times totally without any Grade 2 pressure regulation emergency brake protecting actions encountered. The main lifting equipment operated mainly at 12m/min and the DC motor was operated within ± 200A, the torques at all lifting spots were less than 1000N • m and more than it occasionally, but it was less than 2000N • m.

2. 4 Operation of the computer supervisory control system

The computer supervisory control system has two upper level machines with control programs for automatic and one-step operation. During trial navigation, all control operation flows were in normal working without any control failure arisen. When the shutdown failure occurred to the ♯2 industrial personal computer, automatic switching to the ♯1 industrial personal computer could be realized. The broadcasting system, the fire protection centralized control center and the unit load terminals were all in normal operation.

2. 5 Operation of the boat carrierequipment

Average action time as required for connection of the boat carrier with the upstream devices: 47s for closing the dropping-down door at the boat carrier end, 48s for opening the dropping-down door at the carrier end, 55s for hoisting the anticollision beam, 56s for lowering the anticollision beam, 41s for putting the tightening mechanism into the right place, and 68s for withdrawing the tightening mechanism in place. When the boat carrier is connected with the downstream devices, the action time is 58s for closing the dropping-down door at the carrier end, 57s for opening the dropping-down door at the carrier end, 56s for rising the anticollision beam and 55s for lowering the anticollision beam.

2. 6 Operation of the lower lock-head equipment

The lower lock-head was connected with the boat carrier for over 280 times. The average action time of all the linkage devices were as follows: 116s for connection of the linking device, 61s for removal of connection, 133s for filling interstitial water, 135s for discharging interstitial water, 69s for opening the dropping-down door, 70s for closing the dropping-down door, 13min for boat ingoing and 5min for boat outgoing. The great impact of the peak and frequency regulation of the units on the downstream water level resulted in water depth regulation of the boat carrier for 159 times with average time consumption of 7min/ time.

The prototype observation test and test-run indicate that in all operation modes, the hydraulic conditions and main technical performances of all equipment of the ship-lift and the boat carrier conform to the design requirements and the technical measures taken has effectively improved the operational safety of the ship-lift.

3. Application and appraisal of the achievements

The technical practice in construction and operation of large vertical ship-lift is of great significance in providing solutions for navigation on rivers with high-head power stations and high dams in China's hydropower engineering sector. In combination with construction of the Shuikou ship-lift, the extensive technical research on the key technologies regarding design, test, manufacture, construction, installation and testing, operational and management was also conducted and technological breakthroughs were achieved, which provided good results.

In order to better popularize the application of the technical achievements obtained in construction and operation of the Shuikou ship-lift, the Science and Technology Department of Fujian Province organized scientific and technical appraisal for the ship-lift project. The participants from different units engaged in the project provided complete, detailed and reliable technical data and research outputs to the technical appraisal committee. The spot test and check on site by the testing team of the appraisal committee indicated that the spot test results were basically in conformity with the test values shown in the testing reports for appraisal. The scientific and technical results are detailed as follows.

(1) The technical feasibility to select a ship-lift of wet full-balanced wire rope-hoisted type is proved by the successful test-run of the Shuikou ship-lift. All main technical parameters decided in the light of the specific conditions of Shuikou Hydropower Station are reasonable. The prototype observation test and test-run indicate that in all operational modes, the hydraulic conditions and main technical performances of the boat carrier conform to the design requirements. All technical measures taken to ensure safe operation effectively improve the operational safety of the ship-lift of this type.

(2) The safety locking device developed for the boat carrier of the ship lock (the national patent technology) is verified correct in the witness of 50% water loss test and the device is well designed in regarding of the mechanism. The boat carrier with excess loss of water can be effectively and reliably locked through automatic control of the computer supervisory control system.

(3) The folded-type sealed installation for connection (the national patent technology) and the scheme of setting the connection device on the lock-head gate can well accommodate to deviations at the boat carrier end, moreover, the sealing effect is favorable, and the connection is reliable.

(4) The asymmetric tower and the slender and aseismic thin-wall structure of the tower (the national patent technology) is designed and successfully adopted by the new technology of sliding sash reverse mould, so that the construction quality of the eight supporting towers with a total height of 79.5m each and two traffic towers reached and exceeded the design requirements.

(5) The manufacture quality of the main lifting mechanical equipment is good, the drum processing accuracy exceeds the design requirements, as a result, the levelness deviation (namely the relative height difference from one end to the other end of the boat carrier with a total length of 123m) of the boat carrier during the rise/fall of 59m is less than the design value (10cm). The hydraulic servomotor closed circuit control scheme was successfully applied for the hydraulic levelness regulation system of the boat carrier for the first time. The levelness regulation system has such characteristics as good anti-interference, and stable and correct levelness regulation operation.

(6) The controllable counterweight device is set in the ship-lift weight counter balance system for the first time, which can effectively enhance the capability of the boat carrier against longitudinal overturning

and facilitate improvement of the operation safety of the wire rope hoist vertical ship-lift.

(7) Design of the braking system of the main lifting equipment satisfies normal operation and requirements for safety protection in case of equipment failure, and the control scheme of braking program for Grade 2 pressure regulation accident adopted for the first time can make the operating boat carrier realize more stable braking and shutdown at the speed of 0. 2m/s.

(8) The schemes of the electrical testing, dragging and computer supervisory control system designed and developed for the wet full-balanced wire rope-hoisted vertical ship-lift are correct with complete equipment and reasonable redundancy configuration. The electric control system has perfect protection functions for equipment and operation troubles. The acting sequence cooperation of the main dragging system with all devices during starting is designed accurate, which can ensure smooth and safe starting, and the steady-state speed is better than 1%. Acting sequence of the main dragging system is designed accurate to all devices during starting to ensure smooth and safe start. The coaxial driving devices' output uniformity of the four electric motors is good, the unbalance current at all spots is less than 3%. Undisturbed switching with speed disturbance less than 5% can be reliably realized in the failure state between the main and secondary stations. The oscillation suppression measures are effectively taken, thus, the boat carrier can be steadily lifted at all operational speeds with positioning error less than ±1cm. The operational flow and frame design of the computer supervisory control system is perfect and reasonable.

(9) The air-cushion buoyancy-increase technology was successfully applied for the boat carrier (123m long, 16. 1m wide and 6. 3m high) of the Shuikou ship-lift, and is for the first time adopted to long distance (100km) marine floating practice; and accordingly the transportation and site installation issues of the ultra-large steel structures were solved.

(10) In the site testing and test-run, a great deal of important demonstration data and experience were obtained, which provide proven reference for the subsequent projects to improve the overall technical performances of the ship-lifts, optimize operation parameters, perfect the design scheme, explore the operation and management systems and regulations of the ship-lift.

The appraisal comments of the expert team are as follows: The scientific and technological achievements obtained in construction and operation of the ship-lift at Shuikou Hydropower Station have reached the international advanced level as a whole, and the technologies such as the safety locking device, the folded connection sealing device and the braking device for emergency in Grade 2 pressure-regulation case of the large-inertia vertical hoisting system developed for the wire rope-hoisted ship-lifts, ranks among the leading international level.

4. Conclusions

It is the first time in China to install 2×500t wet full-balanced wire rope-hoisted vertical ship-lift at a hydropower project, which has been successfully in operation. Facing the reality of multi-disciplinary and-complicated systematic engineering and the difficult conditions, i. e. , lack of specifications, construction experience nor engineering examples, the constructors accepted this austere challenge with a mind of "daring to be the first in the world", the scientific way of research, the spirit of defying hardship and being brave in exploration. After the unremitting efforts of the construction units, the ship-lift at Shuikou Hydropower Station was successfully put into operation in April 2005 and so far it has been in good operation.

The success of the Shuikou ship-lift provides valuable and practical experience for construction of large-scale vertical ship-lift in navigable rivers with high dams and pilots in construction of navigation facilities at high dams in China. With the important breakthrough in ship-lift technical field and several innovative technical achievements, the construction and operation of the Shuikou ship-lift was awarded First Prize for Scientific and Technical Advancement of Fujian Province in 2006 and Second Prize for National Scientific and Technical Advancement in 2007. The construction and operation experience at Shuikou offers a practical reference for China's ship-lift construction industry, such as for the Three Gorges Project, and pushes China's navigation technology for rivers with high dams to the world's advanced level.

Topic IV

Hybrid Model Approach and Combination of Physical Approaches with Numerical Simulation

Advances in Hydraulic Modeling and Instrumentation for Bridge Waterways[*]

Colin D. Rennie[1]   Robert Ettema[2]

[1]*University of Ottawa, Canada*
[2]*University of Wyoming, USA*

Abstract: This lecture focuses on advances in hydraulic modeling and field investigation of rivers, especially at bridge waterways. Bridge waterways commonly involve turbulent flow; erosion, failure, and transport of boundary material; and, loads caused by accumulation of materials (notably, ice and woody debris). Long-standing problems include scour at piers and abutments, channel shifting, and loads against piers. Scaled physical models continue to be used extensively for investigating design and operation issues in hydraulic engineering, including the hydraulics of bridge waterways. Developments in lab and field instrumentation are increasing the technical sophistication of physical model and field investigations, such that highly resolved spatiotemporal distributions of water and sediment flux can be measured. Furthermore, scaling of turbulent structures in physical models has recently been considered. Physical modeling, however, continues to be plagued by imperfect scaling of relevant forces. Hydraulic modeling thus continues to require engineering judgment in model design, use, and interpretation. In the present lecture it is argued that hybrid modeling, i. e., the combined use of scaled physical models, numerical models, and field investigations, is the most effective approach.

Key words: bridge waterway; physical model; hybird modeling

---

[*] keynote lecture

Study on River Restoration Method with Large Stones

Masaki Fukushima   Noriaki Hakoishi

River and Dam Hydraulic Research Team, Hydraulic Engineering Research Group, Incorporated Administrative Agency, Public Works Research Institute, 1-6 Minamihara, Tsukuba City, Ibaraki Prefecture, 305-8516, Japan

Abstract: In most Japanese rivers, it is a principal problem of maintaining and improving river channel where gravel-bar area have decreased with forestation at flood channel, which is generally known being caused by construction of cross river structures, gravel mining and so on. As a result of bed degradation, bed rock or hardpan is exposed in several rivers. Recently, large stones are used as a countermeasure for the problem to keep sand and gravel on bed rock or hardpan. However, it is not clear that what rate of large stones and how they should be set in the river. So, effect of large stones on sedimentation was studied in a flume, where cobble was used as a large stone. As one conclusion of some, it was clarified that sediment discharge decreased by 20 to 30 percent of sediment discharge in case of no large stone after setting several large stones on river bed with its half height covered with original riverbed materials.

Key words: experimental Study; bed rock; hardpan; sedimentation; deposition

1. Introduction

Many rivers in Japan have been afflicted with the decline of gravel bars as well as in-channel tree growth mainly because of the construction of cross-river structures and gravel excavation[1]. Because of bed degradation arising in connection with these problems, channel sections with exposed bedrock or soft rock caused by the loss of unconsolidated deposits, which are typically found in mountain stream sections, began to appear in some sections of alluvial river channels. The Tedori River, whose channel is so stony that Ishikawa (which literally means "stone river" in Japanese) Prefecture was given its name, and the Tama River and the Kinu River, both of which are known to have had large gravel bars, are among the rivers where exposed bedrock or other rock surfaces are found today. These exposed rock sections cause various flood control and river environment problems. For the purposes of this study, riverbed restoration is defined as the act of making sand and gravel deposit on such river beds and restoring alluvial river channels with continuous layers of unconsolidated deposits.

In recent years, experimental attempts have been made in some rivers with exposed rock channels to restore unconsolidated deposit layers by using large stones. For example, a field test began in 2008 in the Tedori River, and the ongoing project for the Makomanai River[2], a tributary of the Toyohira River in the Ishikari River System, can be regarded as a pioneering effort. The project aims to restore a layer of unconsolidated deposits by placing groins consisting of interconnected boulders so as to deposit the sand and gravel transported from upstream. In order to design the groins, a hydraulic model test was conducted. A number of groin configurations were field-tested, and their effectiveness is gradually being verified. In view of the continuity with the sediments in the upstream and downstream channels, there is concern about possible adverse effects such as the degradation of the downstream channel bed. Such adverse effects, however, can be mitigated if sediment transport can be controlled appropriately by finding appropriate arrangement patterns of large stones. This approach is also thought to be effective for rivers where a thin layer of gravel is deposited on a hardpan such as the Tama River and the Asa River. When considering what to do with an exposed-rock channel, how to fix large stones themselves is also an important consideration, and riverbed restoration may be very difficult to achieve. In a river channel with a gravel layer, however, large half-embedded stones can be expected to be effective to some extent. Such large-stone arrangements may also be effective in preventing rock exposure. Although "large stone" is not a clearly defined engineering term, it is deemed for the purposes of this study to refer to bed material that is not easily moved by ordinary flood flows.

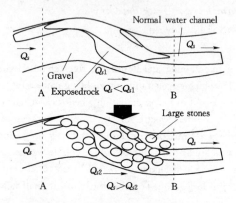

Figure 1   Concept of riverbed restoration
by use of large stones

Figure 1 illustrates the concept of riverbed restoration. The idea is to control the amount of sediment being transported and promote the deposition of gravel in a channel section with a thin layer of unconsolidated deposits such as an exposed-rock bed section. The function is similar to that of sedimentation control facilities in the coast. In Figure 1, Q_s, Q_{s1} and Q_{s2} are sediment discharges in a channel section with a gravel layer, an exposed-rock channel section and a channel section where large stones have been placed, respectively, through cross sections A and B during a certain period, for example, during a single flood. Once rock is exposed, the roughness of the riverbed is thought to decrease so that gravel deposition is reduced and Q_{s1} becomes greater than or equal to Q_s. The effect of large stones was verified by Ashida, Takahashi and Mizuyama[3] through flume experiments. They report that there is no significant effect if the percentage by weight of large stones is 5% or so, and that sediment discharge may be reduced if the percentage is 10% or higher. Because a sediment discharge formula applicable to mixed-size materials including large stones did not exist about thirty years ago, Okabe and Himoto[4] proposed a method for calculating sediment discharge in a field where large

stones exist, taking into consideration the occurrence of restricted sediment flow regions due to large stones and their wake zones and changes in effective tractive force due to the form drag generated according to the degree of protrusion of large stones. This method is useful from the viewpoint of sediment transport formulation refinement, but its practical usefulness is limited mainly because the formula includes many parameters. To be more specific, it is difficult, for example, to determine the percentage of large stones needed to achieve the goal of sediment discharge reduction, and it is also difficult to determine the percentage of large stones needed to maintain a very thin layer of unconsolidated deposits.

In this paper, the effects of different percentages of large stones in the surface layer on sediment discharge were investigated experimentally. The rate of sediment transport, the transport velocity of gravel and the thickness of the transported sediment layer were also measured in a different condition. Following is an organization of the paper. Section 2 briefly describes the experiment, Section 3 reports the experimental results, and Section 4 discusses key considerations for riverbed restoration based on the experimental results.

2. Flume experiment on the influence of large stones on sediment transport

The flume used in the experiment is a variable slope flume that is 30 meters long, one meter wide and 0.8 meters deep (Figure 2). Bedload that dropped into the sieving system at the downstream end is transported to the upstream end by a belt conveyor, and it is supplied at the upstream end through a bucket as bed material. Table 1 shows the experiment conditions and the properties of bed material. Case 1 to Case 6 are the bed conditions created by placing large-grained gravel with a diameter of around 100mm in the initial bed material (Case 0) shown in Figure 3 according to the percentages shown in Table 1 as P_{LS} (Figure 4). Hereafter, the large-grained gravel used in the experiment is referred to "cobble". As shown in Figure 5, the initial bed material shown in Figure 3 was laid on the flume bed to a thickness of 20cm, and the lower halves of the cobbles were embedded in the initial bed material. When the cobbles were placed, part of the initial bed material corresponding in weight to the embedded halves of the cobbles was removed. The state of the half-embedded cobbles did not change significantly because variation of the bed height was not so notable. When grain size distribution is calculated, calculation results vary depending on the thickness of the bed material layer from which to take samples. In this study, grain size distribution was calculated for the material to the depth of 100mm, which is the grain size of the cobbles, including the cobbles protruding from the bed surface. When the flow rate was 0.52m³/s and the flume slope was 1/100, which are the conditions under which tractive force is maximum as shown is Table 1, little movement of the cobble was observed. The bed material used in Case 0 was prepared by mixing together uniform material with grain sizes of 6.4mm, 9.4mm, 15mm, 25mm and 40mm.

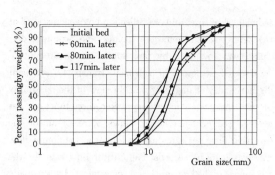

Figure 3　Grain size distribution of initial bed material and transported sediment (Case 0)

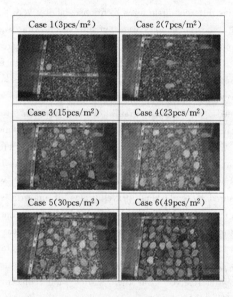

Figure 4　Placement of the cobbles
(number of particles in parentheses)

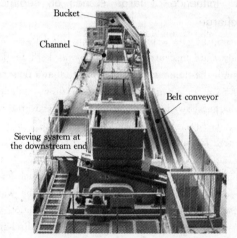

Figure 2　Variable slope flume with sediment circ ulation system

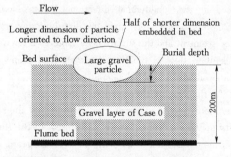

Figure 5　Cobble placed on flume bed

In the experiment, the percentage of cobble at the bed surface was varied, and changes mainly in 1) sediment discharge, 2) the transport velocity of grains of different sizes, and 3) the thickness of the transported sediment layer were examined. For the measurement of

Talbe 1　　Experiment condition

Case	Channel slope	Dishcarge (m³/s)	Bed material		
			D_{50} (mm)	D_{84}/D_{16}	P_{LS} (%)
0	1/100	0.18-0.52	15.0	3.3	0.0
1	1/100	0.2-0.52	20.1	3.4	2.3
2	1/100	0.2-0.52	22.5	3.7	5.2
3	1/100	0.2-0.52	27.0	4.0	10.7
4	1/100	0.2-0.52	31.2	6.7	15.9
5	1/100	0.2-0.52	34.6	12.0	20.1
6	1/100	0.2-0.52	43.0	11.7	30.2

sediment discharge (1), a collection box was installed at the belt conveyor connection point, and sediment discharge was measured by sampling a certain amount of transported sediment. In sediment transport, sediment discharge may change over time. In view of the possibility of such changes, samples were taken three times so that average sediment discharge values were obtained. The sampling time per sampling cycle at the maximum flow rate was 3 to 10 seconds. The critical flow rate for the movement of bed material was determined by gradually increasing the flow rate. For the measurement of transport velocity (2), 25 colored bed material samples of the five different grain sizes mentioned before were prepared, and the samples were put into the flowing water at Point No. 13, which is located 13m from the upstream end of the flume (hereafter, other points are also expressed in a similar manner). After that, water was kept to flow for about five minutes. After the water flow was stopped, the colored grains of bed material were searched in the flume. Then, their transport distances were measured, and their locations in the channel cross section and the state of embedment was observed. The thickness of the transported sediment layer (3) was measured by embedding gravel grains of uniform size with an average grain size of 4mm in the bed material in the form of a cylinder (diameter: about 5cm, depth: about 15cm) in view of the state of the colored gravel grains and measuring the height of the replaced material (hereafter referred to as the "colored sand replacement method"). Measurements were taken at five points aligned in the transverse direction (5cm, 25cm, 50cm, 75cm and 95cm from the left bank wall) and three points aligned in the longitudinal direction (Points No. 15, No. 20 and No. 25). The grain size distribution of transported sediment and the surface layer material after water was kept to flow in the flume, water depth, and bed height were also measured as appropriate.

At the end of this section, applicability of the experiment results to a real river channel and the similarity of the experiment were considered. In the flume experiment, bed material with an average grain size of about 15mm and a maximum grain size of about 45 mm was used, and grain of roughly uniform grain size with an average grain size of about 100mm was used as "large stones". If it is assumed, in view of the application to a real riverbed, that the experiment is conducted at a scale of 1 to 1/10, the experiment result can be applied to a gravel-bed channel with an

average grain size of 15mm to 150mm. Experiment results, however, may be affected by such factors as the ratio between the grain size of in-situ bed material and that of large stones and the grain size distribution of the in-situ bed material. Applicability to a real river channel, therefore, should be considered, keeping in mind that the experiment results described later in this paper were obtained under limited conditions. In the experiment, the Froude number is around 0.9, the water depth-grain size ratio is 7 to 20, and the Reynolds number for sand grains is of the order of 10^4. It is assumed that similarity of sediment transport was maintained.

3. Experimental results

3.1 Influence of large stones on sediment discharge

Figure 6 is a plot of sediment discharges in a steady state measured after water was kept to flow through the flume with a slope of 1/100 at a flow rate of 0.52m³/s. When water began to flow, sediment discharge showed a tendency to decrease over time, but it became stabilized at a certain value. The time required for such stabilization varied among the cases and ranged from about 20minutes to 90minutes. In the experiments conducted under the condition mentioned above, water depth was about 0.3m and the water surface slope was about 1/100, and the friction velocity was about 0.18m/s because water discharge and slope of the flume were fixed in each case. The sediment discharges shown here and the values shown hereafter are values that do not include voids. Figure 7 shows measured values of sediment discharge corre-

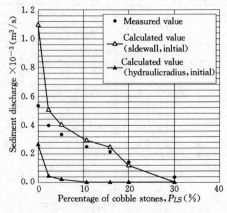

Figure 6   Influence of the percentage of the
cobble stones on sediment discharge

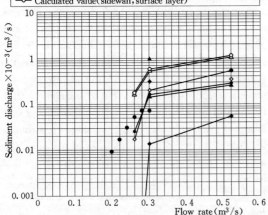

Figure 7　Measured vs. calculated values
of sediment discharge

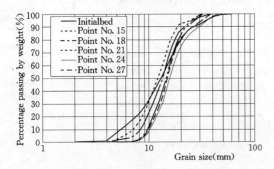

Figure 8　Grain size distribution of initial
and post-flow bed material （Case 0）

sponding to different flow rates under the grain size distribution conditions in Case 0 （$P_{LS} = 0\%$）. Figure 7 also shows calculated values obtained from Ashida-Michiue formula[5] for sand and gravel mixtures often used for bed evolution analysis. Critical tractive force for sediment movement was calculated by using the Egiazaroff formula[6]. Friction velocity was calculated, taking effective tractive force into consideration, by using two types of hydraulic radius, namely, the hydraulic radius determined by dividing the cross-sectional area of flow by wetted perimeter length （indicated as "hydraulic radius" in the legend） and the hydraulic radius calculated in view of the roughness （$0.011 \text{s/m}^{1/3}$ as Manning's coefficient） of the sidewalls[7] （indicated as "sidewall" in the legend）. The

results obtained by using the grain size distributions of the initial bed, the transported sediment and the post-flow surface layer are also shown, respectively. As for water depth and the energy gradient, the water depth and the water surface gradient measured during the sediment discharge measurement were used. The grain size distribution of the initial bed shown in Figure 3 and the grain size distribution at Point No. 15 shown in Figure 8 were used as the transported-sediment and post-flow grain size distributions, respectively. Figure 7 shows the post-flow grain size distributions of the bed material sampled from the transported sediment layer at Point No. 15 to No. 27. As observed during the water flow process, 6.4mm and 9.4mm gravel gradually fell into the voids in the bed material. In 117minutes, the quantity of 10mm or smaller material decreased by half, and the percentage of 15mm or larger material increased. This grain size distribution of the transported sediment is close to the average post-flow riverbed grain size distribution. This indicates that it is good practice to use the average riverbed grain size distribution for calculation. From Figure 7, the calculated values show good agreement with the measured values. It can be reconfirmed, however, that the measured values range widely from 0.1 to 2 times the measured values depending on the hydraulic radius used for friction velocity calculation and the grain size distributions adopted. Figure 6 shows two sets of sediment discharge values obtained by using the initial bed material and two types of hydraulic radius. To incorporate the influence of the cobbles into the sediment discharge formula, the cobbles were regarded as single grain size classes to give a grain size distribution. It was thought that it might be a good way to give a grain size distribution of transported sediment or average riverbed material. The grain size distribution of the initial riverbed material, however, was used partly because transported sediment did not include the cobbles and partly because a material survey paying attention to the cobbles was not conducted after the water flow process. Because water was kept to flow to a depth of 0.3m through a glassy side-walled flume with a width of one meter, it was thought appropriate to estimate sediment discharge by using hydraulic radius considering sidewall roughness. The application of the conventional sediment discharge formula to sediment containing large stones may be considered problematic, but the results indicate that the conventional formula is applicable if the grain size distribution is within the ranges shown in Table 1.

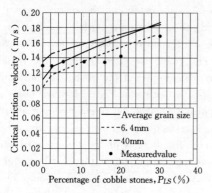

Figure 9　Influence of the cobbles on critical fricti on velocity

Figure 9 is a plot of the critical friction velocities for sediment movement calculated from observation results. The calculated values are the results obtained through evaluation made by using the modified Egiazaroff formula proposed by Ashida and Michiue[5], taking into consideration the differences in grain size distribution.

The observation results have shown that 40mm grains began to move as soon as 6. 4mm (i. e., smallest) grains began to move. At $P_{LS} = 0\%$, all grains actually began to move at or below the friction velocity of 0. 135m/s, at which in theory the largest grains begin to move. Because D_{84}/D_{16} was 3. 3 and because the degree of uniformity increased after the water flow process as shown in Figure 3, it is thought that the behavior observed in Case 0 was similar to that of uniform bed material[8]. According to Egiazaroff formula, the critical friction velocity gradually increases as P_{LS} increases. However, it was observed that the critical friction velocity did not change so much at $P_{LS} = 0$-15. 9%, and the critical friction velocity gradually increased as P_{LS} increased to reach 20. 1% or higher. This is thought to be one of the causes of low reliability of sediment discharge evaluation around the critical friction velocity as shown in Figure 6.

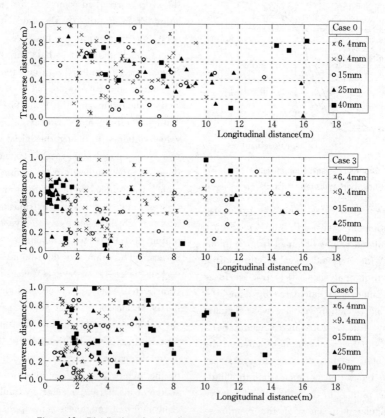

Figure 10　Distribution of colored gravel particles after being transported

3.2 Influence of large stones on sediment transport velocity

Figure 9 shows the results of a study on the movement of colored gravel in Case 0, Case 3 and Case 6. The experiment conditions were a flow rate of 0.52m³/s and a flume slope of 1/100. The longitudinal distance of zero corresponds to the distance of 13.25m from the upstream end of the flume where the colored gravel is estimated to have landed on the riverbed. The figure shows the two-dimensional distribution of colored gravel after water was kept to flow through the flume for five minutes after the gravel was put into the flume. Although the data show variance, it can be seen that the amount of transport in the downstream direction decreases as P_{LS} increases. Figure 11 shows the transport velocities averaged for different grain sizes after calculating the transport velocity of individual grains. As can be seen from the variance shown by the sediment transport

data of Figure 10, it cannot be generalized that as P_{LS} increases, the transport velocity of grains decreases. It seems, however, that with a small number of exceptions, such tendency may be pointed out. Conspicuous exceptions are the grain size of 6.4mm in Case 3, 15mm, 25mm and 40mm in Case 4 and 15mm in Case 5.

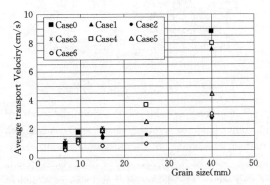

Figure 11   Average transport velocity by grain size

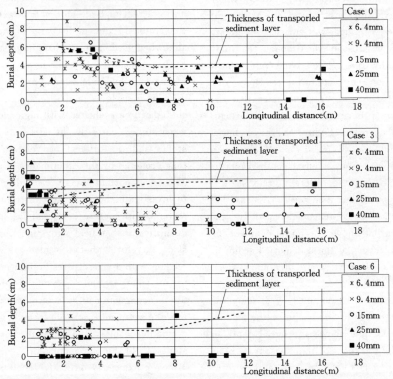

Figure 12   Burial depth of colored gravel and thickness of transported sediment layer

3.3 Influence of large stones on thickness of transported sediment layer

Figure 12 shows the burial depth results corre-

sponding to the results shown in Figure 9. "Burial depth" refers to the depth from the riverbed surface to the lower edge of a colored gravel particle at the location of a colored gravel particle found after the wa-

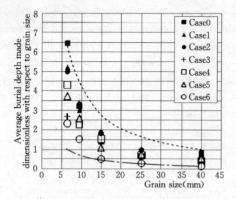

Figure 13　Average burial depth by grain size

ter flow process（see Figure 5）. From the results of a study conducted by the colored sand replacement method, the average of the thicknesses of the transported sediment layer measured at three central points is also shown as "thickness of transported sediment layer." Because the study results obtained at a distance of 5cm from the wall surface included extremely small sediment layer thickness values, it was thought that the influence of the sidewall surface was great, and those results were removed. Although the study results obtained by the colored sand replacement method do not show any significant tendency, the way the colored gravel is buried indicates that the number of shallowly buried colored gravel particles increases as P_{LS} increases. Figure 13 shows burial depth results in all cases averaged for different grain sizes. The vertical axis shows burial depth made dimensionless with respect to different grain sizes. If this burial depth is smaller than one, it means that the gravel particle has remained partially exposed. The broken line shows the burial depth in the case where the thickness of the transported sediment layer corresponds to the particle size of 40mm, which is the largest size of particles that can be transported. This line shows values obtained by dividing the maximum grain size by the grain size of the colored gravel. The dash-dot line shows the burial depth in the case where the thickness of the transported sediment layer corresponds to the smallest grain size of 6.4mm. This line shows values obtained by dividing the smallest grain size by the grain size of the colored gravel. As shown, all data fall between the broken line and the dash-dot line. In Case 0, for example, as often mentioned when discussing the bed load layer, the thickness of the transported sediment layer was

roughly equal to the largest grain size. Paying attention to the grain size of 6.4mm, we can see that the burial depth decreases as P_{LS} increases except in Case 3, and same tendency can be seen in the other grain sizes.

4. Key consideration in riverbed restoration by use of large stones

The preceding chapter looked at the experiment results related to the influence of large stones on sediment transport. As factors contributing to such phenomena, previous studies[4] have pointed out that the movement of large-grained gravel and the bed material in their wake zones is restricted and that the form drag of large gravel particles reduces the effective tractive force acting on the bed material other than large gravel particles. In this chapter, design consideration in applying the experimental results on riverbed restoration was summarized.

One consideration is the sediment discharge reducing effect of large stones. It is generally said that conventional sediment discharge formulas are not reliable when used in cases where large stones are included. It has been confirmed, however, that except in cases where hydraulic conditions are close to the critical conditions for sediment movement, sediment discharge can be evaluated with high accuracy as in "no large stone" cases if the size ratio between large stones and the original bed material is within the range covered by the experiment. As shown in Figure 7, it is important for sediment discharge evaluation to investigate the grain size distribution of the bed material with high accuracy and evaluate the hydraulic quantities in that field with high accuracy. It can be said, therefore, that it is good practice, when proceeding with a riverbed restoration project, to monitor the transport and deposition of sand and gravel after the completion of the project, keeping in mind that the sediment discharge formula gives very rough evaluation results, and adjust the percentage of large stones. Field measurements of sediment transport rates are extremely difficult, but it is thought to be enough to estimate sediment transport rates indirectly, for example, with cross-surveying results. Additionally, in order to clarify the effectiveness of a project, it is also good practice to place large stones so that the percentage of large stones becomes slightly higher than the target value. As stated by Okabe and Himoto[4], the effectiveness of large stones varies depending on the surface roughness of the riverbed, and the sediment discharge reducing effect decreases if the surface roughness of the riverbed decreases because of the deposition of sand and gravel on large stones. This is an important consideration in permitting overdesign.

A second consideration is the scour prevention effect of large stones. Large stones protect river banks from deep scouring[9]. Large stones, therefore, are all the more important for a river that lacks a source of large stones. It may be a good idea, therefore, to provide large stones that promote the deposition of sediment in times of a small-scale flood and are washed away in times of a large-scale flood. Large stones transported in times of a large-scale flood can be expected to protect the scoured areas in the downstream channel from further scouring. The results shown in Figure 7 indicate that there is little need to pay attention to the influence of the placement of large stones on the critical friction velocity for the movement of the original bed material. As in the case of the first consideration, however, it should be kept in mind that the experiment was conducted at a fixed ratio between the grain size of immovable bed material and that of movable material. It is therefore important to conduct further study, taking into consideration other results such as field observation results and future experiment results. Though under these conditions, it can be said that there is no significant influence on the critical friction velocity for sediment movement if P_{LS} is 16% or lower. As can be seen in Figure 4, it is unlikely that a riverbed with a P_{LS} higher than 16% is created through a river improvement project from the viewpoint of river scape. It is thought, therefore, that the placement of large stones does not directly reduce the frequency of disturbances, which are important for the maintenance of gravel beds. Instead, the placement of large stones should have favorable effects on bare land formation through local scouring occurring around large stones in times of a small-scale or mesoscale flood.

A third consideration is the necessity of accurate determination of the grain size distribution of bed material. Figure 14 shows the grain size distributions of transported sediment in the cases plotted on Figure 6. The vertical axis shows the percentage of particles retained in a sieve of a certain size. As the transport velocity and burial depth of different materials shown in Figure 11 and Figure 13 varied, the grain size distribution of transported sediment is not simply related with the increase in P_{LS}. For example, when P_{LS} is 2.3 to 5.2%, the transport of 25mm gravel particles increases. When P_{LS} is 15.9 to 20.1%, the transport of 40mm gravel particles increases. Clearly related with P_{LS} is the transport of gravel particles of the sizes shielded by large stones, namely, 6.4mm and 9.4mm, which decreases as P_{LS} increases. Changes in the materials constituting the transported sediment were not analyzed, but it was thought likely that such changes are related to the step length of individual particles. It was assumed that changes in P_{LS} changed the distances between large stones and, when somehow related to the step length of particles, affected the transport of particles. There is much room for further study in this area, but it was thought to be an important consideration that requires careful attention in connection with riverbed restoration.

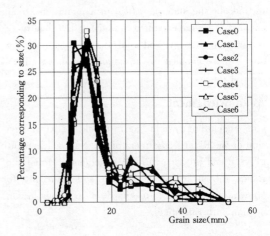

Figure 14   Influence of P_{LS} on grain size distribution of transported sediment

5. Conclusion

This study has focused on methods for forming and maintaining an unconsolidated deposit layer in a river channel with exposed bedrock or hardpan or a channel section where there is concern, in view of the present condition of bed material, about bed degradation due to increases in fluid force acting on the riverbed because of, for example, channel modification to form a compound cross section channel. Mainly the influence of large stones on sediment transport was investigated experimentally, and important considerations in applying the results to real project sites have been identified. The authors are aware that from the analytical point of view, our conclusion is still datable in many ways. It is believed, however, that from the practical point of view, this study has provided useful information.

In view of the growing need to predict riverbed processes occurring in the vicinity of dams, it can be said that the experimental results obtained in this study have yielded important information about riverbed material investigation. It has already been argued that sediment discharge can be reduced by placing large stones on the riverbed surface. In the experiment, sediment discharge was reduced by more than 25% by placing large stones on the channel bed so that the percentage by weight of large stones became 2.3%. It has been pointed out that the grain size of mountain stream bed material tends to vary widely, and many rolling stones are also seen in mountain stream channels. These results suggest that in order to find out the state of a field in a mountain stream channel, it is important to establish a riverbed material investigation method and it is necessary to conduct a more detailed investigation than in an alluvial river channel.

Acknowledgment

The authors would like to thank the River Division of the River Department of the National Institute of Land and Infrastructure Management, Ministry of Land, Infrastructure, Transport and Tourism, for allowing the authors to use its flume facilities in connection with this study.

References

[1] Shimatani Y., Minagawa T.. Present state of alluvial fan rivers in Japan and case studies on environmental conservation (in Japanese). *Proceedings of the International Symposium on River Restoration*, Foundation for Riverfront Improvement and Restoration, 1998: 191-196.

[2] Takahashi H., Maruyama N., Takeuchi K., Watanabe K.. Experimental study on restoration of gravel riverbed of the Makomanai River: Interim report (in Japanese). *Report of Riverfront Research Institute*, 2007 (18): 1-6.

[3] Ashida K, Takahashi T., Mizuyama T.. Study on bedload discharge in mountain streams (in Japanese). *Journal of the Japan Society of Erosion Control Engineering*, 1978 (107): 9-17.

[4] Okabe K., Himoto I.. Study on effective tractive force for sediment transport on mountain stream beds involving large-grained gravel (in Japanese). *Annual Journal of Hydraulic Engineering*, JSCE, 1986, 30: 247-252.

[5] Ashida K., Michiue M.. A basic study on resistance and bedload discharge in flow on movable bed (in Japanese). *Proceedings of the Japan Society of Civil Engineers*, 1972 (206): 59-69.

[6] Egiazaroff. I. V.. Calculation of Nonuniform Sediment Concentration, *Proceedings of the American Society of Civil Engineering*, 1965, 91 (HY4): 225-247.

[7] Einstein H. A.. Formulas for the transportation of bed load. *Trans. ASCE*, 1942, 107 (2140): 561-577.

[8] Yamamoto K.. *Structural alluvial potamology: structural characteristics and dynamics* (in Japanese), Sankaido, 2004: 54-80.

[9] Matsumoto M., Kudo M., Fukuoka S.. Large-scale scouring of the hardpan bed of the Asa River (Tama River System) due to the August 2008 flood and channel management measures" (in Japanese), *Proceedings of River Engineering*, 2009, 15: 285-290.

Observation Study on Water Discharge Structure for Dongping Hydro Junction

Liu Shanyan[1]   Tang Xiangfu[2]   Li Yuanhong   Che Qinquan[1]

[1]*Department of Hydraulics, Yangtze River Scientific Research Institute, Wuhan, 443002*
[2]*Dongping hydropower Limited Liability Company, Enshi, 445000*

Abstract: In this paper the technical method and relevant research results for observing hydrodynamic pressure of stilling pond bottom is introduced when the dam is respectively discharging by middle and surface outlet. The research result indicates that in the observation of working condition the water tongue is fully aerated. After it drops to the stilling pond, the energy cut back fast. The table of discharge due to upstream water-hole high-flow large, As a result of high water level at upstream and large flow, the fluctuating pressure of surface outlets is larger than middle outlets. But the measurement shows that the amplitude fluctuating pressure in the plunge pool is small and the energy is low. Besides the parameter values are within the allowable range in the design. It can be concluded that the water discharge structure discharges smoothly and the stilling pond operates safety.

Key words: water discharge structure; stilling pond; prototype observation; hydrodynamic pressure

1. Project overview

Zhongjian River belongs to Qingjiang river branch in Xuan'en City, Hubei Province. It is 11km from the confluence of Zhongjian River and Qingjiang River. The project mainly consists of the concrete arch dam, surface and middle outlet spillway in dam body, plunge pool behind the dam, the power tunnel at left bank, underground powerhouse, substation and other components. The largest arch dam is 135m, crest elevation is 495.0m, normal water level is 490.0m, total reservoir storage capacity is 343 million m³, it is Ⅱ and large two water conservancy and hydropower engineering.

Flood discharge facilities are surface outlet and middle outlet spillway which are arranged by layer. There are two surface outlets, each outlet's clear width is 11.0m, weir crest elevation is 478.0m, which take differential bucket at two sides. The width of bucket is 3.5m, the middle slope width is 4.0m, the slope gradient is 1 : 2.5 and it is straight bucket which the gradient is 1 : 4.0. The angle between the surface outlets axis and the arch center line is 3 degree. There are three middle outlets, there of two locate at left and right (1♯, 3♯ hole), the cross section size is 5.0m×6.5m, the bottom sill elevation is

430.0m and the other is in the middle (2♯ hole), the cross section size is 5.0m×6.0m, bottom sill elevation is 420.0m. The maximum total discharge flow rate of five outlets is 5970m³/s. The surface and middle outlets in arch dam body are used for flood energy dissipation. The stilling pond which is auxiliary energy dissipater form is set behind the dam bed, the stilling pond length is 175m, bottom elevation is 366m, the top pool elevation of two sides is 405m, the top elevation of tail bucket is 379.0m.

Due to high head and huge energy, the great impact and high velocity may lead to structural damage and it may affect the normal power generation and power station building safety spillway when Dongping dam discharges. To ensure safety and inspection dissipater state, twelve detection points which measure pressure are set on bottom plate of stilling pond, there are DPt11, DPt12, DPt13 (at stilling pond 0 + 68m), DPt19, DPt20, DPt21 (at stilling pond 0 + 88m), DPt13, DPt14, DPt15 (at stilling pond 0 + 108m), DPt16, DPt17, DPt18 (at stilling pond 0 + 128m) respectively. The nine detection points on trial bucket are DPt1, DPt2, DPt3 (1♯ trial bucket), DPt4, DPt5, DPt6 (2♯ trial bucket), DPt7, DPt8, DPt9 (3♯ trial bucket) respectively. This article introduces the special monitoring results about hydraulic.

2. Observation techniques and equipments

2.1 Observation method on hydrodynamic pressure

Hydrodynamic pressure is the pressure which overflowing water acts on the flow section, including both mean pressure and pulse pressure. Hydrodynamic pressure is the main reasons which cause structural vibration and erosion of solid wall. Monitoring aim is to grasp the basic characteristics of water pressure acting on building, determine the flow condition which is not conducive for the building to safely oper-

ate, propose technical measures or operating restrictions conditions to avoid the risk.

Pulse pressure reflects the main hydraulic characteristics of related sites, and it is a important parameter to analyze the hydrodynamic loads and cavitations issues. Using the pressure sensor measures mean pressure and pulse pressure at the same time, after amplification the signal is recorded and analyzed by a specialized computer acquisition system. The pressure signals are all analyzed as the mean value, standard deviation and power spectral, the frequency is not less than 100Hz. The analysis measurement system and observation methods are in Figure 1.

Figure 1   The diagram for hydraulics observation system

2.2 Selection and inspection for test equipment

2.2.1 *Selection for test equipment.* The sensors for the fluctuating pressure monitoring are mainly the series of SZ osmometer which are produced by Nanjing Electric Power Automation Equipment Factory. The sensor selection is based on the following considerations: (1) Because sensors are buried in the stilling pond, the observed cable are laid on the basis of the stilling pond gallery which is high humidity, the sensor and the secondary instrument requirements for the line degree of insulation is relatively low, so it facilitates access to observations. (2) The past practice shows that the main characteristics of the sensor is stable, long durability, and it can meet the observation needs for the mechanical pulse pressure. Obser-

vation instruments are YD28-A strain gauge which are produced by East China Electronic Instrument Factory and DSAP data acquisition and analysis system which are developed by the Beijing Oriental Vibration and Noise Technology Institute.

2.2.2 *Inspection for test equipment.* Device inspection method is below. The conversion characteristic for system which composes of the fluctuating pressure sensor and the second instrument can be determined according to the observing input and output (known), then the corresponding input which cause the output by the system characteristics and the output can be inferred during observing. The process of verification shows in Figure 2.

Figure 2   Block diagram for pulse pressure sensor test

3. Observations results and data analysis

3.1 Observation condition

The hydraulics of the stilling pond under four operating conditions are observed, the hub operating conditions shown in Table 1.

3.2 Discharging only by middle outlet

The trail bucket of stilling pond has not yet fully formed when discharging only by middle outlet, but the downstream cofferdam is not removed, so the water cushion thickness is deepen and length of

Table 1   The hydraulics observation condition for the stilling pond of Dongping project

Group	Running condition	Upstream water level H (m)	Output flow Q (m³/s)
1	The first time passing water for stilling pond	430	267
2	No. 2 middle outlet discharging	423	67
3	Only No. 1 and No. 2middle outlet opened	485-486	530-1000
4	Only No. 1 and No. 2middle outlet opened	486.5-488	1170

plunge pool is extended relative to the original design stilling pond. It is beneficial for energy dissipation. The main features are required by the acquisition and analysis of data from the installed pressure sensor, and they are shown in Table 2. We can see from the table that. pressure fluctuation of the measurement points on the stilling pond floor and tail bucket are small under the observation condition, the amplitude of instantaneous maximum fluctuation is only 0. 90 (9. 81kPa), the amplitude of instantaneous minimum fluctuation is only −1. 02 (9. 81kPa), the standard deviation is 0. 028 to 0. 45 (9. 81kPa), the maximum power spectrum is 0. 20 (9. 81kPa × 9. 81kPa), of which the pulse pressure on the trail bucket is greater than on bottom plate. It is because that the fluctuating pressure measurement points are on the flow direction and 2# trail block not cast to the end of the design elevation. Pulse pressure frequency rate is between 0. 20Hz and 0. 59Hz, while the natural frequency of concrete is generally around 8 Hz. The observational results suggest that the power for pressure fluctuations is small, the frequency is low and it will not cause damage to stilling pond.

Table 2   **Characteristics values of fluctuating pressure of stilling pond**

Measuring points number	Maximum value (9. 81kPa)	Minimum value (9. 81kPa)	Variance (9. 81kPa)	Standard deviation (9. 81kPa)	Dominating frequency (Hz)	Power (9. 81kPa× 9. 81kPa)
DPT-2	0. 90	−1. 02	0. 1992	0. 4463	0. 59	0. 00043
DPT-7	0. 51	−0. 78	0. 0295	0. 1717	0. 59	0. 00061
DPT-8	0. 29	−0. 17	0. 0048	0. 0690	0. 59	0. 00021
DPT-9	0. 22	−0. 30	0. 0083	0. 0911	0. 39	0. 00008
DPT-10	0. 08	−0. 17	0. 0026	0. 0508	0. 05	0. 00009
DPT-11	0. 13	−0. 13	0. 0019	0. 0430	0. 59	0. 00005
DPT-12	0. 10	−0. 15	0. 0026	0. 0508	0. 59	0. 00011
DPT-13	0. 17	−0. 08	0. 0008	0. 0279	0. 59	0. 00002
DPT-14	0. 09	−0. 08	0. 0012	0. 0343	0. 20	0. 00002
DPT-15	0. 13	−0. 30	0. 0016	0. 0402	0. 59	0. 00005
DPT-16	0. 14	−0. 14	0. 0012	0. 0340	0. 20	0. 00004
DPT-17	0. 10	−0. 10	0. 0010	0. 0309	0. 59	0. 00004
DPT-18	0. 11	−0. 11	0. 0015	0. 0392	0. 39	0. 00006
DPT-19	0. 35	−0. 20	0. 0037	0. 0608	0. 39	0. 00006
DPT-20	0. 20	−0. 19	0. 0033	0. 0571	0. 59	0. 19917
DPT-21	0. 24	−0. 13	0. 0025	0. 0503	0. 39	0. 00007

3. 3   Only discharging by surface outlet

The fluctuating pressure of measurement points which are on the bottom of stilling pond and trail bucket are collected, calculated and analyzed under the conditions which the flow is 530m^3/s to 1000m^3/s and 1170m^3/s, the values of its main features are obtained as shown in Table 3.

Table 3   Fluctuation pressure of measuring point under the condition of discharging by surface outlet

Characteristics / Measuring point number	Maximum value (9.81kPa)		Minimum value (9.81kPa)		Standard deviation (9.81kPa)		Dominating frequency (Hz)		Power (9.81kPa× 9.81kPa)	
Values flow (m³/s)	530-1000	1170	530-1000	1170	530-1000	1170	530-1000	1170	530-1000	1170
DPt-1	1.21	2.38	−1.42	−1.93	0.50	0.59	1.56	0.59	0.24	0.34
DPt-2	1.84	4.81	−1.14	−3.43	0.42	1.19	1.17	0.59	0.09	0.12
DPt-3	1.52	1.98	−1.37	−2.33	0.45	0.60	1.17	0.59	0.23	0.50
DPt-4	2.29	2.59	−0.76	−1.52	0.49	0.67	1.17	0.97	0.31	0.17
DPt-5	1.61	2.46	−1.28	−2.03	0.42	0.64	1.17	0.97	0.11	0.40
DPt-6	7.34	7.43	−4.86	−8.52	1.75	2.71	0.97	0.97	2.78	2.50
DPt-7	1.66	1.75	−1.93	−0.74	0.64	0.38	1.17	1.37	0.46	0.004
DPt-8	2.28	2.82	−1.13	−1.45	0.50	0.78	1.56	1.17	0.04	0.23
DPt-9	0.90	1.55	−1.49	−1.08	0.45	0.41	1.37	1.37	0.008	0.008
DPt-10	4.48	10.11	−2.98	−8.52	1.16	2.04	0.78	0.78	0.04	0.35
DPt-11	9.66	11.02	−10.14	−8.39	2.06	2.37	1.37	2.54	1.20	1.34
DPt-12	9.46	7.74	−8.53	−11.81	2.18	1.90	0.98	1.37	1.01	1.06
DPt-14	4.05	15.45	−3.68	−8.18	0.94	3.18	0.98	0.39	0.17	0.31
DPt-15	4.65	14.31	−2.61	−5.25	1.06	2.82	1.17	14.31	0.10	0.35
DPt-16	1.64	0.74	−0.52	−0.91	0.29	0.27	1.17	1.17	0.0001	0.0001
DPt-17	1.48	0.34	−0.91	−1.17	0.38	0.25	1.17	0.98	0.10	0.01
DPt-19	2.57	1.82	−3.73	−1.06	0.87	0.46	1.56	1.37	0.24	4.12
DPt-20	10.11	11.67	−9.58	−9.21	2.63	3.03	0.39	1.76	1.18	2.69
DPt-21	12.29	12.68	−11.11	−9.46	2.22	2.97	1.36	0.98	1.53	4.51

3.3.1 *Fluctuation pressure on bottom plate of stilling pond*

From pressure distribution of the measuring point, the amplitude of instantaneous maximum fluctuation on the bottom plate is only 15.45 (9.81kPa) (DPt14 measuring point), the amplitude of instantaneous minimum fluctuation is −11.81 (9.81kPa) (DPt12 measuring point) the standard deviation is 0.25 to 3.18 (9.81kPa) under the condition which the flow is 1170m³/s. They suggest that the impact district of the water tongue is in this region and the impact fluctuating pressure is greater than the region outside. The maximum amplitude of the amplitude appears near the impact zone and is not in impact area of water tongue, the pressure fluctuations reduce quickly.

3.3.2 *Fluctuation pressure on trail bucket of stilling pond*

Fluctuation pressure on trail bucket is generally within 4.81 and − 3.43 (9.81kPa), the maximum pressure of DPt6 which is at the center of 2 # trail bucket reaches to 7.43 (9.81kPa), the minimum pressure reaches to −8.52 (9.81kPa), the standard deviation of pressure fluctuations is 0.41 to 2.71 (9.81kPa), which the order of magnitude smaller. The fluctuating pressure of the measuring points on the trail bucket consist of pulse pressure and velocity head. The observations show that the fluctuation pressure of DPt-6 is relatively larger and the water is at submerged bottom floor state, the mainstream is at de middle of the trail bucket, it is benefit for stability of downstream slope on two sides.

3. 4   Locale survey

We also make inspections on the discharging construction of dam and the foundation gallery of stilling pond during Dongping dam discharging, as the flow is small and the upstream water level is low, it find no abnormalities when it discharge only by middle outlet. It is able to clearly hear friction and impact voice from bottom plate of stilling pond in the foundation gallery and it sound great in the impact zone. From the measured fluctuation pressure data we also can see that the largest transient pulse pressure appears in the impact zone and it may have a hard object crashing pressure sensor. It is suggested that clean up the stilling pond at the appropriate time and prevent large stones dropping into the stilling pond when running in the future to reduce wearing the bottom plate of stilling pond from the foreign body.

4. Conclusions

(1) The upstream water level is 486. 5m to 488. 0m, the measuring points mean pressure on stilling pond downstream are about 16. 0 (9. 81kPa), the amplitude of instantaneous maximum fluctuation is 15. 45 (9. 81kPa), the amplitude of instantaneous minimum fluctuation is -11.81 (9. 81kPa) and the measuring point does not appear negative pressure when discharging by two surface outlets. Compared with the result discharging by middle outlet, the result discharging by two surface outlets is that as the leakage flux increases and the high water level the mean pressure and pulse pressure increase but the order of magnitude is smaller, it will not cause hazards to the building.

(2) The analysis on fluctuating pressure data and inspection examination reveal that there are hard thing impacting the bottom plate. It is suggested that clean up the stilling pond at the appropriate time so as to avoid or reduce large stones wearing the bottom plate of stilling pond.

References

[1]   Chen Yuanqing, etc.. Hydraulics Prototype Observation Report on Stilling Pool of Geheyan Dam, Yichang Yangtze River Academy of Sciences Research Institute, December 1992.

[2]   Wang Caihuan, etc.. Hydraulics Prototype Observation Report on Geheyan Engineering in 1998. The Yangtze River Academy of Sciences, January 1999.

[3]   Liu Yunkun, etc.. Impoundment Safety Assessment Report on Dongping Hydropower Project in Xuan'en City Hubei Province, China Hydropower Engineering Consulting Group, January 2005.

[4]   Tang Xiangfu, etc.. Hydraulics Equipment Installation and Stage Observation Report on Stilling Pond of Dongping Hydropower Project in Xuan'en City Hubei Province. the Yangtze River Academy of Sciences, October 2005.

Measurements of Instantaneous Bed Pressure in a Backward-facing Step Flow

Kouki Onitsuka[1]   Juichiro Akiyama[1]   Yoshitake Zoshi[2]   Daisuke Mori[3]

[1] Department of Civil Engineering, Kyushu Institute of
Technology, Kitakyushu, 804-8550, Japan
[2] Kyudenko Co., Inc., Minami-ku, Fukuoka, Japan
[3] Toda Corporation, Chuo-ku, Tokyo, Japan

Abstract: A revetment block which is located just downstream of a weir is sometimes destroyed during the flood. Nakagawa et al. (1987) pointed out that the one of the reason why the revetment is destroyed is instantaneous pressure fluctuations near the bed. The drag and lift forces acting on the bed materials above the bed are measured by many researchers. Turbulent structure in open-channel flows behind a backward-facing step is also investigated by making use of a PIV and LDA. In contrast, pressure fluctuations on the bed have not been measured, due to its difficulty. In this study, simultaneously measurements of turbulent structure, pressure on the bed wall and water surface fluctuations in a backward-facing step flow were conducted with a PIV, PTV and a supersonic wave gage. It was found that the instantaneous pressure increases when the K-H instability vortex, which is generated behind the backward-facing step, is falling toward the bed wall. The instantaneous re-attachment point is moving to upward and downward, alternately. This instantaneous deviation value from the time averaged re-attachment length is about 40%. The instantaneous pressure on the bed is not so controlled by the water surface fluctuations, i.e., 3%-10%.

Key words: backward-facing step; PIV; PTV; instantaneous pressure; simultaneously measurements

1. Introduction

Revetment blocks which are located just downstream of a weir are sometimes destroyed during the flood. Nakagawa et al. (1987) suggested that those destroy is caused by the following four factors: 1) river bed degradation and a local scour at arevetment block or at downstream region of a dam apron, 2) a suction of the sand from the gap of a revetment block, 3) piping, 4) a water fall and the direct impacts of rolling stones.

Suzuki et al. (1982) measured a time change of the scour and a maximum scour depth. Kanda et al. (1997) measured the scour at the downstream of revetment blocks.

The hydrodynamic forces in a backward-facing step flow were also measured by some researchers. Kawaguchi et al. (2002) found out that the hydraulic pressure on the bed at the downstream of revetment blocks decreases than the gravitational pressure about 15%. Uchida et al. (2007) measured the hydrodynamic force on gravel bed

at the downstream side of a backward-facing step and suggested the possibility that revetment blocks are broken by the instantaneous hydrodynamic force in the undular hydraulic jump. Onitsuka et al. (2007) measured the pressure near bed in roller of weak hydraulic jump generated on horizontal bed and obtained the conclusion similar to the one of Uchida et al. (2007). As mentioned above, it is thought that measurements of instantaneous hydrodynamic forces are important as well as measurements of time-averaged hydrodynamic forces.

In contrast, studies on the instantaneous turbulent structure in a backward-facing step flow are conducted by many researchers. Eaton & Johnston (1982) implanted thermal tuft probe into the bed behind the step and predicted that the phenomenon of reattachment had not been generated periodically but randomly by the spectrum analyzing. Nezu & nakagawa (1987) measured the backward-facing step flow and found out that the reattachment length decreases with an increase of Reynolds number and also that the location of the reattachment point fluctuated to the upstream and downstream about 50% with time varying. In addition, they pointed out the existence of kolk-boil vortex which was generated near the bed when the flow reattached to the bed. Watanabe et al. (1998) visualized the flow in streamwise-, cross- and horizontal-sections, respectively, by using dye injection technique. As a result, they succeeded in taking pictures of the kolk-boil vortex which was generated near the bed when the flow reattached to the bed. Ohmoto et al. (2002) visualized stepped flow where backward-facing steps existed two continuously to downstream direction by using an YAG-laser and found out that the value of hydrodynamic pressure becomes minus in the reverse flow region by calculating the hydrodynamic pressure from the value of velocity.

Silveira Neto et al. (1993) calculated closed-channel flow which had a backward-facing step with LES (Large Eddy Simulation) and obtained visual data of three-dimensional structures of the coherent vortex. Le et al. (1997) calculated closed-channel flow which has a backward-facing step with DNS (Direct Numerical Simulation) and found out that the fluctuation of fluid pressure was accompanied by the generation of separated vortex and the advection. Nakayama & Yokoshima (2001) and Yokoshima (2006) calculated a backward-facing step open channel flow with $k - \varepsilon$ and $k - \omega$ and pointed out that the RANS model had the availability.

As mentioned above, there are a lot of studies concerning with the hydrodynamic force and time-averaged flow structures in a backward-facing step flow and near sills. However, there are little example of simultaneously measurements of the instantaneous pressure and the instantaneous velocity. Therefore, the influence that the flow field exerts on the bed and the suction mechanisms of bed materials with the former were not clarified. In this study, simultaneously measurements of instantaneous velocity and instantaneous pressure on the bed wall in a backward-facing step flow are conducted. As a result, the relation of instantaneous velocity and instantaneous pressure was found out.

2. Experimental setup and conditions

2.1 Experimental setup and hydraulic conditions

The experiments were conducted in a 1.9m long, 0.15m wide and 0.08m deep tilting flume as shown in Figure 1. The backward-facing step was set up at 1.3m downstream from the channel entrance. The height of the backward facing step H_s is 0.03m. The experimental channel has eight holes of 6mm diameter. There is one hole at the upstream of the step and two holes at the wall of the step. There are five holes at the downstream of the step. Plastic tubes are attached on those holes at vertical.

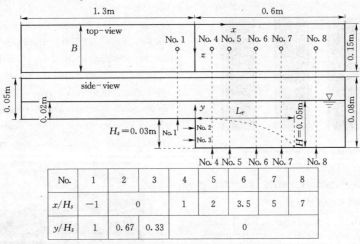

No.	1	2	3	4	5	6	7	8
x/H_s	−1	0		1	2	3.5	5	7
y/H_s	1	0.67	0.33			0		

Figure 1 Experimental setup

$\tilde{u} = U + u$, $\tilde{v} = V + v$ are instantaneous velocities in x, y directions, respectively. U, V are mean velocities and u, v are turbulent velocity fluctuations, respectively. $\tilde{p} = P + p$ is the instantaneous pressure on the wall. P is the

mean pressure and p is the pressure fluctuation. Prime means the RMS value. Table 1 shows the experimental condition. H is the flow depth, U_m is the mean bulk velocity, $Re \equiv HU_m/\nu$ is the Reynolds number, $Fr \equiv$

$U_m / \sqrt{gH}$ is the Froude number, g is the gravity acceleration, ν is the kinematic viscosity.

Table 1   Hydraulic condition

	H (m)	U_m (m/s)	Re	Fr
$x/H_s = 0$	0.02	0.35	6998	0.79
$x/H_s = 10$	0.05	0.14		0.20

2.2 Simultaneous measurements of instantaneous pressure and water surface fluctuations

Manometers which connect to the each hole were set at the right bank wall. The water surfaces in the manometers were illuminated by the halogen light. Those illuminated water surfaces in the manometers were recorded by the high vision camera which has 1440×1080 pixels for 75 seconds every 1/30 seconds. The water surface fluctuations over the hole in open-channel flows was measured by the super sonic wave gauge. The time of both measurement is synchronized by use of electric light. The eclectic light shines and recorded by the video camera. In contrast, this voltage is recorded.

After measurements, the positions of water surfaces in the manometer were calculated by PTV (see Figure 2).

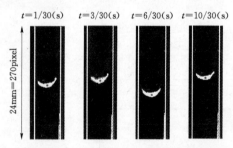

$t=1/30(s)$ $t=3/30(s)$ $t=6/30(s)$ $t=10/30(s)$

24mm = 270pixel

Figure 2   Water surface in manometer

2.3 Simultaneous measurements of instantaneous pressure and velocities

High porous polymers which have 75 to 150μmm diameter with 1.01 specific gravity are dropped in the channel as PIV (Particle Image Velocimentry) tracer particles. A laser beam from the 4W argon-ion laser passed through optical fiber and cylindrical lens was changed to a laser light sheet (LLS). Under such a condition, the area about $0.3m \times 0.3m$ near the step was taken for 75 seconds every 1/60 seconds by a video camera with 640×240 of pixels which was located at the right bank. Pressures from eight holes and PIV were measured simultaneous for 75 seconds every 1/30 seconds.

3. Experimental results and discussions

3.1 Time-averaged values

Figure 3 shows the time averaged velocity of the x component U normalized by the maximum velocity of the x component at the upstream of the backward-facing step U_{maxu}. The shear layer near $y/H_s = 1.0$ diffuses to vertical direction along the downstream direction and velocity gradient dU/dy is shelving. This result is similar to one clarified by Nezu & Nakagawa (1989) with an LDA.

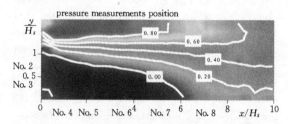

Figure 3   Contour line of mean flow velocity

Nezu & Nakagawa (1989) found out that the reattachment length L_r in a backward-facing step flow decreases with an increase of the Reynolds number. They also found that L_r increases with an increase of the Froude number in subcritical flows ($0.18 \leqslant Fr_d < 1.0$) and decreases with an increase of the Froude number in supercritical flows ($1.0 < Fr_d \leqslant 1.4$). In which Fr_d is the Froude number in streamwise from the step. Therefore, they clarified that the dimensionless reattachment length L_r/H_s is $5 < L_r/H_s < 6$ in a lot of conditions. The stream function ψ is frequently used to calculate the reattachment length.

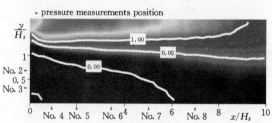

Figure 4   Contour line of stream function

$$\psi \equiv \int_0^y U dy \qquad (1)$$

Figure 4 shows a contour line of $\psi/(U_{maxu}h_u)$. The position in which a line of $\psi/(U_{maxu}h_u)=0$ is connected with the bottom represents the reattachment point. It shows that dimensionless reattachment length L_r/H_s is about 6. This value is similar to the value obtained by Nezu & Nakagawa(1989).

3. 2　Characteristic of pressure near bed

Figure 5 shows time series of the instantaneous pressure $\tilde{p}$ normalized by the time-averaged pressure P. Every instantaneous pressure fluctuates. Amplitude of No. 4 located just downstream of the backward-facing step is relatively small, because No. 4 is located in the dead zone. In contrast, amplitude of No. 8 located far a way from the backward-facing step is relatively large. The value of pressure at No. 8 decreases about 1 second from $t=17.8s$. It is supposed that No. 8 is located near the reattachment point.

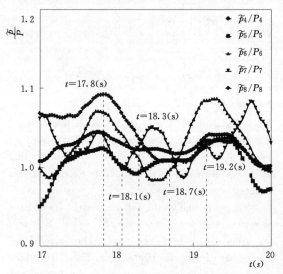

Figure 5　Time series of instantaneous pressure

The normal distribution is written as follows.

$$f\left(\frac{\tilde{p}_i}{P_i}\right)=\frac{1}{\sqrt{2\pi}(p_i'/P_i)}\exp\left[-\frac{(\tilde{p}_i-P_i)^2}{2p_i'^2}\right] \quad (2)$$

Figure 6 shows the histograms of the instantaneous pressures normalized by the time-averaged pressure P. In which N is the amount number of data, n is the number of data in each histogram. The amplitudes of pressure fluctuations at No. 4 and No. 5 are small. This is because those points are located in the dead zone. The amplitudes of pressure fluctuations at No. 7 and No. 8 are large.

Figure 7 shows variation of RMS value of pressure fluctuation p' normalized by the time-averaged pressure P along the streamwise direction. p'/P increases along the streamwise direction and does not change so much after attaining the maximum value near $x/H_s=6.0$. The position where p'/P attains maximum value is almost corresponding the reattachment point.

3. 3　Relation between pressure near bed and velocity field

Figure 8 shows the instantaneous velocity vectors at $t=18.1s$ and $18.7s$ when the pressure takes the minimum at No. 7 and No. 8 (see Figure 5), at $t=19.2s$ and $17.8s$ when the pressure takes the maximum and also at $t=18.3s$ when the middle situations between the minimum and maximum is generated. The downward flow can be seen when $t=19.2s$ at No. 7 $(x/H_s=5.0)$. At this time, the pressure attains the maximum. In contrast, the upward flow can be seen when $t=18.1s$ at No. 7 $(x/H_s=5.0)$. At this time, the pressure attains the minimum.

The instantaneous stream function is obtained as follow:

$$\tilde{\psi}=\int_0^y u\mathrm{d}y \quad (3)$$

Figure 9 shows contour lines of $\tilde{\psi}$ $(U_{maxu}H_u)$ at $t=17.8$, 18.1, 18.3, 18.7 and $19.2s$. Reattachment length x/H_s at $t=18.2s$ was about $x/H_s=5.0$. This time corresponds to that the pressure takes maximum. Reattachment length at $t=18.3s$ was shorter than time-averaged reattachment length $x/H_s=6.0$.

Figure 10 shows the time series of the reattachment length $\tilde{x}_r$ normalized by the height of the backward-facing step H_s. The reattachment length changes periodically.

Figure 11 shows the histogram of the instantaneous reattachment length $\tilde{x}_r/H_s$.

$$f\left(\frac{\tilde{x}_r}{H_s}\right)=\frac{1}{\sqrt{2\pi}(x_r'/H_s)}\exp\left(-\frac{(\tilde{x}_r-H_s)^2}{2x_s'^2}\right) \quad (4)$$

It is recognized that the instantaneous minimum and maximum reattachment lengths are $\tilde{x}_r/H_s=4.6$, 7.0. Those values correspond to the 22% and 19% of the time averaged reattachment length.

The correlation coefficient $R_{ij}(\tau)$ between the instantaneous fluctuation w_i and w_j is obtained from the theory as follows:

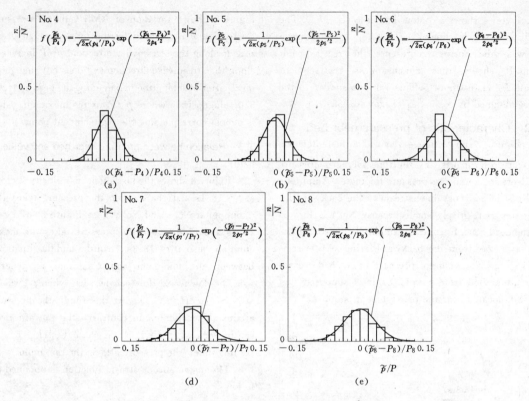

Figure 6　Histogram of instantaneous pressure $\tilde{p}/P$

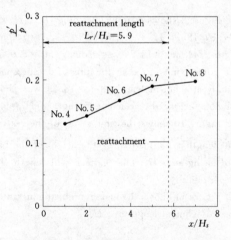

Figure 7　RMS value of pressure fluctuations

$$R_{ij}(\tau) \equiv \frac{\overline{w_i(t) \cdot w_j(t+\tau)}}{w_i' \cdot w_j'} \qquad (5)$$

In which τ is the lag time. Figure 12 shows the self corre-lation coefficients of pressure fluctuation $R_{pi,pi}(\tau)$. $R_{pi,pi}(\tau)$ decreases with the lag time τ and takes the maximum value after taking the minimum value. This maximum value is called the "second mild maximum" and shown by the downward arrow. This lag time

means the period of pressure fluctuation. It was found that the period just downstream of the backward-fa-cing step (No. 4, No. 5) is almost 1.8s and also that the period decreases, i. e., 1.0s, in the downstream direction (No. 7, No. 8).

Figure 13 shows the self correlation coefficients of vertical turbulent fluctuation $R_{ui,ui}(\tau)$ nearest wall bed. Almost periods are about 1.0s. This value is sim-ilar to that of pressure fluctuations. Therefore, there is some relationship between pressure fluctuations and velocity one near the reattach point.

Figure 14 shows the correlation coefficients between the vertical turbulent fluctuation and pressure fluctua-tions $R_{ui,ui}(0)$ under the condition of $\tau=0$ nearest wall bed. $R_{ui,ui}(\tau)$ just downstream of the backward-facing step (No. 4, No. 5) is quite low (0.03). This is be-cause this zone is dead zone. $R_{ui,ui}(\tau)$ increases in the downstream direction and attains 0.1 at No. 8. This im-plies that the vertical velocity fluctuations near bed affects on the pressure fluctuations abut 10%.

3.4　Relation between pressure near bed and flow depth

Figure 15 shows the time series of the instantaneous flow depth $\tilde{h}$ normalized by the time averaged flow depth H. The amplitude of the flow depth just downstream of the backward-facing step (No. 2, No. 3 and No. 4) is small and that at No. 8 is large.

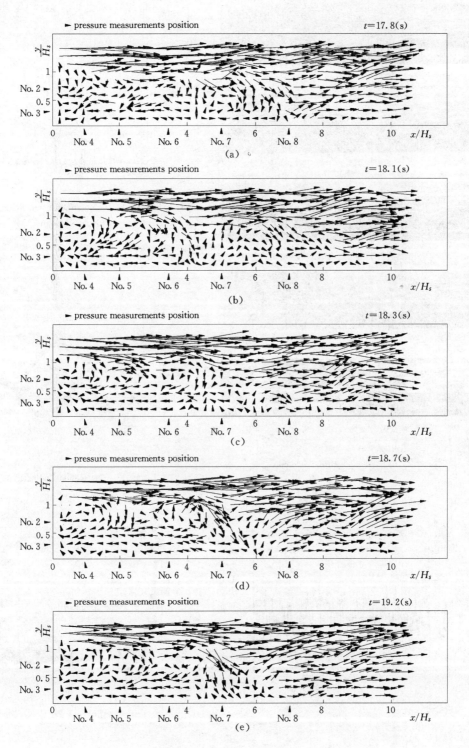

Figure 8   Instantaneous velocity vector

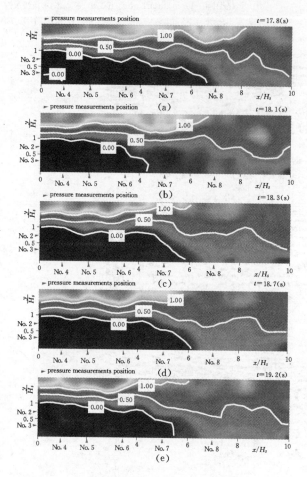

Figure 9   Instantaneous stream function

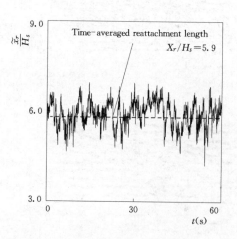

Figure 10   Instantaneous reattachment point

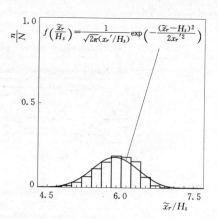

$$f\left(\frac{\widetilde{x}_r}{H_s}\right)=\frac{1}{\sqrt{2\pi}(x_r'/H_s)}\exp\left(-\frac{(\widetilde{x}_r-H_s)^2}{2x_r'^2}\right)$$

Figure 11   Histogram of instantaneous reattachment point

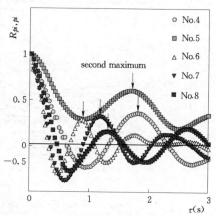

Figure 12   Correlation Coefficients $R_{pi,pi}$ (τ)

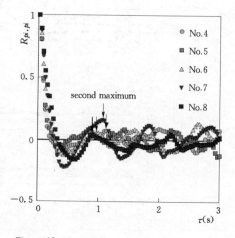

Figure 13   Correlation Coefficients $R_{vi,vi}$ (τ)

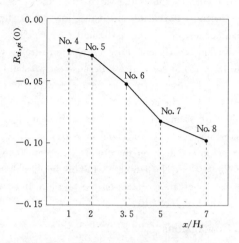

Figure 14　Correlation Coefficients $R_{vi,pi}$ （0）

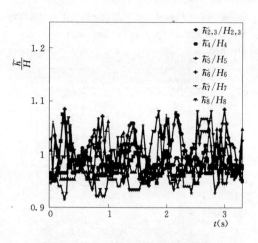

Figure 15　Instantaneous flow depth

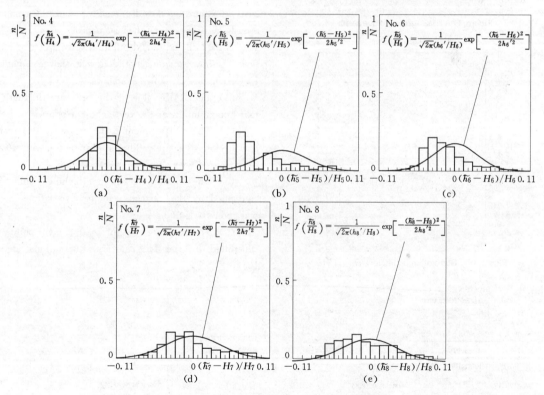

Figure 16　Histogram of Instantaneous flow depth

Figure 16 shows the histogram of the instantaneous flow depth $\tilde{h}$. The histogram is described by the following normal distribution.

$$f\left(\frac{\tilde{h}_i}{H_i}\right) = \frac{1}{\sqrt{2\pi}(h_i'/H_i)} \exp\left(-\frac{(\tilde{h}_i - H_i)^2}{2h_i'^2}\right) \quad (6)$$

The histogram of the instantaneous flow depth $\tilde{h}$ at No. 7 and 8 is well described by the normal distribution. In contrast, that at No. 5 and 6 is not described by the normal distribution.

Figure 17 shows the RMS value of the flow depth fluctuation h' normalized by the time averaged flow depth H. The value of h'/H increases in the downstream direction. This tendency is similar to that of pressure (see Figure 7). This implies that the flow depth fluctuation affects on the pressure fluctuation near the bed.

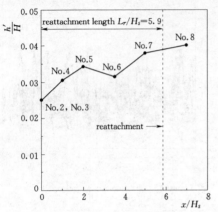

Figure 17   RMS value of flow depth fluctuations h'/H

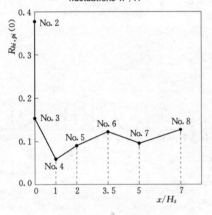

Figure 18   Correlation coefficient $R_{hi,pi}(0)$

Figure 18 shows the variation of the correlation coefficient between instantaneous flow depth and instantaneous pressure $R_{hi,pi}(0)$ under the condition of $\tau = 0$s. The coefficient $R_{hi,pi}(0)$ attains high value (0. 15-0. 40) at No. 2 and 3. Because the pressure hole is near the free surface as compare with the other hole, due to the holes of No. 2 and 3 is located at the backward facing step wall. The coefficient $R_{hi,pi}(0)$ takes low value at No. 4 and increases in the downstream direction lower than the No. 4 point. The value of $R_{hi,pi}(0)$ at No. 8 takes 0. 1. This implies that the surface fluctuation affects on the pressure fluctuation near the bed about 10%.

4. Conclusions

In this study, the relation between instantaneous velocity, pressure and flow depth is investigated by simultaneously measurements of the instantaneous velocity, pressure on the wall and flow depth in a backward-facing step flow. The flow structure in this flow is shown in Figure 19. Conclusions are shown as follows.

(1) The RMS value of the pressure fluctuation takes maximum near the reattachment point.

(2) The reattachment point moves to upstream and stream alternately. The amplitude is about 40% as compared with the time averaged one.

(3) The pressure near the reattachment points is controlled by the vertical velocity fluctuations near bed about 10%.

(4) The pressure near the reattachment points is controlled by the free surface fluctuations about 10%.

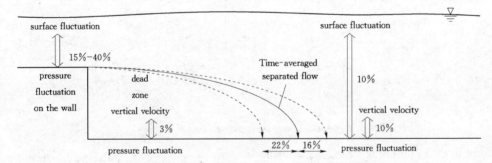

Figure 19   Turbulent structure around backward-facing step open-channel flow

References

[1]   Nakagawa H. , Tsujimoto T. , Shimizu Y. , Murakami S. . Enlargement of the cavity due to sweeping of sands beneath bed protection as a cause of weir disasters, *Proc. Japanese Conference on Hydraulics*, 1987, 31: 359-364.

[2]   Uchida T. , Kawahara Y. , Ikeda M. , Wa-

tanabe A.. Study on hydrodynamic force acting on gravel downstream of a groundsill, *Proc. Japanese Conference on Hydraulics*, 2007, 4: 75-80.

[3] Kawaguchi K, Suwa Y, Takada Y, Suetsugi T. Study on movement of bed particle under bed protection works, unsteady of hydraulic jump and response of its works2. *Manuscript for Advances in River Engineering*, JSCE, 2002, 8: 243-278.

[4] Onitsuka K., Akiyama J., Shigeeda M., Ozeki H., Goto S., Shiraishi T., Fluctuations of free surface and pressure near bed in roller of weak hydraulic jump generated on horizontal bed. *Proc. Japanese Conference on Hydraulics*, 2007, 51: 697-702.

[5] Eaton J. K., Johnston J. P.. "Low frequency unsteadiness of a reattaching turbulent shear layer", *Turbulent Shear Flows*, Springer Verlag, 1982, 3: 162-170.

[6] Nezu I., Nakagawa H.. Experimental investigation on turbulent structure of backward-facing step flow in an open channel. *Journal of Hydraulic Research*, 1987, 25 (1): 67-88.

[7] Nezu I., Nakagawa H.. Turbulent structure of backward-facing step flow and coherent vortex shedding from reattachment in open-channel flows. *Turbulent Shear Flows*, 1989, 6: 313-337.

[8] Silveira Neto A., Grand D., Metais O., Lesieur M.. A numerical investigation of the coherent vortices in turbulent behind a backward-facing step, *J. Fluid Mech.*, 1993, 256: 1-25.

[9] Le H., Moin P., Kim J.. Direct numerical simulation of turbulent flow over a backward-facing step, *J. Fluid Mech.*, 1997, 330: 349-374.

[10] Ohmoto, T., Nariai N., Yakita K.. Turbulent flow structure in a steeped steep open-channel with different ratios of reattachment distance to step length, *Ann. J of Hydraulic Engineering*, 2002, 46: 517-522.

[11] Yokojima T.. A gas-liquid two-phase approach to free-surface turbulence and its application to turbulent open-channel flows, *J Hydraulic, Coastal and Environmental Engineering*, 2006, 62 (4): 419-436.

Effects of Changing Gradient Angle of Notch and Discharge on Migration of Fish

Kouki Onitsuka   Juichiro Akiyama
Koichiro Matsuda   Yusuke Mori   Tsuyoshi Seki

Department of Civil Engineering, Kyushu Institute of Technology,
Kitakyushu, 804-8550, Japan

Abstract: Fishways have been constructed to facilitate migration of fish past dams, waterfalls and rapids. The most of fishway in Japan is pool-and-weir fishway. One of the factors that influences on the migration rate in pool-and-weir fishway is the gradient angle of notch. Wada pointed out that the recommended value of the gradient angle of notch in pool-and-weir fishway is 60 degree. However, there is no measurement at the angle more than 60 degree. In this study, gradient angle of notch in a pool-and-weir fishway was changed in the range from 46 to 82 degree. Moreover discharge was systematically changed from 1 to 13 liter per sec. The migration rates of ayu, Plecoglossus altivelis altivelis, were obtained with the aid of two sets of digital video cameras. The number of used fish is 100. Velocity measurements were conducted with a 3-D electromagnetic current meter. It was found that the migration rates of ayu take the highest value when the gradient angle of notch is from 55 to 64 degree, irrespective of the flow velocity. These results correspond to Wada's point. The migration rates are controlled by the falling velocity from the notch, the distance between the falling flow and orientation area and the angle of fish and falling flow. It was found that there are three factors to cause the migration. The three factors are (1) ayu want to rest close to the falling flow, (2) the angle of fish is close to the angle of falling flow and (3) the falling flow velocity is less than the burst speed and larger than the cruising speed. When one or more of these factors is not filled, the migration becomes difficult.

Key words: pool-and-weir fishway; gradient angle of notch; migration rate

1. Introduction

Fishways have been constructed to facilitate migration of fish past dams, waterfalls and rapids. The most of fishway in Japan is pool-and-weir fishway. However, fishway that is difficult for fish's migration exists. It is necessary to understand appropriate geometrical shape of the fishway. The geometrical shape that has large influence on the migration rate of fish is streamwise pool length, the shape of the notch, the notch ratio, the flow depth and water level difference between the upstream and downstream on pool-and-weir fishway.

Kubota[1] pointed out that the migration rate of iwana is high when streamwise pool length is twice or more the length of iwana. Hayashida[2] et al. pointed out that migration rate is low when the rate of the length of the pool and depth is near one because a high-speed circulation flow is happen in the pool and however, the migration rate is high when the ratio is high more than one be-

cause the low flow velocity occurred in the pool downstream area or the vicinity of the surface of the water. Namihira[3] et al. pointed out that the migration rate is high when the low flow velocity region is secured and the impulse of falling flow is reduced by changing the head between pools or the length of the pool. Wada[4] pointed out that migration rate is high with slope distance of notch shorten. Onitsuka[5] et al. pointed out that migration rate is high when ratio of the notch of width in fishway is small.

On the other hand, the study is few that it is considered the influence that changing angle of notch exerts on migration rate of fish. Wada[4] pointed out that shape of notch which suit for migration of fish is salient type, and round type. In addition, Wada[4] pointed out that angle of notch which suit for migration of fish is 60 degrees by from 15 to 60 degrees and inclination type. However, the experiment is not conducted by angle of notch is 60 degrees or more. It is

not understood whether 60 degrees is the best. Moreover, because flow velocity is almost constant, it is
not understood whether it is possible to apply even at
different flow velocity.

In this study, the influence of the angle of notch
and the discharge on migration rate is investigated by angle of notch and the discharge changes systematically in
pool-and-weir fishway.

2. Experimental set up and hydraulic condi-
tion

Figure 1 shows pool-and-weir fishway that connects
three wooden pools of pool length $L_x = 0.9$m, width of
pool$B = 0.8$m, height from pool bottom to top of
notch$H_t = 0.9$m, height from pool bottom to bottom of
notch$H_u = 0.7$m, thickness of noch$\Delta x = 0.2$m and head
$\Delta y = 0.15$m was used. However, the left sidewall was
made with an acrylic sheet for taking a picture. x axis is
taken in the direction of flowing. y axis is taken perpen-
dicularly and upward. z axis is taken in the direction of
the crossing. Coordinates of bottom of the notch were
provided by using the Eq. (1) and Eq. (2) to change
only angle of notch.

$$y_b = -(x + \Delta x)^n \times \Delta x^{1-n} + H_t \quad (1)$$

$$\tan\theta = \frac{dy_b}{dx}\bigg|_{x=0} = -n \quad (2)$$

Gradient angle of notch is calculated by the Eq. (2) because height of bottom of notchy_b changes and point of
bottom of notch a (0, H_u) and point of top of notch
b ($-\Delta x$, H_t) is fixed by changing n in Figure 1. Figure
2 shows shape of notch in each angle of notch.

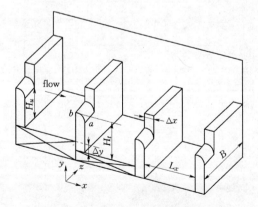

Figure 1　Sketch of experimental channel

Table 1 shows experiment on 30 cases in total that
it is five kinds in the range from 46 to 82 degrees as for
angle of notch and it is six kinds in the range from 1 to

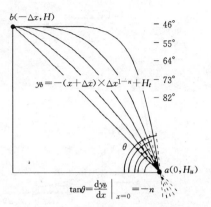

Figure 2　Angle and shape of notch

17L/s as for the discharge. The case name shows that
angle of notch is θ and the discharge is Q. For example,
64Q5 means $\theta = 64°$., $Q = 5$L/s. 100 ayu average length
is 65mm put into the second pool from the upstream and
the flow velocity was gradually increased. Recording was
began with the video camera from the sidewall and the
upper part of the fishway for 30 minutes at the same time
as net for migration prevention was removed after confirming steady flow by watching. Number of migration,
100 ayu swimming positions and fish angle was analyzed
after shooting.

Table 1　Experiment conditions

θ (°) Q (L/s)	46	55	64	73	82
1	46Q1	55Q1	64Q1	73Q1	82Q1
3	46Q3	55Q3	64Q3	73Q3	82Q3
5	46Q5	55Q5	64Q5	73Q5	82Q5
9	46Q9	55Q9	64Q9	73Q9	82Q9
13	46Q13	55Q13	64Q13	73Q13	82Q13
17	46Q17	55Q17	64Q17	73Q17	82Q17

Three components of flow velocity were measured. Average flow velocity (U, V, W) of each direction axis (x, y, z) and synthetic flow velocity
was calculated after measuring. When flow velocity is
measured, the ayu is not put in the fishway.

3. Experimental results and considerations

3.1　Relation between discharge, angle of
notch and migration rate

Migration rate is defined as follows:
The migration rate=

$$\frac{\text{Number of fish that succeeds migration } n}{\text{Number of fish used to experiment } N(=100)} \quad (3)$$

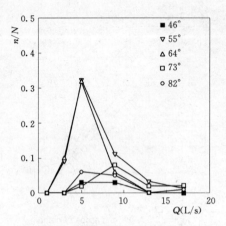

Figure 3　Migration rate with changing discharge in each angle of notch

Figure 3 shows migration rate with changing the discharge in each angle of notch. When angle of notch is 55-64 degrees, high migration rate is shown and when it is larger and is smaller than these angles, migration rate is low in all the discharge in sketch.

Therefore, angle of notch is judged from 55 to 64 degrees is the best value regardless of the discharge. This is almost equal to 60 degrees that is Wada's recommended value.

Moreover, the migration rate becomes the maximum value at the intermediate the discharge (3-9L/s) in all angle of notch. Nakamura has enumerated the best flow velocity and place to rest as a necessary condition for migration. In the following, the factor influencing migration rate is considered.

3.2　Place to rest of ayu in pool

Barycentric position of fish's school is shown in Figure 4 (a) in case when the discharge is 5L/s and angle of notch is changed. Barycentric position of fish's school is shown in Figure 4 (b) case when angle of notch is 55degree and the discharge is changed. h is water depth. A time great change in barycentric position is not seen at each case and the difference between cases is remarkable. Therefore, it is judged time mean swimming position is ayu's place to rest.

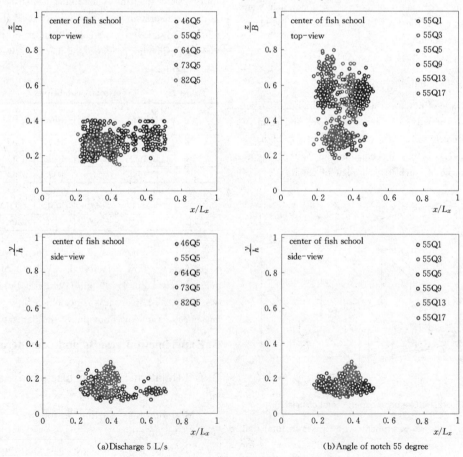

(a)Discharge 5 L/s　　　　　　　　　　(b)Angle of notch 55 degree

Figure 4　Movement situation of barycentric position of fish's school at moment

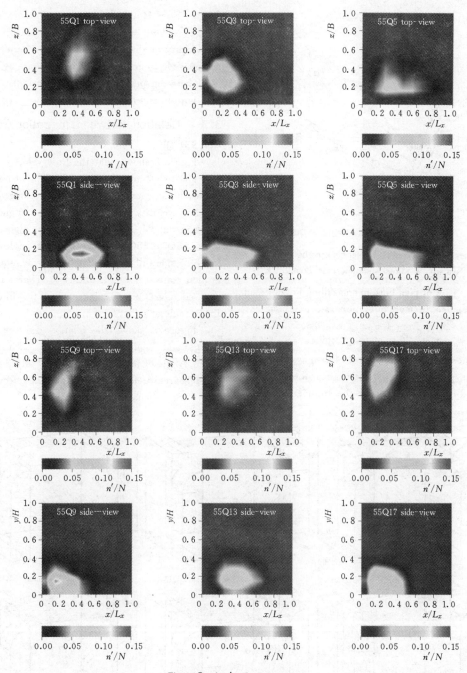

Figure 5   Ayu's place to rest

The time average number n' was calculated by the number counted ayu every ten seconds each 100 meshes separating each the vertical section and the horizontal section in the pool into ten. The contour of the place to rest is shown in angle of notch is 55degree in Figure 5. Ayu's place to rest is the vicinity of the bottom in the vertical cross section regard-less of the discharge changing.

On the other hand, the place to rest leaves from the notch in horizontal section with the discharge increase. It is suggested that ayu's preference place change with the discharge increase.

Figure 6 shows change of horizontal distance between time mean barycentric position of fish and at-

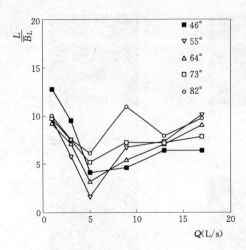

Figure 6　Change of horizontal distance between
ayu's place to rest and attachment point

tachment point of pool bottom and falling flow in each angle of notch. In high migration rate case (55Q5, 64Q5), $L/\overline{B}_L$ is from 1.5 to 3 and ayu is taking a rest comparatively at the vicinity of the attachment point. Therefore, the migration of fish was induced because ayu's swimming position is near the migration route and it is easy to receive the stimulation of the falling flow. However, low migration rate case (46Q5, 46Q9) also exist what ayu taking a rest in the vicinity of the attachment point. Therefore, it is not possible to understand migration rate by only distance between place to rest and attachment point of pool bottom and falling flow.

3.3　Relation between migration rate and fish's angle

Figure 7 shows that the flow velocity vector is blue and fish's angle is red when angle of notch is 55 degree. Fish's angle is random in low migration rate low flow velocity case (55Q1). Fish's angle turn for notch in middle flow velocity case of high migration rate (55Q3, 55Q5, 55Q9). Fish's angle turn for notch or sidewall in high flow velocity case of low migration rate (55Q13, 55Q17). Therefore, it is guessed migration rate is received the influence of fish's angle at taking a rest.

Figure 8 shows the relation between the discharge and angle of falling flow each angle of notch. Angle of falling flow is flat by all flow velocity in 64 degree or less and the angle has decreased by high flow velocity in 73° or more.

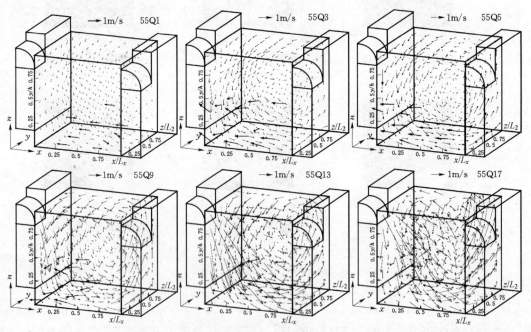

Figure 7　Flow velocity vector (blue) and fish's angle (red)

Figure 9 shows the relation between the discharge and average fish's angle when ayu is taking a rest in each angle of notch. fish's angle becomes the maximum when the discharge is 5L/s in the angle of notch 55 and 64 degree. And fish's angle is the most approaches angle of the falling flow. At this time, migration rate is high as shown in Figure 3. Therefore, this twice case are thought that migration rate became the maximum because fish's angle is the most approaches angle of falling flow and migration became easy. On the other hand, fish's angle becomes the maximum value when the discharge is 17L/s and angle of notch is 46, 73 and 82 degree. Then migration rate is low as shown Figure 3. Therefore, it is not possible to understand the reason influencing migration rite is only the relation between fish's angle and angle of falling flow. In the following, the relation between the falling flow velocity and the migration rate is examined.

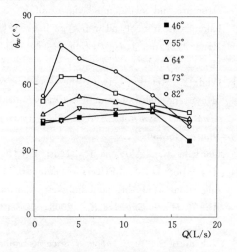

Figure 8   Angle of falling flow

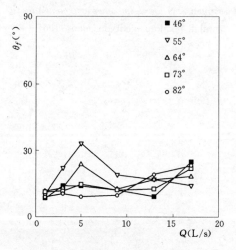

Figure 9   Average fish's angle at taking a rest

Figure 10 shows the relation between the discharge and component falling flow velocity in each angle of notch. Fish's burst speed and cruising speed are shown by

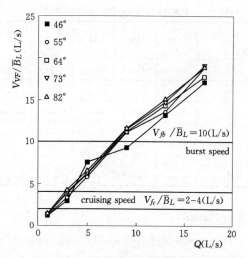

Figure 10   Relation between discharge and component falling flow velocity

the solid line. Composite falling flow velocity increases in all angle of notch with the discharge increases and exceeds fish's burst speed when the discharge is over 8L/s. Therefore, migration rate is low regardless that fish's angle is the most approaches angle of the falling flow when the discharge is 17L/s and angle of notch is 46, 73 and 82 degrees in Figure 8 because the falling flow velocity exceeded the rush speed. On the other hand, composite falling flow velocity is about seven times of fish's body length in high migration rate case that is angle of notch is 55 and 64 degrees and the discharge is 5L/s. It is thought that the fish's migration was induced because falling flow velocity is less than the burst speed and larger than the cruising speed as Koyama[7] and Nakamura[6] have pointed out.

3.4   Factor influencing migration rate

Table 2 shows the factor that the migration was induced and becomes difficult. d of a blue letter shows ayu is taking a rest in the vicinity of the attachment point. a of a blue letter shows fish's angle is near angle of falling flow. v of a blue letter shows the falling flow velocity is less than the burst speed and larger than the cruising speed. The one it was not filled these conditions was shown by red. Cases all are blue letter agree high migration rate case (55Q3, 55Q5, 64Q3) and Cases all are red letter agree low migration rate case shown in Figure 3. Therefore, the migration rate become high when these three conditions are filled and the migration becomes difficult when one or more of these is not filled.

Table 2   Factor influencing migration rate

Q(L/s) \ $\theta(°)$	46	55	64	73	82
1	d,a,v	d,a,v	d,a,v	d,a,v	d,a,v
3	d,a,v	d,a,v	d,a,v	d,a,v	d,a,v
5	d,a,v	d,a,v	d,a,v	d,a,v	d,a,v
9	d,a,v	d,a,v	d,a,v	d,a,v	d,a,v
13	d,a,v	d,a,v	d,a,v	d,a,v	d,a,v
17	d,a,v	d,a,v	d,a,v	d,a,v	d,a,v

4. Conclusions

In this study, it examined influence of angle of notch in pool-and-weir fishway and the discharge on migration rate. Ayu's migration rate is calculated by systematically changing angle of notch and the discharge. As a result, the following have been understood.

(1) Angle of notch is from 55 to 64 degrees is the best value to ayu's migration rates by from 46 to 82 degree, irrespective of the flow velocity.

(2) It was found that there are three factors to cause the migration. The three factors are 1) ayu want to rest close to the falling flow, 2) the angle of fish is close to the angle of falling flow and 3) the falling flow velocity is less than the burst speed and larger than the cruising speed. When these three factors are filled, the migration is induced. However, when one or more of these factors are not filled, the migration becomes difficult.

References

[1] Kubota T.. Fish migration behavior in fishway on Sabo facilities and fishway condition in Situ. *Ann. J. Hydr. Eng. JSCE.*, 1998, 42: 487-492.

[2] Hayashida K., Honda T., Kayaba Y., Shimatani Y.. The flow pattern of the pool-and-fishway in the pool and the swimming behavior of leuciscus hakonensis. *Ann. J. Hydr. Eng. JSCE.*, 2000, 44: 1191-1196.

[3] Namihira A., Goto M., Kobayashi H.. Flow structure and swimming behavior of leuciscus hakonensis at pool-and-weir type fishway with slope 1/5. *Ann. J. Hydr. Eng. JSCE.*, 2008, 52: 1189-1194.

[4] Wada Y.. Relation between the ascending path of ayu and fishway structure. *Proc. of the International Symp. on Fishways '90 in Gifu*, 1990: 445-450.

[5] Onitsuka K., Akiyama J., Mori Y., Iiguni Y., Kobayashi T.. Effects of notch ratio on migration rate in pool-and-weir fishways. *Ann. J. Hydr. Eng. JSCE.*, 2008, 52: 1201-1206.

[6] Nakamura S.. *Topics about fishway*, Sankaidou, 1995.

[7] Koyama N.. *gyodowomegurusyomondai* II kisomigawakakousigentyousadan, 1967.

Relationship Between Flow Conditions of Orientation Area and Behavior behind the Backward-Facing Step

Kouki Onitsuka   Juichiro Akiyama   Tsuyoshi Seki   Yusuke Mori

Department of Civil Engineering, Kyushu Institute of Technology, Kitakyushu, 804-8550, Japan

Abstract: Fishway is a construction which is installed at the side of a large weir for fish migration between the upstream and downstream habitat. It is necessary to facilitate the fish migration that the maximum velocity in the fishway is lower than the burst speed of fish and also offering the orientation area to the fish where fish can rest. There are a lot of investigations on the migration ratio in several type fishway. However, the characteristics of orientation area have not been investigated so much. In this study, the fish behaviors just behind of the backward-facing step were measured and also velocity measurements were conducted with a 3-D electromagnetic current meter, using ayu, Plecoglossus altivelis altivelis. The number of used fish is 20. It was verified that high fish existing probability moves toward the backward-facing step steadily with an increase of height of the backward-facing step. Further, fish existing probability concentrates on particular areas. The areas where fish existing probability is more than 6 percent and is less than 2 percent were defined as orientation area and non-orientation area. Further, comparing the orientation area and non-orientation area, the velocity and turbulence of fish orientation area are lower than one of non-orientation area. Therefore, it was found that the ayu wants to rest in the low velocity and low turbulence area. When there is no location where the ayu wants to rest, the ayu swims in relatively low velocity and low turbulence area. Further, the ayu hates the high shear velocity area, irrespective of low velocity.

Key words: migration; fishway; orientation area; velocity; turbulence; shear

1. Introduction

Fishways are installed at the side of river structures such as dam and weir to facilitate migration of fish. There are many different types of fishways. Nakamura[1] pointed out it requires the condition that maximum flow velocity is less than burst speed, there are rest areas of fish and so on for upstream migration of fish. The migration number was counted in many different types of fishways. However, there is little study which investigated on suitable areas of fish.

Sagou et al[2]. recorded the behavior of ayu in the fishway while changing cobbles density on the bed and pointed out that ayu rest behind the cobble. Wada et al [3]. pointed out that ayu tend to choose area of weak flow. Namihira et al[4]. recorded behavior of ugui, Tribolodin hakonensis, in the pool-and-weir fishway while changing geometrical configuration of the fishway and flow condition. As a result, it was

showed that ayu rest in low flow velocity area of center of pool when the flow belongs to the plunging flow and ayu has positive rheotaxis. Ishikawa et al[5]. recorded behavior of ugui, further, the flow velocity was measured in the vertical-slot fishway. As a result, it was showed that the ground speed is constant. However, the swimming speed varies along with the flow velocity and the swimming direction is consistently opposite to the flow direction. As identified above, it has been shown that fish rests and show positive rheotaxis in low flow velocity area.

There are also studies of rest time length. Wada et al[6]. proposed the modified larinier pass fishway and recorded behavior of juvenile ayu. It was showed that rest time length of juvenile ayu is described by log normal distribution and averaged rest time length is 90s. Izumi et al[7]. recorded behavior of ugui in the hybrid-type fishway and showed that ugui rests in dead water zone and rest time length is almost less than 20s.

The studies that the flow velocity in rest area is quantitatively analyzed have been conducted. Hayashida et al[8]. pointed out that flow velocity in rest area of ugui is 0 to 5 times higher than averaged value of body length. Onitsuka et al[9]. changed the notch location and notch rate in the pool-and-weir fishway and showed that ayu rests in low flow velocity area.

As identified above, there are studies that the rest condition of fish in fishway is verified. However, there is little study which quantitatively analyzed relationship between three dimensional fish behavior, flow condition and rest area. In this study, behavior of ayu was recorded and the flow velocity was measured in the pool-and-weir fishway. Further, behavior of ayu in rest area was closely analyzed and relationship between the flow condition and rest area was investigated.

2. Experimental setup and hydraulic conditions

The experiments were conducted in a 1.8m long, 0.4m width and 0.9m height open-channel as shown in Figure 1. L_x is distance of the backward-facing step to downstream edge, L_y is height of the backward-facing step, h is flow depth, index u and l show downstream side and upstream side of the backward-facing step. The backward-facing step was installed at 0.95m from upstream edge and L_y is 0.85m. In which x, x' and y are the coordinates of streamwise from the backward-facing step, the streamwise from upstream edge and vertical direction, respectively. The depth h_u is 0.1m and L_y/h_l are changed from 0.35 to 0.8 as shown in Table 1 The number of used ayu is 20. Averaged body length $\overline{B_L}$ of ayu is about 7.8cm. Koyama[10] pointed out that positive rheotaxis of ayu depend on the flow velocity and ayu which the body length is range from 6 to 8cm shows the greatest positive rheotaxis when the flow velocity is range from 0.4 to 0.6m/s. Therefore, flow velocity of upstream side of the backward-facing step was 0.6m/s due to averaged body length is 7.8cm in this study. Ayu was released on the location where x is 0.7m and the flow velocity was gradually increased. After it was confirmed that the flow became steady flow and ayu rested, behaviors of 20 ayu were recorded 20 minutes with digital video cameras from the side of the channel. Three components of instantaneous flow velocities, i.e., $\tilde{u}=U+u$ and $\tilde{v}=V+v$ in $x-y$ vertical

cross section were measured at 70 points (10 points in x direction and 7 points in y direction) with a three-dimensional electromagnetic current meter after removing the fish. In which the capital letter denotes the time-averaged flow velocity and the small letter denotes the fluctuation from the time-averaged flow velocity component of x and y direction.

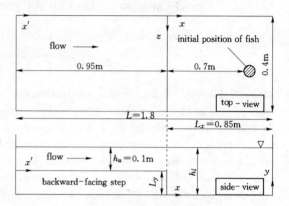

Figure 1 Experimental setup

Table 1　　　Experimental conditions

L_y/h_l	0.35	0.5	0.65	0.8
Case name	C35	C50	C65	C80

3. Experimental results and considerations

3.1. Flow condition

Figure 2 shows the flow velocity vector. The flow velocity of inflow section is high and the vortex is formed bellow the backward-facing step and near the bed. Downstream flow velocity of the vortex is low, comparing flow velocity of inflow section.

Figure 3 (a) and Figure 3 (b) show turbulence intensity of x and y direction i. e., u'/U_{m0} and v'/U_{m0}. It is found turbulence intensity increase in inflow section and decrease with downstream advection. When the flow is uniform flow in the open-channel, ratio of turbulence intensity of x direction u' to one of y direction v'/u' is about 0.55. However, it is found that there is area that v'/U_{m0} is higher than u'/U_{m0} and the flow is complex turbulent flow.

3.2 Definition fish rest area

Figure 4 shows the time-averaged fish existing probability. It is found that high fish existing probability move toward the backward-facing step steadily with an increase of height of the backward-facing step. High fish

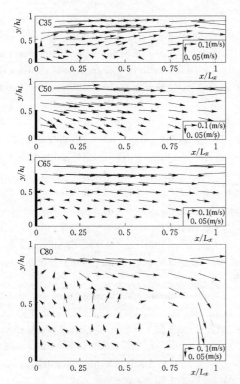

Figure 2   Flow velocity vector

existing probability comparatively locates at near the bed in all cases. However, high fish existing probability of C80 slightly locates at area distant from the bed. Further, fish existing probability concentrates on particular areas. However, comparing other cases, fish existing probability is low. It is suggested that fish discretely swim in C35. It is impossible to understand instantaneous fish behavior due to the time-averaged fish existing probability.

It employs fish school's radius equation of Ishikawa[11] due to determine that ayu randomly swims or behave as fish school.

$$R_x \equiv \sqrt{\sum_{i=1}^{N} \{ x_{fi} - G_x \}^2 / N} \qquad (1)$$

$$R_y \equiv \sqrt{\sum_{i=1}^{N} \{ y_{fi} - G_y \}^2 / N} \qquad (2)$$

In which, R_x, G_x and x_{fi} are fish school's radius, point of fish school's center and point of any fish of x direction, respectively and R_y, G_y and y_{fi} are those of y direction.

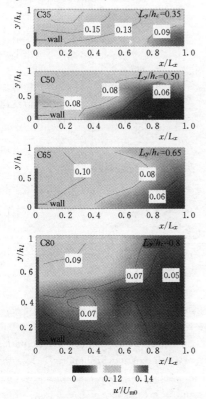
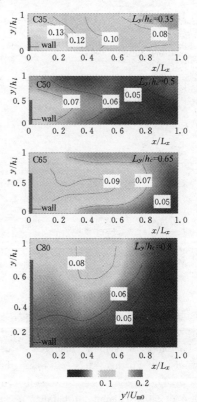

Figure 3   Turbulence intensity

(a) Turbulence intensity of x direction; (b) Turbulence intensity of y direction

587

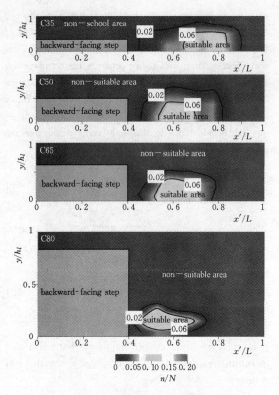

Figure 4　Time-averaged fish existing probability

Figure 5 shows point of fish school's center (G_x, G_y) of 60s per second. Fish school's center variation of x and y direction is few and location of fish school's center fits to location of high fish existing probability. Figure 6 shows fish school's radius of x and y direction and ratio between fish school's radius of x and y direction of time series i. e, $R_x/\overline{B_L}$, $R_y/\overline{B_L}$ and R_x/R_y. Fish school's radius $R_x/\overline{B_L}$ and $R_y/\overline{B_l}$ of all cases except C35 are 1. 0 and 0. 3 and time fluctuation of $R_x/\overline{B_L}$ and $R_y/\overline{B_L}$ is small. Further, R_x/R_y is about 3. Therefore, it is found that fish school is dense. On the other hand, $R_x/\overline{B_L}$ of C35 is fluctuant in the 1. 0 to 2. 0 range and $R_y/\overline{B_L}$ of C35 is smaller than $R_y/\overline{B_L}$ of all cases except C35. Further, R_x/R_y is increasing and maximum value of R_x/R_y is 13.

Figure 7 shows relationship between standard deviation of fish school's radius of x and y direction i. e, $R'_x/\overline{B_L}$ and $R'_y/\overline{B_L}$ and height of the backward-facing step $L_y/\overline{B_L}$. $R'_x/\overline{B_L}$ and $R'_y/\overline{B_L}$ of all cases except C35 is almost constant value regardless of height of the backward-facing step.

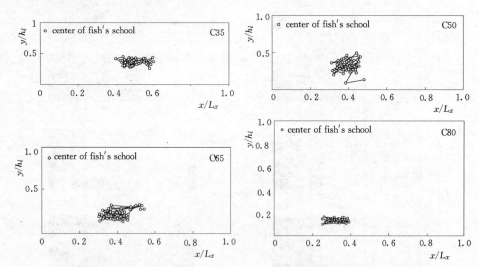

Figure 5　Point of fish school's center

Fish existing probability and fluctuation of fish school's center and radius of all cases except C35 are the similar. Therefore, it is found that ayu choose suitable area and rest. Further, it is considered that ayu swim in a condition close to suitable area in C35.

It is found that fish existing probability in the area of high fish existing probability and location of fish school's center is more than 6 percent. Therefore, the area where fish existing probability is more than 6 percent was defined as orientation area. Further, fish existing probability is less than 2 percent were defined as non-orientation area due to ayu some swim in area where fish existing probability is less than 6 percent.

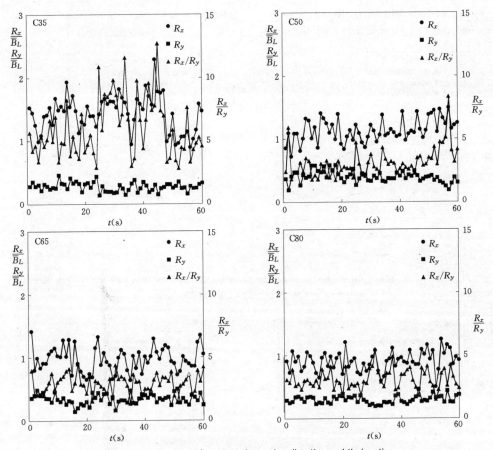

Figure 6　Fish school's radius of x and y direction and their ration

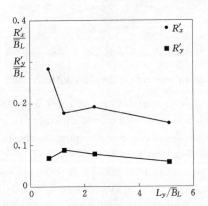

Figure 7　Relationship between standard
deviation of fish school's radius of x
and y direction and height of the
backward-facing step

3.3　Comparing orientation and non-orientation area

Figure 8 (a) and (b) show relationship between the space-averaged flow velocity of x direction $\langle U \rangle$ in orientation area and non-orientation area and height of the backward-facing step L_y and the space-averaged flow velocity of y direction $\langle V \rangle$ and L_y. It is found that the space-averaged flow velocity of x direction decrease with an increase of height of the backward-facing step and there is a significant difference in the space-averaged flow velocity in orientation area and non-orientation area (as shown in Figure 8a). Further, it is understood that the space-averaged flow velocity in orientation area is low as ± 1.0 times higher than averaged value of body length $\overline{B}_L$ and the flow direction is random. The space-averaged flow velocity of y direction $\langle V \rangle$ in orientation area is ± 1.0 times higher than $\overline{B}_L$ and the flow direction is random (as shown in Figure 8b). Figure 9 (a) and Figure 9 (b) show relationship between the space-averaged turbulence intensity of x and y direction $\langle u' \rangle$ and $\langle v' \rangle$ and height of the backward-facing step L_y. It is found that turbulence intensity of x direction $\langle u' \rangle$ is lower than one of y direction $\langle v' \rangle$. Turbulence intensity of x and y direction are less than 0.1 and 0.05 times of averaged value of body length $\overline{B}_L$. It was quantitatively verified that fish favorite area that turbulence intensity is low.

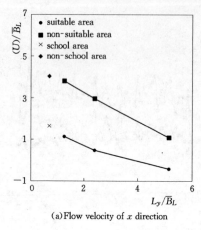

(a) Flow velocity of x direction

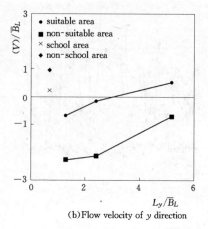

(b) Flow velocity of y direction

Figure 8 Relationship between the space-averaged flow velocity in orientation area and non-orientation area and height of the backward-facing step

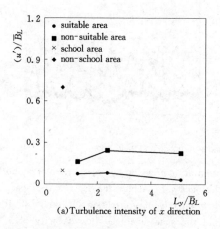

(a) Turbulence intensity of x direction

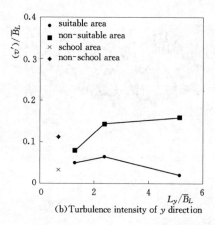

(b) Turbulence intensity of y direction

Figure 9 Relationship between the space-averaged turbulence intensity and height of the backward-facing step

4. Conclusions

In this study, behavior of ayu that swim at downstream side of the backward — facing step was recorded and flow velocity was measured. Further, behavior of resting ayu was analyzed and relationship between floe condition and orientation area was verified. The results of this study are described as follows.

(1) Ayu can choose orientation area and rest when height of the backward-facing step is more than 10cm. However, there is not orientation area when height of the backward-facing step is 5.5cm.

(2) Fluctuation of fish school is few when there is orientation area.

(3) It was verified that ayu that body length is about 7.8cm choose for orientation area to area where flow velocity of x and y direction are about ± 1.0 times higher than averaged value of body length and turbulence intensity of x and y direction are less than 0.1 and 0.05 times of averaged value of body length and the orientation area is independent of the flow direction.

References

[1] Nakamura S.. Topics about fishway. Sankaidou, 1995.

[2] Sagou Z., Honda T., Ooki T., Tanaka N..

A study on upstream migration of Ayu at Otawara barrage fishway. *Ann J of Hydraulic Eng*, JSCE, 1998, 42: 493-498.

[3] Wada K., Azuma N., Nakamura S.. Migratory behavior of Juvenile Ayu related to flow fields in denil and steeppass fishway. *Ann. J. of Hydraulic Eng.*, *JSCE*, 1998, 42: 499-504.

[4] Namihira A., Goto M., Kobayasi H.. Flow and upstream migrating behavior of Leuciscus hakonensis at each pool of pool-and-weir fishway accompanying discharge change. *Ann J of Hydraulic Eng*, JSCE, 2007, 51: 1291-1296.

[5] Ishikawa M., Shirai j., Hu F., Tokai T.. Upstream passage and behavior of Japanese dace school through an experimental verticalslot fishway. *Fisheries Eng*, 2006, 43: 9-20.

[6] Wada K., Koizumi N., Ishikawa M., Nakamura S.. Upstream migrating route of Juvenile Ayu in a modified larinire pass fishway. *Ann. J. of Hydraulic Eng.*, JSCE, 1999,

43: 983-988.

[7] Izumi M., Kudo A., Azuma N., Sato S.. A study on upstream migration of Japanese dace in a pool-type flow of a hybrid-type scale model fishway. *Manuscript for advances in river engineering*, JSCE, 2000, 6: 131-136.

[8] Hayashida K., Honda T., Kayaba Y., Shimatani Y.. The flow pattern of the pool-and-weir fishway in the pool and the swimming behavior of Leuciscus hakonensis. *Ann J of Hydraulic Eng*, JSCE, 2000, 44: 1191-1196.

[9] Onitsuka K., Akiyama J., Kiuchi D., Takahashi Y., Iiguni Y.. Effect of location of notch on migration rate in pool-and-weir fishway. *Ann J of Hydraulic Eng*, JSCE, 2007, 51: 1279-1284.

[10] Koyama N.. Resource investigation committee in Kisomigawa River. 1965.

[11] Ishikawa M.. An experimental study on evaluation of fish schooling behavior properties. *Manuscript for advances in river engineering*, JSCE, 2000, 6: 101-106.

Time-Varying Analysis Model of Monitoring Data in Dam Storage Period

Li Ziyang[1]   Zhou Yongjun[2]   Yuan Hui[1]   Xiang Yan[1]

[1] *State Key Laboratory of Hydrology-Water Resources and Hydraulic Engineering, Nanjing, 210029*
[2] *Water Resources and Hydropower Engineering Bureau, Hangzhou, 310000*

Abstract: Considering the unstable monitoring data and the changed dam behavior in storage period, the forgotten factor is introduced in the traditional statistical model for solving the model parameters, and a forgotten matrix is established to give prominence to the contributions of recent values. Then the time-varying analysis model is set up for monitoring data in dam storage period. Case analysis shows that the model parameters can be periodic updated with the changes of storage state and the accuracy of fitting and forecasting are better than the traditional statistical model, so it is more suitable for analyzing the monitoring data in storage period.

Key words: data analysis; storage period; forgotten matrix; time-varying model

1. Introduction

Dam monitoring data analysis is very important in dam safety evaluation [1]. At present, the methods are mainly single analysis and forecast models including statistical model [2-3], artificial neural network model [4], the gray model [5] etc. and portfolio analysis and forecast model composed of multiple sub-model [6-7]. Based on mechanics of the independent variables and the dependent variable, statistical analysis method has strong physical meaning and good adaptability, so it is the most commonly used models of dam safety monitoring data analysis. However, traditional statistical model has the following problems: model contains statistical properties and anti-jamming ability is poor. When the lack of observation data, larger observation error or less stationary, often we could not obtain desired results. The way of reservoir storage is often raising the water level by stages during initial storage period, and then the monitoring data of dam characteristic has less stationary, so fitting accuracy is often not high when using traditional statistical model. Besides, with the impact of initial storage pressure, the overall state of dam body will change with the rise of water level. That is to say, the model parameters of established monitoring model can not remain the same but should be time-varying and periodic updated with the change of the dam behavior.

Based on above consideration, drawing on the data processing thought of using forgotten factor to highlight the contribution of recent values and reduce it of early values in time series analysis, the forgotten factor is introduced in the traditional statistical model for solving the model parameters in this paper. Through the establishment of forgotten matrix to highlight the contribution of the current and recent values to the model, the time-varying analysis model is set up for monitoring data in dam storage period.

2. Traditional statistical models and parameters solution

With dam theory and mathematical and mechanical principles, the monitoring values of dam state δ are mainly composed of hydraulic components δ_H, temperature components δ_T, rainfall components δ_P, and time components δ_θ [1]. The form is as below

$$\delta(t) = \delta_H(t) + \delta_T(t) + \delta_P(t) + \delta_\theta(t) \qquad (1)$$

where t is time.

For instance, the statistical model of horizontal displacement δ in arch dam is described as following in general,

$$\delta = \delta_H + \delta_T + \delta_\theta$$
$$= \sum_{i=1}^{4} \left[a_1 \left(H_u^i - H_{u0}^i \right) \right]$$
$$+ \sum_{i=1}^{2} \left[b_{1i} \left(\sin \frac{2\pi i T}{365} - \sin \frac{2\pi i T_0}{365} \right) \right.$$

$$+b_{2i}\left(\cos\frac{2\pi iT}{365}-\cos\frac{2\pi iT_0}{365}\right)\Bigg]$$

$$+c_1(\theta-\theta_0)+c_2(\ln\theta-\ln\theta_0)+a_0 \qquad (2)$$

where H_u, H_{u0}—corresponding measured upstream heads of monitoring date and beginning date, which is the difference between measured values of water level and the elevation of dam bottom;

T—cumulative days from beginning date to monitoring date;

T_0—cumulative days from beginning date to the first monitoring date of modeling data series;

θ—T divided by 100;

θ_0—T_0 divided by 100;

a_0—constant.

To facilitate the writing, setting

$$x=[1,H_u^1-H_{u0}^1,H_u^2-H_{u0}^2,H_u^3-H_{u0}^3,H_u^4-H_{u0}^4,$$

$$\sin\left(\frac{2\pi T}{365}\right)-\sin\left(\frac{2\pi T_0}{365}\right),\cos\left(\frac{2\pi T}{365}\right)-$$

$$\cos\left(\frac{2\pi T_0}{365}\right),\sin\left(\frac{4\pi T}{365}\right)-\sin\left(\frac{4\pi T_0}{365}\right),\cos\left(\frac{4\pi T}{365}\right)-$$

$$\cos\left(\frac{4\pi T_0}{365}\right),\theta-\theta_0,\ln\theta-\ln\theta_0]^T$$

$$\theta=[a_0,a_1,a_2,a_3,a_4,b_{11},b_{21},b_{12},b_{22},c_1,c_2]^T$$

Then Eq. (2) can be abbreviated as

$$\delta(t)=x^T(t)\theta \qquad (3)$$

Usually, the solution of θ is using least squares (LS) estimation based on the measured values of $\delta(t)$, $\delta(t-1)$, $\cdots$, $\delta(1)$, that is to say the valuation $\hat{\theta}$ of θ minimize the residual sum of squares

$$J(\theta)=\sum_{i=1}^{t}\varepsilon^2(i)=\sum_{i=1}^{t}[\delta(i)-x^T(i)\theta]^2 \qquad (4)$$

Setting

$$X(t)=\begin{bmatrix}x^T(1)\\ \cdots\\ x^T(t)\end{bmatrix},\Delta(t)=\begin{bmatrix}\delta(1)\\ \cdots\\ \delta(t)\end{bmatrix}$$

So performance index J can be written as

$$J(\theta)=[\Delta(t)-X(t)\theta]^T[\Delta(t)-X(t)\theta] \qquad (5)$$

Setting $\partial J(\theta)/\partial\theta=0$, the matrix solution equation of θ can be obtained by matrix differential computing formula

$$-2X^T(t)[\Delta(t)-X(t)\theta]=0$$

Then the LS valuation $\hat{\theta}$ is gotten based on the measured values of $\delta(t),\delta(t-1),\cdots,\delta(1)$

$$\hat{\theta}=[X^T(t)X(t)]^{-1}X^T(t)\Delta(t) \qquad (6)$$

3. Time-varying parameters solution of traditional statistical models

The solution of statistical model parameters adopts the least-square method (LS), which would lead to large error for time-varying parameters. In order to reflect the time-varying characteristic in dam storage period, the index weighted least squares (IWLS) is adopted to determine model parameters. The following is reasoning and description of this formula.

As the behavior of the dam will change with the rise of water level in storage period, The statistical model of LS algorithm will create large errors with the increase of monitoring values. This is because the new values that reflect the current state will be submerged by a large number of old values. To reflect the time—varying characteristic of model in storage period, the role of new values should be emphasized while gradually forgetting the old data. Based on above consideration, the index weighted coefficient λ^{t-i}, $0<\lambda\leqslant1$ was introduced in performance index J, λ is forgotten factor, then advanced performance index [based on Eq. (4)] is

$$J(\theta)=\sum_{i=1}^{t}\lambda^{t-i}\varepsilon^2(i)=\sum_{i=1}^{t}\lambda^{t-i}[\delta(i)-x^T(i)\theta]^2 \qquad (7)$$

Similar to type (5), the matrix form is

$$J(\theta)=[\Delta(t)-X(t)\theta]^T\lambda(t)[\Delta(t)-X(t)\theta] \qquad (8)$$

Forgotten matrix (or weighted matrix) $\lambda(t)$ is

$$\lambda(t)=\begin{bmatrix}\lambda^{t-1}&&&&\\ &\lambda^{t-2}&&&\\ &&\ddots&&\\ &&&\lambda&\\ &&&&1\end{bmatrix}$$

similarly, setting $\partial J(\theta)/\partial\theta=0$, then

$$-2X^T(t)\lambda(t)[\Delta(t)-X(t)\theta]=0$$

Setting

$$X^T(\lambda t)=X^T(t)\lambda(t)=[x(1),\cdots,x(t)]\begin{bmatrix}\lambda^{t-1}&&\\ &\ddots&\\ &&1\end{bmatrix}$$

$$=[\lambda^{t-1}x(1),\lambda^{t-2}x(2),\cdots,\lambda x(t-1),x(t)]$$

Then similar to Eq. (6), the IWLS valuation formula of $\hat{\theta}$ is gotten based on the measured values of $\delta(t),\delta(t-1),\cdots,\delta(1)$

$$\hat{\theta}=[X^T(\lambda t)X(t)]^{-1}X^T(\lambda t)\Delta(t) \qquad (9)$$

The forgotten factor λ can be determined according to actual situation. If the storage process is faster in analysis series, then $\theta(t)$ is faster time-varying parameter, which should be taken less; On the contrary, $\theta(t)$ should be taken lager.

4. Instance analysis

The radial displacement monitoring values of some dam in storage period is selected as an example to analysis whose field site is on the right 1/4 arch section of 2114m elevation. Considering characteristic of the beginning date of environmental monitoring and the initial storage stages, the selected analysis monitoring series is 1997-1-2 to 2001-5-31.

The process line of water level upstream in storage period in Figure 1 indicates that the changes of water level in initial storage stages has gone through three stages of stepped increases. Therefore, in the solving and fitting of time-varying parameter of statistical models, 7 stages are selected as time-varying parameter updating period and stage 8 as forecasting period, which are shown in Table 1. For instance, the adopted time-varying parameters of the model in 4 fitting period (1999-8-7 to 1999-8-19) is the one calculated with monitoring values of the 4 calculation period of parameters (1997-1-1 to 1999-8-19) by Eq. (9).

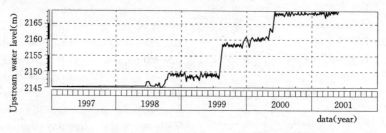

Figure 1　Process line of water level upstream in storage period

Table 1　Divination of calculation period

No.	Fitting/Forecasting period	Calculation period of parameters
1	1997-1-2to1998-9-24	1997-1-1to1998-9-24
2	1998-9-25to1998-10-13	1997-1-1to1998-10-13
3	1998-10-18to1999-8-6	1997-1-1to1999-8-6
4	1999-8-7to1999-8-19	1997-1-1to1999-8-19
5	1999-8-20to2000-5-14	1997-1-1to2000-5-14
6	2000-5-15to2000-6-2	1997-1-1to2000-6-2
7	2000-6-3to2000-12-31	1997-1-1to2000-12-31
8	2001-1-1to2001-5-31	1997-1-1to2000-12-31

Corroding to arch features, displacement statistical model is established as Eq. (2). With the adaptation of stepwise regression analysis, the model parameters are solved respectively by LS method (traditional statistical models) and IWLS method (time-varying models) (as forgotten factor has no experience value, setting $\lambda = 0.985$ under the trial compared). Then regression analysis of measured values of selected point displacement is carried out. the multiple correlation coefficient (R) and the remaining standard deviation (S) of fitting model are shown in Table 2, fitting and forecasting results of two models are shown as Figure 2-Figure 4, comparison of some forecasting values are shown in Table 3.

Table 2　The eigenvalue of fitting model

Parameters ＼ Mmodel	Traditional statistical model	Time-varying model
R	0.960	0.983
S	0.507	0.238

Table 3　Comparison of forecasting values

Date	Measured values (mm)	Forecasting values (mm)	
		Traditional model	Time-varying model
2001-1-3	9.62	8.61	9.45
2001-2-6	9.77	9.44	9.71
2001-2-28	9.49	9.30	9.68
2001-3-21	10.12	9.42	9.69
2001-4-17	10.05	9.43	9.99
2001-5-1	10.31	9.57	10.17
2001-5-23	10.57	10.31	10.48

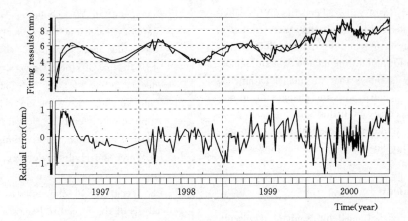

Figure 2　Fitting results of traditional statistical model in fitting period

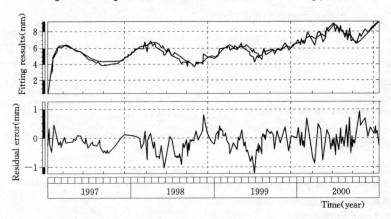

Figure 3　Fitting results of time-varying model in fitting period

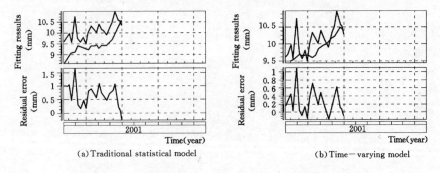

(a) Traditional statistical model　　　　　(b) Time－varying model

Figure 4　Forecasting results of two models in forecasting period

The graph indicates that the accuracy of fitting and forecasting of time-varying model are better than the traditional statistical model，as the model parameters can be periodic updated with the changes of storage state according to the forgotten matrix. Therefore，it is more suitable for reflecting the dam behavior changes and analyzing the monitoring data in storage period.

5. Summary

Considering the shortage of the traditional statistical model with LS parameter estimation period to reflect the behavior of the dam changing with water levels in storage period, the forgotten factor is introduced for solving the model parameters, and a forgotten matrix is established to give prominence to the contributions of recent values. Then the IWLS valuation method is derived to reflect the characteristics of time-varying parameters and the time-varying analysis model is set up for monitoring data in dam storage period. Case analysis shows that the model parameters can be periodic updated with the changes of storage state and the accuracy of fitting and forecasting are better than the traditional statistical model, so it is more suitable for analyzing the monitoring data in storage period.

Acknowledgment

This paper was funded by Nanjiang Hydraulic Research Institute fund (S70901, Y709002, Y710006) and National Nature Science Youth Fund (50909066).

References

[1] Wu Zhongru. Hydraulic Structure Safety Monitoring Theory and Its Applications [M]. Beijing: Higher Education Press, 2003.

[2] Li Ziyang. Gu Chongshi. Analytic Model for Deformation Behavior of Dam Slipmass [J]. Hydropower Automation and Dam Monitoring, 2006, 30 (4): 59-61.

[3] Bao Tengfei, Wu Zhongru, Gu Chong shi. Statistic model and chaos theory-based hybrid forecasting model for dam safety monitoring [J]. Journal of Hohai University (Nature Sciences), 2003, 31 (5): 534-538.

[4] Peng Xinming, Liu Mingjun, Huang Caiyuan. Application of Artificial Neural Network Model in Dam Deformation Observation [J]. China Rural Water and Hydropower, 2006, (11): 99-101.

[5] Bao Tengfei, Xu Zhifeng, Gu Chongshi. Analysis of the Structure Behavior of Dam Using the Grey Process Model [J] . Hongshui River, 2005 (3): 46-48.

[6] Zhou Mingduan, Guo Jiming, Wen Hongyan, etc. Optimized Initial Value-based GM (1, 1) Model and Its Application to Dam Monitoring [J] . Hydropower Automation and Dam Monitoring, 2008, 32 (2): 52-54.

[7] Yan Bin, Zhou Jing, Gao Zhenwei. An Iowga Operator based Dam Safety Monitoring Combination Forecasting Model [J] . Chinese Journal of Rock Mechanics and Engineering, 2007, 26 (Supp. 2): 4074-4078.

Research on Monitoring Method for 3-Dimensional Deformation of Earth Rockfill Dam

Ma Fuheng　Dai Qun

State Key Laboratory of Hydrology-Water Resources and Hydraulic Engineering,
Nanjing Hydraulic Research Institute, Nanjing, 210029

Abstract: The routine methods for monitoring deformation of surface of earth rockfill dam has several defectives, i. e., slow measurement speed, heavy workload, monitoring information can not share with automation system. High precision total station system (TCA2003) can provide solution for research and development of 3-dimensional automation measuring technology of earth rockfill dam, however, atmosphere refraction and other issues affect the measuring precision. In accordance with actual condition of Yanshan Reservoir, the paper analyzes the refraction coefficient, raises that grey relation modification model of atmosphere refraction. As per the actual engineering instance, it can be seen that the model can modify triangle elevation measuring value, afterwards that vertical displacement value can meet precision requirement on safety monitoring of earth rockfill dam.

Key words: earth rockfill dam; Total station system TCA2003; 3-dimensional deformation; atmosphere refraction

1. Precision 3-dimensional automation monitoring apparatus and key issues

1.1　Precision and features of TCA2003 total station system

Tolerance of a single angle at one direction of TCA 2003 total station system is $\pm 0.5''$, when target is a prism, the measuring distance is 1mm + 1ppm $\times D$ (D refers to the horizontal distance)[1], when target is a reflecting piece, the measuring precision is 3mm + 2ppm $\times D$. The apparatus is the best total station system with high precision and performance currently in the whole world. With memory and multiple application programmes. The collimation axis of electronic theodolite, launch axis and reception axis of infra-red range finder is overlapped (three pieces of axis are in concentric).

1.2　Key issues

Showing by Figure 1 for safety monitoring of earth rockfill dam, it can be seen that coordinates of A and B are known, A is designed TCA2003, B is the rear view point, if observing Point C, it can obtain horizontal angle β, horizontal distance D, height angle α, so it can use formula shown by Eq. (1) ~ Eq. (3) to calculate horizontal coordinate and elevation of Point C.

$$x_C = x_A + D\cos\beta \qquad (1)$$
$$y_C = y_A + D\sin\beta \qquad (2)$$
$$H_C = H_A + D\tan\alpha + \frac{1-k}{R}D^2 + H_i - H_v \qquad (3)$$

Where, k is atmosphere refraction coefficient, P is radius of earth, H_i is height of total station system, H_v is height of prism.

TCA2003 total station system can search collimating target automatically when observing, measure horizontal angle, height angle, horizontal distance automatically, and calculate, output final result by in-built software in the system, and then finally work to keep an automatic monitoring on deformation.

By using high precision automatic total station system to establish 3-dimensional automatic monitoring system for earth rockfill dam, the key technology that needs to be solved is the correction issue of atmosphere refraction[2-3]. This kind of system uses single-directional observation when measuring height difference, so actual atmosphere refraction coefficient k cannot be obtained. Within the vertical side, since the complexity of onflow movement of atmosphere, k value shows obvious difference and change at different time at different directions, which causes measurement of height different to be with big tolerance. Therefore, for 3-dimensional monitoring on deformation of dam, the research on change feature of refraction coefficient and efficient model·correction is the base to make 3-dimensional automation monitoring system popular.

2. Analysis on refraction coefficient k

2.1 Meaning of refraction coefficient k

From definition of refraction coefficient k, it can be seen that:

$$k = \frac{R}{R'} \qquad (4)$$

Where, ratio of curvature radius of globe and curvature radius of beam, when passing through air, bend of beam is caused by refractive index gradient of atmosphere. In view of atmosphere refraction formula and onflow theory, and omitting influence amount caused by air pressure and humidity gradient, it can obtain:

$$k = -\frac{81.96R \times P}{T^2 \times 10^6}\left(-0.0342 - \frac{dT}{dh}\right)\cos\alpha \qquad (5)$$

Where: R, T, P, α refers to radius, air temperature, air pressure, view height angle of globe, respectivehy.

Although Formula (5) has certain value in view of theory research, actual atmosphere is quite complex, different height will have different air temperature, air pressure and temperature gradient, when measuring, it is impossible to obtain the feature val-

ues said at each place of beam, but it reflects conno tation of refraction coefficient k, expresses that the influence of T, P, dT/dh, α equivalent value, i. e., function of k value at each direction can be expressed as:

$$k = f\left(T, P, \frac{dT}{dh}\right) \qquad (6)$$

2.2 Calculation model of refraction coefficient k

Evaluating k value by integrating precision geometry level and one-way triangle elevation, it can obtain higher precision. k value obtained by this method represents the overall reflection of influence caused by atmosphere refraction to beam path, reflects authentically influence of atmosphere refraction when measuring. For instance, A, B in Figure 1, in triangle elevation, if Point A is designed the system to observe Point B, it can measure height angle α and horizontal distance D, height different between A and B point is h_A.

Additionally, with precision geometry level, it can measure height different h_w between A and B is:

$$h_A = D\tan\alpha + \frac{1-k}{2R}D^2 + i - v = h_w \qquad (7)$$

$$k = \left[D\tan\alpha + i - v - h_w\right]\frac{2R}{D^2} + 1 \qquad (8)$$

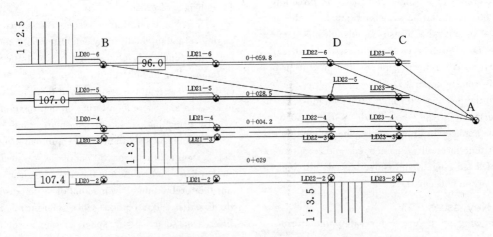

Figure 1　Layout of monitoring station and monitoring points of earth rockfill dam

For differential of Formula (8), it is converted into intermediate tolerance, i. e.,

$$m_k = \pm\frac{2R}{D^2}\sqrt{\tan^2\alpha\, m_D^2 + m_i^2 + m_v^2 + m_w^2 + D^2\sec^4\alpha\frac{m_\alpha^2}{\rho^2}} \qquad (9)$$

If $m_\alpha = \pm 0.5''$, $D = 300m$, $\alpha = 10°$,

$m_D = 1 + 1 \times 10^{-6}D(mm)$, $m_i = m_v = \pm 0.2mm$,

$m_w = \pm 2.0\sqrt{D}\,mm$, it can obtain $m_k = \pm 0.24$, i. e., tolerance for k value by integrating precision geometry level and triangle elevation is ± 0.24 roughly. It can cause a tolerance of ± 2.0mm to calculated height difference, therefore, measuring of triangle elevation within the scope must be available refraction coefficient being less than 0.24. After making refraction correction, which can guarantee vertical displacement of earth rockfill dam to be with a precision of 2mm.

3. Grey relation correction model of atmosphere refraction

3.1 Atmosphere refraction grey relation model GM (1, N)

Monitoring place for deformation of dam is within a very small partial range, atmosphere refraction of each point reflects well the relationship within the partial range and feature that fits for smooth random process. Therefore, variation law of refraction coefficient between measuring points designed out of dam belong to the uncertain information, which can use grey system to establish model. For sequence with few observation data, grey forecast is an efficient tool.

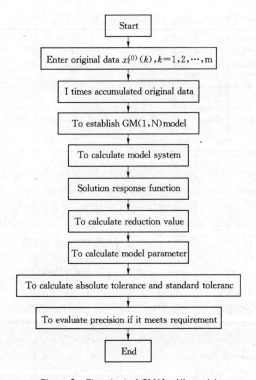

Figure 2　Flowchart of GM(1, N) model

For dam safety monitoring shown by Figure 1, TCA2003 Total station system is designed at base A,

it will be used to observe triangle elevation on measuring point, height different between each base point should be measured firstly by precision level. After multiple observations, each data can be calculated for atmosphere refraction coefficient k_{ij} of each point by using Formula (8) (i refers to number of measuring point, j refers to number of observation period). In view of grey relation analysis theory, the paper takes Point C as an example to establish a varying grey relation model for refraction coefficient between B and C and refraction coefficient of Point D, i.e., GM (1, 3)

$$\frac{\mathrm{d}k_6^{(1)}}{\mathrm{d}t}+ak_6^{(1)}=b_1 k_B^{(1)}+b_2 k_C^{(1)} \qquad (10)$$

Disbanding the above formula, and then using LS to evaluate, then it can obtain $\hat{a}$、$\hat{b}_1$、$\hat{b}_2$, put them into differential equation, it can obtain response function, i.e.,

$$
\hat{k}_6^{(1)}(i+1)=\left[k_6^{(1)}(1)-\frac{b_1}{a}k_B^{(1)}(i+1)\right.
$$
$$
\left.-\frac{b_2}{a}k_C^{(1)}(i+1)\right]\mathrm{e}^{-ai}
$$
$$
+\frac{b_1}{a}k_B^{(1)}(i+1)+\frac{b_2}{a}k_C^{(1)}(i+1) \quad (11)
$$

its reduction value is:

$$\hat{k}_6^{(0)}(i+1)=\hat{k}_6^{(1)}(i+1)-\hat{k}_6^{(1)}(i) \qquad (12)$$

$\hat{k}_6^{(0)}(i+1)$ above represents the forecast value under the condition of main behavior factor $i+1$, it can be calculated by behavior factor k_B and one-time accumulation formed value of k_B at $i+1$ time and a、b_1 and b_2.

3.2 Calculation flowchart of grey relation model GM(1, N)

In accordance with above model-establishing principal and corresponding method, it will use MatLab 6.5[5] to prepare corresponding analysis programme, as per related grey model procedure flowchart, see Figure 2.

The above calculation flow is compiled into related programme, and is processed by computer, which can modify refraction automatically to observation value at each direction.

4. Analysis on engineering example

Yanshan Reservoir is located at Ganjiang River, the upper reaches of Lihe, main branch of Shayinghe River of Huaihe drainage areas, covering a drainage area of 1169km². In accordance with requirement planned for Huaihe drainage area, the development task of Yanshan Reservoir is mainly for flood control, but with packing of the purpose to supply water, irrigate and generate power. With total reservoir volume of 925,000,000 m^3 for engineering, Yanshan Reservoir is Ⅱ class type. The inlet of main buildings of dam, spillway tunnel, flood discharge tunnel and water diversion tunnel is 2nd level, interior building 3rd level. In order to monitor safety of dam, it needs to observe its deformation. As per layout of monitoring point for surface deformation, see Figure 1.

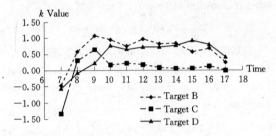

Figure 3　Variation process of k Value

4.1　Test result

On April 18th, 2008, by taking base A at the left bank of Yansha Reservoir to observe B, C and D, observation was lasted from 7am to 17pm and then calculated k value by Formula (8). The horizontal distance between measuring points is D_{AB}, D_{AC} and D_{AD}, and then used 2nd level to measure height different of h_{AB}, h_{AC} and h_{AD}. As per process line for varying k value at each point, see Figure 3.

4.2　Grey relation model of k value

Making analysis on measured data, it can be known that atmosphere refraction changes complicatedly before 9am, so it is not appropriate to observe within such time section. When studying atmosphere refraction correction model, it will select 8 efficient time sections between 10am and 17pm for analysis. By taking k_3 as the main behavior factor, then it will use k_1、k_2 to form a grey relation model GM(1, 3) of k_3.

From the calculation process aforesaid grey relation model GM(1, N), and the integration of precision geometry and triangle elevation, it can obtain the comparison between k and k_0 value, see Table 1 hereinafter.

Table 1　Comparison table between k value obtained from grey relation model and k_0

k ＼ Time	10am	11am	12am	13pm	14pm	15pm	16pm	17pm
k_0	0.78	0.66	0.74	0.74	0.78	0.92	0.80	0.42
Module calculates k		0.54	1.05	1.05	1.03	0.76	0.89	0.33
$\Delta = k - k_0$		−0.12	0.31	0.31	0.25	−0.16	0.09	−0.09
m_k					0.22			

From Table 1, it can be seen that the tolerance between k_3 calculated by following grey relation model with k_1, k_2 in experiment and precision k value is rather small, i.e., $m_k = \pm 0.22$, which means a higher precision. Using k value calculated by model to correct measured value of triangle elevation, then expected tolerance of height will be less than $\pm 2 - \pm 3mm$, which can meet requirement of safety monitoring of earth rockfill dam.

5. Conclusion

(1) Currently, observation methods for surface of earth rockfill dam have several defectives, such as slow measurement speed, heavy workload, hard to guarantee precision, hard to complete within a short time and monitoring information has no way to share automation system interactively. With the use of high precision automatic total station system (such as TCA2003 type total station system) and others, it creates beneficial conditions for research and development of 3-dimensional automation monitoring technology for earth rockfill dam.

(2) By aiming to the issues that affect observation precision, such as atmosphere refraction when observing by total station system, the paper analyzes refraction coefficient, and raises grey relation modification model for atmosphere refraction. As per the actual engineering instance, it can be seen that the model can modify triangle elevation measuring value, afterwards that vertical displacement value can meet precision requirement on safety monitoring of earth rockfill dam.

Acknowledgment

This paper was funded by Nanjiang Hydraulic Research Institute fund (S70901, Y709002, Y710006) and National Nature Science Youth Fund (50909066) .

References

[1] Yu Xingwan, Cheng Mingjian, Xu Zhongyan, et al. Application of TCA2003 for monitoring deformation of Gangkouwan Reservoir [J], Hydropower Automation & Dam's monitoring, 2003, 27 (5): 48-50.

[2] Xu Hui. Discussion on measurement and atmosphere refraction [J], Mapping Report, 1995(7).

[3] Zhang Chunyan. 3-dimensional horizontal difference of total station system and analysis of atmosphere refraction [J]. Railway survey, 2004 (2): 5-8.

[4] Shi Yimin. Control measurement of modern land [M]. Shanghai: Tongji University Press, 2002.

[5] Feisi Technology Product Development center. Analysis and design of Matlab 6. 5 auxiliary nerve network [M]. Beijing: Electronics Industry Press, 2003.

Integration of Local Sediment Transport Rating Curves to Derive Mass Budgets—The Urumqi River

Liu Youcun[1,4] Métivier François[2] , Ye Baisheng[3] , Narteau Clément[2] ,

Lajeunesse Eric[2] , Han Tianding[3]

[1] Key Laboratory of Water Environment and Resource , Tianjin Normal
University , Tianjin , 300387

[2] Laboratory of geological fluids Dynamics , Institute of Earth Physics of Paris and
Paris Diderot University , Paris , 75013 , France

[3] Cold and Arid Regions Environmental and Engineering Research Institute , Chinese
Academy of Sciences , Lanzhou , 730000 , China

[4] Institute of Geochemistry , Chinese Academy of Sciences , Guiyang , 550002 , China

Abstract: We surveyed flow dynamics and sediment transport in a mountain stream of North East Tianshan, the Urumqi River, during three flow seasons (2004 – 2006) . Bedload and suspended load was measured using hand held samplers. Flow velocity was measured using OTT current meters. This survey enables us to derive a simple local rating curve for bedload transport in the river that is both robust and site independent. Using this rating curve we calculate the solid sediment budget at the hero bridge station just before the river lives the mountain ranges and enters the semi-desert piedmont. This budget proves useful first to calculate sediment yield at the catchment scale and second to compare between the measured and the calculated of bed loads in a stream that is similar to other streams in the region.

Key words: sediment transport; bedload transport; grave-bed river; Urumqi River

1. Introduction

Compared with plain sand bed streams, mountain gravel bed streams have different appearances in stream gradients, bed material grain size distributions, cross-sectional channel shape, and stream morphology. And sediment transport in mountain streams differs significantly from those occurs in plain streams. In middle, high-latitude mountain, most rivers are fed by glacial melt water, the streams run along a special channel activated by gravity and carry the surface particle, then form the sediment transport (Ni J. R. and Ma A. N. , 1998; Meunier P. et al. , 2006; Liu Y. , 2008) . It is affected by many factors including slope erosion processes (i. e. , sediment yield), stream hydrodynamics, grain size distribution of transported and bed material, and channel and bed form morphology (Lenzi et al. , 1999) .

In mountainous regions with active uplift, rivers flow on beds covered with coarse materials which they

transport in the form of bedload that then accumulates in the Piedmont. Although bedload constitutes the bulk of preserved sediments in piedmonts of active ranges, it is only grossly included in mass budgets. Despite its unanimously recognized importance and the number of studies that are devoted to it, bedload remains poorly understood and stands as a hurdle to rigorous quantitative studies of alluvial dynamics. The mechanics of bedload seems obscured by sites effects. The fact is, that in the numerous field studies aimed at developing empirical models or testing theoretical laws, emphasis is placed on the problems linked to measurement difficulties, their uncertainties, and especially to their extreme variabilities [Brownlie, 1981; Pitlick, 1988; Garcia et al. , 2000; Bunte and Abt, 2005, 2008; Habersack et al. , 2008]. All of these drawbacks seem to prevent the validation of theoretical laws and render hazardous a move from short timescales to longer ones. Based on data from a three year survey of a gravel bed braided stream we demonstrate that: (1) it is possible to develop a simple power law for bulk transport of bedload; (2) when bulk transport is considered, the threshold of motion seems to have very little influence on sediment transport in such transport limited streams; (3) conventional measurement protocols are at least in part responsible for the inability to derive universal transport relationships from field surveys.

There are numerous transport laws for bedload in gravel bed streams (Garcia, 2008; Parker, 2008) and they can be grouped into two categories. The first concerns total transport of material on the bed, herein referred to as the bulk (Garcia, 2008). The second group of bedload transport relationship deals in some detail with transport of the different size fractions on the river bed [Parker, 2008]. The most famous and probably most often used transport relationship developed for the bulk is the equation of Meyer-Peter and Müller (Graf and Altinakar, 1996; Wohl, 2000; Garcia, 2008).

We hereafter present the dataset acquired on the Urumqi He during three years. We discuss the hydrology of the stream and the dynamics of sediment transport as a Bulk. Our database is then used to show how it is possible to describe Bedload as the sum of an average bedload transport law and a highly fluctuating component. It averaging out the fluctuating part is then possible to extrapolate the balance measured during the maximum flow season to the entire flow season and compare it to measured daily measurements of suspended fluxes at the outlet of the range. Integration of our results at longer timescales is then used to discuss characteristic timescales of formation of the mountain piedmont [Liu et al., 2008].

In this paper, we will treat the variability of bedload measurements. We will outline a simple tool for averaging data and show that a coherent transport relationship for the bulk can be found to be applicable to a gravel bed braided stream. The study site is presented first, followed by the data acquisition protocol and the technique used to filter out variability. Results are then analysed and discussed.

2. Field site and data acquisition

2.1 The Urumqi River

The Urumqi River is a high mountain stream. It is located in the north-eastern part of the Tianshan mountain range. The Tianshan range is an actively growing mountain region absorbing a small albeit non negligible portion of the India-Asia convergence (Avouac et al. , 1993; Molnar et al. , 1994; Métivier and Gaudemer, 1997; Li et al, 2001). Recent tectonic history of the Tianshan has been the subject of numerous studies in is relatively well constrained by analyses both of the deformation of piedmonts in the Tarim basin, south of the range, and in the Dzungar basin, north of the range.

Two gorges where the river flows on rock separate alluvial sections where measurements where performed. The river has a drainage area of 925 km^2, a length of 60 km from source (3900 m A.S.L.) to mountain range front (1700 m) where it is deeply entranced in a piedmont fan (figure1A). The fan itself is mostly made up of quaternary sediments from the Xiyu formation of very coarse and gravelly facies (Zhou et al, 2002; Zhao et al, 2006). Stratigraphy indicates that the river build the fan was pretty much the same as the present day river. Hydrology of the stream is controlled by both precipitations and glacial melting (Huo et al, 2007; Liu et al, 2007), Glacial melting still accounts for about 10% of the daily discharge downstream. In its upper course in the glacial valley of the Tiangger peak the river flows from May to October. It is perennial downstream of the range. Average monthly discharge at the outlet of the range indicates a very modest stream with peak flows in July (mean discharge of 25 cms). At its minimum in winter the river discharge is of about 1 cms.

2.2 Sites description

The river was surveyed in three alluvial portions. Figure1 shows the sites where measurements where performed. Site 1 (i.e. Glacial valley) is located in the upper reach of the river in a glacial valley. Measurements were performed at two different sub sites approximately 100 m apart. Site 1-1, where measurements were made during the three years of survey, is located downstream of a confluence scour. Site 1-2 is located under a small iron bridge that was constructed in 2006 on a straight reach of the river just upstream of site 1-1 (figure1E). We therefore have a double series of measurements in this area in 2006. Figure1B shows the comparison of the gagings performed. The black line is the 1 : 1 line. Results are in very good agreement. Site 2 is located at the outlet of the first gorge at the beginning of the Houxia basin a small intermontane gravelly basin where several tributaries join the main stem before the second gorge (figure1D). Site 3 is located just at the outlet of the second gorge at the entrance of the Urumqi alluvial fan (figure1C). As is the case for all the rivers of North Tianshan, the river is entranced in its fan. Maximum entrenchment reaches about 200 meters at the range front. At all sites the morphology of the stream is that of a more or less pronounced braided stream. For simplicity measurements where performed, when possible, on sections where all the braids joined into one main channel. Surface grain size distribution was studied using Wolman counting. We performed three types of measurements during the years 2004, 2005, and 2006: sedimentologic, hydrologic, and flux measurements. We studied the grain size distribution using both surface counts and volumetric sampling. Velocity profiles were made with a mechanical OTT propeller (Métivier et al., 2004; Meunier et al., 2006; Liu et al., 2008). For each velocity profile there was a coupled bedload taken using a hand held sampler. Between 900 and 1000 samples were measured for each site. The grain size distribution does not vary very much with D_{50} that are very similar and of 3 to 4 cm for the three survey sites.

Figure 1　Map of the Urumqi river drainage basin and location of Survey sites. A: Topographic map of Urumqi River upstream drainage and location of survey site, B: Along stream profile of the river and location of survey sites, C: Photo of site 3, 62 km downstream of glacier No1, D: Photo of site 2, 34 km downstream of glacier No1, E: Photo of site 1, 8 km downstream of glacier No1, the location of the two sub sites is marked

2.3　Measurement and sampling techniques

Gaging was performed using mechanical propeller velocimeters of OTT type or using the Chinese velocimeters in use in hydrology stations. Gaging was performed through the leveling of individual velocity profiles spaced approximately 1 m to 2 m apart across a give section of the stream. The profiles where then integrated according to

$$Q = \int_0^W \left(\int_0^h v(y,z) \mathrm{d}z \right) \mathrm{d}y \qquad (1)$$

the second integral on the right hand side defines the discharge per unit with of the stream,

$$q = \int_0^h v(z) \mathrm{d}z \qquad (2)$$

This parameter will be discussed and used extensively in what follows.

Integration was performed using a simple quadrature formula adapted for our purposes. Gaging was performed this way because we wanted to derive velocity profiles that could be integrated to calculate discharge per unit width and derive a correct average velocity (first moment of the velocity distribution along depth). We further wanted to do this in order not to make any assumption for the calculation of the average velocity of the discharge per unit width (be it uniform flow or a given velocity profile model that would allow for a single point measurement of the velocity). This was done at the price of coarser resolution in the direction perpendicular to the flow.

Bedload was leveled for each velocity profile with a custom made Sampler (hereafter named the Chinese sample with entrance of 30x15 cm (inner dimension) intake and an expansion ratio of 1.4). A polyester mesh sample bag is attached near the nozzle assembly. The mesh sample bag was used with 0.25 mm. Measurements were done at the same place as velocity measurement, and bedload was collected just after velocity measurements. Given these dimensions, our sampler should have the same properties as a Toutle river sampler (Diplas et al., 2008). These samplers were devised following discussions on the problems associated with using samplers with large pressure differences such as the Helley-Smith sampler (Hubbell, 1987; Thomas and Lewis, 1993; Diplas et al., 2008). Sampling efficiency of the Toutle river sampler ranges between 80%-116% (Diplas et al., 2008) so that the measurements obtained are on average likely to be good estimates of the true fluxes. Sampling duration was on average 120s per sample. During the years 2004 and 2005 only the samples with mass in excess of 100 gr were dried and sieved. In 2006 we decided to dry and sieve all the samples in order to get some idea of the fractions transported at low flows. All together more than 1000 samples were dried and sieved in the laboratory. Table 1 summarizes the data used in this study.

Table 1   Bedload data summary

Site	Year	Number of measurements
1	2004	183
1	2005	430
1 (1)	2006	371
1 (2)	2006	369
2	2004	25
2	2006	134
3	2006	106

Water samples were taken with a USDH48 depth integrating sampler. One sample per gaging was taken. Water was filtrated on 0.2 microns filter. Chemical analyses were performed at the Geochemistry Laboratory of IPGP (Institut de Physique du Globe de Paris).

3. Data acquisition

3.1 Hydraulics of flow and relevant scales

There has been a considerable debate on how to describe the flow regime in mountain gravel bed streams. This has been described as an important issue as a correct description of the flow is argued to be a prerequisite for the calculation of the shear stress at the bed (Detert et al, 2005). Recently, on average the flow could be considered as uniform had been shown (Liu et al, 2008). For calculating the shear velocity they also argued that the flow was probably not logarithmic or that the shear stress obtained using the shear velocity computed from velocity profiles was too uncertain to be useful.

The exact form of the flow is not the key point we address here. Rather we believe the most important problem concerns the definition of relevant quantities we can use to describe the flow. Lift and drag forces at the bed although ultimately the physically relevant variables are hard to measure. It is worth asking whether they should be defined at all if only with very high uncertainties. Use of an acoustic Doppler velocimeter on the field remains problematic both for price issues, for comparison purposes and for practical purposes in gravel bed streams where high transport destroys even resistant propellers.

Furthermore when long term averages are concerned, modeling of fluvial transport always assumes that the flow is uniform and can be described by some Chézy or Manning relationships. The average velocity is then parameterized accordingly used with conservation of mass and a transport relation (be it of Bagnold or Meyer-peter and Mueller type) for closure. Hence defining what are the relevant first order variables is therefore an important question as these quantities will be the ones used later to define sediment transport relationships.

As stated before the first variable that can be studied is the local 2D discharge or discharge per unit width as it is the integral along depth of a velocity profile [Eq. (2)]. By virtue of its definition, the average velocity at a given point is nothing but $<v> = q/h$. Figure 2 shows the striking correlation between q and h. An average trend is very well defined that remains valid wherever and whenever the measurements were performed. This lends a very strong credence on the idea that pertinent averages can be obtained from a sufficiently large set of data. In the case of the Urumqi River the unit discharge discloses a power law relationship with depth of the form

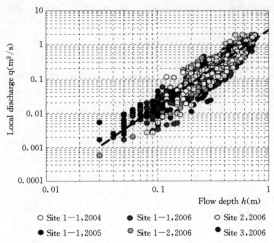

Figure 2   Discharge per unit width $q = \int_0^h v(z)\,dz$ function of local depth h. The trend is remarkable and remains the same for all three years of measurements at each one of the three sites

$$q = (0.4 \pm 0.01)h^{2.23 \pm 0.02} \leftrightarrow q \approx 0.4h^{2.2} \qquad (3)$$

Eq. (3) has two strong implications. The first one is that the average velocity also scales with the flow depth as

$$<v> \approx 0.88h^{1.2} \approx 0.9h^{1.2} \qquad (4)$$

This implies that basically using any one of the three variables h, $<v>$, q is equivalent.

To summarize the main conclusions that can be drawn from our flow measurements are:

(1) On average over successive measurements of velocity profiles spanning the main flow season, the flow of the Urumqi river averages around a very well defined trend that may correspond to uniform flow conditions.

(2) It is useless to try to derive a shear velocity using a given type of velocity profiles.

(3) The relevant quantities namely average velocity (first order moment of the velocity distribution), flow depth and discharge per unit width are equivalent as there exist as a very well defined scaling relationship between these three variables.

3.2   Bedload data

The data gathered is presented in Figure 3. The plot presented here is that of a unit bedload (g/m/s) versus a unit discharge ($q = UH$ in m²/s) obtained from velocity profiling of the river average depth of flow in m). Numerous values are plotted at 0.01 g/m/s. These values are in fact zero fluxes. The use of a unit discharge is arbitrary and it is not the purpose

of this paper to derive any new bedload relationship or to discuss previous ones. Nevertheless let us note that under normal flow conditions

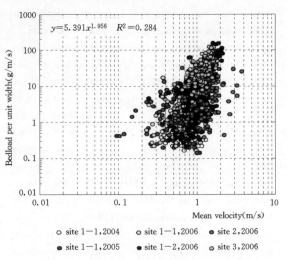

Figure 3   Individual catches of bedload (g/m/s) as a function of unit discharge (m²/s) derived from velocity profiles. Zeros catches are represented at the 10^{-2} g/m/s in this log-log plots so that they can be represented and analyzed.

$$qS = UHS \sim U^3 \sim \tau_b^{3/2} \qquad (5)$$

Where S is the bed slope and τ_b is the bottom shear stress. As most formulas described in the literature define the flux of bedload as a function of $\tau_b^{3/2}$, data should then locally be a function of unit discharge.

The three datasets of the Urumqi River correspond to each other very well. As the three years were similar in discharge and weather it is an infrequent example of reproducible field experiment. Also note that the data for 2006 was leveled at two different sites yet no different trends are noticeable.

A significant difference with classical procedure was that we decided to keep each individual sample at each station. The reasons and significance had been discussed by Liu et al. (2008).

4.   Analysis and results

4.1   Velocity and bedload transport

Several researches have confirmed the shear velocity, far from being usable, just increases uncertainties to a potential transport law for lack of meaning. The only valuable relation is obtained by plotting the transport rate per width as a function of the average velocity v (Meunier et al., 2006; Liu et al,

2008). We computed v by a slide average method. From the correlation on Figure 3, a power law relationship ($q_b = a\,\bar{v}^b$) can be fitted and the fitted results are shown. Table 2 shows the spatio-temporal variety of a and b. We can find that the correlation coefficient value varies from 0.3 to 0.64, depending on time and place. At glacier valley (including site 1-1 and 1-2), the power law relationship is significant. The exponent value varies from 2.3 to 3.3 and the average is approaching 2.8, which is close to the result of Meunier's (exponent as ~3) in "the Torrent de St Pierre" of French Alps (Meunier et al, 2006). At site 3 (Nantaizi), a power law relationship was shown with an exponent of 1.2. However, there is no significant correlation between them at site 2 (Houxia).

Table 2   **The value of a and b in function $q_b = a\,\bar{v}^b$**

Site/Year	1-1/04	1-1/05	1-1/06	1-2/06	2/06	3/06
a	4.2±0.6	12.4±1.2	13.5±1.3	7.3±1.1	7.8±1.1	32.1±2.1
b	2.9±0.5	2.3±0.2	2.5±0.2	3.3±0.3	0.72±0.2	1.23±0.2
R	0.47	0.64	0.63	0.62	0.3	0.55

The obtained relation at glacier valley is close to the Bagnold formulation for the bed load transport ((Raudkivi, 1990; Meunier et al., 2006; Liu Y., 2008).

$$\frac{Q_S}{W} \propto (\tau_* - \tau_{c*})^{3/2} \propto u_0^3 \qquad (6)$$

where τ_* and τ_{c*} are, respectively the shear stress and the critical shear stress required for the induction of movement.

Then the shear velocity was replaced by the average velocity:

$$\frac{Q_S}{W} \propto \bar{v}^3 \qquad (7)$$

Starting from the expression of the discharge per unit width:

$$\frac{Q}{W} = \bar{v}H \qquad (8)$$

where H is the river depth.

This result is in good agreement with field measured results in "the Torrent de St Pierre" of French Alps and bed load descriptions provided by experimental results on micro-scaled braided rivers (Meunier et al, 2000; Métivier et al, 2003; Meunier et al, 2006). These formulae derive from Meunier' literature and similar conclusion was obtained. In fact, the disadvantage of using the average velocity is that the strong turbulence effects are filtered out. However, the information contained in dispersion may be useful too (Meunier et al, 2006). As turbulence effects are related to bed grain size and bedload grain size, all of our bed load samples have been returned to the laboratory and analyzed. A study is currently in progress to quantify the bed load transport per grain size and relate it to velocity variations. A closer relation with laboratory studies cited above could be established by carrying a differential fluxes measurement. They suggest that the bedload flux measured at a given cross section is a function of its value at the inlet of the braiding plain and the distance between the cross section and the inlet. Our field work on the Urumqi River at different locations along the alluvial plain is to enforce that idea.

For site 2, the absence of any correlation cannot be explained by the duration of the sampling time. Actually, we have shown that this sampling duration (2 minutes) is sufficient to draw uniform flow relationships from the data and that the measured average velocity is relevant. This sampling time is very long with regard to high frequency fluctuations of the velocity field, also corresponds to the sampling time of bed load, and hence, to the velocity of the flow that shears the bed during sampling. Indeed, a part of bed load may enter the sampler under the influence of the shock caused by the sampler on the bed when it is laid down and kept to stability. Meanwhile, a part of bedload may exit the sampler in the process of taking it out, because the sampler is ponderous (20 kg or so) cannot keep its balance in rapid flow even when the measurement was cooperatively performed by two or three persons. The relative importance of this error may become higher when river flux is turbulent, especially in Houxia.

Compared with continuous measuring at glacier valley and Houxia, the measurement is incomplete at site 3. It was preformed only at the final stage of flood period (in late Aug). Different with Houxia, the relatively smaller flow velocity eases the bedload sampling at site 3. The results are more accurate. As for the exponent difference, it cannot be explained well. The sample number is less than 100, not sufficient enough and the point distributes discretely. Meanwhile, the sampling process of bedload is under the effect of shocking, and the effect is significant especially for close threshold motion particles. Certainly, we cannot exclude the validity of measured data. The high ratio between bedload transport rate and mean velocity was caused by river turbulence, because there are some small or medium boulders at the right side of the cross section. A study on this point of view will be done.

4. 2 Discharge and bedload transport

Despite the problem of evaluating equations developed initially for different ranges of flow and sediment conditions, extensive tests have been carried out (e. g. Task Committee, 1971; White et al., 1973; Batalla and Martin-vide, 2001; Meunier et al., 2006; Liu et al., 2008). A widely accepted theory has been obtained. Simple relationship of the forms

$$Q_b = kQ^n \tag{9}$$

Where Q_b is bedload transport rate, Q is discharge; k & n are respectively the constant and exponent. Figure 4 shows the relationship between bedload transport rate and discharge in Urumqi River upstream. Although the scatter in the data points is inherent to many uncertainties linked with bedload measurement, a power law relationship can be fitted with a function as

$$Q_b = 7.183Q^{1.726} \approx 7Q^2 \tag{10}$$

where the unit of Q_b and Q is respectively g/s and m^3/s. A problem appears in the figure that bedload transport rate increases while the discharge remains the same when the value of discharge is less than 1. The only possible explanation is that the bed load transport measurement at low velocity ($<1 m/s$) is wrong as a result of intrinsic errors of Helley-Smith sampler. In fact, a part of bed load entering the sampler is possible under the effect of the shock caused

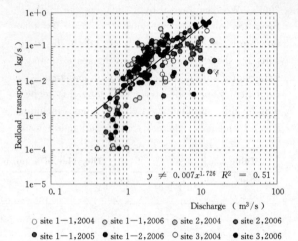

Figure 4　Bedload transport rate (kg/s) correlated with discharge (m³/s). A power law can be fitted with an exponent near to2.

by the sampler on the bed. The relative importance of this error may become higher when bed load rate is low.

After the simple power law relationship between bedload transport rate and discharge was obtained, we are seeking for a local bedload transport relationship (Liu et al., 2008). It is an average bedload transport rate plus random fluctuations around that average

$$q_b(q) = \overline{q_b}(\overline{q}) + q'_b, (q, D_{50}, \cdots) \tag{11}$$

where q_b is the local bedload transport rate ($g/m/s$) (bedload transport per unit width), and $q = \overline{u}h = \int_0^h v(z)dz$ is the local discharge per unit width (m^2/s) calculated through integration of each individual velocity profile. Figure 5 shows the bedload transport rate per unit width fitted with discharge per unit width.

As the sample is of sufficiently high frequency, it is possible to average over the random fluctuations by defining the bedload integral function as

$$BI = \int_0^q q_b(\eta)d\eta = \int_0^q \overline{q_b}(\eta)d\eta + \int_0^q q'_b(\eta)d\eta \tag{12}$$

If the random distribution of the second item on the right of Eq. (12) vanishes

$$BI = \int_0^q \overline{q_b}(\eta)d\eta \rightarrow \overline{q_b} = \frac{dBI}{dq} \tag{13}$$

Figure 6 confirms that all the datasets gathered during three years on an approximately 60 km long

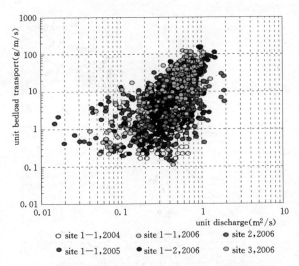

unit discharge(m²/s)

○ site 1−1,2004　　⊙ site 1−1,2006　　● site 2,2006
● site 1−1,2005　　● site 1−2,2006　　◐ site 3,2006

Figure 5　Bedload transport per unit width
fitted with unit discharge.

segment of the river are similar in upstream of Urumqi River. A single composite sample is made from the entire dataset.

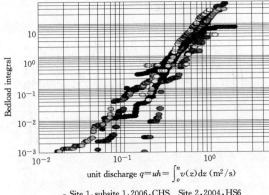

unit discharge $q = uh = \int_o^n v(z)dz$ (m²/s)

■ Site 1,subsite 1,2006,CHS　　Site 2,2004,HS6
■ Site 1,subslte 2,2006,　Site 2,2006,CHS
◦ Site 1,subsite 1,2005,CHS　　Site 3,2004,HS6
◦ Site 1,subsite 1,2004,HS6　　Site 3,2006,CHS

Figure 6　Plot of bedload integral $\int_0^q q_b(\eta)d\eta$ with unit

discharge $\int_0^h v(z)dz$ (m²/s)

Figure 7 shows the bedload integral BI fitted with unit discharge. It displays a very clear trend which can be fitted by a very simple power law of the form

$$BI = 10^{1.23}q^{3.08} \sim 20q^3 \qquad (14)$$

The constant is an independent unit and here only valid for a bedload transport rate given in g/m/s.

The average bedload transport rate for the Urumqi River can then be obtained by derivation of Eq.

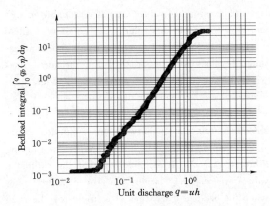

Unit discharge $q = uh$

Figure 7　The plot of bedload integral BI
with unit discharge

(14) according to Eq. (13), hence

$$q_b \simeq 60q^2 \qquad (15)$$

Here q_b is in (g/m/s) and q in (m²/s) .

It is a very simple transport equation, valid on average. It probably remains valid for all the reaches of the river that are transport limited (alluvial) as in the case of all our sampling sites. All the sites are mildly to strongly braided and not armored so that basically all fractions are more or less available here and there. It is also constituted with a basically constant size distribution of grains along the river.

Assuming a very realistic, rectangular channel this procedure can be used to derive bedload rates from hydrographs provided the width of the channel is known

$$q_b \sim 60\left(\frac{Q}{W}\right)^2 \rightarrow Q_b \sim Wq_b \sim 60Q^2W^{-1} \qquad (16)$$

This model can then be applied downstream using daily discharge measurements made at the Hero Bridge Station by the Chinese Hydrology Service and compared to suspended load measurements. As the hydraulic cross section is close to U-shaped, the width that we choose for calculating bedload transport rate varies from 17m to 18.5m according to the discharge variation.

Figure 8 shows the daily discharge variation, comparison between calculation bedload transport rate 〔By Eq. (16)〕and measured bedload transport rate in site 1-1 during sampling periods from 2004 to 2006. No values of measured bedload transport rate are plotted as 0.01 g/s. From the figure, we can see generally the estimated values keep pace with daily variations of measured ones, yet with some limitations. The estimated values are not reasonable enough and tend to be obviously large at the time of small discharge. Taking site 1 as an example, almost all estimated values are higher than the measured ones when the discharge is less than 1 m³/s while estimated values are smaller than the measured ones when the discharge is more than 3m³/s. Figure 9 shows the relationship between the estimated values and measured values of bedload transport rate during sampling periods. It also implies that the estimated values are relatively larger than the measured ones when the bedload discharges at a low rate; the ratio of the estimated values to the measured ones approaches 1 ∶ 1 in all stations except site 2. Big errors are encountered here at site 2, which might be caused by the low accuracy of bedload data. In summary, although with deviations to a certain extent, overall, this formula can be widely applied to the calculations on bedload transport rate in Mountain River basin and small watersheds (braided and no armored) in particular in mountainous regions.

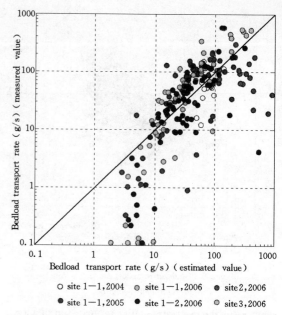

○ site 1—1, 2004   ◉ site 1—1, 2006   ● site 2, 2006
● site 1—1, 2005   ● site 1—2, 2006   ◌ site 3, 2006

Figure 8   Comparison between the estimated value of bedload transport rate and the measured value of bedload transport rate at different sites during sampling periods in upstream of Urumqi River.

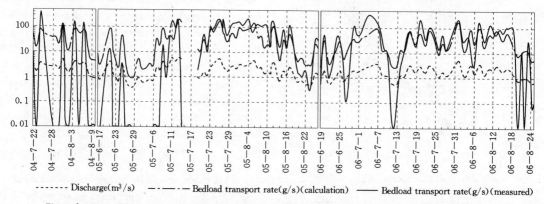

------ Discharge(m³/s)   ─ ─ ─ Bedload transport rate(g/s)(calculation)   ── Bedload transport rate(g/s)(measured)

Figure 9   Daily discharge, comparison of calculation bedload transport rate [By Eq. (16)] and measured bedload transport rate in site 1-1 during sampling periods from 2004 to 2006.
No values of measured bedload transport rate are plotted as 0. 01 g/s.

Figure 10 shows daily discharge, bedload transport rate and suspended load transport rate in Hero Bridge Station from 2002 to 2006, in which the discharge and suspended load transport rate is measured data, and the bedload transport rate data is estimated. The result shows that the bedload transport rate may be slightly higher in dry seasons and lower in flood periods.

5. Summary and conclusions

To study the sediment transport in gravel-bed

river, a series of field sampling and measurements were performed in upstream of Urumqi River. A great deal of field data was obtained and a consistent relation is proposed between the three load types and water discharge. The summary is as follows:

1) It is possible to average over the very large fluctuations recorded by hand-held bedload samplers and derive well constrained average bedload transport relationships.

2) For bedload transport, the relationships between velocity and bedload transport rate, discharge

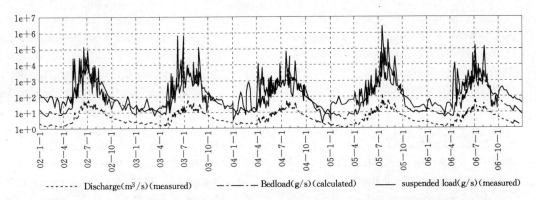

------ Discharge(m³/s)(measured)   — · — · — Bedload(g/s)(calculated)   —— suspended load(g/s)(measured)

Figure 10   Daily discharge, bedload transport rate and suspended load transport rate in upstream of
Urumqi River. Discharge and suspended load transport rate are measured data by Hero bridge station;
bedload transport rate is calculation data by Eq. (16)

and bedload transport rate are analyzed. There exists a power law relationship between the ratio of bedload transport to the width of channel and average velocity, as predicted by the most commonly used sediment transport relationships. This result is in good agreement with field measured results in "the Torrent de St Pierre" (Meunier and Métivier, 2006).

3) The absence of any correlation at site 2 and a non-significant correlation at site 3 were shown. This may be caused by the shock of sampler or intrinsic errors of Helley-Smith sampler, or measuring difficulties with large discharge.

4) By fitter between bedload transport rate and discharge, a simple function $Q_b \approx 7Q^2$ is proposed. Meanwhile, a function of bedload transport rate (Q_b) with discharge (Q) and width (W) is derived, that is, $Q_b = 60Q^2/W$. The latter equation has been tested by the measured data. It shows that the estimated is consistent with measured ones, yet with some limitations. Above equations are applied for calculating the sediment transport of the Urumqi River upstream.

Acknowledgments

The research program was funded by the French Programme de Recherches Avancées (PRA-T05 grant to F. Métivier), by the French ministry of foreign affairs (PhD thesis grant to Y. Liu) and Talent Introduction Item of Tianjin Normal University (5RL085). We are indebted to P. Meunier who helped us in field work.

References

[1] Avouac J., Tapponnier P., Bai M., You H. and Wang G.. Active thrusting and folding along the northern Tien Shan and late Cenozoic rotation of the Tarim relative to Dzungaria and Kazakhstan. *J. Geophys. Res.*, 1993, 98 (4): 6655-6804.

[2] Batalla RJ and Martin-Vide JP. Thresholds of particle entrainment in a poorly sorted sandy gravel-bed river. *Catena*, 2001, 44 (3): 223-243.

[3] Brownlie W.. Prediction of flow depth and sediment discharge in open channels. *Technical report*, W. M. Keck laboratory of hydraulics and water resources, California Institute of Technology, 1981.

[4] Bunte K. and Abt S.. Effect of sampling time on measured gravel bed load transport rates in a coarse-bedded stream. *Water. Resour. Res.*, 2005, 41 (W11405).

[5] Bunte K., Abt S., Potyondy J. and Swingle K.. A comparison of coarse bedload transport measured with bedload traps and Helley-smith samplers. *Geodinamica Acta*, 2008, 21 (1-2): 53-66.

[6] Detert M., Klar M., Wenka T. and Jirka G. H.. Pressure-and Velocity-Measurements Above and Within a Porous Gravel Bed at the Threshold of Stability. *Proceedings of 6th Gravel Bed Rivers Conference*, 5-9/9/2005, St. Jakob/Austria.

[7] Diplas P., Kuhnle R., Gray J., Glysson D. and Edwards T.. Sediment transport meas-

urements. In *M. Garcia*, editor, *Sedimentation engineering: processes, management, modeling, and practice*, ASCE Manuals and Reports on Engineering Practice, 2008, 110: 307-353. ASCE.

[8] Ferguson R.. Flow resistance equations for graveland boulder-bed streams. Water Resour. Res. 2007, 43 (W05427).

[9] Garcia C., Laronne J. and Sala M.. Continuous monitoring of bedload flux in a mountain gravel-bed river. *Geomorphology*, 2000, 34: 23-31.

[10] Garcia M.. Sediment transport and morphodynamics. In *M. Garcia*, editor, *Sedimentation engineering: processes, management, modeling, and practice*, ASCE Manuals and Reports on Engineering Practice, 2008, 110: 21-163.

[11] Graf W., Altinakar M.. Hydraulique Fluviale, Ecoulement non Permanent et phenomenes de transport. Eyrolles, Paris, 1996.

[12] Habersack H., Seitz H. and Laronne J.. Spatio-temporal variability of bedload transport rates. *Gedinamica acta*, 2008, 21 (1-2): 67-79.

[13] Hubbell D.. Bed load sampling and analysis. In *C. Thorne, J. Bathurst, and R. Hey*, editors, *Sediment transport in gravel bed rivers*, Jonh Wiley and Sons Ltd., Chichester, 1987: 89-106.

[14] Huo L., Gong J. X. and Wang Z. B.. Runoff variation of Urumqi in Dry season. *Xinjiang water resource*, 2007 (4), 25-28.

[15] Li X. K., Liu G. N. and Cui Z. J.. Glacial valley cross-profile morphology, TianShan Mountains. *Geomorphology*, 2001, 38: 153-166.

[16] Liu Y., Métivier F., Narteau C., Lajeunesse E., Meunier P., Gaillardert J., and Ye B.. Mass transport and hydraulics of flow in a high mountain gravel bed stream, the Urumqi River (Chinese Tianshan). Abstract, 2007 *fall meeting (AGU)*, Session: H51I-0891, 2007.

[17] Liu, Y., Metivier, F., Lajeunesse, E., Clement N. and Meunier, P.. Measuring bed load in gravel-bed mountain rivers: averaging methods and sampling strategies. *Geodinamica Acta*, 2008, 21 (1-2): 81-92.

[18] Meunier, P., Métivier F., Lajeunesse E. and Mériaux A., 2006. Flow pattern and sediment transport in a braided river: The "torrent de St Pierre" (French Alps), J. *Hydrol.*, 2006, 330 (3-4): 496-505.

[19] Ni J. R, Ma A. N.. Stream Dynamic Geomorphology. The Beijing University Press, 1998.

[20] Métivier F. and Gaudemer Y.. Mass transfer between eastern Tien Shan and adjacent basins. Constraints on regional tectonics and topography. *Geophysical J. International*, 1997, 128: 1-17.

[21] Métivier F., Meunier P., Moreira M., Crave A., Chaduteau C., Ye B. and Liu G.. Transport dynamics and morphology of a high mountain stream during the peak flow season: the Urumqi River (Chinese Tianshan. In *M. Greco, A. Carravetta,* and *R. Della Morte* editors, *River Flow*, 2004: 769-776.

[22] Molnar P., Thorson Brown E., Burchfiel B., Deng Q., Feng X., Li J., Raisbeck M., Shi J., Wu Z., You F. and You, H.. Quaternary climate change and the formation of river terraces across growing anticlines on the north flank of the Tien Shan area. *Jour. Geology*, 1994, 102: 583-602.

[23] Paker G. Transport of gravel and sand mixtures. In M. Garcia, editor, Sedimentation engineering: processes, management, modeling and practice, ASCE Manuals and Reports on Engineering Practice, 2008, 110: 165-251.

[24] Pitlick J.. Variability of bed load measurement. *Water Resources Research*, 1988, 24 (1): 173-177.

[25] Raudkivi A. J.. *Loose boundary hydraulics*, Pergamon Press, Inc., New York, 1990.

[26] Thomas R., Lewis J.. A new model for bed load sampler calibration to replace the probability-matching method. *Water Resources Research*, 1993, 29 (3): 583-597.

[27] White W. R. , Milli M. , Crabe A. D. . Sediment transport: an appraisal of available methods. Report Ⅱ 119, *Hydraulics Research*, Wallingford, 1973.

[28] Wohl E. . Mountain rivers. In *Water resources monograph*. American Geophysical Union. Washington, 2000.

[29] Zhao J. D. , Zhou S. Z. , He Y. Q. . ESR dating of glacial tills and glaciations at the head area of the Urumqi River, Tianshan Mt. China. *Quaternary International*, 2006, 144: 61-67.

[30] Zhou S. Z. , Jiao K. Q. , Zhao J. D. , Zhang S. Q. , Cui J. X. , Xu L. B. . Geomorphology of the Urumqi river valley and the uplift of the Tianshan mountain in Quaternary. Science in china (D), 2002, 45 (11): 961-969.

Prototype Observations on Vibration Performance of Drain Grating in Paunglaung Myanmar

Yan Genhua   Chen Fazhan   Zhao Jianping

State Key Laboratory of Hydrology-Water Resources and Hydraulic Engineering,
Nanjing Hydraulic Research Institute, Nanjing, 210029

Abstract: This essay presents the prototype observations results on ground working gate and dynamic characteristics of drain grating in Myanmar Paunglaung Hydropower Station. Through system testing and recognition on the vibration model of main beam of gate and drain grating in downstream and lateral directions, we acquire the characteristics of vibration model in two main directions. By means of analysis we also acquire the influence degree the aqueous medium acting on the vibratory frequency of gate structure. It points out if the aqueous medium is taken into consideration the model vibratory frequency of main beam in lateral direction will be relatively low. Besides, we conduct parametric recognition to the vibration mode of drain grating gate leaf, acquiring the vibration mode parameters of drain grating gate leaf, which provides estimation basis for sluicing vibration. In conclusion interrelated suggestions and requirements are proposed for operation of ground gate and drain grating according to prototype observation results.

Key words: hydropower station; plane gate; drain grating; dynamic characteristics; prototype observation

1. Introduction

Paunglaung hydropower station is located within Myanmar borders, and there're four machine sets with 70MW, which is the largest hydropower project in Myanmar at the present time. The bottom outlets are rebuilt from diversion tunnel. Two accident maintenance gates and two plane working gates are added additionally at the end of scour outlet as flow-controlling structure. The size of orifice of accident gate is $1.4 \times 3.75m^2$, that of orifice of working gate $1.4 \times 3.45m^2$, the water-retaining head of gate 120m, the hydro-dynamic operating water head 80m, and the plane working gate has to be operated with being partially opened.

Two intake towers parallel to each other are set at inlet opening of hydropower station, each of which is commonly used by two machine sets. Two maintenance gate slots for drain grating are set at the intake. There's a bypassing inlet opening at 10m high of upside of intake tower's right orifice. A accident maintenance gate is at the bypassing inlet opening, which is commonly used with the maintenance gate at inlet opening in hydropower station. There is a working sector gate at the outlet. The effects of bypassing prot can be concluded in three aspects: (1) while the power station is in construction, the bypassing prot can adjust and control the water level of reservoir without electricity-generating from machine sets; (2) the water level of reservoir has to be lowered while there's emergencies; (3) the bypassing drainage can be controlled by sector gate with machine sets shutting down.

The supporting beam of drain grating is inbuilt in concrete abutment pier at orifice. The designed hydraulic head between beams and stakes is 10m.

The dam in construction is shown in Figure 1, and the drain grating structure at inlet opening after completion is shown in Figure 2.

Figure 1   The dam and dam side in construction

Figure 2　The drain grating structure
at inlet opening after construction

In light of the special design of inlet opening and complicated operation conditions for Paunglaung hydropower station, there's working conditions combining different opening degrees of electricity-generating and bypassing prot sector gates, where the safety of drain grating structure is very prominent. Besides, the ground working gate needs to be operated with partially opening, so the flow-excited vibration and cavitation corrosion of ground gate need to be paid attention to. This is an outbound project, and it is of great project importance and profound international significance to make sure of the safety operation of all draining buildings. In order to expound and prove the design rationality of ground working gate and drain grating at the inlet opening as well as to formulate reasonable operation specifications, Yunnan Machinery Import and Export Corporation is consigned to conduct observations on flow-excited vibration of the main beams as well as stakes of drain grating at inlet opening and ground working gate to get the vibratory magnitude and spectral density under different working conditions. Through analysis and treatment on experiment results we do safety evaluations and formulate reasonable operation specifications.

2. The dynamic characteristics of gate structure

The characteristics of structure vibration mode reflect the fixed vibration behavior of structure. The measurement and research on dynamic characteristics of ground working gate and drain grating structure are conducted by using modal analysis technology to reveal the internal cause of structural vibration. As to

ground working gate, we respectively make research and observation to the vibratory characteristics of structure with dry mode and fluid-solid coupling in order to more comprehensively obtain the information and data about gate vibratory mode.

2.1 Mode analysis theory about structure dynamic characteristics

The movement of gate structure can be expressed by the following discretization matrix differential equation:

$$[M]\{\ddot{X}\}+[C]\{\ddot{X}\}+[K]\{X\}=\{F(t)\} \quad (1)$$

In the equation, $[M]$, $[C]$ and $[K]$ respectively are quality, damping and rigidity matrix at the order of $N\times N$; $\{\ddot{X}\}$, $\{\ddot{X}\}$ and $\{X\}$ respectively are acceleration, velocity and displacement column matrix of discrete nodal motion of structure; $\{F(t)\}$ is vector of applied force.

In order to clearly and visually reflect the dynamic characteristics of structure, we transform time-domain movement to frequency-domain movement to do research and observation. Make Laplace transformation to Eq. (1) and suppose the initial condition as zero, and get:

$$[B(S)]\{X\{S\}\}=\{F(S)\} \quad (2)$$

Thereamong, $[B(S)]=[M]S^2+[C]S+[K] \quad (3)$

Define $[B(S)]$ as impedance matrix of gate structure, and then transfer function matrix of system is:

$$[H(S)]=[B(S)]^{-1} \quad (4)$$

Apparently, in Eq. (4) includes quality, rigidity, damping and natural frequency of structure and such information about the whole mode. Therefore, once the transfer function matrix of system is obtained, the dynamic characteristics of working gate can be completely and correctly reflected.

All the constituents of the transfer function matrix can be expressed as:

$$[H(S)]=[B(S)]^{-1}=\frac{\text{adj}[B(S)]}{\det[B(S)]} \quad (5)$$

In the equation, adj $[B(S)]$ is an adjoint matrix of $[B(S)]$, which is a multinomial of 2N-2 powers of S; det $[B(S)]$ is a determinant, which is a multinomial of 2N powers of S. Expand Eq. (5) according to culminating points, and get transfer function matrix in residue form:

$$[H(S)]=\sum_{i=1}^{N}\left[\frac{(A_1)}{S-S_1}+\frac{(A_1)^*}{S-S_i^*}\right] \quad (6)$$

In the equation, (A_1) is residue matrix of relevant culminating point S^1, and (A_1) as well as $(A_1)^*$ conjugate each other.

Vector of vibration mode $[\Phi^i]$ is defined as:

$$[B(S^1)][\Phi^i]=\{0\} \tag{7}$$

At the left and right of Eq. (6) $[B(S)]$ and $(S-S^i)$ are multiplied, and make $S=S^i$ and then get:

$$[B(S^1)][A(S^i)]=\{0\} \tag{8}$$

$$[B(S^1)][A(S^i)]^T=\{0\} \tag{9}$$

Then there must be vector of vibration mode at the order of I in every line and every row of residue matrix, so:

$$[A(S^i)]=a^i[\varphi^i][\varphi^i]^T \tag{10}$$

Make $a^i=1$

Then $\quad[A(S^i)]=[\Phi^i][\Phi^i]^T \tag{11}$

The relation between residue in Line or Row q and vector of vibration mode at the order of I is:

$$A^q(S^i)=[\Phi^i][\Phi^i]^T qi$$

And then:

$$[\Phi^i]=\frac{A_q(S_i)}{qi}, \ [\Phi^i]^T=\frac{A_q(S_i)^T}{qi} \tag{12}$$

Substitute Eq. (12) into Eq. (10) and get:

$$[A(S^i)]=A^q(S^i)A^q(S^i)^T : \Phi^q\Phi^q$$

$$=A^q(S^i)A^q(S^i)^T : A^q(S^i) \tag{13}$$

Obviously, residue matrix $[A(S^i)]$ can be constructed by $A^q(S^i)$ in Line q. We can get transfer function matrix of the whole system from Eq. (6). In the same way, it can also constructed by the constituent of residue matrix in some line. Therefore, through measuring response on order-point, vibration on all the nodal points of gate or excitement on order-point, response on all the nodal points of gate we can obtain transfer function matrix of the whole structure. The latter one is adopted in this experiment.

2.2 The equipment for modal experiment of ground working gate and drain grating

In order to simplify testing course and shorten test cycle, we do pulse testing by hammer. The CL-YD-302 piezoelectric type of force sensor with steel caps is set at hammerhead to obtain modal information at a wide frequency range. The vibratory signal can be obtained by YD-1 type of accelerometer unit.

The YE5858A type of multifunction of adaptability amplifier is used as secondary instrument, and the outcoming signals are collected by dedicated data acquisition unit then treated by computer. The test analysis system is shown on Figure 3. The experiment with boundary conditions is conducted according to the limitation under the practical operation of gate and drain grating.

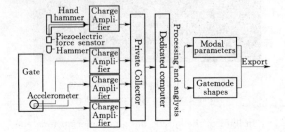

Figure 3　System for measurement and analysis of structure mode

2.3　Mode analysis on dynamic characteristics of ground working gate

2.3.1　Layout for nodal points of ground gate

In this modal experiment of ground working gate the gate structure disperses to 50 nodal points, each of which will be measured in downstream, lateral and vertical direction. The movement of gate structure after discretization is set in cylindrical coordinate to be observed in radial (ρ), tangential (θ) and side (Z) direction. The layout for nodal points of gate is shown on Figure 4. We collect and treat force signals and response signals as well as analyzing the modes in the above three directions at the meantime.

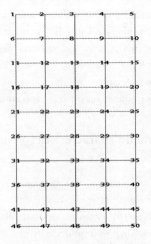

Figure 4　Geometric model of gate structure gate for modal test

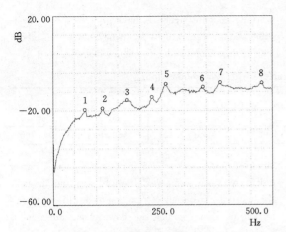

Figure 5　The transfer function of gate
for vibration mode

2.3.2　*Analysis on dry modal gate structure*

The transfer function of vibration modes of gate

structure without water is drawn on Figure 5. The structure mode parameters after modal parameters recognition are listed on Table 1. The experiment results show that the fundamental frequency of gate without water is 75.35Hz, and the frequency at the second order is 115.03Hz. The vibration mode chart of modal vibration of gate structure at the former orders is drawn on Figure 6.

Table 1　The parameters of gate structure modes

Serial Number of Modes	1	2	3	4
Frequency（Hz）	75.35	115.03	172.80	229.53
Damping Ratio（%）	5.38	4.608	5.38	2.897
Serial Number of Modes	5	6	7	8
Frequency（Hz）	254.47	343.84	380.92	473.53
Damping Ratio（%）	2.74	4.94	5.38	2.374

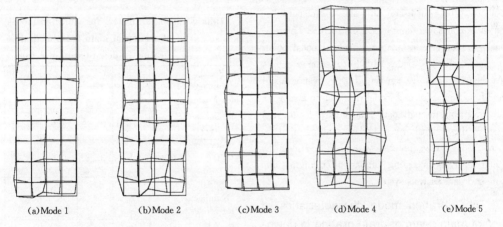

(a)Mode 1　　　(b)Mode 2　　　(c)Mode 3　　　(d)Mode 4　　　(e)Mode 5

Figure 6　The vibration modes of model vibration of gate structure

2.3.3　*Analysis on modes of gate structure with fluid-solid coupling*

（1）Boundary conditions of test. The following aspects need to be taken into consideration for boundary conditions in this modal test: 1）limitation from water-sealing on the top and at the two sides of gate; 2）limitation from cylinder rod of headstock gear; 3）there's water in upper stream; 4）the gate is opened with a height from 10m to 20m. Directions of excitation: downstream direction; direction of response: downstream direction.

（2）The experimental results. The transfer functions of vibration mode of gate structure with flow-solid coupling are drawn on Figure 7. The structure mode parameters after mode parameter recognition are listed on Table 2. The experiment results show that

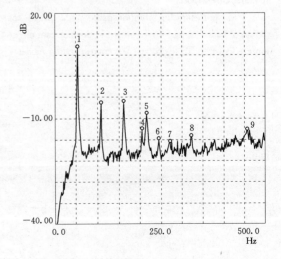

Figure 7　The transfer functions of vibration mode

the fundamental frequency of gate structure with water is 53. 7Hz, and the frequency at the second order is 106. 8Hz. The characteristics of vibration fluid working on gate structure have great effect on first-order mode, the rate of frequency descend is about 30%; the second-order frequency descends to 8%.

Table 2 **The mode parameters of gate structure with flow-solid coupling**

Order	1	2	3	4	5	6	7	8	9
Frequency (Hz)	53. 7	106. 8	164. 9	207. 3	214. 4	247. 4	272. 5	319. 6	466. 2
Damping (%)	0. 100	1. 241	1. 283	5. 380	0. 100	1. 324	1. 860	2. 126	0. 100

3. Analysis on dynamic characteristics mode of drain grating structure

The field test of dynamic characteristics of drain grating structure pays attention to dynamic characteristics of main beam and stakes, and in downstream as well as lateral direction the mode characteristics of main beam are tested in order to completely obtain the vibration characteristics of drain grating structure. When it is in observation and test, the main beam and stakes are dispersed to several space nodes according to space structure. Three-directional vibration sensors are set at critical nodal points and hammer each space nodes with force hammer one by one. At the meantime, we can get vibratory response signals and force signals of each nodal pint and do analog-to-digital conversion through data acquisition system after amplifier of these signals by secondary multiplication instrument. And then we proceed with data treatment by proprietary analysis processing software after entering into computerized data analysis system.

3. 1 The vibration mode characteristics of section of main beam of drain grating in downstream direction

The parameters for vibration mode characteristics of section of main beam of drain grating in downstream direction are listed on Table 3, and the corresponding modal vibration modes at low-level frequency of several former orders are drawn on Figure 8. It can be seen from the table the first-order frequency of vibration mode of main beam's section of drain grating in downstream direction is 6. 27Hz, and the second-order frequency is 10. 34Hz. If aqueous medium is taken into consideration, the first-order fundamental frequency will have 30% decreases and the fundamental frequency under flow action will decrease to 4. 0Hz.

Table 3 **The parameters for vibration modes of section of main beam of drain grating in downstream direction**

Serial Number for Modes	1	2	3	4	5	6
Frequency (Hz)	6. 27	10. 344	17. 69	23. 93	37. 37	42. 83
Damping Ratio (%)	0. 1	5. 38	5. 38	0. 1	2. 74	5. 38
Serial Number for Modes	7	8	9	10	11	12
Frequency (Hz)	49. 43	66. 66	65. 199	133. 43	200. 96	270. 36
Damping Ratio (%)	1. 86	0. 1	0. 98	0. 1	0. 98	0. 1

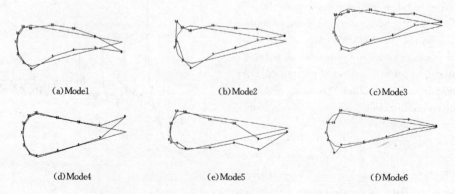

| (a)Mode1 | (b)Mode2 | (c)Mode3 |
| (d)Mode4 | (e)Mode5 | (f)Mode6 |

Figure 8   The modal vibration modes of main beam of drain grating in downstream direction

3.2 The vibration mode characteristics of main beam of drain grating structure in lateral direction

The layout of nodal points for testing vibration mode of main beam of drain grating structure in lateral direction is shown on Figure 9, and the mode parameters are listed on Table 4. The corresponding modal vibration modes at low-level frequency of several former orders are drawn on Figure 10. It can be seen from the table the first-order frequency of vibration mode of main beam of drain grating in lateral direction is 1.817 Hz, and the second-order frequency is 17.45 Hz. If aqueous medium is taken into consideration, the first-order fundamental frequency will have about 30% decreases and the fundamental frequency under flow action will decrease to about 1.0 Hz.

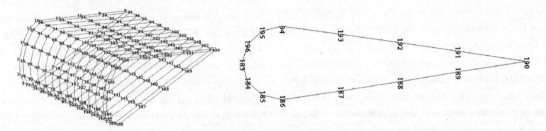

Figure 9　The layout of nodal points of main beam of drain grating in lateral direction

Table 4　　　**The parameters for vibration modes of the whole main beam of drain grating in lateral direction**

Serial Number for Modes	1	2	3	4	5	6
Frequency (Hz)	1.817	17.45	23.71	30.21	42.922	66.07
Damping Ratio (%)	5.38	5.38	3.72	3.62	1.64	0.1
Serial Number for Modes	7	8	9	10	11	12
Frequency (Hz)	135.91	161.86	180.24	192.23	214.41	280.23
Damping Ratio (%)	0.556	0.626	0.326	0.173	0.1	0.1

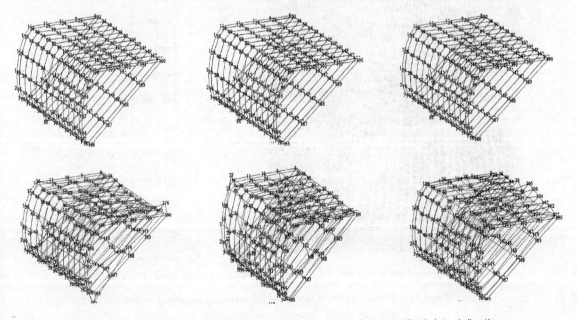

Figure 10　The modal vibration modes of the whole main beam of drain grating in lateral direction

3.3   The vibration mode characteristics of drain grating's cascade

The drain grating structure at inlet opening after construction is shown on Figure 11 (a) .

The layout of nodal points for testing vibration mode of drain grating's cascades is shown on Figure 11 (b) and the vibration mode parameters of cascades after parametric fitting and recognition. The corresponding modal vibration modes at low-level frequency of several former orders are drawn on Figure 12. It can be seen from the table the first-order frequency of vibration mode of cascade of drain grating is 15.19Hz, and the second-order frequency is 28.56Hz. If aqueous medium is taken into consideration, the first-order fundamental frequency will have about 30% decreases and the fundamental frequency under flow action will decrease to about 10.0Hz.

The experiment results of structural vibration modes show that the fundamental vibration frequency of ground working gate without water is 75.35Hz, and the second-order frequency is 115.03Hz. There's significant change on the vibration mode of gate structure with fluid-solid coupling, and the fundamental frequency with water is 53.7Hz and the second-order frequency is 106.8Hz. The experiment results indicate that the characteristics of vibration fluid working on gate structure have great effect on first-order mode, the rate of frequency descend is about 30%; the second-order frequency descends to 8%.

Main beam and cascade constitute drain grating structure. The modal experiment results indicate that the first-order frequency of vibration mode of main beam's section in downstream direction is 6.27Hz, and the second-order frequency is 10.34Hz. If aqueous medium is taken into consideration, the first-order fundamental frequency will have 30% decreases and the fundamental frequency under flow action will decrease to 4.0Hz. The observation data also shows that the first-order frequency of vibration mode of main beam of drain grating in lateral direction is 1.82Hz, and the second-order frequency is 17.45Hz. If aqueous medium is taken into consideration, and the fundamental frequency under flow action will decrease to about 1.3Hz. The vibration mode frequency of main beam in lateral direction is relatively low.

The vibration mode experiment of drain grating's cascades points out that the first-order frequency of vibration mode drain grating's cascades is 15.19Hz, and the second-order frequency is 28.56Hz. If aqueous medium is taken into consideration, the fundamental frequency under flow action will decrease to about 10.0Hz.

(a) Connecting of main beams and cascades of drain grating

(b)Layout for nodal points on drain grating's cascades

Figure 11   Connecting of main beams and cascades of drain grating as well as layout for nodal points on drain grating cascades

Table 5　　　　　　　　**The parameters for cascade modes of drain grating**

Serial Number for Modes	1	2	3	4	5	6
Frequency（Hz）	15. 19	28. 56	42. 38	58. 56	84. 92	186. 18
Damping Ratio（%）	5. 38	5. 38	5. 38	5. 38	5. 38	1. 34
Serial Number for Modes	7	8	9	10	11	12
Frequency（Hz）	256. 82	281. 11	336. 44	391. 69	440. 95	472. 29
Damping Ratio（%）	0. 98	1. 42	0. 98	1. 02	0. 81	1. 86

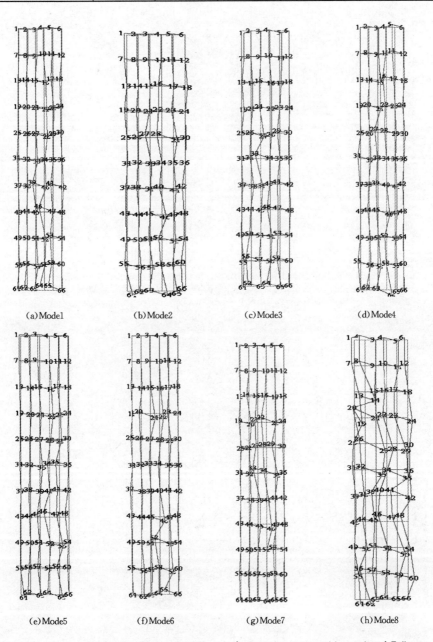

(a)Mode1　　　　　　(b)Mode2　　　　　　(c)Mode3　　　　　　(d)Mode4

(e)Mode5　　　　　　(f)Mode6　　　　　　(g)Mode7　　　　　　(h)Mode8

Figure 12　The modal vibration modes of drain grating's cascades at several low orders 4 Epilogues

References

[1] Yan Genhua. Analysis on Experimental Modes of Dynamic Characteristics of Hydraulic Sector Gate, Water Conservancy and Transportation Scientific Research, 1990, 1.

[2] Yan Genhua. Research on Vibration Characteristics and Anti-design of Drain Grating for Hydropower Station, Hydroelectric Generation, October, 2006.

Advance of Hydraulic Tests Concerning about Fishes

Yang Yu[1]　Gao Yong[2]　Sun Feng[1]　Han Yeqin[3]

[1] State Key Laboratory of Hydrology-Water Resources and Hydraulic Engineering,
Nanjing Hydraulic Research Institute, Nanjing, 210029
[2] Institute of Chinese Sturgeon, China Three Gorges Corporation Yichang, 443002
[3] Nanjing Foreign Language School, Nanjing, 210008

Abstract: With the fast development of economy and water projects in China, ecologic and environmental issues have attracted more and more attention. And the protection of fishes against the impacts of water project construction is increasingly urgent. Test studies related with fishes in the designs and scientific researches of water projects are summarized and called Hydraulic Tests Concerning about Fishes (HTCF) in this paper. The definition of HTCF is given, and its research contents and methods are elaborated in classification. The development of HTCF is predicted as a conclusion of this paper.

Key words: fish; hydraulics test; fish ethology

1. Foreword

Impacts of water project construction on a river ecosystem can not be ignored. A river ecosystem covers a wide range of animals, plants and microorganisms, which interact on each other and are of complex relationship. It is extremely difficult to research the impacts of water projects on a river ecosystem.

The primary difference between a river ecosystem and a terrestrial ecosystem lies in the different dynamic environments where organisms are. Therefore, the specific evolutions of organisms are obviously induced by the space media (i. e. water) in a river ecosystem. These evolutions exert a far-reaching influence on the shapes and behaviors of organisms in the river ecosystem.

Fishes as a group of higher aquatic organisms in the river ecosystem occupy the top of the food chain in the river ecosystem. The evolution of fishes is widely and deeply influenced by the river hydrodynamic characteristics. Therefore, it is an important means to carry out hydraulic tests concerning about fishes (HTCF) to gain relevant information. The aim of this paper is to sum up current HTCF and generalize the study range of HTCF in order to lay the research foundation of hydraulics concerning about fishes.

2. Hydraulic tests concerning about fishes (HTCF)

HTCF in this paper refer to all hydraulic tests to gain fish demands, rehabilitate the habitats of fishes, analyze the impacts of hydrology and water environment alternation on fishes, and assess the impacts of water projects on fishes. According to different research levels, the research objectives can be classified into four layers as following: fish demands (fish demands on water environment), physiological tolerance of fishes (in view of individual scale), ecological tolerance of fishes (in view of species scale) and characteristics and restoration of fish habitats.

Field measurement and investigation, prototype tests, laboratory tests and numerical simulation are the test methods of HTCF.

2.1　Field measurement and investigation

2.1.1　*Field measurement and investigation (Direct observation)*

Field measurement and investigation are the direct methods to study the fish demands on water environment and the important means to validate lab test results. They have the advantage of good reflection of the reality and the disadvantage that observation results are affected by the complex environmental factors in the field. Besides, field measurement usually needs various expensive and accurate instruments, such as flow meter, water quality analyzer, diving equipment, survey ship, etc. , resulting in a high cost.

2.1.2 *Method of catches* (*Indirect observation*)

Catch method was a method to analyze the stock and spatial and temporal distribution of fishes by catching a target fish. After further development, the catch method combined with accurate spatial orientation is used in the habitat preference presumption of target species[1,2].

2.2 Prototype tests

2.2.1 *Tests of fishes' passing through non-fish passage*

The purpose of relevant hydraulic tests of fishes' passing through spillways or turbines is to assess the extent of injuries of fishes in the process of downstream migration through spillways or turbines so as to provide data support for eco-friendly water projects[3]. This kind of tests is generally realized by the recapture of marked target fishes[4].

2.2.2 *Tests of fishes' passing through fish-ways*

The existing fish passage facilities are the research objects in this kind of tests. The test purpose is to assess the efficiency of fish passage facilities and the working status of all parts of the facilities[5]. Generally, fishes passing over dams are counted with the help of fish-way monitoring room or the target species passing through fish-ways are caught to analyze the efficiency of fish-ways.

2.3 Laboratory tests

2.3.1 *Flume tests*

Flume test is a normal method to study the demands of fishes, usually conducted in a laboratory under controlled conditions. It has the advantages of simple quantitative observation, easy obtainment of test results in a short time and reappearance of test results after repeated tests and the disadvantages of small test scale in a flume, demands on special lab instruments and control of experimental environment in order to study the impacts of a single factor on fishes. Test on group fish needs 3 fish at least. Besides, the same fish shall not be used in a repeated experiment to prevent the fish adaptation to experiments.

2.3.2 *Model tests*

It is quite common to study a water project by the utilization of a small scale model. Generally, the model test of a water project should fulfill the requirement of Froude Number similarity. In a scaled model, the target fish cannot fulfill the similarity requirement. Therefore, model tests are commonly used for the simulation of hydraulic characteristics. For example, a fish-way model is used to analyze the velocity distribution of different parts of the fish-way.

2.4 Numerical simulation

2.4.1 *Numerical simulation of hydro-environment*

Numerical simulation of hydraulics and hydro-environment is comparatively mature. Based on Navier-Stokes Equations and integrated with diffusion equation and relevant biological growth equations, the method has been widely used with the tools of difference and finite element method[6]. The common commercial software of computational fluid dynamics (CFD) has the coupling computation function of water environment.

2.4.2 *Numerical simulation coupling fish behavior with ecological succession*

Above all, the principal factors that induce fish behaviors shall be identified and a numerical model of fish behaviors shall be built[7-8]. A coupling fish distribution model at a reach scale is finally set up by coupling the fish behavior model with the hydraulic or water environment model[9]. The difficulty in the coupling fish model lies in the establishment of a fish behavior model. Usually, the stimulating factors of fish behaviors can be gained by filed survey[10], prototype test, laboratory test or catch method.

3. Interdiscipline in HTCF

Actually, HTCF is the test that intersects hydraulics and ichthyology. Both the hydraulic environment and fish demands should be concerned in HTCF. Relevant disciplines include fish ethology, fish physiology and fish ecology, etc. .

3.1 Hydraulic tests in fish ethology

The closest discipline to HTCF is fish ethology[11-13]. Fish ethology is to study the internal and external reactions of fishes to stimulation, including reaction type, pattern, mechanism, etc. (see Figure 1). It belongs to animal ethology and has close relationship with fish physiology and fish ecology. Individual and group behaviors of fishes both under natural and experimental conditions are the study objects of fish ethology.

Stimulation is a prerequisite for animal behaviors. In most cases, external environmental changes can be stimulations. For example, water temperature change can drive fishes to migrate. Internal environment changes act as stimulation too. For example, hunger drives fishes to find food, and gonadal maturation drives fishes to reproduce.

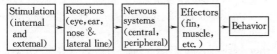

Figure 1　Stimulations to animal behaviors

The research objects of fish ethology include behavior pattern, evolution, genetics, development and physiological mechanism. The main research content of HTCF is the impacts of different flows on fish behaviors. In other words, researchers choose a species, proceed with accurate, detailed and repeated observations on fish behaviors in a special flow, and record and analyze its corresponding behavior patterns. As a further research, some researchers conducted profound simulation studies on group fish behaviors in different water environments based on the research of individual fish behaviors[7].

3.1.1　*Fish rheotaxis*

Fish rheotaxis is a critical juncture of fish ethology and hydraulics. Rheotaxis is an important fish tropism in addition to phototaxis, thigmotaxis, geotaxis, etc.. Fishes can adjust their swimming direction and speed according to flow direction and velocity to keep themselves contranatant or staying in a specific place over a long time. Specific indexes for the study on fish rehotaxis include perceived velocity, preferred velocity, limit velocity, etc..

Perceived velocity is a velocity that fishes just start to react to, which can be also called start velocity. Preferred velocity is an optimum velocity range which fishes can get over. Fishes like to swim upstream at this velocity range. Limit velocity is a max. velocity that fishes can get over and which is also called max. velocity.

Figure 2　Flume for rheotaxis tests

3.1.2　*Flume tests of characteristic velocities*

Flume is a normal device for the research of fish rheotaxis. (see Figure 2) The inlet and outlet in a flume can control the flow velocity and rate in it. The velocity is measured by velocity meter or computed by flowmeter (see Figure 3).

Researchers use rheotaxis test devices to gain some general rules of fish rheotaxis as following. Fishes of spindle body possess a strong swimming capacity, while fishes of flat body are usually weak in swimming. The relationship between fish limit velocity and morphology is similar to that between fish swimming capacity and morphology. The rheotaxis of sweetfish and orfe in streams is stronger than that of German carp and crucian carp in lakes with slow or no flow. Both fish limit velocity and preferred velocity increase with fish body length[15-16]. Big fishes are of stronger rheotaxis while the juvenile are of weaker rheotaxis, drifted with flow in general[12].

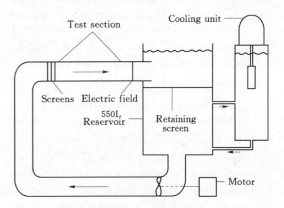

Figure 3　Pipes for rheotaxis tests[14]

3.1.3　*Field tests of characteristic velocities*

Large fishes are not suitable for flume tests. Their preferred velocity and water depth can be obtained by the hydraulic investigation of their habitats. Chinese National Fisheries Administration conducted the flume tests of rheotaxis for juvenile Chinese sturgeon at Nanjing Hydraulic Research Institute in 1979. At the same time, they conducted a field investigation of the swimming capability of adult Chinese sturgeon in the upper and middle reaches of the Yangtze River[17]. YANG Yu and et al. (2004) investigated the habitat of Chinese sturgeon in detail, and the preferred velocity and depth of Chinese sturgeon were gained with the integration of fish detector data[18].

3.1.4 *Mechanism of rheotaxis*

The general view of fish rheotaxis is that fish rheotaxis is induced by synthesized sensations including visual sensation, touch sensation and flow sensation. Hofer (1908) considered that fish lateral line was a sense organ that controlled fish rheotaxis. Lyon (1904) and Loeb (1990) considered fish determination of flow direction depended on the touch sensation by the friction from bottom on the one hand, and on the sight of surrounding fixed targets on the other hand. Many researchers considered fish rheotaxis was decided by its habitude of target maintenance to keep a certain target at a special place of its retina[12].

3.2 Hydraulic tests in fish physiology

Hydraulic tests in fish physiology are used to analyze fish hydrodynamic response and hurts by current. Fish hydrodynamic response means how fishes perceive water wave or flow. Hurts by current include sports injuries by swimming fatigue, physical injuries by a complex flow field with high shear force and high fluctuation when fishes pass through water projects and gas bubble trauma by dissolved gas supersaturation induced by discharge of high dams.

3.2.1 *Receptivity capacity of lateral line*

Fish lateral line has a function of feeling flow. The sensory receptor top of lateral line will incline and the lower sensory cells will send nerve impulses when flow impacts a fish body. So the fish has a feeling on water flow. Hofer (1908) confirmed the sensory capacity of lateral line using pike as a test target. Dijkgraaf (1933) conducted a further research of lateral line using experimental biologic method. He considered the main function of lateral line was to perceive bait, enemy or company at a certain distance by feeling flow disturbance. Research results show that fish lateral line plays a certain role in fish feeding, migration, clustering, reproduction and defense.

3.2.2 *Physiological injuries by swimming*

When fishes pass through fish-ways in hydraulic projects, they keep swimming without rest, sometimes resulting in poisoning death due to excessive anaerobic metabolism and lactic acid accumulation in vivo. Therefore, the max. sustainable swimming speed (MSSS) of the target species should be measured. The max. sustainable swimming speed can also

be called critical swimming speed (CSS), which can be measured by constant flow method and incremental flow method. Beamish (1996) considered that the swimming speed a fish can keep for a continuous swimming of 240 min. was MSSS.

3.2.3 *Gas bubble trauma*

Dissolved gas supersaturation (DGS) and gas bubble trauma (GBT) of fishes reflect the relationship between physical cause and biological effect, which has gradually drawn the attention of environmental scientists in the past several decades[19]. DGS is a widespread phenomenon in natural and man-made water bodies in the world. It occurs when the sum of partial pressures of gas in solution exceeds the atmospheric pressure. Many study findings demonstrated DGS resulting from high dam discharge could produce a variety of physiological signs which were harmful or fatal to fishes and other aquatic and marine organisms [20-24] (see Figure 4) (Renfro, 1963; Stroud et al., 1976; Weitkamp et al., 1980; Cornacchia et al., 1984; Johnson et al., 1984; Gray et al., 1985; Fidler, 1988; White et al., 1991).

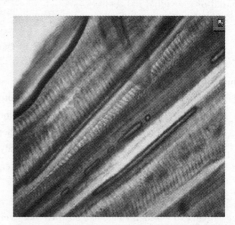

Figure 4　Intra-corporeal bubbles in
rainbow trout lamella[24]

In China, there is little study on GBT of fishes in natural rivers. It was found that only Chinese carps suffered from GBT in many spawns downstream of the Gezhouba Dam (2006 Report supported by NSFC, Project No. 30490230). According to the data from Water Environment Monitoring Center at the Three Gorges, the dissolved oxygen saturation (DOS) was more than 120% between the Huanglingmiao section and the Dongyuemiao section from August to September 2003. The max. DOS was 130% in

September 2003. Form June to October 2004, DOS at the Huanglingmiao section was more than 110%. The middle reaches of the Yangtze River is a natural breeding site of many fishes. Therefore, analysis of DGS caused by the discharge of the Three Gorges Project and measures to protect rare fishes and their reproduction will have great economic and social benefits[19,25-29]. Nanjing Hydraulic Research Institute is building a fish test device with high pressure, which will be used to research GBT under laboratory condition[30].

3.2.4 *Injuries by complex flow regime*

Many studies show that stronger turbulence and vortex and higher shear stress will generate injuries to fishes. The difference of injuries mainly depends on different fish species and growth stages[31-32] and different turbulence amplitudes (see Figure 5).

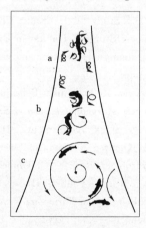

Figure 5   Turbulence amplitude
compared with fish body

Coutant and Whitney found that migratory fishes used river turbulence to improve their swimming speed. If the conclusion by Coutant is correct, there will be certain turbulent intensity which is the most appropriate to migratory fish survival. Stronger turbulences will hurt fishes, while weaker turbulences will deplete fish energy. Morgant, et al., used a barrel to simulate vortex impacts on fish eggs and juveniles. He got the result that the mortality rate of many fishes was higher under the condition of stronger turbulences. Wang Dexiang (2007) with Hohai University used oscillating-grid turbulence generator to test turbulence hurts to crucian carp (see Figure 6). His conclusion showed the tested fish would be injured or die when the turbulent intensity was higher than 9.09cm/s[33].

Figure 6   Crucian carp in oscillating-grid
turbulence generator

3.2.5 *Injuries by fishes' passing through hydraulic projects*

There are 3 passages for fishes to pass through hydraulic projects, i.e. turbine, spillway and fishway. Complex current movement in turbines and spillways as abnormal passages often produce grievous injuries to fishes. Turnpenny (1992) used submerged jet to simulate turbine shear force and observe fish injury rate and mortality. Test results showed that the fish injury rate and short-term mortality increased with the increase of jet velocity and shear force.

Heisey (2000) summarized 10 relevant research reports, and concluded the probability of fish balance loss was 0.4%-4.1% after passing through turbines. At same time, 38 tests were conducted in 6 hydropower projects in the Columbia River Basin, and the fish mortality and probability of balance loss were 0-7.0% and 0.1%-4.7% respectively in the downstream spillways.

GAO Yong, et al. (2008), carried out the tests of fishes' passing through turbines with a result the survival rate of fishes after passing through run-of-river turbines with low head could be up to 94.7%.

3.3   Hydraulic tests in fish ecology

Fish ecology is a discipline to study fish life pattern and interaction between fishes and environment. The main contents of hydraulic tests in fish ecology are to acquire, simulate and restore the hydraulic characteristics of fish habitats.

With the development of positioning technology and ultrasonics, it becomes more convenient to acquire the hydraulic characteristics of habitats[18,34-35]. Acoustic Doppler Current Profile (ADCP) can be used to study the hydraulic characteristics of a research area. High-resolution fish detector (such as DIDSON) can be used to identify fish species from images.

Simulation of fish habitats includes physical and numerical models. There is little progress in physical models due to scale effects[36-37]. There is a relatively great progress in 1D, 2D or even 3D numerical simulation of hydraulic environment of fish habitats. And such ecological simulation software as PHABSIM[38], IFIM[39] and WESP-TOOL[40] have been developed.

3.4 Hydraulic tests of fish passage facilities

Hydraulics tests of fish passage facilities started comparatively early, and well develop at the same time[41-44]. At present, it becomes an important content of engineer hydraulic tests.

The tests simulate the flow inside the model and the flow at fish inlet by the model establishment of fish passage facilities such as fish way, fish ladder and fish elevator. Velocity distribution in the facilities is studied to guarantee the safe and quick passage of target fishes[7]. Many studies show that the flow regime and velocity at fish inlet is a key point to attract fishes. With the development of computed fluid dynamics (CFD), the velocity distribution of more and more fish passage facilities can be gained by numerical simulation. Meanwhile, the nature-like engineering of fish passage facilities is gradually extended due to the further development of habitat restoration theory in eco-hydraulics[43].

3.5 Hydraulic tests in bionics

Bionics researchers focus on the study on fish body types and propulsion of fish locomotorium (such as pectoral fin, dorsal fin and caudal fin)[45-46] to analyze the energy efficiency, power and motility of fishes in different shapes. Such studies can be done in laboratory under the control conditions, based on fish ethology and physiology. Researchers obtain relevant data of optimum body type, swimming pattern and body surface mucus by the analysis of fish energy consumption under different hydrodynamic conditions in order to provide instructions for ship design.

4. Development of HTCF

In conclusion, the development of HTCF lies in a more in-depth systematic interdisciplinary study coupling fish ecology and ethology with hydraulics. Based on fish ecology theory, action rules and mechanism of flow in fish ecology and behaviors can be further explored with the study contents of fish ethology and by the utilization of hydraulic research method. Fish eco-hydraulic theory aiming at fish ecosystem protection is going to be set up, based on the frame of fish ecology and integrated with fish ethology and physiology, to provide theoretic guidance to further development of hydraulics of fish passage facilities and fish habitat restoration.

Acknowledgment

This research is supported by NSFC for Youth (No. 50909064) and Ministry of Water Resources for Nonprofit Research (No. 200801105). Hou yiqun and Wang dexiang have worked for this research.

References

[1] Yu Yang. Study on Hydraulic Characteristics of Chinese Sturgeon Habitat in Yangtze River [D], Nanjing, Hohai University Hydro-power Engineer, 2007: 120.

[2] Qi Weiwei. Reproductive Behavioral Ecology of Chinese sturgeon (*Acipenser Sinesis Gray*) with its Stock Assessment [D], Wuhan, China, Chinese Academy of Science Institute of Hydrobiology, 2003 (in Chinese): 47-78.

[3] Diplis Mathur, Paul G. Heisey. Ask Young Clupeids if Kaplan Turine are Revolving Doors or Blenders [C]. Hydraulic Engineering, 1993: 1332-1337.

[4] Yong Gao, Yu Yang, Yiqun Hou. Assessment of survival rate of fish passing through the turbine of Gezhouba Dam [C]. Nanjing, International Symposium on Hydraulic Physical Modeling and Field Investigation, 2010.

[5] Yong Gao. Fish Passing Capacity and Design of Fish Passage Facilities in Run-of-River Hydraulic Project with Low Head [R]. China Three Gorges Corporation, Yichang, 2008: 100.

[6] Qiuwen Chen, Ouyang Zhiyun. Coupling Model of Ecology and Hydraulics and Its Application [J]. *Journal of Hydraulic Engineering*, 2005: 1273-1279.

[7] R. A. Goodwin, J. M. Nestler, J. J. Anderson

et al. . Forecasting 3-D fish movement behavior using a Eulerian-Lagrangian-agent method (ELAM) [J] . *Ecological Modelling* , 2006: 197-223.

[8] D. J. Booker. Hydraulic modelling of fish habitat in urban rivers during high flows [J]. *Hydrological Processes* , 2003: 577-599.

[9] M. Leclerc. Ecohydraulics: A New Interdisciplinary Frontier for CFD. 2005: 429-460.

[10] T. Vehanen, A. Huusko, T. Yrjana, et al. . Habitat preference by grayling (Thymallus thymallus) in an artificially modified, hydropeaking riverbed: a contribution to understand the effectiveness of habitat enhancement measures [J], *Journal of Applied Ichthyology* , 2003: 15-20.

[11] Xiaorong Huang, Zhuang Ping. Research of Fish Ethology and Its Application in Practice [J], *Freshwater Fisheries* , 2002: 53-56.

[12] Daren He, Cai Houcai. Fish Ethology [M]. Xiamen, Xiamen University Press, 1998: 338.

[13] Yi Chai, Xie Congxin, Wei Qiwei et al. . Researches of Fish Ethology [J]. *Reservoir Fisheries* , 2006: 1-2.

[14] Robert G. Otto Ian S. Hartwell. Critical Swimming Capacity of the Atlantic Silverside, Menidia menidia L. [J] . *Estuaries* , 1991: 218-221.

[15] C. Tudorache, P. Viaene, R. Blust, et al. . A comparison of swimming capacity and energy use in seven European freshwater fish species [J] . *Ecology of Freshwater Fish* , 2008: 284-291.

[16] Eric D. Tytell. The hydrodynamics of eel swimming II. Effect of swimming speed [J]. *J. Exp. Biol.* , 2004: 3265-3279.

[17] Test Team of Chinese Sturgeon Swimming Capacity of National Fisheries Administration, Test Report on Chinese Sturgeon Swimming Capacity [R], Yangtze River Fisheries Research Institute, Nanjing, 1979: 26.

[18] Yu Yang, Zhongmin Yan, Jianbo Chang. Assessing Chinese Sturgeon Spawning Ground by Using Acoustic Doppler Current Profiler to Measure River Velocities [C] . Beijing, China, International Symposium on Flood Forecasting and Water Resources Assessment for IAHS-PUB, 2006.

[19] Decai Tan, Ni Zhaohui, Zheng Yonghua. Dissolved Gas Supersaturation Induced by High Dam and Its Impacts on Fishes [J]. *Freshwater Fisheries* , 2006: .35-38.

[20] D. A. Geldert, J. S. Gulliver, S. C. Wilhelms. Modeling dissolved gas supersaturation below spillway plunge pools [J] . *Journal of Hydraulic Engineering* , 1998: 513-521.

[21] Donna Schulze Lutz. Gas Supersaturation and Gas Bubble Trauma in Fish Downstream from a Midwestern Reservoir [J] . *Transactions of the American Fisheries Society* , 1995: 423-436.

[22] J. J Orlins, J. S Gulliver. Dissolved gas supersaturation downstreamof a spillway. II : Computational model [J] . *Journal of Hydraulic Research* , 2000: 151-159.

[23] D. E. Weitkamp. Total dissolved gas supersaturation in the natural river environment [R]. Chelan County Public Utility District No. 1, Washington, 2000: 12-13.

[24] L. E. Fidler, S. B. Miller. British Columbia Water Quality Guidelines for Dissolved Gas Supersaturation [R] . Aspen Applied Sciences Ltd. , 1994: 46.

[25] Daqing Chen, Summarization of Project Supported by National Natural Science Foundation of China [R] . Beijing, 2006, Project No. 30490230.

[26] Xiangju Cheng, Chen Yongcan, Gao Qianhong. Analysis of Reoxygenation Supersaturation Induced by Discharge from Three-Gorge Reservoir [J] . *Journal of Hydroelectric Engineering* , 2005: 17-20.

[27] Fufeng Peng. Cause and Prevention of Gas Bubble Trauma of Fish Fry [J] . *Scientific Fish Farming* , 1998: 25.

[28] State Environmental Protection Administration. Bulletin on Ecological and Environmental Monitoring Results of the Three Gorges Project in 2003 [R] . 2004: 1.

[29] Chenggen Wu. Gas Bubble Trauma of Rainbow Trout [J]. *China Fisheries*, 1994: 12.

[30] Yu Yang, Hu Ya'an, Han Changhai. Test Device for Gas Bubble Trauma of Fishes [P]. China, 2010.

[31] Eric D. Tytell. Median fin function in bluegill sunfish Lepomis macrochirus: streamwise vortex structure during steady swimming [J]. J. *Exp. Biol.*, 2006: 1516-1534.

[32] James C. Liao, David N. Beal, George V. Lauder et al.. The Karman gait: novel body kinematics of rainbow trout swimming in a vortex street [J]. *J. Exp. Biol.*, 2003: 1059-1073.

[33] Dexiang Wang. Experimental Study on the Effects on Fish of Flow turbulent [D]. Nnajing, Hohai University College of Water Conservancy and Hydrapower Engineering, 2007: 69.

[34] Fdouglas Jr Shields, Charles M Cooper, Scott S Knight. Experiment in stream restoration [J]. *Journal of Hydraulic Engineering*, 1995: 494-502.

[35] Elizabeth A. Nystrom, Kevin A. Oberg, Chris R. Rehmann. Measurement of Turbulence with Acoustic Doppler Current Profilers-Sources of Error and Laboratory Results [C]. Estes Park, Colorado, USA, Hydraulic Measurements and Experimental Methods Conference 2002, 2002.

[36] David W. Crowder. Reproducing and Quantifying Spatial Flow Patterns of Ecological Importance with Two-Dimensional Hydraulic Models [D]. Blacksburg, VA, Virginia Polytechnic Institute and State University Civil Engineering, 2002: 170.

[37] David W. Crowder, Panayiotis Diplas. Evaluating spatially explicit metrics of stream energy gradients using hydrodynamic model simulations [J]. *Canadian Journal of Fisheries and Aquatic Sciences*, 2000: 1497-1507.

[38] Midcontinent Ecological Science Center. PHABSIM for Windows User's Manual and Exercises [R]. U. S. Department of the Interior U. S. Geological Survey, 2001: 299.

[39] Clair Stalnaker, Berton L. Lamb, Jim Henriksen et al.. The Instream Flow Incremental Methodology A Primer for IFIM [R]. U. S. Department of the Interior National Biological Service, Washington, D. C., 1995: 49.

[40] Helmut Lorek, Michael Sonnenschein. Modelling and simulation software to support individual-based ecological modelling Elsevier Science, 1999: 199-216.

[41] C. H. Clay. Design of fishways and other fish facilities [M]. Boca Raton, Lewis Publishers, 1995: 15.

[42] John Cocks. Fishway on the Nokomai [J]. *New Zealand Engineering*, 1993: 18.

[43] A. Laine, T. Jokivirta, C. Katopodis, Atlantic salmon, Salmo salar L., and sea trout, Salmo trutta L.. Passage in a regulated northern river -fishway efficiency, fish entrance and environmental factors [J]. *Fisheries Management and Ecology*, 2002: 65-77.

[44] Nanjing Hydraulic Research Institute. Fishway [M], Nanjing, Electric Power Industry Press, 1982: 278.

[45] C. D. Wilga, G. V. Lauder. Function of the heterocercal tail in sharks: quantitative wake dynamics during steady horizontal swimming and vertical maneuvering [J]. *J. Exp. Biol.*, 2002: 2365-2374.

[46] E. D. Tytell, E. M. Standen, G. V. Lauder. Escaping Flatland: three-dimensional kinematics and hydrodynamics of median fins in fishes [J]. *Journal of Experimental Biology*, 2008: 187.

Prototype Debug Study on the Hydraulic Characteristics of the Filling and Emptying System of Qiaogong Ship Lock on Hongshui River

Hu Ya'an   Yan Xiujun   Quan Qiang   Xue shu

*State Key Laboratory of Hydrdogy-Water Resources and Hydranlic Engineering,
Nanjing Hydraulic Research Institute, Nanjing, 210029*

Abstract: The transportation valve of Qiaogong adopts the Flat—gate type, its factual working head is higher than other projects which have completed at home and abroad, this ship lock breaks records of the adoption of Flat-gate type. The working conditions of this type of valve directly affect the running of the ship lock such as cavitation and vibration, etc. This paper systematic introduces the achievement of this ship lock prototype observation, including hydraulic characteristics, the process of lifting force of tumble gate when opening and closing, and the process of uplifting force of the miter gate. The fruits of this prototype debug have provided the important scientific basis for the safety navigation of ship lock.

Key words: Qiaogong; ship lock; valve; cavitation; prototype debug; hydraulic characteristics

1. Introduction

Qiaogong hydropower station is the ninth step of the 10 steps cascade hydropower station of Hongshui River. Its location is about 1km away from Qianjiang, Laibing City. Qiaogong hydropower station is a large scale hydro-junction which is mainly designed for generating electricity and also can be used for transportation, irrigation, etc. The scale of the lock is designed for barge of 500t. The design navigation volume of the lock is 2,500,000 tons per year. One-way capacity is 2,000,000 tons per year.

The highest and lowest upstream navigable water stage of ship lock are 84.00m and 82.00m respectively and the highest and lowest downstream navigable water stage of ship lock are 71.15m and 59.35m respectively. The maximum design head is 24.65m. The amplitude of downstream navigable water stage is 11.8m. The effective dimension of ship lock is 120m ×12.0m× 3.0m (length × width × sill depth). The Multiport Filling and Emptying System uses the open ditch energy dissipation of the conveyance culvert under the lock. for General layout, see Figure1. The layout of the open ditch energy dissipation see Figure 2. Layout characteristics of filling and emptying system see Table 1. The valve of water transmission

adopts plain batten door, ship lock has a new culvert type with "top-abrupt expansion + bottom-abrupt expansion".

In May-June 2008 (construction of navigation) and in November to December 2009 (formal navigation), the hydraulic characteristics in Qiaogong ship lock has been conducted twice. At the first prototype debug (construction of navigation), because the head is low, the control procedure has not completed totally and just adopt temporary control procedure to make the gate up and down. the hydraulic characteristics mainly through the methods of reading the water level, recording the flow pattern with video and the analysis of calculating. during the second time of prototype debug we have carried out the comprehensive hydraulics monitor of the ship lock when debugged the ship lock with and without water. This paper introduces the achievement of observation of hydraulic characteristics of water transportation system.

2. The characteristic of filling and emptying system when two sides of valve continued open

In the overall model experiment of scale ratio 1:25 Qiaogong ship lock has investigated the hydraulic characteristics of water transportation and the

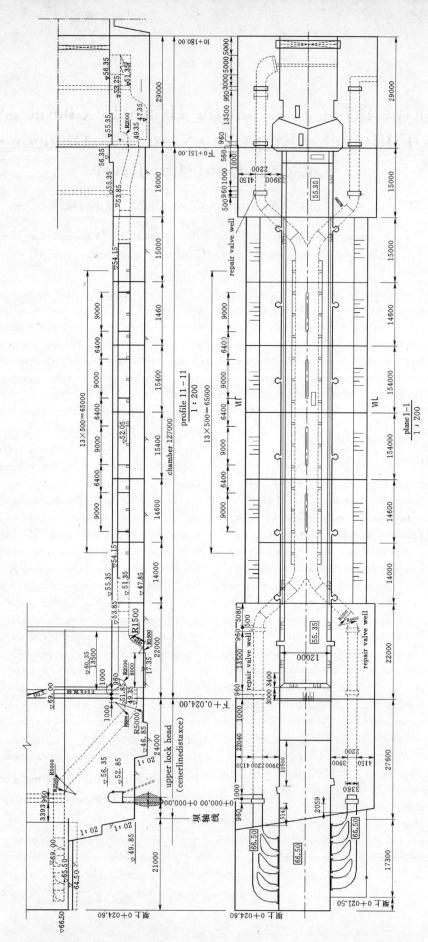

Figure 1 Overall layout of filling and emptying system of Qiaogong ship lock

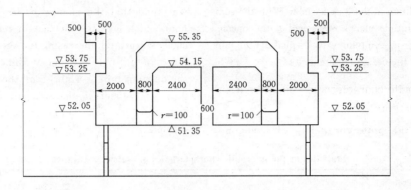

(a) The lock chamber profile

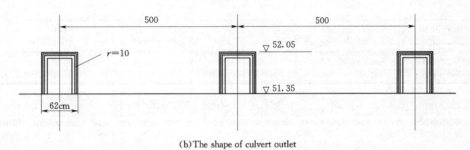

(b) The shape of culvert outlet

Figure 2　The layout of outlet in the chamber

Table 1   **Layout characteristics of filling and emptying system**

Position	Description	Area (m²)	Area ratio with valve
Upstream inlet	4 branches holes is vertical with lock wall on the upstream approach channel	50. 0	4. 45
Conveyance valve culvert	The height of top valve in the two sides culvert is 51. 85m, submerged depth is 7. 5m	11. 0	1. 00
Outlet culvert	Two sides main culvert converge at the culvert of outlet	13. 44	1. 22
Outlet branches holes	From up and down stream: totally 14 holes, every hole size is 0. 60×0. 70m, the interval of outlet hole is 5. 0m, the total length of culvert in the outlet is 65. 0m, the length ratio with effective length of lock chamber is 0. 54	11. 76	1. 07
Open ditch energy dissipation	Cross-section is 2. 00m×4. 00m, lay a retain-sill of 0. 50m× 0. 5m in the ditch	—	—
Downstream outlet	(1) the area of culvert outlet increase the double; (2) the branches holes'size: 4. 4m × 0. 5m; in the grid type of energy dissipation lay two deflecting flow sill	24. 2	2. 00 4. 40
Side empting culvert	Connect with the tail water of the hydropower station at downstream	9. 0	1. 80

mooring condition in the lock basin[2]. With the comparison of multiple plans, we select t_v-the opening time of filling and emptying valve, equal to 6 minutes which fulfills the design requirements.

2.1 Hydraulic characteristics of water transportation system

During the prototype debug of Qiaogong, we found that the quality of the construction is poor which lead the increase of resistance coefficient. so through increasing the speed of the opening time to satisfy the time of filling and empting, the opening time is 4 min, and the closing time is 3.3 min. Observed characteristic values of water transportation under all kinds of water head are in Table 2.

Table 2　　　　　　　**Statistics of the hydraulic characteristics of water conveyance**

condition	Water head (m)	Equal level (m)	Valve Open time (s)	Flow time (s)	Maximum flow rate (m³/s)	Over filling (empting) (m)	Maximum velocity of water lever (m/min)	Average velocity of water lever (m/min)
Two sides filling	23.52	83.6	240	724	94.0	0.08	3.67	1.95
Two sides filling	24.2	83.9	240	734	95.3	0.04	3.61	1.98
Two sides empting	23.9	59.6	200	670	85.22	0.08	3.23	2.14
Two sides empting	24.2	59.6	200	682	108.75	0.10	4.12	2.13

Combined with Table 2, we derive the following conclusions from the synthesized analysis of above observation.

(1) The designed opening time of filling and emptying valve is $t_v = 6$min. In the real case, the left and right valves' opening time $t_v = 240$s when filling water both sides. The left and right valves'opening time $t_v = 200$s when emptying water both sides. The opening rate of valve is bigger than designed.

(2) With water level combination between 83.9m and 59.6m and valve working water head is 24.2m which is closed to design water head (24.65m), the water filling time is 734s (12.31min) which is close to water filling time 11.36min as designed when $t_v = 6$min. That means it fulfill the requirement. The maximum flow capacity in the water filling process is $Q = 95.3$m³/s and maximum water level elevation speed is 3.61m/min while the average speed is 1.98m/min. The hydraulic indexes of water filling system are high.

In the frequent range (water level combination from 83.62m to 60.1m and valve working water head 23.52m) of working water head, the filling water time is 724s ie 12min. The maximum flow capacity in the water filling process is $Q = 95.3$m³/s and maximum water level elevation speed is 3.61m/min.

(3) The water emptying time is 640s (10.67min) when valves are open in both sides and under the condition that water level combination between 83.8m and 59.6m and valve working water head is 24.2m which is close to design water head (24.65m). That means it fulfill the requirement. The maximum flow capacity in the water filling process is $Q = 108.75$m³/s and maximum water level descending speed is 4.12m/min while the average speed is 2.27m/min.

According to the actual gate water level and downstream access door water level, we calculate the drag coefficient and discharge coefficient of water transportation system when filling and emptying water and compare them with the experimental results of whole scale ship lock model. The results are in Table 3.

Table 3　　　　　　　**Resistance coefficients in prototype and model**

Operating condition		Resistance coefficient	Flow coefficient
Both sides filling	prototype	3.26	0.554
	model	3.01	0.577
Both sides empting	prototype	3.19	0.56
	model	2.79	0.599

It turns out that prototype drag coefficients are both larger than that in experiment and discharge coefficient of prototype and experiment are basically close.

2. 2 The flow state of upstream inlet

The bottom elevation of upstream approach channel at entrance area and the region about 800m behind it is 79m, and then the elevation gradually reduce to the 69. 0m until the inlet, so satisfy the submerged depth of inlet culvert be enough. In addition, the upper head entrance of the culvert used vertical multi—port in the wall. To ensure the flow rate uniform of every branch hole. Such layout makes sure that the water surface is smooth enough when filling in the lock chamber and without the phenomenon of swirl.

Because the upstream approach channel is designed longer, the observation have proved that water surface wave is little so that the ship sail in the approach channel is safe without dangerous.

2. 3 Water surface fluid state in lock basin

2. 3. 1 *Water filling fluid state in lock basin*

By selecting a small βvalue (0. 88) which is the ratio of total area of outlet hole against the area of main culvert to keep flow capacity uniformly distributed in weir direction in lock basin both filling and emptying water and water flow fluid state steady be-

sides we place open ditch energy dissipation in both sides of outlet hole in the weir direction of main culvert at the bottom of lock basin. From the model experiment, we know the arrangement of outlet hole section of water transportation culvert in the lock basin has improved the mooring condition in the lock basin as expected. Water flow in the lock basin is rather steady and has small turbulence. The mooring force of fleet of $2 \times 500t$ and single ship of 500t is well below the value in the manual.

With 24. 2m water head which is close to the designed value, the starting water depth is 4. 2m. The fluid state in the lock basin can be seen from Figure 3 when filling water both sides and when opening the valve.

From the photos, water comes out of lock basin uniformly at the beginning of opening the valve and local turbulence of water is small. To 90-120s, the effluent of left part is more than right of first half of lock basin and it is contrary in the second half. The intensity of water surface turbulence is big respectively. Although water flow of lock basin reaches its maximum when valve is fully open, the fluid state trends to be steady for submergence become 7m deeper than beginning. There is little bubble obviously escaping have explained that the amount of aeration at top sealing sill and drop sill is suitable without any disadvantageous effects.

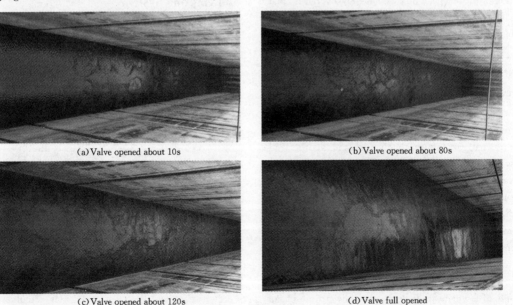

(a) Valve opened about 10s

(b) Valve opened about 80s

(c) Valve opened about 120s

(d) Valve full opened

Figure 3　The flow state of lock chamber when two sides filling

2.3.2 *Water emptying flow state in lock chamber*

Similar to most ship lock, water flow fluid state is very steady in the lock basin in the water emptying under all kinds of working water head and continuous open of water emptying valve with $t_v = 200$s. There is no vortex flow or obvious flow in weir direction. Though the filling and empting observation process indicate that the ship is stable in the lock chamber and the cleat is loose, so the force of water wave have produced is small.

2.4 The flow state at downstream approach channel and the outlet

The filling and emptying system of Qiaogong ship lock adopts that left side culvert emptying the water to the downstream approach and right side to the downstream of hydropower station. The left side outlet utilizes the grid type energy dissipation and the energy dissipation room set two deflecting flow sills. So that uniforms the flow. The outlet of right side is single hole emptying. When emptying the water in the lock chamber, the outlet flow state of downstream approach channel is very power and strong. but when water flow to the guide wall, the water surface is smooth. At the same time, because the length of downstream approach channel is short (640m) and only one side emptying, its water wave is weak at the pier structure downstream. In the course of prototype debug we have discovered no matter left side or right side outlet, there are a lot of bubbles while emptying.

3. The measures of eliminating the water level difference of the miter gate at the downstream

Qiaogong ship lock adopt another side empting system, one side of outlet at the downstream flow in the downstream approach channel and another side flow in the right side of river which at the downstream of hydro — power station. The water level at the right side of river is higher than the left side of approach channel, so at the end of empting the water from river of right side will conversely flow in the chamber lead the result of water level difference of the miter gate. In the prototype observation, the largest water level differential is 0.7m around the first MITR of lower head. It both extends the time of water transportation and analyzes valve hoist loads of MITR with and without water head which is presen-

ted in Figure 4.

Figure 4　The flow state of
lock chamber when two sides filling

In the prototype observation, the largest water level differential is 0.7m around the first MITR of lower head. the flow pattern of different water level see at Figure 4.

In this situation, it is vary dangerous to the miter gate, will break the hoisting equipment of miter gate. faced with the situation, we suggest that the empting valve should be closed in advance when the water head is 1.0m between the chamber and the downstream channel so as to prevent the water flow in the chamber in turn.

4. The mathematical model calculated and predicted hydraulic characteristics of filling and empting System

As the water level in the prototype debugging limitations, difficult to achieve the maximum navigable level design portfolio, in order to correctly guide the prototype debug require a combination of mathematical model calculations. According to Bernoulli equation can be written description of a single—stage ship lock when the unsteady flow equations, namely:

$$H_1 - H = (\zeta_1 + \zeta_{v1}) \frac{Q_1 |Q_1|}{2g\omega_1^2} + \frac{L_1}{g\omega_1} \frac{dQ_1}{dt} \quad (1)$$

$$H - H_2 = (\zeta_2 + \zeta_{v2}) \frac{Q_2 |Q_2|}{2g\omega_2^2} + \frac{L_2}{g\omega_2} \cdot \frac{dQ_2}{dt} \quad (2)$$

$$Q_1(t) = S \frac{dH(t)}{dt} \quad (3)$$

$$Q_2(t) = -S \frac{dH(t)}{dt} \quad (4)$$

Where H_1, H, H_2 respectively upstream water level, lock chamber water level and water level (m); ζ, ζ_v respectively the drag coefficient of culvert and valve resistance coefficient; ω is the area of the culvert of water transmission valve (m²); Q is the water flow of chamber (m³/s); S is the water area of the chamber (m²); L is the length of corridor conversion (m); subscript 1, 2 are filling and empting, respectiely.

Mathematical model to predict the hydraulic characteristics when filling (empting) valve open time is $t_v = 4$min, 3. 3min, the water level combinations: 84. 0-59. 46m (the largest head of permanent navigation), charge, discharge flow of the process line shown in Figure 5 and Figure 6. The results of calculation show that the design is expected in the design of the largest navigable head 24.54m (highest navigable water level upstream of the downstream minimum navigable level combinations), chamber filling, emptying time was 690s and 755s, respectively, corresponding to the maximum flow rate 97. 0m³/ s and 127. 2m³/ s.

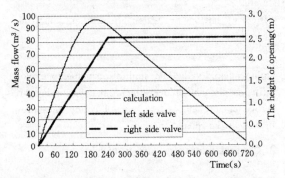

Figure 5   the mass flow rate hydrograph of filling

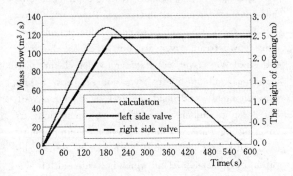

Figure 6   the mass flow rate hydrograph of empting

5. Summery

(1) Two sides filling conditions close to the design head, valves are continuously opened with $t_v = 4$min rate, the filling time is 12min, is the same as the design requirements 12min. After the valve fully opened, the flow coefficient of distribution system are 0. 554 (filling), almost equal with model values. When two sides filling, the flow pattern is stable in the lock chamber. Vertical flow rate is very small, there is some lateral velocity, and mooring rope is slack, the force of flow is small. Design of the ship when fully loaded, as long as in accordance with the specification, the ship fastened double cable, and dual cable is outside the "八" or of "八" shape, mooring line forces to meet regulatory requirements.

(2) The conditions of both sides empting, due to adoption of another side flow into the downstream of hydropower station, because the downstream of river has not renovation. so the water level of right side river is higher than the downstream approach channel obviously. at the end of empting the water from river of right side will conversely flow in the chamber lead the result of water level difference of the miter gate, so adjust the manner of right side valve closing advance. Through the method involved above solve these issues effectively.

(3) According to Bernoulli equation establish the mathematical model to calculate the maximum designed head 24. 54m and supply the hydraulic characteristics, providing the safe basis in operation of high head condition for Qiaogong ship lock.

References

[1]   Hu Yaan, Ling Guozeng, Yan Xiujun. Valve Hydraulic Model Test of Qiaogong Ship Lock on Hongshui River [R] . Nanjing: Nanjing Hydraulic Research Institution, 2005.

[2]   Xuan Guoxiang, Huang-Yue. Filling and Empting System Hydraulic Model Test of Qiaogong Ship Lock on Hongshui River [R]. Nanjing: Nanjing Hydraulic Research Institution, 2005.

[3]   Hu Yaan, Li Jun, Zong Muwei. Prototype debug study of letan ship lock on hongshui river [R]. Nanjing: Nanjing Hydraulic Research Institution, 2006.

[4]   Hu Yaan, Li Jun, Li Zhonghua. Prototype debug study of dahua ship lock on Hongshui River [J] . Nanjing: Nanjing Hydro Science and Engineering. Published in March, 2008: 88-92.

Deep-thickness Covering Layer Riverbed Closure Difficulty and Safety Control Countermeasures

Li Xuehai[1,2]   Huang Guobing[2]   Cheng Zibing[2]

[1]*College of Water Resources and Hydroelectric Engineering,*
Wuhan University, Wuhan, 430072
[2]*Department of Hydraulics Research, Changjiang River Scientific*
Research Institutde, Wuhan, 430010

Abstract: According to project examples, two type deep-thickness covering layer riverbed formation and its effect on river closure safety are analyzed in detail and new understanding to covering layer riverbed closure difficulty is provided. The safety control countermeasures of deep-thickness covering layer riverbed closure is explored from the view of lowering closure difficulty methods, bottom protection effectiveness, dike collapsing prevention and closure material stability computation method etc.

Key words: deep-thickness covering layer, river closure difficulty; bottom protection effectiveness; embankment collapsing; stability computation; sand protection and leakage prevention

1. Introduction

In China, deep-thickness covering layer riverbed is very common in the main rivers[1]. For example, covering layer is more than 100m thick in Xinshizheng-Yibing river section. As the maximal depth covering layer found in hydropower projects, the maximal covering layer depth is more than 420m at dam site of Yele hydropower station on Nanya River, which is a branch river of Dadu river. Around the world, deep thickness covering layer is distributed in many rivers. For example, sand-cobble covering layer is 230m thick at dam base of Tarbela earth-rockfill dam in Pakistan. The covering layer depth is 170m at dam base of Aswan high dam in Egypt. In France, the covering layer depth is 110m in dam site of Serrepon-con earth-rockfill dam. In Canada, the covering layer depth of Manic-3 dam site zone is 130m[2]. For river closure in deep-thickness covering layer riverbed mentioned above, due to greater discharge, water head, gap flow velocity and low anti-scour ability of covering layer, unreasonable protection will cause scour damage, leakage damage, bottom protection instability and embankment collapsing, which may threat the safety of construction staff and machine and prolong closure difficulty time. Closure material shortage may cause project failure. Therefore, it has high academic value and application value to explore difficulty elements in deep-thickness covering layer riverbed closure and safety control countermeasures, to perfect the related theory.

2. Project examples and analysis

2.1 Classic Project examples

Example 1: Changzhou hydraulic project's outside river course closure in Guangxi Province[3].

Changzhou hydraulic project is located in Xunjiang downstream river section which is the main stream of West river water system. When closure project is carried out in its river course, flow passes through middle and inside natural river course. With 3.7-8.8m ground elevation of closure gap section, its covering layer is 3.0-15m thick and made up of pelitic sand and pelitic sand gravel. With 22.5m ground elevation of Sihuazhou island which is located at left side of closure gap section, its covering layer is 23.6-27m thick and made up of floured clay, floured earth and pelitic sand gravel. Due to left side revetment being too short in the first closure process, flow rounds reinforcing steel bar cages to scour bank slope, resulting in scour damage and foundation collapsing. Due to low revetment height, covering layer is scoured away by water flow, the revetment can not fall on base rock and is pushed to downstream section. Due to bad closure material quality, the material's wash away quantity increases greatly. Due to great lacking of special material like concrete tetrahedroid and abatis etc. covering layer is scoured seriously, which causes the first closure to be interrupted. After adopting improvement measures, closure project is completed successfully.

Example 2: Xiangjiaba hydraulic project's main river course closure in Jinshajiang river[5].

Xiangjiaba hydraulic project is located in downstream river section of Jinshajiang river. Adopting 6 diversion bottom outlets to release flood and two-direction end-dumping advancing with single dike scheme, main river course closure has the features of great flow discharge, high flow velocity, 61m thick covering layer, slow diversion ratio increasing, near 20h difficult period of time and related with navigation closely. In river closure zone, riverbed covering layer is made up of egged gravel layer, sand layer and egged gravel layer filled with rock block from top to bottom in vertical direction. The covering layer is hardened and coarsened after long-term flow scour in flood season. Due to containing abundant sand and smaller pebble, the covering layer's conveyable velocity is less than 3m/s. Due to no bottom protection project carried out preliminarily, the covering layer becomes conveyable in large-scale during difficult period of time, local scour pits appear in gap section, with increased closure material running away and dike head collapsing. By increasing dumping intensity in three hours, open channel diversion ability starts to increase, flow velocity begins to decrease and difficult closure period is completed successfully.

2.2 Covering layer formation and its effect on closure safety

The first example mentioned above belongs to deep-thickness covering layer closure project with soft foundation. Its covering layer formation is due to sediment deposition in placid terrain or deeply cut river valley and it has bad anti-scour ability. Due to greater discharge, water head, gap flow velocity and low anti-scour ability of covering layer, unreasonable protection will cause scour damage, leakage damage, bottom protection instability and embankment collapsing, which may threat the safety of construction staff and machine and prolong closure difficulty time. Closure material shortage may cause project failure.

The second example mentioned above belongs to accumulated covering layer closure project in west mountain zone. Its covering layer formation is due to large-scale old-time landslide accumulation caused by geology movement, river erosion and river valley being deeply cut. Due to its rough composition, this kind of covering layer's anti-scour velocity is greater than that of soft-based covering layer. However, due to narrow river valley and turbulent flow in west mountain zone, covering layer scour may still cause dike head collapsing, which brings troubles to closure safety control.

3. Closure safety control countermeasures

For riverbed closure projects of soft-based covering layer or accumulated covering layer, the reasonable safety control countermeasures should be determined according to covering layer property, river course form and other conditions.

3.1 Understanding covering layer riverbed closure difficulty

Closure difficulty involves dumped stone stability, closure scale and closure safety factor[6]. Deep-thickness covering layer increases closure difficulty in three aspects mentioned above.

The past research shows that when covering layer exists, dumped stone stability depends on the stability of covering layer, viz. covering layer's existence reduces dumped stone stability. Covering layer scour prolongs closure difficult period and increases closure difficulty and scale. In the condition of no bottom protection or bottom protection failure, covering layer scour may cause dike head collapsing. In the case of soft base, covering layer scour may cause leakage damage. Therefore, covering layer's existence reduces closure safety factor.

Covering layer riverbed closure difficulty is closely related with its thickness, composition and graduation. Small thickness covering layer has little effect on closure difficulty. Greater thickness covering layer increases difficult time and quantities, also may cause dike head collapsing and dangers the safety of construction staff and machine. Different covering layer composition and graduation mean different anti-scour ability. Smaller anti-scour ability of covering layer has greater effect on closure safety factor and closure difficulty.

3. 2 The coutermeasures to reduce closure difficulty

Reducing closure difficulty can reduce bottom protection cost, covering layer scour and increases closure safety factor, is a effective method for closure safety control. The countermeasures to reduce closure difficulty are explored in the following three aspects.

(1) There are many closure related hydraulic elements such as discharge Q_0, final closure head Z_{max}, the maximal gap flow velocity V_{max} and water depth H etc. By adopting reasonable closure standard, enhancing hydrology forecast and adopting upstream step reservoir regulation, closure discharge can be reduced. By optimizing diversion structure form and diversion ability, lowering the demolition elevation of upstream and downstream embankments reasonably, installing wing dam at diversion tunnel outlet, gap hydraulic index can be declined. For mild water surface slope ratio river course, double or more dikes can share water head. For rapid slope river course, wide dike can reduce closure difficulty. For closure projects in backwater zone of their downstream step project, upraising upstream reservoir level can reduce closure head.

(2) The closure solid boundary involves dumped material's physical and mechanical property, foundation composition and relative roughness Δ/d etc. With greater volumetric weight, weight or graduation, relative roughness, dumped material can be equipped with greater anti-scour ability. Therefore, when hydraulic index is lowered and local stone material can not still meet the need of dike advancing, series connected stone and steel bar cage filled with stones can increase dumped material stability. There are other methods such as increasing bottom plate roughness, utilizing stone barrier and stone screen to increase dumped material stability. In addition, reducing water depth can prevent dike head collapsing. Due to its good accommodation in covering layer riverbed, concrete tetrahedrons can be prepared for the sake of additional safety.

(3) In river closure project, there are some subjective elements such as dumping intensity, dumping method, step reservoir regulation, closure scheme, diversion structures etc. Increasing dumping intensity can prevent closure material running off ef-

fectively in gap section and is a good method for deep-thickness covering layer riverbed closure due to its inability for bottom protection. Increased dumping intensity can be achieved by optimizing material factory distance, full material preparation, improving traffic, increasing machine loading capacity etc. For dumping method, it is recommended to adopt bigger stone material dumped in upstream dike head zone. Due to flow trajectory effect in upstream dike head zone, flow velocity will be reduced in downstream dike section and on dike axis line, it is in favor of later smaller stone material dumping.

3. 3 The countermeasures to increase bottom protection effective

For soft base covering layer riverbed, bottom protection can increase riverbed anti-scour ability, reduce later project cost, increase riverbed roughness and later dumped material stability, reduce material run off. The effective of bottom protection involves material choice, protection range and depth. In navigation river course, it is very easy to utilizing bottom opened barge or ship hanging-up system to carry out bottom protection project. Due to narrow river valley and turbulent flow, it is very difficult to carry out bottom protection operation in non navigation river course. It is recommended to adopt buoyancy pontoon with double reversible plates to dump stone cage under water, like Dachaoshan closure project, as shown in Figure 1.

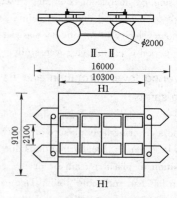

Figure 1 The schematic illustration of buoyancy pontoon with double reversible plates

The choice of bottom protection material should base on material stability computation. The bottom protection width may adopt the gap width in the condition of large-scale covering layer starting to move and should also be determined according to closure procedure, gap hydraulic condition in different stages and covering layer's anti-scour ability. The bottom protection length can be computed based on vertical velocity distribution along gap central line. The bottom protection thickness should not less than double material thickness, viz. double-layer protection should be adopted. Due to small relative roughness Δ/d, single layer protection has bad stability. Double-layer or multi-layer protection can increase relative roughness, press down bottom material, which is in favor of their stability. Due to covering layer scour also existing in bottom protection downstream zone, if protection zone's thickness and length are not enough, upstream-downstream protection material collapsing or running off will cause the bottom protection system's stability failure and threat dike stability. Therefore, bottom protection scheme should be determined with the full consideration of all elements mentioned above and keep certain safety redundancy.

3.4 The countermeasures to prevent dike collapsing

In general, the main dike collapsing problem refers to dike head collapsing in closure project, which is related with water depth, flow action, dumped material, submerged effect, machine load, terrain elements, covering layer thickness, dumping method and intensity etc. For deep-thickness soft base covering layer riverbed closure, large scale dike collapsing appears due to covering layer scour. In addition, during dike advancing process, piping effect will happen to sand base layer under high head, which causes downstream dike slope sliding and integral dike instability. In this case, except bottom protection, sand base protection and dike leakage prevention should be carried out. It is recommended that geotextile is paved under gravel protection layer to prevent piping effect and increase sand base stability. During dike advancing process, such material as rock ballast and clay can be filled to increase dike leakage prevention ability and its stability.

3.5 Closure material graduation computation

Isbash formula[8] [Formula (1)] developed by the former USSR is the most authoritative formula for closure material graduation computation.

$$V = K \sqrt{2g \frac{r'-r}{r} d} \qquad (1)$$

In this formula, K value is adopted as 0.86 in anti-slide condition, as 1.20 in anti-tilting motion condition. Based on end-dumping closure, professor Xiao Huanxiong from Wuhan University corrects K value to 0.89 for single stone material, 1.07 for well-proportioned material collective and 0.93 for mixed material collective.

In the practice, it is found out that closure material stability is also related with water depth H, relative roughness Δ/d, and vertical velocity distribution coefficient α etc. According to closure research of Gezhouba project and Three Gorges Project, Wang Dingyang from CRSRI develops practical computation formula in end-dumping and movement stop condition[10]:

$$V = \left(\frac{H}{d}\right)^{\alpha} \sqrt{2g \frac{r'-r}{r} d} \left[0.65 + 0.35 \left(\frac{\Delta}{d}\right)^{0.5}\right] \qquad (2)$$

During difficult period, rectangle velocity distribution in vertical direction appears in gap zone along dike axis line. Due to gap riverbed filled with protection material or earlier dumped material, when the same or nearly same material is dumped on it, Δ/d value is 1 nearly, Formula (2) is uniform with Formula (1). Therefore, Isbash formula is suited for material stability computation in difficult period, which is proved by many projects' practice. For non gap section closure and scour pit appearing in gap covering layer section, due to non rectangle velocity distribution in vertical direction, the computation result by Formula (1) can not match with project practical condition.

For deep-thickness covering layer riverbed closure, during initial stage, river course type velocity distribution in vertical direction appears in gap zone. Due to low flow velocity and slight covering layer scour, sediment movement formulas, for example sediment movement velocity formula [Formula (3)] developed by professor Lu Jinyou from CRSRI[11], can be adopted for covering layer stability computation. Its application conditions include $V = 1.2$-3.94m/s, $H = 2.92$-37.0m and $D = 12$-255mm.

$$V = 0.95 \left(\frac{H}{D}\right)^{1/6} \sqrt{\frac{r'-r}{r} gD} \qquad (3)$$

With gap width reducing, flow velocity increasing and covering layer scour depth increasing, vertical velocity distribution changes. So Formula (3) can be used to estimate scour depth firstly, then Formula (2) is adopted to compute material graduation. In the case of gap covering layer moving in large scale, bottom protection should carry out for soft base riverbed. Due to bottom protection material scoured by water flow in gap section, for the sake of project safety, Formula (1) can be adopted to compute its graduation in dike area. According to vertical velocity distribution, Formula (2) can be used as reference for bottom protection material choice in upstream and downstream dike area. For accumulated covering layer riverbed in narrow river valley, other measures should carry out to reduce closure difficulty according to covering layer's unique features firstly, then Formula (2) can be used to compute dumped material graduation according to changed gap hydraulic condition.

4. Conclusion

River closure difficulty of deep-thickness covering layer riverbed is related with covering layer's depth, composition and graduation. The river closure safety control countermeasures include adopting water-earth-person related measures to reduce closure difficulty, adopting effective bottom protection measures and dike collapsing prevention to eliminate all kinds of possible damage and ensure successful later procedures and preparing enough closure material, whose graduation and quantity can be computed according to design scheme, covering layer condition, bottom protection scheme and hydraulic condition. By these safety control methods, deep-thickness covering layer riverbed closure difficulty can be reduced, difficult period of time can be cut down and river closure construction safety extent can be raised. In addition, local stone material can be utilized to reduce project cost scientifically.

Acknowledge

This paper is sponsored by National Eleventh Five-Year Plan Supporting Project (2008BAB29B02-02).

References

[1] Xu Qiang. New Views on Forming Mechanism of Deep Overburden on River Bed in Southwest of China, *Advances in Earth Sciences*, 2008 (5): 448-454.

[2] Lu Shenwu, Ren Dechang. A review on "Research on Building Dam on Thick Overburden" [J]. *Sichuan Water Power*, 1986 (4): 11-21.

[3] Li Faxiao. Preliminary exploration on closure technology in deep-thickness soft base riverbed. *Sichuan Water Power*, 2006 (12): 39-45.

[4] Sun Zhiyu. Understanding river closure technology in Xiangjiaba hydraulic project. *China Three Gorges Construction*, 2009 (1): 26-28.

[5] Bo Ling. Understanding end-dumping closure difficulty and improvement measures. *Journal of Yangtze River Scientific Research*, 2001 (8): 7-8.

[6] Long Dehai. River closure design and construction in Dachaoshan hydropower station. *Water power*, 1998 (9): 50-54.

[7] Huang Zhiming. Experimental study on Xixi river closure and practice in Chaozhou water supply project. *Journal of Yangtze River Scientific Research*, 2003 (10): 21-24.

[8] Isbash C B. River closure hydraulics. *Chinese industry publishing house*, Beijing, 1964.

[9] Xiao Huanxiong. Construction hydraulics. *China WaterPower Press*, Beijing, 1992.

[10] Wang Dingyang. Practical hydraulics computation of End-dumping Closure. *Journal of Hydraulic Engineering*, 1983 (9): 11-18.

[11] Lu Jinyou. Sediment motion velocity formula exploration in Changjiang river. *Journal of Yangtze River Scientific Research*, 1991 (12): 57-64.

Evolution of a Typical Meandering and Bifurcated Channel in the Middle Yangtze River

Liqin Zuo   Chengwei Xu

State Key Laboratory of Hydrobogy-Water Resources and Hydraulic Engineering,
Nanjing Hydraulic Research Institute, *Nanjing*, 210029

Abstract: With the impoundment of the Three Gorges project (TGP), water and sediment conditions, as well as the evolution process, in the downstream of the dam has been changed. In this paper, the Yaojian reach in the middle Yangtze River was chosen as a typical example of meandering and bifurcated channels. A 2D flow-sediment mathematical model of the Yaojian reach was developed. The measured data in different periods, including seasons of the low water levels, the middle water levels and the flood levels from 2003 to 2007 (after the impoundment of TGP), were used to verify this model. The bed erosion and deposition in the following 20 years after the impoundment of TGP was calculated and analyzed according to the historical data. The results showed that, the Yaojian reach will be scoured, and the main stream still keeps in the right branch. Although the main tendency is erosion, the low water level in bends will still decrease by 0. 1 to 1. 2m. Therefore, the navigation depth in the inlet of the right branch will also decrease. A series of projects such as groins and fishbone dams are going to be built to improve the navigation conditions. The 2D model was used to study the effect of these proposed projects.

Key words: river bed evolution; meandering and bifurcated channel; numerical model; Yangtze River

1. Introduction

With the impoundment of the Three Gorges Project (TGP), water and sediment conditions, as well as the evolution process, in the down stream of the dam has been changed. Many researchers have conducted relative investigations (Han, et al. , 2003; Li, et al. , 2004; Lu, et al. , 2006) . Generally speaking, sediment will be kept in the Three Gorges reservoir, so the sediment concentration in its down stream will be unsaturated, leading to erosion in a long reach for a long time. The TGP impounded in 2003 and its water stage will rise to 175m gradually, thus have great impacts on flood control and waterway. From the point of channel navigation, it is in urgent need to study the evolution trend of the watercourse and possibility of navigation obstruction, for the further study of channel regulation projects. In this paper, the Yaojian reach, a typical meandering and bifurcated reach in the lower Jingjiang reach of the middle Yangtze River, was taken as an example. A 2D flow-sediment mathematical model of the Yaojian reach was developed, and the measured data

after the impoundment of TGP were used to verify it. Then the evolution trend and navigation obstruction of the Yaojian reach in 20 years (2008 to 2027) was analyzed and the channel regulation projects were studied.

2. Yaojian reach and its water and sediment characteristics

Yaojian channel belongs to the lower Jinjiang reach, middle Yangtze River. It is a typical meandering and bifurcated channel (Figure 1) . The banks of lower Jingjiang reach are mainly formed by sediments of the floodplain which deposited in modern times. Constructed by silty loam, silty clay etc. , those sediments have rich water content and loose texture. It has a dual structure of soil and sand—the medium fine sand forms its bottom and 2-15cm's sticky floodplain soil covers its surface, so it's vulnerable to erosion (Li et al. , 2004) . The main stream of Yaojian reach flows along the right bank of Egongtu, through Tashiyi and then is diverted into two streams: the left and right branch separated by the Wuguizhou Island , in which the right one contains

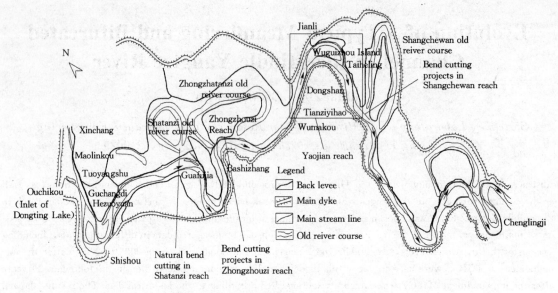

Figure 1   Sketch of Yaojian reach and Jinjiang reach

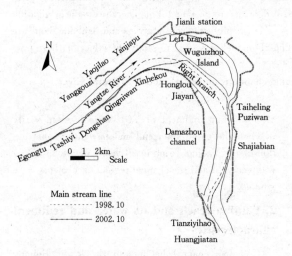

Figure 2   River regimes of Yaojian reach

90 percent. They meet near Taiheling and keeps down along the left bank to the Damazhou channel, then flow to the opposite bank and flow down along Tianziyihao reach (Figure 2). In the history, its river bed changed greatly, for the main branch and the sub-branch switch frequently, and the deposition and erosion in the shoal and groove was complex. Since 2000, its navigation channel condition has been worsened rapidly, with the reasons as follows: the right branch of Wuguizhou Island is wide and shallow; its natural depth is seriously inadequate for navigation; the entrance of the right branch varies greatly; there is serious back siltation after dred-

ging. So it becomes one of the worst channels with navigation obstruction in the middle Yangtze River.

The average measured runoff of Jianli station in Yaojian reach is 360 billion m³ per year. During 1955-2002, its average discharge was 11400m³/s. After the impounding of TGP, its runoff changed barely while during 2003-2006, its average discharge is 11200m³/s. Before the impounding of TGP, the average sediment concentration for many years in Jianli station was 0.996kg/m³. From 2003 to 2006, after the impounding, this data decreased greatly to 0.197kg/m³. The average sediment discharge has been reduced to 231million ton in 2003, while in 1998, this data was 407milllion ton (Wu et al., 2005).

3. Riverbed evolution

3.1   Evolution before the impoundment of the TGP

Yaojian reach is a part of the meandering reach of lower Jinjiang reach in middle Yangtze River. Its evolution is restricted by the longitudinal sediment transport equilibrium, as well as the dual structure boundary conditions (Yu, 2006). The evolutions of the Jinjiang reach and the Yaojian channel have been studied by many scholars (Han et al., 2003; Li et al., 2004; Peng et al., 2006).

In 1960s, Yaojian reach is a single bend called Jianli bend. In 1912, a relatively large shoal formed in the center of Jianli bend. Then the sinuosity increased gradually, and the evolution of the Jianli bend had a periodic law: a new right branch forms→the section enlarges→the talweg shifts left→the sinuosity grows larger→the right branch declines→a new right branch forms (Peng et al., 2006). Such process repeated, but the bifurcation feature had never changed. Since 1995, the main stream has always kept in the right branch (Li et al., 2004). Before 1960s, affected little by the human activity, the evolution of Yaojian reach generally agrees with the natural water and sediment conditions. However, after 1960s, its evolution has been affected by many factors, especially the human activity, which generally includes the following points: bend cutting of lower Jinjiang River, flood in 1996 and 1998 and so on.

The bend cutting projects were operated in 1967, in Zhongzhouzi, and 1969, in Shangchewan. Besides, in 1972, bend cutting happened naturally in Shatanzi. All these had shortened the Jingjiang River by 78km, increased its gradient, and caused the retrogressive erosion of Yaojian reach. Moreover, the incoming sediment was relatively less, caused less deposition in the side beach of Wuguizhou Island and low elevation. During long flood period, the right branch is scoured and being main stream.

Generally speaking, 1996 is a year with mean water and relatively little sediment from upstream. But the flood period lasted long and greatly affected by the Dongting Lake, Yaojian reach had large deposition. In particular, in the left branch of Wuguizhou Island as well as its entrance, the deposition was so great that in some place the river bed was even higher than the lowest navigation stage. 1998 was a year with large water and sediment. Besides, the flood in Jingjiang reach met the flow in Dongting Lake inlet, caused slow velocity and significant sediment settlement. In particular, with the overbank flow, heavy deposition happened on the river beach.

3.2 Evolution after the impoundment of the TGP

With the impoundment of the TGP, water and sediment conditions in the middle and downstream of Yangtze River have been changed. Flow with unsaturated sediment concentration released to the downstream of the dam will inevitably form a new riverbed, which will cause the alongshore erosion on the riverbed in downstream of TGP.

At the first stage of the impoundment, from Jun. 2003, to Sep. 2004, the sediment washed from upper Yaojian reach met the backwater action and morphological resistance of the Jianli bifurcated bend, as well as the flood jacking of the outlet water from Dongting Lake. So in Yaojian reach great deposition happened. At Yaojilao beach, Xinhekou shoal and the inlet of the right branch of Wuguizhou Island, the deposition depth was more than 2m. So the depth of the dredged channel became shallow (Figure 3).

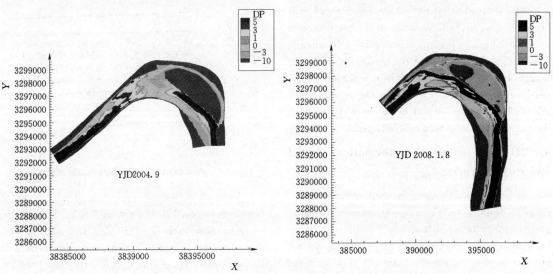

Figure 3   Navigation depth in 2004 and 2008 of Yaojian reach (m)

From Sep. 2004 to Sep. 2005, there were both e-rosion and siltation in Yaojian reach. In the bending reach, it is obvious of erosion in beach and siltation in groove. Erosion in beach ahead of Wuguizhou Island is about 2-5m, and the beach's volume was smaller than it was in 2004. The inlet of the right branch of Wuguizhou Island has erosion, but still no stable deep navigation groove.

From Sep. 2005 to Jan. 2006, it was in a period of emersion and erosion, and Yaojian reach was in a scoring condition. The erosion region mainly lied in the deep groove of the right branch of Yaojilao, the inlet of right branch of Wuguizhou Island, and the left branch. Compared with Sep. 2005, in Jan. 2006, water depth of the inlet of right branch is deeper. But there was still no stable channel (Figure 3).

From Jan. 2006 to Jan. 2007, the groove in the inlet of right branch was deepened, while the middle reach deposited, and in the downstream of the branch, the beach deposited and deep groove scoured. The navigation depth was more than 3.5m (Figure 3).

From Jan. 2007 to Jan. 2008, deposition in the inlet of right branch was about 2-5m. The navigation condition was so bad that the navigation depth was less than 3m. There were two deep groves in the inlet of right branch, the upper one near the beach ahead of Wuguizhou and the lower one near the Xinhekou (Figure 3). The two grooves share flows, which was bad for scouring the navigation channel in middle and low water season.

The Wuguizhou Island and the beach ahead of it belong to a whole part. Variation of the beach is close to the stability of the right branch groove. Its integrity will affect the stability of the entrance of the right branch. Horizontal gradient might cause inlets in the middle and lower part of the beach. During this time, the beach was relatively stable, so the deep groove in the right branch inlet was relatively stable too.

4. 2D flow-sediment mathematical model and its verification

To overcome difficulties in computation for natural water bodies, such as irregular boundary figures, great disparity between length and width of a calculated area, etc., the boundary-fitting orthogonal coordinate system is employed in 2D flow-sediment mathematical model in this paper. Basic equations include the flow continuity equation, flow momentum equation, the non-equilibrium transport equations of suspended load, the non-equilibrium transport equations of bed load, the gradation equation of bed materials and the bed deformation equations. About this model, its basic governing equations, numerical solutions, solution on key problems, initial boundary conditions, and movable boundary technique could be found in the reference (Lu et al., 2005).

4.1 Calculated region and orthogonal curvilinear grid

The calculated region covers a length of 25km, from Tashiyi to Huangjiatan. There are 165×81 grid points in this area. After orthogonal calculation, a orthogonal curvilinear grid is got, which is 30-300 m in length and 15-30m in width (Lu et al., 2008).

4.2 Verification of water level, velocity distribution and diversion ratio

According to the hydrometrics data in Sep., 2003, Jan. 2006, Jul., 2006, Oct., 2006, Jan., 2007, and Jul., 2007, their corresponding discharge are 30500m³/s, 5280m³/s、16050m³/s、10059m³/s、4452m³/s and 28900m³/s. Water surface lines in right and left bank and velocity distributions in 6 hydrologic sections along this reach are verified. The calculated water level has less than 0.1m diversion with the measured data (Lu et al., 2008). Figure 4 shows the verification of water surface line in right bank.

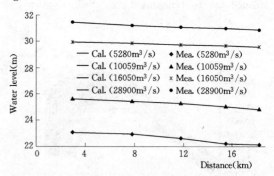

Figure 4　Verification of water surface line in right bank

Figure 5 shows the verification of velocity distribution of section 1♯ and 4♯ in Jan. 2006, with the corresponding discharge 5280m³/s. It is shown that, the calculated velocities are in good agreement with the measured data. Figure 6 shows the flow filed under discharge of 5280m³/s. It could be seen that this model has high accuracy in simulating the flow field.

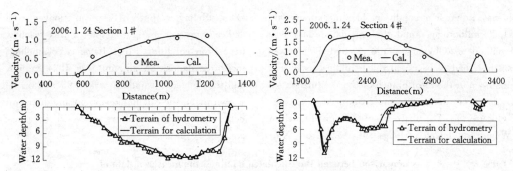

Figure 5　Verification of velocity profile at discharge of 5280m³/s

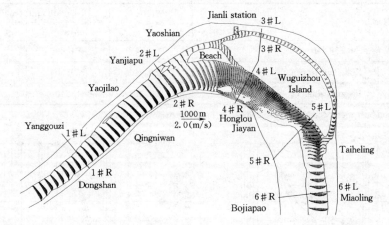

Figure 6　Flow field of Yaojian reach at discharge of 5280m³/s and layout of vertical lines

Table 1 shows comparison between the calculated and measured diversion ratios of the left branch of the Wuguizhou Island. It is shown that the divided flow in the left branch is relatively small, with only 3%-4% in dry season and less than 10% in middle flood season. Compared with the measured data, the calculated ratio has a diversion of less than 1.2%.

Table 1　Verification of diversion ratio of the left branch of Wuguizhou Island (%)

Q (m³/s)	Date	The left branch		
		Mea.	Cal.	Diversion
5280	2006.01	3.80	3.50	−0.30
10059	2006.10	5.56	4.62	−0.94
28900	2007.07	8.96	7.76	−1.19
30500	2003.09	8.12	8.64	0.52

4.3　Verification on river bed deformation

River bed deformations of Jan., 2007 to Jan., 2008 were calculated, with a daily mean discharge. Processes of measured sediment concentration in Jianli station were employed. Calculation of suspended load and bed material load was conducted for 6 groups of grain size relatively, and percentage of various grain sizes employs the annual mean value. In Yaojian reach, the bed material load is fine and its distribution is relatively uniform. The diameter size is mainly about 0.125-0.355mm, and the median diameter is about 0.20mm. The median diameter of suspended load is 0.15-0.15mm in dry season of Jan. 2006, Oct., 2006, Jan. 2007, under the discharge of 4452-10059m³/s, and 0.06mm in flood season of Jul. 2006, under the discharge of 16050m³/s. It is shown that, the suspended load is relatively coarse in dry season (Figure 7).

Table 2 shows the comparison between the calculated data and the measured data of the erosion and deposition amount in Yaojian reach from Jan. 2007 to Jan. 2008. Figure 8 shows the verification of the erosion and deposition distribution. From Yanjiapu to Wuguizhou Island, there were both erosion and deposition: the upstream of Island was eroded, and one side of Xinhekou was silted. But in general, this reach was in erosion condition. The measured data was 4.89 million m³, and the calculated data was 2.41 million m³. Obviously, the calculated data was too small. It had to be noted that, in this reach and the upstream of the right branch of the Wuguizhou Island, they were dredged to meet the need of navigation (Figure 8). The upstream of the right branch of the Wuguizhou Island includes the reach from Xin-

hekou to Jiayan, both of its deep grooves were eroded to some extent. The middle of this reach was in deposition condition. From Jiayan to Taiheling, the measured deposition amount was 1.24 million m³, and the calculated amount was 0.96 million m³. It's shown that, the calculated amount and distribution of deposition and erosion are close to the measured ones.

Table 2   Comparison between the calculated data and the measured data of
the erosion and deposition amount in Yaojian reach from Jan., 2007 to Jan., 2008

Reaches		Length (km)	Width (m)	Amount (million m³)	
				Mea.	Cal.
Yanjiapu-Wuguizhou Island		2.2	1429.4	−4.890	−2.410
The right branch	Xinhekou-Jiayan	2.7	780.5	−0.004	−0.045
	Jiayan-Taiheling	1.1	591.4	1.240	0.960

Note Positive numbers mean deposition, and negative numbers mean erosion.

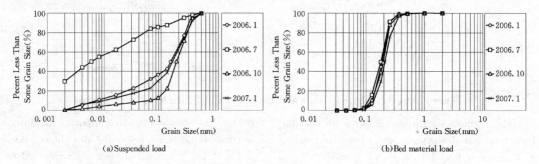

(a) Suspended load   (b) Bed material load

Figure 7   Gradient curve lines of suspended load and bed material load in Yaojian reach

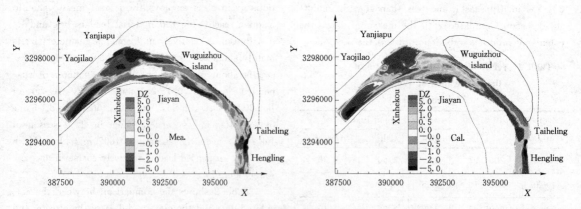

Figure 8   Comparison between calculated and measured distributions of
erosion and deposition from Jan. 2007 to Jan. 2008

5. Evolution prediction of Yaojian reach after the operation of the TGP

According to the water and sediment boundary condition from 2008 to 2027 after operation of the TGP provided by a 1D mathematical model (Huang, 2007), the erosion and deposition of the Yaojian reach is predicted. The condition of the flowing water and sediment is based on the water and sediment series from 1991 to 2000. From 2008 to 2027, the mean discharge is 11601m³/s, the mean sediment concentration is 0.22kg/m³, and the annual sediment discharge is 149 million ton (Talbe 3). At the first stage of the operation of the TGP, most of the sediment was deposited in the reservoir, the sediment flow out were few and fine. The median diameter of the suspended load was 0.01-0.02mm. The sediment carrying capacity of the outlet flow is far greater than the sediment concentration, so the flow has to scour the river bed to satisfy its sediment transport capacity. As time passing, the diameter of the suspended load tends to be finer (Huang, 2007).

Table 3 Water and sediment characteristics during diffident periods (Huang, 2007)

Periods	Mean discharge (m³/s)	Mean suspended load Concentration (kg/m³)	Annual sediment runoff (million ton)
2008-2012	11802	0.24	165
2013-2017	11224	0.22	138
2008-2017	11516	0.23	151
2018-2027	11680	0.21	146
2008-2027	11601	0.22	149

According to the data of discharge, sediment concentration as well as its gradation of Tashiyi section (inlet) and water level hydrograph of Huangjiatan (outlet) (Figure 9), the deposition and erosion process of Yaojian reach from 2008 to 2027 were calculated.

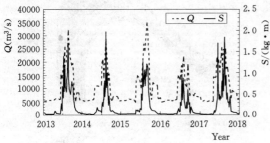

Figure 9   Discharge and sediment concentration process of inlet section and water level hydrograph of outlet section

5.1 Calculated region and orthogonal curvilinear grid

Figure 10 shows the changing process of the amount of deposition and erosion with time. It could be seen that: 1) By 2012, from Dongshan to Huanjiatan, this 22.4km long reach will be scoured by 17.5 million m³; by 2017, this data will become 31.4 million m³. Deposition happens in flood season and erosion happens in dry season. So erosion and deposition alternate through the whole reach. 2) During 2008 to 2027, from Dongshan to Huangjiatan, the erosion a-

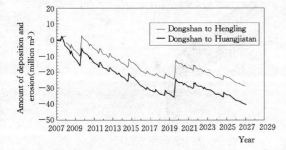

Figure 10   Erosion and deposition amount from 2008 to 2027 in Yaojian reach

mount is about 40.3 million m³ in this 22.4km long reach, and the average scouring area in the cross section is 1799.7m². By 2027, the Yaojian reach is still in erosion.

5.2 The fall of water level in dry season

Table 4 shows the changes of the design water level in Tashiyi, Jianli and Taiheling. It could be seen that at the discharge of 5000-5500m³/s, compared with 2007, in 2010 the water level in Tashiyi falls by 0.09m, in 2012, by 0.17m, in 2017, by 0.40m, and in 2027, by 0.86m. As for Jianli station, the data is 0.08m in 2010, 0.16m in 2012, 0.34m in 2017 and 1.16m in 2027. The decline of water level in Taiheling is close to that in Jianli station.

Table 4   The fall of low water level at discharge of 5000-5500m³/s (m)

Year	Tashiyi	Jianli	Taiheling
2010	−0.09	−0.07	−0.08
2012	−0.17	−0.16	−0.20
2017	−0.40	−0.34	−0.36
2022	−0.57	−0.88	−0.87
2027	−0.86	−1.16	−1.14

5.3 Distribution of erosion and deposition and changes of the navigation depth

With the impoundment of the TGP, the Yanggouzi side beach, which also called the transitional reach of Yaojilao, its left side beaches deposited and it gradually connect with the deposition area in Wuguizhou Island. It helps in restricting the development of the left branch of Wuguizhou Island, and keeps the main stream at the right branch. In the inlet of the right branch, the beach ahead of Wuguizhou Island will be scoured in its head and silted in lower side, and the beach will move to left and down-stream. There will be deposition in the centre of channel at the upper section of right branch. Then, staggered shoal will form in the inlet of the right branch, and water depth will be less than 0m under the design discharge. To get a stable waterway in this area, projects are needed. There will be little erosion in left branch of Wuguizhou Island. By 2027, the Yaojian reach will be scoured, the main stream still keeps in the right branch and the river regime remain unchanged. Although the main tendency is erosion, the low water level in bends will decrease by 0.1 to 1.2m, so the navigation depth in the inlet of right branch will also decrease (Figure 11).

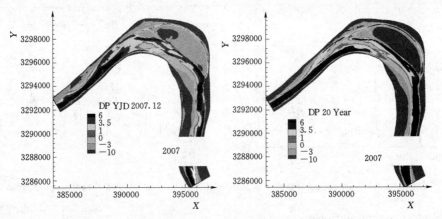

Figure 11　Navigation depth in 2007 and 2027 in Yaojian reach

6. Study of navigation regulation projects

6.1 Scheme of navigation regulation

From the study on evolution tendency of Yaojian reach, it could be known that the navigation depth in the inlet of right branch will decrease. So regulation projects are needed to keep the navigation depth. A series of projects have been studied (Lu et al., 2008), and the final scheme is shown in Figure 12. Six parts form this scheme, including beach protection of

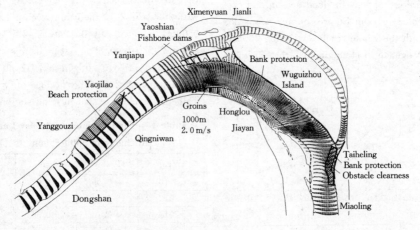

Figure 12　Flow field after accomplishment of the regulation scheme（$Q = 7000 \text{m}^3/\text{s}$）

Yanggouzi; fishbone dams in beach ahead of Wu-guizhou Island; bank protection in Wuguizhou Island's head, as well as it's right bank and tail; groins in side beach of Xinhekou; bank protection and obstacle clearness in Taiheling.

6.2 Velocity changes

Figure 12 gives the flow field after the scheme under discharge of 7000m³/s. It shows that, fishbone dams and groins will narrow the inlet of right branch of Wuguizhou Island, the flow will concentrate and the velocity will increase. Under design and regulation discharge, the velocity is increased by 0.20-1.00m/s in the regulation channel. Among which, there is 1.00-1.30m/s at front area of groin head, 0.05-0.15m/s at upper regime of right branch of Wu-guizhou Island. Velocity in groin field and downstream area is decreased by 0.20-1.30m/s.

6.3 Distribution of erosion and deposition and change of navigation depth

Flow and sediment process from 2005 to 2007 were employed to calculate erosion and deposition results after the regulation scheme. The annual dischar-ges from 2005 to 2007 were 12799m³/s, 8627m³/s and 11585m³/s, and the annual suspended load con-centrations were 0.225kg/m³, 0.118kg/m³ and 0.165kg/m³. The results showed that, navigation obstacle area (inlet of the right branch of Wuguizhou Island) is narrowed by the fishbone dams and groins, so water depth is enlarged. Erosion in regulation channel is mainly 2-5m after 3 years. The required navigation depth of 3.5m will be satisfied in 1 or 2 years. There is a river regime of concave bank in the inlet of right branch. But the Xinhekou groove still develops in low sediment concentration years.

7. Conclusions

Yaojian channel belongs to the lower Jingjiang reach, middle Yangtze River. It is a typical meander-ing and bifurcated channel, and it is one of the worst obstructed navigation channels. A 2D flow-sediment mathematical model of the Yaojian reach was devel-oped. Based on verification, its bed erosion and depo-sition in the following 20 years after the impoundment of TGP was calculated and analyzed. The results showed that, the Yaojian reach will be scoured, and the main stream will still keep in the right branch. After accomplishment of Yaojian reach's reg-ulation scheme, fishbone dams and groins will nar-row the inlet of right branch. the flow will concentrate and the velocity will increase, which is good for keeping navigation depth. The required navigation depth of 3.5m will be satisfied in 1 or 2 years. There is a river regime of concave bank in the inlet of right branch.

Acknowledgment

This work was financially supported by the Na-tional Natural Science Foundation of China (Grant No. 50879047, No. 50779037 and No. 50909065) and China's "Eleventh Five-Year" National Science and Technology Support Program (Grant No. 2008BAB29B08).

References

[1] Han Qiwei, Yang Kecheng. The tendency of river pattern variation in the lower Jingjiang River after completion of the Three Gorges Pro-ject. *Journal of Sediment Research*, 2003 (3): 1-11. (in Chinese)

[2] Huang Yue. Study on water and sediment chan-ges in Yaojian reach after impoundment of the TGP by 1D mathematical model. Wuhan China: Yangtze River Scientific Research Institute, 2007. (in Chinese)

[3] Li Xianzhong, Lu Yongjun, Liu Huaihan. In-fluences on waterway of Jingjiang River after impoundment of Three-gorges Junction and re-search on countermeasures. *Port & Waterway Engineering*, 2004, 367 (8): 55-59. (in Chinese)

[4] Lu Jinyou, Huang Yue, Gong Ping. Scouring and silting variation in middle and lower channel of the Yangtze river after TGP. *Yangtze River*, 2006, 37 (9): 55-57. (in Chinese)

[5] Lu Yongjun, Wang Zhaoyin and Zuo Liqin. 2D numerical simulation of flood and fluvial process in the meandering and island-braided middle Yangtze River. *International Journal of Sedi-ment Research*, 2005, 20 (4): 333-349.

[6] Lu Yongjun, Zuo Liqin, JI Rongyao, et al. 2D mathematical model for navigation regulation

projects of Yaojian reach in middle Yangtze River. Nanjing China: Nanjing Hydraulic Research Institute, 2008. (in Chinese)

[7] Peng Yuming, Gao Zhibin. Analysis of evolution of Jianli reach in Yangtze River. *Yangtze River*, 2006, 30 (12): 78-81. (in Chinese)

[8] Wu Wensheng, He Guagnshui. Study on evolution pattern of lower Jingjiang River. *Journal of Hunan Water Conservancy and Hydroelectric*, 2005 (6): 33-36. (in Chinese)

[9] Yu Wenchou. Preliminary study on forming condition of lower Jingjiang meandering channels of middle Yangtze River. *Journal of Yangtze River Scientific Research Institute*, 2006 (6): 9-13. (in Chinese)

Study on the Effect of Streambed Structures in Spillways of Landslide Dams

Huaixiang Liu[1]   Zhaoyin Wang[2]   Chengwei Xu[1]   Guoan Yu[2]

[1] *State Key Laboratory of Hydrology-Water Resources and Hydraulic engineering,*
Nanjing Hydraulic Research Institute, Nanjing, 210029
[2] *Department of Hydraulic and Hydropower Engineering, Tsinghua*
University, Beijing, 100084

Abstract: An 8.0 Ms earthquake occurred at Wenchuan (31°01′16″N, 103°22′01″E), China on May 12, 2008. The earthquake-hit area was mainly mountainous district lying on the margin of Tibetan plateau and therefore numerous disasters such as avalanches and landslides were triggered. These mass movements created many landslide dams and quake lakes along the earthquake belt. Some of them were regarded dangerous because massive amounts of the lake water might break the dam and potentially endangering the downstream people. Thus their stability is very important in the security control. In this paper, the influences of streambed structures on the landslide dams stability were studied through the field measurement work in 5/12 Wenchuan earthquake-hit area. The results indicated that streambed structures were developed with huge stones or clusters of big stones in many landslide dam spillways during water scour, stabilizing the dams due to their high resistance to erosion. And with the enhancement of streambed structure intensity Sp, erosion was impeded and local geomorphology features such as longitudinal profiles was altered. The preservation proportion of landslide dams are in direct ratio to streambed structure intensity. And coarser grain size of dams leaded to larger intensity. Therefore It was revealed that the development of streambed structures and finally the preservation proportion of dams could be estimated according to the grain size distribution, so that the stability of these dams could be assessed.

Key words: Wenchuan earthquake; landslide dams; streambed structures; structure intensity; preservation proportion

1. Introduction

Large scale landslides may dam rivers and form landslide dams or quake lakes. For instance, at Tianshan Mountain, which is shared by China, Kazakhstan, Uzbekistan, Kyrgyzstan, and Tajikistan, the Sarez earthquake triggered the Usoi rockslide with a volume of 2.2 billion m^3 and blocked the Murgab River with the highest landslide dam in the world (500-600m), forming the Sarez Lake, in 1911 (Gaziev, 1984). In 1996, a landslide dammed Halden Creek, British Columbia, Canada (Geertsema et al., 2006). A great earthquake measured at 8.0 Ms according to the China Seismological Bureau occurred at Wenchuan (31°01′16″N, 103°22′01″E) at 14:28 May 12, 2008 (Beijing time), 80 km northwest of Chengdu, the capital of Sichuan, with a depth of 19 km. According to a report by the state government of

China on July 31, 2008, the earthquake killed 68, 197 people, injured 374, 176 people and left 18, 222 missing. The Wenchuan earthquake triggered landslides created more than 100 quake lakes, 35 of which have been identified as dangerous.

The management of landslide dams and quake lakes is a challenge for river engineers. Naturally, quake lakes are preserved or flooded. In many cases the risk is high because 1) the dam break floods occur infrequently and local people usually do not expect them, 2) the warning time is very short, which limits the potential for evacuation, and 3) the peak discharge may be many times greater than any normal rainfall flood, potentially placing "normally safe" infrastructure and lives at risk (Becke et al., 2007). There have been many examples of large numbers of casualties caused by quake-lake break floods. On October 6, 1999, a large rock avalanche from Mount

Adams on the west coast of the South Island, New Zealand, fell into the Poerua Valley. The landslide dammed the Poerua River and created a large lake. The potential for overtopping and failure of the landslide dam presented a potential dam-break flood hazard that was assessed as posing a serious danger to Poerua Valley residents located downstream. The dam eventually failed 6 days after it formed (Becke et al. , 2007). In China, the Kangding earthquake in 1786 and the Diexi earthquake in 1933 also created huge landslide dams. And the death toll caused by dam failures was several times higher than the one directly caused by earthquakes.

From an emergency management perspective, it is important to ensure that quake lakes are safe enough and the risk of dam failure is reduced to a minimum. The most precarious of all the wenchuan quake lakes is the Tangjiashan quake lake, which located in extremely difficult terrain in Qingchuan County, accessible only by foot or air. The quake lake was formed by a huge landslide from Tangjiashan Mountain. The volume of the sliding body is about 20.37 million m^3 and the landslide dam is 612m long (across the river), 803 m wide (along the river), and 82-124m high. It consists mainly of quaternary deposit, eluvial soil and clastation rocks. The total storage capacity was about 316million m^3. There was a high risk of dam-break flooding when 200million m^3 of water had stored in the lake. An artificial spillway was first dug and then the water began releasing through the spillway. To help the water scour the spillway bed to a low level, large boulders were removed or exploded. The channel inlet bottom elevation was cut down from 740m to about 714m, and the channel bed was cut wider from about 10m to 100m. Then the lake water was reduced from 246 million to 86 million m^3. More than 160 million m^3 of lake water was drained out of the lake. Although the peak discharge rate was as high at 6420 m^3/s during the course of channel cutting and lake-water draining, no casualties or damage were caused by the draining flood.

On the other hand some quake lakes were still preserved even without human intervention. Generally a landslide-dammed lake that persists for more than 10 years (usually several decades) is assigned the "stable" status class. There are many landslide dams and quake lakes in New Zealand (Costa et al. , 1988; Perrin et al. , 1992). Statistics for an inventory of 232 landslide dams and quake lakes showed that only 37% of all landslide dams appear to have failed (Korup, 2004). This result indicates that the preservation of quake lakes is not only possible but may also become a major strategy for quake lake management. And the degree of dam preservation and dam stability under natural conditions should be studied for that.

The formation and failure of landslide dams have been studied by some researchers. It is believed that the longevity of the quake lakes are related with stiffness of the dam itself, the incision of the spillway scoured inside the dam and other factors. For example, some scholars argued that the most important reason should be the size of the initial dam, the proportion of huge stones and the geometrical characteristics of the river valley (Costa et al. , 1991). Nie (2004) divided the landslide dams into three main types: ephemeral, dangerous and stable. Yin (2008) gave a suggestion of how to evaluate the failure possibility of landslide dams according to his survey in wenchuan earthquake, that several rankings including "very high risk", "high risk", "medium risk", "low risk" were used. The parameters he put into consideration were: 1) the dam height; 2) the maximum capacity of the quake lakes; 3) the dam configurations. Most of the results were relatively ambiguous classification as shown above. Only several categories which reflect the differences in dam stability were derived from the statistical data. This paper tried to study the dam stability quantitatively from a new aspect, that is, the streambed structures development in the dam spillways. And the analysis were mainly based on the field work in 5/12 Wenchuan earthquake-hit area.

2. Streambed Structures

Streambed structures are certain bed arrangements which are formed by coarse sediment particles (Figure 1). They have been reported from a wide range of humid and arid environments, and analogous forms have even been observed in supraglacial streams (Chin, 2002). Generally they develop in river beds where severe scour exists and so that these structures are more stable under erosion (Chin et al. , 2007). Otherwise the structures can not sustain for a long time. Erosion and incision are more common in mountain streams compared with other rivers. Therefore streambed structures in these mountain streams are more obvious and larger in scale. In morphology the streambed structures can be divided into several categories (Wang, 2008): step-pool system, ribbing structure, star-studded boulders, bank stones, pebble clusters and so on.

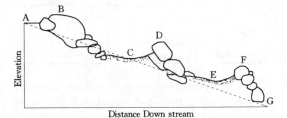

Figure 1　The longitudinal profile of a channel with streambed structures development

Streambed structures function in resisting channel incision and stabilizing the channel bed. The large boulders in them act as a framework tight interlocking the structure and resulting in considerable stability. The additional mechanical strength of the streambed structures is envisaged as deriving from three sources: 1) grain-to-grain contact involving intergranular friction; 2) particle interlock; and 3) shelter, especially of the particles in the wake tail. The structures reduce the lift and drag forces acting on the particles in the lee side of the structures (Reid, 1992).

Streambed structures also acts an important role in landslide dam stability. Many landslide dams consist of mainly rock debris and there are many big stones. Once the quake lake reaches its full capacity and the flow over the landslide dam deposits begins, erosion-resistant boulders overlap and construct streambed structures naturally (Figure 2). This process reduces the original landslide dam material to a smaller mass composed of boulders and cobbles, stabilizing the dam and protecting the top of the initial deposit from further erosion. Finally, the spillway becomes a narrow and steep reach cutting through the landslide dam but with streambed structures to consume the flow energy. In this way the dams can be preserved. However, if the spillway cannot be stabilized before the flood season, or the structures are not strong enough to resist a big flood, the landslide dam may fail at high flood and cause great flood disaster.

The most important hydraulic feature of streambed structures is the extremely high bed rough-

Figure 2　The streambed structures in spillways of a small landslide dam

ness, which maximizes the resistance and reduces the flow velocity. To represent the bed roughness of the streambed structures in a certain channel length, a parameter S_P called structure intensity is introduced, which may be used to describe the development degree of streambed structures. As shown in Eq. (1) and Figure 1, the parameter SP is defined as the ratio of the length of curve ABCDEFG to the length of straight line AG minus one, i.e.,

$$S_P = \frac{(\overset{\frown}{AB} + \overset{\frown}{BCD} + \overset{\frown}{DEF} + \overset{\frown}{FG})}{\overline{AG}} - 1 \qquad (1)$$

It is obvious that for a flat bed without any streambed structures, $S_P = 0$. And the larger S_P value represent a rougher bed and in turn a more stable spill way in the landslide dam.

3. Landslide dams survey and results

The quake-hit area of the great Wenchuan earthquake is about 440,000km². Numerous avalanches and landslides were triggered during the earth shaking. After the earthquake, investigations were conducted using Beijing No. 1 and IKONOS satellite images with identification scales of 32 and 1 m, respectively (Di, 2008). The earthquake caused in total 11,700 avalanches and landslides on the steep bank slopes of numerous streams in the quake-hit area. The total area of avalanches and landslides is 2260km², which changed this district from green vegetation cover into naked rock and soil. It is estimated that the average thickness of soil and rock mobilized by the avalanches and landslides is about 2-3m. Thus, the total mobilized sediment volume is about 4.5-6.8 billion m³. The direct reason of this earthquake was the faults moving on the margin of the Qinghai-Tibet plateau. Thus, after the earthquake more than 100 landslide dams (and quake lakes) were created along the faults stretching in the northeast-southwest direction. In this study, some typical landslide dams were taken as samples. The distribution of them on the NE-SW faults belt was relatively even so that their representatives could be guaranteed (Figure 3).

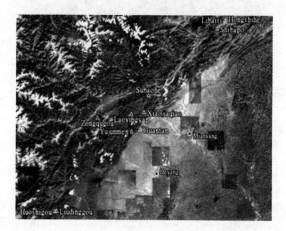

Figure 3   The landslide dams sampling
(Triangle dots)

3. 1   Structures development in the spillways

The appearance of landslide dams changed the local river geomorphology and streambed structures development to a great extent. The Xiaojiaqiao landslide dam (31°38′48″N, 104°16′41″E) was formed by the right bank landslide in the Chaping river, which locates in the Anxian County, Mianyang City. The dam length was about 370m along the river and 200m across the river, with its total volume reaching 2 million m³. After the water scour many big stones remained in the incised dam spillway, making the bed much coarser than the nearby natural river section without any influences of landslide dams. According to the statistic data in Table 1, the Xiaojiaqiao dam increased the proportion of large boulders and cobbles in the channel bed materials. Therefore the streambed structures development was enhanced and in turn larger S_P value was observed. Given the same discharge, larger structure intensity meant larger energy consumption and steeper channel slope S. That is, a higher gradient channel could be sustained due to structures development in the dam spillway.

Table 1   The influence of streambed structures development (Xiaojiaqiao)

Location	Slope S	S_P	D_{max} (m)	Size proportion in bed materials (%)		
				>1m	>0.5m	>0.1m
Dam spillway	0.022	0.22	3	5	15	45
No dam	0.006	0.13	0.5	0	0	30

The profiles of original dams and scoured spillways within them at several typical dam sites were plotted in Figure 4. It is revealed that most spillways showed convex shapes in their longitudinal profiles when relatively stable. The incision depth, which is defined as the depth between the original dam top height and the present spillway bed height, could be dozens of meters larger in downstream than in upstream. That is to say, the main scouring mode of the spillways was retrogressive erosion. The downstream channel incised earlier and deeper, so that the structures intensity S_P there was larger and could sustain steeper slope (given the same discharge). On the contrary, the upstream channel was scoured much slighter, leading to smaller structures intensity S_P and milder slope. If the upper quake lakes were also put into consideration, then in a large scale the whole spillway can be regarded as a steep slope compared with the almost flat lake section. The remaining dam will act as a knick point (a geomorphologic break point in slope) and will control the further evolution of surrounding channel sections. In that way the local landform and landscape will be altered greatly due to the development of streambed structures.

3. 2   Preservation proportion of landslide dams

The field survey and the results of Figure 4 both suggested that there were great differences between dams after erosion of water flows. Some landslide dams were almost cut through by the channel, such as the Xiaojiaqiao dam in Figure 4 (c). While some landslide dams were only very slightly destroyed, mainly in the top surface layer, such as the Laoyingyan dam in Figure 4 (b). That means the preservation proportion of dams were totally different. One of the most important reasons was the differences in streambed structures which developed within the dam spillways, and another reason being, of course, the

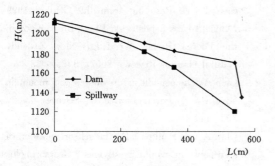

(a) Huoshigou

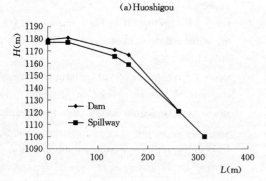

(b) Laoyingyan

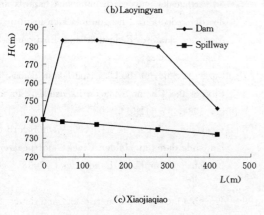

(c) Xiaojiaqiao

Figure 4   The profiles of spillways and original dams

tensity S_P was, the larger the dam preservation proportion would be. For the dams where super strong structures developed ($S_P > 0.45$), they could be almost totally preserved. So that their dam preservation proportions were approaching 1; on the contrary, for weak-structures dams ($S_P < 0.15$), few parts could still survive. That is, their dam preservation proportions were approaching 0.

The influences of discharge between different rivers can not be ignored and thus should be discussed. Even for the same structures intensity S_P, dams under the scour of larger discharge would be eroded severer, resulting in a smaller preservation proportion. However, we found that this effect would be weakened along with the increase of S_P, especially for those super strong structures. A river called wasihe has been examined before. It is rich in hydraulic power resources since the slope reaches 0.1 and the unit-width discharge reaches $3m^2/s$, the erosive power of which is larger than most mountain streams in our field survey and the structures of S_P 0.4-0.5 could still hold on under such circumstance. Therefore for most landslide dams, they can still be almost fully preserved if structures intensity in their spillways can exceed 0.4 or larger.

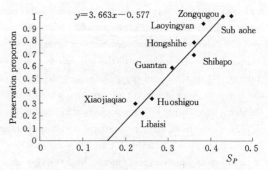

Figure 5   Dam preservation proportions and structures intensity in the spillways

river discharge. If strong structures could develop in spillways, then the erosion and incision will be impeded and in turn the original landslide dams could be preserved to the greatest extent. If not, then the particles of the loose landslide dams' bodies could be flushed away in a short time without the protection of streambed structures.

Figure 5 explained the relationship between the dam preservation proportions and the structures intensity S_P in their spillways. The preservation proportion here was exactly defined as the ratio of remaining dam height to the original dam height. An approximate direct ratio could be found in Figure 5. The larger the streambed structures in-

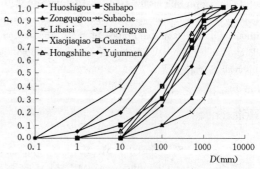

Figure 6   Size distributions of the landslide dams' bodies

The possible intensity of structures developed in the spillways was mainly controlled by the size distributions of the landslide dams. Figure 6 listed the size distributions of corresponding landslide dams in Figure 7, and the connection between the two figures can be easily recognized. The component ratio of coarse grain were evidently large in the Subaohe and Zongqugou landslide dams, thus their S_P value and preservation proportions were both larger. While the Libaisi, Xiaojiaqiao and Huoshigou landslide dams were mainly consist of fine particles, leading to very small S_P value and preservation proportions.

4. Conclusions

The development degree of streambed structures in the spillways has great effect on the stability of landslide dams. And the geomorphology of near-dam river section is also under the influences of these structures. According to the field survey results of landslide dams in the 5/12 Wenchuan earthquake-hit area, the following conclusions can be drawn:

(1) The proportion of coarse particles and the streambed structures intensity S_P in landslide dams were evidently higher than nearby reaches of the same river. Thus high gradient spillways could be sustained and knick points in longitudinal river profiles could be formed.

(2) The preservation proportions of landslide dams were in linear relationship with the streambed structures intensity S_P in their spillways. The larger the S_P was, the larger the dam preservation proportion would be. For the dams where super strong structures developed ($S_P >$ 0. 45), their preservation proportions were approaching 1; on the contrary, for weak-structures dams ($S_P <$ 0. 15), their preservation proportions were approaching 0.

(3) The possible intensity of structures developed in the spillways was mainly controlled by the size distributions of the landslide dams. Large S_P values could be obtained if the component ratios of huge stones in the dams were very high. While only fine particles would lead to small S_P values.

References

[1] Becker J S, Johnston D M, Paton D, Hancox G T, Davies T R, McSaveney M J and Manville V R. Response to Landslide Dam Failure Emergencies: Issues Resulting from the October 1999 Mount Adams Landslide and Dam-Break Flood in the Poerua River. Westland, New Zealand, Natural Hazards Review, 2007, 8 (2): 35-42.

[2] Chin A. The periodic nature of step-pool mountain streams, American Journal of Science, 2002, 302: 144-167.

[3] Chin A. and Phillips J D. The self-organization of step-pools in mountain streams. Geomorphology, 2007, 83: 346-358.

[4] Costa J E, Schuster R L. The formation and failure of natural dams. Geol Soc Amer Bull, 1988, 100: 1054-1068.

[5] Costa J E, Schuster R L. Documented historical landslide dams from around the world. U. S. Geological Survey Open-File Report, 1991, 91-239, 486.

[6] Di B F. Investigation of sediment erosion induced by the May 12 Wenchuan earthquake. Symposium on earthquake induced mountainous hazards and disaster reduction, Chengdu Research Institute of Mountain Hazard Reduction and Environment, Aug. 1-3, 2008.

[7] Gaziev E. Study of the Usoi landslide in Pamir, in Proc. of the 4th Int. Symp. on Landslides, Toronto, 1984, 1: 511-514.

[8] Geertsema M. Clague J J. 1,000-year record of landslide dams at Halden Creek, northeastern British Columbia. Landslides, 2006, 3: 217-227.

[9] Korup O. Geomorphometric characteristics of New Zealand landslide dams. Engineering Geology, 2004, 73: 13-35.

[10] Nie G Z, Gao J G, Deng Y. Preliminary study on earthquake-induced dammed lake. Quaternary Sciences, 2004, 24 (3): 293-301. (in Chinese)

[11] Perrin N D, Hancox G T. Landslide-dammed lakes in New Zealand preliminary studies on their distribution, causes and effects. In: Bell, D. H. (Ed.), Landslides. Glissements de Terrain. Proceedings of the Sixth International Symposium 10-14 February 1992, Christchurch. Balkema, Rotterdam, 1992: 1457-1466.

[12] Reid J B. The Owens River as a tiltmeter for

Long Valley Caldera, California. Journal of Geology, 1992, 100: 353-363.

[13] Wang Z Y, Melching C S, Duan X H, et al. Ecological and hydraulic studies of step-pool system. Journal of Hydraulic Engineering, ASCE, 2008, 130 (7): 792-800.

[14] Yin Y P. Researches on the geohazards triggered by wenchuan earthquake, sichuan. Journal of engineering geology, 2008, 16 (4): 433-444. (in Chinese)

Experimental Research of Microcystis Aeruginosa as Tracer Particle in Particle Image Velocimetry

Yang Li[1,2]   Xu Shikai[3]   Wu Shiqiang[3]   Lu Ming[1]   Wang Yong[3]

[1] *College of Environment, HoHai University, Nanjing, 210089*
[2] *College of Environment, Nanjing University of Technology, Nanjing, 210009*
[3] *State Key Laboratory of Hydrology-Water Resources and Hydraulic Engineering, Nanjing Hydraulic Research Institute, Nanjing, 210029*

Abstract: It was verified by experimental research of the feasibility of microcystis aeruginosa as tracer particle in water flow particle image velocimetry (PIV). It was concluded by the experiment that microcystis aeruginosa has the advantage of outstanding tracing and light scattering characteristics in water flow. Besides, owing to the characteristics, microcystis aeruginosa as tracer particle in water flow showed excellent tracing effect and the entire flow field images exported by PIV system displayed high definition. In addition, microcystis aeruginosa is more affordable and easier to obtain than the traditional solid tracer particles. It can be concluded that microcystis aeruginosa as tracer particle in PIV system is an applicable substitute for the traditional solid tracer particles.

Key words: particle image velocimetry; tracer particle; microcystis aeruginosa; experimental research

1. Introduction

The implementation of PIV technology is spreading the appropriate tracer particles in the flow field, obtaining the tracer particle images by using imaging system, then processing and analyzing the images, obtaining the distribution of flow field at last[1]. PIV technology not only has the adequate accuracy and resolving power of single point measurement technology, but also can obtain the overall configuration and the transient images of flow. PIV technology is an effective method for quantitative measurement of complex flow field. It can obtain the transient structure of some flow field which cannot observed by the other traditional measurement technology. Therefore, PIV technology has been developed and promoted deeply in recent years and become the most mature measurement technology of the entire flow field velocimetry.

As the movement of tracer particles in the flow field is used to reflect the movement condition of flow, the characteristics of tracer particles have a great influence on the final measurement results of the PIV system[2]. If the tracer particles have outstanding follow capability with flow, light scattering characteristics and uniform distribution, the movement condition of flow can be truly reflected by the movement of tracer particles. Otherwise, it will appear a distinct difference between the particle images and the actual flow.

Solid particles or air bubbles are generally used as tracer particles in water flow[3]. The characteristic of solid particles is generally undissolved in water, non-toxic, non-corrosive and chemically stable. According to the particularity of PIV measurement, the researchers and manufacturers in China and abroad have developed many kinds of solid tracer particles which have better tracing and quality, such as pollen, silver hollow glass ball, fluorescence particle, nylon particle and Al_3O_2 powder, etc. However, the prices of all of the tracer particles above are so high that they cannot be attainable easily. The hollow glass bead coating by silver is currently used as tracer particle in PIV water flow measurement. The density of the tracer particles is about $1g/cm^3$ which is close to the density of water. In addition, the tracer particle diameter is $5\text{-}20\mu m$. The foregoing outstanding characteristics result in the excellent tracing trait of hollow glass bead coating by silver in water flow. However, producing such fine, hollow glass beads with uniform density is a complex process and the tracer particle cannot be made in China at present. Therefore, the high price problem limits the application of tracer particle in measurement of water flow by PIV system. Especially when some common experiments need a large amount of tracer particles to spread into water dynamics flow field in PIV system and the tracer particles cannot be recycled, the expense on tracer particle is so high as not to be carried out successfully. Consequently, it is necessary to find low cost, easy to obtain and excellent tracing traits PIV tracer particles to substitute the traditional particles, thus can make PIV technology to be really popular in the study of water movement.

In this paper, it is introduced that the characteristics of microcystis aeruginosa and the comparison experiments have been conducted in water hydrodynamic flow between microcystis aeruginosa and hollow glass balls used as tracer particles. The results of the experiments verified the feasibility of microcystis aeruginosa as tracer particle. Besides, microcystis aeruginosa as tracer particle in water flow has not been reported in China.

2. Characteristics of microcystis aeruginosa

Microcystis aeruginosa, as a kind of single-celled phytoplankton cyanobacteria, is a kind of prokaryote. It is spherical with hollow interior (also known as air bags) configuration. Due to the configuration, it has good suspending trait and its density is close to the density of water. In the protoplasm of microcystis aeruginosa cell, there are many granules with obvious reflective capability. Its dimension is so small that a single cell is about 2-10 μm or so. In addition to meet the PIV system requirements of non-toxic, non-corrosion, wearable and clean, microcystis aeruginosa as tracer particles also has the excellent characteristics as follows:

(1) Because the dimension of microcystis aeruginosa is so appropriate that its density is close to water and it can flow with water fluently, the difference between microcystis aeruginosa tracer particles and actual flow is small and the fluid motion can be truly reflected.

(2) Due to its spherical and uniform particle size, the shape of microcystis aeruginosa with good light scattering capability. The light scattering trait will not change in all directions and it is helpful to imaging process of the PIV system.

(3) As a single-cell phytoplankton in the labora-

tory, when microcystis aeruginosa is applied as tracer particle in the measurement of the PIV system, it cannot result in the overlapping trouble in the imaging process and high quality simulation images can be obtained accordingly.

All in all, microcystis aeruginosa meets the basic requirements of tracer particles of PIV system. Besides, compared with traditional tracer particles, it is cheaper and easier to acquire than the other tracer particles because it can be enlarge-cultured in the laboratory during a short time.

3. Experiment and discussion

3.1　PIV system

The technical principle of PIV system is that after tracer particles are spread into water flow and flow field is illuminated by the pulse laser light, the images of the particles are recorded in the film or the CCD camera when exposal occur twice or more times. Then, PIV negatives or images recorded by CCD are point-by-point processed by using Young's method, autocorrelation method or cross-correlation method. Thus the simulation images of the flow velocity distribution can be obtained[1,4]. Essentially, it is the velocity of tracer particles obtained which can represent the actual flow velocity at the same location of the particles in flow field.

As the schematic drawing of PIV imaging shows (see Figure 1), PIV system usually consists of four major parts: tracer particle, optical lighting system, image recording system and image processing system[5].

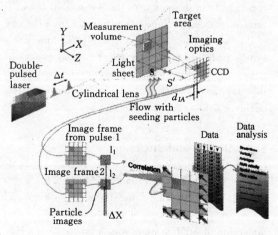

Figure 1　Schematic drawing of PIV system[6]

Tracer particle: display the details of flow, obtain the information of the flow field by means of measuring the movements of tracer particle.

Optical lighting system: provide light source, illuminating the flow field (generally laser light is used).

Image recording system: record the particle images.

Image processing system: obtain the velocity data via image processing and velocity extraction algorithm.

In this experiments, the PIV system is produced by TSI corporation and the model of PIV is ULTRA-II.

3.2　Experimental setup

In order to verify the measurement accuracy of flow field using PIV system of microcystis aeruginosa tracer particles, the comparison experiments have been conducted in water hydrodynamic flow between microcystis aeruginosa and hollow glass ball used as tracer particles. The experiments must be conducted under three conditions according to various velocities respectively. Then, the simulation images of the flow velocity distribution obtained by experiments can be analyzed and draw a comparison of the difference between microcystis aeruginosa and hollow glass ball as tracer particles.

The experiments were carried out in an open flume system, which is a water-circulating system. After the tracer particles mentioned above are spread into the water at the middle of the flume, the circulation pump was launched thus the water was flow at a certain velocity in the flume. The light sheets emit vertically from the bottom of the flume while the CCD camera recorded the images horizontally from the side of the flume. The recording mode of CCD camera was synchronous and continuous and the finally image obtained should be averaged out over many images. In this way, a velocity field of a water interface in water-circulating flow could be acquired by using of a PIV system.

3.3　Results and discussion

In the experiments, the water-circulating flume is long enough to allow uniform flow. The width and the water depth of the flume are 25cm and 19.50cm, respectively. The frequency of control apparatus of PIV system was adjusted to1.4Hz, 2.1Hz, and 3.2Hz and the corresponding flow ve-

locity in the flume was 0.07m/s, 0.14m/s and 0.24m/s at respective frequency. Through the PIV measurement system, the velocity data could be acquired. The velocity data is acquired via image analysis, velocity extraction algorithm process and display software of PIV system. The results are shown from Figure 2 to Figure 7. In the figures as following, X-coordinate represents the width direction of the interface and Z-axis represents the depth direction of the interface.

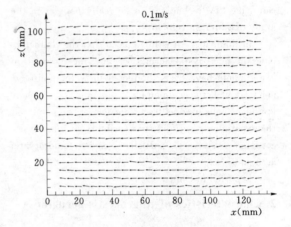

Figure 2　The velocity flow of microcystis aeruginosa at 0.07m/s

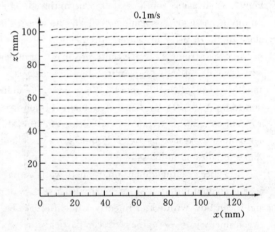

Figure 3　The velocity flow of hollow glass ball at 0.07m/s

The difference of the measurements between microcystis aeruginosa and hollow glass balls is small. It was verified microcystis aeruginosa as tracer particle can meet the requirements of measurement precision of PIV system.

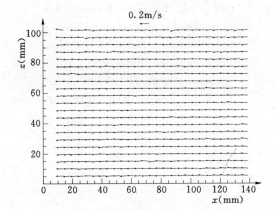

Figure 4　The velocity flow of microcystis
aeruginosa at 0. 14m/s

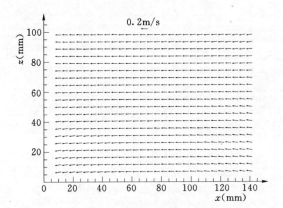

Figure 5　The velocity flow of hollow
glass ball at 0. 14m/s

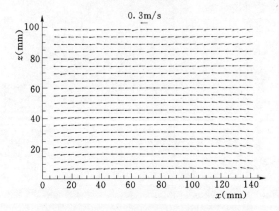

Figure 6　The velocity flow of microcystis
aeruginosa at 0. 24m/s

4. Conclusions

It was verified the feasibility of microcystis

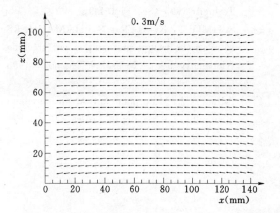

Figure 7　The velocity flow of hollow
glass ball at 0. 24m/s

aeruginosa as tracer particle in water flow PIV sytem via experimental study. The experimental results demonstrated that as follows:

(1) Microcystis aeruginosa as tracer particle in the experiments showed outstanding tracing and light scattering characteristics in water flow and thereby the high-definition images can be obtained after data processing.

(2) When measuring the velocity of water flow, it is only a small number of microcystis aeruginosa as tracer particles to be needed to add to the flow. The distribution and concentration of a small dosage of microcystis aeruginosa can meet the imaging requirements of PIV system.

Consequently, it was verified by experiments that microcystis aeruginosa as tracer particle in PIV system is applicable and economical.

Acknowledgement

This work was supported by the Nanjing Hydraulic Research Institute initiative scitific research program (Grant No. 200801001) and Nanjing University of Technology scitific research program (Grant No. 39727007).

References

[1]　Adrian R J. Particle-imaging techniques for experimental fluid mechanics [J]. Annu. Rev. Fluid Mech. 1991 (23): 261-304.

[2]　MELLING A. Tracer particles and seeding for particle image velocimetry [J]. Meas. Sci.

Technol, 1997 (8): 1406-1416.

[3] SCHMIT T, KOSTER J N, HAMACHERS H. Particle design for displacement tracking velocimetry [J]. Meas. Sci. Technol, 1995 (6): 682-689.

[4] Adrian R J. Multi-point optical measurement of simultaneous vectors in unsteady flow-a Re-view [J]. Int. J. Heat&Fluid Flow. 1986 (7): 127-145.

[5] Tang Hongwu. Modern Flow Measurement Technology and Application [M]. Beijing: Science Press, 2009: 202-203.

[6] Extracted from Marktec Technology Limited (http: //www. marktec-technology. com).

Experimental Research on Structure Optimization of Energy Dissipation of Myitsone Hydropower Station

Li Jing   Jiang Bole

Changjiang River Scientific Research Institute of Changjiang Water Resources Commission, Wuhan, 430010

Abstract: On weir surface and middle orifice local physical model with geometry scale of 1 : 60, energy dissipation capacity, velocity and flow pattern, pressure distribution and related hydraulics problem of Myitsone hydropower station discharging structure layout were tested. Basic practicable discharging structure pattern and stilling pool pattern were put forward, though comparison and optimization of more than ten schemes. The study result shows that the surface orifice adapt proper pattern of flaring gate pier was favorable to increase the energy dissipation efficiency and shorten the length of stilling pool, but had no effect on the phenomenon of water wing hitting both sides of sidewall caused by flaring gate pier, the schemes of canceling flaring gate pier of surface orifice and adapting the declined pattern stilling pool could decrease the stilling pool base slab velocity of under flow energy dissipation efficiently, at the same time diminish the phenomenon of water wing hitting sidewall of both sides caused by flaring gate pier and obtain certain satisfying energy dissipation effect. The more than ten energy dissipation research schemes in this paper have reference value for all similar engineering project design and science research.

Key words: myitsone; surface orifice; under flow; deflecting flow; energy dissipation; stilling pool

1. Introduction

Burmese Myitsone hydropower station locates on main stream of Ayeyarwady River. The main development task of the hydropower station is electricity generating, to some extent, playing the role of flood control for Myitkyina City on downstream. After the construction of this engineering, it has the synthesis utilization effect of increasing the dry period flow of downstream; creating condition to develop navigation of reservoir area, additionally.

Myitsone hydropower station has the characteristics of large discharge, many estuaries and so on. Moreover, its geological condition is complicated, and it is subtropical marine monsoon climate area. According to the plan, the installed capacity of this hydropower station is 6000MW, the peak flood discharge is 65700m³/s, the check water level is 248.37m, the normal water level is 245m, and the maximum dam height is 130.5m. The weir structure layout for three levels, the top level has 11 overflow surface orifices, the dimension of the orifice is 14m×

23m, the outlet takes the pattern of flaring gate pier, the flow width of surface contracts from 14m to 5.6m; the contractive ratio of the orifice is 0.4. There are 10 mid orifices on the middle level, the mid orifice is pressured short pipe, the top and side of the inlet section takes ellipse curve, the elevation of floor plate is 165m, the dimension of the orifice is 5m×8m (width× height). There are 11 guide flow bottom orifices; the floor plate with the elevation of 134m, the dimension of orifice is 9m×14.5m. The elevation of stilling pool floor plate is 127m, the length of the pool was 150m, and the width of the pool is 264m. The end of the stilling pool is connected to concrete apron with the length of 45m. The open channel is connect to apron by 1 : 5 slop and layout horizontal, the elevation of open channel is 130m, and the length is 1800m. The surface orifice and bottom orifice take under flow energy dissipation pattern, the mid orifice takes the energy dissipation pattern of deflecting flow. The layout of the hydro project is shown in Figure 1, the profile of mid orifice is shown in Figure 2, the profile of surface orifice and bottom orifice is shown in Figure 3.

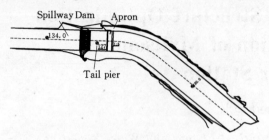

Figure 1   Flood discharge structure layout plan

In permanent operating period, under operating condition of surface orifice and mid orifice combined discharge, the flow pattern in energy dissipation area was complicated. Therefore, more than ten schemes experiments were taken out on local hydraulic physical model of surface orifice and mid orifice (the geometrical scale was 60). Analysis of hydraulic characteristics of flow pattern, dynamic water pressure and so on, research of dimension of stilling pool, dynamic load of stilling pool floor plate and the pressure characteristic in discharging orifice, the optimization of the energy dissipation pattern and the structure pattern of the discharge orifice were carried out through the experiments. The experimental operating condition was shown in Table 1.

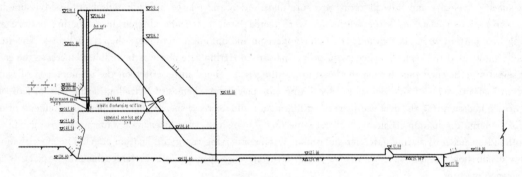

Figure 2   Mid orifice profile

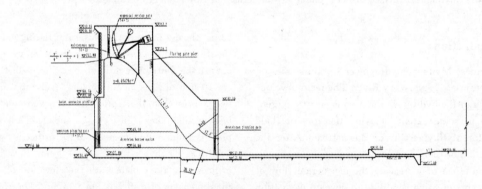

Figure 3   Surface and bottom orifice profile

Table 1   **Experimental Conditions**

Operating condition	1	2	5	6
Frequency	$P=0.01\%$	$P=0.1\%$	$P=2\%$	Mid orifice and power plant combined discharge
Peak flood discharge (m³/s)	65700	51000	—	—
Maximum water level in front of the dam (m)	248.37	245	—	—
Surface orifice discharge (m³/s)	43190	34663	34670	—
Mid orifice discharge (m³/s)	13840	13524	—	13530

Operating condition	1	2	5	6
Power plant discharge (m³/s)	0	2800	4480	4480
The total discharge of the hub (m³/s)	57030	51000	39150	18010
Corresponded downstream water level (m)	164. 73	162. 57	158. 13	149. 23

2. Experiment of original scheme

(1) The outlet took the structure pattern of flaring gate pier, the flaring gate pier made the flow width of surface orifice contracting from 14m to 5. 6m, the contractive ratio of the flaring gate pier was 0. 4, the contract angle was 16. 7°, the contracted section was 14m long.

Under operating condition 1, the starting point of water wing of flaring gate pier was near Pile No. 0 +48m, with the elevation of 220m, the peak point of water wing located Pile No. 0+60m, with the elevation of 225-228m, the drop down point of water wing was between Pile No. 0 + 125m and 0 + 132m. The water wing intermittent swung left and right, and hit the top of both sides of the sidewall (near Pile No. 0 + 115m), the spray splashed around. The water wing started contacting with the sidewall at Pile No. 0+63m, elevation 224-228m. In stilling pool the hydraulic jump head located at Pile NO. 0 + 84m-0 + 93m, was submerged hydraulic jump, the hydraulic jump end did not exceed the stilling pool. The mid orifice reflecting flow reflected at bottom platform of 150m, and then reflected to the stilling pool. The entry point of outer edge of water tongue located at 0+124m-0+132m. Under operating condition 2, the starting point of water wing of flaring gate pier was near Pile No. 0+51m, with the elevation of 214m, the peak point of water wing located Pile No. 0+60m, the with the elevation of 217-222m. The water wing intermittent swung left and right, and hit the top of both sides of the sidewall, the spray splashed around. The water wing started contacting with the sidewall at Pile No. 0+72m, elevation 210-216m. The drop down point of water wing located near Pile No. 0 + 140m-0 + 144m. In stilling pool the hydraulic jump head located Pile No. 0 + 85m-0 + 96m, was submerged hydraulic jump, the

hydraulic jump end did not exceed the stilling pool. The reflecting flow of mid orifice did not reflect out successfully, the flow reflected at bottom 150m platform, then reflected to the stilling pool. The entry point of outer edge of water tongue located at (0+126) m.

At Pile No. 0 + 264m in front of the tail-pier, and Pile No. 0+276m at back of the tail-pier, velocity measuring points were set. Under operating condition 1, when surface orifice and mid orifice combined discharge flow 57030m³/s, the maximum bottom velocity in front of the tail-pier was 6. 42m/s, the maximum bottom velocity at back of the tail-pier was 2. 30m/s.

(2) According to the flow pattern of the above conditions, the water wing was not stable, and swung left and right hit the top of the both sides of the sidewall, the flaring gate pier was cancelled to verify the energy dissipation effect without flaring gate pier.

Under each operating condition, upstream flow was stable and fluent; there was no particular flow pattern. Under operating condition 1, in stilling pool, hydraulic jump head located at Pile No. 0+174m-0+ 184m, was far drive hydraulic jump, the hydraulic jump end exceeded stilling pool. The mid orifice reflecting flow reflected at the 150m platform then reflected into the stilling pool, the peak point of water tongue located at Pile No. 0 + 136m-0 + 138m, with the elevation of 159-164m, the thickness of the water tongue approximately 21. 6m, the width of the water tongue was about 24m. The entry point of outer edge of water tongue located at Pile No. 0 + 180m-0 + 184m in stilling pool, the entry point of inner edge of water tongue located at Pile No. 0 + 156m. Because flow of the adjacent two surface orifices proliferated and intersected at the end of the pier, and at Pile No. 0+132m, began generating rather strong water wing, water wing was discrete form intensive water, the peak point of water reflected by water wing located at Pile No. 0+151m, with the elevation of 141m.

The velocity measuring points were set at Pile No. 0+137.4m, 0+192m, 0+264m, 0+276m, 0+324m section. From the test result, it showed that the velocity of the surface orifice flow entering the stilling pool was all about 40m/s, the velocity distribution before the tail-pier basically got the maximum value at bottom of the section; the velocity distribution at back of the tail-pier got the maximum value at middle of the section. The velocity distribution of section at Pile No. 0+324m had the form of the surface having large velocity value; the bottom having small velocity value. Under operating condition1, the maximum bottom velocity was 4.45m/s.

The test result of this scheme showed that the water flow can not reflect out of mid orifice, additionally, without the flaring gate pier, if the stilling pool pattern did not reformed, the hydraulic jump in the stilling pool was far drive hydraulic jump, the bottom velocity in stilling pool was large, the length of the stilling pool was far from enough.

3. Research on flaring gate pier

(1) To the unstable and intermittent swinging left and right water wing at the back of flaring gate pier, however if the flaring gate pier was cancelled, the energy dissipation effect was not satisfying in original scheme, in Scheme 1, compared and refer to the flaring gate pier parameter of other similar hydraulic project, the contractive ratio of flaring gate pier was modified to 0.6, the contractive angle was modified to 11.31°, the contract section length was modified to 14m. Meanwhile, took into account of the original scheme that the flow of mid orifice can not deflect out completely, the mid orifice was extended to the direction of downstream of 25m, with deflecting angle of 15°, radius of 40m. In this scheme, the energy dissipation was not satisfying, and the flaring gate pier had no obvious energy effect.

(2) In Scheme 2, the contractive ratio of flaring gate pier in the above scheme was modified to 0.4, the contractive angle was modified to 11.87°, the contractive section length was modified to 20m, and the reflecting angle was modified to 0°. To the upstream entry flow was not so fluent, perfected the strut of frusta more round. In Scheme 3, the flaring

gate pier of Scheme 2 was shifted to downstream of 17m. In Scheme 4, the flaring gate pier in Scheme 3 was modified to bottom hatched pattern flaring gate pier, the impact pressure intensity on stilling pool floor plate could be reduced maximal nearly 30% compared to normal flaring gate pier.

(3) In Scheme 5, the flaring gate pier was modified to X-shape flaring gate pier, with the contractive ratio of 0.4, contractive angle of 16.7°, and contractive section length of 14m.

With the X-shape flaring gate pier, a part of the water tongue cling to the dam surface and extend breadth wise, another part withdraw in vertical direction through narrow slot of the flaring gate pier, moreover, the bottom extend breadth wise flow support partly water mat effect for the longitudinal water tongue. In this scheme, the start upstream water level that water flow can reflect from mid orifice was 220m, and the reflecting water tongue drop down in stilling pool.

Under operating condition 1, the total water tongue cross section displays "工" shape, the bottom of water tongue extend breadth wise along the dam surface, the middle of the water tongue forms withdraw water tongue in vertical direction through the narrow slot of flaring gate pier, on top of the flaring gate pier, the water tongue extended breadth wise and reflected across the flaring gate pier. At back of the flaring gate pier the happened point of water wing that generated by the conjunction water of impact wave located at Pile No. 0+72m, with the elevation of 207m. In most time, the water wing located in the middle of both sidewalls, the crest of the water sometimes hit the top of right sidewall at the range of Pile No. 0+112m-0+115m, the peak point of the water wing located at Pile No. 0+84m, with the elevation of 204m. In stilling pool, the hydraulic jump head located at Pile NO. 0+108m-0+120m, was critical hydraulic jump. The reflecting flow from the mid orifice dropped down in the hydraulic jump area in stilling pool, and aerated, tumbling violently, the water surface fluctuates up and down, and the fluctuation of water surface on the tail-pier was approximately 4m. The water flow began comparatively fluent at the back of Pile No. 0+444m. Under operating condition 2, the water wing mainly located in the middle of both of sidewalls, sometimes swung left and right hit the top of the sidewall. In stilling pool, the hydraulic jump head located Pile No. 0＋96m-0＋108m, was submerged hydraulic jump, the hydraulic jump tail did not exceed the stilling pool. The mainstream whirlpool still existed until in front of the tail-pier, the wave on the tail-pier was comparatively large, the wave height was about 3m. The reflecting flow from the mid orifice drop down in the hydraulic jump area, the flow began fluent after Pile No. 0＋384m.

Under suitable flaring gate pier pattern and contractive ratio, the energy dissipation effect was comparatively good, but the phenomenon that water wing hitting both sidewalls can not eliminate completely.

4. Research on schemes without flaring gate pier

(1) In Scheme 6, the flaring gate pier was cancelled, the mid orifice still took the extension of 10m, reflecting angle of 15°, radius of 40m.

Under operating condition 1, the hydraulic jump head in stilling pool located at Pile No. 0＋156m-0＋180m, was far drive hydraulic jump. The entry point of reflecting flow from the mid orifice located at Pile No. 0＋192m. The peak point of the outer edge of water tongue located at Pile No. 0＋96m, with the elevation of 176m. In stilling pool the flow surged and rolling, the wave on tail-pier was very large, the wave height was about 6.5m. At the back of the tail-pier, the secondary hydraulic jump generated at Pile No. 0＋348m-0＋360m. Under operating condition 5, the hydraulic jump head in stilling pool located at Pile No. 0＋180m-0＋192m was far drive hydraulic jump, the hydraulic jump tail exceeded stilling pool. Water wing was generated at the end of the frusta of surface orifice. The starting point of water wing located at Pile No. 0＋132m, the peak point of water wing located at Pile No. 0＋180m, with the elevation of 144m. The flow above the tail-pier rolled, the wave height was about 6m.

Under operating condition 6, the width of the water tongue was about 18m, after the flow entered, hydraulic jump was generated, the hydraulic jump head in stilling pool located at Pile No. 0＋186m-0＋190m, the hydraulic jump tail did not exceed stilling pool. There were small bubbles on water flow surface above the tail-pier; the wave height was about 2.5m. The flow began comparatively fluent near Pile No. 0＋366m.

(2) Because the energy dissipation effect in scheme 6 was not satisfying, far drive hydraulic jump happened; optimized measures were considered in Scheme 7, the length of stilling pool was extended of 50m, at the same time the height of tail-pier was raised to 140m. In scheme 8, the length of stilling pool was extended to downstream of additional 30m, at the same time the height of tail-pier was reduced of 3m to 137m.

The experimental results indicated that there were little difference in these two schemes and Scheme 6. Without flaring gate pier, the phenomenon of water wing hitting both sidewalls was eliminated, but the entry velocity of surface orifice flow exceeded 40m/s. Only relied on increasing pool length and tail-pier height, it was difficult to solve the problem of high stilling pool entry velocity and the drop water generated easily behind the tail-pier.

(3) Aimed at eliminating the above phenomenon, in scheme 9, the stilling pool was dug down of 9m, and was extended of 75m based on original scheme, the stilling pool length was 225m, and the elevation of tail-pier was 134m.

Under operating condition 1, the hydraulic jump head in stilling pool located at Pile No. 0＋90m-0＋96m, was submerged hydraulic jump, the hydraulic jump tail did not exceed stilling pool. Clockwise whirlpool was generated under the falling-sill that at back of surface orifice, the flow pattern was complicated. The entry point of outer edge of mid orifice reflecting flow located at Pile No. 0＋162m-0＋168m, the peak point located at Pile No. 0＋96m, with the elevation of 176m. The flow at back of tail-pier was somewhat fluent. Under operating condition 2, the hydraulic jump head in the stilling pool located at Pile No. 0＋85m-0＋96m, was submerged hydraulic jump. Clockwise whirlpool was generated under the falling-sill that at back of surface orifice, the flow pattern was comparatively complicated. The entry point of outer edge of mid orifice reflecting water tongue located at Pile No. 0＋162m-0＋168m, the entry point of inner edge located at Pile No. 0＋150m-0＋156m. There were bubbles assembled on the water flow surface in front of the tail-pier, the flow began fluent at Pile No. 0＋420m. Under operating condition 5, the hydraulic jump head in stilling pool located at Pile No. 0＋84m-0＋98m, was submerged hydraulic jump. There was a little amount of hydraulic jump water rushed into 150m platform of mid orifice intermittently. In front of tail-pier there were a bit of bubbles on the water surface, at back of the tail-pier, the water flow were fluent. Under operating condition 6, the entry point of outer edge of mid orifice reflecting water tongue located at Pile No. 0 + 186m-0 + 190m, the entry point of inner edge located at Pile No. 0+165m-0+168m, in front of the tail-pier, the flow was somewhat fluent, at back of the tail-pier, there were only small amount of bubble on water surface.

(4) In Scheme 10, considering the demand of stilling pool maintenance in the future, the tail-pier was raised to elevation of 144m.

Under operating condition 1, hydraulic jump head in stilling pool located at Pile No. 0 + 80m-0 + 86m, was submerged hydraulic jump, the hydraulic jump tail did not exceed stilling pool. The entry point of the outer edge of mid orifice reflecting flow water tongue located at Pile No. 0 + 156m, the peak point located at Pile No. 0+96m, the entry point of inner edge of water tongue located at Pile No. 0+137m-0+144m. The water level that leaving the stilling pool had large difference in height with the water level of downstream, at the exit of stilling pool, comparatively large water drop was generated, and at Pile No. 0+396m downstream of the exit of stilling pool, undulant form hydraulic jump was generated. In downstream watercourse somewhat large surface wave was generated. Under operating condition 2, the hydraulic jump head in stilling pool located at Pile No. 0 + 81m-0 + 86m, was submerged hydraulic jump, the entry point of outer edge of mid orifice reflecting water tongue located at Pile No. 0+158m-0+162m, the entry point of inner edge of water tongue located at Pile No. 0+144m-0+132m. At back of tail-pier, the secondary hydraulic jump was generated by drop water at Pile No. 0+396m-0+406m. Under operating condition 5, the hydraulic jump head in stilling pool located at Pile No. 0 + 80m-0 + 90m, was submerged hydraulic jump. In front of the tail-pier, bubbles assembled on water surface, at back of the tail-pier, the secondary hydraulic jump was generated by drop water at Pile No. 0 + 390m-0 + 396m. Under operating condition 6, the entry point of outer edge of mid orifice reflecting flow located at Pile No. 0 + 178m-0+180m, the entry point of inner edge located

at Pile No. 0 + 156m-0 + 166m. At back of the tail-pier, the secondary hydraulic jump was generated by drop water at Pile No. 0+366m-0+372m.

(5) Considering of the structure safety of constructions, the stilling pool floor plate was raised to elevation of 122m, the elevation of tail-pier was raised to 144m.

Under operating condition 1, the hydraulic jump head in stilling pool located at Pile No. 0 + 80m-0 + 86m, was submerged hydraulic jump, clockwise whirlpool was generated under the falling-sill at back of the surface orifice. The entry point of mid orifice reflecting flow located at Pile No. 0+156m, the peak point located at Pile No. 0+96m, with the elevation of 176m. The entry point of inner edge of water tongue located at Pile No. 0 + 137m-0 + 144m. The secondary hydraulic jump was generated at Pile No. 0 + 396m, at back of the tail-pier in stilling pool. Under operating condition 2, the hydraulic jump head in stilling pool located at Pile No. 0+81m-0+86m, was submerged hydraulic jump, clockwise whirlpool was generated under the falling-sill at back of the surface orifice. The entry point of mid orifice deflecting flow located at Pile No. 0 + 81m-0 + 86m, was submerged hydraulic jump, clockwise whirlpool was generated under the falling-sill at back of the surface orifice, the entry point of mid orifice reflecting located at Pile No. 0 + 158m-0 + 162m, the entry point of inner edge of water tongue located at Pile No. 0+144m-0+132m. The secondary hydraulic jump was generated at Pile No. 0 + 396m-0 + 406m at back of the tail-pier in stilling pool. Under operating condition 5, the hydraulic jump head in stilling pool located Pile No. 0 + 80m-0 + 90m, was submerged hydraulic jump, clockwise whirlpool was generated under the falling-sill at back of the surface orifice. Bubbles were assembled on the flow surface in front of the tail-pier. The secondary hydraulic jump was generated at Pile No. 0 + 390m-0 + 396m at back of the tail-pier in stilling pool. Under operating condition 6, the entry point of outer edge of mid orifice reflecting flow water tongue located at Pile No. 0 + 178m-0+180m, the entry point on inner edge located at Pile No. 0+156m-0+166m. The secondary hydraulic jump was generated at Pile No. 0+366m-0+372m at back of the tail-pier in stilling pool.

(6) In Scheme 12, aimed at eliminating the problem of drop water at back of the tail-pier and cre-

ating secondary hydraulic jump, the elevation of tail-pier was reduced to 134m.

Under operating condition 1, the hydraulic jump head in stilling pool located at Pile No. 0+86m-0+94m, was submerged hydraulic jump. The entry point of outer edge of mid orifice reflecting flow located at Pile No. 0+168m, the peak point located at Pile No. 0+96m, with the elevation of 176m, the entry point of inner edge of water tongue located at Pile No. 0+156m. At the place where the mid orifice flow entering the stilling pool, the spray splashed, near Pile No. 0+168m-0+216m, the water drop intermittent splashed and exceeded elevation of 193m. Near Pile No. 0+406m back of tail-pier, sometimes weak undulant hydraulic jump was generated. Under operating condition 2, the hydraulic jump head in stilling pool located at Pile No. 0+92m-0+98m, was submerged hydraulic jump, the entry point of outer edge of mid orifice reflecting water tongue located at Pile No. 0+168m, the inner edge of water tongue located at Pile No. 0+156m. At the place where the mid orifice flow entered the stilling pool, the water drop intermittent splashed and did not exceed elevation of 193m. The bubble generated by rolled upward water, floated downstream with flow. Under operating condition 5, the hydraulic jump head in stilling pool located at Pile No. 0+90m-0+104m, was submerged hydraulic jump, near Pile No. 0+144m, there were water drop intermittent splashed on water surface. At Pile No. 0+204m of upstream, there was circumfluence on water surface, on downstream of the Pile point there was fluent flow on water surface. Under operating condition 6, the entry point of outer edge of mid orifice reflecting water tongue located at Pile No. 0+186m-0+189m, the entry point of inner edge located at Pile No. 0+162m. At back of Pile No. 0+366m, the flow was fluent, there was small amount of bubbles on water surface.

The test results of Scheme 9 to Scheme 12 indicated that, the energy dissipation effect of digging down stilling pool were all satisfying, the entry velocity of stilling pool near the bottom were about 20m/s. In the range that the design of structure could accept, and at the same elevation of stilling pool floor plate, the higher the elevation of tail-pier was, the nearer the hydraulic jump head from the upstream. The problem of the generation of secondary hydraulic jump at back of tail-pier could be released by further optimizing the elevation of tail-pier combined with structure design.

5. Conclusions

(1) The water wing generated at back of flaring gate pier all hit both sidewalls in different degree. For normal flaring gate pier, bottom hatched pattern flaring gate pier, X shape flaring gate pier, as well as the same pattern of flaring gate pier with different contractive ratio, the flow pattern at back of flaring gate pier all had different characteristics, among them, the X-shape flaring gate pier had the best flow pattern.

(2) Without flaring gate pier, the energy dissipation effect in stilling pool was rather poor, the entry velocity of water flow exceeded 40m/s. If combined with the engineering task amount of construction and the structure safety of hydraulic structures, increasing the thickness of stilling pool properly, at the same time took the optimizing measures of extending the length of stilling pool, raising the height of tail-pier, the energy dissipation effect could be improved, the entry velocity could be reduced to 20m/s. To be considered synthetically, Scheme 11 could be taken as the recommendable scheme of this project.

Acknowledgment

The authors would like to thank Changjiang River Scientific Research Institute of Changjiang River Water Resources Commission and Changjiang Institute of Survey, Planning, Design and Research of Changjiang Water Resources Commission for the assistance and collaboration which was essential in making this study possible. Also, the authors would like to thank the two anonymous reviewers for their comments and suggestions, thus improving this paper.

References

[1] Yin Jinbu, Liang Zongxiang, Gong Honglin. Experimental on application & development of X type flaring gate piers. Journal of hydroelectric engineering. 2007, 26 (4): 35-40.

[2] Xu Zhengfan. Hydrodynamics [M] . Senior

Education Press, 1986.

[3] Li Jing, Jiang Bole. Surface and mid orifice discharge local hydraulic physical modal experimental report of Myitsone Hydropower Station of Hydropower Project on the Upstream of Ayeyarwady River [R]. Changjiang River Scientific Research Institute of Changjiang Water Resources Commission, 2009.

[4] Liang Zongxiang, Yin Jinbu. Study On the method for style designing Of X type flaring gate piers. Journal of Northwest A & F University (Natural Science Edition). 2008, 36 (5): 206-210.

[5] Yu Zidan. Application and understanding of flaring gate pier joint energy dissipater with low-middle water head. Advances in science and technology of water resources. 2009, 29 (2): 36-39.

[6] Li Zhongshu, Pan Yanhua, Han Lianchao. Hydraulics of stilling basin with flaring gate piers. Advances in water resources. 2000, 11 (1): 82-87.

[7] Mo Zhengyu, Wu Chao, Lu Hong. Relationship of location of flaring gate piers, froude number at section of starting flare and weir head. Journal of Sichuan University (engineering science edition) . 2007, 39 (4): 26-30.

[8] Nan Xiaohong, Liang Zongxiang, Liu Hansheng. New type of flaring pier for improving energy dissipation of stepped surface overflow dam. Journal of hydraulic engineering. 2003 (8): 49-57.

[9] Zhang Ting, Wu Chao, Mo Zhengyu. Comparison on hydraulic characteristics of X-shape and Y-shape flaring gate piers. Journal of hydraulic engineering. 2007, 38 (10): 1207-1213.

[10] Sun Shuangke, Liu Haitao, Xia Qingfu. Study on stilling basin with step-down floor for energy dissipation of of hydraulic jump in high dams. Journal of hydraulic engineering. 2005, 36 (10): 1188-1193.

[11] Wang Haijun, Zhao Wei, Yang Hongxuan. Experimental research on hydraulic characteristics of underflow for energy dissipator with step-down floor. Water resources and hydropower engineering, 2007, 38 (10): 39-41.

[12] Wang Haijun, Zhang Qiang, Tang Tao. Energy dissipation mechanism and hydraulic calculation of falling-sill bottom-flow energy dissipator. Water resources and hydropower engineering. 2008, 39 (4): 46-52.

Fluctuant Characteristics of Stilling Basin Floor of YinPan Hydropower Station

Li Jing   Jiang Bole

ChangJiang River Scientific Research Institute of ChangJiang Water Resources Commission, Wuhan, 430010

Abstract: Stilling basin floor is light structure, its structural vibration, impact, stability and so on, directly related to the flow fluctuation intensity of hydraulic jump area of the floor. Using ODYSSEY high-frequency high-capacity data acquisition system to test and study fluctuant characteristics of stilling basin floor of YinPan hydropower station, get the hydraulic jump fluctuating load, fluctuating frequency distribution along the way, as well as the probability distribution law of fluctuating amplitude, and quantitatively analyses the worst dynamic load of stilling basin floor. The results could be reference for similar projects.

Key words: YinPan; stilling basin; pressure fluctuations; amplitude; frequency

1. Introduction

YinPan Hydropower Station is the first 11 steps of hydropower develop planning in the main stream of Wu River. It is large-scale water utilities, mainly for power generation, with due consideration to shipping and other utilization. YinPan hydropower station mainly consists of water retaining structures, flood discharging structures, power plants and navigation components structures and so on. It is concrete gravity dam, the normal water level is 215.00m, corresponding capacity is 320000000m³, the maximum dam height is 78.5m, installed 4 power stations, with total capacity of 600MW, and navigation structure is single-stage 500t class ship lock.

Water retaining structure designed for 100-year flood recurrence, the corresponding peak discharge is 27100³/s, 1000-year flood recurrence is check flood condition, the corresponding peak discharge is 35600³/s. Energy dissipation and erosion control designed for 50-year flood recurrence, the corresponding peak discharge is 24400³/s. The main characteristics of the hub are:

(1) The hydropower station has large flood discharge capacity, high downstream water level with wide variable range, the upstream and downstream water level difference variable in the range of 31 – 32m, under 100-year flood recurrence condition, downstream water level raise up to 217m, beyond normal water level.

(2) YinPan hydropower station has no flood control task; in order to reduce submerging influence on upstream Pengshui county in the main flood season, on the condition of 20-year recurrence flood (20800m³/s), the upstream water level in front of dam required less than 213.5m.

(3) The project construction is divided into three phases, it is needed to set up more discharging holes to meet the requirement of the third-phase river closure.

(4) Flood discharging in the flow of less than 5500m³/s, flow condition in upstream and downstream approach channel entrance area meets the requirement of navigation.

The hub layout pattern is: power station structure locates on the left side of the river, flood discharging structures locates on the right side of the river, the ship lock locates on the right bank. In which, flood discharging structure uses 10-hole surface hole to form discharge overflow, and is divided into left, central and right 3 regions, left region has 4 holes (the downstream energy dissipation type is surface flow energy dissipation), central region has 4 holes (the downstream energy dissipation type is underflow energy dissipation), the right region has 2 holes (the downstream energy dissipation type is underflow energy dissipation), the regions are separated with sidewall or cofferdam, detailed layout see Figure 1.

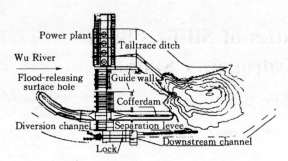

Figure 1   Wu River YinPan hydropower
overall layout diagram

2. Flow turbulent theory

By Renault test, we can see that viscosity of liquid and fluctuation of flow layer lead to a lot of different size of vortex caused in water flow, the vortex generates pressure difference between upper and lower vortex body, forming lift-P acting on vortex.

When lift P is greater than the vortex's resistance, the vortex flow-out of the original layer, into a new layer; when lift P is less than the vortex's resistance, despite the vortex body cannot be divorced from the original layer, into a new layer, in the surface hole, with the increase of velocity and volatile fluctuation of flow layer, the vortex still migrate downstream and breakdown, this makes the internal turbulence of flow intensified, the surface water fluctuating, particle trajectories in the flow twists and turns. When the border changes, superimposed with the shock waves generated by water flow, such turbulence becomes more evident, at this time each point in the flow, its velocity and pressure all random change with time, in view of the random turbulence, when research on turbulent flow, we must at the same time apply the methods of mechanics and mathematical statistics.

Transient pressure P of turbulent flow can be viewed as consists of both time-average pressure $\overline{P}$ and pressure fluctuation P'. For time-average pressure, use mechanical methods to study, for the pressure fluctuation, adopt the methods of mathematical statistics to research. The fluctuating phenomenon of turbulence has significant impact on time-average movement (mainly in the flow energy), also has impact on hydraulic structure's load, vibration and cavitation. Experiments indicate that the fluctuating with uniform isotropic turbulence is in line with the normal distribution.

Stilling basin floor is light structure, pressure fluctuation of hydraulic jump area directly related to structural vibration, impact, stability and so on. With cyclical changes in pressure fluctuation, pressure sometimes big, sometimes small, back and forth impact on the structure, which may cause the light structure generate strong vibration, when the dominant frequency of hydrodynamic pressure fluctuation and the natural frequency of the hydraulic structures are identical or similar, may also cause resonance and thus create a serious threat to the safety of structures. Therefore, It is need to study on hydraulic jump fluctuating load and frequency on the stilling basin floor of dam, through the physical model, provide test data for the stilling basin floor structure design, combined with estimation of stilling basin floor natural frequency, analyze the possibility of resonance of stilling basin, so as to ensure the safety of the dam energy dissipation facilities.

3. Physical model test

According to a large number of tests and research of the predecessors and prototype observation, the low-frequency turbulence of hydraulic jump area mainly generated by the large-scale vortex and the surface fluctuations, the force acting on stilling basin floor to be considered only has significant low-frequency fluctuation, in which gravity and inertia force play a leading role. As a result, this model designed according to the gravity similar guidelines, geometric scale $\lambda_L = 50$, the pressure fluctuating amplitude, the frequency of scale are considered in line with gravity similar criteria. Take 2 complete surface holes, establish local normal hydraulic model. The simulated range of model: on the direction of dam axis, from the left sideline of the 5th section to the right sideline of sluice pier of the 7th section; on the direction of flow, from the surface hole to the downstream apron.

Along the central line of the stilling basin floor lays total 12 points for pressure fluctuation test. In order to get the changes of pressure fluctuation along the direction of flow, along the central line of hole on the direction of flow layout 7 points for pressure fluctuation test. In order to analyze the area pressure fluctuation, on the apron board layout 5 points for pressure fluctuation test, the central point locates on the central line of hole. Furthermore, for the analysis of pressure fluctuation of the sidewall, along the direction of flow, on the sidewall sets 7 points for pressure fluctuation test on three different elevation.

The amplitude and dominant frequency of points pressure fluctuation, respectively, obtained by using statistical analysis and spectrum analysis. And uses points pressure fluctuation correlation analysis to compute the area fluctuating load of the observation area.

The United States NICOLET's ODYSSEY high-frequency high-capacity data acquisition system, ZHP (0-40) kPa pressure sensors, YD-28-type dynamic resistance strain gauge are used for data acquisition and processing, according to the low-frequency fluctuant characteristics of the pressure fluctuation of hydraulic jump, use 100Hz sample frequency, sampling time is 300s.

As the pressure fluctuation is random, measured extreme value has relationship with record length of time, in fact the accurate value of the largest possible amplitude is difficult to determine, the test can only get extreme value under condition of a certain probability.

Its methods for analysis based on random function theory, which are statistical data analysis and spectrum analysis method, the results got RMS (root mean square) σ (standard deviation, the unit in kPa) and the dominant frequency f_k [corresponding to the peak frequency of power density of $G(f)$, unit in Hz] are taken as main parameters for the analysis of points pressure fluctuation characteristics, at the same time get $P95$ and $P99$ of the amplitude as parameter (in kPa) corresponding to probability of 95% and 99%.

According to different construction stages of hydropower stations, on different upstream water levels and sluice gate open operating condition (e), when the upstream water level reaches 215m measures 6 different operating conditions, when the upstream water level reaches 210.5m measures 8 different operating conditions, on open discharge condition measures 3 operating conditions. For pressure fluctuation test results, see Figure 2 and Figure 3.

3.1 Points pressure fluctuation

3.1.1 *Time domain characteristics*

Pressure fluctuation on the region of hydraulic jump can be viewed as generated by strong separation

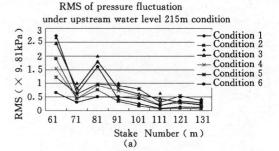

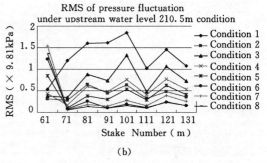

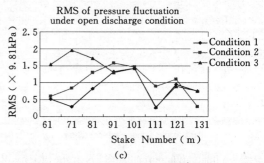

Figure 2   RMS of pressure fluctuation test results under each condition

and mixing, caused by the random movement of vortex. The first half of the Stilling basin is strong turbulent region, the floor flow fluctuating is stationary random process; pressure fluctuation probability distribution is approximate to normal distribution. Generally, for flat and width-equaled stilling basin, the pressure fluctuation on the hydraulic jump area floor depends on discharge per width, water head above floor, the submerging coefficient of hydraulic jump and distance from the measuring point to the contracted section, the bigger the discharge per width and water head above floor is, the greater the submerging coefficient of hydraulic jump is, and the smaller the distance from measuring point to the contracted section is, the greater the value of the floor pressure fluctuation is, so the measured maximum value of the

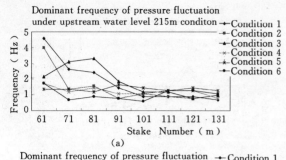

(a)

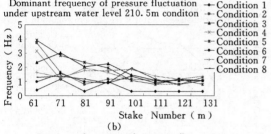

(b)

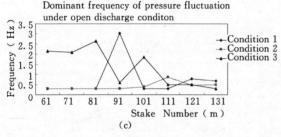

Figure 3   Dominant frequencies of pressure fluctuation
test results under each condition

pressure fluctuation normally occurs in the vicinity of hydraulic jump head. Test results show that flow pressure fluctuation in the vicinity of the most turbulent region of hydraulic jump area, has maximum intensity, followed by a gradual decay process along, but in the general trend of decay, on some conditions there are slight fluctuations; the point that has the maximum fluctuating value whose position related to the length of the hydraulic jump L_j, the distance from hydraulic jump X, the water depth of hydraulic jump head h_1 and so on. From open discharge test results, we can see that in this projects the maximum pressure fluctuation intensity of the stilling basin locates at $X/L_j = 0.33-0.47$. Under each flow condition, maximum RMS of pressure fluctuation on apron floor locates in the vicinity of hydraulic jump head. Under each upstream water level condition, the RMS of fluctuation on sidewall along the way reduce, along the direction of the elevation, the lower the position is, the larger the value is, vice versa.

Under operating conditions of 215m upstream water level, the apron floor pressure fluctuation are $(0.06-2.73) \times 9.81$kPa, on the apron, the largest RMS in the flow of 5570m³/s (6-holes controlled flow discharge, $e = 0.2$), at $0 + 61$m Pile No. points, elevation is 175.1m, RMS is 2.73×9.81kPa, corresponding to 99% probability, amplitude is 7.56×9.81kPa, observed flow pattern shows the hydraulic jump head in the vicinity of Pile No. $0 + 39$m-$0 + 40$m. According to the layout of measuring points, Pile No. $0 + 61$m is the measuring point nearest to the hydraulic jump head.

Pressure fluctuation on sidewall are $(0.35-1.41) \times 9.81$kPa, the largest RMS in the flow of 8560m³/s (6-holes controlled flow discharge, $e = 0.4$), at Pile No. $0 + 81$m, elevation is 178m, RMS is 1.41×9.81kPa, corresponding to 99% probability, amplitude is 3.36×9.81kPa, flow pattern observed show the hydraulic jump head in the vicinity of Pile No. $0 + 34$m-$0 + 35$m.

Under operating conditions of 210.5m upstream water level, the apron floor pressure fluctuation are $(0.06-1.85) \times 9.81$kPa, on the apron, the largest RMS in the flow of 8320m³/s (6-holes controlled flow discharge, $e = 0.6$) condition, at $0 + 61$m Pile No., elevation is 175.1m, RMS is 1.85×9.81kPa, corresponding to 99% probability, amplitude is 4.73×9.81kPa, flow pattern observed shows the hydraulic jump head in the vicinity of Pile No. $0 + 34$m-$0 + 35$m, according to measuring point setting, measuring point of Pile No. $0 + 61$m is the measuring point nearest to the hydraulic jump head, operating condition 1 has far drive hydraulic jump, the pressure fluctuation can be found gradually increase along, when reach the hydraulic jump head get the maximum value, and then decreases along the way. In the general trend of reduce, there are small fluctuations.

Pressure fluctuation on sidewall are $0.07-1.14 \times 9.81$kPa, the largest RMS in flow of 10200m³/s (6-holes controlled flow discharge, $e = 0.8$) condition, locating on the measuring point Pile No. $0 + 81$m, elevation 178m, RMS is 1.14×9.8kPa, corresponding to 99% probability, amplitude is 2.66×9.81kPa, flow pattern observed shows the hydraulic jump head locates in the vicinity of Pile No. $0 + 28$m-$0 + 38$m.

Under operating conditions of open discharge flow of 4000m³/s, 5500m³/s, 7000m³/s, the pressure fluctuations of apron floor are $(0.26-1.95) \times$

9. 81kPa, the largest RMS σ reach 1. 95 $\times$ 9. 81kPa, for the flow of 7000m³/s condition (H_{upstream} = 209. 03m), the No. 2 measuring point, Pile No. 0 + 71m, the flow pattern observed shows the hydraulic jump head locates at 0 + 50m - 0 + 60m, elevation is 175m, corresponding to 99% probability amplitude is 5. 28 $\times$ 9. 81kPa. On sidewall, pressure fluctuation are (0. 24-3. 19) $\times$ 9. 81kPa, the largest RMS σ reaches 3. 19 $\times$ 9. 81kPa, on flow of 5500m³/s condition (H_{upstream} = 206. 86m), Pile No. 0 + 81m, the flow pattern observed shows the hydraulic jump head locates at Pile No. 0 + 70m-0 + 77m, elevation is 178m, the corresponding probability of 99%, amplitude is 12. 05 $\times$ 9. 81kPa.

3. 1. 2 *Frequency domain characteristics*

Test results show that under each condition, pressure fluctuation on apron and sidewall is low-frequency fluctuating, and its dominant frequency in general below 4. 59Hz. In the vicinity of hydraulic jump head, the pressure fluctuation value is the maximum, and descending along the process. Distribution of frequency and the fluctuating value are basically the same, from the hydraulic jump head section rise gradually, after reaching the maximum value, coming down. Judging from the frequency characteristics, supercritical flow region has high fluctuating frequency; sub-critical flow region has low frequency; hydraulic jump area is in the range of the two, from the hydraulic jump head to the hydraulic jump tail, the trend is decreasing. In short, stilling basin fluctuating frequency is the same with amplitude, related to the location of the measuring point, energy head on contraction section, discharge per width, downstream water level and other relevant factors. Dominant frequency of hydraulic jump head and hydraulic jump tail are very prominent, in the middle section of the hydraulic jump area, relative proportion of high-frequency energy increased, major frequency moves to high frequency. All these show from the frequency domain, too, that the roller and rotating region of hydraulic jump area is the strongest turbulent region, the appearance of a large number of large and small sizes of vortex make low-frequency and high frequency energy increase significantly. Therefore, we can see the water flow has higher fluctuating energy in the central hydraulic jump area.

Under operating conditions the upstream water level reaches 215m, flow fluctuating frequency of stilling basin is not high, its dominant frequency are 0. 06-4. 59Hz, frequency of inside edge are 0. 10-2. 54Hz.

Under operating conditions the upstream water level reaches 210. 5m, flow fluctuating frequency of stilling basin is not high, its dominant frequency is 0. 29-3. 71Hz, frequency of inside edge are 0. 29-2. 73Hz.

Under operating conditions of open discharge flow of 4000m³/s, 5500m³/s and 7000m³/s, flow fluctuating frequency of stilling basin is not high, its dominant frequency are 0. 29-3. 03Hz, frequency of inside edge is 0. 29-2. 54Hz.

3. 2　Area pressure fluctuation

Compared the area average pressure fluctuation with the points pressure fluctuation, the ratio in addition to related to the structure parts (such as the spillway and apron board is not the same), but also relevant with the size of sub-block region.

In this experiment, according to the law of pressure fluctuation along the direction of flow and the sub-block of apron board, on the apron board (10 $\times$ 11. 5m²) sets 5 pressure fluctuation sensors, while the No. 2 measuring point locates at board center, so as to analyze and observe the area average pressure fluctuation.

During the test, under operating condition of upstream water level reaches 215m, the largest area fluctuating load occurs in flow of 8560m³/s condition (6 holes-controlled flow discharge, $e = 0. 4$), the load is 95. 7 tons; when the upstream water level reaches 210. 5m, the largest area fluctuating load occurs in flow of 8320m³/s (6 holes controlled flow discharge $e = 0. 6$) operating condition, the load is 96. 3 tons; under open discharge flow of 4000m³/s, 5500m³/s, 7000m³/s conditions, the largest area fluctuating load is in flow of 7000m³/s operating condition, the load is 90. 9 tons.

4. Discussion

(1) The pressure fluctuation intensity on stilling basin floor, from the hydraulic jump head gradually increase and then decrease along the way, the maximum value occurs in the maximum turbulent intensity region of hydraulic jump area, and related to the length of hydraulic jump, the distance from the hydraulic jump head, water depth of hydraulic jump head. Under each upstream water level, the RMS of

pressure fluctuation on the sidewall reduce along the way, along the direction of the elevation, the lower the location is, the higher the value is, vice versa.

(2) Flow fluctuating frequency of stilling basin are 0.06-4.59Hz, belonging to low-frequency fluctuating, as stilling basin floor natural frequencies is normally larger, it will not cause harmful resonance.

(3) If estimate the maximum fluctuating amplitude using pressure fluctuation RMS σ, the flow pressure fluctuation probability distribution is in compliance with the normal distribution, using $A_{max} = 3\sigma$ to estimate flow pressure fluctuation maximum value, could get 99.7 percent qualified probability, it is enough to meet engineering design requirements.

Acknowledgment

We would like to thank Changjiang Scientific Research Institute of Changjiang River Water Resources Commission and Changjiang Institute of Survey, Planning, Design and Research of Changjiang Water Resources Commission for the assistance and collaboration which was essential in making this study possible. Also, we would like to thank the two anonymous reviewers for their comments and suggestions, thus improving this paper.

References

[1] Tsinghua University Hydraulics Research Group. Hydraulics [M] .Beijing: People's Education Press, 1981.

[2] Zhang Shengming. Yangtze River Academy of Sciences. Fluctuating pressure flow studies of progress in a number of issues Summary [Z]. 1992.

[3] Tang Xiangfu, Ling Ting-jun, Chen Yuanqing. Elementary introduction to calculation of area fluctuating pressures in hydraulics. Yangtze River Academy of Sciences [J]. 2008, 17 (4): 12-15.

[4] Lu Guanping, Wang Xiaoquan. Fluctuation load on the hydraulic structures. Journal of Hefei University of Technology [J]. 2002, 25 (3): 23-26.

[5] Lu Fangchun, Jiang Ying, Bao Zhongjin. Research on stilling basin hydraulic jump fluctuation pressure characteristics of Caoe River sluice. Rural China Water Conservancy and Hydropower [J]. 2005 (6): 34-38.

[6] Nie Mengxi, Li Linlin, Duan Bin. Fluctuant characteristics analysis of sidewall and bottom behind sudden lateral enlargement and vertical drop form. Journal of Tsinghua University (Natural Science) [J]. 2007, 47 (3): 45-49.

Experimental Study on Cavitation Characteristics of Middle Discharging Orifice Inlet Section Structure of Goupitan Hydropower Station

Jiang Bole[1] Xiang Guanghong[2]

[1] *Changjiang Scientific Research Institute of CWRC, Wuhan*, 430010
[2] *Changjiang Institute of Survey, Planning, Design and Research of CWRC, Wuhan*, 430010

Abstract: In Goupitan Hydropower Station, the layout type of middle discharging orifice was classified into flat-floored type and up-bended type, the maximum operational water head was 95. 36m, and the emergency maintenance gate used upstream inclined slanting gate slot, the velocity of gate slot region was beyond 30m/s. From the result of vacuum tank experiment, the steam type cavitation intensity in gate slot region of up-bended type middle orifice did not surpass firstborn stage. The structure of this middle orifice basically belonged to cavitation free type. The cavitation intensity in gate slot zone of flat-floored type middle orifice was severe. The steam type cavitation reached developing stage. According to the comparison of structure optimal schemes, adapting measure of reducing the area of pressure section outlet or enlarging the dimension of pressure section could solve the above problem effectively. Therefore, when slanting gate slot was laid in pressure section of discharging structure, the mean velocity of gate slot section should be controlled under feasible value; otherwise, the rather serious cavitation would happen more easily in slanting gate slot, compared to straight gate slot.

Key words: Middle orifice; slanting gate slot; steam cavitation; gas cavitation; Goupitan hydropower station

1. Introduction

Goupitan hydropower station river dam belongs to concrete double curvature arch dam, the highest dam height is 232. 5m. The dam spillway segment consists of 6 surface orifices and 7 middle orifices. The 7 discharging middle orifices adopts outlet type of flat-floored type and up-bended type separately. The inlet bottom elevation of up-bended middle orifice (2#, 4#&6#) is 543m. The end of pressure segment adopts outlet type of up-bended type. The inlet bottom elevation of flat-floored middle orifice (1#, 3#, 5#&7#) is 550m. The elliptic curve equation of inlet top is $x^2/82 + y^2/3. 292 = 1$, the equation of side ellipse is $x^2/3. 62 + y^2/1. 22 = 1$. There is a maintenance gate slot with the tilt angle of 85° arranged in inlet segment, the Width/Depth ratio of gate slot is $W/D = 1. 51/0. 9 = 1. 68$; the height of mean straight section of orifice is 8m, and the width is 6m; the outlet section of pressure section diffuses in both sides in plane (the diffusion angle is 2. 38°),

the top shrinks with the slop of 1 : 6. The top platen is connected to the anterior horizontal segment by an arc with angle of 9° and radius of 30m. The top elevation of outlet top is 556m, the control cross-section dimension of pressure outlet is 6m × 7m (height × width). The structure layout of flat-floored middle orifice is shown in Figure 1.

Because the inlet elevation of up-bended middle orifice is 7m lower than that of flat-floored middle orifice, the flood discharging capacity of them are basically equal, the flow mean velocity of gate slot cross-section is equal, the time-averaged pressure of up-bended middle orifice is 7m larger than that of flat-floored middle orifice in front of the gate slot , accordingly, the gate slot referencing flow cavitation number is a bit large, the steam type cavitation intensity in gate slot region of up-bended type middle orifice did not surpass firstborn stage, The structure of this middle orifice basically belonged to cavitation free type. Therefore , in this paper the inlet segment structure of 1# middle orifice (flat-floored) was discussed, especially the experimental result of gate slot zone.

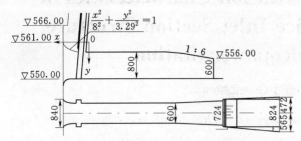

Figure 1   Layout of Goupitan Hydropower Station's middle discharging orifices (Flat base type)

2. Hydraulic model

The model was designed according to the gravity similar criterion, the model length scale is $Lr = 30$. The single orifice model was constructed of intact 1# middle orifice.

The hydraulic model was completely made by PMMA, the surface roughness coefficient of PMMA is 0.008, the roughness coefficient scale is calculated by $n_r = Lr^{1/6}$, which converted to prototype roughness coefficient is $n_p = 0.014$, that is similar with steel mould concrete wall roughness coefficient. The selection of model material could meet the similar requirement of prototype and hydraulic model.

The decompression model experiment conformed to the Froude similar criterion, in addition, it should conform to cavitation similar criterion, that is the flow cavitation number of prototype and hydraulic model should be equal ($\sigma_p = \sigma_m$).

3. Methodology of cavitation experimental analysis

The noise that created by flow cavitation could be taken as proof to judge the different stage of cavitation development, therefore, use cavitation noise detecting technology to study the problem of flow cavitation is an efficient technology way[1-4]. Normally, the noise power spectrum level different method was used to analyze the cavitation characteristic. the noise spectrum level different value $\Delta SPL = SPL_f - SPL_o$, in which SPL_f is the total noise power spectrum in flow under similar air pressure (indicated of line 1 in spectrum figure plot). SPL_o is the background noise spectrum level (indicated of line 2 in spectrum figure plot) that was tested when no cavita-

tion happened, it was the environmental noise in experiment process and had the characteristic of low frequency (normally under 20kHz). Under background condition, the vacuum degree in vacuum tank is relatively low (approximately 70% of vacuum degree under similar condition). The gas nucleus content in flow is higher that of similar vacuum degree, so noise spectrum level plot ordinarily indicated that in low frequency section SPL_o is higher than SPL_f. Based on large amount of hydraulic experiment and prototype observation references, Guo Jun li et al (2000) suggested that whether ΔSPL reaches 5dB can be taken as the judge index of cavitation firstborn. When certain cavitation type happened apparently, ΔSPL value will surpass 10dB, when cavitation is weak and the spectrum level difference ΔSPL is between 5 dB and 10dB, it can be taken as cavitation firstborn stage.

Based on the flow cavitation creation organism, the instability of gas nuclear in flow creates cavitation bubble (that is hollow bubble). The cavitation that cavitation bubble mainly including gas is called gas cavitation, mainly including steam (liquid boil away, hollow bubble exploding collapse) is called steam cavitation. The acoustic energy radiation of cavitation consists of low frequency hollow bubble oscillation component and high frequency sharp pulse component. The low frequency component of noise radiation prevails in gas cavitation, however, high frequency sharp pulse component of noise radiation is predominant in steam cavitation, thus the cavitation type could be estimated[6]. In general, the gas cavitation normally has no cavitation erosion destructive power, but if steam cavitation reached curtain strength, it has cavitation destructive power.

4. Experimental result

4.1   Original scheme flow pattern

Under operation condition of water level 630m (normal water level), 632.89m (design water level), and 638.36m (check water level), the upstream flow were fluent and stable, there were no obvious vortex in front of orifice inlet, and no significant eddy current in maintenance gate slot, the flow pattern in other part of orifice was normal. Under similar air pressure condition, near the junction corner of maintenance gate slot and the end of top elliptic curve, bearded cloudy cavitating flow appeared now and then, and it began more significant with increasing of upstream water level.

4.2 Original Scheme underwater noise and cavitation characteristics

4.2.1 *Inlet top elliptic curve*

Under the three upstream water level conditions, in the rather wide band range of 31.5-200kHz of underwater noise spectrum level, there were significant cavitation noise signals, furthermore, gas cavitation was stronger than stream-typed cavitation. Under normal water level of 630m and above conditions, the cavitation noise spectrum level difference ΔSPL_{max} all surpassed 10dB, in high frequency band of 80 – 200kHz, and the steam cavitation strength was on developing stage.

4.2.2 *Inlet side elliptic curve*

Under the three upstream water level conditions, in the middle and low frequency band the noise spectrum level ΔSPL value was apparently larger than that of high frequency band, which indicated that gas cavitation is stronger than steam cavitation, the noise spectrum level difference ΔSPL_{max} value that representing the steam cavitation strength was 12-18dB, the steam cavitation strength all belonged to developing stage.

4.2.3 *Maintenance gate slot zone*

Under the three upstream water level conditions, the whole analysis band (3.15-200kHz) of spectrum level all had obvious cavitation noise signal, the gas cavitation and steam cavitation all strong, and both cavitation type all getting stronger with increasing of upstream water level, the spectrum level difference of high band ΔSPL_{max} was 25-35dB, steam cavitation strength belonged to developing stage. The underwater noise spectrum figure of maintenance gate slot was shown in Figure 2.

4.3 Modified scheme experimental result

Based on original result, there were strong cavitation happed in inlet top, side curve and maintenance gate slot zone; the structure should be modified and optimized. Combined with the analysis of flow pattern, time average pressure and underwater noise, the main reason of cavitation was that the flow cavitation number was too low along the flow way. Therefore, two methods were considered, one way is to reduce the height of outlet control cross-section of pressure orifice, increase the pressure in orifice, but the

$H_{upstream} = 638.36m$

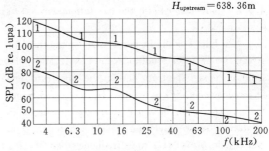

(a) Upstream water level 632.89m

$H_{upstream} = 630.00m$

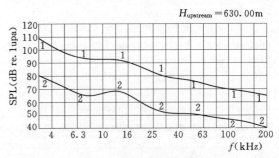

(b) Upstream water level 630.00m

Figure 2　Spectrum level of underwater noise in area of emergency service gate slots for original scheme

flood discharging capacity of mid orifice would be decreased, another way is to keep the area of outlet cross-section of pressure orifice, increase the area of inlet section and the cross-section area along the flow way in orifice, decrease the velocity along the flow way to increase the flow cavitation number, the flood discharging capacity of mid orifice basically had no change. In order to keep the flood discharging capacity, the second way was adapted as modified scheme.

The specific modifying measure was that the inlet top elliptic curve equation was changed to $x^2/92 + y^2/32 = 1$, the top elevation of mean straight segment at back of maintenance well was 559.00m (rising 1m compared to that of original scheme).

4.3.1 *Flow pattern*

Under the three characteristic upstream water level conditions, the flow through the orifice all belonged to pressure flow, there were no obvious eddy current created in maintenance gate slot and nearby, the pressure outlet posterior diffusion flow impacted the support pivots of left side curved gate frusta apparently. Under decompression condition of similar gas pressure, there was no "cloudy cavitating flow" phenomenon found in orifice by visual measurement.

4.3.2 *Underwater noise and cavitation characteristics*

(1) Inlet top elliptic curve. Under the three upstream water level conditions, in 63kHz and below band of noise spectrum level, the spectrum level difference value$\Delta SPL_{max} \approx 8dB$ ($H_{upstream} = 638.63m$), the gas cavitation belonged to firstborn stage, the cavitation noise spectrum level difference value ΔSPL_{max} in 80-200kHz high frequency band was all within 3dB, which indicated that no steam cavitation happened.

(2) Inlet side elliptic curve. Under the three upstream water level conditions, in 63kHz and below band of noise, spectrum level, the spectrum level difference value$\Delta SPL_{max} \approx 7$-10dB, and the ΔSPL_{max} value increased a bit with increasing of upstream water level, but did not surpass firstborn stage, the cavitation noise spectrum level difference value $\Delta SPL_{max} \approx 8$-9dB in 80-200kHz high frequency band, but ΔSPL_{max} value changed little with the change of efficient water head, there was steam cavitation on firstborn stage happened.

(3) Maintenance gate slot zone. Under the three upstream water level conditions, the cavitation noise spectrum level difference value ΔSPL_{max} all below 5dB in 80-200kHz high frequency band, in 63kHz and below band of noise, spectrum level, the spectrum level difference value $\Delta SPL_{max} \approx 9dB$ ($H_{upstream} = 638.63m$), ΔSPL_{max} value decreased with decreasing of upstream water level, it indicated that there was gas cavitation on firstborn stage created in maintenance gate slot zone. The underwater noise spectrum figure of maintenance gate slot zone was shown in Figure 3.

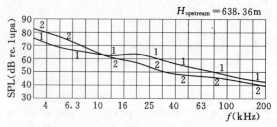

(a) Upstream water level 632.89m

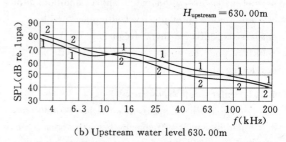

(b) Upstream water level 630.00m

Figure 3　Spectrum level of underwater noise in area of emergency service gate slots for modified scheme

4.4 gate slot flow cavitation number

The factors that influence flow cavitation is rather more and complicated, but the main decisional condition is flow velocity and pressure. The flow cavitation number is a dimensionless parameter that inflects the joint influence of pressure and velocity. The gate slot flow cavitation number was calculated by pressure on gauge point that hug the gate slot upstream wall and the mean velocity of the cross-section where the point located. The specific result was shown in Table 1.

Table 1　　　　Reference flow cavitation number of emergency service gate slots

Scheme	Upstream water level (m)	Pressure of referencing point (mH$_2$O)	Referencing mean velocity (m/s)	Cavitation number σ
2 # Mid orifice	638.36	39.14	31.92	0.937
	632.89	37.22	30.79	0.967
	630.00	36.26	30.17	0.986
	627.30	35.09	29.60	0.998
1 # mid orifice (original scheme)	638.36	32.72	31.83	0.817
	632.89	30.62	30.69	0.836
	630.00	29.29	30.08	0.841
	627.30	28.00	29.52	0.844

Scheme	Upstream water level (m)	Pressure of referencing point (mH$_2$O)	Referencing mean velocity (m/s)	Cavitation number σ
1 # mid orifice (modified scheme)	638. 36	40. 53	28. 57	1. 219
	632. 89	38. 13	27. 39	1. 245
	630. 00	36. 63	26. 87	1. 253
	627. 30	35. 13	26. 39	1. 266

Based on decompression experimental results, the firstborn cavitation number of steam cavitation of mid orifice maintenance gate slot was about 0. 967. Judged from above table, 2 # mid orifice structure cavitation would be created in gate slot zone under 638. 36m, 632. 89m two level of upstream water levels, in original scheme 1 # mid orifice structure under all experimental upstream water level conditions, in gate slot zone cavitation would be created, in modified scheme 1 # mid orifice structure under all experimental upstream water level conditions, the referencing flow cavitation number in gate slot improved a lot compared to original scheme, no steam cavitation would created in gate slot zone.

The above experimental results was consistent with decompression experimental result (Under 638. 36m, 632. 89m two level upstream water level conditions, with 2 # mid orifice structure would create steam cavitation, its cavitation strength did not surpass firstborn stage, under each level upstream water level condition, with 1 # mid orifice original scheme structure, steam cavitation with cavitation strength surpassed developing stage would be created in gate slot zone, with 1 # mid orifice modified scheme, no steam cavitation created in gate slot zone).

5. Conclusions

The gate slot zone of high water head discharging orifice is a part prone to create cavitation, because when water flow through gate slot, eddy current would create in gate slot, when pressure of vortex center below certain degree, cavitation would happen. Furthermore, the lift pier effect of slot gate downstream wall is also easy to cause flow division so as to create cavitation. So when inclined slot gate is need in structure design layout of high water level

discharging orifice, it should be aware that the area adoption of the cross-section where the slot gate located should be suitable, the mean velocity of its control cross-section should be no higher than 30m/s, otherwise the slot gate cavitation erosion destruction risk is quite high.

Acknowledgment

We would like to thank Changjiang Scientific Research Institute of Changjiang River Water Resources Commission and Changjiang Institute of Survey, Planning, Design and Research of Changjiang Water Resources Commission for the assistance and collaboration which were essential in making this study possible. Also, we would like to thank the two anonymous reviewers for their comments and suggestions, thus improving this paper.

References

[1] Huang Jitang. Principle and Application of Cavitation and Cavitation Erosion, Tsinghua University Press, 1991: 77-79.

[2] Bark, G.. Prediction of propeller cavitation noise from model tests and its comparison with full scale data, J. of Fluids Engineering, 1985, 107 (1): 112-120.

[3] Leggat, L. J. and Sponagle, N. C.. The study of propeller cavitation noise using cross-correlation methods, J. of Fluids Engineering, 1985, 107 (1): 127-133.

[4] Higuchi, H., Arndt, R. E. A. and Rogers, M. F.. Characteristics of tip vortex cavitation noise, J. of Fluids Engineering, 1989, 111 (4): 495-501.

[5]   Guo Junli, Lu Junying. Cavitation study on surface spillway of Three Gorges Project, Journal of Yangtze River Scientific Research Institute, 2000, 17 (4): 12-14.

[6]   Mao Chunpu. Cavitation erosion and cavitation noise. Shanghai Jiaotong University, 1988.

Experimental study on inverted siphon channel in South-to-North water diversion project

Cheng Zibing[1]   Wang Feng[2]   Yan Wei[1]

[1] *Department of Hydraulics Research, Changjiang River Scientific Research Institute, Wuhan, 430010*

[2] *Department of River Research, Changjiang River Scientific Research Institute, Wuhan, 430010*

Abstract: There are 13 inverted siphon channels located in Taocha-south Shahe section in the middle line project of South-to-North water diversion. By hydraulics model test, the hydraulic characteristics such as inlet flow pattern and head loss etc. are studied systematically for inverted siphon channel in Huangjin river. Due to this project being representative, according to the model test result, other inverted siphon channel projects'head loss is computed. In order to reduce inlet local head loss and improve inlet whirlpool pattern, it is recommended to adopt circle arc jointed by titled slope for inverted siphon inlet form and adopt half circle head form for diversion pier.

Key words: middle line project of South-to-North water diversion; inverted siphon channel; inlet flow pattern; head loss coefficient; intlet submerged depth

1. Introduction

In the middle line project of South-to-North Water Diversion, as the first section of the main water transmission channel, Taocha-south Shahe section is 239km long, passing through 31 intersected rivers in Changjiang River drainage basin and Huihe River drainage basin. In Taocha head channel, average annual water diversion capacity is 9.5 billion cubic meters, with design discharge of 320-350m³/s. The total water head of Taocha-south Shahe section is 15m, including intersected building design distribution head of 5.8m and channel distribution head of 9.2m along its way. Due to the main channel intersected with river, ditch, channel, public road and railway, adding control projects, there are total 350 buildings, including 30 river-channel intersected structures and 100 left bank draining buildings. In these river-channel intersected structures, there are 6 categories of intersected form such as beam aqueduct, culvert aqueduct, inverted siphon channel, inverted siphon river course, drainage culvert and drainage aqueduct.

2. Choosing research object and model design

There are 13 inverted siphon channel projects in Taocha-south Shahe section, including 11 river-channel intersected structures. These inverted siphon channel projects have almost the same design inlet-outlet form, with inlet submerged depth of 2.82-3.68m and inner flow velocity of 1.94-3.08m/s under increased discharge of 380-410m³/s. Due to the main concerned hydraulic characteristics being inlet flow pattern and local head loss, Huangjin river inverted siphon channel project is representative to be chosen as research object. In this project, there are 4 inverted siphon pipes with single pipe dimension of 6.85m×6.85m. Its bottom plate slope gradient of upstream channel and downstream channel is 1/25000 respectively. The form detail can be found out in Figure 1.

In model design, according to roughness factor value of 0.014 for prototype inverted siphon pipe, adopting synthetic glass simulating inverted siphon pipe, the geometry scale of normal hydraulic model is determined as 1 : 18.5 based on gravity similarity rule. In general, in order to ensure whirlpool simulating similarity, flow Reynolds number should be more than 3×10^4 or Weber number should be more than 3×10^4 in the model. According to chosen model scale, model flow Reynolds number is computed to be $3.5 \times 10^4 - 4.4 \times 10^4$ and meets the basic need of whirlpool simulating similarity.

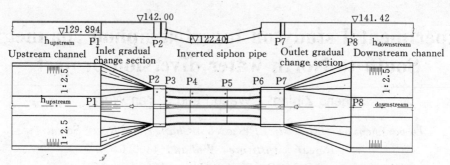

Figure 1　The schematic illustration of Huangjin river inverted siphon channel project

3. Original design scheme

3.1　Inlet whirlpool

For inverted siphon pressure inlet, due to limited submerged depth and unsymmetrical incoming flow pattern, surface funnel whirlpool formation is a common hydraulic phenomenon. Whirlpool can be classified as air intake type and non-air intake type. There are three kinds of air intake whirlpool such as intermittent air intake whirlpool, intermittent run-through whirlpool, and stable run-through whirlpool. As shown in Figure 2, Example A would happen generally in the condition of high level or low level with no harm, belonging to non-air intake whirlpool level range. Example B would happen appear generally in the condition of middle level with air intake appearing and small harm. In Example C, due to air entering into the pipe by run-through whirlpool intermittently, flow becomes unstable and flow discharge is reduced correspondingly. If air cell forms in the pipe due to air concentration, it may burst in some sections, causing great pressure pulse and endangering operation safety. Due to air intake and air cell bursting randomly, strong flow turbulence may result in cavitation and vibration damage. Therefore, air intake whirlpool, especially run-through whirlpool should be avoided to appear at pressure inlet in water diversion projects.

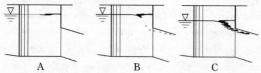

Figure 2　The schematic illustration of inlet whirlpool pattern

In the model test, it is observed that intermittent air intake funnel whirlpool of 0.15-0.90m diameter appears on both inlet sides of inverted siphon pipes. Due to low flow velocity in the pipes, this flow pattern may not bring safety threat to water transmission operation, but has some bad effect on discharge capacity and causes head loss. In addition, due to diversion pier adopting rectangular head form, obvious circumferential motion and water surface rising phenomenon can be observed around it.

3.2　Head loss and head loss coefficient

Head loss or head loss coefficient is one of important parameters concerned in inverted siphon channel design. In general, linear loss can be compute according to channel's roughness factor. Due to local head loss related with flow pattern, building boundary form and flow velocity etc. it is usually checked by hydraulic model test for large-scale projects.

According to the arrangement characteristics of Huangjin River project, 10 measuring points are installed in inlet section, inverted siphon pipe and outlet section to obtain these sections' head loss respectively, as shown in Figure 1. $P_1 - P_3$ measuring points are installed in inlet section (including inlet gradual change channel section, inlet descent section and inlet gradual change pipe section). $P_4 - P_5$ measuring points are installed in horizontal straight pipe section. $P_6 - P_8$ measuring points are installed in outlet section (including outlet gradual change pipe section, outlet ascent section and outlet gradual change channel section). The design scheme's head line is shown in Figure 3.

According to Figure 3, in 330m³/s discharge condition, inverted siphon pipe inlet's head loss h_{inlet} is 0.053m, 44.9% of total head loss, with integrative inlet local head loss coefficient ξ_{inlet} of 0.332. Pipe body's head loss h_{pipe} is 0.014m, 11.9% of total head loss. Inverted siphon pipe outlet's head loss h_{outlet} is 0.051m, 43.2% of total head loss, with integrative

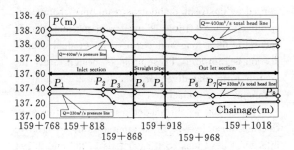

Figure 3   Original design scheme's
pressure line and total head line

inlet local head loss coefficient ξ_{inlet} of 0.320.

In $400m^3/s$ discharge condition, inverted siphon pipe inlet's head loss h_{inlet} is 0.075m, 44.9% of total head loss, with integrative inlet local head loss coefficient ξ_{inlet} of 0.319. Pipe body's head loss h_{pipe} is 0.020m, 12% of total head loss. Inverted siphon pipe outlet's head loss h_{outlet} is 0.072m, 43.1% of total head loss, with integrative inlet local head loss coefficient ξ_{inlet} of 0.306.

4. Optimized scheme

4.1   Form optimizing

Inlet whirlpool formation is related with submerged depth, inlet form, flow velocity and incoming flow symmetry. In practical projects, the following methods are usually adopted to reduce whirlpool intensity or eliminate whirlpools.

(1) Increase submerged depth, or decrease flow velocity, but resulting in increased project investment.

(2) Increase upstream guiding channel length to reduce flow circulation around inlet and ensure incoming flow symmetry, but also resulting in increased project investment.

(3) Adopt diversion pier with streamline shape to avoid backflow formation near inlet.

(4) Adopt inlet top form of circle arc, or increase inlet top tilt extent.

Inlet top form and diversion pier form are optimized based on prototype project condition. For inlet top form, circle arc of 4m radius jointed by 1 : 5 tilted slope replaces design scheme's 1 : 5 tilted slope with vertical breast wall and its top point elevation is close to 138.15m. For diversion pier form, 1/4 circle arc of 2m radius replaces design scheme's rectangular head.

4.2   Flow pattern

After optimizing inlet top form and diversion pier form, only small surface whirlpool appears with no intake. This A type whirlpool does not bring air into inverted siphon pipe and will not cause any bad effect. In addition, water surface rising phenomenon is improved around inlet diversion pier with rising value reduced from 5-15cm to 3-8cm.

In the inverted siphon pipes, flow moves smoothly and no bad flow pattern appears. Submerged flow forms in the outlet.

4.3   Head loss and head loss coefficient

For the optimized scheme, in $330m^3/s$ discharge condition, inverted siphon pipe inlet's head loss h_{inlet} is 0.034m, 33.7% of total head loss, with integrative inlet local head loss coefficient ξ_{inlet} of 0.253. Pipe body's head loss h_{pipe} is 0.013m, 12.9% of total head loss. Inverted siphon pipe outlet's head loss h_{outlet} is 0.054m, 53.4% of total head loss, with integrative inlet local head loss coefficient ξ_{outlet} of 0.304.

In $400m^3/s$ discharge condition, inverted siphon pipe inlet's head loss h_{inlet} is 0.057m, 39.6% of total head loss, with integrative inlet local head loss coefficient ξ_{inlet} of 0.243. Pipe body's head loss h_{pipe} is 0.019m, 13.2% of total head loss. Inverted siphon pipe outlet's head loss h_{outlet} is 0.068m, 47.2% of total head loss, with integrative inlet local head loss coefficient ξ_{outlet} of 0.291.

4.4   The possibility to upraise inlet bottom plate

In general, pipe inlet's submerged depth should be greater than critical submerged depth in inverted siphon project design. Gordon J. L formula is always used to compute critical submerged depth. Adopting this formula, inverted siphon pipe inlet's critical submerged depth is computed to be 4.07m and 3.36m respectively in $400m^3/s$ discharge condition and $330m^3/s$ discharge condition. In the model test, measured inlet submerged depth is 3.39m and 2.58m respectively, less than computed critical submerged depth. Considering Gordon J. L formula mainly used for hydropower water intake inlet, due to low flow velocity at inverted siphon pipe inlet, it is unreasonable to adopt this formula to determine its critical submerged depth.

In the model test, the possibility to upraise inverted siphon pipe inlet bottom plate is explored by lowering downstream water level (reducing inlet submerged extent). For optimized scheme, in 400m³/s discharge condition, by reducing downstream water level to 137.21-137.97m and inlet submerged depth to 2.58-3.34m accordingly, it is observed that inlet flow pattern change is not obvious with small whirlpool appearing occasionally and no air intake happening. Therefore, it is possible to upraise inverted siphon pipe inlet's bottom plate.

5. Application to other projects

In 13 inverted siphon channel projects of Taocha-south Shahe section, the basic design form of water transmission buildings adopts inverted siphon pipe jointed by upstream and downstream open channel respectively, most projects' upstream gradual change section is 60m long and downstream gradual change section is 75m long. These projects' design form, inlet top form and diversion pier form are similar to Huangjin river project.

According to the model test result, after subtracting linear loss of titled pipe section, integrative inlet local head loss coefficient ξ_1 is 0.283 and 0.194 respectively for design scheme and optimized scheme. Integrative outlet local head loss coefficient ξ_2 is 0.271. So other projects' head loss can be computed initiatively by integrative local head loss coefficient, as shown in Table 1.

Table 1   **Some projects' computed head loss**

Project		Xizhaohe	Baihe	Qinghe	Feihe
Discharge (m³/s)		410	400	400	380
Upstream level (m)		144.33	141.27	138.97	133.37
Pipes		4	4	6	4
Dimension (m)		6.5×6.5	6.8×6.8	5.7×5.7	7.0×7.0
Velocity (m/s)		2.43	2.16	3.08	1.94
Inlet form	Top	1:5 slope	circle arc	circle arc	circle arc
	Pier	half circle	half circle	rectangular head	half circle
Pipe length (m)		140	652	150	254
ξ_{inlet}		0.283	0.194	0.283	0.194
ξ_{outlet}		0.271	0.271	0.271	0.271
Head loss (m)	Inlet	0.085	0.067	0.137	0.037
	outlet	0.082	0.064	0.131	0.052
	total	0.252	0.426	0.442	0.178

6 Conclusion

(1) By optimizing inverted siphon pipe's inlet top form and diversion pier form, inlet flow pattern is improved and total head loss is reduced 13.8% and 14.4% respectively in 400m³/s discharge condition and 330m³/s discharge condition. For similar inverted siphon pipe projects, it is recommended to adopt circle arc jointed by titled slope for inverted siphon inlet form and adopt half circle head form for diversion pier.

(2) Due to Huangjin river project's representative, according to the model test results, other inverted siphon channel projects' head loss is compu-

ted. The computation result can be used as initiative reference for the design staff.

(3) The model test results show that the bottom plate elevation of inverted siphon pipe inlet can be up-raised modestly in Huangjin river project and it is recommended to carry out further hydraulic model test to determine.

References

[1]   Wu Chigong. Hydraulics. Beijing: Higher Education Press, 2003.

[2]   Fu Zhiyuan. Design specification for intake of hydraulic and hydroelectric engineering. Beijing: China Waterpower Press, 2003.

Large Eddy Simulation of Water Flow over Series of Dunes

Jun Lu   Lingling Wang   Hai Zhu   Zhenzhen Yu

The College of Water Conservancy and Hydropower, Hohai University, Nanjing, 210098

Abstract: Large eddy simulation is used to investigate the spatial development of open channel flow over a train of twelve dunes. The three dimensional filtered N-S equations are numerically solved by fractional-step method in sigma-coordinate. The subgrid turbulent stress is modeled by a dynamic coherent eddy model proposed by the authors. The computed velocity profiles are in good agreement to the available experimental results. The mean velocity and the turbulent Reynolds stresses and so on affected by a series of dune shaped structures on are compared and analyzed. The variations of turbulence statistics along the flow affected by the wavy bottom roughness have been studied.

Key words: LES; Dunes; Turbulent boundary layer; Flow separation

1. Introduction

Dunes are the most common bed form structures in sandy rivers. The presence of dunes in river beds causes flow separation and recirculation, which can alter the overall flow resistance and, consequently, can affect sediment and contamination transport within the river. The flow is characterized by an attached flow upstream of the dune crest, separation at the crest and formation of a recirculation eddy on the crest leeside[1].

The turbulent open channel flow over bottom with dunes has been studies under the assumption of a fixed bed without sediment movement, both experimentally and numerically. Lyn[2] reported the experimental study on the mean flow and turbulence characteristics over artificial space-periodic one-dimensional bed form features using Laser Doppler velocimentry. Wiberg and nelson[3] conducted the experiments under unidirectional flow over asymmetric and symmetric features including high-angle and low angle ripples. Similar experiments have been done by many researchers, such as Nelson and Smith[4], and Ojha and Mazumder[5]. Recently, numerical studies on the developing turbulent flow by a solution of the Reynolds averaged Navier-Stokes equations (RANS) by researchers[6,7], e.g. Peric[8] numerical simulated the flow over a typical dune employing the k-ε turbulence model. The influence of sand grain roughness was taken into account with wall functions approach. Their comparisons with experiments show that the computed the separating and reattaching flow over a dune is well with experimental data. Similar studies were reported by Mendoza and Shen[9], who presented an algebraic-stress model with wall functions in place of the k-ε model. They were able to obtain quite realistic predictions of the detailed pressure, velocity and turbulence profiles. Johns[10] employed one-equation turbulence model, with the turbulence model length scale prescribed from an empirical correlation. The comparisons with experiments show that near wall velocity and turbulence data, particularly the wall shear stress, are not in good agreement with experimental data. Yoon and Patel[11] considered the flow over a fixed dune employing k-ε turbulence model of Wilcox. Comparisons of model predictions with measured velocity, turbulence fields, as well as the pressure and friction distributions along the dune shows good agreement. Lu[12] compared three different k-ε models for the separated flows passing sills. Their computed results show the performance of V2F models among the three models is encouraging. Fourniostis et al[1] used the commercial CFD code FLUENT to compute sub-critical, turbulent, open-channel flow over a bottom with five dunes with free surface treatment method based on the rigid-lid approximation. They reported that the mean velocity and turbulence numerical predictions are in good agreement to available experimental data.

Although the RANS model can predict the realistic predictions of flow features over dunes, they cannot calculate the power spectrum of hydrodynamic turbulence, which is a very interesting parameter for some especially engineering applications, and refined turbulence structures, which is a very important for turbulence research. In recent years, large eddy simulation (LES) has been used to study hydrodynamic turbulence, for examples, the separated flows passing sills by Lu and Wang [13] and so on. The computed results show that the LES is very powerful and encouraging. So it is essential that a detailed numerical simulation open channel flow over a series of dunes using the large eddy simulation technique.

2. Governing equations and numerical methods

2.1 The governing equations in the σ-coordinate and turbulence model

The LES approach under σ-coordinate is used in the paper[14]. The equations for the large-scale motion can be obtained by applying a spatial filter to the Navier-Stokes equations (indicated by an over-bar). With a top-hat filter based on the Boussinesq assumption and according to the principle of chain differentiation, the governing equation for an incompressible fluid in the general coordinate can be written as:

$$\frac{\partial}{\partial \xi^k}(\overline{J^{-1}\,\xi_i^k\overline{u_i}})=0 \qquad (1)$$

$$\frac{\partial(\overline{J^{-1}\overline{u_i}})}{\partial t}+\frac{\partial}{\partial \xi^k}(\overline{J^{-1}\,\xi_i^k\overline{u_i}\,\overline{u_i}})=f_i-\frac{1}{\rho}\frac{\partial}{\partial \xi^k}(\overline{J^{-1}\,\xi_i^k\overline{p}})$$
$$+\upsilon\frac{\partial}{\partial \xi^k}(\overline{J^{-1}\,\xi_i^k\,\xi_i^l\frac{\partial \overline{u_i}}{\partial \xi^l}})-\frac{\partial}{\partial \xi^k}(\overline{J^{-1}\,\xi_i^k\overline{\tau_{ij}}}) \qquad (2)$$

where ξ^k is the coordinate direction in the transformed space, $\xi_i^k=\frac{\partial \xi^k}{\partial x_i}$, x_i is the Cartesian coordinate, J^{-1} is the Jacobin of the transformation, f_i is the gravity force, ρ is the water density, u_i is the Cartesian components of the velocity field, p is the static pressure, υ is the kinematic viscosity and $\overline{\tau_{ij}}$ is the subgrid stress, which must be modeled.

In open channel flow the free surface elevation varies with time and the bottom is uneven. This causes certain difficulty in the discretiztion of domain along the vertical direction. To solve the uneven physical domain, the vertical σ-coordinate transformation is as follows:

$$\xi^1=x_1=x,\xi^2=x_2=y,\xi^3=x_3=\delta=\frac{z+h}{h+\eta} \qquad (3)$$

where η is the surface elevation and h is the static water depth. The subgrid stresses ($\overline{\tau}_{ij}$) can be decomposed into the sum of a trace-free factor ($\overline{\tau}_{ij}'$) and a diagonal tensor ($\overline{\tau}_{kk}'$):

$$\overline{\tau}_{ij}'-\frac{\delta_{ij}}{3}\overline{\tau}_{kk}'=-2\upsilon_t\,\overline{S_{ij}} \qquad (4)$$

$$\overline{S}_{ij}=\frac{1}{2}\left(\frac{\partial \overline{u_i}}{\partial x_j}+\frac{\partial \overline{u_j}}{\partial x_i}\right) \qquad (5)$$

The term $\overline{\tau}_{kk}'/3$ is absorbed in the pressure term. In this study, we model the eddy viscosity (υ_t) by Lu and Wang (2009)'s coherent eddy model[12]:

$$\upsilon_t=(C_s\Delta)^2\,|S_Q| \qquad (6)$$

$$|S_Q|=\alpha 2\,\overline{S_{ij}}\,\overline{S_{ij}}+(1-\alpha)2Q \qquad (7)$$

where $2Q=(\overline{\Omega}_{ij}^2-\overline{S}_{ij}^2)$ and $\overline{\Omega}_{ij}=(\partial \overline{u_i}/\partial x_j-\partial \overline{u_j}/\partial x_i)/2$; α is weighted factor; Δ is a length scale defined here as $(\Delta_x^2+\Delta_y^2+\Delta_z^2)^{1/2}$ (Bardina, 1980), Δ_x, Δ_y and Δ_z are the control volume dimension in the x, y and z directions. C_s is a coefficient.

Here, the dynamic procedure computes the coefficient. The dynamic constant C_s^2 is calculated as follows[15,16]:

$$C_l^2=\frac{L_{ij}M_{ij}}{M_{ij}M_{ij}} \qquad (8)$$

$$L_{ij}=\overline{\overline{u_i}\,\overline{u_j}}^t-\overline{\overline{u_i}}^t\,\overline{\overline{u_j}}^t \qquad (9)$$

$$M_{ij}=-(\overline{\Delta}^t)^2\,|\overline{S}^t|\,2\,\overline{S_{ij}}^t+(\Delta)^2\,\overline{|\overline{S}|\,2\,\overline{S_{ij}}}^t \qquad (10)$$

where the test filter width $\overline{\Delta}^t=k\Delta$ and the coefficient k can be optimized. The coefficient k is taken to be $\sqrt{5}$; α is taken to be 0.5. C_l^2 is the exhibiting long time negative which generates numerical instability. To solve this, here we relax the value C_s^2 by a filter approach in stead of a general averaged time method[14], as follows:

$$C_s^2=\hat{C}_l^2(x,t)=\int_\Omega G(x-\xi)C_l(x,t)\mathrm{d}\xi \qquad (11)$$

where $G(x-\xi)$ is the smooth function. Although the smooth function can be in many forms, a box filter function was used for a convenient in this paper. In addition, the condition $\upsilon_t+\upsilon\geqslant 0$ is imposed. This condition keeps the total resolved dissipation remaining positive or zero.

2.2 Numerical methods and boundary conditions

The splitting operator approach is used to numerically solve the governing equations. At each time interval the momentum equations are split into three steps: advection, diffusion and pressure propagation. A combination of the quadratic backward characteristic method and Lax-Wendroff method is used to solve the flow advection. The central difference method is used to solve the diffusion. The pressure propagation step solves additional source and sink terms which include pressure and gravitational forces. The projection method is used to calculate the pressure and velocity field so that the updated velocity field satisfies the divergence free condition as imposed by the continuity equation. The CGSTAB method is used to solve the above equation.

The governing equations may be solved only when adequate boundary conditions are proposed. Several types of boundary conditions are usually imposed in open channel flow problems. No-slip boundary condition is imposed on the bottom wall and a zero gradient boundary condition is imposed on the two side walls. At inflow boundary, the inflow rate with a predetermined velocity distribution with added Gaussian distribution random signals is specified and the gradient of the water surface elevation is assumed to zero. At the outflow boundary, a convective boundary condition is. A Lagrange-Euler Method is used to locate the free surface elevation. Details of these boundary conditions are been found in Lin [17-18] and Lu and Wang[14].

3. Computed case

Figure 1 is a typical domain of open channel flow over train of twelve identical dunes. The numbers and shape of the dunes shown here correspond to that used in the experiments of Ojha and Mazumder. The experimental channel consists of a length of 10m, a width of 0.5m and a depth of 0.5m. The dunes had mean wavelength, L=32cm and mean height, H_d=3cm at the crest which results in the steepness H_d/L=0.094. The angles of the stoss side and lee side slope of the dunes were 6°and 50°, respectively (seeing Figure 1). Velocity profiles was measured by acoustic Doppler velocimeter (ADV). The mean flow depth h is kept 30cm; the discharge Q=0.04m³/s. The corresponding Reynolds number was found to be $Re=Uh/v$=1.5×10⁵; and the Froude number was $Fr=U/\sqrt{gh}$=0.29, where h is water depth.

The simulation domain is carefully chosen for proper setting up the inflow and outflow boundaries. To ensure the inlet flow is fully developed, the length of the main channel upstream of the first dune is extended to 0.5m. The length of the main channel downstream of the last dune is also extended to 0.5m to avoid the downstream outflow influence. A non-uniform grid of 451 ×11 ×45 nodes in the X direction, the Y direction and the Z direction respectively, was used to discretize the computational domain (5m long, 0.1m wide and 0.3m deep). The expansion ratio of the grid did not exceed 1.01. The time step was 0.0002 s. The grids and time step were sufficiently small enough to obtain grid convergent results. The total computed time was 40s.

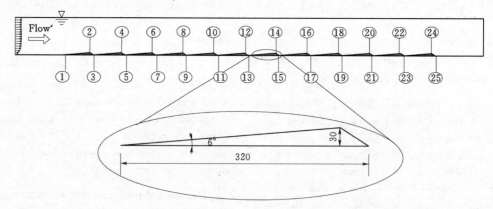

Figue 1　Schematic of computed domain and dunes profiles of Ojha and Mazumder（mm）

4. Results and discussion

Figure 2 shows the computed stream-wise mean velocity profiles at the trough and crest points, respectively, of each dune. Figure 3 shows the vertical mean velocity profiles at the corresponding points. In the figures, odd numbers inside circles represent the positions of velocity profiles in the consecutive trough points starting with 1 at the tail point the first dune and even numbers inside circles represent the positions of the velocity profiles at the crest point starting with 2 at the first dune crest (referring to Figure 1). Profiles of vertical mean velocity, corresponding to those for the horizontal mean velocity in Figure 2 are shown in Figure 3. From Figure 2 and Figure 3, it can be seen that the computed velocity results agree well with the experimental data. The most striking feature is that the horizontal mean velocity near at the trough points is negative except for the trough of the first dune, which means existence of flow reversal at the trough points. The computed results clearly show that developing flow from the leading dune up to the seventh dune and it reaches quasi-steady state.

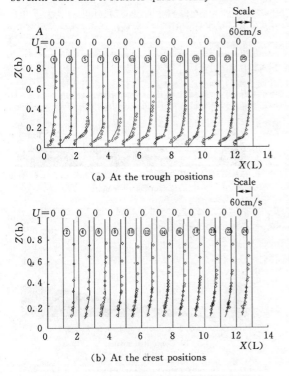

(a) At the trough positions

(b) At the crest positions

Figure 2   Horizontal mean velocity U over dunes covered surface. The solid lines are computed results and the points are experimental data

To keep the figure readable, the velocity field of the first half of the total train of dunes is shown as in Figure 4. As can been seen, the spatial development of turbulent open channel flow over the train of the twelve dunes is characterized by the flow separation at the dune crest and reattachment on the stoss side of the next dune. The similar phenomenon was given by Fourniostis et al in their computed results of five

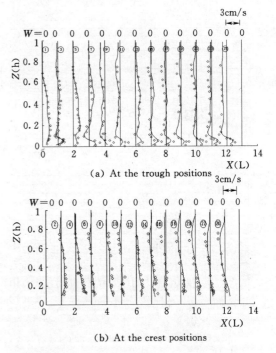

(a) At the trough positions

(b) At the crest positions

Figure 3   Vertical mean velocity W over dunes covered surface. The solid lines are computed results and the points are experimental data

dunes. For a better view of the flow field over a complete trough region, a close up view of velocity vector between the ninth and tenth dune in the fully developed region is given in Figure 7. The figure contains both the experimental and corresponding computational velocity vector field. Agreement is seen to be close within the recirculation zone. According to the flow pattern, three distinct layers are observed: the internal layer, $Z/h \leqslant 0.05$; the advecting and diffusing layer, $Z/h \leqslant 0.15$; the outer flow layer, $Z/h \leqslant 0.05$ (Fourniostis et al.).

Figure 5 shows the spatial development of the Reynolds stress ($-\rho uw$) along the flow over dunes. From Figure 5, it can be seen that the flow characteristics vary up to the seventh dune, and beyond which the entrance effect disappears. Again, this is in qualitative agreement with the experimental observations.

The reattachment point is determined as the wall nearest location at which the mean velocity changes sign. The predicted values at each the reattachment point are shown in Figure 6. From Figure 6, it can be seen that the first two computed reattachment length are much larger than the rest values. The flow reaches the full developing state after the seventh dune,

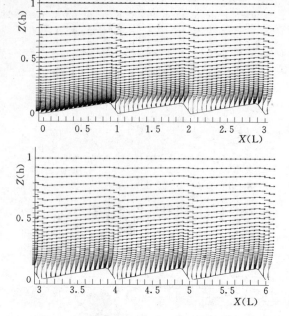

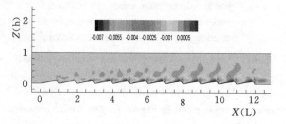

Figure 4　Computed velocity field

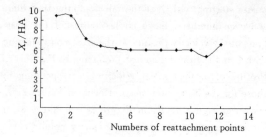

Figure 6　Reattachment length at each dune

are shown in Figure 8. From the figure, in can be seen that the surface elevation reaches the maximum value at the position of $X/L=1$ and $X/L=13$ due to the effect from lee side of the dune, respectively. The surface elevation gradually decreases along the X distance between the fifth dune and eleventh dune.

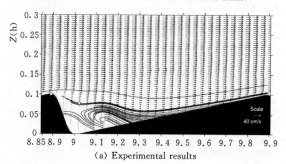

(a) Experimental results

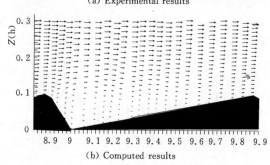

(b) Computed results

Figure 7　Velocity vector field between 9th and 10th dune

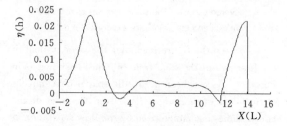

Figure 8　Free surface files over dunes

In order to analysis the head loss causing by

Figure 5　Reynolds shear stress ($-\rho uw$) distribution

which is agreement with the result mentioned above. The predicted reattachment length is at $X_r/H_d=6.0$ at quasi-steady state (as shown in Figure 7), which is considerably larger than the experimental value 5.8 (Fourniostis et al), where X_r is the reattachment length. Kasagi and Matsunaga[19] experiments the flow over a backward-facing step at Reynolds number of 5540, gave a value 6.5. As the reattachment position tend to move upstream with increasing Reynolds number in range $1200<Re<6600$. Dejoan and Leschziner[20] gave the computed value is about 7.0. It seems the experimental result underestimates the location of the reattachment point for purely 2D conditions.

Free surface plays a significant role on the open channel over dunes. The normalized free surface files (η/h, where η is the free surface level) over dunes

dunes, the energy equation between the cross-section between the inflow and outflow can be expressed as:

$$Z_{in} + \frac{P_{in}}{\rho g} + \frac{\alpha_{in} U_{in}}{2g} = Z_{out} + \frac{P_{out}}{\rho g} + \frac{\alpha_{out} U_{out}}{2g} + h_w \quad (12)$$

where Z_{in} is the inflow height of position, Z_{out} is the outflow height of position, α_{in} is the coefficient, α_{out} is the coefficient. In this paper, $\alpha_{in} = 1$ and $\alpha_{out} = 1$.

From the Eq. (12), one found the head loss between the inflow and outflow section is about 0.271m.

5. Conclusion

Large eddy simulation of open channel flow over series of dunes in the sigma coordinate have been carried out, in which the subgrid stress modeled by the dynamic coherent eddy model proposed by the authors. The computed velocity profiles are well agreement with the available experimental data. It found that turbulence does not reach a fully developed state until to the seventh dune. The computed results show that the length of separation zone at fully development region is about 6 H_d, which is little large than those in the experiment. The spatially mean free surface level decreases in the flow direction. The head loss caused by dunes is about 0.271m.

Acknowledgment

The authors gratefully acknowledge the financial support provided by the National Natural Science Foundation of China, with the project number of 50679023, the Ministry of Education Doctoral Foundation Project, No. 20070294012, the Postgraduate Innovation Project of Jiangsu Province and the Outstanding Doctoral Dissertation Incubation Program of Hohai University.

References

[1] N. Th. Fourniostis, et al. Numerical computation of turbulence development in flow over sand dunes. 16th congress of Asia and pacific division of international association of hydraulic engineering and research and 3th IAHR international symposium of hydraulic structures, Nanjing, 2008: 943-848.

[2] D. A. Lyn. Turbulence measurement in open channel flows over artificial bedforms.

J. Hydraulic. Eng, 1993, 119: 306-326.

[3] P. L. Wiberg, J. M. Nelson. Unidirectional flow over asymmetric and symmetric ripples. J Geophys Res, 1992, 97: 12745-12761.

[4] J. M. Nelson, J. D. Smith. Mean flow and turbulence over two dimensional bed forms. water resource research, 1993, 29: 3925-3953.

[5] S. P. Ojha, B. S. Mazumder. Turbulence characteristics of flow region over a series of 2-D dune shaped sructures. advance in water resources, 2008, 31: 561-576.

[6] W. X. Huai, Y. P. Sheng, T. Komatsu. Hybrid finite analytic solutions of shallow water circulation. Applied mathematics and mechanics-English edition. 2003, 24, 1081-1088.

[7] W. X. Huai, S. G. Fang. Numerical simulation of obstructed round buoyant jets in a static uniform ambient. Journal of hydraulics engineering. 2006, 132: 428-431.

[8] M. Peric, M. Ruger, G. Scheuerer. Calculation of the two-dimensional turbulent flow over a sand dune model. Rep. No. SRR-TN-88-02, Univ. of Erlangen, Germany, 1988.

[9] C. Mendoza, H. W Shen. Investigation of turbulent flow over dunes. J. Hydr. Engrg., ASCE, 1990, 116: 459-477.

[10] B Johns, R. L Soulsby, J. Xing. A comparison of numerical model experiments of free surface flow over topography with flume and field observations. J. Hydr. Res., 1993, 31: 215-228.

[11] J. Y. Yoon, V. C. Patel, R. Ettema. Numerical model of flow in ice-covered channels. J. Hydr. Engrg., ASCE, 1995, 122 (1): 19-26.

[12] J. Lu, L. L. Wang. Comparison of several turbulent models for calculating separated flows passing on sill. advance in water science 2009, 20: 255-260.

[13] J. Lu, L. L. Wang. Numerical study of large eddy structures—separated flows passing

sills. 16th congress of Asia and pacific division of international association of hydraulic engineering and research and 3th IAHR international symposium of hydraulic structures, 2008: 1795-1799.

[14] J. Lu, L. L. Wang, F. Peng, A novel dynamic eddy model and its application to LES of turbulent jet with free surface, Science in China Series G (In press).

[15] M. Germano, U, Piomelli, P. Moin, W. H Cabot. A dynamic subgrid-scale eddy viscosity model, Phys. Fluids A, 1991, 3: 1760-1765.

[16] D. K. Lilly. A proposed modification of the germano subgrid scale closure method. Phys. Fluids A, 1992, 4: 633-635.

[17] P. Z. Lin, C. W. Li. A σ-coordinate three-dimensional numerical model for surface wave propagation. International journal for numerical methods in fluids, 2002, 38: 1045-1068.

[18] C. W. Li. , F. X. Ma. Large eddy simulation of diffusion of a buoyancy source in ambient water, Applied mathematical modelling, 2003, 33: 247-2264.

[19] N. Kasagi, A. Matsunaga. Three-dimensional particle tracking velocimetry measurement of turbulence statistics and energy budget in a backward-facing step flow. J fluid and heat flow, 1995, 16: 477-485.

[20] A. Dejoan, M. A. Leschziner. Large eddy simulation of periodically perturbed separated flow over a backward-facing step. J fluid and heat flow, 2004, 25: 581-592.

Research on the Measurement and Controlling System of Tidal Current and Sediment Physical Model Test Applied on the Changjiang Estuary Reach

Xia Yunfeng   Xu Hua   Wu Daowen   Zhang Shizhao
Du Dejun   Wang Xiaojun

State Key Laboratory of Hydrology-Water Rescurces and Hydraulic Engineering,
Nanjing Hydraulic Research Institute, Nanjing, 210029

Abstract: At present, for the earlier researches navigation regulation projects in large estuaries in China, physical model tests are usually employed to solve some key technical problems about hydrodynamics and bed erosion and deposition. By analyzing the measurement and control system of physical model for the estuarine reach of Changjiang River based on communication, frequency conversion techniques, etc., the building characteristics of platforms for tide generating system and tide level and flow field measuring system and the development of the measurement and control techniques under digital conditions are discussed. It may provide reliable and information-based numerical data with high precision. It may improve the general level of model tests.

Key words: physical model; measurement and control system; tide generating system; tide level and flow field measuring technique

1. Introduction

China is not only a continental country with vast territory and dense rivers but also a large marine country with the coastline of about 18,000km. It has great development potential. The evolution of rivers and coasts is closely related to the development of human society. However, we have not understood enough the natural laws of rivers and coasts. The rivers have different lengths and discharges, the geographical and geological environments are different, the sediment transportation is complex, the flows and river beds change and interact with each other. Accordingly, the evolution of the rivers is complex. With the rapid development of hydraulic projects, more and more technical issues need to be understood, studied and solved, Such as port construction, river regulation, bridge and wharf construction, river and sea beach reclamation, hydraulic hinge layout, local anti-scour project around hydraulic structures, sand prevention projects for water intakes as well as researches on flood control and disaster mitigation projects. At present, the physical model, mathematical model and prototype observation are the three main means to study these engineering problems.

Nowadays, the key technical problems in the exploitation and utilization of large rivers and estuarine waterway regulation projects in China such as Changjiang Three Gorges Project and Changjiang deep water channel regulation projects in the estuarine reach of Changjiang River are hydrodynamics and river bed erosion and deposition. They are often solved by the use of physical model tests, in which the quality of test results is dependent upon the precision of measurement and control system.

The lower reach of Changjiang River from Tiansheng Port to Wusong Port is located in the Changjiang River delta with the most advanced economy in China, and owns the highest shipping benefit in the golden waterway of Changjiang River. However, there are three obstructing navigation shallow shoals along the reach: Baimao, Tongzhou and Fujiang, called as "Three Shoals". With regarded to the studies on the Changjiang 12.5 m-deepwater waterway regulation project on Changjiang River extended to Nanjing, commissioned by Changjiang Changjiang River Waterway Bureau, the tidal and sediment physical model tests on the deepwater channel regulation project are performed.

A large model for the estuarine reach of Changjiang River is established to study the hydrodynamical and sediment problems in the channel regulation project so as to provide important technical support for the project design. Considering the high precision requirements, large model range and massive measurement data, a new measurement and control system is developed.

2. Model measurement and control system

The model range from Jiangyin to Wusong is about 270m in length. It is equivalent to 176km in length of the prototype. The horizontal scale is 655, and the vertical one is 100. The twisted waterway to simulate tidal limit connects the measuring weir and the model. The flow control is used in the upstream of model, and the tidal level control is used in the downstream. The model layout is shown in Figure 1.

The model measurement and control system has good real-time effect. Various states of equipments, collected data and controlled output signals can be arbitrarily viewed. The software is programmed with VB language, the density of the data acquisition is large, and the time interval of control and operation is short. The precision of computer acquisition interface is much higher than one-time precision of measuring instruments.

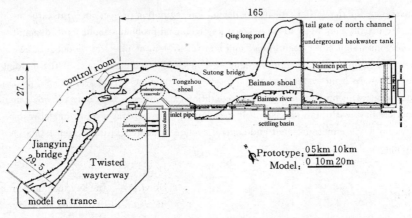

Figure 1　Model layout of estuarine reach of Changjiang River

2.1　Control system

The real-time water level of measuring weir is monitored by grating automatic tracing water level meter, and the incident discharge of measuring weir is controlled by computer adjusting electric valve at the upstream boundary of the model.

The tide generating system for the southern branch in the downstream of the model is mainly consisted of tide box (25m in length, 2m in width, and the volume is about 140m³, see Figure 2), frequency conversion blower, electric servo exhaustion valve, electrical drain valve, automatic tracing water level meter and so on. During flooding, the water in the tide box is extruded to enter into the model through blower air inlet. According to the tidal level collected by the water level meter at the control station, the flow rate of the blower air and the opening of the exhaustion valve are controlled by computer. During ebbing, the pressure is reduced in the tide box by the exhaustion valve, then the water flows into the tide box, at the same time the tide falling process

is also controlled by the electrical drainage valve.

The tide generating system for the northern branch in the downstream is mainly consisted of electric servo tail gate control system, inlet water pipes, automatic tracing water level meter and so on. During flooding or ebbing, according to the tidal level collected by the water level meter at the control station, the opening of tail gate is controlled through the servo motor by computer, and the prototype tide phenomenon is repeatedly shown in the model.

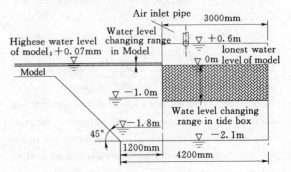

Figure 2　Cross-section layout of tide box

2.2 Data acquisition system

The data acquisition system is mainly consisted of industrial control computer, automatic tracing water level meter, particle surface flow field measurement system, underwater rotor current meter, photoelectric sediment measuring meter and ultrasonic three-dimensional topography meter.

The water level is measured by the grating automatic tracing water level meter (Figure 3). Its resolution is 0.01mm, and its measurement precision can reach 0.05mm, and its tracing speed is not less than 15mm/s. RS 485 serial grids are employed for the transformation parameters between the water level meter and the main control machine, and the transfer interval is 50ms.

The underwater velocity is measured by the photoelectrical rotor current meter. The current meters are fixed on the tracing shelf, and the water level and velocity are regularly collected. The data is processed and saved by computer.

The particle image velocimetry (PIV) technology is a large-scale synchronous velocity measurement system based on the particle tracing measurement technology. By use of the system, a wide range of surface flow fields in steady or unsteady flow tests can be arbitrarily measured. The flow field, velocity distribution and velocity process at a point can be quickly and easily got. The system hardware includes many CCD cameras, video transmission lines, video splitter, video acquisition card and computer, and the system software includes particle data acquisition and processing program.

The sediment concentration is measured by the infrared adjustment reflecting photoelectric sediment measurement sensor which is composed of sensors and instruments, etc. The maximum measured sediment concentration is about $12kg/m^3$.

For the topography, the ultrasonic topography meter (Figure 4) horizontally moves on the measuring bridge by using a measuring car, and the measuring rode controlled by the topography meter vertically moves up and down. The horizontal positioning of the measuring car is completed by controlling the device of measuring bridge. The vertical measurement range is between 0 to 1m, and the vertical measurement error is less than 1mm. The horizontal measurement range is between 0 to 10m, and the horizontal meas-

urement error is 1mm. The horizontal moving speed is 0 to 0.15m/s.

Figure 3　Grating automatic tracing water level meter

Figure 4　Ultrasonic topography meter

3. Application of model measurement and control system‘

Since its establishment in 2005, the model has been employed to study many hydraulic projects such as navigation regulation of Fujiang shoal, navigation regulation of Tongzhou and Baimao shoals, Shanghai-Nantong Railway Bridge on Changjiang River, regulation and exploitation on the west waterway of Zhangjiagang, overall planning on harbor district of Zhangjiagang, integrated regulation and exploitation on the Jiangyin-nantong reach. The precision of the control system is satisfactory, and a lot of reliable numerical data with high precision are obtained.

3.1 Application of control system

The repeatability verification results of the model are shown in Figure 5. The results indicate that the tidal level and velocity will be stable from the third cycle. The repeatability precision satisfies the standard requirements.

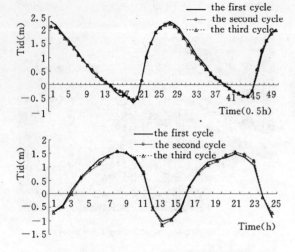

Figure 5 Repeatability verification on tidal level
and velocity course of the model

3.2 Application of measurement and control system

The results of the tidal level course, velocity course, surface flow field and topography change are shown in Figure 6-Figure 9. The test results are visual and convenient. It offers much help for analyzing the effect on the current and river bed evolution after project construction.

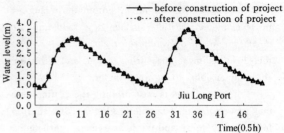

Figure 6 Comparison of tidal level course between
before and after project construction

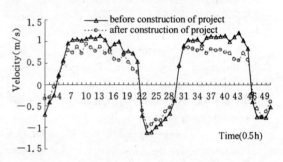

Figure 7 Comparison of velocity course between
before and after project construction

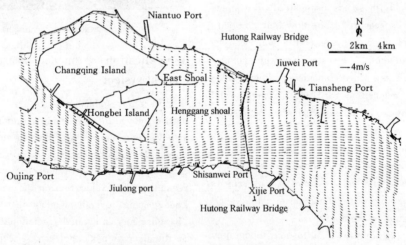

Figure 8 Surface flow field in the study reach

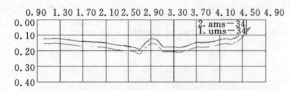

Figure 9 Comparison of topography between
before and after project construction

4. Conclusion

The measurement and control system applied in large-scale model of the estuarine reach in Changjiang River is employed to well realize the measurement and control aim required by the standards by use of technical means of frequency conversion control and 485 communication and so on. The precision of control system is satisfactory, and a large number of numerical data with high precision are obtained. Recently, many large-scale hydraulic projects have been accomplished by using the proposed model, and abundant research results are achieved. At the same time, the measurement and control system is constantly updated and improved during the course of model tests, thus the overall level of model tests is enhanced. The test results provide important technical support for the planning, design and decision-making of projects.

References

[1] Hui Yujia, Wang Guixian. Similarity Theory of River Model [M]. China Water Conservancy and Hydropower Press, 1999.

[2] Li Changhua, Jin Dechun. River Model Tests [M]. China Communications Press, 1981.

[3] Xia Yunfeng. Model test on Deep-water depth navigation regulation project in the Fujiang shoal reach of the Changjiang River [R]. NHRI, 2006.

[4] Qu Bo. Research on controlling tide level of tide box in river model [C]. The 14th Ocean (Coast) Engineering Symposium of China, 2009: 1232-1237.

Author Index

A

A. Ahmadi
A. Duarte
A. Nichols
A. Ricardo
A. Sukhodolov
A. R. Zarrati
Abdolreza Karaminejad
Abul Basar M. Baki
Ahmed Y. Al-Taee
Ali Parvaneh
Alireza Keshavarzi
Amir Reza Zarrati
An Mengxiong
Atsushi Ono
Azza N. Al-Talib

B

B. C. Basak
Benjamin J. Dewals

C

C. C. Sung
C. W. Li
C. Zeng
Cai Ying
Chanseng Phongpachith
Che Qinquan
Chen Airong
Chen Fazhan
Chen Huiling
Chen Huiquan
Chen Wenxue
Cheng Zibing
Chenglu
Chengwei Xu
Chih-Shien Yang
Christos Katopodis
Chunhua Wang
Colin K. C. Wong
Cui Wei

D

D. J. Dimitriou
Dai Qun
Daisuke Mori
Dang Yuanyuan
Dao-Wen Du
Darrienyauseng
David Walker
David Z. Zhu
De-Jun Zhang
Der-Liang Young
Ding Limin
Dinh Cong San
Dong Jieying
Dong Xinglin

E

Ebrahim Jabbari

F

F. Z. Lee
Fan Zi-Wu
Fang Kezhao
Felipe Justo Breton
Feng Huan
Fong-Zuo Lee
Franci Steinman
Fu Hui
Fu Zhimin

G

G. Zhao
GasPer Rak
Gao Xueping
Gao Yong
Gong Jie
Gorazd Novak
Gu Hua
Gu Xingwen
Guo Xiaochen
Guo Xinlei
Guo Yongbin
Guo Yongxin

Guoan Yu

H

H. Golmohammadi
Hai Zhu
Hamed Sarkardeh
Han Changhai
Han Tianding
Han Xijun
Hikaru Takeuchi
Hou Guibing
Hsiang-Kuan Chang
Hu Qulie
Hu Ya' An
Huaixiang Liu
Huang Dong
Huang Guobing
Huang Minghai
Huang Yue
Huei-Tau Ouyang

I

I. Schauder

J

J. D. Demetriou
J. S. Lai
James Ball
James Yang
Javad Farhoudi
Jean-Marie Vuillot
Ji Hongjun
Jiang Bole
Jiang Shuhai
Jiang Yunpeng
Jianjun Zhao
Jie Jin
Jihn-Sung Lai
Joseph H. W. Lee
Juichiro Akiyama
Jun Lu

K

K Attenborough
K. Blanckaert
K. Horoshenkov
Khodadad Safavi
Klaus Schmitt

Koichiro Matsuda
Kouki Onitsuka

L

Lai Saihin
Lajeunesse Eric
Li Chuanqi
Li Fang
Li Jing
Li Jun
Li Xi
Li Xiangfu
Li Xuehai
Li Yuanfa
Li Yuanhong
Li Yun
Li Zhenhai
Li Zhongyi
Li Ziyang
Liangshi Xiong
Lin Jinsong
Lingling Wang
Liqin Zuo
Liu Chengdong
Liu Hansheng
Liu Huojian
Liu Shanyan
Liu Tonghuan
Liu Wenzhong
Liu Youcun
Liu Benqin
Luo Shaoze
Lv Hongxing

M

M. R. Jalili Ghazizadeh
M. T. Pusch
Ma Fuheng
Ma Jiming
Mah
Mahdi Esmaeili Varaki
Malte Cederstrom
Mansour Abolghasemi
Masaki Fukushima
Md. Altaf Hossain
Md. Lutfor Rahman
Md. Showkat Osman
Meng Xiangwei

Métivier Francois
Mi Zhankuan
Michel Pirotton
Ming-Hsi Hsu
Mohammad Reza Jalili Ghazizadeh
Mu Xiangpeng
Mwafaq Y. Mohammed

N

N. Rajaratnam
Narteau Clément
Nathan C. Cox
Nicole Von Lieberman
Nicole Von Lieberman
Ning Bo
Noriaki Hakoishi

O

Oudomsack Philavong
Oulaphone Ongkeo

P

P. J. Visser
Patrik Reasson
Peng Li
Pierre Archambeau

Q

Qi Erong
Qi Liang
Qiao Jianping
Qiu Ruofeng
Qu Zhaosong
Quan Qiang

R

R. Wilkes
Rasool Ghobadian
Ren Mingxuan
Ren Mingxuan
Reza Roshan
Robert Keller
Rozita Hamidi
Ruan Shiping

S

S. Esfiari
S. Shepherd

S. Tait
Saso Santl
Samad Emamgholizadeh
Sébastien Erpicum
Seyed Mahmood Borghei
Shen Dengle
Shizhao Wang
Shun-Chung Tsung
Shuqing Yang
Sima Safarkhani
Song Zhenbo
Steven L. Barfuss
Sun Feng

T

Tang Xiangfu
Taro Uchimura
Teng Juan
Thorsten Albers
Tong Zhongshan
Tsang-Jung Chang
Tsuyoshi Seki

U

V

V. Dugué

W

W. Ottevanger
W. Van Balen
W. S. J. Uijttewaal
Wang Chen
Wang Feng
Wang Lin
Wang Qi
Wang Tao
Wang Xiaogang
Wang Xin
Wang Yong
Wangfeng
Wei Zhangping
Weilin Xu
Wenbin Zhang
William J. Rahmeyer
Wu Caiping
Wu Fusheng
Wu Huabao

Wu Shiqiang
Wu Wei

X

X. -F. Garcia
Xia Lijuan
Xia Yun-Feng
Xiang Guanghong
Xiang Yan
Xiao Chenmeng
Xiao-Jun
Xie Di
Xie Min
Xie Xinghua
Xie Yaqiong
Xing Linghang
Xiong Run' E
Xu Guangming
Xu Hua Wu
Xu Shikai
Xuan Guoxiang
Xue Shu
XuXinmin

Y

Y. G. Tan
Y. J. Huang
Yan Genhua
Yan Wei
Yan Xiujun
Yan Yixin
Yang Kailin
Yang L
Yang Wei
Yang Yanxiong
Yang Yu

Yanqiong Geng
Yao Shiming
Ye Baisheng
Yoshitake Zoshi
Yuan Hui
Yuan Qiang
Yufang Han
Yusuke Mori

Z

Zhang Congjiao
Zhang Hongwei
Zhang Jiabo
Zhang Landing
Zhang Luchen
Zhang Ming
Zhang Ruikai
Zhang Shaochun
Zhang Xiantang
Zhao Jianjun
Zhao Jianping
Zhao Jianping
Zhao Rong
Zhaoyin Wang
Zheng Jianchun
Zheng Jun
Zheng Shuangling
Zhenhua Huang
Zhenzhen Yu
Zhichang Chen
Zhihuang Lu
Zhonghua Feng
Zhou Jie
Zhu Chao
ZongMuwei
Zou Zhili